GPS for Geodesy

Springer

Berlin
Heidelberg
New York
Barcelona
Budapest
Hong Kong
London
Milan
Paris
Singapore
Tokyo

Peter J. G. Teunissen
Alfred Kleusberg (Eds.)

GPS for Geodesy

Second Completely Revised and Extended Edition
with 127 Figures and 32 Tables

PROFESSOR DR. PETER J. G. TEUNISSEN
Delft University of Technology
Faculty of Civil Engineering and Geosciences
Dept. of Mathematical Geodesy and Positioning
Thijsseweg 11
2629 JA Delft
The Netherlands

PROFESSOR DR. ALFRED KLEUSBERG
University of Stuttgart
Institute for Navigation
Geschwister-Scholl-Straße 24D
70174 Stuttgart
Germany

ISBN 3-540-63661-7 2nd ed.Springer-Verlag Berlin Heidelberg New York
ISBN 3-540-60785-4 1st ed.Springer-Verlag Berlin Heidelberg New York

Cataloging-in-Publication Data applied for
GPS for geodesy / Alfred Kleusberg, Peter J.G. Teunissen, [eds.]. -- 2nd ed.
"Contains the edited lecture notes of the International School GPS for Geodesy in Delft, The Netherlands, March 2-8 1997"--Pref. Includes bibliographical references and index
ISBN 3-540-63661-7
1.Global Postioning System--Congresses. 2. Geodesy--Congresses. 3. Surveying--Congresses. I. Kleusberg, Alfred. II. Teunissen, P. J. G.
QB275.G65 1998

Typesetting: Camera-ready by Jasmine van der Bijl
Cover design: design & production, Heidelberg
SPIN:10635190 32/3020-5 4 3 2 1 0 - Printed on acid -free paper

PREFACE TO SECOND EDITION

This second edition contains the edited lecture notes of the 2nd International School **GPS for Geodesy** in Delft, The Netherlands, March 2-8, 1997. The school was again held at the excellent facilities of the DISH Hotel, and organized with the knowledgeable help of Frans Schröder of the Netherlands Geodetic Commission. It attracted 60 geodesists and geophysicists from around the world.

The 2nd school was organized using the successful formula of the first school. Following the rapid developments in GPS geodesy, new topics were added to the original lecturing program. These include GPS quality control, active GPS reference stations, long-distance kinematic GPS and atmospheric models from GPS. The expanded program was presented by the team of experts of the first school and three invited guest lecturers.

The typescript of the lecture notes of this second and enlarged edition was edited by Frank Kenselaar and Jasmine van der Bijl of the Department of Geodetic Engineering of the Delft University of Technology. They succeeded in transforming the original lecture notes to one coherently formatted manuscript.

The 2nd school received the same generous support as the first school. This support is gratefully acknowledged.

January 1998
Delft, The Netherlands
Stuttgart, Germany

Peter Teunissen
Alfred Kleusberg

PREFACE

This monograph contains the revised and edited lecture notes of the International School **GPS for Geodesy** in Delft, The Netherlands, March 26 through April 1, 1995. The objective of the school was to provide the necessary information to understand the potential and the limitations of the Global Positioning System for applications in the field of Geodesy. The school was held in the excellent facilities of the DISH Hotel, and attracted 60 geodesists and geophysicists from America, Asia, Australia and Europe.

The school was organized into lectures and discussion sessions. There were two lecture periods in the morning and two lecture periods in the afternoon, followed by a discussion session in the early morning. A welcome interruption to this regular schedule was a visit to the European Space Research and Technology Centre (ESTEC) in Noordwijk in the afternoon of March 29. A tour of the Noordwijk Space Expo and the ESA satellite test facilities, and presentations by ESTEC personnel of GPS and GNSS related activities at ESTEC, provided a different perspective to space geodesy.

The school had the support of the International Association of Geodesy, the Netherlands Geodetic Commission, the Department of Geodetic Engineering of the Delft University of Technology, the Department of Geodesy and Geomatics Engineering of the University of New Brunswick, and the Survey Department of Rijkswaterstaat. This support is gratefully acknowledged.

The organization of the International School began in early 1994, with the knowledgeable help of Frans Schröder of the Netherlands Geodetic Commission. Throughout the year of preparation and during the school, Frans Schröder looked after student registration and organized facilities, and thereby ensured the success of the school.

The International School **GPS for Geodesy** would not have been possible without a team of dedicated lecturers of international reputation with expertise in GPS geodesy. The lecturers were willing to agree beforehand to a shared responsibility for parts of the school presentation and the preparation of the corresponding lecture notes. All authors tried to adhere to a common notation throughout the chapters of the lecture notes, and avoided unnecessary repetitions.

The typescript of these lecture notes was edited by Wendy Wells of the Department of Geodesy and Geomatics Engineering of the University of New Brunswick. She received expert help on Chapter 8 from Jasmine van der Bijl of the Department of Geodetic Engineering, Delft University of Technology. Ms Wells succeeded in producing a coherently formatted manuscript from bits and pieces created with three different word processors on two different computer platforms.

January 1996
Fredericton, Canada
Delft, The Netherlands

Alfred Kleusberg
Peter Teunissen

TABLE OF CONTENTS

LIST OF CONTRIBUTORS

G. Beutler
Astronomical Institute, University of Bern, Sidlerstrasse 5, CH - 3012 Bern, Switzerland.

Geoffrey Blewitt
Department of Geomatics, University of Newcastle, Newcastle upon Tyne, NE1 7RU, United Kingdom.

Yehuda Bock
Cecil H. and Ida M. Green, Institute of Geophysics and Planetary Physics, Scripps Institution of Oceanography, University of California, San Diego, 9500 Gilman Drive, La Jolla, California, 92093-0225 USA.

Oscar L. Colombo
University of Maryland/NASA's Goddard S.F.C., Code 926, Greenbelt, Maryland 20771, U.S.A.

Clyde C. Goad
Department of Geodetic Science and Surveying, The Ohio State University, 1958 Neil Avenue, Columbus OH 43210-1247 U.S.A.

Alfred Kleusberg
Institut für Navigation, Universität Stuttgart, Geschwister-Scholl-Str. 24D, 70174 Stuttgart, Germany.

Richard B. Langley
Geodetic Research Laboratory, Department of Geodesy and Geomatics Engineering, University of New Brunswick, P.O. Box 4400, Fredericton, N.B., Canada E3B 5A3.

Hans van der Marel
Faculty of Civil Engineering and Geosciences, Department of Mathematical Geodesy and Positioning, Thijsseweg 11, 2629 JA Delft, The Netherlands.

Peter J.G. Teunissen
Faculty of Civil Engineering and Geosciences, Department of Mathematical Geodesy and Positioning, Thijsseweg 11, 2629 JA Delft, The Netherlands.

INTRODUCTION

The topic of these lecture notes of the International School **GPS for Geodesy** is the description of the use of the Global Positioning System (GPS) measurements for geodetic applications. The term geodetic applications is used in the sense that it covers the determination of precise coordinates for positions in a well defined reference system, and the monitoring of temporal changes of these coordinates.

These lecture notes are organized in sixteen chapters, each of which begins with the full address of the author(s), and a section introducing the theme of the chapter. After the main body of text, each chapter is concluded by a summary section and a list of references. The individual chapters have been written independently, and they can also be read and studied independently. Their sequence, however, has been arranged to provide a logical and coherent coverage of the topic **GPS for Geodesy**.

Chapter 1 introduces global reference systems for Cartesian and ellipsoidal coordinates and local reference frames, and their basic relation to the GPS measurements. Transformations between and motions of the Celestial and Terrestrial Reference Frames are described. Time systems are introduced to provide an independent variable for the description of motion and earth deformation. The concepts and realizations of Conventional Reference Systems are explained.

The topic of Chapter 2 is the description of the computation of GPS satellite orbits, and the dissemination of GPS satellite ephemerides. Starting from the equations of motion for satellites, first the Keplerian orbit is introduced and then generalized to include the pertubations resulting from non-central forces. Various sources of orbital information to GPS end users are described, and the chapter is concluded with a brief introduction of the effect of unmodelled orbit errors on positions determined from GPS measurements.

Chapter 3 introduces the GPS signal, its components, and its generation in the satellites' circuitry. The aspects of signal propagation from the satellite to the GPS receiver are described, including the effects of refraction, multipath, and scattering. Chapter 4 begins with an introduction to the basic building blocks of a GPS receiver, and shows how pseudoranges and carrier phases are being measured in the receiver circuits. The chapter is concluded with a discussion of the measurements errors in these two observables.

Chapter 5 starts from the complete non-linear observation equations for pseudoranges and carrier phases and introduces a number of different linear combinations, in order to eliminate, reduce, and/or emphasize parts of the equations. Following this exploratory analysis, the observation equations are linearized with respect to the parameters to be determined. Basic properties of the linearized equations in the context of single point positioning and relative positioning are discussed, with particular emphasis on parameter estimability.

Chapters 6 through 8 present details on the adjustment and testing of GPS data. Practical and methodological aspects of GPS data processing are discussed in Chapter 6, and methods for GPS quality control are presented in Chapter 7. The quality control procedures are developed for both batch solutions and recursive solutions. The process of finding and validating the correct integer carrier phase ambiguities is discussed in Chapter 8. This chapter introduces the theory of integer least-squares and outlines various ambiguity search strategies, including the process of ambiguity decorrelation.

Chapters 9 through 16 present details on the use of GPS measurements for the spectrum of geodetic applications. There might be a number of ways of structuring these applications; we have chosen to start with single-receiver applications followed by multi-receiver applications for which the network scale is used as the criterion. Accordingly we begin with two chapters, Chapter 9 and Chapter 10, on the use of a single receiver. The first chapter discusses the components of an active GPS reference station, together with its potential on-line and off-line applications. Chapter 10 proceeds to discuss single site solutions through linearization of the observation equations. Also included is a presentation of the direct solution of pseudorange equations without the requirement of a priori information.

Chapter 11 discusses short distance GPS models. In the context of these Lecture Notes, "short" means that atmospheric and orbital errors do not significantly affect the accuracy of the positioning results, and do not have to be included explicitly. In order to retain the full capability of GPS in networks of larger scale (typically between 50 km and 1000 km), the atmospheric refraction effects and inaccuracies of the GPS satellite orbits need to be explicitly included. The corresponding mathematical models for the GPS measurements and procedures for the estimation of geodetic parameters are outlined in Chapters 12 and 13. The first of these two chapters considers large scale network solutions with applications for monitoring crustal deformation, while the second chapter concentrates on the positioning of moving platforms over long distances so as to assist large-area surveys with remote-sensing instruments on board ships or aircraft.

The last three chapters of this book discuss the Global Positioning System for applications on a global scale. Chapter 14 discusses the estimation of various parameters of interest for geodynamics applications, together with the necessary link to the determination and maintenance of global reference systems. Chapter 15 discusses the contribution of GPS to atmospheric modelling. In this scenario, the ionospheric and tropospheric delays are seen as remotely sensed data related to atmospheric parameters, with the possibility to recover some or all of these parameters through proper GPS data analysis. Finally, the last chapter of this book puts GPS in perspective by discussing its role in space geodesy, now and in the future.

1. REFERENCE SYSTEMS

Yehuda Bock
Cecil H. and Ida M. Green, Institute of Geophysics and Planetary Physics,
Scripps Institution of Oceanography, University of California, San Diego,
9500 Gilman Drive, La Jolla, California, 92093-0225 USA

1.1 INTRODUCTION

Of fundamental importance in space geodetic positioning is the precise definition and realization of terrestrial and celestial reference systems. It is appropriate then that this topic be covered in the first chapter of these notes on the Global Positioning System (GPS).

As its name implies, the purpose of GPS is to provide a global absolute positioning capability with respect to a consistent terrestrial reference frame. The original intention was instantaneous three-dimensional position with 1–2 meter precision for preferred users; ordinary users would be allowed 1–2 orders of magnitude less precision.

Geodesists realized, at least 15 years ago, that GPS could be used in a differential mode, much like very long baseline interferometry (VLBI), to obtain much more accurate relative positions. Relative positioning with 1 mm precision was demonstrated in the early 1980's using single-frequency geodetic receivers over short distances (100's of meters). Precision decreased, however, in approximate proportion to intersite distance, about 1–2x10^{-6} (1–2 ppm) *circa* 1983, primarily due to satellite orbital errors and ionospheric refraction. Since then, geodesists have been able to attain about 3 orders of magnitude improvement in horizontal precision (Table 1.1). Vertical precision has also improved but has always been about 3–4 times less precise than horizontal precision. These dramatic improvements were achieved by advancements in geodetic receiver technology, full implementation of the GPS satellite constellation, and the establishment of a worldwide tracking network to estimate precise GPS satellite orbits. Today GPS can be considered an “absolute” global geodetic positioning system (GGPS) providing nearly instantaneous three-dimensional positions at the 10–20 mm level with respect to a consistent global terrestrial reference system. Uncertainties in the positions and velocities of the global tracking stations, i.e., in the global reference frame, still limit the accuracy of GGPS (Table 1.1).

Table 1.1: Improvements in Positioning and Limiting Error Sources

Year	b (ppm)	Source of Improvement	Primary Error Sources
~ 1983	1	geodetic receivers (carrier phase measurements)	atmospheric refraction, orbital accuracy
~ 1986	0.1	dual-frequency carrier phase measurements	tropospheric refraction, orbital accuracy
~ 1989	0.01	global tracking stations (CIGNET)	tropospheric refraction, orbital accuracy
~ 1992	0.005	enhanced global tracking (IGS)	troposphere, orbits, antenna phase center errors
1997	0.001	improved orbit, troposphere, antenna modeling	global reference frame, site specific errors, atmospheric gradients
$s^2_{Horizontal}(mm^2) = [0.1\text{–}1.0\ mm]^2 + [2bS_{ij}\ (km)]^2$ S_{ij} is the distance between sites i and j			

1.1.1 Basic GPS Model

The geometric term of the model for GPS carrier phase can be expressed in simple terms as a function of the (scalar) range ρ_i^k such that

$$\phi_i^k(t) = \frac{t_0}{c}[\rho_i^k(t, t - \tau_i^k(t))] = \frac{t_0}{c}\mathbf{r}^k(t - \tau_i^k(t)) - \mathbf{r}_i(t) \quad (1.1)$$

where t is the travel time of the radio signal, $\mathbf{r}_i$ is the geocentric vector for station i at reception time t, $\mathbf{r}^k$ is the geocentric vector for satellite k at satellite transmission time (t-τ), f_0 is the nominal signal frequency, and c is the speed of light. The station position vector is given in a geocentric Cartesian reference frame as

$$\mathbf{r}_i(t) = \begin{bmatrix} X_i(t) \\ Y_i(t) \\ Z_i(t) \end{bmatrix} \quad (1.2)$$

The equations of motion of a satellite can be expressed by six first-order differential equations, three for position and three for velocity,

$$\frac{d}{dt}(\mathbf{r}^k) = \dot{\mathbf{r}}^k \tag{1.3}$$

$$\frac{d}{dt}(\dot{\mathbf{r}}^k) = \frac{GM}{r^3}\,\mathbf{r}^k + \ddot{\mathbf{r}}^k_{\mathrm{Perturbing}} \tag{1.4}$$

where G is the universal constant of attraction and M is the mass of the Earth. The first term on the right-hand side of (1.4) contains the spherical part of the Earth's gravitational field. The second term represents the perturbing accelerations acting on the satellite (e.g., non-spherical part of the Earth's gravity field, luni-solar effects and solar radiation pressure).

In order to difference the station vector and satellite vector in (1.1), both must be expressed in the same reference frame. The station positions are conveniently represented in a terrestrial (Earth-fixed) reference frame, one that is rotating in some well defined way with the Earth. Solving the equations of motion of the GPS satellites (i.e., orbit determination) requires a geocentric celestial (inertial) reference frame. In order to compute (1.1), either the station position vector needs to be transformed into the celestial frame or the satellite position vector needs to be transformed into the terrestrial reference frame. Furthermore, fundamental concepts of time epoch, time interval, and frequency must be rigorously described, and fundamental constants (e.g., speed of light and the GM value[1]) must be defined.

1.1.2 The Fundamental Polyhedron

As we shall see in this chapter, the orientation of the Earth in space is a complicated function of time which can be represented to first order as a combination of time varying rotation, polar motion, a nutation, and a precession. The realization of celestial and terrestrial reference systems are quite involved because of the complexity of the Earth's composition, its interaction with the atmosphere, and its mutual gravitational attraction with the Moon and the Sun. The definition of the terrestrial reference system is complicated by geophysical processes that make the Earth's crust deform at global, regional, and local scales, at a magnitude greater than the precision of present-day space geodetic measurements. The definition of the celestial reference system is complicated by the fact that stellar objects have proper motions (or structure) and are not truly point sources.

The realization of a reference system is by means of a reference frame, i.e., by a catalogue of positions which implicitly define spatial coordinate axes. The

[1] IERS [1996] adopted values: c=299792458 m/s and GM = 3986004.4356 x 10^8 m^3/s^2.

celestial reference system is realized by a catalogue of celestial coordinates of extragalactic radio sources determined from astrometric (VLBI) observations. These coordinates define, at an arbitrary fundamental epoch, a celestial reference frame (CRF). The terrestrial reference system is realized through a catalogue of Cartesian station positions at an arbitrary fundamental epoch, t_0, i.e.,

$$[\mathbf{r}(t)]_0 = [X(t), Y(t), X(t)]_0 \qquad (1.5)$$

determined from a variety of space geodetic observations, including satellite laser ranging (SLR), VLBI, and GPS. These positions define a *fundamental polyhedron*. Implicit in the coordinates of its vertices are conventional spatial Cartesian axes that define the terrestrial reference frame (TRF). Maintaining the reference frame means relating the rotated, translated, and deformed polyhedron at a later epoch to the fundamental polyhedron. The deformations of the polyhedron are by definition those motions that do not contain any rotations or translations. Deformations are accommodated, at least to first order, by supplementing the station catalogue with station velocities derived from global plate models and/or long-term geodetic measurements. The reference frame does not change (unless a new one is defined, of course). It is fixed to the station positions at t_0 (the fundamental polyhedron) and consists of a set of spatial Cartesian axes with a particular origin and orientation. It is the reference system that is changing and moving with the polyhedron. Therefore, the terrestrial system and its frame coincide, in principle, only at the initial epoch. The terrestrial system is not only a set of changing positions. Its definition includes descriptions of anything that influence these positions (e.g., the initial station positions (1.5), plate motion models, gravity models, fundamental constants, precession models, nutation models, etc.). The purpose of the terrestrial system is to make the reference frame accessible to the user who can then determine time-tagged positions on the Earth's surface. The connection between the fundamental polyhedron and the celestial system is given by nutation, precession, and the Earth orientation parameters (EOP).

Positioning, therefore, is intricately linked to a reference system. The reference system is realized by the reference frame which is in turn defined by station positions. Thus, space geodetic positioning is a bootstrapping process of incremental improvements in station positions, physical models, reference frames, and reference systems. Any factor that affects station and satellite positions affects the reference frame, and *vice versa*. Any change in adopted physical models affects the reference system and therefore the reference frame.

Today, the TRF and CRF are maintained through international cooperation under the umbrella of the International Earth Rotation Service (IERS) (section 1.6). The IERS is also responsible for maintaining continuity with earlier data collected by optical instruments. Global GPS activities are coordinated by the International GPS Service for Geodynamics (IGS) (section 1.7), in collaboration with the IERS. These international efforts are under the umbrella of the

International Association of Geodesy (IAG) with important links to the International Astronomical Union (IAU).

1.2 TRANSFORMATION BETWEEN THE CELESTIAL AND TERRESTRIAL FRAMES

It is useful to consider the transformation of the position of a stellar object in the CRF into TRF coordinates as two rotations such that

$$\mathbf{r}_{TRF} = \mathbf{T}_{3x3}\,\mathbf{U}_{3x3}\,\mathbf{r}_{CRF} \tag{1.6}$$

where **U** includes the rotations of the Earth caused by external torques, and **T** the rotations to which the Earth would be subjected to if all external torques would be removed. This formulation is approximately realized in practice by the well known transformation consisting of 9 rotations

$$\begin{aligned}\mathbf{r}_{TRF} &= \mathbf{R}_2(-x_P)\,\mathbf{R}_1(-y_P)\,\mathbf{R}_3(GAST)\,\bullet \\ &\quad \bullet\,\mathbf{R}_1(-\varepsilon-\Delta\varepsilon)\,\mathbf{R}_3(-\Delta\psi)\,\mathbf{R}_1(\varepsilon)\,\mathbf{R}_3(-z_A)\mathbf{R}_2(\theta_A)\mathbf{R}_3(-\zeta_A)\,\mathbf{r}_{CRF} \qquad (1.7) \\ &= \mathbf{S}\,\mathbf{N}\,\mathbf{P}\,\mathbf{r}_{CRF}\end{aligned}$$

whose elements are described in the sections below such that

$$\mathbf{S} = \mathbf{R}_2(-x_P)\,\mathbf{R}_1(-y_P)\,\mathbf{R}_3(GAST) \cong \mathbf{T} \tag{1.8}$$

$$[\mathbf{N}][\mathbf{P}] = [\mathbf{R}_1(-\varepsilon-\Delta\varepsilon)\,\mathbf{R}_3(-\Delta\psi)\,\mathbf{R}_1(\varepsilon)][\mathbf{R}_3(-z_A)\mathbf{R}_2(\theta_A)\mathbf{R}_3(-\zeta_A)] \cong \mathbf{U} \tag{1.9}$$

$\mathbf{R}_i$ represents a single right-handed rotation about the i axis with a positive rotation being counterclockwise when looking toward the origin from the positive axis. The elements of $\mathbf{R}_i$ may be computed, see Kaula [1966] by $j = i$ (modulo 3) + 1; $k = j$ (modulo 3) + 1

$$r_{ii} = 1;\ r_{ij} = r_{ji} = r_{ik} = r_{ki} = 0$$

$r_{jj} = r_{kk} = \cos\alpha;\ r_{jk} = \sin\alpha;\ r_{kj} = -\sin\alpha$

For example for i=3, j=1, and k=2 so that

$$\mathbf{R}_3(\alpha) = \begin{bmatrix} \cos\alpha & \sin\alpha & 0 \\ -\sin\alpha & \cos\alpha & 0 \\ 0 & 0 & 1 \end{bmatrix}$$

$\mathbf{R}_i$ is an orthogonal matrix with the properties

$$\mathbf{R}_i^{-1} = \mathbf{R}_i^T;\ \mathbf{R}_i\mathbf{R}_i^T = \mathbf{R}_i^T\mathbf{R}_i = \mathbf{I}$$

where **I** is the (3x3) identity matrix.

The rotation matrix **S** (1.8), including the transformation for polar motion and Earth rotation, approximates **T**. The matrix product **NP** (1.9) of the rotation matrices for nutation (**N**) and precession (**P**) approximates **U**. The tidal response of the Earth prevents the separation (1.6) from being perfectly realized (see section 1.6.2); by definition, **S** also contains the tidally induced nearly-diurnal forced terms of polar motion and free nutation terms, or equivalently, **NP** does not contain the tidally induced free nutation terms.

The transformation (1.7), then, describes the rotation of position from the CRF to the TRF. We see now that (1.1) should be more properly expressed as

$$\phi_i^k(t) = \frac{f_0}{c}[\rho_i^k(t,t-\tau_i^k(t))] = \frac{f_0}{c}\left\|\mathbf{SNPr}^k(t-\tau_i^k(t)) - \mathbf{r}_i(t)\right\|_{\text{Terrestrial}} \equiv \frac{f_0}{c}\left\|\mathbf{r}^k(t-\tau_i^k(t)) - \mathbf{P}^T\mathbf{N}^T\mathbf{S}^T\mathbf{r}_i(t)\right\|_{\text{Inertial}} \quad (1.10)$$

The elements of this transformation include the EOP, and the conventional precession and nutation models.

It is only by convention that the net celestial motion of the Earth's pole of rotation is split into precession and nutation. It could just as well have been defined by three rotations (rather than six rotations). Although more complicated, the six rotation representation is still in use because it is geometrically and physically intuitive, it allows continuity with earlier astronomic measurements, and it is convenient for intercomparison of space geodetic techniques. For example, elements of the current adopted expressions for **P** and **N** are still being updated according to improved Earth models based on discrepancies detected by the analysis of space geodetic measurements. The EOP elements of **S** include the unpredictable part of the Earth's rotation and are determined empirically from space geodetic observations.

1.3 TIME SYSTEMS

Space geodesy essentially measures travel times of extraterrestrial signals. Precise definition of time is therefore fundamental. Two aspects of time are required, the epoch and the interval. The epoch defines the moment of occurrence and the interval is the time elapsed between two epochs measured in units of some time scale. Two time systems are in use today, atomic time and dynamical time. GPS receivers tag phase and pseudorange measurements in atomic time (UTC or GPS time) which is the basis of modern civilian timekeeping. The equations of motion of the GPS satellites are expressed in dynamical time.

Prior to advent of atomic time, the civilian time system was based on the Earth's diurnal rotation and was termed universal (or sidereal) time. This is no longer the case although atomic time (UTC) is made to keep rough track of the Sun's motion for civil convenience. Nevertheless, it is necessary to retain the terminology of sidereal and universal "time" since the primary rotation angle between the CRF and the TRF is given as a sidereal angle (the Greenwich Apparent Sidereal Time — GAST). In addition, variations in the Earth's rotation are expressed as differences between universal time (UT1) and atomic time (UTC).

The interrelationships between atomic (TAI, UTC) and dynamic (TDT) *times* and the sidereal (GMST, GAST) and universal (UT1) *angles* can be visualized in this expression for GAST,

$$\begin{aligned}\mathrm{GAST} = \mathrm{GMST}_0 + \frac{d(\mathrm{GMST})}{dt}\,[&\mathrm{TDT} - (\mathrm{TDT} - \mathrm{TAI}) \\ &- (\mathrm{TAI} - \mathrm{UTC}) - (\mathrm{UTC} - \mathrm{UT1})] + \mathrm{Eq.\,E}\end{aligned} \tag{1.11}$$

where the individual terms are discussed below.

1.3.1 Atomic Time

Atomic time is the basis of a uniform time scale on the Earth and is kept by atomic clocks. The fundamental time scale is International Atomic Time (Temps Atomique International — TAI) based on atomic clocks operated by various national agencies. It is kept by the International Earth Rotation Service (IERS) and the Bureau International des Poids et Mesures (BIPM) in Paris who are responsible for the dissemination of standard time and EOP. TAI is a continuous time scale, related by definition to TDT by

$$\mathrm{TDT} = \mathrm{TAI} + 32.184 \text{ seconds} \tag{1.12}$$

Its point of origin was established to agree with universal time (UT) at midnight on 1 January, 1958[2].

The fundamental interval unit of TAI is one SI second. The SI second was defined at the 13th general conference of the International Committee of Weights and Measures in 1967, as the "duration of 9,192,631,770 periods of the radiation corresponding to the transition between the two hyperfine levels of the ground state of the cesium 133 atom." The SI day is defined as 86,400 seconds and the Julian century as 36,525 days. The time epoch denoted by the Julian date (JD) is expressed by a certain number of days and fraction of a day after a fundamental epoch sufficiently in the past to precede the historical record, chosen to be at 12^h UT on January 1, 4713 BCE. The Julian day number denotes a day in this continuous count, or the length of time that has elapsed at 12^h UT on the day designated since this epoch. The JD of the standard epoch of UT is called J2000.0 where

$$\mathrm{J2000.0} = \mathrm{JD}\ 2{,}451{,}545.0 = 2000 \text{ January } 1^d.5 \text{ UT} \tag{1.13}$$

[3]

All time arguments denoted by T are measured in Julian centuries relative to the epoch J2000.0 such that

$$T = (\mathrm{JD} - 2451545.0)/36525 \tag{1.14}$$

Because TAI is a continuous time scale, it does not maintain synchronization with the solar day (universal time) since the Earth's rotation rate is slowing. This

---

[2]At this date, universal and sidereal times (section 1.3.3) ceased effectively to function as time systems.

[3]The astronomic year commences at 0^h UT on December 31 of the previous year so that 2000 January $1^d.5$ UT = 2000 January 1 12^h UT. JD which is a large number is often replaced by the Modified Julian Date (MJD) where MJD = JD - 2,400,000.5, so that J2000.0 = MJD 51,444.5.

problem is solved by defining Universal Coordinated Time (UTC) which runs at the same rate as TAI but is incremented by leap seconds periodically[4].

The time signals broadcast by the GPS satellites are synchronized with the atomic clock at the GPS Master Control Station in Colorado. Global Positioning System Time (GPST) was set to 0^h UTC on 6 January 1980 but is not incremented by UTC leap seconds. Therefore, there is an offset of 19 seconds between GPST and TAI such that

$$\text{GPST} + 19 \text{ seconds} = \text{TAI} \tag{1.15}$$

At the time of this writing (June, 1997), there have been a total of 11 leap seconds since 6 January 1980 so that currently, GPST = UTC + 11 seconds.

1.3.2 Dynamical Time

Dynamical time is the independent variable in the equations of motion of bodies in a gravitational field, according to the theory of General Relativity. The most nearly inertial reference frame to which we have access through General Relativity is located at the solar system barycenter (center of mass). Dynamical time measured in this system is called Barycentric Dynamical Time (Temps Dynamique Barycentrique — TDB). An Earth based clock will exhibit periodic variations as large as 1.6 milliseconds with respect to TDB due to the motion of the Earth in the Sun's gravitational field. TDB is important in VLBI where Earth observatories record extragalactic radio signals. For describing the equations of motion of an Earth satellite, it is sufficient to use Terrestrial Dynamical Time (Temps Dynamique Terrestre — TDT) which represents a uniform time scale for motion in the Earth's gravity field. It has the same rate (by definition) as that of an atomic clock on Earth.

According to the latest conventions of the IAU, Kaplan [1981]

$$\text{TDB} = \text{TDT} + 0.^s 001658 \sin (g + 0.0167 \sin g) \tag{1.16}$$

where

$$g = (357^0.528 + 35999^0.050\, T)\left(\frac{\pi}{180^0}\right) \tag{1.17}$$

[4]Leap seconds are introduced by the IERS so that UTC does not vary from UT1 (universal time) by more than 0.9s. First preference is given to the end of June and December, and second preference to the end of March and September. DUT1 is the difference UT1-UTC broadcast with time signals to a precision of ±0.1s.

and T is given by (1.14) in Julian centuries of TDB.

1.3.3 Sidereal and Universal Time

Prior to the advent of atomic clocks, the Earth's diurnal rotation was used to measure time. Two interchangeable time systems were employed, sidereal and universal time (not to be confused with UTC which is atomic time). Their practical importance today is not as time systems (they are too irregular compared to atomic time) but as an angular measurement used in the transformation between the celestial and terrestrial reference frames.

The local hour angle (the angle between the observer's local meridian and the point on the celestial sphere) of the "true" vernal equinox (corrected for precession and nutation) is called the apparent sidereal time (AST). When the hour angle is referred to the Greenwich "mean" astronomic meridian, it is called Greenwich apparent sidereal time (GAST). Similarly, MST and GMST refer to the "mean" vernal equinox (corrected only for precession). The equation of the equinoxes, due to nutation, relates AST and MST

$$\text{Eq. E} = \text{GAST} - \text{GMST} = \text{AST} - \text{MST} = \Delta\psi \cos(\varepsilon + \Delta\varepsilon) \qquad (1.18a)$$

which varies with short periods with a maximum value of about 1 second of arc. McCarthy [1996] recommends, starting on 1 January, 1997, the following expression

$$\text{Eq. E} = (\Delta\psi + \Delta_{\text{Pls}}) \cos\varepsilon + 0.''00264 \sin\Omega + 0.''000063 \sin 2\Omega + \Delta_{\text{Ppl}} \qquad (1.18b)$$

adding two terms which involve the longitude of the lunar node, accounting for the shift in the position of the Celestial Ephemeris Pole (see section 1.6.2) due to the changes of the precession constant, and where Δ_{Pls} and Δ_{Ppl} are shifts in luni-solar and planetary precession constants, respectively. See section 1.4.1 for definition of the nutation terms on the right hand sides of (1.18a) and (1.18b).

Since the apparent revolution of the Sun about the Earth is non-uniform (this follows from Kepler's second law), a fictitious mean sun is defined which moves along the equator with uniform velocity. The hour angle of this fictitious sun is called universal time (UT). UT1 is universal time corrected for polar motion. It represents the true angular rotation of the Earth and is therefore useful as an angular measurement though no longer as a time system[5].

The conversion between sidereal and universal time is rigorously defined in terms of the IAU (1976) system of constants, Kaplan [1981]; Aoki et al. [1982]

---

[5]The instability of TAI is about six orders of magnitude smaller than that of UT1.

$$GMST = UT1 + 6^h\,41^m\,50^s.54841 + 8640184^s.812866\,T_u + 0^s.093104\,T_u^2 - 6^s.2 \times 10^{-6}\,T_u^3 \tag{1.19}$$

in fractions of a Julian century

$$T_u = \frac{(\text{Julian UT1 date} - 2451545.0)}{36525} \tag{1.20}$$

The first three terms in (1.19) are conventional and come from the historical relationship

$$UT = h_m + 12^h = GMST - \alpha_m + 12^h \tag{1.21}$$

where h_m and α_m are the hour angle and right ascension of the fictitious sun, respectively. The last term is an empirical one to account for irregular variations in the Earth's rotation. The relationship between the universal and sidereal time interval is given by[6]

$$\frac{d(GMST)}{dt} = 1.002737909350795 + 5.9006 \times 10^{-11}\,T_u - 5.9 \times 10^{-15}\,T_u^2 \tag{1.22}$$

such that

$$GMST = GMST_0 + \frac{d(GMST)}{dt}\,UT1 \tag{1.23}$$

In order to maintain a uniform civilian time system, national and international time services compute and distribute TAI-UTC (leap seconds) where UTC is made to keep rough track of the Sun's motion for civil convenience Thus, UT1-UTC establishes the relationship between atomic time and the universal (sidereal) angle, and describes the irregular variations of the Earth's rotation. These variations are best determined by analysis of VLBI data which provides long-term stability and the connection between the celestial and terrestrial reference systems. GPS now supplements VLBI by providing near real-time values.

[6]This relationship shows why the GPS satellites (with orbital periods of 12 hours) appear nearly 4 minutes earlier each day.

1.4 MOTION OF THE EARTH'S ROTATION AXIS

The Earth's rotation axis is not fixed with respect to inertial space, nor is it fixed with respect to its figure. As the positions of the Sun and Moon change relative to the Earth, the gradients of their gravitational forces, the tidal forces, change on the Earth. These can be predicted with high accuracy since the orbits and masses of these bodies are well known. The main motion of the rotation axis in inertial space is a precession primarily due to luni-solar attraction on the Earth's equatorial bulge. In addition, there are small motions of the rotation axis called nutation. The motion of the Earth's rotation axis with respect to its crust (in the terrestrial system) is called polar motion. Nutation and polar motion are due to both external torques (forced motion) and free motion. Nutation represents primarily the forced response of the Earth, see e.g. Gwinn et al. [1986]; Herring et al. [1986]; polar motion represents the forced and free response in almost equal parts. Currently, only the forced response of the nutation can be well predicted from available geophysical and orbital models, supplemented by space geodetic measurements (VLBI). The free response of nutation and polar motion can only be determined by space geodesy (by VLBI and increasingly by GPS). Knowledge of the motions of the Earth's rotation axis are essential for GPS positioning and are described in detail in this section.

1.4.1 Motion in Celestial System

The pole of rotation of the Earth is not fixed in space but rotates about the pole of the ecliptic. This motion is a composite of two components, precession and nutation, see e.g., Mueller [1971], and is primarily due to the torques exerted by the gravitational fields of the Moon and Sun on the Earth's equatorial bulge, tending to turn the equatorial plane into the plane of the ecliptic. Luni-solar precession is the slow circular motion of the celestial pole with a period of 25,800 years, and an amplitude equal to the obliquity of the ecliptic, about 23°.5, resulting in a westerly motion of the equinox on the equator of about 50".3 per year. Planetary precession consists of a slow (0°.5 per year) rotation of the ecliptic about a slowly moving axis of rotation resulting in an easterly motion of the equinox by about 12".5 per century and a decrease in the obliquity of the ecliptic by about 47" per century. The combined effect of luni-solar and planetary precession is termed general precession or just precession. Nutation is the relatively short periodic motion of the pole of rotation, superimposed on the precession, with oscillations of 1 day to 18.6 years (the main period), and a maximum amplitude of 9".2.

By convention, the celestial reference frame is defined by the 1976 IAU conventions (i.e., the precession model - see below) as a geocentric, equatorial frame with the mean equator and equinox of epoch J2000.0, see e.g., Lieske et al [1977]. This definition is supplemented by the 1980 nutation series which defines the transformation from the mean equinox and equator to the true or instantaneous equinox and equator. In 1996 the International Earth Rotation Service published a modified theory of precession/nutation, McCarthy [1996]. It is expected that the IAU will soon adopt these changes.

In practice the best approximation to a truly inertial reference frame is a reference frame defined kinematically by the celestial coordinates of a number of extragalactic radio sources observed by VLBI, assumed to have no net proper motion. Their mean coordinates (right ascensions and declinations) at epoch J2000.0 define the Celestial Reference Frame (CRF). GPS satellite computations are performed with respect to the CRF by adopting the models for precession and nutation. See section 1.6 for more details.

Precession Transformation. The transformation of stellar coordinates from the mean equator and equinox of date at epoch t_i to the mean equator and equinox at another epoch t_j is performed by the precession matrix composed of three successive rotations

$$\mathbf{P} = \mathbf{R}_3(-z_A)\mathbf{R}_2(\theta_A)\mathbf{R}_3(-\zeta_A) \tag{1.24}$$

The processional elements are defined by the IAU (1976) system of constants as

$$\begin{aligned}\zeta_A = {} & (2306''.2181 + 1''.39656T_u - 0''.000139T_u^2)t \\ & + (0''.30188 - 0''.000344T_u)t^2 + 0''.017998t^3\end{aligned} \tag{1.25}$$

$$\begin{aligned}z_A = {} & (2306''.2181 + 1''.39656T_u - 0''.000139T_u^2)t \\ & + (1''.09468 - 0''.000066T_u)t^2 + 0''.018203t^3\end{aligned} \tag{1.26}$$

$$\begin{aligned}\theta_A = {} & (2004''.3109 - 0''.85330T_u - 0''.000217T_u^2)t \\ & - (0''.42665 - 0''.000217T_u)t^2 - 0''.041833t^3\end{aligned} \tag{1.27}$$

where again

$$T_u = (JD - 2451545.0)/36525$$

and t is the interval in Julian centuries of TDB between epochs t_i and t_j.

The IERS 1996 theory of nutation/precession, see McCarthy [1996] includes corrections to precession rates of the equator in longitude (-2.99 mas/yr) and obliquity (-0.24 mas/yr) obtained from VLBI and LLR observations, Charlot et al. [1995]; Walter and Sovers [1996].

Nutation Transformation. The transformation of stellar positions at some epoch from the mean to the true equator and equinox of date is performed by multiplying the position vector by the nutation matrix composed of three successive rotations

$$\mathbf{N} = \mathbf{R}_1(-\varepsilon - \Delta\varepsilon)\,\mathbf{R}_3(-\Delta\psi)\,\mathbf{R}_1(\varepsilon) \tag{1.28}$$

where ε is the mean obliquity of date, $\Delta\varepsilon$ is the nutation in obliquity, and $\Delta\psi$ is the nutation in longitude.

1980 IAU Model. The 1980 IAU nutation model is used to compute the values for $\Delta\psi$ and $\Delta\varepsilon$. It is based on the nutation series derived from an Earth model with a liquid core and an elastic mantle developed by Wahr [1979]. The mean obliquity is given by

$$\begin{aligned}\varepsilon &= (84381''.448 - 46''.8150T_u + 0''.00059T_u^2 + 0''.001813T_u^3)\\ &+ (-46''.8150 - 0''.00177T_u + 0''.005439T_u^2)t\\ &+ (-0''.00059 + 0''.005439T_u)t^2 + 0''.00181t^3\end{aligned} \tag{1.29}$$

The nutation in longitude and in obliquity can be represented by a series expansion

$$\Delta\psi = \sum_{j=1}^{N} [(A_{0j} + A_{1j}T)\sin(\sum_{i=1}^{5} k_{ji}\alpha_i(T))] \tag{1.30}$$

$$\Delta\varepsilon = \sum_{j=1}^{N} [(B_{0j} + B_{1j}T)\cos(\sum_{i=1}^{5} k_{ji}\alpha_i(T))] \tag{1.31}$$

of the sines and cosines of linear combinations of five fundamental arguments of the motions of the Sun and Moon, Kaplan [1981]:

(1) the mean anomaly of the Moon

$$\alpha_1 = l = 485866''.733 + (1325^r + 715922''.633)\,T + 31''.310T^2 + 0''.064T^3 \tag{1.32}$$

(2) the mean anomaly of the Sun

$$\alpha_2 = l' = 1287009''.804 + (99^r + 1292581''.224)\,T - 0''.577T^2 - 0''.012T^3 \quad (1.33)$$

(3) the mean argument of latitude of the Moon

$$\alpha_3 = F = 335778''.877 + (1342^r + 295263''.137)\,T - 13''.257\,T^2 + 0''.011\,T^3 \quad (1.34)$$

(4) the mean elongation of the Moon from the Sun

$$\alpha_4 = D = 1072261''.307 + (1236^r + 1105601''.328)\,T - 6''.891\,T^2 + 0''.019\,T^3 \quad (1.35)$$

(5) the mean longitude of the ascending lunar node

$$\alpha_5 = \Omega = 450160''.280 - (5^r + 482890''.539)\,T + 7''.455\,T^2 + 0''.008\,T^3 \quad (1.36)$$

where $1^r = 360° = 1296000''$. The coefficients in (1.30–1.31) are given by the standard 1980 IAU series, see e.g., Sovers and Jacobs [1996], Appendix C, Table XXIV.

Improvements on the 1980 IAU Model. The 1980 IAU tabular values for $\Delta\psi$ and $\Delta\varepsilon$ have been improved by several investigators, including additional terms (free core nutations and out-of-phase nutations). Zhu and Groten [1989] and Zhu et al. [1990] have refined the IAU 1980 model by reexamining the underlying Earth model and by incorporating experimental results. Herring [1991] has extended the work of Zhu et al. and used geophysical parameters from Mathews et al. [1991] to generate the ZMOA 1990-2 nutation series (ZMOA is an acronym for Zhu, Mathews, Oceans and Anelasticity). The 1980 IAU lunisolar tidal arguments (1.32)–(1.36) have been modified based on Simon et al. [1994]. All numerical coefficients can be found in the software (KSV_1996_3.f) [McCarthy, 1996, electronic version]. See also Sovers and Jacobs, [1996], Appendix C, for a convenient compilation of nutation models.

1.4.2 Motion in Terrestrial System

If all external torques on the Earth were eliminated, its rotation axis would still vary with respect to its figure primarily due to its elastic properties and to exchange of angular momentum between the solid Earth, the oceans and the atmosphere. Polar motion is the rotation of the true celestial pole as defined by the precession and nutation models presented in section 1.4.1 with respect to the pole (Z-axis) of a conventionally selected terrestrial reference frame. Its free component (all external torques eliminated) has a somewhat counterclockwise circular motion with a main period of about 430 days (the Chandler period) and an amplitude of 3–6 m. Its forced component (due to tidal forces) is about an order of magnitude smaller, with nearly diurnal periods (hence, termed diurnal polar motion), whereas its forced annual component due to atmospheric excitation is nearly as large as the Chandler motion.

Polar motion is not adequately determined by the most sophisticated Earth models available today so as to have a negligible effect on space geodetic positioning. Therefore polar motion is determined empirically, i.e., by space geodetic measurements. Its observed accuracy today is 0.2–0.5 milliseconds of arc which is equivalent to 6–15 mm on the Earth's surface. Polar motion values are tabulated at one day intervals by the IERS based on VLBI, SLR, and GPS observations. The latter is playing an increasingly important role in this determination because of the expansion of the global GPS tracking network and the implementation of the full GPS satellite constellation (section 1.7).

Earth Orientation Transformation. The Earth orientation transformation was introduced in section 1.2 as a sequence of three rotations, one for Earth rotation and two for polar motion

$$\mathbf{S} = \mathbf{R}_2(-x_P)\,\mathbf{R}_1(-y_P)\,\mathbf{R}_3(\mathrm{GAST}) \qquad (1.8)$$

Earth Rotation Transformation. The transformation from the true vernal equinox of date to the zero meridian (X-axis) of the TRF, the 1903.0 Greenwich meridian of zero longitude[7], is given by a rotation about the instantaneous (true) rotation axis (actually the CEP axis — see section 1.62) such that

$$\mathbf{R}_3(\theta) = \begin{bmatrix} \cos\theta & \sin\theta & 0 \\ -\sin\theta & \cos\theta & 0 \\ 0 & 0 & 1 \end{bmatrix} ; \theta = \mathrm{GAST} \qquad (1.37)$$

where GAST is given by

$$\mathrm{GAST} = \mathrm{GMST}_0 + \frac{d(\mathrm{GMST})}{dt}\,[\mathrm{UTC} - (\mathrm{UTC} - \mathrm{UT1})] + \mathrm{Eq.\ E} \qquad (1.38)$$

[7]Referred to today as the IERS Reference Meridian (IRM).

such that GMST is given by (1.19), the time derivative of GMST is given by (1.22), UTC-UT1 is interpolated from IERS tables, and Eq. E is given by (1.18).

Polar Motion Transformation. The polar motion rotations complete the transformation between the CRF and TRF. Polar motion is defined in a left-handed sense by a pair of angles (x_P, y_P). The first is the angle between the mean direction of the pole during the period 1900.0–1906.0 (the mean pole[8] of 1903.0 — see section 1.6.3) and the true rotation axis. It is defined positive in the direction of the X-axis of the TRF. The second is the angle positive in the direction of the 270° meridian (the negative Y-axis). Recognizing that these angles are small, the polar motion transformation can be approximated by

$$\mathbf{R}_2(-x_P)\,\mathbf{R}_1(-y_P) \cong \begin{bmatrix} 1 & 0 & x_p \\ 0 & 1 & -y_p \\ -x_p & y_p & 1 \end{bmatrix} \tag{1.39}$$

where the two angles are interpolated from IERS tables at the epoch of observation.

Tidal Variations in Polar Motion and Earth Rotation. Tidal forces effect mass redistributions in the solid Earth, i.e., changes in the Earth's moment of inertia tensor. This causes changes in the Earth's rotation vector in order to conserve a constant angular momentum.

Solid Earth Tidal Effects on UT1. Yoder et al. [1981] computed the effects of solid Earth tides and some ocean effects on UT1, represented by

$$\Delta UT1 = \sum_{i=1}^{41} [A_i^R \sin(\sum_{j=1}^{5} k_{ij}\alpha_j)] \tag{1.40a}$$

and including all terms with periods from 5 to 35 days.

S. R. Dickman reconsidered the effects of the oceans and added 21 new terms with longer periods (90 days to 18.6 years) such that McCarthy [1996]

$$\Delta UT1 = \sum_{i=1}^{62} [A_i^S \sin(\sum_{j=1}^{5} k_{ij}\alpha_j) + B_i^S \cos(\sum_{j=1}^{5} k_{ij}\alpha_j)] \tag{1.40b}$$

The values and periods for the A_i's, B_i's and k_{ij}'s are tabulated by Sovers and Jacobs [1996, Appendix A, Table XIV], and a_j for (j=1,5) are the fundamental

[8]Referred to today as the IERS Reference Pole (IRF).

arguments for the nutation series (1.32)–(1.36). Yoder's expressions are referred to as UT1-UT1R and Dickman's expressions by UT1-UT1S, explaining the R and S superscripts in (1.40).

Ocean Tidal Effects. The dominant effects on polar motion and UT1 are diurnal, semidiurnal, fortnightly, monthly and semiannual. The ocean tidal effects can be written compactly as

$$\Delta\Theta_l = \sum_{i=1}^{N} \{A_{il} \cos[\sum_{j=1}^{5} k_{ij}\alpha_j + n_i(\theta + \pi)] + B_{il} \sin[\sum_{j=1}^{5} k_{ij}\alpha_j + n_i(\theta + \pi)]\} \quad (1.41)$$

for l = 1,2,3 (polar motion and UT1 respectively) and q = GMST (1.19). The cosine and sine amplitudes A and B can be calculated from theoretical tidal models, see Brosche et al. [1989; 1991]; Gross [1993]; Ray et al. [1994]; Seiler and Wunsch [1995] or from space geodetic data, Herring [1992]; Sovers et al. [1993]; Herring and Dong [1994]; Watkins and Eanes [1994]; Chao et al. [1996]. The argument coefficients (for i=1,8) for these models are tabulated for polar motion and UT1 by Sovers and Jacobs [1996; Appendix B, Tables XV–XXIII].

Estimation of Earth Orientation Parameters (EOP). In GPS analysis Earth orientation parameters (Chapter 14) are typically estimated as corrections to tabulated values of UTC-UT1, x_P and y_P, and their time derivatives. For example, we can model changes in UT1 by

$$\begin{aligned}(UTC - UT1)_{t_0} \\ = (UTC - UT1)_{tabulated(t_0)} + \Delta(UTC - UT1)_{t_0} + \frac{d(UTC - UT1)}{dt}(t - t_0)\end{aligned} \quad (1.42)$$

where the time derivative of UTC-UT1 is often expressed as changes in length of day ("lod"). See, for example, Ray [1996].

The effect of small errors in EOP (in radians) on a GPS baseline can be computed by

$$\begin{bmatrix} \delta X \\ \delta Y \\ \delta Z \end{bmatrix} = \begin{bmatrix} 0 & \delta\theta & -\delta x_P \\ \delta\theta & 0 & \delta y_P \\ \delta x_P & -\delta y_P & 0 \end{bmatrix} \begin{bmatrix} \Delta X \\ \Delta Y \\ \Delta Z \end{bmatrix}$$

1.5 EARTH DEFORMATION

1.5.1 Rotation vs. Deformation

The time derivative of the position vector for a station fixed to the Earth's surface is given by

$$\left[\frac{d\mathbf{r}}{dt}\right]_I = \left[\frac{d\mathbf{r}}{dt}\right]_T + \omega \times \mathbf{r} \tag{1.43}$$

or

$$\mathbf{v}_I = \mathbf{v}_T + \omega \times \mathbf{r} \tag{1.44}$$

where I and T indicate differentiation with respect to an inertial and terrestrial reference frame, respectively, and ω is the Earth's rotation vector. For a rigid Earth, $\mathbf{v}_T = \mathbf{0}$ since there are no changes in the station position vector $\mathbf{r}$ with respect to the terrestrial frame. For a deformable Earth, station positions are being displaced so that the rotation vector may be different for each station. However, deviations in the rotation vector are small and geophysicists have defined an instantaneous mean rotation vector such that

$$\iiint (\mathbf{v}_T \cdot \mathbf{v}_T)\rho \, dE = \text{minimum} \tag{1.45}$$

where the integration is taken over the entire Earth and ρ denotes density. This condition defines the Tisserand mean axes of body. If the integration is evaluated over the Earth's outer layer (the lithosphere) then this condition defines the Tisserand mean axes of crust. Since geodetic stations are only a few in number, the discrete analog of (1.45) is

$$\sum_{i=1}^{P} m_i \, (\mathbf{v}_{T_i} \cdot \mathbf{v}_{T_i}) = \text{minimum} \tag{1.46}$$

Now let us consider a polyhedron of geodetic stations with internal motion (deformation) and rotating in space with the Earth. Its angular momentum vector $\mathbf{H}$ is related to the torques $\mathbf{L}$ exerted on the Earth by Euler's equation

$$\mathbf{L} = \left[\frac{d\mathbf{H}}{dt}\right]_I = \left[\frac{d\mathbf{H}}{dt}\right]_T + \omega \times \mathbf{H} \tag{1.47}$$

The total angular momentum is given by

$$\mathbf{H} = \sum_{i=1}^{P} m_i\,(\mathbf{r}_i \times \mathbf{v}_i) \tag{1.48}$$

From (1.44)

$$\begin{aligned}\mathbf{H} &= \sum_{i=1}^{P} m_i\,[\mathbf{r}_i \times (\omega \times \mathbf{r}_i + \mathbf{v}_{T_i})] \\ &= \sum_{i=1}^{P} m_i\,[\mathbf{r}_i \times (\omega \times \mathbf{r}_i)] + \sum_{i=1}^{P} m_i\,[\mathbf{r}_i \times \mathbf{v}_{T_i}]\end{aligned} \tag{1.49}$$

$$= \mathbf{I} \cdot \omega + \mathbf{h} = \mathbf{H}_R + \mathbf{h} \tag{1.50}$$

where **I** is the inertia tensor. The angular momentum is given by a rigid body term $\mathbf{H}_R$ and a relative angular momentum vector **h**. Now returning to (1.46)

$$\begin{aligned}T &= \sum_{i=1}^{P} m_i\,(\mathbf{v}_{T_i} \cdot \mathbf{v}_{T_i}) \\ &= \sum_{i=1}^{P} m_i\,(\mathbf{v}_I - \omega \times \mathbf{r}_i) \cdot (\mathbf{v}_I - \omega \times \mathbf{r}_i)\end{aligned} \tag{1.51}$$

Minimizing T with respect to the three components of ω, i.e.,

$$\frac{\partial T}{\partial \omega_1} = \frac{\partial T}{\omega_2} = \frac{\partial T}{\partial \omega_3} = 0$$

yields in matrix form

$$\sum_{i=1}^{P} m_i \begin{bmatrix} Y^2+Z^2 & -XY & -XZ \\ -XY & X^2+Z^2 & -YZ \\ -XZ & -YZ & Y^2+X^2 \end{bmatrix}_i \begin{bmatrix} \omega_1 \\ \omega_2 \\ \omega_3 \end{bmatrix} = \sum_{i=1}^{P} m_i \begin{bmatrix} YV_Z - ZV_Y \\ XV_Z - ZV_X \\ YV_X - XV_Y \end{bmatrix}_i \tag{1.52}$$

or

$$\mathbf{I} \cdot \omega = \mathbf{H}_R \tag{1.53}$$

implying that $\mathbf{h} = \mathbf{0}$ from (1.50) or

$$\mathbf{h} = \sum_{i=1}^{P} m_i \begin{bmatrix} 0 & -Z & Y \\ Z & 0 & -X \\ -Y & X & 0 \end{bmatrix}_i \begin{bmatrix} V_X \\ V_Y \\ V_Z \end{bmatrix}_i = \mathbf{0} = \sum_{i=1}^{P} m_i \begin{bmatrix} 0 & -Z & Y \\ Z & 0 & -X \\ -Y & X & 0 \end{bmatrix}_i \begin{bmatrix} dX \\ dY \\ dZ \end{bmatrix}_i \tag{1.54}$$

in terms of differential station displacements (or deformation of the polyhedron with respect to the fundamental polyhedron — see section 1.1.2).

As pointed out by Munk and McDonald [1975] only the motions of the Tisserand axes are defined by the above constraints; the origin and orientation are arbitrary. Therefore the constraints (1.46) and equivalently (1.54) can be used to maintain the orientation of the fundamental polyhedron at a later epoch, i.e., a no net rotation constraint. An origin constraint can take the form of

$$\sum_{i=1} m_i \begin{bmatrix} 1 & 0 & 0 \\ 0 & 1 & 0 \\ 0 & 0 & 1 \end{bmatrix} \begin{bmatrix} dX \\ dY \\ dZ \end{bmatrix}_i = \mathbf{0} \tag{1.55}$$

where the mass elements in the equations above can be interpreted as station weights, e.g., Bock [1982].

1.5.2 Global Plate Motion

The NNR-NUVEL1 (NNR — no net rotation) plate tectonic model by Argus et al. [1994] describes the angular velocities of the 14 major tectonic plates defined by a no net rotation constraint. Fixing any plate (the Pacific Plate is usually chosen) to zero velocity will yield velocities in the NUVEL-1 relative plate motion model which are derived from paleomagnetic data, transform fault azimuths, and earthquake slip vectors, DeMets et al. [1990]. Note that a recent revision of the paleomagnetic time scale has led to a rescaling of the angular rates by a factor of 0.9562 defining the newer models NUVEL-1A and NNR-NUVEL-1A, see DeMets et al. [1994], which have been adopted by the IERS, McCarthy [1996].

The velocity of station i on plate j in the NNR (or other) frame is given on a spherical Earth as a function of spherical latitude, longitude and radius $(\phi, \lambda, R)_S$ by

$$\mathbf{v}_{ij} = \Omega_j \times \mathbf{r}_i \tag{1.56}$$

$$= R\,\omega_j \begin{bmatrix} \cos\phi_j \sin\phi_i \sin\lambda_j - \sin\phi_j \cos\phi_i \sin\lambda_i \\ \sin\phi_j \cos\phi_i \cos\lambda_i - \cos\phi_j \sin\phi_i \cos\lambda_j \\ \cos\phi_j \cos\phi_i \sin(\lambda_i - \lambda_j) \end{bmatrix} \tag{1.57}$$

where the angular velocity of plate j is

$$\Omega_j = \begin{bmatrix} \omega_x \\ \omega_y \\ \omega_z \end{bmatrix} = \omega_j \begin{bmatrix} \cos\phi_j \cos\lambda_j \\ \cos\phi_j \sin\lambda_j \\ \sin\phi_j \end{bmatrix} \tag{1.58}$$

with rate of rotation ω_j and pole of rotation (ϕ_j, λ_j), and

$$\mathbf{r}_i = \begin{bmatrix} X_i \\ Y_i \\ Z_i \end{bmatrix} = R \begin{bmatrix} \cos\phi_i \cos\lambda_i \\ \cos\phi_i \sin\lambda_i \\ \sin\phi_i \end{bmatrix} \tag{1.59}$$

is the geocentric coordinate vector of station i. Station coordinate corrections for global plate motion are then given by

$$\mathbf{r}_{ij}(t) = \mathbf{r}_{ij}(t_0) + (\Omega_j \times \mathbf{r}_i)(t - t_0) \tag{1.60}$$

1.5.3 Tidal Effects

The gravitational attractions of the Sun and Moon induce tidal deformations in the solid Earth, Melchior [1966]. The effect is that instantaneous station coordinates will vary periodically. The amplitude and period of these variations and the location of the station will determine the effect on station position. For GPS measurements, the penalty for ignoring tidal effects will generally be more severe as baseline length increases.

In principle, Earth tides models need to be defined as part of the definition of the terrestrial reference system. To first order, Earth tide deformation is given by the familiar solid Earth tides. Three other secondary tidal affects may need to be considered; ocean loading, atmospheric loading, and the pole tide. Their descriptions are extracted from the excellent summary of Sovers and Jacobs [1996].

Solid Earth Tides. The tidal potential for the phase-shifted station vector $\mathbf{r}_s$, due to a perturbing object at $\mathbf{R}_P$ is given by

$$\begin{aligned} U_{tidal} &= \frac{GM_P}{R_P}\left[\left(\frac{r_s}{R_P}\right)^2 P_2(\cos\theta) + \left(\frac{r_s}{R_P}\right)^3 P_3(\cos\theta)\right] \\ &= U_2 + U_3 \end{aligned} \tag{1.61}$$

where G is the universal gravitational constant, M_p is the mass of the perturbing object, P_2 and P_3 are the 2nd and 3rd degree Legendre polynomials, and θ is the angle between $\mathbf{r}_S$ and $\mathbf{R}_P$. To allow a phase shift ψ of the tidal effects from its nominal value of 0, the phase-shifted station vector is calculated by applying the lag (right-handed rotation) matrix $\mathbf{L}$ about the Z-axis of date

$$\mathbf{r}_s = \mathbf{L}\,\mathbf{r}_0 = \mathbf{R}_3(\psi)\,\mathbf{r}_0 \tag{1.62}$$

The tidal displacement vector on a spherical Earth expressed in a topocentric system is

$$\delta = \sum_i [g_1^{(i)}, g_2^{(i)}, g_3^{(i)}]^T \tag{1.63}$$

where $g_j^{(i)}(i = 2,3)$ are the quadrupole and octupole displacements. The components of δ are obtained from the tidal potential as

$$\delta_1^{(i)} = \frac{h_i U_i}{g} \tag{1.64}$$

$$\delta_2^{(i)} = \frac{l_i \cos\phi_s \left(\frac{\partial U_i}{\partial \lambda_s}\right)}{g} \tag{1.65}$$

$$\delta_3^{(i)} = \frac{l_i \left(\frac{\partial U_i}{\partial \phi_s}\right)}{g} \tag{1.66}$$

where h_i (i = 2,3) are the vertical (quadrupole and octupole) Love numbers, l_i (i = 2,3) are the corresponding horizontal Love numbers, and g is the gravity acceleration

$$g = \frac{GM_E}{r_s^2} \tag{1.67}$$

In this formulation, the Love numbers are independent of the frequency of the tide-generating potential. A more sophisticated treatment involves harmonic expansions of (1.65) and (1.66) and different vertical and horizontal Love numbers for each frequency. Currently the first six largest nearly diurnal components are allowed to have frequency-dependent Love numbers, see McCarthy [1992])[9].

There is a permanent deformation of the solid Earth due to the average gradient of the luni-solar attraction, given approximately in meters as a function the geodetic latitude (see section 1.6.6) by

$$\begin{aligned} \Delta W &= -0.12083\left(\tfrac{3}{2}\sin^2\phi - \tfrac{1}{2}\right) \\ \Delta U &= -0.05071 \cos\phi \sin\phi \end{aligned} \tag{1.68}$$

in the up and north directions, respectively.

Ocean Loading. Ocean loading is the elastic response of the Earth's crust to ocean tides. Displacements can reach tens of millimeters for stations near continental shelves. The model of Scherneck [1983, 1991] includes vertical and horizontal displacements. All eleven tidal components have been adopted for the IERS standards, McCarthy [1992]. Corrections for ocean tide displacements take the form of

$$\delta_j = \sum_{i=1}^{N} \xi_i^j \cos(\omega_i t + V_i - \delta_i^j) \tag{1.69}$$

where ω_i is the frequency of tidal constituent i, V_i is the astronomical argument, and ξ_i^j and δ_i^j are the amplitude and phase lag of each tidal component j determined from a particular ocean loading model. The first two quantities can be computed from the Goad algorithm, Goad [1983]. The eleven tidal components include K_2, S_2, M_2, N_2 (with about 12-hour periods); K_1, P_1, O_1, Q_1 (24 hour periods); M_f (14 day periods); M_m (monthly periods); S_{sa} (semiannual periods).

[9]For tidal computations the following physical constants have been recommended by the IERS Standards [1996]: h_2=0.609, l_2=0.0852, h_3=0.292, l_3=0.0151; GM_E = 3986004.356 x 10^8 m^3/s^2 (Earth); GM_S = 1.32712440 x 10^{20} m^3/s^2 (Sun); M_E/M_M = 81.300585 (Earth/Moon mass ratio)

Atmospheric Loading. Atmospheric loading is the elastic response of the Earth's crust to a time-varying atmospheric pressure distribution. Recent studies have shown that this effect can have a magnitude of several mm in vertical station displacement. Unlike the case of ocean loading, however, it does not have a well-understood periodic driving force. A simplified model proposed by Rabbel and Schuh [1986] requires a knowledge of the instantaneous pressure at the site and an average pressure over a circular region of radius R=2000 km surrounding the site. The expression for vertical displacement (in mm) is

$$\Delta W = -0.35(p - p_{STD}) - 0.55\bar{p};\ p_{STD} = 1013.25\, e^{-(h/8.567)}\ (\text{mbar}) \tag{1.70}$$

where p is the local pressure reading, p_{STD} is the standard pressure of 1013.25 mbar, $\bar{p}$ is the pressure anomaly within the 2000 km region, and h is the ellipsoidal height of the site (Eq. 1.75). The reference point is the site location at its standard (sea level) pressure.

Empirical estimates of atmospheric loading coefficients have been determined by Manabe et al. [1991], Van Dam and Herring [1994], MacMillan and Gipson [1994], and Van Dam et al. [1994].

Pole Tide. The pole tide is the elastic response of the Earth's crust to shifts in the pole of rotation. An expression for pole tide displacement in terms of unit vectors in the direction of geocentric spherical latitude, longitude and radius $(\phi, \lambda, R)_S$ is given by Wahr [1985]

$$\begin{aligned}\delta = -\frac{\omega^2 R}{g}[\, & \sin\phi\cos\phi\,(x_p\cos\lambda + y_p\sin\lambda)\, h_2\, \hat{R} \\ & + \cos 2\phi\,(x_p\cos\lambda + y_p\sin\lambda)\, l_2\, \hat{\phi} \\ & + \sin\phi\,(-x_p\sin\lambda + y_p\cos\lambda)\, l_2\, \hat{\lambda}\,]\end{aligned} \tag{1.71}$$

where ω is the rotation rate of the Earth, (x_p, y_p) represent displacements from the mean pole, g is the surface acceleration due to gravity, and *h* and *l* are the vertical and horizontal quadrupole Love numbers[10]. Considering that the polar motion components are on the order of 10 m or less, the maximum displacement is 10–20 mm.

---

[10] IERS (1996) Standards include R=6378.1363 km, w = 7.2921151467 x 10^{-5}rad/s, g=9.80665 m/s^2.

1.5.4 Regional and Local Effects

Other significant deformation of the Earth's crust are caused by a variety of regional and local phenomena, including:
(1) diffuse tectonic plate boundary (interseismic) deformation, with magnitudes up to 100–150 mm/yr, Genrich et al. [1996];
(2) coseismic and postseismic deformation with magnitudes up to several meters, and several mm/day, respectively, for major earthquakes, see e.g., Wdowinski et al. [1997];
(3) postglacial rebound (mm/yr level in the vertical) in the higher latitudes, e.g., Mitrovca et al. [1993]; Peltier [1995]; Argus [1996];
(4) monument instability due to varying local conditions, Wyatt [1982, 1989;] Langbein et al. [1995]; Langbein and Johnson [1997].

1.5.5 Non-Physical Effects

Site survey errors are not due to deformation *per se* but contribute nevertheless to station position error. For example, a GPS antenna may be displaced from its surveyed location, not oriented properly, and have its height above the monument erroneously recorded or a tie error may be made when surveying the offset between a VLBI reference point and a GPS reference point. Surprisingly, site survey errors of the latter type are one of the largest error sources remaining today in defining the terrestrial reference frame from a combination of space geodetic techniques.

A similar error is due to differing phase center characteristics between unlike (and like) GPS geodetic antennas. In general, for highest precision, referencing the phase center to the monument position requires careful antenna calibration, see e.g., Schupler et al. [1994]; Elosegui et al. [1995]. Switching antennas at a particular site may result in an apparent change of position (primarily in the vertical, but horizontal offsets are also a possibility, Bock et al. [1997]).

1.6 CONVENTIONAL REFERENCE SYSTEMS

1.6.1 International Earth Rotation Service (IERS)

Present day reference systems are maintained through international cooperation by the International Earth Rotation Service (IERS)[11] under the umbrella of the

[11] IERS information is provided through Internet from the IERS Central Bureau located at the Paris Observatory [E-mail: iers@obspm.fr] and the IERS Sub-Bureau for Rapid Service and Predictions located at the U.S. Naval Observatory, Washington, D.C. [E-mail: eop@ usno01.usno.navy.mil; anonymous ftp: maia.usno.navy.mil or 192.5.41.22; NEOS Bulletin Board (202 653 0597)].

International Association of Geodesy (IAG) and with links to the International Astronomical Union (IAU) [IERS, 1995]. There are IERS Analysis Centers for each of the different space geodetic methods including VLBI, SLR, LLR (lunar laser ranging), and GPS. The Central Bureau combines the results, disseminates information on the Earth's orientation, and maintains the IERS Celestial Reference Frame (ICRF) and the IERS Terrestrial Reference Frame (ITRF).

The IERS Reference System is composed of the IERS standards, see McCarthy [1989, 1992, 1996], the ICRF, and the ITRF. The IERS standards are a set of constants and models used by the analysis centers. The standards are based on the state of the art in space geodetic analysis and Earth models and may differ from the IAG and IAU adopted standards, e.g., precession and nutation parameters. The ICRF is realized by a catalogue of compact extragalactic radio sources, the ITRF by a catalogue of station coordinates and velocities.

1.6.2 Celestial Reference System

Definition. The small motions of the Earth's rotation axis can be described as the sum of two components: (1) astronomical nutation with respect to a celestial (inertial) coordinate system as described in section 1.4.1, and (2) polar motion with respect to a terrestrial reference system as described in section 1.4.2 . We indicated earlier that free polar motion is not adequately modeled analytically and must be determined from space geodetic measurements. Luni-solar effects can be predicted much better in both (free) nutation and (forced) polar motion, although improvements are also being made in these models (see 1.4.1). Therefore, it is reasonable to compute precession and nutation for the angular momentum axis whose small motions are not affected by nearly diurnal (forced) polar motion as viewed from the terrestrial frame, and by nearly diurnal (free) nutation as viewed from the inertial frame. This axis is called the Celestial Ephemeris Pole (CEP), i.e., the one defined by the theory of nutation and precession. It differs from the Earth's instantaneous rotation axis by quasi-diurnal terms with amplitudes under 0".01, Seidelmann [1982].

The IERS Celestial Reference System (ICRS) [Arias et al., 1995] is defined by convention to be coincident with the mean equator and equinox at 12 TDB on 1 January 2000 (Julian date 2451545.0, designated J2000.0), with origin at the solar system barycenter.

Realization. The ICRS is realized by the ICRF defined by a catalogue of adopted equatorial coordinates (right ascensions and declinations) of more than 200 compact extragalactic radio sources at epoch J2000.0 [IERS, 1995], computed to have no net proper motion. The ICRF catalogue implicitly define the direction of the frame axes. Coordinates of radio sources are computed annually by several IERS Analysis Centers and independent VLBI groups. The IERS produces weighted coordinates with increasing precision as more data are available, while maintaining the initial definition of the axes to within ±0.0001 arcseconds, McCarthy [1996]. The origin is realized by modeling observations in the framework of General Relativity. The transformation from the CRF to the true of date frame (with third axis in the direction of the CEP) is given by the precession and nutation transformations as described in section 1.4.1.

1.6.3 Terrestrial Reference System

Definition. The Celestial Ephemeris Pole also moves with respect to the Earth itself. The IERS Terrestrial Reference System (ITRS) is defined with origin at the Earth's geocenter (center of mass including oceans and atmosphere), and pole at the 1903.0 Conventional International Origin (CIO) frame adopted by the IAU and IAG in 1967. The X-axis is oriented towards the 1903.0 meridian of Greenwich (called the IERS Reference Meridian - IRM), the Z axis is towards the CIO pole (called the IERS Reference Pole — IRP), and the Y-axis forms a right-handed coordinate system. The CIO pole is the mean direction of the pole determined by measurements of the five International Latitude Service (ILS) stations during the period 1900.0 to 1906.0. Although this definition is somewhat cumbersome it helps to preserve continuity with the long record of optical polar motion determinations which began formally in 1899 with the establishment of the ILS. The scale of the ITRS is defined in a geocentric frame, according to the relativistic theory of gravitation. Its orientation is constrained to have no residual global rotation with respect to the Earth's crust (see section 1.5.1).

Realization. The ITRS is realized by the ITRF, which is defined by the adopted geocentric[12] Cartesian coordinates and velocities of global tracking stations derived from the analysis of VLBI, SLR, and GPS data. The ITRF coordinates implicitly define the axes of the frame (orientation and origin). The unit of length is the SI meter. Coordinates are given in a frame in which all tidal effects are removed, McCarthy [1996]. The latest in a series of annual ITRF frames is ITRF94, Boucher et al. [1996]. Its definition includes all IERS data collected through 1994. Also included are station velocities computed by the IERS from a combination of the adopted NNR-NUVEL1-A model (section 1.5.2) and long-term space geodetic measurements. Annual refinements of the ITRF are to be expected at up to the 10 mm level in position and several mm/yr in velocity, with a gradual increase in the number of defining stations (mainly GPS).

1.6.4 Transformation between ICRF and ITRF

The IERS Earth orientation parameters provide the tie between the ICRF and the ITRF, in conjunction with the precession and nutation models. The EOP describe the orientation of the CEP in the terrestrial and celestial systems (pole coordinates x_p, y_p; nutation offsets $d\psi, d\varepsilon$, and the orientation of the Earth around this axis (UT1-UTC or UT1-TAI), as a function of time, McCarthy [1996]. The pole

[12]The origin is located at the Earth's center of mass (±5 cm).

coordinates are the displacements of the CEP relative to the IRP. UT1-UTC (see section 1.3.3) provides access to the direction of the IRM in the ICRF, reckoned around the CEP axis.

Two IERS bulletins [e.g., IERS, 1994] provide Earth orientation information in the IERS Reference System, including UT1-UTC, polar motion, and celestial pole offsets. Bulletin A gives advanced daily solutions and is issued weekly by the Sub-Bureau for Rapid Service and Predictions. Bulletin B gives the standard solution and is issued at the beginning of each month by the Central Bureau. An Annual Report is issued six months after the end of each year. It includes the technical details of how the products are determined, and revised solutions for earlier years. The IERS is also responsible for maintaining continuity with earlier data collected by optical instruments[13]. Long term homogeneous series including polar motion (from 1846), UT1 (from 1962) and nutation parameters (from 1981) are also available.

1.6.5 WGS 84

The terrestrial reference system used by the U.S. Department of Defense (DoD) for GPS positioning is the World Geodetic System 1984 (WGS 84). The GPS navigation message includes Earth-fixed satellite ephemerides expressed in this system. WGS 84 is a global geocentric coordinate system defined originally by DoD based on Doppler observations of the TRANSIT satellite system (the predecessor of GPS). WGS 84 was first determined by aligning as closely as possible, using a similarity transformation (see section 1.6.7), the DoD reference frame NSWC-9Z2, and the Bureau International de l'Heure (BIH) Conventional Terrestrial System (BTS) at the epoch 1984.0 (BIH is the predecessor of the IERS, and BTS is the predecessor of ITRF). It was realized by the adopted coordinates of a globally distributed set of tracking stations with an estimated accuracy of 1–2 meters (compare to the 10–20 mm accuracy of ITRF). In January 1987, the U.S. Defense Mapping Agency (DMA) began using WGS 84 in their computation of precise ephemerides for the TRANSIT satellites. These ephemerides were used to point position using Doppler tracking the coordinates of the ten DoD GPS monitoring stations. GPS tracking data from these stations were used until recently to generate the GPS broadcast orbits, fixing the Doppler derived coordinates (tectonic plate motions were ignored). Abusali et al. [1995] computed the similarity transformation between SLR/VLBI and WGS 84 reference frames.

In an attempt to align WGS 84 with the more accurate ITRF, the DoD has recoordinated the ten GPS tracking stations at the epoch 1994.0 using GPS data collected at these stations and a subset of the IGS tracking stations whose ITRF91 coordinates were held fixed in the process, Malys and Slater [1994]. This refined WGS 84 frame has been designated WGS 84 (G730). The 'G' is short for GPS derived, and '730' is the GPS week number when these modifications were

---

[13]The IERS Reference Pole (IRP) and Reference Meridian (IRM) are consistent with the earlier BIH Terrestrial System (BTS) (±0.005") and the Conventional International Origin (CIO) (±0.03").

implemented by DMA in their orbit processing (the first day of this week corresponds to 2 January 1994). In addition, the original WGS 84 GM value was replaced by the IERS 1992 standard value of 3986004.418 x 10^8 m^3/s^2 in order to remove a 1.3 m bias in DoD orbit fits. Swift [1994] and Malys and Slater [1994] estimate that the level of coincidence between ITRF91, ITRF92, and WGS 84 (G730) is now of the order of 0.1 m. The Air Force Space Command implemented the WGS 84 (G730) coordinates on 29 June, 1994, with plans to implement the new GM value as well.

The change in WGS 84 has resulted in a more precise ephemeris in the GPS broadcast message (approximately 1 part in 10^7).

1.6.6 Ellipsoidal and Local Frames

Although the geocentric Cartesian frame is conceptually simple, other frames are more convenient for making certain model corrections, in particular tidal corrections and site eccentricity computations.

Geodetic Coordinates $(\phi, \lambda, h)_G$. For an ellipsoid with semi-major axis a and eccentricity e the geocentric Cartesian coordinates can be computed in closed form from geodetic coordinates (geodetic latitude, geodetic longitude and height above the ellipsoid) by

$$\begin{aligned} X &= [N + h] \cos\phi \cos\lambda \\ Y &= [N + h] \cos\phi \sin\lambda \\ Z &= [N(1 - e^2) + h] \sin\phi \end{aligned} \tag{1.72}$$

where

$$N = \frac{a}{\sqrt{1 - e^2 \sin^2\phi}} \tag{1.73}$$

is the radius of curvature of the ellipsoid in the prime vertical. The reverse transformation can be computed by Heiskanen and Moritz [1967]

$$\tan\lambda = \frac{Y}{X} \tag{1.74}$$

and solving the following two equations iteratively for h and ϕ

$$h = \frac{p}{\cos \phi} - N \tag{1.75}$$

$$\tan \phi = \frac{Z}{p}(1 - e^2 \frac{N}{N+h})^{-1} \tag{1.76}$$

where

$$p = (X^2 + Y^2)^{1/2} \ (= (N + h) \cos \phi) \tag{1.77}$$

Topocentric coordinate frame (U,V,W). The conversion from (right handed) geocentric Cartesian coordinates to a left-handed topocentric system (U-axis positive towards north, V-axis positive to the east, and W-axis positive up along the ellipsoidal normal) by

$$\begin{bmatrix} U \\ V \\ W \end{bmatrix} = \mathbf{P}_2 \, \mathbf{R}_2(\phi - 90°) \, \mathbf{R}_3(\lambda - 180°) \begin{bmatrix} X \\ Y \\ Z \end{bmatrix} \tag{1.78}$$

$$\mathbf{P}_2 = \begin{bmatrix} 1 & 0 & 0 \\ 0 & -1 & 0 \\ 0 & 0 & 1 \end{bmatrix} \tag{1.79}$$

This transformation is useful for reducing GPS antenna height to geodetic mark, expressing baseline vectors in terms of horizontal and vertical components, and correcting for site eccentricities.

1.6.7 Similarity Transformation

A seven-parameter (three-translations, three rotations and scale) similarity transformation (sometimes referred to as a "Helmert transformation") is often used to relate two terrestrial reference frames

$$\mathbf{r}_2 = s\mathbf{R}\mathbf{r}_1 + \mathbf{t}_{12} \tag{1.80}$$

where

$$\mathbf{R} = \mathbf{R}_1(\varepsilon)\,\mathbf{R}_2(\psi)\,\mathbf{R}_3(\omega) \tag{1.81}$$

For infinitesimal rotations, (1.80) can be written as

$$\begin{bmatrix} X_2 \\ Y_2 \\ Z_2 \end{bmatrix} = (1+\Delta s)\begin{bmatrix} 1 & \omega & -\psi \\ -\omega & 1 & \varepsilon \\ \psi & -\varepsilon & 1 \end{bmatrix}\begin{bmatrix} X_1 \\ Y_1 \\ Z_1 \end{bmatrix} + \begin{bmatrix} \Delta X_{12} \\ \Delta Y_{12} \\ \Delta Z_{12} \end{bmatrix} \tag{1.82}$$

1.7 The IGS

The International GPS Service for Geodynamics (IGS) contributes essential data to the IERS Reference System, including precise geocentric Cartesian station positions and velocities (the global polyhedron) and Earth orientation parameters, Beutler and Brockmann [1993]. The IGS was established in 1993 by the International Association of Geodesy (IAG) to consolidate worldwide permanent GPS tracking networks under a single organization. Essentially two major global networks, the Cooperative International GPS Network (CIGNET) spearheaded by the U.S. National Oceanic and Atmospheric Administration (NOAA) and Fiducial Laboratories for an International Natural science Network (FLINN) led by the U.S. National Aeronautics and Space Administration (NASA), were merged with several continental-scale networks in North America, Western Europe and Australia, Chin [1989]; Minster et al. [1989, 1992]. A highly successful proof of concept and pilot phase was initiated in June 1992, and formal operations began in January 1994, Beutler et al. [1994].

The current operational and planned stations of the IGS network are shown in Figure 1.1 [IGS, 1995]. The IGS collects, distributes, analyzes, and archives GPS data of geodetic quality (dual frequency phase and pseudorange) from these stations, see e.g., Kouba [1993]. The data are exchanged and stored in the Receiver Independent Exchange Format (RINEX), Gurtner [1994]. The primary IGS products includes high-quality GPS orbits, satellite clock information, Earth orientation parameters, and ITRF station positions and velocities. The coordinate and EOP information are provided to the IERS. The IGS supports worldwide geodetic positioning with respect to the International Terrestrial Reference Frame. Approximate accuracies of IGS products are given in Table 1.2.

The organization of the IGS is shown in Figure 1.2, Zumberge et al. [1994]. It includes three Global Data Centers, five Operational or Regional Data Centers, eight Analysis Centers, an Analysis Center coordinator, a Central Bureau[14], and an International Governing Board. Currently more than 50 institutions and organizations contribute to the IGS.

Table 1.2: Approximate Accuracy of IGS Products

IGS Products	Accuracy
Polar Motion (Daily)	0.2–0.5 mas
UT1-UTC rate (Daily)	0.1–0.5 ms/day
Station Coordinates (Annual)[15]	3–10 mm
GPS Orbits	50–100 mm
GPS Clocks	0.5–5 nsec

[14]IGS information is provided through the Central Bureau located at the Jet Propulsion Laboratory in Pasadena, California, through Internet (E-mail: igscb@igscb.jpl.nasa.gov; anonymous ftp: igscb.jpl.nasa.gov, directory igscb), and the World Wide Web (http://igscb.jpl.nasa.gov).

[15]Station coordinate and full covariance information are computed on a weekly basis and distributed in the Software Independent Exchange Format (SINEX) format (see Chapter 12).

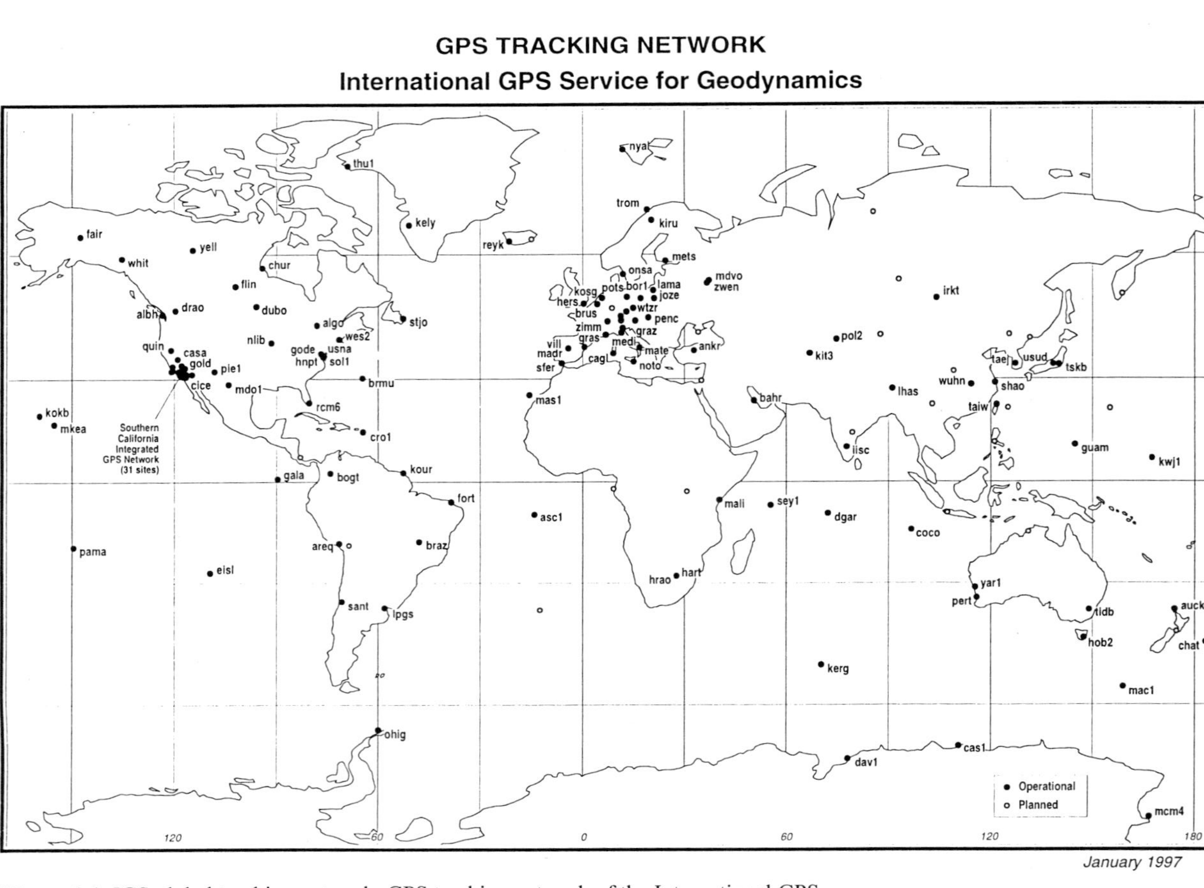

Figure 1.1. IGS global tracking network. GPS tracking network of the International GPS Service for Geodynamics: Operational and planned stations.

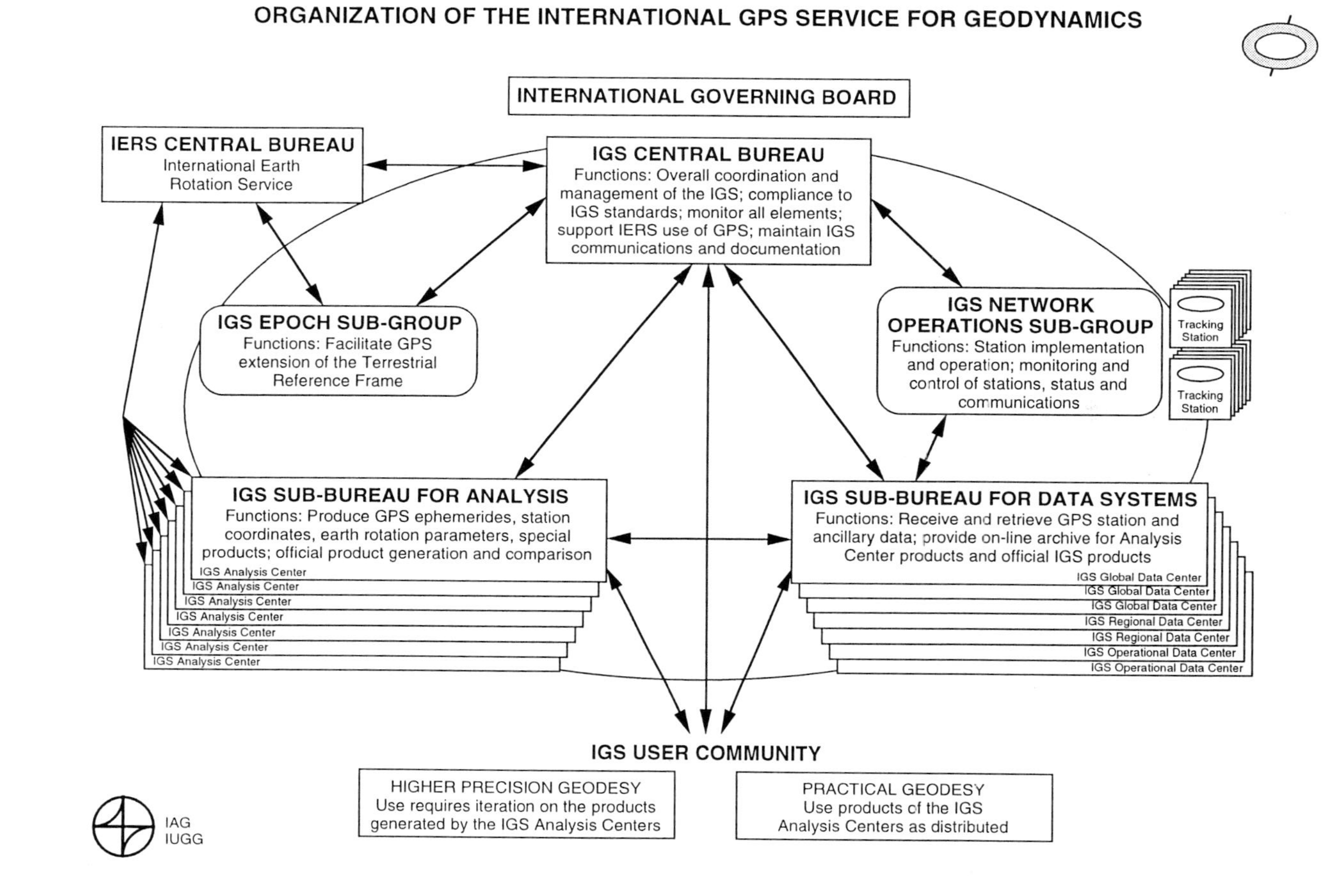

Figure 1.2. Organization of the International GPS Service for Geodynamics

1.8 SUMMARY

We have described the fundamental importance of terrestrial and celestial reference systems in GPS positioning. A reference system is realized through the definition of a reference frame at a fundamental epoch and all the physical models and constants that are used in the determination of coordinates at an arbitrary epoch in time. The celestial reference system is realized through a catalogue of coordinates of extragalactic radio sources. The right ascensions and declinations of these radio sources at epoch J2000.0 define the IERS celestial reference frame (ICRF). The terrestrial reference system is realized through the station coordinates of a global space geodetic tracking network defining the vertices of a deforming terrestrial polyhedron. The coordinates of these stations at a specified epoch define the IERS terrestrial reference frame (currently ITRF94), and the fundamental polyhedron.

The transformation from the ICRF to the ITRF includes a sequence of rotations including precession, nutation, Earth rotation and polar motion, as well as precise definitions of time systems. These are described in sections 1.2–1.4. Maintenance of the terrestrial reference system requires a knowledge of how the terrestrial polyhedron is deforming in time. The different phenomena that cause the Earth to deform are presented in section 1.5. The celestial and terrestrial reference systems in use today are described in section 1.6. In section 1.7, the International GPS Service for Geodynamics (IGS) is discussed.

Acknowledgments

I would like to thank Bob King for his critical comments on an earlier version of this chapter and for his support over the last 14 years, Ivan Mueller for his training and inspiration, Shimon Wdowinski for providing a comfortable writing environment in Tel Aviv, and Peter Teunissen and Alfred Kleusberg for organizing the GPS school in Delft. The assistance of Hans van der Marel, Paul de Jonge, and Frans Schroder in Delft is appreciated. Rob Liu and Jim Zumberge provided IGS material from the Central Bureau. Significant material for this chapter have been extracted from Sovers and Jacobs [1996], Moritz and Mueller [1987], and King et al. [1985] listed in the references. Supported by the U.S. National Science Foundation grant (EAR 92 08447) to the Scripps Orbit and Permanent Array Center (SOPAC).

References

Abusali, P. A. M., B. E. Schutz (1995), B. D. Tapley and M. Bevis, Transformation between SLR/VLBI and WGS-84 reference frames, *Bull. Géodesique*, *69*, 61-72.

Aoki, S., B. Guinot, G. H. Kaplan, H. Kinoshita, D. D. McCarthy and P. K. Seidelmann (1982), The new definition of Universal time, *Astron. Astrophys.*, *105*, 359-361.

Argus, D. F. and R. G. Gordon (1991), No-Net-Rotation model of current plate velocities incorporating plate rotation model NUVEL-1, *Geophys. Res. Lett.*, *18*, 2039-2042.

Argus, D. F. (1996), Postglacial rebound from VLBI geodesy: On establishing vertical reference, *Geophys. Res. Lett. 23*, 973-976.

Arias, E. F., P. Charlot, M. Feissel, J.-F. Lestrade (1995), The Extragalactic reference system of the International Earth Rotation Service, ICRS, *Astron. Astrophys.*, *303*, 604-608.

Beutler, G. and E. Brockmann (eds.) (1993), *Proceedings of the 1993 IGS Workshop*, International Association of Geodesy, Druckerei der Universität Bern.

Beutler, G., I. I. Mueller, and R. E. Neilan (1994), The International GPS Service for Geodynamics (IGS): Development and start of official service on January 1, 1994, *Bull. Géodesique* , *68*, 39-70.

Bock, Y. (1982), The use of baseline measurements and geophysical models for the estimation of crustal deformations and the terrestrial reference system, *Department of Geodetic Science and Surveying Report No. 337*, The Ohio State University.

Bock, Y., et al. (1997), Southern California Permanent GPS Geodetic Array: Continuous measurements of regional crustal deformation between the 1992 Landers and 1994 Northridge earthquakes, J. Geophys. Res., in press.

Boucher, C., Z. Altamimi, M. Feissel, and P. Sillard (1996), Results and Analysis of the ITRF94, *IERS Technical Note 20*, Observatoire de Paris.

Brosche, P., U. Seiler, J. Sundermann, and J. Wünsch (1989), Periodic changes in Earth's rotation due to oceanic tides, *Astron. Astrophys.*, *220*, 318-320.

Brosche, P., J. Wünsch, J. Campbell, and H. Schuh (1991), Ocean tide effects in universal time detected by VLBI, *Astron. Astrophys.*, *245*, 676-682.

Chao, B. F., R. D. Ray, J. M. Gipson, G. D. Egbert, and C. Ma (1996), Diurnal/semidiurnal polar motion excited by oceanic tidal angular momentum, *J. Geophys. Res.*, *101*, 20,151-20,163.

Charlot, P., O. J. Sovers, J. G. Williams, and X. X. Newhall (1995), Precession and nutation from joint analysis of radio interferometric and lunar laser ranging observations, *Astron. J.*, *109*, 418-427.

Chin, M. (Ed.) (1989), *GPS Bulletin 2*, CSTG Subcommission., National Geodetic Survey, Silver Spring, Md.

DeMets, C., R. G. Gordon, D. Argus and S. Stein (1990), Current Plate Motions, *Geophys. J. Int.*, *101*, 425-478.

DeMets, C., R. G. Gordon, D. Argus and S. Stein (1994), Effects of revisions to the geomagnetic reversal time scale on estimates of current plate motions, *Geophys. Res. Lett.*, *21*, 2191-2194.

Elosegui, P., J. L. Davis, R. T. K. Jaldehag, J. M. Johansson, A. E. Niell and I. I. Shapiro (1995), Geodesy using the Global Positioning System: The effects of signal scattering on estimates of site position, *J. Geophys. Res.*, *100*, 9921-9934.

Genrich, J. F., Y. Bock, R. McCaffrey, E. Calais, C. W. Stevens, and C. Subarya (1996), Accretion of the southern Banda arc to the Australian plate margin determined by Global Positioning System measurements, *Tectonics*, *15*, 288-295.

Goad, C.C. (1983), in IAU, IUGG Joint Working Group on the Rotation of the Earth, Project MERIT Standards, *U.S. Naval Observatory Circular No. 167*, A7-1 to A7-25, USNO, Washington, D.C.

Gross, R. S. (1993), The effect of ocean tides on the Earth's rotation as predicted by the results of an ocean tide model, *Geophys. Res. Lett.*, *20*, 293-296.

Gurtner, W. (1994), RINEX-The Receiver Independent Exchange Format, *GPS World*, *5*, July, 48-52.

Gwinn, C. R., T. A. Herring and I. I. Shapiro (1986), Geodesy by radio interferometry, Studies of the forced nutations of the Earth 2. Interpretation, *J. Geophys. Res.*, *91*, 4755-4765.

Heiskanen, W. A. and H. Moritz (1967), *Physical Geodesy*, W. H. Freeman and Company, San Francisco.

Herring, T. A., C. R. Gwinn and I. I. Shapiro (1986), Geodesy by radio interferometry: Studies of the forced nutations of the Earth, Part I: Data analysis, *J. Geophys. Res.*, *91*, 4745-4754.

Herring T. A. and D. N. Dong (1991), Current and future accuracy of Earth rotation measurements, Proceedings AGU Chapman Conference on geodetic VLBI: Monitoring global change, *NOAA Tech. Rept. NOS 137 NGS 49*, 306-324, Rockville, MD.

Herring, T. A. (1992), Modeling atmospheric delays in the analysis of space geodetic data, in Refraction of Transatmospheric Signals in Geodesy, J. C. DeMunck and T. A. Th Spoelstra, *Netherlands Geodetic Commission*, Delft, The Netherlands.

Herring, T. A. (1993), Diurnal and semidiurnal variations in Earth rotation, in The orientation of the planet Earth as observed by modern space techniques, *Advances in Space Research*, Pergamon Press, New York, 147-156.

Herring T. A. and D. N. Dong (1994), Measurement of diurnal and semidiurnal rotational variations and tidal parameters, *J. Geophys. Res.*, *99*, 18,051-18,071.

International GPS Service for Geodynamics (1995), *Resource Information*, Int. Assoc. of Geodesy, May.

International Earth Rotation Service (1994), *Explanatory Supplement to IERS Bulletins A and B*, IERS, March.

International Earth Rotation Service (1995), *IERS: Missions and Goals for 2000*, Bureau Central de l'IERS, Paris, May.

International Earth Rotation Service (1995), *1994 Annual Report*, Observatoire de Paris, July.

Kaula, W. M. (1966), *Theory of Satellite Geodesy*, Blaisdell Publishing Company.

Kaplan, G. H. (1981), The IAU resolutions of astronomical constants, time scales, and the fundamental reference frame, *United States Naval Observatory Circular No. 163*, U.S. Naval Observatory, Washington, D.C.

King, R. W., E.G. Masters, C. Rizos, A. Stolz and J. Collins (1985), *Surveying with GPS*, School of Surveying Monograph No. 9, The University of New South Wales, Australia.

Kouba, J., (ed.) (1993), Proceedings of the IGS Analysis Center Workshop, Oct. 12-14, Ottawa, Canada.

Lambeck, K. (1988), *Geophysical Geodesy*, Clarendon Press, Oxford.

Langbein, J. O., F. Wyatt, H. Johnson, D. Hamann, and P. Zimmer (1995), Improved stability of a deeply anchored geodetic monument for deformation monitoring, *Geophys. Res. Lett.*, *22*, 3533-3536.

Langbein, J. O., and H. Johnson (1997), Correlated errors in geodetic time series: Implications for time-dependent deformation, *J. Geophys. Res.*, *102*, 591-604.

Leick A. (1990), *GPS Satellite Surveying*, John Wiley and Sons, New York.

Lieske, J. H., T. Lederle, W. Fricke and B. Morando (1977), Expressions for the precession quantities based upon the IAU (1976) system of astronomical constants, *Astron. Astrophys.*, *58*, 1-16.

MacMillan, D. S., and J. M. Gipson (1994), Atmospheric pressure loading parameters from very long baseline interferometry observations, J. Geophys. Res., 99, 18,081-18,087.

Malys, S. and J. Slater (1994), Maintenance and enhancement of the World Geodetic System 1984, *J. Institute of Navigation*, *41*, 17-24.

Manabe, S., T. Sato, S. Sakai, and K. Yokoyama (1991), Atmospheric loading effect on VLBI observations, Proceedings of the AGU Chapman Conference on Geodetic VLBI: monitoring Global Change, 111-122, *NOAA Tech. Rep. NOS 137*, *NGS 49*.

Mathews, P.M., B. A. Buffett, T. A. Herring, and I. I. Shapiro (1991), Forced nutations of the Earth, Influence of inner core dynamics: 1. Theory, *J. Geophys. Res.*, *96B*, 8219-8242.

Mathews, P. M., B. A. Buffett, and I. I. Shapiro (1995), Love numbers for a rotating spheroidal Earth: New definitions and numerical values, *Geophys. Res. Lett.*, *22*, 579-582.

McCarthy, D. D. (1989), International Earth Rotation Service Standards, *IERS Technical Note 3*, Observatoire de Paris.

McCarthy, D. D. (1992), International Earth Rotation Service Standards, *IERS Technical Note 13*, Observatoire de Paris.

McCarthy, D. D. (1996), International Earth Rotation Service Conventions 1996, *IERS Tech. Note 21*, Observatoire de Paris.

Melbourne, W., R. Anderle, M. Feissel, R. King, D. McCarthy, D. Smith, B. Tapley and R. Vicente (1983), Project MERIT Standards, *U.S. Naval Observatory Circular No. 167*, A7-1 to A7-25, USNO, Washington, D.C.

Melchior, P. (1966), *The Earth Tides*, Pergamon Press, New York.

Minster, B., W. H. Prescott, L. Royden, Y. Bock, K. Kastens, M. McNutt, G. Peltzer, R. Reilinger, J., Rundle, J. Sauber, J. Scheid and M. Zuber (1989), *Report of the Plate Motion and Deformation Panel*, NASA Coolfont Workshop, August.

Minster, J. B., B. H. Hager, W. H. Prescott and R. E. Schutz (1991), *International global network of fiducial stations*, U.S. National Research Council Report, National Academy Press, Washington, D.C.

Mitrovica, J. X., J. L. Davis, and I. I. Shapiro (1993), Constraining proposed combinations of ice history and earth rheology using VLBI determined baseline rates in North America, *Geophys. Res. Lett.*, *20*, 2387-2390.

Moritz, H. and I. I. Mueller (1987), *Earth Rotation, Theory and Observation*, Ungar Publishing Company, New York.

Mueller, I. I. (1971), *Spherical and Practical Astronomy as Applied to Geodesy*, Ungar, New York.

Mueller, I. I., S. Y. Zhu, and Y. Bock (1982), Reference frame requirements and the MERIT campaign, *Department of Geodetic Science and Surveying Report No. 329*, The Ohio State University.

Mueller, I. and S. Zerbini (eds.) (1989), The interdisciplinary role of space geodesy, *Lecture Notes in Earth Sciences, Vol. 22*, Springer Verlag, Berlin.

Munk, W. H. and G. J. F. MacDonald (1975), *The Rotation of the Earth*, Cambridge Univ. Press, U.K.

Peltier, W. R. (1995), VLBI baseline variations from ICE-4G model of postglacial rebound, *Geophys. Res. Lett.*, *22*, 465-468.

Rabbel, W. and H. Schuh (1986), The influence of atmospheric loading on VLBI experiments, *J. Geophys.*, *59*, 164-170.

Ray, R. D., D. J. Steinberg, B. F. Chao, and D. E. Cartwright (1994), Diurnal and semidiurnal variations in the Earth's rotation rate induced by ocean tides, *Science*, *264*, 830-832.

Ray, J. R. (1996), Measurements of length of day using the Global Positioning System, *J. Geophys. Res.*, *101*, 20,141-20,149.

Scherneck, H. G. (1983), Crustal loading affecting VLBI sites, University of Uppsala, Institute of Geophysics, *Dept. of Geodesy Report No. 20*, Uppsala, Sweden.

Scherneck, H. G. (1991), A parameterised solid Earth tide model and ocean tide loading effects for global geodetic baseline measurements, *Geophys. J. Int.*, *106*, 677-694.

Schupler, B. R., R. L. Allshouse and T. A. Clark (1994), Signal characteristics of GPS user antennas, *J. Inst. Navigation*, *41*, 277-295.

Seidelmann, P. K. (1982), The 1980 theory of nutation: the final report of the IAU Working Group on Nutation, *Celestial Mechanics*, *27*, 79-106.

Seidelmann, P. K., (ed.)(1992), *Explanatory Supplement to the Astronomical Almanac*, University Science Books, Mill Valley, California.

Seiler, U., and J. Wunsch (1995), A refined model for the influence of ocean tides on UT1 and polar motion, *Astron. Nachr., 316*, 419-423.

Simon, J. L., P. Bretagnon, J. Chapront, M. Chapront-Touze, G. Francou, and J. Laskar (1994), Numerical expressions for precession formulae and mean elements for the Moon and Planets, *Astron. Astrophys. 282*, 663-683.

Sovers O. J., C. S. Jacobs and R. S. Gross (1993), Measuring rapid ocean tidal Earth orientation variations with VLBI, *J. Geophys. Res.*, *98*, 19,959-19,9971.

Sovers, O. J. and C. S. Jacobs (1996), Observation models and parameter partials for the JPL VLBI parameter estimation software "MODEST"—1996, *JPL Publication 83-89*, Rev. 6.

Swift, E. (1994), Improved WGS 84 coordinates for the DMA and Air Force GPS tracking sites, *J. Institute of Navigation*, *41*, 285-291.

Van Dam, T. M. and T. A. Herring (1994), Detection of atmospheric pressure loading using very long baseline interferometry, *J. Geophys. Res.*, *99*, 4505-4517.

Van Dam, T. M., G. Blewitt, and M. B. Heflin (1994), Atmospheric pressure loading effects on Global Positioning System coordinate determinations, *J. Geophys. Res.*, *99*, 23,939-23,950.

Wahr, J. M. (1979), The tidal motions of a rotating, elliptical, elastic and oceanless Earth, PhD thesis, Dept. of Physics, University of Colorado, Boulder.

Wahr, J. M., Deformation induced by polar motion, *J. Geophys. Res.*, *90*, 9363-9368, 1985.

Walter, H. and O. J. Sovers, Precession and nutation from the analysis of positions of extragalactic radio sources, *Astrophys. J., 308*, 1001-1008, 1996.

Watkins, M. M. and R. J. Eanes (1994), Diurnal and semidiurnal variations in Earth orientation determined from LAGEOS laser ranging, *J. Geophys. Res.*, *90*, 18,073-18,079.

Wdowinski, S., Y. Bock, J. Zhang, P. Fang, and J. Genrich (1997), Southern California Permanent GPS Geodetic Array: Spatial Filtering of Daily Positions for Estimating Coseismic and postseismic displacements Induced by the 1992 Landers earthquake, *J. Geophys. Res.*, 102, 18,057-18,070.

Wyatt, F. (1982), Displacements of surface monuments: horizontal motion, *J. Geophys. Res.*, *87*, 979-989.

Wyatt, F. (1989), Displacements of surface monuments: vertical motion, *J. Geophys. Res.*, *94*, 1655-1664.

Yoder, C.F., J. G. Williams and M. E. Parke (1981), Tidal variations of Earth rotation, *J. Geophys. Res.*, *86*, 881-891.

Zhang, J., Y. Bock, H. Johnson, P. Fang, J. Genrich, S. Williams, S. Wdowinski and J. Behr (1997), Southern California Permanent GPS Geodetic Array: Error analysis of daily position estimates and site velocities, *J. Geophys. Res*., 102, 18,035-18,055.

Zhu, S. Y. and E. Groten (1989), Various aspects of numerical determination of nutation constants. I. Improvement of rigid-Earth nutation, *Astron. J.*, *98*, 1104-1111.

Zhu, S. Y., E. Groten and C.. Reigber (1990), Various aspects of numerical determination of nutation constants. II. An improved nutation series for the deformable Earth, *Astron. J.*, *99*, 1024-1044.

Zumberge, J. F., R. E. Neilan, G. Beutler and W. Gurtner (1994), The International GPS Service for Geodynamics- benefits to users, Institute of Navigation, Proceedings ION GPS-94.

2. GPS SATELLITE ORBITS

G. Beutler, R. Weber, U. Hugentobler, M. Rothacher and A. Verdun
Astronomical Institute, University of Berne, Sidlerstrasse 5, CH - 3012 Berne, Switzerland

2.1 INTRODUCTON

Nominally the Global Positioning System (GPS) consists of 24 satellites (21 + 3 active spares). The satellites are in almost circular orbits approximately 20 000 km above the surface of the Earth. The siderial revolution period is almost precisely half a siderial day (11^h 58^m). All GPS satellites, therefore, are in deep 2:1 resonance with the rotation of the Earth with respect to inertial space. This particular characteristic gives rise to perturbations to be discussed in section 2.3.3. Thanks to this particular revolution period essentially the same satellite configuration is observed at a given point on the surface of the Earth at the same time of the day on consecutive days (the constellation repeats itself almost perfectly after 23^h 56^m UT).

The first GPS satellite, PRN 4, was launched on 22 February 1978. PRN 4 was the first in a series of 11 so-called Block I satellites. The orbital planes of the Block I satellites have an inclination of about 63 degrees with respect to the Earth's equator. The test configuration was optimized for the region of North America in the sense that four or more satellites could be observed for a considerable fraction of the day there. The test configuration was not optimal in other parts of the world.

In February 1989 the first of the Block II (or production) satellites was launched. The Block II satellites are arranged in six orbital planes (numbered A, B, C, D, E, and F), separated by about 60 degrees on the equator, and inclined by about 55 degrees with respect to the Earth's equator. Twenty-four Block II satellites are operational today. Figure 2.1 gives an overview of the arrangement of the satellites in the orbital planes, Figure 2.2 contains a drawing of a Block I, a Block II, and a Block IIR satellite (taken from Fliegel et al. [1992]). Figure 2.3 gives an impression of the orbital planes around the Earth in space as seen from a point in 35 degrees latitude, and as seen from the pole (North or South). The philosophy behind the 21+3 active spare satellites may be found in Green et al. [1989].

The present constellation allows for a simultaneous observation of at least four GPS satellites from (almost) every point on the surface of the Earth at (almost) every time of the day. Eight or more satellites may be observed at particular times and places. Figure 2.3 shows that the constellation is problematic in the Arctic

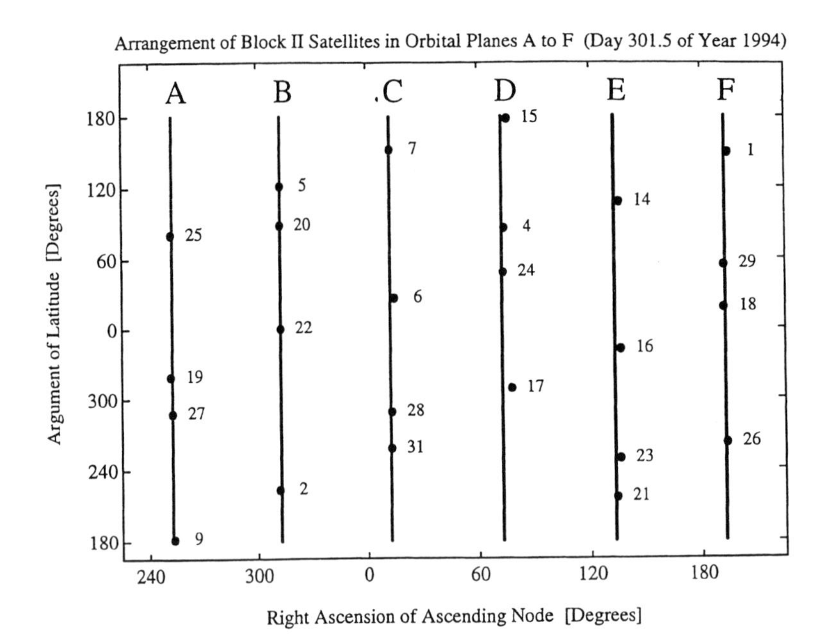

Figure 2.1. Arrangement of the GPS satellites in the orbital planes A-F.

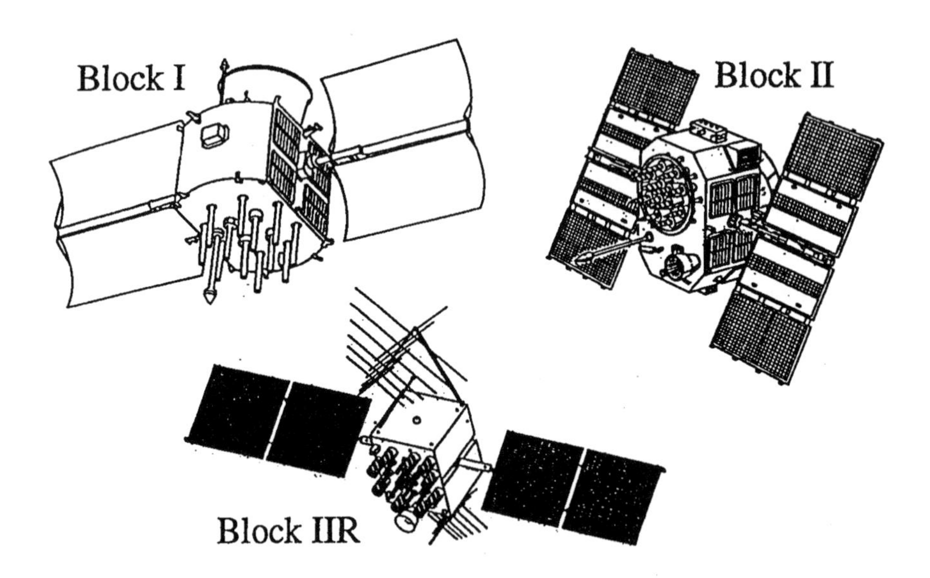

Figure 2.2. (a) Block I satellite, (b) Block II satellite, (c) Block II R.

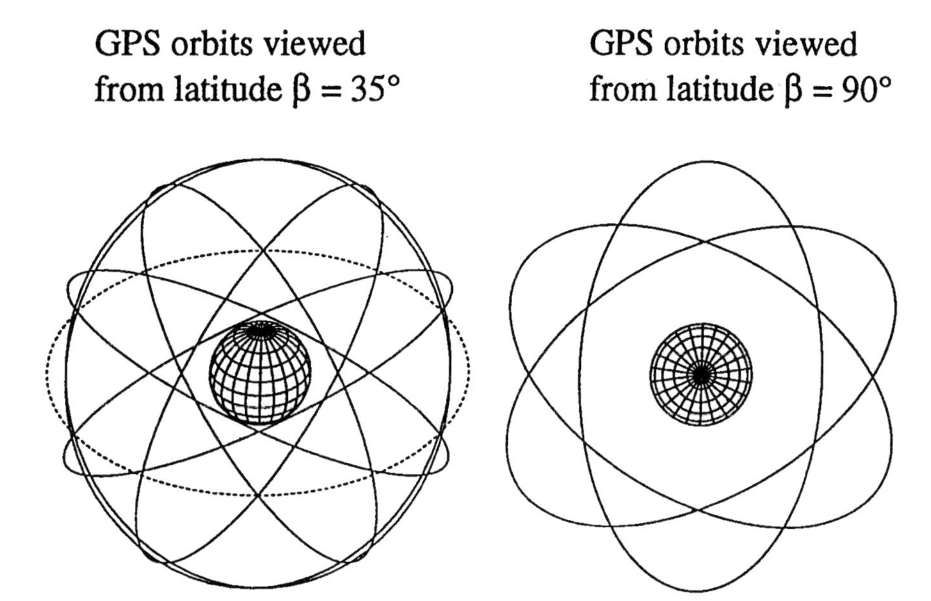

Figure 2.3. The GPS as seen from the outside of the system (Earth and orbital planes in scale).

regions: The maximum elevation for the satellites is 55 degrees only. In view of the fact that tropospheric refraction is roughly growing with $1/\cos(z)$, z = zenith distance, this may be considered as a disadvantage of the system. On the other hand, a receiver set up at the pole will be able to see simultaneously all six orbital planes which implies that a fair number of satellites will always be visible simultaneously at the poles!

Let give us an overview of the sections of Chapter 2: In section 2.2 we will present and discuss the equations of motion for an artificial Earth satellite. We will introduce the Keplerian elements as the solution of the two body (or one body) problem, and introduce the concept of osculating elements in the presence of perturbing forces. Subsequently we will present and discuss the so-called perturbation equations, first-order differential equations for the time development of the osculating elements. We will make the distinction between osculating and mean elements to get an overview of the long-term evolution of the GPS orbits. Most of the perturbing accelerations may be considered as known from earlier investigations in satellite geodesy. This is true in particular for the Earth's gravity field — with the possible exception of some resonance terms — and for the gravitational effects of Sun and Moon (including tidal variations). Due to the bulkiness of the satellites the same is not true for the radiation pressure. If highest orbital accuracy is aimed at, we have to solve for parameters of the radiation pressure acting on the satellites in addition to the initial conditions (position and velocity components at an initial epoch) with respect to the osculating elements at the same epoch. Thus, in general, each arc of a GPS satellite is described by more than six parameters. We have to define one possible set of such parameters. We will also briefly review numerical integration techniques as the general method to solve the so-called initial value problem in satellite geodesy.

In section 2.3 we will analyze the perturbing forces (with respect to accelerations) in the case of GPS satellites. We will in particular look at radiation pressure and at the resonance terms of the Earth's gravity field. The section will be concluded by studying the development of the GPS since mid-1992. This includes the detected manoeuvres of GPS satellites.

In section 2.4 we will present the two most commonly used types of orbits; namely, the broadcast and the IGS orbits. We will give some indication of the accuracies achieved and achievable today.

The chapter will be concluded by a summary (section 2.5) and by a bibliography for the topic covered here.

2.2 EQUATIONS OF MOTION FOR GPS

2.2.1 The Keplerian Elements

In 1609 Johannes Kepler published his first two laws of planetary motion in his fundamental work *Astronomia Nova,* Kepler [1609]. The third law was published ten years later in *Harmonices Mundi Libri V,* Kepler [1619].

The Keplerian Laws , Danby [1989]:

1. The orbit of each planet is an ellipse; with the Sun at one of the foci.
2. Each planet revolves so that the line joining it to the Sun sweeps out equal areas in equal intervals of time (*law of areas*).
3. The squares of the periods of any two planets are in the same proportion as the cubes of their mean distances to the Sun.

These laws — to a first order — are also valid for the revolution of (natural and) artificial Earth satellites around the Earth. We just have to replace the terms "Sun" resp. *Planet* by *Earth* resp. *Satellite* in the laws above. The parametrization of orbits is essentially still the same as that given by Kepler:

Keplerian Elements (for an artificial Earth satellite):

a : semi-major axis of the ellipse
e : numerical eccentricity (or just eccentricity)
i : inclination of the orbit with respect to the reference plane, the mean Earth's equatorial plane referring to a standard epoch
Ω : right ascension of the ascending node
ω : argument of perigee (angle between the perigee and the ascending node, measured in the orbital plane in the direction of motion)
T_0 : perigee passing time.

(2.1)

Figure 2.4 shows the Keplerian elements, which are very easy to understand. This probably is the reason for their popularity.

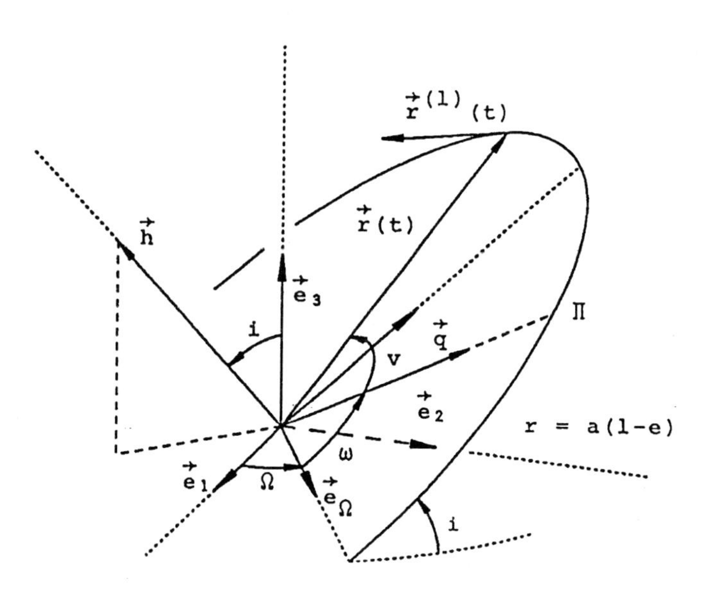

Figure 2.4. The Keplerian elements.

Kepler also solved the problem of computing the position of the celestial body at an arbitrary time t using the above set of elements. To be honest, he had to know *in addition* the revolution period U of the planet (the satellite in our case) to solve this problem. From this revolution period U he computed what he called the mean motion n. In radians we may write:

$$n = \frac{2 \cdot \pi}{U} \tag{2.2}$$

Obviously n is the mean angular velocity of the celestial body in its orbital plane around the Sun. In order to solve the problem of computing the position and velocity vector for any given point in time he introduced the so-called mean anomaly M and the eccentric anomaly E. M is a linear function of time, namely

$$M = n \cdot (t - T_0) \tag{2.3}$$

Often, not the perigee passing time T_0 but the mean anomaly

$$\sigma = M(t_0)$$

at an initial epoch t_0 is used as the sixth of the Keplerian elements. In this case the mean anomaly at time t is computed as:

$$M = \sigma + n \cdot (t - t_0) \tag{2.4}$$

The eccentric anomaly E is the angle (in the orbital plane) between the line of apsides (center of ellipse to perigee) and the line from the center of the ellipse to the projection P' (normal to the semi-major axis) of the satellite P on the circle of radius a around the ellipse. Figure 2.5 illustrates the situation. In the same figure we also find the true anomaly v. From Kepler's second law (by applying it to the time intervals (T_0, t) and $(T_0, T_0 + U)$ and by using Figure 2.5) it is easy to come up with *Kepler's Equation*:

$$E = M + e \cdot \sin E \tag{2.5}$$

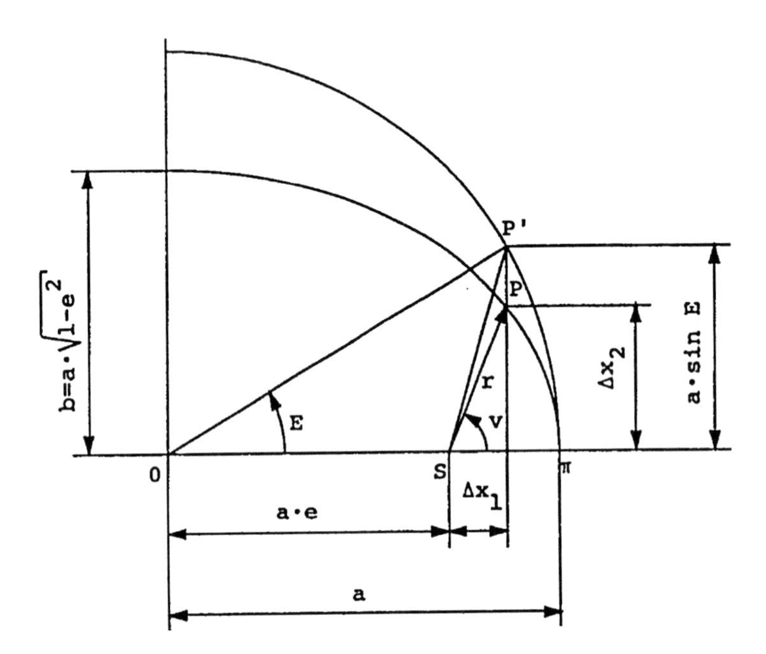

Figure 2.5. Eccentric and true anomalies E and v plus other useful relationships in the ellipse geometry.

This equation may be used to compute the eccentric anomaly E as a function of the mean anomaly M (and the orbital element e of course). Introducing a coordinate system with the orbital plane as reference plane and, with the line of apsides as first coordinate axis, we may compute the coordinates x, y, z of the satellite in this particular system:

$$\begin{aligned} x &= a \cdot (\cos E - e) \\ y &= a \cdot \sqrt{(1-e^2)} \cdot \sin E \\ z &= 0 \end{aligned} \tag{2.6}$$

From these equations we conclude that:

$$z = \sqrt{x^2 + y^2 + z^2} = a \cdot (1 - e \cdot \cos E) \tag{2.7}$$

Denoting by $R_i(w)$ the 3x3 matrix describing a rotation about angle w around axis i, we may compute the coordinates x', y', z' in the equatorial system as:

$$\begin{pmatrix} x' \\ y' \\ z' \end{pmatrix} = R_3(-\Omega) \cdot R_1(-i) \cdot R_3(-\omega) \cdot \begin{pmatrix} x \\ y \\ z \end{pmatrix} \tag{2.8}$$

The same type of transformation must be applied for the computation of the velocity components u', v', w' in the equatorial system as a function of the components u, v, w in the orbital system:

$$\begin{pmatrix} u' \\ v' \\ w' \end{pmatrix} = R_3(-\Omega) \cdot R_1(-i) \cdot R_3(-\omega) \cdot \begin{pmatrix} u \\ v \\ w \end{pmatrix} \tag{2.9}$$

where u, v, w are obtained by taking the first derivatives of eqns. (2.6) with respect to time t (using Kepler's equation):

$$\begin{aligned} u &= -a \cdot \sin E \cdot \dot{E} &&= -n \cdot \frac{a^2}{r} \cdot \sin E \\ v &= a \cdot \sqrt{(1-e^2)} \cdot \cos E \cdot \dot{E} &&= n \cdot \frac{a^2}{r} \cdot \sqrt{(1-e^2)} \cdot \cos E \\ w &= 0 &&= 0 \end{aligned} \tag{2.10}$$

where we have used that

$$\dot{E} = \frac{n}{(1 - e \cdot \cos E)} = n \cdot \frac{a}{r} \tag{2.11}$$

a result which is obtained by taking the first time derivation of Kepler's equation (2.5). We have thus given — in the Keplerian approximation — the algorithms to compute the rectangular coordinates of the position and the velocity vectors at any instant of time t using the Keplerian elements as input. We have thus shown that the position and the velocity vectors are a function of the Keplerian elements (and of time t).

2.2.2 Equations of Motion in Rectangular Coordinates

Sir Isaac Newton (1643-1727) published his *Philosophiae naturalis principia mathematica* in 1687, Newton [1687]. His well known *laws of motion*, but also his famous *law of universal gravitation* are written down in this outstanding book. Newton could show that Kepler's laws are a consequence of his more general laws of motion and the law of universal gravitation. He could also show that Kepler's laws are only valid if two (spherically symmetric) bodies are involved.

Newton's laws of motion, Danby [1989]:

1. Every particle continues in a state of rest or uniform motion in a straight line unless it is compelled by some external force to change that state.
2. The rate of change of the linear momentum of a particle is proportional to the force applied to the particle and takes place in the same direction as that force.
3. The mutual actions of any two bodies are always equal and oppositely directed.

It was Leonhard Euler (1707-1783) who for the first time transformed these laws into a modern mathematical language and formulated what we now call the *Newton-Euler equations of motion,* Euler [1749]. These are differential equations of second order in time: the momentum (in law no. 2) is the first derivative of the product *mass· velocity* of a particle, the term *change of momentum* has to be interpreted as the time derivative of the mentioned product. This obviously involves a first derivative of the velocity vector, thus a second derivative of the position vector. Newton's laws also imply the concept of a *force* acting on the bodies of a system. Assuming that the mass of our particle is constant in time Euler concluded from law number 2:

$$m \cdot \ddot{\vec{r}} = \vec{F} \tag{2.12}$$

where: m is the (constant) mass of the particle,
$\vec{r}$ its position vector in inertial space,
$\vec{F}$ the force acting on particle with mass m.

Actually, $\vec{F}$ should be understood as the vectorial sum of all forces acting on the particle resp. the satellite.

Newton's law of gravitation states that between two particles of masses M and there is an attracting force $\vec{F}$ of magnitude $F = |\vec{F}|$

$$F = G \cdot \frac{m \cdot M}{r^2} \tag{2.13}$$

where: G is the Newtonian gravitational constant

r is the distance between the two bodies.

It is assumed that either the (linear) dimensions of the two particles are very small (*infinitesimal*) compared to the distance r between the two bodies or that the mass distribution within the bodies is spherically symmetric.

Assuming that M is the total mass of the Earth, that the mass distribution within the Earth is spherically symmetric, interpreting m as the mass of an artificial Earth satellite, and neglecting all other forces that might act on this satellite, we obtain the *equations of motion* for an artificial Earth satellite in their simplest form:

$$m \cdot \ddot{\vec{r}} = -G \cdot \frac{m \cdot M}{r^2} \cdot \frac{\vec{r}}{r}$$

or

$$\ddot{\vec{r}} = -GM \cdot \frac{\vec{r}}{r^3} \tag{2.14}$$

where: $GM =$ $398.600415 \cdot 10^{12} \mathrm{m}^3 \cdot \mathrm{s}^{-2}$ is the product of the gravitational constant G and the Earth's mass M (value taken from the IERS Standards, McCarthy [1992]).

One easily verifies that the vector defined by its components (2.6) is a solution of the above equations of motion (2.14) *provided* we adopt the relationship:

$$n^2 \cdot a^3 = GM \tag{2.15}$$

This is in fact the equivalent to Kepler's law no. 3 in the Newtonian (Eulerian) formulation. It is *true* if the mass m of our test particle may be neglected. If this is not the case (e.g., for the Moon) the right-hand side of eqn. (2.15) must be replaced by $G \cdot (M+m)$. In the case of an artificial Earth satellite we may always neglect m.

It is relatively easy to verify Kepler's laws starting from the equations of motion (2.14). We may, e.g., multiply eqn. (2.14) by $\vec{r}\ \times$ (vector product) and obtain:

$$\vec{r} \times \ddot{\vec{r}} = \vec{0}$$

which implies that the vector product of $\vec{r}$ and $\dot{\vec{r}}$ is constant in time, which in turn proves that the motion is taking place in a plane (the so-called *orbital plane*, where $\vec{h}$ is a vector normal to the orbital plane):

$$\vec{r} \times \dot{\vec{r}} = \vec{h} \tag{2.16}$$

If we denote by h_1, h_2, h_3 the components of $\vec{h}$ in the equatorial system we may immediately compute the right ascension of the ascending node Ω and the inclination i with respect to the equatorial plane as:

$$\Omega = \arctan(h_1 / (-h_2)) \tag{2.16a}$$

$$i = \arctan\left(\sqrt{h_1^2 + h_2^2} \Big/ h_3\right) \tag{2.16b}$$

We have thus demonstrated that two of the Keplerian elements may be written as a function of the (components of) position vector $\vec{r}(t)$ and the velocity vector $\dot{\vec{r}}(t) = \vec{v}(t)$. This actually is a characteristic of all six Keplerian elements: *each* of the elements may be written as a function of the position and velocity vectors at one and the same (arbitrary) time t. Without proof we include these relationships:

$$\frac{1}{a} = \frac{2}{r} - \frac{v^2}{GM} \tag{2.16c}$$

$$e^2 = 1 - \frac{h^2}{GM \cdot a} \tag{2.16d}$$

$$\omega = u - \arctan\left\{\sqrt{\frac{a(1-e^2)}{GM}} \cdot \frac{\vec{r} \cdot \vec{v}}{r} \Big/ \left(\frac{a(1-e^2)}{r} - 1\right)\right\} \tag{2.16e}$$

$$E = 2 \cdot \arctan\left(\sqrt{\frac{1-e}{1+e}} \cdot \tan\left(\frac{u-\omega}{2}\right)\right)$$

$$\sigma(t) = E - e \cdot \sin E \tag{2.16f}$$

$$T_0 = t - \sigma / n \tag{2.16g}$$

where u is the *argument of latitude* of the satellite at time t, i.e., the angle in the orbital plane measured from the ascending node to the position of the satellite at time t. The above formulae, together with the formulae for the position and the velocity components (2.6) resp. (2.10), prove that there is a one to one correspondence between the position and velocity vector at time t on one hand and the Keplerian elements on the other hand. This fact is of importance for our subsequent developments.

Let us now generalize the equations of motions (2.14). We have to take into account that the mass distribution within the Earth is not spherically symmetric and that the gravitational attractions on our artificial satellite are stemming from the Moon and the Sun; moreover we allow for non-gravitational forces (like the radiation pressure). We first have to write down equations of motion for the satellite and for the center of mass of the Earth with respect to an (arbitrary) inertial system. Figure 2.6 illustrates the situation.

The equations of motion for the satellite (in the inertial system) may be written in the following form (direct consequence of Newton's laws of motion and Newton's law of gravitation):

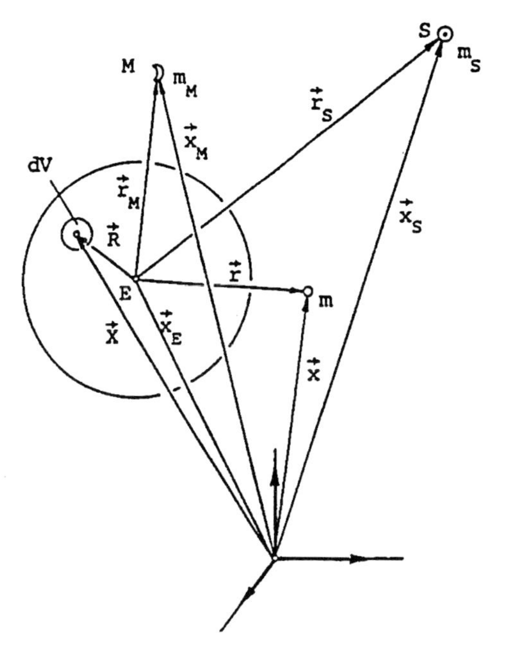

Figure 2.6. Center of mass of the Earth E, Sun S, Moon M, a volume element dV in the Earth's interior, their position vectors $\vec{x}_E, \vec{x}_S, \vec{x}_M, \vec{X}$ with respect to the origin of the inertial system, and their geocentric position vectors $0, \vec{r}_S, \vec{r}_M$, and $\vec{R}$.

$$m \cdot \ddot{\vec{x}} = -G \cdot m \cdot \int_{Vol} \frac{\vec{x} - \vec{X}}{|x - X|^3} \cdot \rho(\vec{X}) \cdot dV - G \cdot m \cdot m_M \cdot \frac{\vec{x} - \vec{x}_M}{|\vec{x} - \vec{x}_M|^3} -$$

$$- G \cdot m_S \cdot \frac{\vec{x} - \vec{x}_S}{|\vec{x} - \vec{x}_S|^3} + \vec{F}_{NG} \qquad (2.17)$$

where: m_M, m_S are the masses of Moon and Sun respectively
F_{NG} is the sum of all non-gravitational forces
$\rho(X)$ is the mass density at point $\vec{X}$ of the Earth's interior.

The equations of motion for the center of mass of the Earth may be written down in the following form:

$$M \cdot \ddot{\vec{x}}_E = -G \cdot M \cdot m_M \cdot \frac{\vec{x}_E - \vec{x}_M}{|\vec{x}_E - \vec{x}_M|^3} - G \cdot M \cdot m_S \cdot \frac{\vec{x}_E - \vec{x}_S}{|\vec{x}_E - \vec{x}_S|^3} \qquad (2.18)$$

Dividing eqn. (2.17) by the mass m of the satellite, dividing eqn. (2.18) by the mass M of the Earth and forming the difference of the two resulting equations we obtain the *equations of motion for the geocentric motion* $\vec{r}(t)$ *of the satellite*:

$$\ddot{\vec{r}} = -G \cdot \int_{Vol} \frac{\vec{r} - \vec{R}}{\left|\vec{r} - \vec{R}\right|^3} \cdot \sigma(\vec{R}) \cdot dV - G \cdot m_M \cdot \left\{ \frac{\vec{r} - \vec{r}_M}{\left|r - r_M\right|^3} + \frac{\vec{r}_M}{r_M^3} \right\} -$$

$$- G \cdot m_S \cdot \left\{ \frac{\vec{r} - \vec{r}_S}{\left|\vec{r} - \vec{r}_S\right|^3} + \frac{\vec{r}_S}{r_S^3} \right\} + \vec{F}'_{NG} \tag{2.19}$$

where F'_{NG} is the sum of all non-gravitational accelerations, $F'_{NG} = F_{NG} / m$.

In order to solve the equations of motion (2.19) we have to introduce a coordinate system. In this chapter we will select the equatorial system referring to a reference epoch; we may think, e.g., of using the system J2000. The geocentric system underlying eqn. (2.19) is *not* an inertial system (because of the motion of the Earth's center of mass around the Sun), but it is at any time parallel to the inertial system underlying the original equations (2.17). Due to the rotation of the Earth, the mass density $\rho(R)$ is a function of time. This time dependence may be taken out of the integral, if we formulate the equations of motion in the Cartesian coordinates referring to the equatorial system.

Let:

$$\mathbf{r} = \begin{pmatrix} r_1 \\ r_2 \\ r_3 \end{pmatrix} \tag{2.20a}$$

the Cartesian coordinates of $\mathbf{r}$ in the equatorial system referring to a standard epoch,

$$\mathbf{r}'' = \begin{pmatrix} r_1'' \\ r_2'' \\ r_3'' \end{pmatrix} \tag{2.20b}$$

the Cartesian coordinates of $\mathbf{r}''$ in an Earth-fixed system. Let furthermore the transformation matrix between the two systems be described by the following sequence of rotation matrices (orthonormal matrices):

$$\mathbf{r}'' = R_2(-x) \cdot R_1(-y) \cdot R_3(\Theta) \cdot N(t) \cdot P(t) \cdot \mathbf{r} =: R(t) \cdot \mathbf{r} \tag{2.21}$$

where: $R_i(w)$ characterizes a rotation around axis i and about angle w.
x, y are the components of polar motion,
Θ is the true Greenwich siderial time,

$N(t)$, $P(t)$ are the resp. nutation precession matrices.

For a detailed discussion of the transition between the Celestial and the Terrestrial Reference Frames we refer to the IERS Standards (1992), McCarthy [1992].

Using the abbreviated form of eqns. (2.21) we may write eqns. (2.19) in coordinate form as follows:

$$\ddot{\mathbf{r}} = -G \cdot R(t) \cdot \int_{Vol} \frac{\mathbf{r}'' - \mathbf{R}''}{|\mathbf{r} - \mathbf{R}|^3} \cdot \rho(\mathbf{R}'') \cdot dV - G \cdot m_M \cdot \left\{ \frac{\mathbf{r} - \mathbf{r}_M}{|\mathbf{r} - \mathbf{r}_M|^3} + \frac{\mathbf{r}_M}{r_M^3} \right\} - G \cdot m_S \cdot \left\{ \frac{\mathbf{r} - \mathbf{r}_S}{|\mathbf{r} - \mathbf{r}_S|^3} + \frac{\mathbf{r}_S}{r_S^3} \right\} + F'_{NG} \tag{2.22}$$

Assuming that the Earth is a rigid body, the mass distribution $\rho(\mathbf{R}'')$ in the Earth-fixed system is no longer time dependent.

The integral in equation (2.22) may be written as the gradient ∇ of the so-called Earth potential V, a scalar function of the coordinates of the satellite position:

$$-G \cdot \int_{Vol} \frac{\mathbf{r}'' - \mathbf{R}''}{|\mathbf{r} - \mathbf{R}|^3} \cdot \rho(\mathbf{R}'') \cdot dV = G \cdot \nabla \int_{Vol} \frac{\rho(\mathbf{R}'')}{|\mathbf{r} - \mathbf{R}|} \cdot dV =: \nabla V \tag{2.23}$$

We follow the usual procedure and develop $V(\mathbf{r}'')$ into a series of normalized Legendre functions, see e.g., Heiskanen and Moritz [1967]. Using the polar coordinates **r** (length of geocenric radius vector), λ (geocentric longitude), and β (geocentric latitude) instead of the Cartesian coordinates we may write:

$$V(r,\lambda,\beta) = \frac{G \cdot M}{r} \left\{ 1 + \sum_{n=1}^{\infty} \left(\frac{a_E}{r} \right)^n \cdot \sum_{m=0}^{n} P_n^m(\sin\beta) \cdot \left(C_{nm} \cdot \cos(m \cdot \lambda) + S_{nm} \cdot \sin(m \cdot \lambda) \right) \right\} \tag{2.24}$$

where: $P_n^m(\sin\beta)$ are the (fully normalized) associated Legendre functions, defined, e.g., in Heiskanen and Moritz [1967], a_E is the equatorial radius of the Earth, C_{nm} and S_{nm} are the coefficients.

In general, if we are working in the center-of-mass-system, where the terms with $n = 1$ and C_{21}, S_{21} are all equal to zero. For the numerical values of the coefficients we again refer to the IERS Standards [McCarthy, 1992], where the references to the more important gravity models may be found.

We distinguish between the *zonal terms* (where $m = 0$), which depend only on latitude, the *sectorial terms* and (where $n = m$), which only depend on longitude, and the *tesseral terms* (n and m arbitrary), which depend on both, latitude and longitude. Examples may be found in Figures 2.7a, 2.7b, and 2.7c.

Figure 2.7a. Zonal harmonics - zones of equal sign (n=6, m=0).

Figure 2.7b. Sectorial harmonics - sectors of equal sign (n=m=7).

Figure 2.7c. Tesseral harmonics - regions of equal sign ($n = 13$, $m = 7$).

Let us briefly summarize this section. We started from Newton's laws of motion and wrote down the equations of motion in their simplest form (2.14). We stated that each of the Keplerian elements may be written as a function of the position and velocity vectors. We then wrote down the general equations of motion for an artificial

Earth satellite, first referring to an inertial system (eqns. (2.17)), then referring to a geocentric system (2.19). After that we wrote the equations of motion for the Cartesian coordinates (eqns. (2.22)), which allowed us to compute the gravitational attraction stemming from the Earth in an Earth-fixed coordinate system. The Earth's gravitational potential (2.24) was introduced, where we made (as usual) the distinction between zonal, sectorial, and tesseral terms.

Let us conclude the section with the remark that the effects due to the elastic properties of the Earth may still be taken into account by the development (2.24), if we allow the coefficients C_{nm}, S_{nm} to be functions of time. Only the lowest terms must be taken into account usually. This is, e.g., the case for the solid Earth tides, McCarthy [1992, chapter 7].

Let us mention that major parts of this section were extracted from Beutler and Verdun [1992], where more details concerning the development of the Earth's gravitational potential may be found.

2.2.3 The Perturbation Equations in the Elements

The equations of motion (2.22) may be written in the form

$$\ddot{\mathbf{r}} = -GM \cdot \frac{\mathbf{r}}{r^3} + \mathbf{P}(\mathbf{r}, \dot{\mathbf{r}}, t) \tag{2.25}$$

where the first term is the Keplerian term (eqn. (2.14)), the second the perturbation term. $\mathbf{P}(\mathbf{r}, \dot{\mathbf{r}}, t)$ contains all but the main term stemming from the Earth's gravitational potential (2.24), the gravitational attractions by Sun and Moon, and, last but not least, all non-gravitational terms $\mathbf{F'}_{NG}$. Usually we may assume that

$$\frac{GM}{r^2} \gg \left|\mathbf{P}(\mathbf{r}, \dot{\mathbf{r}}, t)\right| \tag{2.26}$$

This relation certainly holds for GPS satellites. The solution of the unperturbed equation (2.14) thus is a relatively good approximation of the equation (2.22) if the same initial conditions are used in both cases (at an initial epoch t_0) – at least in the vicinity of this initial epoch. It thus makes sense to speak, e.g., of an orbital plane which evolves in time. It also makes sense to introduce an *instantaneous* or *osculating* ellipse, and to study the semi-major axis and the eccentricity as a function of time.

In the preceding section we said that there is a one-to-one correspondence between the Keplerian elements on one hand and the components of the position and the velocity vector on the other hand *in the case of the Keplerian motion.* Let us assume that $\mathbf{r}(t)$ and $\dot{\mathbf{r}}(t)$ for each time argument t are the *true* position and velocity vectors as they emerge from the solutions of the equations of motion (2.22) (corresponding to one and the same set of initial conditions $\mathbf{r}(t_0)$, $\mathbf{v}(t_0)$). We now define the osculating elements at time t as the Keplerian elements computed from $\mathbf{r}(t)$, $\mathbf{v}(t)$ using the relationships (2.16a-g) of the unperturbed two body problem. Through this procedure we define time series of osculating elements $a(t)$, $e(t)$, $i(t)$, $\Omega(t)$, $\omega(t)$, $\sigma(t)$ (or $T_0(t)$) associated with the perturbed motion. The Keplerian orbit corresponding to the osculating elements at time t are by design tangential to the perturbed orbit (at time t) because the two orbits (perturbed and unperturbed) share the same position and velocity vectors.

The celestial mechanic is used to think in terms of these osculating elements. They are the ideal quantities to study the evolution of an orbit. Of course we should keep in mind that, in principle, each orbit is completely specified by one set of osculating elements (e.g., at time t) and by the perturbation equations (2.25).

It is possible to introduce (and solve) differential equations *not* for the rectangular coordinates of the position vector, *but* directly for the osculating elements. It is very instructive to study the perturbation equations for the osculating elements. Let us first introduce the following notation:

$$\{K_1(t), K_2(t), K_3(t), K_4(t), K_5(t), K_6(t)\} := \{a(t), e(t), i(t), \Omega(t), \omega(t), \sigma(t)\} \quad (2.26a)$$

Furthermore, let

$$K_i(t) \in \{K_1(t),\ i = 1,...6\} \quad (2.26b)$$

In view of the relationships we gave in the previous section we may write:

$$K_i(t) = K_i(\mathbf{r}(t), \mathbf{v}(t)) = K_i(\mathbf{r}(t), \dot{\mathbf{r}}(t)) \quad (2.26c)$$

i.e., the time dependence of K_i is only given through the vectors $\mathbf{r}(t)$, $\mathbf{v}(t)$. Let us now take the first derivative of eqn. (2.26c)

$$\dot{K}_i = \sum_{j=1}^{3} \frac{\partial K_i}{\partial r_j} \cdot \dot{r}_j + \sum_{j=1}^{3} \frac{\partial K_i}{\partial v_j} \cdot \ddot{r}_j \quad (2.26d)$$

Replacing the second time derivative in eqn. (2.26d) by the right-hand side of the equations of motion, and taking into account that K_i is constant for the unperturbed motion, we obtain the following simple relation:

$$\dot{K}_i = \sum_{j=1}^{3} \frac{\partial K_i}{\partial v_j} \cdot P_j(\boldsymbol{r}, \boldsymbol{v}, t) =: \nabla_v(K_i) \cdot \boldsymbol{P}(\boldsymbol{r}, \boldsymbol{v}, t) \quad (2.26e)$$

Keeping in mind that:

$$\mathbf{r} = \mathbf{r}(K_1, K_2, ..., K_6, t), \quad \mathbf{v} = \mathbf{v}(K_1, K_2, ..., K_6, t) \quad (2.26f)$$

we have thus shown that the set of osculating elements may be described by a first-order differential equation system in time t. Instead of one system of second order of type (2.25) we may thus consider one first-order differential equation system of six equations:

$$\dot{K}_i = \nabla_{\mathbf{v}}(K_i) \cdot \mathbf{P}(K_1, K_2, ..., K_6, t) \qquad i = 1, ..., 6 \quad (2.27)$$

The differential equation systems (2.25) and (2.27) are equivalent in the sense that the same orbit will result, provided the same initial conditions are used to produce particular solutions. Equations (2.25) and (2.27) are different mathematical formulations of one and the same problem.

Apart from the fact that eqns. (2.27) are of first, whereas eqns. (2.25) are of second order, eqns. (2.27) are by no means of simpler structure than eqns. (2.25).

However, eqns. (2.27) allow for relatively simple approximate solutions. Let us first state that for

$$\mathbf{P}(K_1, K_2, \ldots, K_6, t) = 0 \tag{2.28a}$$

we have

$$\dot{K}_i = 0 \quad \text{or} \quad K_i = K_{i0} = K_i(t_0) \tag{2.28b}$$

Equations (2.28b) simply repeat that in the case of the unperturbed motion the Keplerian elements are constants of integrations. In view of inequality (2.26) the elements K_i may nevertheless be considered as an approximate solution for equations (2.27). *First-order perturbation theory* gives us a much better (but not yet the correct) solution:

$$\dot{K}_i = \nabla(K_i) \cdot \mathbf{P}(K_{10}, K_{20}, \ldots, K_{60}, t) \qquad i = 1, \ldots, 6 \tag{2.29a}$$

That is, we compute the perturbing acceleration **P** (and the gradient) using the Keplerian approximation for the orbit. Equations (2.29a) are much easier to solve than the original equations (2.27), because the unknown functions are no longer present on the right-hand sides. As a matter of fact equations (2.29a) consist of six uncoupled integrals only which may be solved easily:

$$K_i(t) = K_{i0} + \int_{t_0}^{t} \nabla_{\mathbf{v}}(K_{i0}) \cdot \mathbf{P}(K_{10}, K_{20}, \ldots, K_{60}, t') \cdot dt' \quad i = 1, \ldots, 6 \tag{2.29b}$$

In first-order perturbation theory it is therefore possible *to study the perturbation of each orbit element independently of the others*. This is a remarkable advantage over the formulation (2.25) which does not allow for a similar structuring of the problem.

Let us go back to the perturbation equations (2.27) and derive the perturbation equation for the semi-major axis a. Equation (2.16c) gives a as a function of position and velocity:

$$\frac{1}{a} = \frac{2}{r} - \frac{v^2}{GM}$$

From this equation we conclude

$$\nabla_{\mathbf{v}} a = \frac{2 \cdot a^2}{GM} \cdot \mathbf{v}$$

and therefore:

$$\dot{a} = \frac{2}{n^2 a} \cdot (\mathbf{v} \cdot \mathbf{P}) \tag{2.29c}$$

We are of course free to choose any coordinate system to compute the scalar product ($\mathbf{v} \cdot \mathbf{P}$) in the above equation. We may even select different coordinate systems at different instants of time t; we only have to use the same coordinate system for $\mathbf{v}$ and $\mathbf{P}$ at time t (!). Several coordinate systems are actually used in celestial mechanics. Subsequently we will decompose $\mathbf{P}$ into the components R, S, and W, the unit vectors $\vec{e}_R, \vec{e}_S$, and $\vec{e}_w$ in R, S, and W directions from a right-handed coordinate system, where $\vec{e}_R$ points in the radial direction, $\vec{e}_w$ is normal to the orbital plane, and $\vec{e}_S$ lies in the instantaneous orbital plane and points approximately into the direction of motion. Figure 2.8 illustrates the decomposition.

Without proof we give the perturbation equations using the R, S, W decomposion of the perturbing accelerations. For the derivation of these equations we refer to Beutler and Verdun [1992].

$$\dot{a} = \sqrt{\frac{p}{GM}} \cdot \frac{2a}{1-e^2} \cdot \left\{ e \cdot \sin v \cdot R + \frac{p}{r} \cdot S \right\} \tag{2.30a}$$

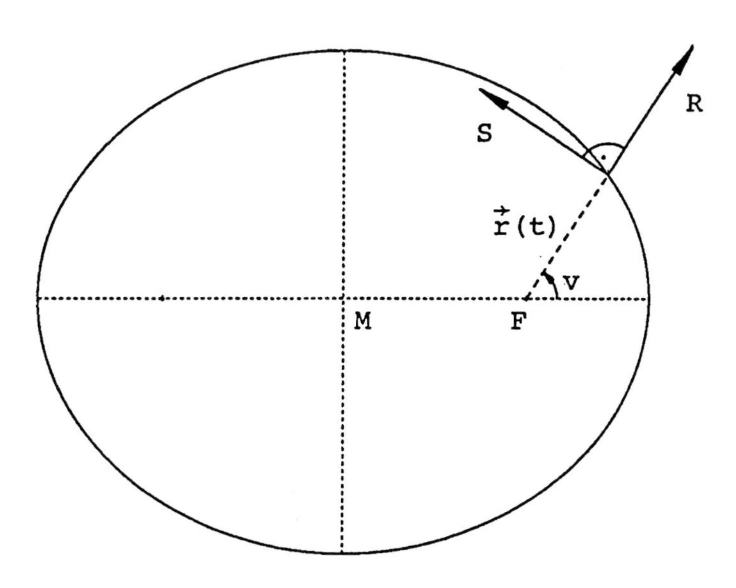

Figure 2.8. The decomposition of the perturbing acceleration into the components R, S, and W. (Only first two components drawn.)

$$\dot{e} = \sqrt{\frac{p}{GM}} \cdot \{\sin v \cdot R + (\cos v + \cos E) \cdot S\} \tag{2.30b}$$

$$i^{(1)} = \frac{r\cos(\omega + v)}{n \cdot a^2 \cdot (1 - e^2)^{1/2}} \cdot W \tag{2.30c}$$

$$\dot{W} = \frac{r \cdot \sin(\omega + v)}{n \cdot a^2 \cdot (1 - e^2)^{1/2} \cdot \sin i} \cdot W \tag{2.30d}$$

$$\dot{\omega} = \frac{1}{e} \cdot \sqrt{\frac{p}{GM}} \cdot \left\{ -\cos v \cdot R + \left(1 + \frac{r}{p}\right) \cdot \sin v \cdot S \right\} - \cos i \cdot \dot{\Omega} \tag{2.30e}$$

$$\dot{\sigma} = \frac{1}{na} \cdot \frac{1 - e^2}{e} \cdot \left\{ \left(\cos v - 2 \cdot e \cdot \frac{r}{p} \right) \cdot R - \left(1 + \frac{r}{p}\right) \cdot \sin v \cdot S \right\} + \frac{3}{2} \cdot \frac{n}{a} \cdot (t - t_0) \cdot \dot{a} \tag{2.30f}$$

where $p = a \cdot (1 - e^2)$ is the parameter of the ellipse, and v denotes the true anomaly.

Equations (2.30) are convenient in the sense that we may discuss the influence of the components R, S, and W separately. We can, e.g., see at once that only the component W perpendicular to the orbital plane is capable of changing the position of the orbital plane (elements i and Ω). We can also see that for $e \ll 1$ it is mainly the acceleration S which will change the semi-major axis. Actually eqn. (2.29c) tells us that in this particular case a may only be influenced by an acceleration in tangential direction (direction of velocity). This will, e.g., be of importance when considering satellite manoeuvres.

If we want to study the influence of any perturbing acceleration, we have to compute the components R, S, and W of this acceleration, and we have to integrate eqns. (2.30). If we are satisfied with first-order perturbation theory, we may use the osculating elements of the initial epoch on the right-hand sides of eqns. (2.30). In this case the problem of solving a coupled system of non-linear differential equations is reduced to the solution of six definite integrals.

Let us conclude this section by two types of examples:

(1) we outline an approximate solution for the elements $a(t)$, $i(t)$, $\Omega(t)$, and for the term C_{20} of the Earth's gravitational potential using the characteristics of GPS orbits.

(2) we give the osculating elements as a function of time for a time period of 3 days using the *complete* force field for one GPS satellite.

The perturbing acceleration due to the term C_{20} may be written in the equatorial coordinate system as Beutler and Verdun [1992, eqn. (8.25)]:

$$\mathbf{P} = -\frac{3}{2} \cdot GM \cdot a_E^2 \cdot J_{20} \cdot \frac{1}{r^5} \cdot \begin{pmatrix} r_1 \cdot \left(1 - 5 \cdot r_3^2 / r^2\right) \\ r_2 \cdot \left(1 - 5 \cdot r_3^2 / r^2\right) \\ r_3 \cdot \left(3 - 5 \cdot r_3^2 / r^2\right) \end{pmatrix} \tag{2.31a}$$

where $J_{20} = 1082.6 \cdot 10^{-6}$ (2.31b)

The R, S, W components may easily be computed (by a series of transformations, Beutler and Verdun [1992, eqns. (8.29), (8.30)]):

$$\begin{pmatrix} R \\ S \\ W \end{pmatrix} = -\frac{3}{2} \cdot GM \cdot a_E^2 \cdot J_{20} \cdot \frac{1}{r^4} \cdot \begin{pmatrix} 1 - 3 \cdot \sin^2 i \cdot \sin^2 u \\ \sin^2 i \cdot \sin(2u) \\ \sin(2i) \cdot \sin u \end{pmatrix} \tag{2.31c}$$

where u is the argument of latitude at time t, i.e. $u = \omega + v(t)$, and v is the true anomaly.

If we are only interested in a crude approximation we may neglect the terms of order 1 or higher in e in the perturbation equations, because the GPS orbits are almost circular. This means that we may replace $\mathbf{r}$ and $\mathbf{P}$ by the semi-major axis a in the perturbation equations. Moreover we do of course use first-order perturbation theory.

With these simplifying assumptions the perturbation equation for the semi-major axis a reads as:

$$\dot{a} = \frac{2}{n} \cdot S$$

Replacing the component S in the above equation according to eqn. (2.31c), where we again use the approximation $r = a$ we obtain:

$$\dot{a} = -3 \cdot n \cdot a \cdot \left(\frac{a_E}{a}\right)^2 \cdot J_{20} \cdot \sin^2 i \cdot \sin(2u)$$

In view of the fact that in our approximation we may write $u(t) = \omega + n \cdot (t - T_0)$ this equation may easily be integrated to yield:

$$a(t) = \frac{3}{2} \cdot a \cdot \left(\frac{a_E}{a}\right)^2 \cdot J_{20} \cdot \sin^2 i \cdot \cos(2u) + C \tag{2.32a}$$

where the integration constant is of no interest to us here. We see that the main effect in the semi-major axis due to the oblateness of the Earth is a *short periodic perturbation* (period = half a revolution ≈ 6 hours for GPS satellites). The amplitude A is:

$$A = \frac{3}{2} \cdot a \cdot \left(\frac{a_E}{a}\right)^2 \cdot J_{20} \cdot \sin^2 i = 1.67 \text{ km} \tag{2.32b}$$

using the values $a = 26'500$ km, $a_E = 6'378$ km, $i = 55°$, $J_{20} = 1082.6 \cdot 10^{-6}$.

In the same approximation and with $e = 0$ the equation for the right ascension of the ascending node has the form:

$$\dot{\Omega} = -\frac{3}{2} \cdot \left(\frac{a_E}{a}\right)^2 \cdot J_{20} \cdot \cos i \cdot n \cdot (1 - \cos(2u)) \tag{2.32c}$$

where we have made use of the formula

$$\sin^2 u = \frac{1}{2} \cdot (1 - \cos(2u)) .$$

Equation (2.32) might be solved easily, but we already see the essential properties: there is a regression (backwards motion) of the node with an average rate of:

$$\dot{\Omega}_{\text{mean}} = -\frac{3}{2} \cdot \left(\frac{\mathrm{a_E}}{\mathrm{a}}\right)^2 \cdot \mathrm{J}_{20} \cdot \cos \mathrm{i} \cdot \mathrm{n} = -0.039 \left[°/\text{day}\right] = 14.2 \left[°/\text{year}\right] \tag{2.32d}$$

where we used the same numerical values as above. We also mention that twice per revolution, for $u = 0°$ and $u = 180°$ (i.e., in the nodes) the instantaneous regression vanishes, the maximum backwards motion is expected for $u = 90°$ and $u = 270°$ (i.e., at maximum distances form the equatorial plane). We thus expect that the nodes of all GPS satellites are performing a rotation of 360° in about 25 years on the equator. How does the inclination behave? Using the same approximations we obtain:

$$i^{(1)} = -\frac{3}{4} \cdot n \cdot \left(\frac{a_E}{a}\right)^2 \cdot J_{20} \cdot \sin(2i) \cdot \sin(2u)$$

$$i(t) = \frac{3}{8} \cdot \left(\frac{a_E}{a}\right)^2 \cdot J_{20} \cdot \sin(2i) \cdot \cos(2u) \tag{2.32e}$$

which means that there is *no* secular effect on the inclination i due to the oblateness perturbation term. There is, however, a short period term with the period of half a revolution. We thus expect the normal vectors to the orbital planes to perform essentially a complete revolution on a latitude circle of 35° in about 25 years.

Let us now reproduce in Figures 2.9a-f the osculating elements for a particular GPS satellite (PRN 14) over three days (in November 1994). The figures were based on a numerical integration for the orbit of PRN 14, where the entire force field (to be introduced in section 2.3) was included. PRN 14 was *not* in an eclipse season at that time, it is meant to be an *average* GPS satellite for the time interval considered.

Let us mention a few aspects:

- We easily see that our crude approximation (2.32a-e) is not too far away from the *truth*. We see in particular that the dominating effect in the semi-major axis a actually is an oscillation with an amplitude of about 1.7 km (Figure 2.9a), and that the node is moving backwards with an average speed of about 0.04°. We also see that the backwards motion is zero twice per revolution as mentioned above. That *real life* is so close to our crude approximations is due to the fact that actually the GPS satellites are low-eccentricity satellites ($e \approx 0.003$ in that time period for PRN 14) and that the term C_{20} is the dominant perturbation term.
- We can also see that there are long-period variations on top of the short period variations which we did not expect from our crude analysis. This is true in particular for the inclination i, where in addition to of the short period variation of an amplitude we would expect from eqn. (2.32e) there is a long-period variation which we did not explain above. This variation is not caused by the oblateness.
- We can also see that the osculating argument of perigee and the mean anomaly at time t0, the starting time of the arc, show rather big short period variations, but that they are highly correlated; would we compute the sum of the two terms (corresponding more or less (?) to the argument of latitude at time t0), the variations would be much smaller. This behavior just reflects the fact that the argument of perigee is not well defined for low eccentricity orbits. We should thus avoid to use the argument of perigee and the mean anomaly at an initial time as orbit parameters in an adjustment process.

The osculating elements are *not* well suited to study the long-term evolution of the satellite system. Small changes – well below the amplitudes of the short-period perturbations – are not easily detected in Figures 2.9a-f. This is the motivation for the next section.

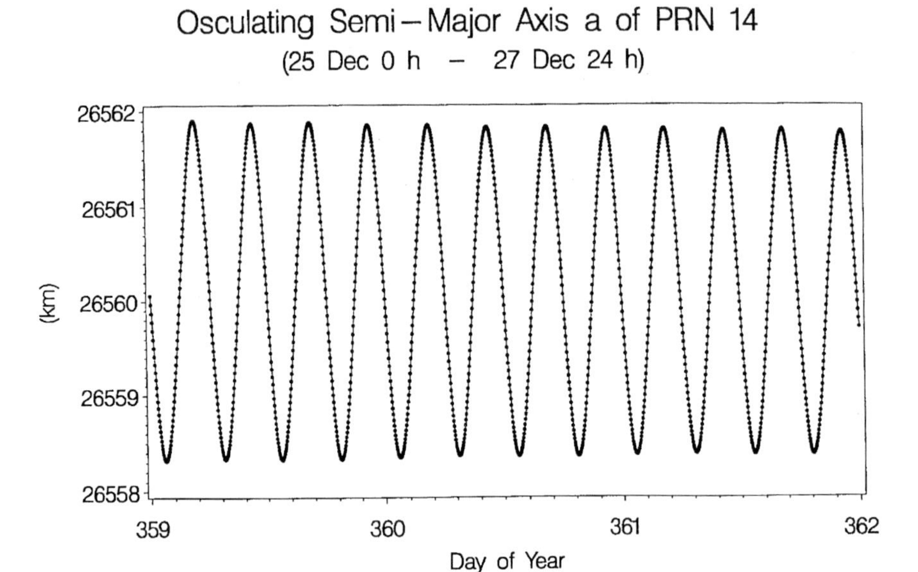

Figure 2.9a. Osculating semi-major axis a of PRN 14 (25 Dec 0 h - 27 Dec 24 h).

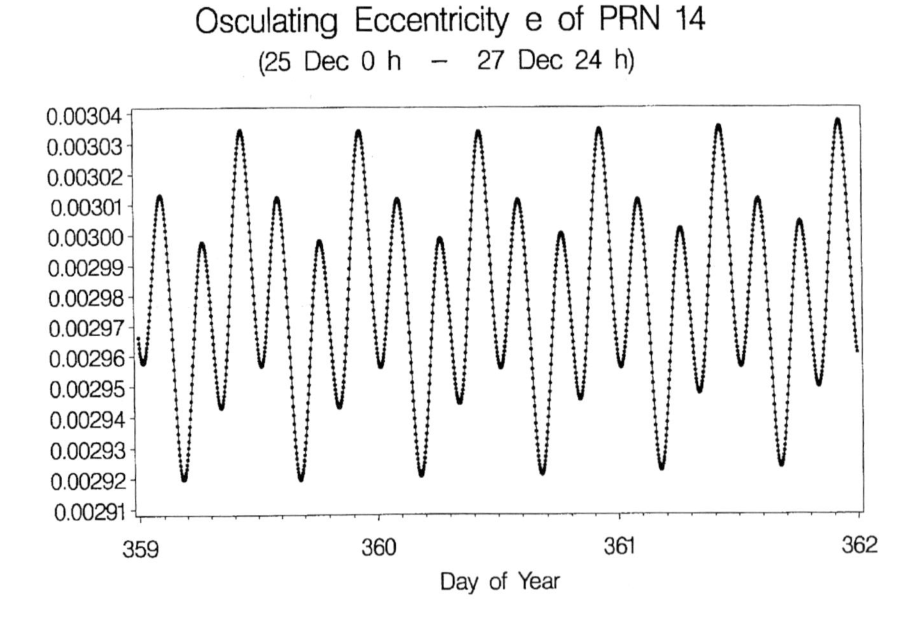

Figure 2.9b. Osculating eccentricity e of PRN 14 (25 Dec 0 h - 27 Dec 24 h).

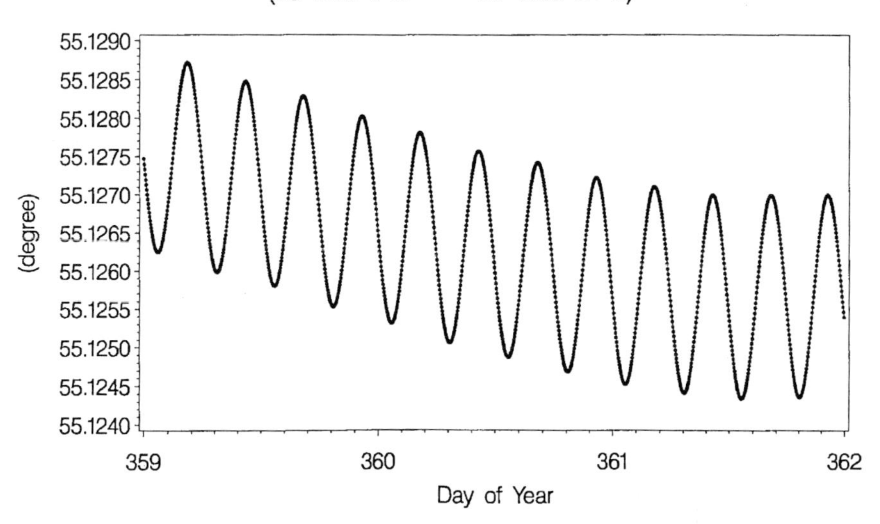

Figure 2.9c. Osculating inclination i of PRN 14 (25 Dec 0 h - 27 Dec 24 h).

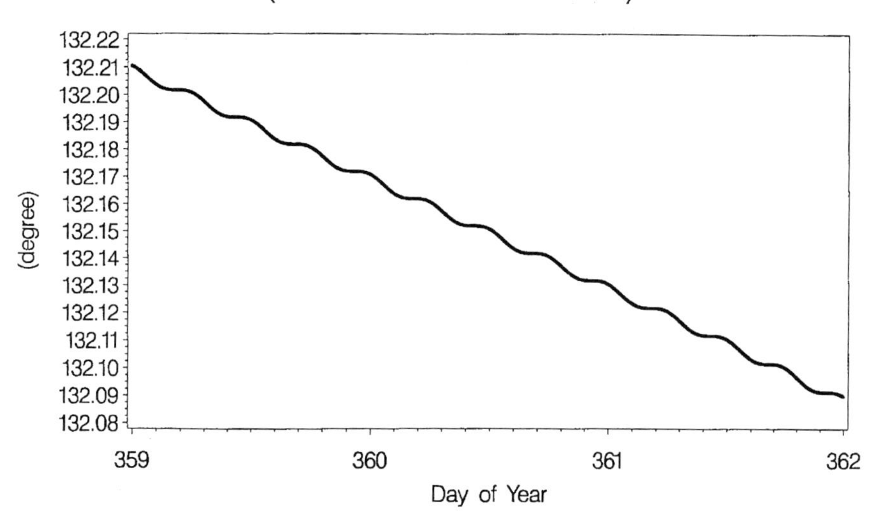

Figure 2.9d. Osculating right ascension of the ascending node Ω of PRN 14 (25 Dec 0 h - 27 Dec 24 h).

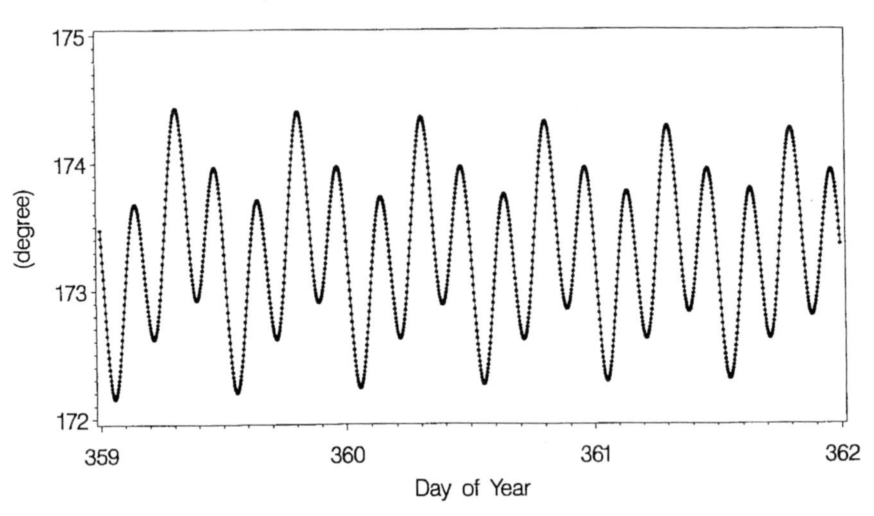

Figure 2.9e. Osculating argument of perigee ω of PRN 14 (25 Dec 0 h - 27 Dec 24 h)

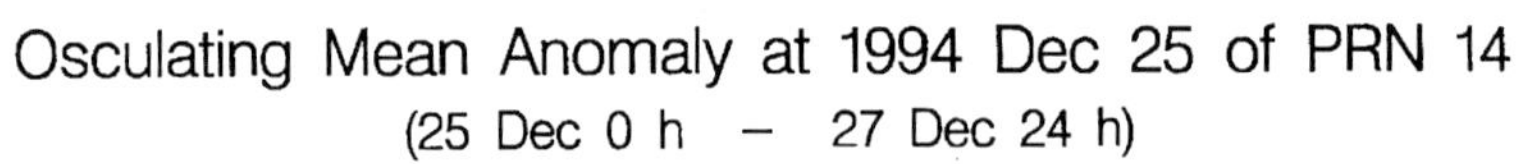

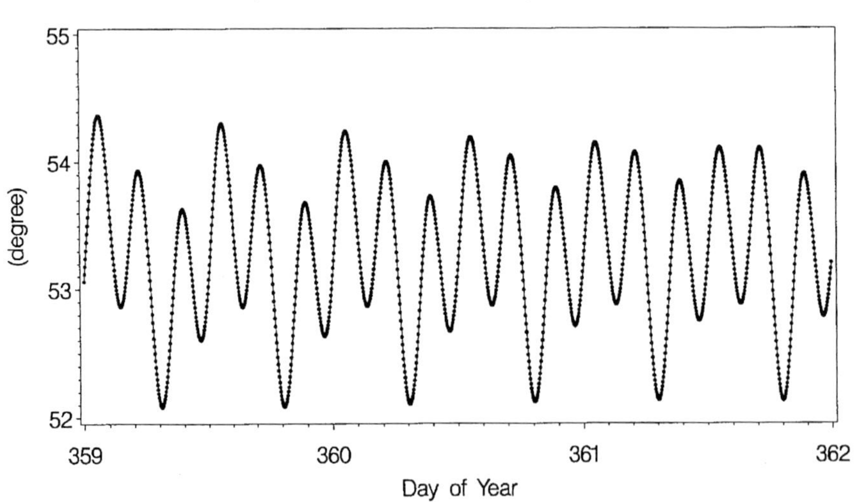

Figure 2.9f. Osculating mean anomaly σ at 1994 Dec 25 of PRN 14 (25 Dec 0 h - 27 Dec 24 h)

2.2.4 Mean Elements

There are many different ways to define mean orbital elements starting from a series of osculating elements. The purpose is the same, however, in all cases: one would like to remove the higher frequency part of the spectrum in the time series of the elements. There are subtle differences between different definitions of mean elements, but they are not relevant for our purpose. Here we just want to use mean elements to give an overview over the development of the GPS in time periods stretching from weeks to years.

Let us use the notation introduced in equations (2.26a,b). Starting from the osculating element $K_i(t)$ we define the *mean element* $\bar{K}_i(t)$ using the following definition:

$$\bar{K}_i(t) := \frac{1}{U(t)} \cdot \int_{t-U/2}^{t+U/2} K_i(t') \cdot dt' \tag{2.33}$$

where $U(t)$ essentially is the sidereal revolution period as computed from the osculating elements at time t.

Let us point out that neither eqn. (2.33) is the only possible definition, nor is it necessarily the best possible. In view of the importance the argument of latitude plays in the short period perturbations due to the oblateness of the Earth, it might have been better to use *not* the siderial revolution period, *but* the draconic revolution period in eqn. (2.33) (i.e., the revolution period from one pass through the ascending node to the next). The differences between different definitions of mean elements are of second order in the differences (osculating-mean elements).

Be this as it may: our mean elements will all be based on the definition equation (2.33), and Figures 2.10a-e show the development of the mean elements in the time interval mid 1992 till end of 1994 for the same satellite which PRN 14 we already used in Figures 2.9a-f. We do not include a figure for the mean anomaly at an initial time t_0 because, due to obvious reasons, such data are not readily available. The result would not be very instructive anyway: In essence we would see the difference between the mean perturbed motion and the mean Keplerian motion multiplied by the time interval $(t-t_0)$ – in a figure this would be a straight line.

A comparison of Figures 2.10a-2.10e with the corresponding Figures 2.9a-2.9d reveals that the short period perturbations were indeed removed successfully. Of course we have to take into account that the time interval is much longer in Figures 2.10 than in Figures 2.9. Interesting facts show up when looking at the mean elements!

Let us first look at mean semi-major axis of PRN 14 in Figure 2.10a: We clearly see an average drift of about 7 m/day and two manoeuvres setting back the mean semi-major axis by about 2.5 km resp. 2.9 km. The manoeuvres were necessary because of that drift in order to keep PRN 14 from overtaking the space vehicles in the same orbital plane in front of it. As a matter of fact PRN 14 was

110° ahead of PRN 21 in November 1994 – no reason to worry at present. Let us see how this will change, now.

We compute the change in the mean motion n associated with a change in the semi-major axis using Kepler's law no. 2 eqn. (2.15):

$$dn = -\frac{3}{2} \cdot \frac{n}{a} \cdot da$$

where in our case da is a linear function of time:

$$da = \dot{a} \cdot (t - t_0) \quad .$$

Because the mean anomaly M is a linear function of time, too (at least if the mean motion n is constant), $M(t) = n \cdot (t - T_0)$, the change dM in the mean anomaly associated with the above drift must be computed in the following way:

$$dM = \int_{t_0}^{t} dn(t') \cdot dt' \approx -\frac{3}{2} \cdot \frac{n}{a} \cdot \dot{a} \cdot \int_{t_0}^{t} (t' - t_0) \cdot dt' \approx -\frac{3}{4} \cdot \frac{n}{a} \cdot \dot{a} \cdot (t - t_0)^2$$

The above formula gives dM in radian, $(t\text{-}t_0)$ has to be express in seconds, the drift $\dot{a}$ in units of m/sec; t_0 is an arbitrary initial epoch. It is more convenient to have a formula which gives dM in degrees, where the time argument is expressed in years, and $\dot{a}$ in m/day. The following relation (derived by simple scaling operations from the above equation) may be used for this purpose:

$$dM[°] = -\frac{3}{4} \cdot \frac{180}{\pi} \cdot \frac{n}{a} \cdot (86400 \cdot 365.25)^2 / 86400 \cdot (dT)^2 \cdot \dot{a} \approx -2.7° \cdot (dT)^2 \cdot \dot{a} \qquad (2.34)$$

where dT is the time difference expressed in years, $\dot{a}$ must be given in [m/day]; the values $a = 26'500'000$ m and $n = (4 \cdot \pi / 86400)$ were used to establish the numerical values in eqn. (2.34). According to eqn. (2.34) PRN 14 changes its nominal position in orbit by about 19° per year, by about 86° per two years. Corrective manoeuvres are therefore unavoidable about once per year for such a satellite (assuming that the other satellites in the orbital plane do not show a similar drift). Figure 2.10a shows that such manoeuvres actually took place.

We should *not* conclude from Figure 2.10a that PRN 14 is on its way *home* down to Earth. The perturbation actually is periodic, but the period is very long (even compared to our time basis of about 2.5 years). We refer to section 2.3.3 for more information.

Figure 2.10b shows that the eccentricity decreases linearly with time (very much like the semi-major axis). The reason again has to be sought in the resonance terms. We also see an expressed annual term (which we attribute mainly to radiation pressure) and perturbations of shorter period (caused by the

Moon). We also see that the eccentricity was slightly changed at the times of the manoeuvres.

Figures 2.10c again contains a perturbation of very long period. The reason again is resonance. The semi-annual perturbation with an amplitude of about 0.03° is caused by the gravitational attraction due to the Sun, the semi-monthly terms by lunar attraction. There is no trace of a manoeuvre in the element i, which indicates that the impulse change took place in the osculating orbital plane.

Figure 2.10d is really nice! We just see the backwards motion of the node (of about 14.5°/year). If we remove the linear drift, periodic variations show up, too, of course. Figure 2.11 gives the result. Figures 2.11 and 2.10c are of particular interest for people who want to estimate UT1-UTC or the nutation in longitude (actually the first time derivatives of these quantities). Frequencies present in these Figures might also be found in UT1-UTC curves or in nutation drift curves derived by GPS, because the definition of the nodes is crucial for the estimation of these terms. Apart from that it is fair to state that Figure 2.10c (mean inclination) and Figure 2.11 look quite similar. There are prominent semi-annual perturbations in both cases, there is a semi-monthly term in both cases, and there is a resonance effect of very long period on top of that. In view of the close relationship of the corresponding perturbation equations (2.30c,d) this does not amaze us.

Figure 2.10e shows the mean argument of perigee as a function of time. Again we have to point out that, in view of the small eccentricity, the motion of the node is not so dramatic for the orbit of the satellite: most of the effect would be counter-balanced by the perturbation in the mean anomaly.

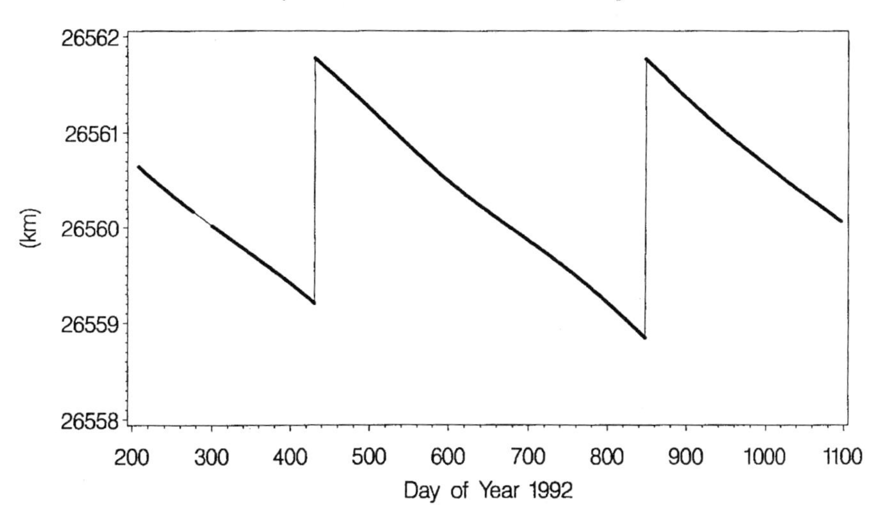

Figure 2.10a. Mean semi-major axis a of PRN 14 (Mid 1992 - End of 1994).

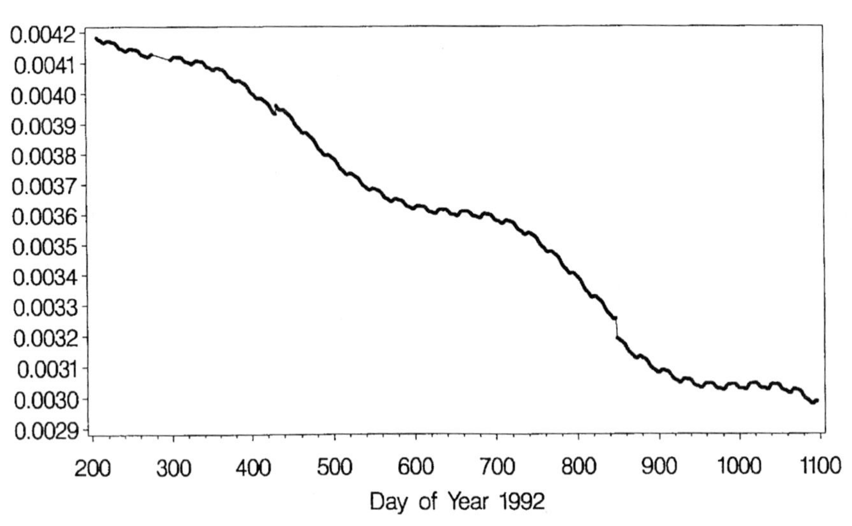

Figure 2.10b. Mean eccentricity *e* of PRN 14 (Mid 1992 - End of 1994).

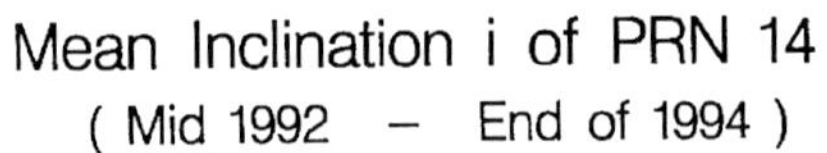

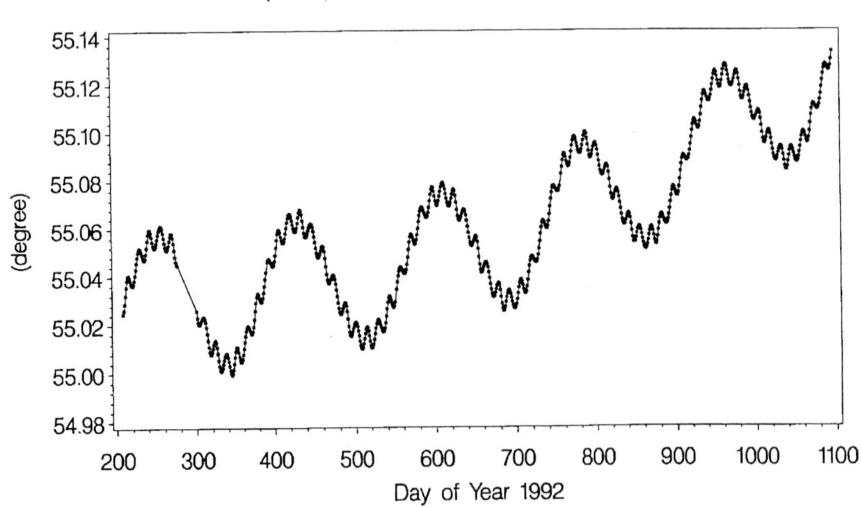

Figure 2.10c. Mean inclination *i* of PRN 14 (Mid 1992 - End of 1994).

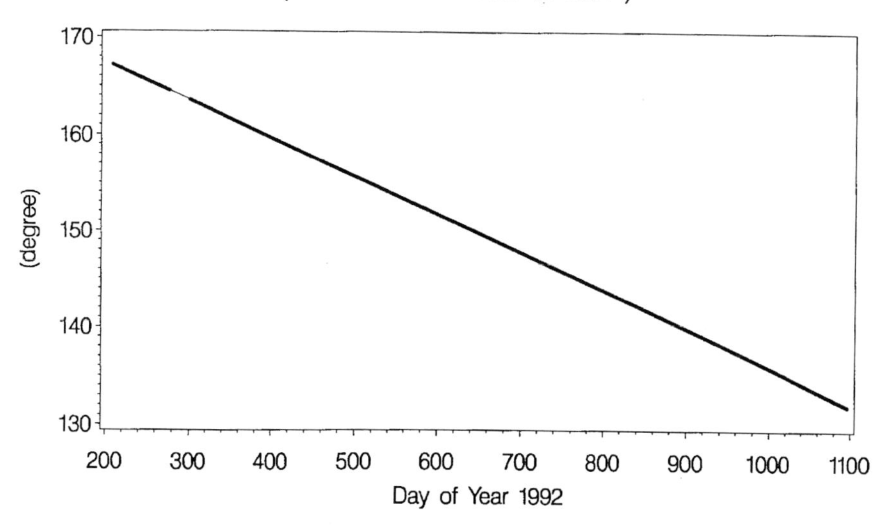

Figure 2.10d. Mean right ascension of the ascending node Ω of PRN 14 (Mid 1992 - End of 1994).

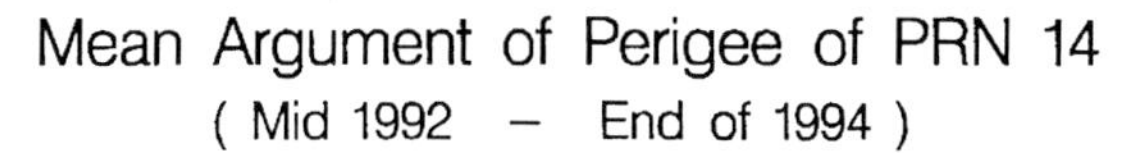

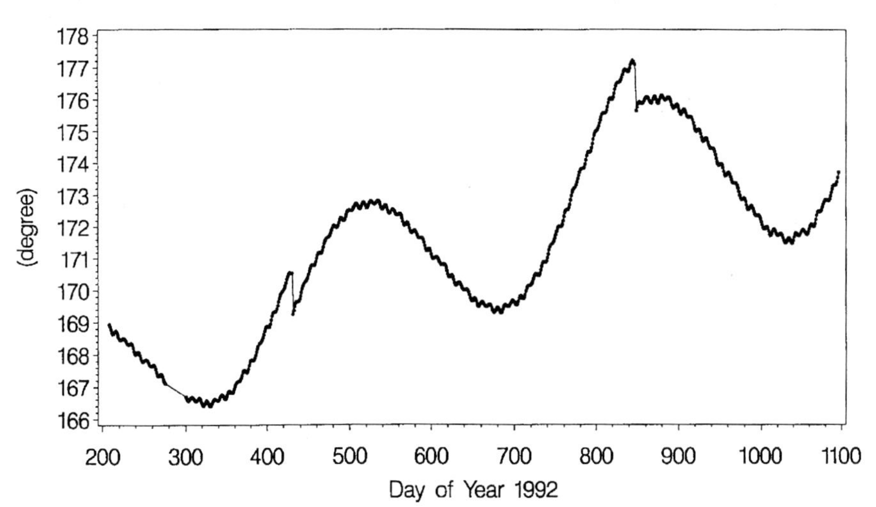

Figure 2.10e. Mean argument of perigee ω of PRN 14 (Mid 1992 - End of 1994).

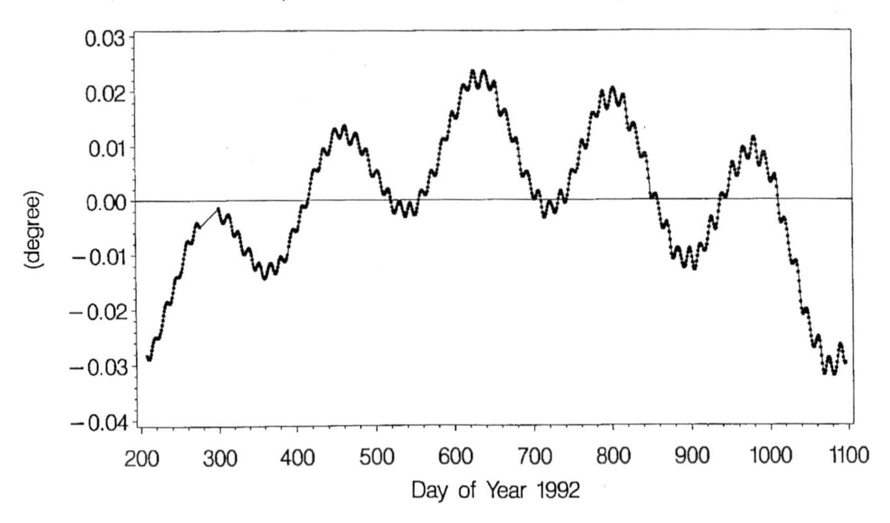

Figure 2.11. Mean R.A. of the ascending node of PRN 14 after removal of linear drift (Mid 1992 - End of 1994).

2.2.5 The Parametrization of Satellite Orbits, Linearization of the Orbit Determination Problem

The developments and considerations in this section are based on the equations of motion of type (2.25). Let us mention, however, that we might as well use equations (2.30) as an alternative formulation of the equations of motion.

In *orbit determination* we are never discussing the general solution of eqns. (2.25) but we are interested in a so-called *particular solution* of eqns. (2.25). Such a solution is uniquely defined, if, in addition to eqns. (2.25), *initial conditions* are given, as well. We are thus considering an *initial value problem* of the following kind (see eqns. (2.25), (2.26a)):

$$\ddot{\mathbf{r}} = -GM \cdot \frac{\mathbf{r}}{r^3} + \mathbf{P}(\mathbf{r}, \mathbf{v}, q_1, q_2, \ldots, q_d, t) \tag{2.35a}$$

$$\mathbf{r}(t_0) = \mathbf{r}(\mathbf{K}_1, \mathbf{K}_2, \ldots, \mathbf{K}_6, t_0),\ \dot{\mathbf{r}}(t_0) = \mathbf{v}(\mathbf{K}_1, \mathbf{K}_2, \ldots, \mathbf{K}_6, t_0) \tag{2.35b}$$

where we assume that the parameters $K_1, K_2, \ldots, K_6$ are the six osculating orbital elements at time t_0.

In the equations of motion (2.35a) (in rectangular coordinates) we have assumed that d *dynamical parameters* $q_1, q_2, \ldots, q_d$ are unknown. In equations

(2.35b) we described the initial position and velocity vectors (actually the component matrices) by the six *osculating elements at time* t_0.

Our orbit determination problem has thus the following 6+d *unknown parameters* $p_1, p_2, \ldots, p_n$, $n = 6+d$:

$$\{p_1, p_2, \ldots, p_n\} = \{K_1, K_2, \ldots, K_6, q_1, q_2, \ldots, q_d\} \quad , \quad n = 6 + d \tag{2.36}$$

These parameters have to be determined using the observations made in a certain time interval $I = (t_0, t_1)$ by a network of tracking stations on ground. (Space borne GPS receivers might be used in addition.)

If we are only considering the solution of the initial value problem (2.35a,b) in the time interval $I = (t_0, t_1)$ we are also speaking of a *satellite arc* with an *arc length* $\ell = |t_1 - t_0|$.

Usually in celestial mechanics the term *orbit determination* is used in a more restricted sense. One assumes that all dynamical parameters are known, i.e., that $d = 0$. This is the case, e.g., in the problem of *first orbit determination* in the planetary system, where from a short (few weeks) series of astrometric observations of a minor planet or a comet we have to derive a particular solution of the equations of motion *without* having any a priori information (other than the observations).

Let us mention that alternatives to the formulation as an initial value problem are possible, too. As a matter of fact the most successful algorithm of first orbit determination is due to C.F. Gauss (1777-1855), who based his considerations on a *boundary value problem*, i.e., he replaces the equation for the velocity in eqn. (2.35b) by an equation for the position at a time $t_1 \neq t_0$, Gauss [1809]. We also refer to Beutler [1983] for more details.

In the case of the GPS we may assume that *most* of the parameters of the force field on the right-hand side of eqns. (2.35a) are *known*. It should be formally acknowledged at this point that in GPS we are making extensive use of the information acquired by other techniques in satellite geodesy, in particular in *SLR* (*S*atellite *L*aser *R*anging). Modeling techniques and the coefficients of the Earth's gravity field are essentially those established in *SLR*. On the other hand, when processing GPS observations, it is never possible to assume that *all* parameters are known. At present usually $d = 2$ dynamical parameters are determined by the IGS processing centers. They are both related to radiation pressure.

We have thus seen that in the case of processing GPS data we have to solve a *generalized orbit determination problem* as compared to the standard problem in celestial mechanics. It is only fair to acknowledge that our orbit determination problem is simpler than the standard problem in the sense that we never have a problem to find a good a priori orbit. So, in principle, we should speak of *orbit improvement* (at times it might be even wise to speak of orbit modification).

We may thus assume that we know an a priori orbit $\mathbf{r}_0(t)$ which *must* be a solution of the following initial value problem:

$$\ddot{\mathbf{r}}_0 = -GM \cdot \frac{\mathbf{r}_0}{r_0^3} + \mathbf{P}(\mathbf{r}_0, \dot{\mathbf{r}}_0, q_{10}, q_{20}, ..., q_{d0}, t) =: \mathbf{f} \qquad (2.37a)$$

$$\mathbf{r}_0(t_0) = \mathbf{r}(K_{10}, K_{20}, ..., K_{60}, t_0), \quad \dot{\mathbf{r}}_0(t_0) = \mathbf{v}(K_{10}, K_{20}, ..., K_{60}, t_0) \qquad (2.37b)$$

Let us now linearize the orbit determination problem: We assume that the unknown orbit **r**(t) may be written as a Taylor series development about the known orbit $\mathbf{r}_0(t)$. As usual we truncate the series after the terms of order 1:

$$\mathbf{r}(t) = \mathbf{r}_0(t) + \sum_{i=1}^{n} \frac{\partial r(t)}{\partial p_i} \cdot (p_i - p_{i0}) \qquad (2.38)$$

Equation (2.38) is the basic equation for the orbit determination process. It gives the unknown orbit as a *linear function* of the unknown orbit parameters p_i, $i=1,2,...,n$.

We know that $\mathbf{r}_0(t)$ is the solution of the initial value problem (2.37a,b). We will now show that the partial derivatives of the approximate orbit are solutions of a linear initial value problem. We just have to take the derivative of eqns. (2.37a,b) for that purpose. But let us first introduce the following symbol for the partial derivatives of the orbit with respect to one orbit parameter $p \in \{p_1, p_2, ..., p_n\}$:

$$\mathrm{z}(t) := \frac{\partial \mathbf{r}_0(t)}{\partial p} \qquad (2.39)$$

Taking the derivative of eqn. (2.37a) with respect to parameter p gives the following differential equation for **z**(t):

$$\ddot{\mathbf{z}} = A_0 \cdot \mathbf{z} + A_1 \cdot \dot{\mathbf{z}} + \mathbf{P}_p \qquad (2.40)$$

where A_0 and A_1 are 3x3 matrices the elements of which are defined in the following way:

$$A_{0,ik} = \frac{\partial \mathbf{f}_i}{\partial r_{0,k}} \quad , \quad i,k = 1,2,3 \qquad (2.40a)$$

$$A_{1,ik} = \frac{\partial \mathbf{f}_i}{\partial \dot{r}_{0,k}} \quad , \quad i,k = 1,2,3 \qquad (2.40b)$$

$$\mathbf{P}_p = \frac{\partial \mathbf{P}}{\partial p} \tag{2.40c}$$

where all the partials have to taken at the known orbit position; $\mathbf{f}_i$ is the component no i of $\mathbf{f}$. Equations. (2.40) are called the *variational equations* associated with the original equations of motion, which are also called *primary equations* in this context. The initial conditions associated with eqns. (2.40) result by taking the derivative of the initial conditions (2.37b) of the primary equations:

$$\mathbf{z}(t_0) = \frac{\partial \mathbf{r}_0}{\partial p} \quad , \quad \dot{z}(t_0) = \frac{\partial \mathbf{v}_0}{\partial p} \tag{2.41}$$

We can thus see that all of the partials in eqn. (2.38) are solutions of the same type of variational equations (2.40) but that the initial conditions are different.

If $p \in (K_1, K_2, ..., K_6)$ we have (2.42a)

$$\mathbf{P}_p = 0 \text{ and } \mathbf{z}(t_0) \neq 0 \quad , \quad \dot{\mathbf{z}}(t_0) \neq 0 \tag{2.42b}$$

whereas *for* $p \in \{q_1, q_2, ..., q_d\}$ we have (2.43a)

$$\mathbf{P}_p \neq 0 \text{ and } \mathbf{z}(t_0) = \mathbf{0} \quad , \quad \dot{\mathbf{z}}(t_0) = \mathbf{0} \tag{2.43b}$$

In the case (2.42a,b) the variational equations (2.40) are even homogeneous. Let us also mention that for GPS orbits we may assume that $A_1 = 0$ because at present no velocity-dependent forces are modelled, to our knowledge.

2.2.6 Numerical Integration

In the present section we have to discuss methods to solve initial value problems of type (2.37a,b) and of type (2.40,41). The corresponding differential equations (2.37a) and (2.40) are *ordinary differential equations*, the latter is even *linear*.

We know the *true* solution of eqn. (2.37a) in terms of trigonometric and elementary functions in the case of the two body problem. Unfortunately this is not true in the general case. Our initial value problems thus have to be solved approximately. In general one makes the distinction between

a) *analytical methods*
b) *numerical methods* .

In case (a) the perturbation equations (2.30) are used to describe the primary equations. The right-hand side of these equations has to be developed into a series of functions of the orbital elements and of time which may be *formally integrated.*

Approximations are unavoidable in this process. The tools are those developed in perturbation theory: In first-order perturbation theory the orbital elements are considered as constant (time independent) on the right-hand side of the perturbation equations, in second-order theory the solutions of the first order are used on the right-hand side, in the theory of order n the solutions of order n-1 are used on the right-hand sides. In section 2.2.3 we have given very simple approximate solutions for some of the orbital elements using first order theory for the term C_{20} of the Earth's gravity field. It is a very important characteristic of these solution methods that the problem of *finding the solution of a differential equation system* is reduced to *quadrature*, i.e. to the problem of formally integrating known basic functions. It is most attractive that the solutions are eventually available as (linear) combinations of known basic functions and that function values may be computed (relatively) easily for any time argument t (even if t is far away from the initial epoch t_0). Another nice characteristic has to be seen in the circumstance that – because the solution is available as a known function not only of time but also of the osculating elements (at time t_0) and of the dynamical parameters $q_1, q_2, \ldots, q_d$ – the partial derivatives with respect to the orbit parameters may be computed in a straightforward way, too. Thus, if we have solved the primary equations with analytical methods, we may also claim to have solved all variational equations associated with them. Analytical methods played an important role in the first phase of satellite geodesy. The first models for the Earth's gravity field stemming from satellite geodesy (SAO Standard Earth I, II, III) were all based on analytical developments of pioneers like I.G. Izaak, Y. Kozai, W.G. Kaula, see Lunquist and Veis [1966].

Analytical solution methods still play an important role in many domains of celestial mechanics and satellite geodesy. In particular, these methods are well suited for understanding phenomena like resonance, see e.g., Hugentobler et al. [1994], Kaula [1966]. Analytical theories completely disappeared from the field of routine orbit determination, however. The reason may be seen in the extreme complexity of the method (every new force type asks for new developments), the difficulty to model phenomena like radiation pressure in the case of eclipsing satellites, and, mainly because of the growing precision requirements. Today we have to ask (at least) for 1 cm orbit consistency even for very long arcs (think, e.g., of multiple months Lageos arcs). This accuracy requirement would ask for a very complex analysis indeed (many terms would have to be taken into account, high-order perturbation theory should be used). We will not consider analytical solution methods from here onwards.

Our discussion of numerical solution methods (b) of the initial value problem is based on Beutler [1990] and on Rothacher [1992]. We should point out that the numerical solution of the initial value problem in satellite geodesy might serve itself as a topic for a one week series of lectures. We therefore have to limit the discussion to the basic facts.

Let us fist discuss the solution of the initial value problem (2.37a,b). For the purpose of this section we may use the following simple notation:

$$\ddot{\mathbf{r}} = \mathbf{f}(\mathbf{r}, \dot{\mathbf{r}}, t) \tag{2.44a}$$

$$\mathbf{r}(t_0) = \mathbf{r_0}, \quad \dot{\mathbf{r}}(t_0) = \mathbf{v_0} \tag{2.44b}$$

Let us briefly characterize the following *different* techniques to solve the initial value problem (2.44a,b):

a) *The Euler Method*
b) *Direct Taylor Series Methods*
c) *Multistep Methods*
d) *Runge-Kutta Methods*
e) *Collocation Methods*

We assume that we want to have the solution (position and velocity vector) available at time t which is *far away* from the initial epoch t_0 (by *far away* we understand that it will not be possible to bridge the time interval I = (t_0, t) with a truncated Taylor series of order ≤ 10).

The Euler Method. This method in a certain sense may be considered as the common *origin* for all other methods to be discussed afterwards. Most of the important aspects of numerical integration already show up in the Euler method , Euler [1768]. As a matter of fact the Euler method plays an essential role in the existence and uniqueness theorems for the solutions of ordinary differential equations in pure mathematics.

Euler divides the interval I in – let us say – m subintervals (Figure 2.12). In each of the subintervals he *defines* an initial value problem with the left interval boundary as initial epoch. In each of the subinterval he approximates $\mathbf{r}(t)$ by a Taylor series of order 2:

```
     t1        t2       ...                          tm-1

|----x--------x-------x--------x---------x------------x---x---x--|
t0                                                               t
```

Figure 2.12. Subdivision of time interval I = (t_0, t).

$$\mathbf{r}_i(t) = \mathbf{r}_{i0} + (t - t_{i-1}) \cdot \mathbf{v}_{i0} + \frac{1}{2} \cdot (t - t_{i-1})^2 \cdot \mathbf{f}(\mathbf{r}_{i0}, \dot{\mathbf{r}}_{i0}, t_0) \tag{2.45a}$$

The corresponding formula for the velocity follows by taking the time derivative of eqn. (2.45a):

$$\dot{\mathbf{r}}_i(t) = \mathbf{v}_{i0} + \left(t - t_{i-1}\right) \cdot \mathbf{f}(\mathbf{r}_{i0}, \dot{\mathbf{r}}_{i0}, t_0) \qquad i = 1, \ldots, m \tag{2.45b}$$

Equations (2.45a,b) define the approximate solution in the subinterval i. The initial conditions at the left initial boundaries are defined in the following way:

$$i = 1: \mathbf{r}_{i0} = \mathbf{r}_0, \quad \mathbf{v}_{i0} = \mathbf{v}_0 \tag{2.45c}$$

$$i \geq 1: \mathbf{r}_{i0} = \mathbf{r}_{i-1}(t_{i-1}), \quad \mathbf{v}_{i0} = \dot{\mathbf{r}}_{i-1}(t_{i-1}) \tag{2.45d}$$

Equation (2.45b) tells us that the errors in the velocities will be of the second order in the lengths of the subintervals (first omitted term in the Taylor series development). Without proof we mention that the error will also be of second order in $\mathbf{r}(t)$ for large $|t - t_0|$. Therefore, by dividing each of the subintervals of I into two subinterval of equal length and by applying Euler's method to the grid with $2 \cdot m$ subintervals, we will get a solution which is four times more accurate than the solution corresponding to m subintervals. This procedure of finer and finer subdivisions of the interval I is in essence the procedure which is also used in the existence and uniqueness theorems.

The idea of dividing the interval I into finer and finer subintervals and of defining subsidiary initial value problems at the left subinterval boundaries will be the same for all methods to be defined below. (The same procedure may be followed for a backwards integration; we just have to replace the left by the right interval boundary.)

Therefore, from now on we only have to consider one of the subintervals (i.e., one of the initial value problems) in a small environment of the initial epoch. For the sake of simplicity of the formalism we will always consider the original initial value problem (2.44a,b).

Direct Taylor Series Methods. This method may be very efficient if it is relatively easy to compute analytically the time derivatives of the function $\mathbf{f}(\mathbf{r}, \dot{\mathbf{r}}, t)$. For the applications we have in mind this is an (almost) hopeless affair for higher than the first derivative of f(..). The first derivative may easily be computed if the matrices $A_0(t)$, $A_1(t)$ (see definitions (2.40a,b)) used to set up the variational equations are available:

$$\dot{\mathbf{f}}(\mathbf{r}, \dot{\mathbf{r}}, t) = A_0(t) \cdot \dot{r} + A_1(t) \cdot \mathbf{f} + \frac{\partial \mathbf{f}}{\partial t}$$

For GPS satellites we may even write (no velocity dependent forces)

$$\dot{\mathbf{f}}(\mathbf{r}, t) = A_0(t) \cdot \dot{\mathbf{r}} + \frac{\partial \mathbf{f}}{\partial \mathrm{t}}$$

where the partial derivative with respect to t might, e.g., be computed numerically. The next derivative would require the computation of the first time derivative of $A_0(t)$ – a lost case!

The direct Taylor series method would just add the terms of higher order in eqns. (2.45a,b). The advantage over the Euler method resides in the fact that the error is no longer proportional to the square but to higher orders of the lengths of partial intervals.

Multistep Methods. Historically these methods were developed in the environment of interpolation theory. It is generally assumed that a series of q-1 *error-free* function values

$$\mathbf{f}\left(\mathbf{r}(t_k),\dot{\mathbf{r}}(t_k),t_k\right),\quad k=-(q-2),-(q-3),\ldots,0 \tag{2.46a}$$

is available initially; no two time arguments are allowed to be identical. We may, e.g., imagine that such a series was established using the Euler method with a very fine partition of the intervals. The approximating function $\mathbf{r}'(t)$ of the true solution is now defined as a polynomial of degree q:

$$\mathbf{r}'(t):=\sum_{i=1}^{q}\mathbf{a}_i\cdot(t-t_0)^i \tag{2.46b}$$

where the coefficients $\mathbf{a}_i$, i=1,...,q+1 are defined as follows:

$$\ddot{\mathbf{r}}''(t_k):=\sum_{i=2}^{q}i\cdot(i-1)\cdot\mathbf{a}_i\cdot(t_k-t_0)^{i-2}=\mathbf{f}\left(t_k,\mathbf{r}(t_k),\dot{\mathbf{r}}(t_k)\right) \tag{2.46c}$$
$$k=-(q-2),-(q-3),\ldots,0$$

$$\mathbf{a}_0=\mathbf{r}_0,\quad \mathbf{a}_1=\mathbf{v}_0 \tag{2.46d}$$

Obviously $\mathbf{r}'(t)$ has the same values as $\mathbf{r}(t)$ at time t_0 (the same is true for the first derivatives at epoch t_0). The second derivative of $\mathbf{r}'(t)$, a polynomial of degree q-1, is the interpolation polynomial of the function values (2.46a). The coefficients $\mathbf{a}_i$, i=2,3,...,q are obtained by solving the systems of linear equations (2.46c). There is one such system of linear equations for each of the three components of $\mathbf{r}(t)$. As we see from eqns. (2.46c) the same coefficient matrix results for each component.

Once the coefficients $\mathbf{a}_i$, i=(0),(1),2,...,q are determined we may use eqn. (2.46b) to compute the values $\mathbf{r}'(t_1),\dot{\mathbf{r}}'(t_1)$. This allows us to compute the right-hand side of eqn. (2.44a) for the time argument t_1. This already closes the loop for the simplest multistep method (an Adams-type method). Accepting

$$\mathbf{f}\left(t_1,\mathbf{r}(t_1),\dot{\mathbf{r}}(t_1)\right)=\mathbf{f}\left(t_1,\mathbf{r}'(t_1),\dot{\mathbf{r}}'(t_1)\right)$$

as the final function value for $\mathbf{f}$(...) at time t_1, we may shift the entire integration scheme by one partial interval and solve the initial value problem at time t_1. If we do that we have used a pure *predictor integration procedure*. We also may refine the function value f(...) at time t_1 by using the interpolation polynomial at times $t_{-(q-3)},\ldots,t_1$ to solve the initial value problem at time t_0. In this case we would speak of a *predictor-corrector procedure*.

Multistep methods may be very efficient. This is true in particular if we use *constant step size*, i.e., if all partial intervals are of the same length. This actually implies that the coefficient matrix of the system of linear equations (2.46c) is identical in every subinterval. In celestial mechanics constant step size is attractive in the case of low-eccentricity orbits, i.e., for eccentricities $e \le 0.02$. This actually is the case for GPS satellites.

Runge-Kutta Methods. Sometimes these methods are also called *single step methods* in particular when compared to the *multi-step methods* discussed above. Runge-Kutta methods are very attractive from the theoretical point of view and they are very simple to use, too. This is the main reason for their popularity.

As opposed to the other methods discussed here Runge-Kutta methods never try to give a local approximation of the initial value problem in the entire environment of the initial epoch t_0. *Their goal is to give an approximation for the solution of the initial value problem for exactly one time argument t_0+h.*

Runge-Kutta methods are equivalent to a Taylor series development up to a certain order q. Runge-Kutta algorithms usually are given for first-order differential equations systems – which do not pose any problems because it is always possible to transform a higher-order differential equation system into a first-order system.

The original Runge-Kutta method is a method of order 4, i.e., the integration error is of order 5. This means that by reducing the step-size h by a factor of 2 the integration error is reduced by a factor of $2^5 = 32$. Runge-Kutta methods were generalized for higher integration orders, too. Moreover, the method was adapted for special second-order systems (no velocity dependent forces, Fehlberg [1972]).

Let us comment the classical Runge-Kutta method of order 4. It approximates the following initial value problem:

$$\dot{\mathbf{y}} = f(\mathbf{y},t) \tag{2.47a}$$

$$\mathbf{y}(t_0) = \mathbf{y}_0 \tag{2.47b}$$

The algorithm

$$\mathbf{y}(t_0 + h) := \mathbf{y}_0 + \frac{1}{6}\cdot(\mathbf{k}_1 + 2\cdot\mathbf{k}_2 + 2\cdot\mathbf{k}_3 + \mathbf{k}_4) \tag{2.47c}$$

where:

$$\begin{aligned}\mathbf{k}_1 &= h\cdot\mathbf{f}(t_0,\mathbf{y}_0)\\ \mathbf{k}_2 &= h\cdot\mathbf{f}(t_0 + h/2, \mathbf{y}_0 + \mathbf{k}_1/2)\end{aligned}$$

$$\mathbf{k}_3 = h \cdot \mathbf{f}\left(t_0 + h/2, \mathbf{y}_0 + \mathbf{k}_2/2\right)$$
$$\mathbf{k}_4 = h \cdot \mathbf{f}\left(t_0 + h, \mathbf{y}_0 + \mathbf{k}_3\right) \tag{2.47d}$$

The solution of the initial value problem for time t_0+h is thus given as a linear combination of function values **f**(...) in the environment of the point $(t_0,\mathbf{y}_0)$. Obviously the function values have to be computed in the order $\mathbf{k}_1,\mathbf{k}_2,\mathbf{k}_3$ and then $\mathbf{k}_4$ because it is necessary to have the function values $\mathbf{k}_1,\mathbf{k}_2,...,\mathbf{k}_{i-1}$ available to compute $\mathbf{k}_i$.

The algorithm (2.47c,d) is a special case of the general Runge-Kutta formula of order 4:

$$\dot{\mathbf{y}}\left(t_0 + h\right) := \mathbf{y}_0 + a_1 \cdot \mathbf{k}_1 + a_2 \cdot \mathbf{k}_2 + a_3 \cdot \mathbf{k}_3 + a_4 \cdot \mathbf{k}_4 \tag{2.48a}$$

where:

$$\mathbf{k}_1 = h \cdot \mathbf{f}\left(t_0, \mathbf{y}_0\right)$$
$$\mathbf{k}_2 = h \cdot \mathbf{f}\left(t_0 + \alpha_2 \cdot h, \mathbf{y}_0 + \beta_1 \cdot \mathbf{k}_1\right)$$
$$\mathbf{k}_3 = h \cdot \mathbf{f}\left(t_0 + \alpha_3 \cdot h, \mathbf{y}_0 + \beta_2 \cdot \mathbf{k}_1 + \gamma_2 \cdot \mathbf{k}_2\right) \tag{2.48b}$$
$$\mathbf{k}_4 = h \cdot \mathbf{f}\left(\mathrm{t}_0 + \alpha_4 \cdot h, \mathbf{y}_0 + \beta_3 \cdot \mathbf{k}_1 + \gamma_3 \cdot \mathbf{k}_2 + \delta_3 \cdot \mathbf{k}_3\right)$$

where the coefficients a...,α...,β...,γ..., and δ... have to be selected in such a way that the Taylor series development of $\mathbf{y}(t_0+h)$ is identical with the series development of eqn. (2.48) up to terms of order 4 in $(t-t_0)$. It is thus easy to understand the principle of the Runge-Kutta method. It is also trivial to write down the equations corresponding to eqns. (2.48b) for higher than fourth order. The actual determination of the coefficients, however, is not trivial at all: The conditions to be met are non-linear, moreover the solutions are not unique. For more information we refer to Fehlberg [1972].

Let us again point out that Runge-Kutta methods are *not* efficient when compared to multistep methods or collocation methods, *but they are extremely robust and simple to use*. As opposed to all other methods they do only give the solution vector at one point in the environment of the initial epoch.

Collocation Methods. They are closely related to multistep methods. They are more general in the following respect:

(1) It is not necessary to know an initial series of function values **f**(...).
(2) Multistep methods may be considered as a special case of collocation methods.

Collocation methods may be used like single step methods in practice. In addition the function values and all derivatives may be computed easily after the integration.

As in the case of multistep methods the approximating function (for each component) of the solution vector is assumed to be a polynomial of degree q:

$$\mathbf{r}'(t) := \sum_{i=1}^{q} \mathbf{a}_i \cdot (t - t_0)^i \tag{2.49a}$$

Here, the coefficients are determined by asking that the approximating functions has the same initial values at time t_0 as the true solution, *and that the approximating function is a solution of the differential equation at* q-1 *different instants of time* t_k:

$$\ddot{\mathbf{r}}(t_k) := \sum_{i=2}^{q} i \cdot (i-1) \, \mathbf{a}_i \cdot (t_k - t_0)^{i-2} = \mathbf{f}\left(t_k, \mathbf{r}'(t_k), \dot{\mathbf{r}}'(t_k)\right) \qquad k = 0,1,\ldots,q-1 \tag{2.49b}$$

$$\mathbf{a}_0 = \mathbf{r}_0, \qquad \mathbf{a}_1 = \mathbf{v}_0 \tag{2.49c}$$

The system of condition equations (2.49b) is a *non-linear* system of equations for the determination of the coefficients $\mathbf{a}_i$. It may be solved iteratively, e.g., in the following way:

$$\ddot{\mathbf{r}}'^{I+1}(t_k) := \sum_{i=2}^{q} i \cdot (i-1) \cdot \mathbf{a}_i^{I+1} \cdot (t_k - k_0)^i = \mathbf{f}\left(t_k, \mathbf{r}'^I(t_k), \dot{\mathbf{r}}'^I(t_k)\right)$$

where I stands for the I-th iteration step.

In the first iteration step we might, e.g., use the Keplerian approximation for $\mathbf{r}$(t) and $\dot{\mathbf{r}}(t)$ on the right-hand side, or, more in the tradition of numerical integration, the Euler approximation. The efficiency of the method depends of course to a large extent on the number of iteration steps. In the case of GPS orbits time intervals of one to two hours may be bridged in one iteration step essentially by using first approximation stemming from first-order perturbation theory and methods of order q=10,11,12.

As we pointed out at the beginning of this section we have to limit the discussion of numerical integration to the key issues. There are of course many more aspects which should be covered here (e.g., the distinction between stiff/non-stiff equations). Let us conclude this section with a few remarks concerning the problem of *automatic step size control* and the *integration of the variational equations*.

Automatic Step-Size Control. The problem is very difficult to handle for a broad class of differential equations, because, in general, we do not know the propagation of an error made at time t_i to a time t_j, $j >> i$.

In satellite geodesy we are in a much better position. We know that in the one-body problem the mean motion is a function of the semi-major axis only. The error analysis given by Brouwer [1937] is still the best reference. It thus makes sense to ask that the change $d(a)$ in the semi-major axis a associated with the numerical solution of the initial value problem at time t_i should be controlled in the following sense:

$$|d(a)| = \left|\sum \frac{\partial a}{\partial r_k} \cdot dr_k + \sum \frac{\partial a}{\partial v_k} \cdot dv_k\right| \approx \left|\sum \frac{\partial a}{\partial v_k} \cdot dv_k\right| \leq e \tag{2.50}$$

where ε is a user defined small positive value. For multistep and collocation methods we may use the last term (term with index q of the series (2.46b) resp. (2.49a)) to have a (pessimistic) estimation for the error dv_k at time t_i. In practice automatic step sign control would lead to much smaller step sizes near the perigee than near the apogee. For GPS satellites step size control is not of vital importance because the orbits are almost circular. We refer to Shampine and Gordon [1975] for a more general discussion.

Integration of the Variational Equations. In principle we might skip this paragraph with the remark that each of the methods presented for the solution of the primary equations (2.44a,b) may also be used for the integration of the variational equations (2.40,41). This actually is often done in practice. Let us mention, however, that there are very efficient algorithms making use of the linearity of equations (2.40). The system of condition equations (2.49b) becomes, e.g., linear for such problems. It is thus not necessary to solve this system iteratively, it may be solved in one step.

Let us add one more remark: whereas highest accuracy is required in the integration of the primary equations, the requirements are much less stringent for the variational equations. In principle we only have to guarantee that the terms

$$\left|\frac{\partial r}{\partial p_i} \cdot dp_i\right| = \left|\mathbf{z}_i(t) \cdot dp_i\right|, \quad dp_i = (p_i - p_{i0}) \tag{2.51}$$

are small compared to the orbit accuracy we are aiming at (p_i is one of the orbit parameters, p_{i0} is the known a priori value). Because the quantities dp_i are becoming small when the orbit determination is performed iteratively, rather crude approximations for the partials $\mathbf{z}_i(t)$ are sufficient. In the case of the GPS even the Keplerian approximation is sufficient for the partials with respect to the initial conditions for arc length up to a few days! The partial derivatives for the two body problem may be found in Beutler et al. [1995].

2.3 THE PERTURBING FORCES ACTING ON GPS SATELLITES

2.3.1 Overview

The discussions in the overview section are based on Rothacher [1992]. Here we consider essentially the same forces as Rothacher [1992] and as Landau [1988] before him. We left out ocean tides, albedo radiation pressure, and relativistic effects from our considerations because these effects are very small for arc lengths up to three days. Both authors present the (typical) accelerations for these terms and the orbit error after one day if the effect is neglected and identical initial conditions are used.

Table 2.1 gives an overview of the important perturbing accelerations and the effect of neglecting these terms after one day of orbit integration. In addition we include in Table 2.1 the rms error of an orbit determination based on 1 day, resp. 3 days of *pseudo-observations* (geocentric x, y, and z positions of the satellite every 20 minutes) if the respective terms are *not* included in the force model. As in the routine environment we characterize each orbit by 8 parameters (six for the initial conditions, two for the radiation pressure).

From Table 2.1 we conclude that all perturbations with the exception of radiation pressure should be known well enough from satellite geodesy using low Earth orbiters. In view of the fact that all gravitational parameters are known with a relative precision of about 10^{-6} we conclude that it does not make sense to solve for such parameters in an orbit determination step in the case of GPS satellites.

Due to the shape of the satellite and due to the fact that attitude control never can be done without an error the same is not true for the radiation pressure terms. We always have to include or model these effects when dealing with a particular satellite arc.

Let us now consider *resonance* and *radiation pressure* in more detail.

2.3.2 The Radiation Pressure Models

For a satellite absorbing the entire solar radiation, the perturbing acceleration due to radiation pressure may be written as Rothacher [1992]:

$$\mathbf{a}_d = \mu \cdot \left\{ P_s \cdot C_r \cdot \frac{A}{m} \cdot a_s^2 \cdot \frac{\mathbf{r} - \mathbf{r}_s}{|\mathbf{r} - \mathbf{r}_s|^3} \right\} \tag{2.52}$$

where:

$\mathbf{a}_d$ is the acceleration due to the direct radiation pressure,

μ is the eclipse factor (= 1 if the satellite is in sunlight, = 0 if the satellite is in the Earth shadow),

A/m is the cross-section area of the satellite as seen from the Sun divided by its mass,

a_s is the astronomical unit (AU),

$P_s = S/c$ is the radiation pressure for a completely absorbing object with $A/m = 1$ at the distance of one astronomical unit. (S is the solar constant, c the velocity of light),

C_r is a reflection coefficient,

$\mathbf{r}, \mathbf{r}_S$ are the geocentric coordinates of satellite and Sun respectively.

Table 2.1. Perturbing Acceleration and their effect on satellite orbits: Net effect when used/left out in the equations of motion and after an orbit determination using one resp. three days of data.

Perturbation	Acceleration [m/s²]	Orbit error after 1 day [m]	rms of orbit determination [m] using 1 day of observations	rms of orbit determination [m] using 3 days of observations
Kepler term of Earth potential	0.59	∞	∞	∞
Term C_{20}	$5 \cdot 10^{-5}$	10'000	1'700	5'200
Other terms of Earth potential	$3 \cdot 10^{-7}$	200	15	50
Attraction by the Moon	$5 \cdot 10^{-6}$	3'000	100	300
Attraction by the Sun	$2 \cdot 10^{-6}$	800	45	150
Fixed body tides	$1 \cdot 10^{-9}$	0.30	0.03	0.08
Direct radiation pressure	$9 \cdot 10^{-8}$	200	0.0	0.0
y-bias	$6 \cdot 10^{-10}$	1.5	0.0	0.0

The same formula is valid for a perfectly spherical satellite even if we allow for absorption and reflection of solar radiation. The difference would only consist of different numerical values for the reflection coefficient C_r.

The perturbing acceleration due to radiation pressure (we also speak of *direct* radiation pressure in this context) always points into the direction Sun → satellite in model (2.52). For spherical satellites the ratio A/m may be assumed as constant.

For GPS satellites the cross section area A as seen from the Sun is attitude dependent. This cross section area thus will be variable over one revolution, there also will be variations over the year due to the changing angle between the normal to the orbital plane and the unit vector pointing to the Sun.

Moreover one has to take the reflective properties of the satellite into account. As soon as we allow for reflection there also are acceleration components perpendicular to the direction Sun → satellite. The most commonly used radiation pressure models for GPS satellites may be found in Fliegel et al. [1992]. The authors give relatively simple formulae for the radiation pressure in a spacecraft fixed coordinate system (see Figure 2.13, taken from Rothacher [1992]):

Assuming perfect attitude control Fliegel et al. [1992] show that the resulting force always lies in the (x, z) plane. They give simple algorithms to compute the force components in x- and z- directions as a function of one parameter ß only. ß is the angle between the positive z axis and the direction from the Sun to the satellite. The models are called Rock4 (for Block I satellites) and Rock42 models (for Block II satellites). They are recommended in the IERS standards, McCarthy [1992]. The distinction is made between the standard model or S-model (no longer recommended by the authors) and the T-model which includes thermal re-radiation of the satellite.

It is worthwhile pointing out that in practice the differences between the two Rock models (S or T) and the much simpler model assuming a constant acceleration in the direction Sun → satellite are very small, *provided* either a direct radiation pressure parameter, or (what is equivalent), a scaling parameter for the Rock – models is estimated. The differences between the three models are of the order of 2 % of the total radiation pressure only (i.e., of the order of the y-bias, see Table 2.1 and the discussion below).

So far we assumed that the GPS attitude control is perfect. In theory the y-axis of the satellite (Figure 2.13) should always be perpendicular to the direction Sun → satellite. The attitude control is based on a feedback loop using solar sensors, it is performed by momentum wheels. These momentum wheels rotate about the x-axis (with the goal that the z-axis is always pointing to the Earth) and about the z-axis (with the goal to have the y-axis perpendicular to the direction to the Sun). If the solar panels axes are perfectly normal to the direction to the Sun, there is no y-bias. In all other cases there will be a net force in the direction of the y-axis. It proved to be essential to solve for one so-called *y-bias* for each satellite arc of one day or longer.

The perturbing acceleration due to the y-bias p_2 has the following form:

$$\mathbf{a}_y = \mu \cdot p_2 \cdot \mathbf{e}_y \tag{2.53}$$

where $\mathbf{a}_y$ is the acceleration in the inertial space, $\mathbf{e}_y$ is the unit vector of the solar panels' axis in inertial space, and μ is the eclipse factor.

Let us point out that the integration has to be initialized at the light-shadow boundaries in order to avoid numerical instabilities.

In the section 2.3.4 we will present values for the direct radiation pressure and for the y-bias based on 2.5 years of results gathered at the CODE processing center.

2.3.3 Resonance Effects in GPS Satellite Orbits

All terms with $m = 2,4,...$, $n = 2,3,4,...$ of the Earth's gravitational potential (2.24) are candidates to create resonance effects because identical perturbing accelerations result after each revolution for these terms (at least in the Keplerian

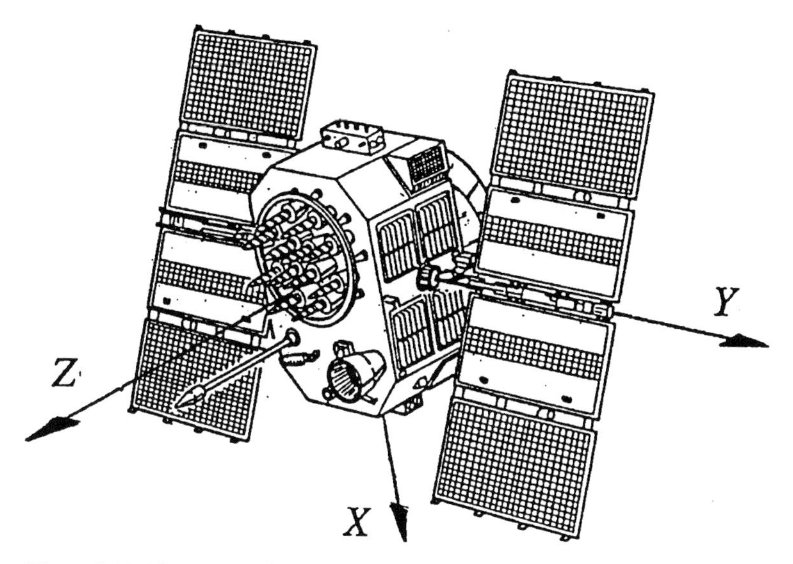

Figure 2.13. The spacecraft-fixed coordinate system (x, y, z).

approximation). A full discussion of the resonance problem is rather complex. Below we present a geometrical discussion only for the semi-major axis a and only for two terms ($n = 2$, $m = 2$ and $n = 2$, $m = 3$). For a more complete treatment of the problem we refer to Hugentobler and Beutler [1994] and to Hugentobler [1996].

The contributions due to the terms ($n = 2$, $m = 2$) and ($n = 3, m = 2$) may be written as follows (we are only interested in the mathematical structure of the terms, not in the numerical values of the coefficients) :

$$V_{22} = cs_{22} \cdot r^{-3} \cdot \cos\beta \cdot \cos(2\lambda + \Theta_{22}) \tag{2.54a}$$

$$V_{32} = cs_{32} \cdot r^{-4} \cdot \cos\beta \cdot \sin(2\beta) \cdot \cos(2\lambda + \Theta_{32}) \tag{2.54b}$$

Considering only circular orbits (i.e., using the approximation $e = 0$) we may write down the following simplified version of the perturbation equation for a (see eqn. (2.30a)):

$$\dot{a} = \frac{2}{n} \cdot S \tag{2.54c}$$

where n is the mean motion of the satellite, S is the pertubing acceleration in tangential direction (circular motion). *Resonance will only show up if the mean value of S over one revolution will significantly differ from zero.* The mean value of $\dot{a}$ over one resultion simply may be approximated by

$$\bar{\dot{a}} = \frac{2}{n} \cdot \bar{S} \tag{2.54d}$$

In order to compute this mean value $\bar{S}$ we have to take the derivatives of the expressions (2.54a,b) with respect to r, λ, and ß. We conclude right away that the derivative with respect to r is of no importance in this context (the resulting acceleration is by definition normal to the along-track component). We are thus left with the partials with respect to λ and ß as contributors to S. Let us first compute the accelerations a_λ and a_β parallel and normal to the equator due to the potential terms (2.54a,b) (again we are not interested in the numerical values of the coefficients):

$$\begin{aligned}
&n = 2,\ m = 2: \\
&a_\lambda = cs'_{22} \cdot \cos\beta \cdot \sin(2\lambda + \Theta_{22}) \\
&a_\beta = cs''_{22} \cdot \sin\beta \cdot \cos(2\lambda + \Theta_{22}) \\
&n = 3,\ m = 2 \\
&a_\lambda = cs'_{32} \cdot \sin 2\beta \cdot \sin(2\lambda + \Theta_{32}) \\
&a_\beta = cs''_{32} \cdot \cos\beta \cdot (3\cos 2\beta - 1) \cdot \cos(2\lambda + \Theta_{32})
\end{aligned} \tag{2.54e}$$

These accelerations have to be projected into the orbital plane. Let us look at the geometry in an arbitrary point P (corresponding to a time argument t) of the orbit. Let us furthermore introduce in point P the angle γ between the velocity vector at time t and the tangent to the sphere with radius a in the meridian plane and pointing towards the north pole (see Figure 2.14).

The S component in point P is computed as

$$S = a_\beta \cdot \cos\gamma + a_\lambda \cdot \sin\gamma$$

Let us assume that at time t_0 the satellite is in the ascending node and that the geocentric longitude of the node is λ_0 at time t_0. It is relatively easy to prove that $\overline{S} = 0$ for the term $(n = m = 2)$. For the term $(n = 2, m = 3)$ we have

$$\overline{S} = \frac{1}{4} \cdot cs''_{32} \cdot \sin i \cdot \left(1 - 2\cos - 3\cos^2 i\right) \cdot \cos\left(2 \cdot \lambda_0 + \phi_{32}\right)$$

$$= \frac{15}{8} \cdot GM \cdot \left(\frac{a_e}{a}\right)^3 \cdot \frac{1}{a^2} \cdot J_2 \cdot \sin i \cdot \left(1 - 2 \cdot \cos i - 3\cos^2 i\right) \cdot \cos\left(2\lambda_0 + \phi_{32}\right) \quad (2.54f)$$

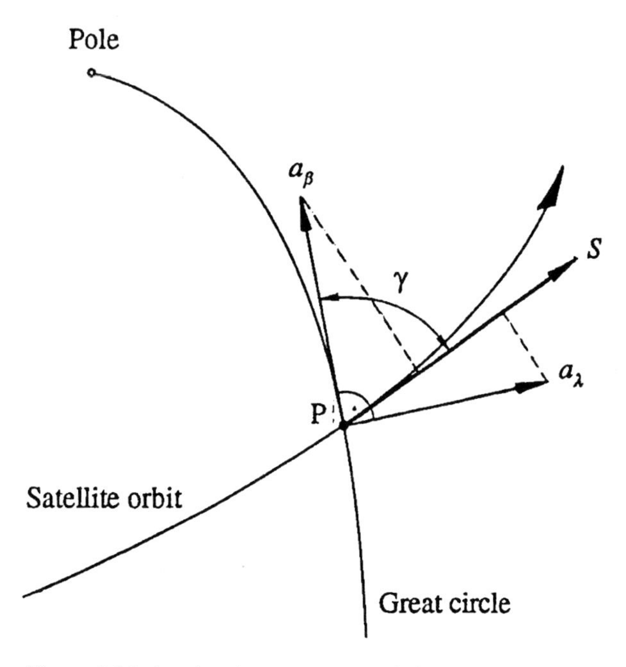

Figure 2.14. Accelerations a_λ, a_β and S at an arbitrary point P of the orbit. γ is the angle between the velocity vector and the tangent vector in P pointing to the north pole.

Relation (2.54f) reveals that the satellites in one and the same orbital plane actually will have significantly different drifts in the mean semi-major axis. As a matter of fact these drifts must be significantly different if the satellites are separated by 120° nominally. Examples may be found in the next section.

Let us conclude this section with a few remarks going beyond our geometrical treatment of the problem:

- Hugentobler and Beutler, [1994] show that the term with $n = 2$, $m = 3$ actually is the dominant contributor to resonance for GPS satellites.

- Two other terms, $n = m = 2$ and $n = m = 4$ also give significant contributions (about a factor of 5 smaller than the term $n = 2$, $m = 3$). Why did we not discover the term $n = m = 2$? Simply because we did not consider the radial perturbation component R in the perturbation equation (use of eqn. (2.54c) instead of eqn. (2.30a)). The term $n = m = 2$ gives rise to a term of first order in the eccentricity e, whereas the other two terms are of order zero in e.
- In Figure 2.10a we get the impression that the mean drift in a stays more or less constant over a time period of 2.5 years. The impression is correct, but we have to point out that the mean drift over long time intervals (let us say over 25 years) must average out to zero. As a matter of fact PRN 14 is artificially kept at this extremely high drift rate because of the manoeuvres! These manoeuvres prevent the satellite from significantly changing the longitude λ_0 of the ascending node!
- The actual periods for the periodic changes of a are different for different satellites. Typically these periods range between 8 and 25 years, Hugentobler [1996].
- Figure 2.15 shows the development of the semi-major axis for PRN 12. This spacecraft, an *old* Block I satellite, was *not* manoeuvred in the time period considered due to a lack of fuel. We see a significant change of the drift in a from 2.1 m/day to 3.6 m/day. Figure 2.15 illustrates that the changes in the semi-major axes due to resonance are not secular, but of long periods (if the satellites are no longer manoeuvred).
- Resonance phenomena also exist in the other orbital elements (see section 2.2.4) but they are not important for the arrangement of the satellites in the orbital planes.

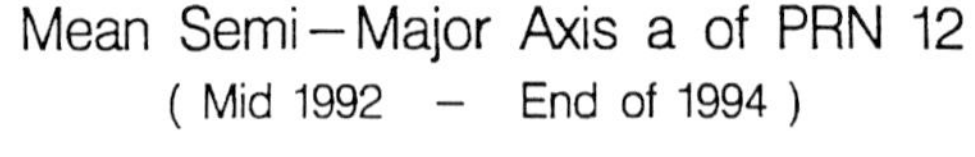

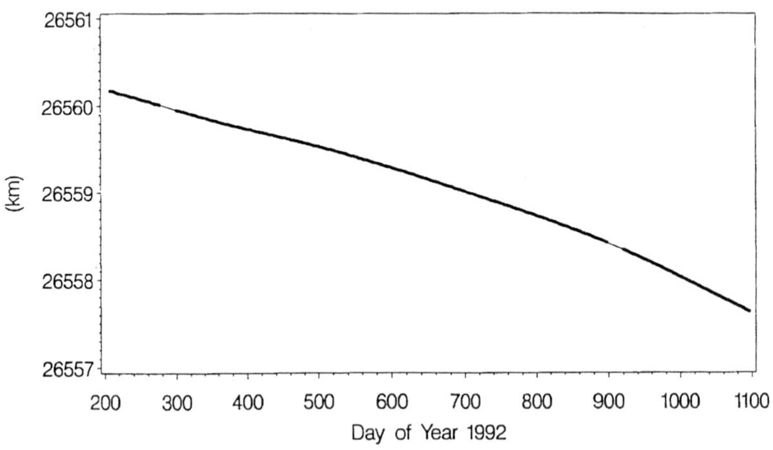

Figure 2.15. Development of the semi-major axis a for PRN 12 (mid 1992 to end of 1994).

2.3.4 Development of the Satellite Orbits Since mid 1992

The material presented in this section is extracted from results produced by the CODE Analysis Center of the IGS, in particular from the annual reports for 1992, 1993, and 1994. Let us start this overview with Table 2.3 containing the essential elements of the satellite constellation (Block II satellites only) on day 301 of year 1994.

We see that in general the six orbital planes are very well defined (inclination i and right ascension of the ascending node Ω). The distribution of the satellites within the orbital planes is identical with that given in Figure 2.1. The longitude of the ascending node is of course no orbital element. It was added to Table 2.2 for later use in this section.

Table 2.3 tells us that manoeuvres are rather frequent events in the life of the satellites (about one per year on the average). In general the semi-major axis is changed by 1.5 to 3 km by such a manoeuvre. In general only the semi-major axis and the argument of perigee are dramatically changed by a manoeuvre. Perturbation equation (2.30a) tells us that the impulse change must take place in the orbital plane, more specifically in S-direction to achieve that. Since we know the masses of the satellites we even might calculate the impulse change involved in the individual manoeuvres, and, with some knowledge of chemistry (horrible dictu) even the fuel which was used during such events.

The mean values for $\dot{a}$ or the Block II satellites were pretty stable during the time period considered. In view of the discussions in the preceding section we might be tempted to display the mean drifts in a as a function of the argument 2λ . If actually what we said at the end of the preceding paragraph is true, we would expect a sinusoidal change of a as a function of the mentioned argument. Figure 2.16 shows that the behavior is as expected. Taking the coefficients C_{32} and S_{32} of the potential field (2.25) and computing the phase angle ϕ_{32} we even are able to verify that the maxima, minima, and zero crossings occur roughly at the correct values of the argument.

Radiation pressure parameters (rpr-parameters) are estimated by all IGS processing centers. Each center has uninterrupted series of daily rpr-parameters available for each GPS satellite since the start of the 1992 IGS test campaign on 21 June 1992. These parameters are (a) scale factors for the Rock models used or, alternatively, direct radiation pressure parameters on top of the Rock models, and (b) y-biases for each satellite and for each day since the start of the 1992 IGS test campaign.

It was mentioned before that in practice, in consideration of the short arcs (1-3 days) usually produced, the differences between different models for the direct radiation pressure (Rock4/42, versions S or T, or no a priori model) are small. It is even possible to reconstruct from one series of results based on a priori model A the parameters for a series based on a priori model B *without* actually reprocessing the entire series.

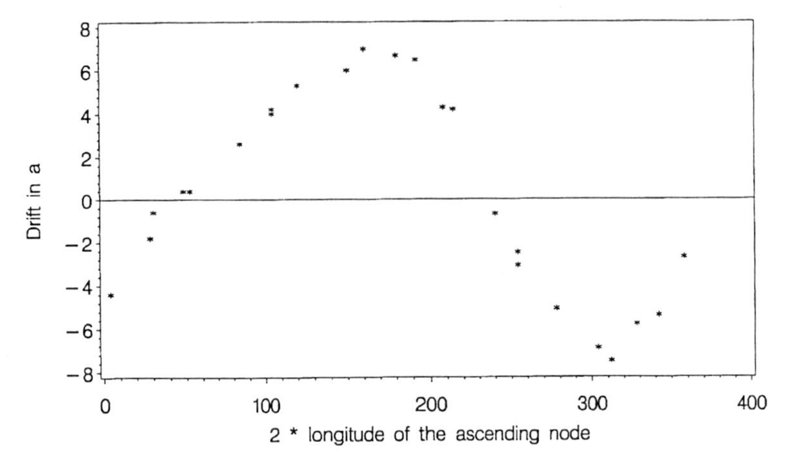

Figure 2.16. Drifts in a as a function of $2 \cdot \lambda_0$, (λ_0 = longitude of ascending node).

Table 2.2. Mean orbit elements of Block II satellites on day 301 of year 1994.

MEAN ELEMENTS FOR ALL SATELLITES: Day 301 of Year 1994

SAT PL	a [m]	e	i [deg]	Ω [deg]	ω [deg]	U_0 [deg]	Longitude of ascending node [deg]
7 C	26561622	0.00685	55.2	12.1	208.5	151.2	52
31 C	26560679	0.00507	55.2	12.2	37.1	256.5	104
28 C	26560215	0.00486	55.6	12.6	170.2	287.6	120
24 D	26561289	0.00580	55.8	72.7	235.9	47.1	60
4 D	26560292	0.00312	55.2	73.1	288.2	84.9	80
15 D	26560152	0.00699	55.5	75.0	103.4	177.0	127
17 D	26560561	0.00788	55.6	77.0	114.8	307.8	195
14 E	26560461	0.00302	55.1	134.5	171.5	107.0	152
21 E	26560589	0.01152	54.7	132.7	164.4	215.2	236
23 E	26560573	0.00870	54.9	134.7	225.1	248.0	222
16 E	26560643	0.00064	54.9	135.1	285.1	341.4	270
18 F	26560649	0.00589	54.0	191.2	77.8	16.7	164
29 F	26561616	0.00487	54.7	191.2	254.5	53.6	182
1 F	26559252	0.00345	54.7	193.7	290.0	147.8	232
26 F	26560998	0.00834	54.9	192.5	307.4	260.7	287
25 A	26560728	0.00567	54.1	251.3	171.27	79.6	255
9 A	26560768	0.00315	54.5	252.8	332.68	180.9	307
27 A	26560634	0.01092	54.3	252.1	142.84	286.4	359
19 A	26559916	0.00044	53.5	251.2	201.00	318.1	14
20 B	26561320	0.00462	54.9	311.6	81.21	87.7	319
5 B	26561417	0.00223	54.7	311.9	236.93	120.8	336
2 B	26560263	0.01364	54.6	311.0	210.81	221.9	26
22 B	26560566	0.00759	54.6	311.9	347.62	359.4	96

Table 2.3. Satellite events since mid 1992, including the manoeuvres as they were detected at CODE processing center, the change in the semi-major axis associated with the manoeuvres, and the mean rate of change of **a** over the time period mid 1992 to end of 1994.

"n" : New satellite included into the CODE processing } Flags F
"+" : Old satellite excluded from the CODE processing } Flags F

PRN	Plane	Processed since	until	F	#	Manoeuvre Epochs		*da*	*da/dt*
09	A	1993 7 25	1994 12 31	n	1	1994	420	2113 m	-3.1 m/d
19	A	1992 7 26	1994 12 31		2	1993	116	1318 m	-1.8 m/d
						1994	1215	1467 m	
27	A	1992 9 30	1994 12 31	n	1	1994	3 3	1701 m	-2.7 m/d
25	A	1992 7 26	1994 12 31		2	1993	325	-2334 m	6.0 m/d
						1994	317	-2121 m	
02	B	1992 7 27	1994 12 31		1	1993	830	-572 m	0.4 m/d
05	B	1993 9 28	1994 12 31	n	1	1994	9 2	2980 m	-7.5 m/d
20	B	1992 7 26	1994 12 31		2	1993	413	2402 m	-5.1 m/d
						1994	816	2755 m	
22	B	1993 4 7	1994 12 31	n	2	1993	527	526 m	6.5 m/d
						1994	2 9	-3025 m	
06	C	1994 3 27	1994 12 31	n	2	1994	411	53462 m	-5.4 m/d
						1994	416	31744 m	
07	C	1993 6 18	1994 12 31	n	2	1993	1216	594 m	4.2 m/d
						1994	1110	-2386 m	
28	C	1992 7 26	1994 12 31		1	1992	1216	788 m	-0.7 m/d
31	C	1993 4 29	1994 12 31	n	1	1993	111	-2020 m	4.3 m/d
04	D	1993 11 21	1994 12 31	n	1	1994	328	-2695 m	7.0 m/d
15	D	1992 7 26	1994 12 31		1	1993	8 2	1730 m	-2.5 m/d
17	D	1992 7 26	1994 12 31		1	1994	120	720 m	-0.6 m/d
24	D	1992 7 26	1994 12 31		2	1993	927	-2539 m	5.3 m/d
						1994	1129	-2334 m	
14	E	1992 7 26	1994 12 31		2	1993	3 5	2579 m	-6.9 m/d
						1994	427	2938 m	
16	E	1992 7 26	1994 12 31		2	1992	124	-2660 m	6.7 m/d
						1994	2 2	-3044 m	
23	E	1992 7 26	1994 12 31		1	1993	920	-1678 m	2.6 m/d
21	E	1992 7 26	1994 12 31		0				0.4 m/d
01	F	1992 12 7	1994 12 31	n	1				4.0 m/d
						1994	1011	-2257 m	
18	F	1992 7 26	1994 12 31		2	1993	317	2569 m	-5.8 m/d
						1994	5 6	2425 m	
26	F	1992 7 26	1994 12 31		1	1993	812	-2381 m	4.2 m/d
29	F	1993 1 4	1994 12 31	n	3	1993	520	1914 m	-4.4 m/d
						1993	9 7	-1161 m	
						1993	114	1528 m	
						1994	1028	2006 m	

Block II Satellites						
03	-	1992 7 26	1994 04 07	+	0	0.2 m/d
11	-	1992 7 26	1993 5 4	+	0	-0.1 m/d
12	-	1992 7 26	1994 12 31		0	-2.9 m/d
13	-	1992 7 26	1993 12 31	+	0	1.5 m/d

Figure 2.17a contains the result of such a reconstruction for PRN 19, a Block II satellite. Based on the results of the CODE processing center, which uses Rock4/42 (Type S) as a priori rpr model, the direct radiation pressure parameters corresponding to the *no* (zero) radiation pressure model were computed.

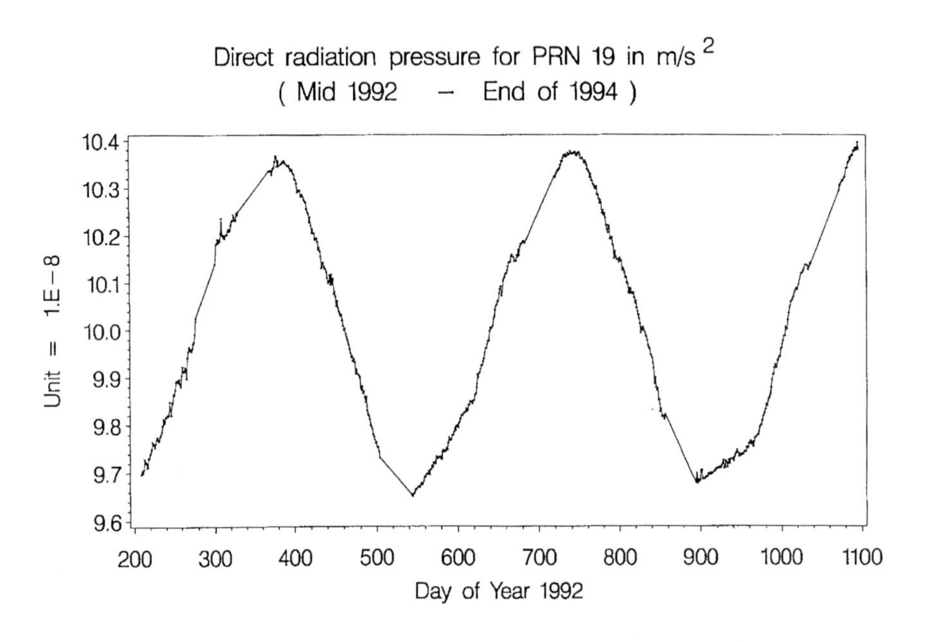

Figure 2.17a. Direct radiation pressure for PRN 19 in m/s².

The mean value for the acceleration due to direct solar radiation is about $1 \cdot 10^{-7}$m/s². Figure 1.17a thus shows the result corresponding to radiation pressure model (2.52) (where the variation due to the ellipticity of the Earth's orbit around the Sun was *not* taken into account). The annual oscillation actually is caused by the ellipticity of the Earth's orbit: the maximum is in January, the minimum in June, the expected variation is

$$\frac{rpr(\max) - rpr((\min)}{\frac{1}{2} \cdot (rpr(\max) + rpr(\min))} = \frac{1}{(1-e)^2} - \frac{1}{(1+e)^2} = 4 \cdot e \approx 0.067 \tag{2.55}$$

So, we have just rediscovered the ellipticity of the orbit of the Earth ... ! This signature may of course be taken into account as indicated by eqn. (2.52). Let us mention that the rpr values gathered during eclipse seasons were taken out in Figures 2.17 – the results were somewhat noisier.

Figure 2.17b shows that the dominant characteristic after removing the annual variation is roughly semi-annual (solid line, best fitting curve $p \cdot a^2/r(t)^2$ subtracted). The residuals are clearly correlated with the angle $2 \cdot \gamma$, where γ is the angle between the normal to the orbital plane and the direction from the Earth to the Sun. The dotted line shows the residuals after taking out in addition the semi-annual term (best fitting trigonometric series truncated after the terms of order 2 in the argument $2 \cdot \tau$).

Figure 2.17b demonstrates that the direct radiation pressure is constant in time over rather long time intervals. The semi-annual variations are of the order of a few units in $10^{-10} m/s^2$. We are thus allowed to conclude that direct radiation pressure may be predicted in a quite reliable way. This is also underlined by Figure 2.18 which shows the mean values for the direct rpr-parameters over the time interval of 2.5 years for all GPS satellites processed in this time by the CODE processing center. We see in particular that the rpr parameters are quite consistent within the classes of Block II and Block IIA spacecrafts. PRN 23 is an exception: it seems that the solar panels are not fully deployed - the result is a somewhat smaller value for the direct radiation pressure. We should mention that the orbit of PRN 23 is particularly difficult to model.

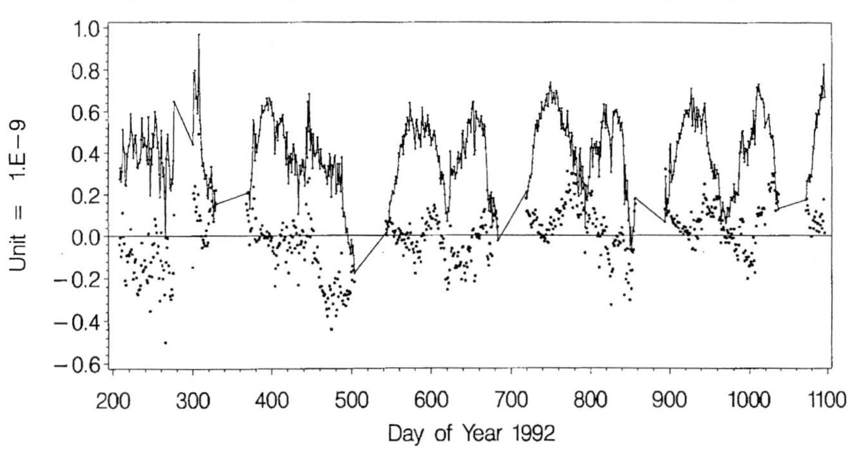

Figure 2.17b. Direct radiation pressure (a) after removing annual term of the form $p_0 \cdot (a_E^2 / r^2)$ (solid line), (b) after removing in addition the semi-annual variation (dotted line).

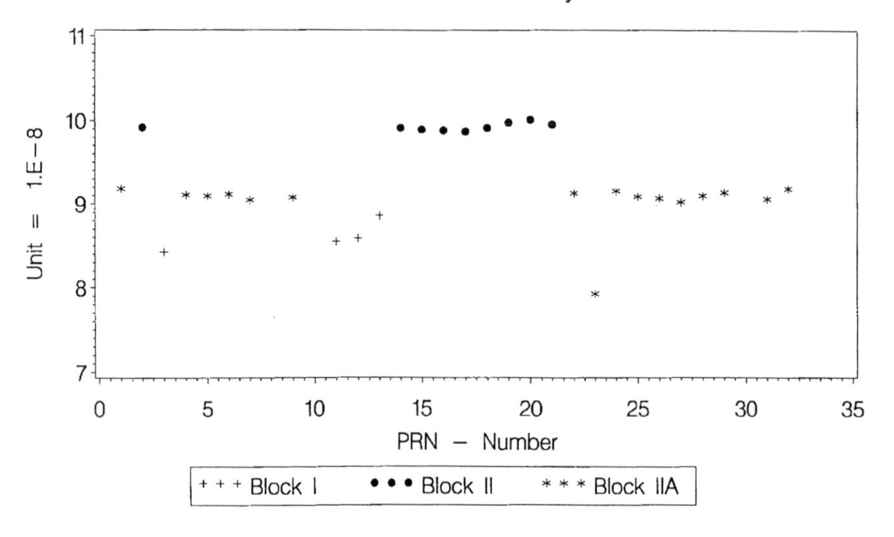

Figure 2.18. Mean values for direct radiation pressure parameters over a time interval of 2.5 years.

Let us have a look at the y-biases. We start with Figure 2.19 corresponding to Figure 2.17a, which gives the y-biases as a function of time for PRN 19. First of all we see that the y-biases are much smaller in absolute value than the direct radiation pressure parameter (about a factor of 200). Nevertheless the y-bias is an important perturbing acceleration because the mean value of the S-component of the perturbing acceleration is *not* zero over one revolution.

The y-biases seem to be quite consistent until mid 1994. Afterwards the mean values are quite different before and after the eclipse seasons. This behavior might be caused by the change in the attitude control of the satellites, Bar Sever et al [1994]. Other satellites show a similar behavior.

All y-biases seem to be slightly negative, the absolute values are of the order of a few units of 10^{-10} m/s². It is of little value to reproduce a figure with the mean values for the y-biases for all GPS satellites. There are significant changes in time (like, e.g., those in Figure 2.19) and for some satellites there are also significant differences for the cases $\gamma < 90°$ and $\gamma > 90°$. These results have to be analyzed in more detail before coming up with useful predictions.

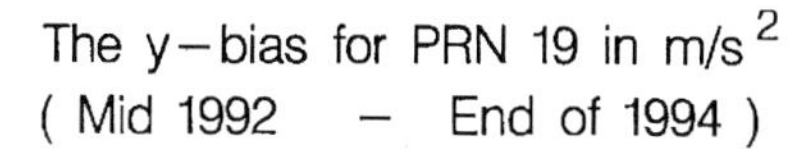

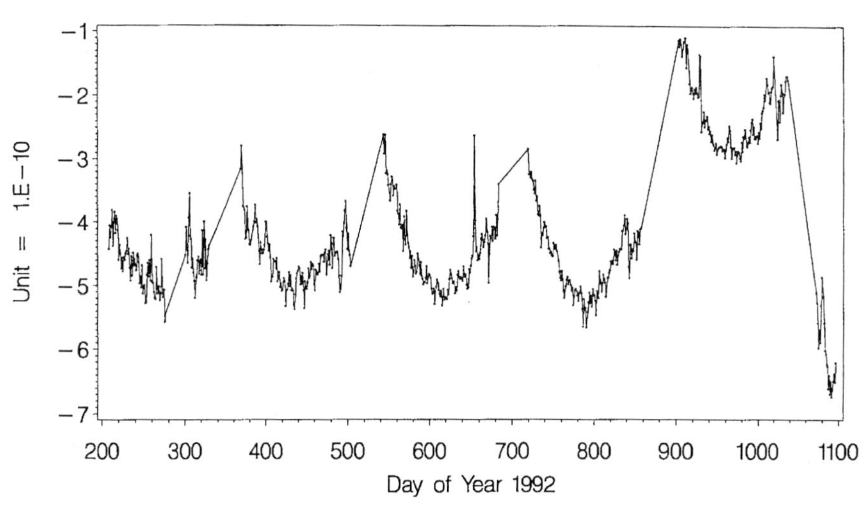

Figure 2.19. The *y*-bias for PRN 19 mid 1992 - end of 1994 in m/s².

2.4 GPS ORBIT TYPES

The information concerning Broadcast and Precise Orbits presented in this section was extracted from Rothacher [1992], Hofmann-Wellenhof [1994], van Dierendonck et al. [1978], the information concerning the IGS stems from papers listed in the references of this chapter. We refer to these references for more information. Below we have to confine ourselves to a short outline of the principles underlying the following orbit types:

- *Broadcast orbits* which are available in real time, their name indicates that they are transmitted by the satellites,
- *Precise Orbits* produced by the Naval Surface Warfare Center together with the DMA, available upon request about 4-8 weeks after the observations.
- *IGS orbits*, produced by the International GPS Service for Geodynamics (IGS), available to the scientific world about 2 weeks after the observations.

2.4.1 Broadcast and Precise Orbits

The Operational Control System (OCS) for the GPS became operational in September 1985. The Master Control Station, situated at Colorado Springs, is responsible for satellite control, the determination, prediction and dissemination of satellite ephemerides and clocks information. Five monitor stations, at Colorado

Springs, Hawaii (Pacific Ocean), Ascension Islands (Atlantic Ocean), Diego Garcia (Indian Ocean), and Kwajalein (Pacific Ocean, near Indonesia) are tracking the GPS satellites. Their recorded pseudorange data (not the phase data) are used for routine orbit and satellite clock determination and prediction.

The Naval Surface Warfare Center (NSWC) together with the Defence Mapping Agency (DMA) generate the so-called *Precise Orbits* about two months after the actual observations. In addition to the five stations mentioned data from Quito (Ecuador), Buenos Aires (Argentina), Smithfield (Australia), Hermitage (England), and Bahrein are used for this routinely performed analysis. Relatively long spans of pseudorange data (8 days) are analyzed to produce long arcs.

In practice broadcast orbits are of much greater importance than precise orbits, because the former are available in real time. When we compare broadcast orbits with the high precision orbits of the IGS (see section 2.4.3) we should keep in mind that broadcast orbits are *predicted*, the prediction period being somewhere between 12 and 36 hours. In view of this fact and in view of the small number of only five tracking stations used for orbit determination it must be admitted that broadcast orbits are of an amazing and remarkable quality.

Broadcast orbits are based on a numerically integrated orbit. The orbits are made available in the so-called B*roadcast Navigation Message,* van Dierendonck et al. [1978]. Instead of just transmitting an initial state and velocity vector, or, alternatively, a list of geocentric satellite coordinates, pseudo-Keplerian orbit elements including the time derivatives for some of the elements are transmitted. The orbit parameters are determined to fit the numerically integrated orbit in the relevant time interval. New broadcast elements are transmitted every two hours. Broadcast orbits refer to the WGS-84 (World Geodetic System-84). For more information we refer to van Dierendonck et al. [1978] and to Hofmann-Wellenhof [1994].

2.4.2 The IGS Orbits

IGS orbits are produced by the Analysis Centers of the International GPS Service for Geodynamics (IGS) since the start of the 1992 IGS Test Campaign on 21 June 1992. One of the IGS Analysis Centers, Scripps Institution of Oceanography (SIO), started its orbit determination activities even about one year earlier. Today there are seven active IGS Analysis Centers (see Chapter 1).

As opposed to broadcast orbits (or precise orbits) the IGS orbits mainly rely on the phase observations gathered by a relatively dense global network of precision P-code receivers (Figure 2.20). The IGS Analysis Centers produce daily orbit files containing rectangular geocentric satellite coordinates in the ITRF (IERS Terrestrial Reference Frame) and, in some cases, GPS clock information every 15 minutes for the entire satellite system tracked. The information is made available in the so-called SP3-Format, Remondi [1989].

Since 1 November 1992, the start of the *IGS Pilot Service*, the daily orbit series of first six, then seven IGS Analysis Centers were compared every week. This comparison was performed by the IGS Analysis Center Coordinator [Goad, 1993].

GPS TRACKING NETWORK OF THE INTERNATIONAL GPS SERVICE FOR GEODYNAMICS
OPERATIONAL AND PLANNED STATIONS

Figure 2.20. The IGS network of tracking stations in Spring 1995.

The comparison consisted of seven parameter Helmert transformations between the orbit files of all possible combinations of IGS Analysis Centers (with seven centers $7 \cdot 6/2 = 21$ combinations were possible). From the rms values (per satellite coordinate) after the transformations it was possible to extract an estimation for the orbit quality of individual analysis centers. Figure 2.21 from Beutler et al. [1994a,b]) shows the development of the orbit quality for all centers since September 1992 till end of December 1993.

Figure 2.21 demonstrates that already in the initial phase the consistency of estimates of different processing centers was below 50 cm. This led to the idea of producing a combined, official IGS orbit based on a weighted average of the individual orbit series. At the IGS Analysis Center Workshop in Ottawa, see Kouba [1993] it was decided to produce the IGS combined orbit based on the paper, Beutler et al. [1995b]; the original version of the paper may be found in the proceedings of the Ottawa workshop.

Since the start of the official IGS on January 1, 1994, daily IGS orbit files are made available by the new IGS Analysis Center Coordinator in weekly packages. They are distributed through the IGS Central Bureau Information System (IGS CBIS) and through the Global and Regional IGS Data Centers. The IGS combined orbits are available within 10 days after the end of the week. They proved to be *extremely reliable* and they have an accuracy comparable with that of the best individual contributions. The statistical information associated with the IGS orbits is made available each week in the IGS Report Series, see, e.g., Kouba et al. [1995].

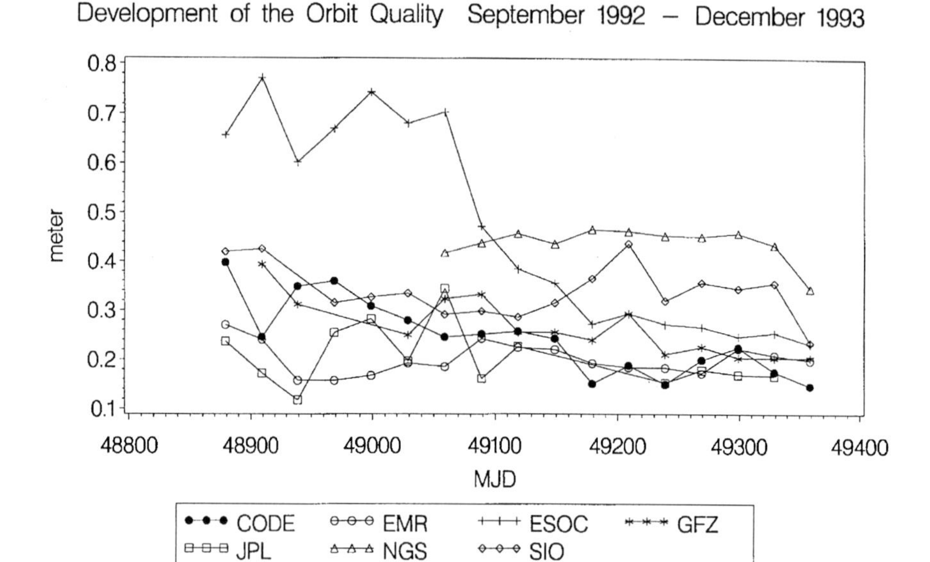

Figure 2.21. Development of the orbit quality November 1992 - December 1993.

The quality of the individual contributions is monitored (a) through the rms of the individual centers with respect to the combined orbit and (b) through the rms of a long arc analysis performed separately with the weekly data of each analysis center (one week arc using an improved radiation pressure model, Beutler et al. [1994c], see also Chapter 10, section 10.5).

Figures 2.22a and 2.22b show for each IGS Analysis Center the development of the weekly mean values of the daily rms values produced by the IGS Analysis Center Coordinator (a) based on the weighted average of the daily orbit solutions (b) based on the long-arc analysis. These figures underline that the best individual contributions are of the order of 10 cm rms per satellite coordinate today, a value which all IGS Analysis Centers seem to reach asymptotically. This allows us to conclude that the combined IGS orbits today are of (sub-)decimeter accuracy. We also see a high degree of consistency of Figures 2.22a and 2.22b.

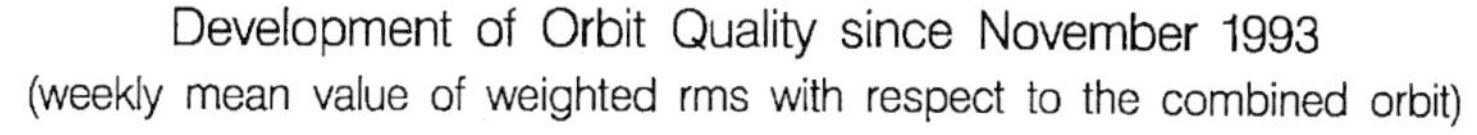

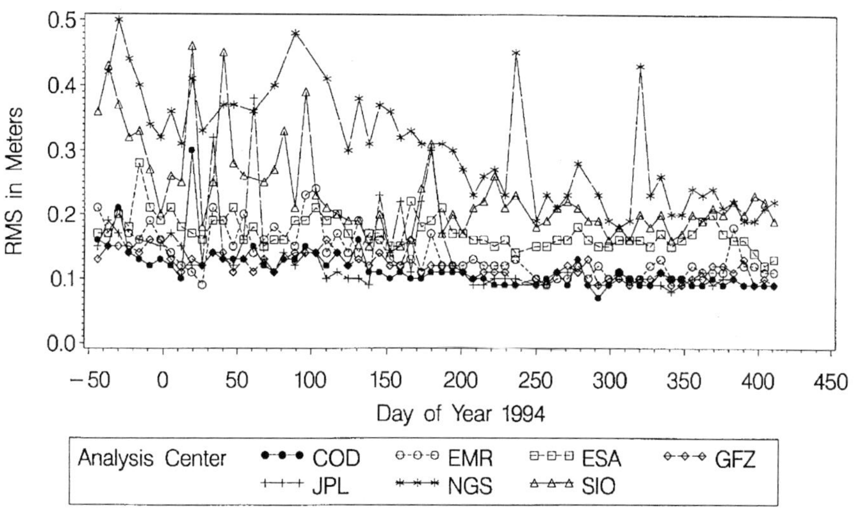

Figure 2.22a. Development of orbit quality since November 1993 (weekly mean value of weighted rms with respect to the combined orbit).

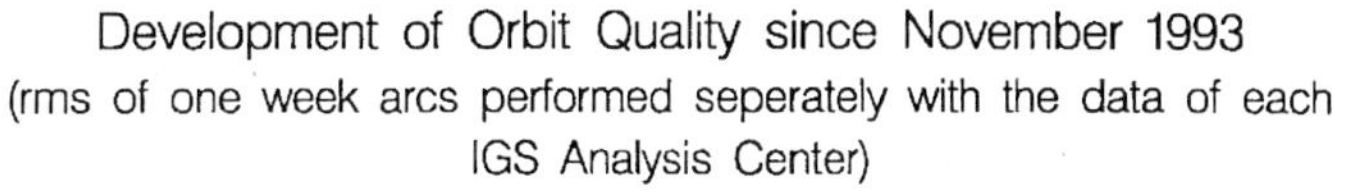

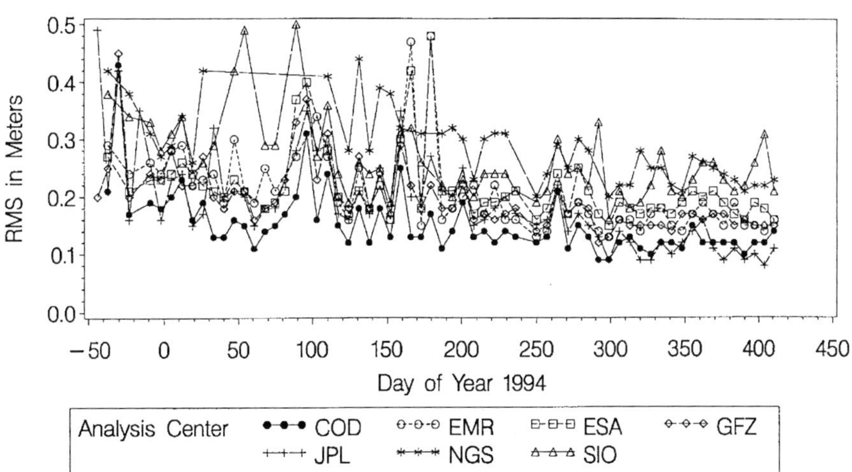

Figure 2.22b. Development of orbit quality since November 1993 (rms of one week arcs performed separately with the data of each IGS Analysis Center).

2.4.3 Propagation of Orbit Errors into Baselines and Networks

Bauersima [1983, eqn. 84] states that errors **dr** in the coordinates of a satellite orbit propagate into errors **db** in the coordinates of a baseline of length b according to the following rule:

$$\left|\frac{\mathbf{db}}{b}\right| = \left|\frac{\mathbf{dr}}{r}\right| \tag{2.56a}$$

where r is the mean distance between station and satellite. Zielinski [1988], using statistical methods, argues that this value is too pessimistic. He comes up with the following rule:

$$\left|\frac{\mathbf{db}}{b}\right| = \left|\frac{\mathbf{dr}}{k \cdot r}\right|, \quad 4 \langle k \langle 10 \tag{2.56b}$$

There is a slight difference in eqns. (2.56a,b), however. Whereas we are looking at errors **db** of the baseline length in eqn. (2.56b), we are looking at errors **db** in the components (latitude, longitude, height) of the baseline in eqn. (2.56a). In practice (2.56b) actually seems to be a fair rule for the propagation of orbit errors into the baseline lengths, whereas eqn. (2.56a) seems to be adequate for the propagation of the orbit errors into the height component. We should mention, however, that the height determination is also contaminated by the necessity to estimate tropospheric scale parameters for all stations, a fact which is not taken into account by either of the above formulae. Additional work is required in this area.

What kind of results are achieved in practice? Figures 2.23a and 2.23b give the residuals of daily estimates of the baseline Onsala-Graz (length about 1000 km) relative to the average coordinate solution over a time interval of about three months using broadcast orbits (Figure 2.23a) resp. IGS orbits (Figure 2.23b).

The repeatabilities of the horizontal components are clearly of the order of a few millimeters rms only – which we would expect according to both of our rules presented above. We even might argue that the rms of the horizontal components is no longer driven by the orbits. The results are unfortunately not as good in height. Here the rms is of the order of 1 cm, which would let us expect an orbit accuracy of about 25 cm according to eqn. (2.56a), an orbit accuracy of about 1 m if we trust eqn. (2.56b). Because we have reason to believe that the IGS orbits actually are accurate to about 10 cm we conclude that in practice eqn. (2.56a) is quite useful as a rule of thumb for the propagation of orbit errors into the height component of the baseline.

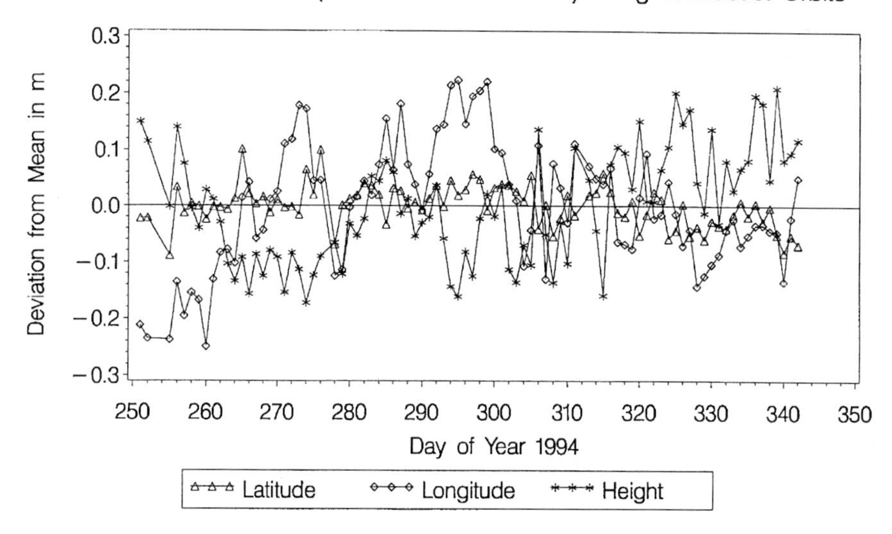

Figure 2.23a. Daily repeatabilities of latitude, longitude, height of the baseline Onsala-Graz (from 8 Sept 94 to 8 Dec 94) using broadcast orbits.

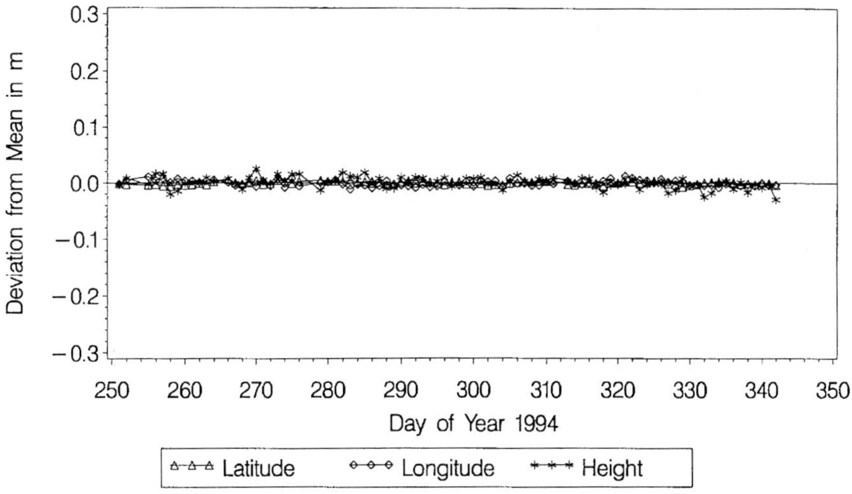

Figure 2.23b. Daily repeatabilities of latitude, longitude, height of the baseline Onsala-Graz (from 8 Sept 94 to 8 Dec 94) using IGS orbits.

2.5 SUMMARY AND CONCLUSIONS

Chapter 2 was devoted to the orbits of the GPS satellites. We first presented a few facts concerning the entire GPS system, which is fully operational today (section 2.1).

In section 2.2.1 we introduced the Keplerian elements and we developed the equations of motion for an artificial Earth satellite in rectangular geocentric coordinates in section 2.2.2 (eqn.(2.22)). We stated that there is a one-to-one correspondence between the osculating Keplerian elements at time t and the geocentric position- and velocity- vectors at the same time. This fact allowed us to derive the perturbation equations, first-order differential equations for the osculating elements (eqns. (2.30), section 2.2.3). We showed that there are simple approximate solutions of the perturbation equations giving us some insight into the structure of different perturbations. In section 2.2.4 we introduced the concept of mean elements. In section 2.2.5 we introduced the generalized orbit determination problem we have to solve when analyzing GPS data. Section 2.2 was concluded with some remarks concerning numerical integration as an universal tool in orbit determination (section 2.2.6). We made the distinction between the solution of the equations of motion and the variational equations associated with them.

In section 2.3 we studied the perturbing accelerations actually acting on the GPS satellites. In particular we discussed the problem of modeling radiation pressure (section 2.3.2) and looked at the effects of the deep 2:1 resonance of the entire GPS with the Earth's rotation. The section was concluded with an overview of the development of the GPS between mid 1992 and the end of 1994.

In section 2.4 we studied different orbit types available for the GPS. In particular we introduced the *Broadcast Orbits*, as the only information available in real time, and the *IGS Orbits* as orbits of highest accuracy, which are available to the scientific community about two weeks after the observations. Working with IGS orbits and including the observations of one or more permanent IGS tracking sites (using the coordinates and the site information made available through the IGS) is a guarantee that the results of a GPS survey automatically refer to the ITRF.

We tried to present in this chapter the orbit information which is relevant for the user of the GPS. We conclude with the remark that, thanks to the existence of the IGS, it is no longer necessary that groups working in regional geodynamics produce their own orbits. It is much safer to rely on the IGS orbits.

Acknowledgements

The author wishes to express his gratitude to Dr. Robert Weber and Dipl.-Astr. Andreas Verdun for their invaluable assistance in editing and writing this chapter. Many of the examples presented are extracted from results acquired by the CODE processing center of the IGS. Let me cordially thank Dr. Markus Rothacher, who is the head of CODE, for making these results available and for many valuable suggestions. CODE stands for Center for Orbit Determination in Europe, a joint venture of four European institutions (Astronomical Institute University of Bern (Switzerland); Federal Office of Topography (Switzerland); Institute for Applied Geodesy (Germany); Institute Geographique National (France)); IGS stands for International GPS Service for Geodynamics.

Section 2.3.3 dealing with resonance effects in GPS orbits is based on research work performed by Dipl.-Phys. Urs Hugentobler in the context of his Ph.D. thesis.

Last but not least, I would like to thank Ms. Christine Gurtner for the actual typing of the manuscript. Her contribution was essential for the timely completion of the manuscript.

References

Bauersima, I. (1983), "Navstar/Global Positioning System (GPS) (II)", *Mitteilung No. 10 der Satellitenbeobachtungsstation Zimmerwald*, Druckerei der Universität Bern.

Beutler, G. (1990), "Numerische Integration gewöhnlicher Differentialgleichungssysteme: Prinzipien und Algorithmen", *Mitteilung Nr. 23 der Satellitenbeobachtungsstation Zimmerwald.*

Beutler, G., A. Verdun (1992), "Himmelsmechanik II: Der erdnahe Raum" *Mitteilung No. 28 der Satellitenbeobachtungsstation Zimmerwald*, Druckerei der Universität Bern.

Beutler, G., I.I. Mueller, R.E. Neilan, R. Weber (1994a), "IGS - Der Internationale GPS-Dienst für Geodynamik" *Zeitschrift für Vermessungswesen, Deutscher Verein für Vermessungswesen (DVW),* Jahrgang: 119, May, Heft 5, S. 221-232.

Beutler, G., I.I. Mueller, R.E. Neilan (1994b), "The International GPS Service for Geodynamics (IGS): Development and Start of Official Service on January 1, 1994", *Bulletin Géodésique*, Vol. 68, 1, pp. 39-70.

Beutler, G., E. Brockmann, W. Gurtner, U. Hugentobler, L. Mervart, M. Rothacher, A. Verdun (1994c), "Extended Orbit Modelling Techniques at the CODE Processing Center of the IGS: Theory and Initial Results", *Manuscripta Geodaetica*, Vol. 19, pp. 367-386.

Beutler, G., E. Brockmann, U. Hugentobler, L. Mervart, M. Rothacher, R. Weber (1995a), "Combining Consecutive Short Arcs into Long Arcs for Precise and Efficient GPS Orbit Determination", *Journal of Geodesy*, Vol. 70, pp. 287-299, 1996.

Beutler, G. J. Kouba, T. Springer (1995b), "Combining the Orbits of the Analysis Centers", *Bulletin Géodésique*, Vol. 69, pp. 200-222, 1995.

Brouwer, D. (1937), "On the accumulation of errors in numerical integration", *Astronomical Journal*, Volume 46, No 1072, p. 149 ff.

Danby, J.M.A. (1989), "*Fundamentals of Celestial Mechanics*", Willmann-Bell, INC., Richmond, Va., Second Edition, Second Printing, ISBN 0-943396-20-4.

Euler, L. (1749), "Recherches sur le mouvement des corps célestes en général", *Mémoires de l'académie des sciences de Berlin* [3] (1747), pp. 93-143.

Euler, L. (1768), "*Institutio calculi integralis*", Volumen primum, sectio secunda: De integratione aequatiorum differentialium per approximationem. Caput VII. Petropoli, Academiae Imperialis Scientiarum.

Fehlberg, E. (1972), "Classical Eighth and Lower Order Runge-Kutta-Nystrom Formulas with Stepsize Control for Special Second Order Differential Equations", *NASA Technical Report* TR-R-381.

Fliegel, H.F., T.E. Gallini, E.R. Swift (1992), "Global Positioning System Radiation Force Model for Geodetic Applications", *Journal of Geophysical Research*, Vol. 97, No. B1, pp.559-568.

Gauss, C.F. (1809), "*Theoria motus corporum coelestium in sectionibus conicis solem ambientum*", Hamburgi, Perthes & Besser.

Goad, C. (1993), "IGS Orbit Comparisons", *Proceedings of 1993 IGS Workshop*", pp. 218-225, Druckerei der Universität Bern.

Green, G.B., P.D. Massatt, N.W. Rhodus (1989), "The GPS 21 Primary Satellite Constellation" Navigation, *Journal of the Institute of Navigation*, Vol 36, No.1, pp. 9-24.

Heiskanen, W.A., H. Moritz (1967), "*Physical Geodesy*", W.H. Freeman and Company, San Francisco and London.

Hofmann-Wellenhof, B., H. Lichtenegger, J. Collins. (1994), "*GPS Theory and Practice*", Third revised edition, Springer-Verlag Wien, New York.

Hugentobler, U., G. Beutler (1994), "Resonance Phenomena in the Global Positioning System", In:: Kurzynska, K./Barlier F./Seidelman, P.K./Wytrzyszczak, I. (eds.) "Dynamics and Astrometry of Natural and Artificial Celestial Bodies", *Proceedings of the Conference on Astrometry and Celestial Mechanics*, Poznan, Poland, September 13-17, 1993. Published by Astronomical Observatory of A. Mickiewicz University, Poznan, 1994, p. 347-352.

Hugentobler, U. (1996),"*Astrometry and Satellite Orbits: Theoretical Considerations and Typical Applications*", Ph.D. Thesis, University of Bern.

Kaula, W.M. (1966), "*Theory of Satellite Geodesy*", Blaisdell Publication Cie., Waltham.

Kepler, J. (1609), *"Astronomia nova de motibus stellae Martis ex observationibus Tychonis Brahe*", Pragae.

Kepler, J. (1619), "*Harmonices mundi libri* V", Lincii.

Kouba, J. (1993), "*Proceedings of the IGS Analysis Center Workshop"*, October 12-14, 1993, Geodetic Survey Division, Surveys, Mapping, and Remote Sensing Sector, NR Can, Ottawa, Canada.

Kouba, J., Y. Mireault, F. Lahaye (1995), "Rapid Service IGS Orbit Combination - Week 0787", *IGS Report No 1578*, IGS Central Bureau Information System.

Landau, H. (1988), "*Zur Nutzung des Global Positioning Systems in Geodäsie und Geodynamik: Modellbildung, Software-Entwicklung und Analyse*", Ph.D. Thesis, Studiengang Vermessungswesen, Universität der Bundeswehr München, Neubiberg.

Lundquist, C.A., G. Veis (1966), "Geodetic Parameters for a 1966 Smithsonian Institution Standard Earth", Volumes I-III, Smithsonian Astrophysical Observatory, *Special Report 200.*

McCarthy, D.D. (1992), "IERS Standards (1992)", *IERS Technical Note* No. 13, Observatoire de Paris.

Newton, I. (1687), "*Philosophiae naturalis principia mathematica*", Joseph Streater, Londini.

Remondi, B.W. (1989), "Extending the National Geodetic Survey Standard GPS Orbit Formats", *NOAA Technical Report* NOS 133 NGS 46, Rockville, MD.

Rothacher, M. (1992), "Orbits of Satellite Systems in Space Geodesy*", Geodätisch-geophysikalische Arbeiten in der Schweiz*, Vol 46, 253 pages, Schweizerische Geodätische Kommission.

Seeber, G. (1993), "*Satellite Geodesy*", Walter de Gruyter, Berlin/New York.

Shampine, L.F., M.K. Gordon (1975), *"Computer solution of ordinary differential equations, the Initial value problem"*, Freedman & Cie.

Van Dierendonck, A., S. Russell, E. Kopitzke, M. Birnbaum (1978), "The GPS Navigation Message", *Journal of the Institute of Navigation*, Vol. 25, No. 2, pp. 147-165, Washington.

Zielinski, J.B. (1988), "Covariances in 3D Network Resulting From Orbital Errors". *Proceedings of the International GPS-Workshop,* Darmstadt, April 10-13, published *in Lecture Notes in Earth Sciences*, GPS-Techniques Applied to Geodesy and Surveying, Springer Verlag, Berlin, pp. 504-514.

3. PROPAGATION OF THE GPS SIGNALS

Richard B. Langley
Geodetic Research Laboratory, Department of Geodesy and Geomatics Engineering, University of New Brunswick, P.O. Box 4400, Fredericton, N.B., Canada E3B 5A3

3.1 INTRODUCTION

The Global Positioning System is a one-way ranging system. The GPS satellites emit signals — complex modulated radio waves — which propagate through space to receivers on or near the earth's surface.[1] From the signals it intercepts, a receiver measures the ranges between its antenna and the satellites. In this chapter, we will examine the nature of the GPS signals. After a brief review of the fundamentals of electromagnetic radiation, we will describe the structure of the GPS signals. Since the signals, in propagating to a receiver, must travel through the ionosphere and the neutral atmosphere, we will examine the effect these media have on the signals. Finally, we will look at the propagation phenomena of multipath and scattering and the effects they have on the measurements made by a GPS receiver.

3.2 ELECTROMAGNETIC WAVES

The infinitely wide electromagnetic spectrum stretches from below the extremely low radio frequencies — 30 Hz to 3 kHz with an equivalent wavelength of 10,000 to 100 kilometres — used by the U.S. Navy in communications tests with submerged submarines to the frequencies characteristic of gamma rays: about 3 x 10^{19} Hz and beyond with corresponding wavelengths shorter than 10 picometres (10^{-11} metres)! The radio part of the spectrum extends to frequencies of about 300 GHz, but the distinction between millimetre radio waves and long infrared light waves is a little blurry (see Figure 3.1).

An electromagnetic wave is a self-propagating wave with both electric and magnetic field components generated by the rapid oscillation of a charged particle. The characteristics of the wave, and in fact the possibility for the actual

[1] GPS receivers can also be used on low-earth-orbiting spacecraft.

existence of electromagnetic waves, is given by Maxwell's equations[2] (see, for example, Lorrain and Corson [1970] or Feynman et al. [1964]):

$$
\begin{aligned}
\nabla \cdot \mathbf{E} &= \frac{\rho}{\varepsilon_0} \\
\nabla \cdot \mathbf{B} &= 0 \\
\nabla \times \mathbf{E} &= -\frac{\partial \mathbf{B}}{\partial t} \\
\nabla \times \mathbf{B} - \varepsilon_0 \mu_0 \frac{\partial \mathbf{E}}{\partial t} &= \mu_0 \mathbf{J}_m
\end{aligned}
\tag{3.1}
$$

where $\mathbf{E}$ is the electric field intensity, $\mathbf{B}$ is the magnetic induction (or magnetic flux density), $\mathbf{J}_m$ is the current density due to the flow of charges in matter, ρ is the total electric charge density, ε_0 is the permittivity of free space, and m_0 is the permeability of free space. ε_0 = 8.854 187 818 x 10^{-12} F m^{-1} and μ_0 = 1.256 637 062 x 10^{-6} H m^{-1} are fundamental constants that relate electric charge to the Coulomb or electrostatic force and current flow to the magnetic force respectively.

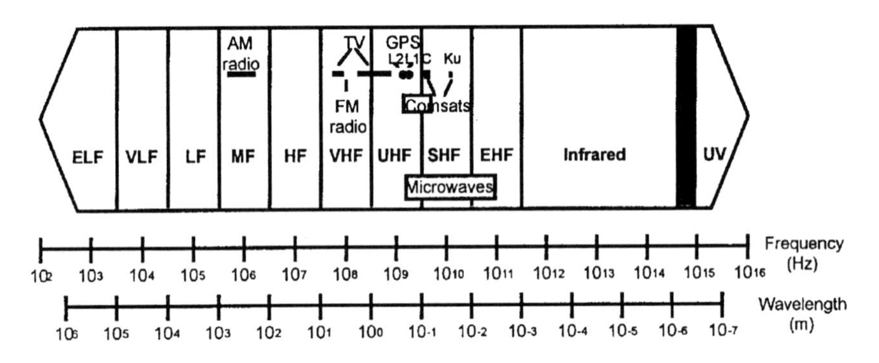

Figure 3.1. The radio and light portions of the electromagnetic spectrum.

If the charged particle generating the wave oscillates in a sinusoidal fashion, then in free space ($\rho = 0$, $\mathbf{J}_m = \mathbf{0}$), Maxwell's equations (in phasor form) reduce to

[2] The first of these equations has an indirect link to geodesy. It is the general form of Gauss' law — named after Johann Karl Friedrich Gauss, the eighteenth century polymath and father of modern geodesy.

$$\nabla \cdot \mathbf{E} = 0$$

$$\nabla \cdot \mathbf{H} = 0 \qquad (3.2)$$

$$\nabla \times \mathbf{E} + i\omega\mu_0\mathbf{H} = 0$$

$$\nabla \times \mathbf{H} - i\omega\varepsilon_0\mathbf{E} = 0$$

where **H** is the magnetic field ($= \mathbf{B}/\mu_0$), $i^2 = 1$, and w is the angular frequency of oscillation of the particle. The solution of these equations (obtained by taking the curls of the third and fourth equations) yields the following pair of differential equations:

$$\nabla^2\mathbf{E} + \varepsilon_0\mu_0\omega^2\mathbf{E} = 0$$

$$\nabla^2\mathbf{H} + \varepsilon_0\mu_0\omega^2\mathbf{H} = 0. \qquad (3.3)$$

These equations are those of an unattenuated wave propagating with a speed

$$c = \frac{1}{\sqrt{\varepsilon_0\mu_0}} \qquad (3.4)$$

which is the speed of light — 2.997 924 58 x 10^8 metres per second. It can be shown that in free space, or in any homogeneous, isotropic, linear, and stationary medium, the electric and magnetic fields are transverse to the direction of propagation and the fields are mutually perpendicular.

For a plane wave propagating in the direction of the positive z-axis (equation (3.3) also can yield spherical waves), the **E** vector of the wave can be written as (the **H** vector can be similarly expressed)

$$\mathbf{E} = \mathbf{E}_0 e^{i\omega\left(t - \frac{z}{c}\right)} \qquad (3.5)$$

where $\mathbf{E}_0$ gives the amplitude and the direction of polarisation of the wave. $\mathbf{E}_0$ can be decomposed into two orthogonal vectors: $\mathbf{E}_{0,x}$, parallel to the positive x-axis and $\mathbf{E}_{0,y}$, parallel to the positive y-axis. If $\mathbf{E}_{0,x}$ and $\mathbf{E}_{0,y}$ have the same phase (or an integer multiple of π), the wave is linearly polarised (**E** is always directed along a line). If $\mathbf{E}_{0,x}$ and $\mathbf{E}_{0,y}$ differ in phase, their sum describes an ellipse about the z-axis. This is an elliptically polarised wave. If $\mathbf{E}_{0,x}$ and $\mathbf{E}_{0,y}$ have the same amplitude but are $\pi/2$ (or an odd multiple of $\pi/2$) out of phase, the ellipse becomes a circle and the wave is said to be circularly polarised. If **E** and **H** rotate clockwise (counterclockwise) for an observer looking towards the source of the wave, the polarisation is right-handed (left-handed). For a good description and conceptual illustration of linear, eliptical, and circular polarisation, see Kraus (1950).

Using the relationship

$$f\lambda = \frac{\omega}{2\pi}\lambda = \frac{\omega}{2\pi}\frac{2\pi}{k} = \frac{\omega}{k} = c \tag{3.6}$$

where f is the frequency of the wave in cycles per second, λ is the wavelength, and where k is called the propagation wave number, equation (3.5) may be written as

$$\mathbf{E} = \mathbf{E}_0 e^{i(\omega t - kz)}. \tag{3.7}$$

More generally, for a wave traveling in direction **k** (the magnitude of **k** is the wave number), the field at some point defined by vector **r** is

$$\mathbf{E} = \mathbf{E}_0 e^{i(\omega t - \mathbf{k} \cdot \mathbf{r})}. \tag{3.8}$$

At a fixed point in space, the electric field intensity may be written as

$$\mathbf{E} = \mathbf{E}_0 e^{i(\omega t - \phi)} \tag{3.9}$$

where $(\omega t - \phi)$ is the phase or phase angle of **E** and ϕ is a phase bias or constant.

The concept of a plane electromagnetic wave is somewhat artificial. A plane wave is one that travels in some particular direction and whose intensity and phase are constant over any plane normal to the direction of propagation. Such plane electromagnetic waves do not actually exist in nature. Far from a transmitter of electromagnetic waves, the surface of constant phase is a sphere. So the electromagnetic waves typically encountered in practice are spherical rather than plane. However, at a sufficiently large distance from the transmitter, a portion of the surface of the sphere may be approximated by a plane, and therefore far from the transmitter, a spherical wave behaves very much like a plane wave.

An electromagnetic wave may be generally characterized by four parameters: amplitude, frequency, phase, and polarization. If one of these parameters is varied in some controlled fashion — or modulated — then an electromagnetic wave can convey information. Amplitude modulation (AM) is commonly used, for example, for long wave, medium wave, and short wave radio broadcasting, and for most aeronautical communications; frequency modulation (FM) is used for very high frequency high fidelity broadcasts; and phase modulation (PM) is typically used for data transmissions. The modulating signal may either be continuously varying (analogue) or have a fixed number of levels (digital) — two in the case of binary modulation.

3.3 THE GPS SIGNALS

The radio signals transmitted by the GPS satellites are amazingly complex. This complexity was designed into the system in order to give GPS its versatility. GPS is required to work with one-way measurements (receive only); serve an unlimited number of both military and civilian users; provide accurate, unambiguous, real-time range measurements; provide accurate Doppler-shift measurements; provide accurate carrier-phase measurements; provide a broadcast message; provide ionospheric delay correction; allow simultaneous measurements from many satellites; have interference protection; and have multipath tolerance. The GPS signals contain a number of components in order to meet these requirements. The official description of the GPS signals is contained in the Interface Control Document, ICD-GPS-200 [ARINC, 1991]. Spilker [1978, 1980] is also a primary reference for details of GPS signal structure. The condensed description given here, much of which has been published previously by Langley [1990], has been based, in large measure, on those documents.

3.3.1 The Carriers

Each GPS satellite transmits signals centred on two microwave radio frequencies, 1575.42 MHz, referred to as Link 1 or simply L1, and 1227.60 MHz, referred to as L2[3]. These channels lie in a band of frequencies known as the L band (1 to 2 GHZ, see Fig. 3.1). Within the L band, the International Telecommunications Union, the radio regulation arm of the United Nations, has set aside special sub-bands for satellite-based positioning systems. The L1 and L2 frequencies lie within these bands.

Such high frequencies are used for several reasons. The signals, as we have said, consist of a number of components. A bandwidth of about 20 MHz is required to transmit these components. This bandwidth is equal to the whole very high frequency (VHF) FM broadcast band! So a high, relatively uncluttered part of the radio spectrum is required for GPS-type signals. The GPS signals must provide a means for determining not only high accuracy positions in real-time, but also velocities. Velocities are determined by measuring the slight shift in the frequency of the received signals due to the Doppler effect — essentially the same phenomenon, albeit for sound waves, that gives rise to the change in pitch of a locomotive's whistle as a train passes in front of you at a level crossing. In order to achieve velocities with centimetre-per-second accuracies, centimetre wavelength (microwave) signals are required.

A further reason for requiring such high frequencies is to reduce the effect of the ionosphere. As we will see later in this chapter, the ionosphere affects the

[3] The GPS satellites also transmit an L3 signal at 1381.05 MHz associated with their dual role as a nuclear burst detection satellite as well as S-band telemetry signals.

speed of propagation of radio signals. The range between a satellite and a receiver derived from measured signal travel times, assuming the vacuum speed of light, will therefore be in error. The size of this error gets smaller as higher frequencies are used. But at the L1 frequency it can still amount to 30 metres, or so, for a signal arriving from directly overhead. For some applications, an error of this size is tolerable. However there are applications, such as geodetic positioning, that require much higher accuracies. This is why GPS satellites transmit on two frequencies. As we will see, if measurements made simultaneously on two well-spaced frequencies are combined, it is possible to remove almost all of the ionosphere's effect.

Although high frequencies are desirable for the reasons just given, it is important that they not be too high. For a given transmitter power, a received satellite signal becomes weaker the higher the frequency used[4]. The L band frequencies used by GPS are therefore a good compromise between this so-called *space loss* and the perturbing effect of the ionosphere.

GPS signals, like most radio signals, start out in the satellites as pure sinusoidal waves or *carriers*. But pure sinusoids cannot be readily used to determine positions in real-time. Although the phase of a particular cycle of a carrier wave can be measured very accurately, each cycle in the wave looks like the next so it is difficult to know exactly how many cycles lie between the satellite and the receiver.

In order for a user to obtain positions independently in real-time, the signals must be modulated; that is, the pure sinusoid must be altered in a fashion that time delay measurements can be made. This is achieved by modulating the carriers with *pseudorandom noise (PRN) codes*.

These PRN codes consist of sequences of binary values (zeros and ones) that at first sight appear to have been randomly chosen. But a truly random sequence can only arise from unpredictable causes which, of course, we would have no control over and could not duplicate. However, using a mathematical algorithm or special hardware devices called *tapped feedback registers*, we can generate sequences which do not repeat until after some chosen interval of time. Such sequences are termed *pseudo*random. The apparent randomness of these sequences makes them indistinguishable from certain kinds of noise such as the hiss heard when a radio is tuned between stations or the "snow" seen on the screen of a television when tuned to an unoccupied channel (some radios and televisions sense the lack of a signal and blank out the noise). Although noise in a communications device is generally unwanted, in this case the noise is very beneficial.

Exactly the same code sequences are independently replicated in a GPS receiver. By aligning the replicated sequence with the received one and knowing the instant of time the signal was transmitted by the satellite, the travel time, and hence the range can be computed. Each satellite generates its own unique codes, so it is easy for a GPS receiver to identify which signal is coming from which satellite even when signals from several satellites arrive at its antenna

---

[4]Expressed in dB, the loss is given by $32.5 + 20 \log_{10}\rho + 20 \log_{10} f$ where ρ is the distance between the satellite and the receiving station in km and f is the operating frequency in MHz, Roddy and Coolen [1984].

simultaneously — a communications technique known as code division multiple access (CDMA).

3.3.2 The Codes

The C/A-code. Two different PRN codes are transmitted by each satellite: the C/A or coarse/acquisition code and the P or precision code. The C/A-code is a sequence of 1,023 binary digits or *chips* which is repeated every millisecond. This means that the chips are generated at a rate of 1.023 million per second and that a chip has a duration of about 1 microsecond. Each chip, riding on the carrier wave, travels through space at the speed of light. We can therefore convert a time interval to a unit of distance by multiplying it by this speed. So one microsecond translates to approximately 300 metres. This is the *wavelength* of the C/A-code.

Because the C/A-code is repeated every millisecond, a GPS receiver can quickly lock onto the signal and begin matching the received code with the one it generates.

Each satellite is assigned a unique C/A-code. There are a total of 32 codes available for the satellites. An additional four unique C/A-codes are available for other uses such as ground transmitters.

The P-code. The precision of a range measurement is determined in part by the wavelength of the chips in the PRN code. Higher precisions can be obtained with shorter wavelengths. To get higher precisions than are afforded by the C/A-code, GPS satellites also transmit the P-code. The wavelength of the P-code chips is only 30 metres, one-tenth the wavelength of the C/A-code chips; the rate at which the chips are generated is correspondingly 10 times as fast: 10.23 million per second. The P-code is an extremely long sequence. The pattern of chips does not repeat until after 266 days or about 2.35×10^{14} chips! Each satellite is assigned a unique one-week segment of this code which is re-initialized at Saturday/Sunday midnight each week.

The Y-code. As part of a procedure known as Anti-spoofing (AS), the U.S. Department of Defense has encrypted the P-code by combining it with a secret W-code. AS was formally activated at 00:00 UTC on 31 January 1994 and now is in continuous operation on Block II satellites.

Other Properties. The GPS PRN codes have additional useful properties. When a receiver is processing the signals from one satellite, it is important that the signals received simultaneously from other satellites not interfere. The GPS PRN codes have been specially chosen to be resistant to such mutual interference. Also the use of PRN codes results in a signal that has a certain degree of immunity to unintentional or deliberate jamming from other radio signals.

At the present time, the C/A-code is modulated onto the L1 carrier whereas the encrypted P-code is transmitted on both L1 and L2. This means that users with dual frequency GPS receivers can correct the measured ranges for the effect of the ionosphere. Users of single frequency receivers must resort to models of the ionosphere which typically account for only a portion of the effect (see section 3.5.2). It is access to the lower accuracy C/A-code which is provided in the GPS *Standard Positioning Service* (SPS), the level of service authorized for civilian users. The *Precise Positioning Service* (PPS) provides access to both the C/A-code and the encrypted P-code and is designed (primarily) for military users. The SPS incorporates a further intentional degradation of accuracy, called *Selective Availability* (SA). SA is effected through satellite clock dithering (the so-called "delta-process") and broadcast orbit ephemeris degradation (the "epsilon-process"). Reports indicate that currently SA primarily uses the delta-process. The clock dithering affects all pseudorange and phase measurements. Different levels of SA are possible; the level that is presently used is one which yields the current SPS horizontal position accuracy of 100 m 2-d.r.m.s (twice the distance root mean square). As with AS, authorized users employ a cryptographic key to overcome SA. Almost all of the effect of SA can also be removed by the use of differential techniques (see Chapter 5). SA had been enabled on Block II satellites during part of 1990. SA was turned off between about 10 August 1990 and 1 July 1991 due to Gulf crisis. The standard level was re-implemented on 15 November 1991. Since then, SA has been temporarily turned off for different purposes as has AS. There have been calls from the civilian community for the reducation and eventual removal of SA. Currently, two of the Block II satellites appear to have little or no SA imposed. The sole remaining operational Block I satellite is free of both SA and AS.

3.3.3 The Broadcast Message

In order to convert the measured ranges between the receiver and the satellites to a position, the receiver must know where the satellites are. To do this easily in real-time requires that the satellites broadcast this information. Accordingly, there is a message superimposed on both the L1 and L2 carriers along with the PRN codes. Each satellite broadcasts its own message which consists of orbital information (the *ephemeris*) to be used in the position computation (see Chapter 2), the offset of its clock from GPS System Time (see section 3.3.5), and information on the health of the satellite and the expected accuracy of the range measurements. The message also contains *almanac* data for the other satellites in the GPS constellation as well as their health status and other information. The almanac data, a crude description of the satellite orbit, is used by the receiver to determine where every satellite is. It uses this information to quickly acquire the signals from satellites that are above the horizon but are not yet being tracked. So once one satellite is being tracked and its message is decoded, acquisition of the signals from other satellites is quite rapid. For further details of the structure and content of the message, see ARINC [1991] or Van Dierendonck et al. [1978, 1980].

The broadcast message (also referred to as the navigation message) contains another very important piece of information for receivers that track the P-code. As we mentioned, the P-code segment assigned to each satellite is 7 days long. A GPS receiver with an initially unsynchronized clock has to search through its generated P-code sequence to try to match the incoming signal. It would take many hours to search through just one second of the code, so the receiver needs some help. It gets this help from a special word in the message called the *hand-over word* (HOW) which tells it where in the P-code to start searching.

The GPS broadcast message is sent at a relatively slow rate of 50 bits per second, taking 12.5 minutes for all the information in the message to be transmitted. To minimize the delay for a receiver to obtain an initial position, the ephemeris and satellite clock offset data are repeated every 30 seconds.

The C/A-code and encrypted P-code chip streams are separately combined with the message bits using *modulo 2 addition*[5]. This is just the binary addition that computers and digital electronics do so well. If the code chip and the message bit have the same value (both 0 or both 1) the result is 0. If the chip and bit values are different, the result is 1. The carriers are then modulated by the code and message composite signal. This is readily done with the L2 channel as it only carries the encrypted P-code. But the L1 channel has to carry both the encrypted P-code and the C/A-code. This is achieved by a clever technique known as *phase quadrature.* The encrypted P-code signal is superimposed on the L1 carrier in the same way as for the L2 carrier. To get the C/A-code signal onto the L1 carrier, the unmodulated carrier is tapped off and this tapped carrier is shifted in phase by 90°. This quadrature carrier component is mixed with the C/A-code signal and then combined with the encrypted P-code modulated in-phase component before being transmitted by the spacecraft antenna.

3.3.4 Binary Biphase Modulation

As mentioned in section 3.2, carrier waves can be modulated in a number of ways. Phase modulation is the approach used for the GPS signals. Because the PRN codes and the message are binary streams, there must be two states of the phase modulation. These two states are the normal state, representing a binary 0, and the mirror image state, representing a binary 1. The normal state leaves the carrier unchanged. The mirror image state results in the unmodulated carrier being multiplied by -1. Therefore a code transition from 0 to 1 (normal to mirror image) or from 1 to 0 (mirror image to normal) each involves a phase reversal or a phase shift of 180°. This technique is known as *binary biphase modulation.* An interesting property of binary biphase modulation was exploited by one of the first commercially available GPS receivers, the Macrometer™. By electronically squaring the received signal, all of the modulation is removed leaving a pure carrier. The phase of the carrier could then be measured to give ambiguous range measurements (this is discussed further in Chapter 4). Of course, the broadcast

---

[5] Modulo 2 addition of the P-code and the encryption W-code is used to produce the Y-code, Ashjaee and Lorenz [1992].

message was lost in the process and so orbit data had to be obtained from an alternate source.

3.3.5 The GPS Satellite Clocks and Time

The timing and frequency for the carriers, the PRN codes, and the message are all coherently derived from an atomic oscillator on board the satellite running at 10.23 MHz (and compensated for most of the relativistic frequency shift). The L1 frequency, 1575.42 MHz = 154 x 10.23 MHz; the L2 frequency, 1227.6 MHz = 120 x 10.23 MHz. Each satellite carriers four oscillators (two cesiums and two rubidiums in the Block II satellites), any one of which may be commanded on by the GPS Master Control Station.

The GPS signals are referenced to GPS (System) Time, which until June 1990 was the time kept by a single atomic clock at one of the U.S. Air Force GPS monitor stations. However, GPS Time is now derived from a composite or "paper" clock consisting of all monitor stations and the operational satellite clocks.

GPS Time is steered over the long run to keep it within about 1 microsecond of UTC, ignoring leap seconds. So unlike UTC, GPS Time has no leap second jumps. At the integer second level, GPS Time equaled UTC in 1980, but currently, due to the leap seconds that have been inserted into UTC, it is ahead of UTC by 10 seconds plus a fraction of a microsecond that varies day to day.

A particular epoch is identified in GPS Time as the number of seconds that have elapsed since the previous Saturday/Sunday midnight. Such a time measure is, of course, ambiguous, so one must also indicate in which week the epoch is. GPS weeks start with week 0 on 6 January 1980, and are numbered consecutively.

3.3.6 Polarization

The signals transmitted by the GPS satellites are right-hand circularly polarized (RHCP). Circular polarization is commonly used for signals transmitted from spacecraft in order to combat the fading problem associated with Faraday rotation of the plane of polarization due to the earth's magnetic field. For a RHCP signal to provide maximum signal strength to a receiver, a RHCP antenna must be used. This subject is discussed further in section 3.6 and in Chapter 4.

3.3.7 Putting it all Together

The composite GPS signal transmitted by a GPS satellite consists then of carriers modulated by the PRN C/A and encrypted P-codes and the broadcast message. The combining of these different components is illustrated in Figure 3.2. The composite signal is transmitted from the shaped-beam antenna array on the nadir-facing side of the satellite. The transmitted power levels are +23.8 dBW and +19.7 dBW for the encrypted P-code signal on L1 and L2 respectively and +28.8 dBW for the L1 C/A-code signal, Nieuwejaar [1988]. The array radiates near-uniform power to users on or near the earth's surface of at least -163 dBW and -166 dBW for the L1 and L2 encrypted P-code signals respectively and -160 dBW for the L1 C/A-code signal. Actual received signal levels may be larger than these values for a variety of reasons including satellite transmitter power output variations. Maximum received signal levels are not expected to exceed -155.5 and -158.0 dBW for the L1 and L2 encrypted P-code signals respectively and -153.0 dBW for the L1 C/A-code signal.

Forgetting for a moment that GPS is a ranging system, we could consider the satellites to be simply broadcasting a message in an encoded form. The bits of the

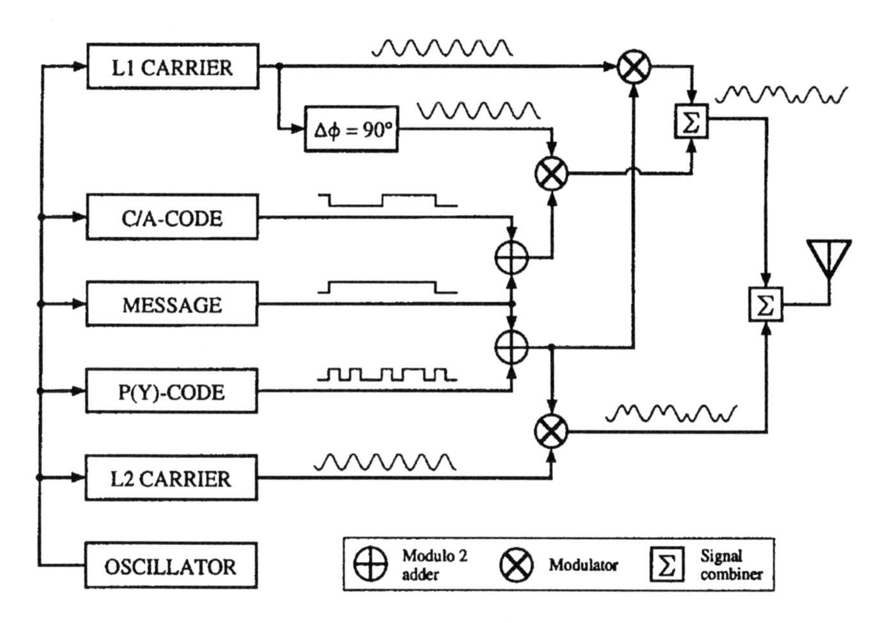

Figure 3.2. How the components of the GPS signal are combined. Note that the various waveforms are not to scale.

message have been camouflaged by the PRN code chips. The effect of this camouflaging is to increase the bandwidth of the signal. Instead of occupying only a fraction of one kiloHertz, the signal has been spread out over 20 MHz. Inside a GPS receiver, the code matching operation de-spreads the signal allowing the message to be recovered. Clearly this can only be done if the receiver knows the correct codes. The de-spreading operation conversely spreads out any interfering signal considerably reducing its effect. This is a common technique, especially in military circles, for ensuring security and combating interference and is known as *direct sequence spread spectrum communication*. Spread spectrum signals have the additional property of limiting the interference from signals reflected off nearby objects *(multipath)*.

The L1 signal transmitted by a GPS satellite can be represented in equation form as

$$S_{L1_i}(t) = A_p P_i(t) W_i(t) D_i(t) \cos(\omega_1 t + \phi_{n,L1,i}) + A_c C_i(t) D_i(t) \sin(\omega_1 t + \phi_{n,L1,i}) \tag{3.10}$$

where

A_p and A_c	represent the amplitudes of the encrypted P and C/A-code components respectively.
$P_i(t)$	represents the P-code of satellite i.
$W_i(t)$	represents the encryption code. $Y_i(t) = P_i(t)W_i(t)$.
$C_i(t)$	represents the C/A-code of satellite i.
$D_i(t)$	represents the data transmitted by satellite i in the broadcast (navigation) message.
ω_1	is the L1 frequency.
$\phi_{n,L1,i}$	represents a small phase noise and oscillator drift component.

Similarly, the L2 signal transmitted by satellite i can be represented as

$$S_{L2_i}(t) = B_p P_i(t) W_i(t) D_i(t) \cos(\omega_2 t + \phi_{n,L2,i}) \tag{3.11}$$

where

B_p represents the amplitude of the L2 signal.

3.4 PROPAGATION OF SIGNALS IN REFRACTIVE MEDIA

Of critical importance to any ranging system is the speed of propagation of the signals. It is this speed when multiplied by the measured propagation time interval that provides a measure of the range. If an electromagnetic signal propagates in a vacuum, then the speed of propagation is the vacuum speed of light — valid for all frequencies. However, in the case of the signals transmitted by the GPS satellites, the signals must pass through the earth's atmosphere on their way to receivers on or near the earth's surface. The signals interact with the constituent charged particles and neutral atoms and molecules of the atmosphere with the result that their speed and direction of propagation are changed — the signals are *refracted*.

Before discussing the effects of the propagation media on the GPS signals, we will first define some basic characteristics of signals propagating in a refractive medium.

3.4.1 Refractive Index

The speed of propagation of an electromagnetic wave (a pure carrier) in a medium is given by an equation analogous to equation (3.4):

$$v = \frac{1}{\sqrt{\varepsilon\mu}} \tag{3.12}$$

where ε is the permittivity of the medium and μ is its permeability. The ratio of the speed of propagation in a vacuum to the speed in the medium is known as the refractive index of the medium:

$$n = \frac{c}{v}. \tag{3.13}$$

In a medium, the speed of propagation of a pure (unmodulated) wave, referred to as the phase velocity (we should really call it the phase *speed* as we are not specifying a direction of the motion but the term phase velocity is quite pervasive), is related to the angular frequency of the wave, ω, and the wave number, k:

$$v = \frac{\omega}{k}. \tag{3.14}$$

A medium may be dispersive, in which case the phase velocity and wave number are functions of the frequency of the wave. A plot of frequency vs. wave number yields the dispersion curve of the medium. At any point on the dispersion curve, the slope of the line joining that point to the origin is the phase velocity.

A signal, or modulated carrier wave, can be considered to result from the superposition of a group of waves of different frequencies centred on the carrier frequency. If the medium is dispersive, the modulation of the signal will propagate with a different speed from that of the carrier; this is called the group velocity. The group velocity is given by

$$\begin{aligned} v_g &= \frac{d\omega}{dk} \\ &= v + k\frac{dv}{dk} \end{aligned} \tag{3.15}$$

which is the local tangent slope at a point on the dispersion curve. Corresponding to the phase refractive index, n, we can define a group refractive index, n_g:

$$n_g = \frac{c}{v_g} \tag{3.16}$$

$$= n + f\frac{dn}{df}.$$

In general, a medium will not be homogeneous, in which case, n and n_g will be functions of position in the medium.

At the interface between two media of different refractive indices (or within a medium of varying refractive index), bending of the signal's ray path (as given by vector **k**) will occur as described by Snell's law. Snell's law states that

$$n_1 \sin\theta_i = n_2 \sin\theta_t \tag{3.17}$$

where n_1 is the refractive index in the first media, θ_i is the angle of incidence (between the direction of the incident signal and the normal to the surface between the media), n_2 is the refractive index of the second medium, and θ_t is the transmitted angle (between the direction of the transmitted signal and the normal to the surface). The path bending is a direct consequence of Fermat's principle (of least time) that states that out of all possible paths that it might take, light (and other electromagnetic waves) takes the path that requires the shortest time. It is, in fact, possible to derive Snell's law from Fermat's principle (left as an exercise for the student).

3.4.2 Phase Delay and Group Delay

Due to the fact that the speed of propagation of a carrier wave in a non-ionized medium is less than that in a vacuum, the arrival of a particular phase of the carrier will be delayed in comparison to a wave traveling in a vacuum. This phase delay is given by

$$\tau = \int_S \frac{1}{v} dS - \int_{S'} \frac{1}{c} dS' \tag{3.18}$$

where the integrations are carried out along the refracted path, S, and the non-refracted or rectilinear path S'. The delay may be expressed in units of distance as

$$d_\phi = c\tau = \int_S n dS - \int_{S'} dS' = \int_{S'} (n-1) dS' + \left[\int_S n dS - \int_{S'} n dS' \right]. \quad (3.19)$$

The bracketed integrals account for the bending of the path followed by the wave. Typically, the bending contributes only a small amount to the delay.

Similarly, the modulation of a signal is delayed by

$$d_g = \int_{S'} (n_g - 1) dS' + \left[\int_S n_g dS - \int_{S'} n_g dS' \right]. \quad (3.20)$$

3.5 ATMOSPHERIC REFRACTION

When describing the effects of atmospheric refraction on radio waves, it is convenient to separate the effects of neutral atoms and molecules, the bulk of which are contained in the troposphere, from those of charged particles, primarily contained in the ionosphere. We will look at the effects of both of these media on GPS signals in turn. There is an extensive bibliography on the effects of the troposphere and ionosphere on space geodetic systems. A useful report on the state of the art (circa 1992) in understanding and modeling atmospheric effects on these systems is the *Proceedings of the Symposium on Refraction of Transatmospheric Signals in Geodesy* which was held in The Hague, The Netherlands, in May 1992 by de Munck and Spoelstra [1992]. Brunner [1988] documented significant advances in several aspects of refraction effects on space measurements as of 1988 and subsequently to 1991 by Brunner [1991]. Brunner and Welsch [1993] have authored a tutorial on the effect of the troposphere on GPS measurements and Yunck [1993] has discussed the effects of both the ionosphere and troposphere on ground-level and satellite GPS positioning and how to cope with them. Continued interest in studying the effects of the troposphere is evidenced by the convening of a special session entitled "Applications of GPS Meteorology" at the American Geophysical Union Fall Meeting in December 1994 [AGU, 1994]. A bibliography of the literature on tropospheric propagation delay, both recent and historical, has been put together by Langley et al. [1995].

Much of the following discussion was previously presented by Langley [1992] but appears here for the first time in published form.

3.5.1 Troposphere

The troposphere is the lower part of the earth's atmosphere (see Figure 3.3) where temperature decreases with an increase in altitude. The thickness of the troposphere is not everywhere the same. It extends to a height of less than 9 km over the poles and in excess of 16 km over the equator, see Lutgens and Tarbuck [1989]. Figure 3.4 illustrates the temperature structure of the atmosphere as given by example standard atmospheres. Shown are the temperature profiles of the U.S. Standard Atmosphere, 1976 (identical to the International Civil Aviation Organization Standard Atmosphere up to 32 km), see NOAA/NASA/USAF [1976] and the U.S. Standard Atmosphere Supplements, 1966 for the tropical and polar (summer and winter) regions, see ESSA/NASA/USAF [1966]. The slight kink in the profile for the tropical region between 2 and 3 km reflects the trade wind inversion over ocean areas.

The presence of neutral atoms and molecules in the troposphere affects the propagation of electromagnetic signals. Atoms and molecules in the stratosphere also exist in sufficient numbers to affect the propagation of signals. However, since the bulk of the neutral atmosphere lies within the troposphere, the whole neutral atmosphere is often loosely referred to as the troposphere.

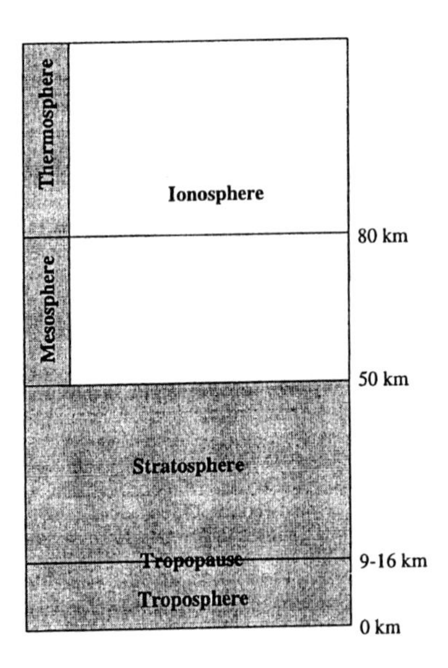

Figure 3.3. The structure of the earth's atmosphere. Note that the thermosphere ranges to a height of 500 km or so and the ionosphere to more than 1,000 km.

Refractivity of Air. The refractivity (or refractive modulus) of a parcel of air, $N = 10^6 (n - 1)$, is a function its temperature (T) and the partial pressures of the dry gases (P_d) and the water vapour (e):

$$N = K_1\left(\frac{P_d}{T}\right)Z_d^{-1} + \left[K_2\left(\frac{e}{T}\right) + K_3\left(\frac{e}{T^2}\right)\right]Z_w^{-1} \tag{3.21}$$

where K_1, K_2, and K_3 are empirically determined coefficients and Z_d is the compressibility factor for dry air and Z_w is the compressibility factor for water vapour. The compressibility factors are corrections to account for the departure of the air behaviour from that of a perfect gas (one for which $P/T = R\rho$ where R is the appropriate gas constant and ρ is the density of the gas, see Owens [1967]). For typical conditions in the earth's atmosphere, Z_d and Z_w depart from unity by less than 1 part in 10^3. The first and second terms in equation (3.21) are due to ultraviolet electronic transitions of the induced dipole type for dry air molecules and water vapour respectively, and the third term is due to the permanent dipole infrared rotational transitions of water vapour.

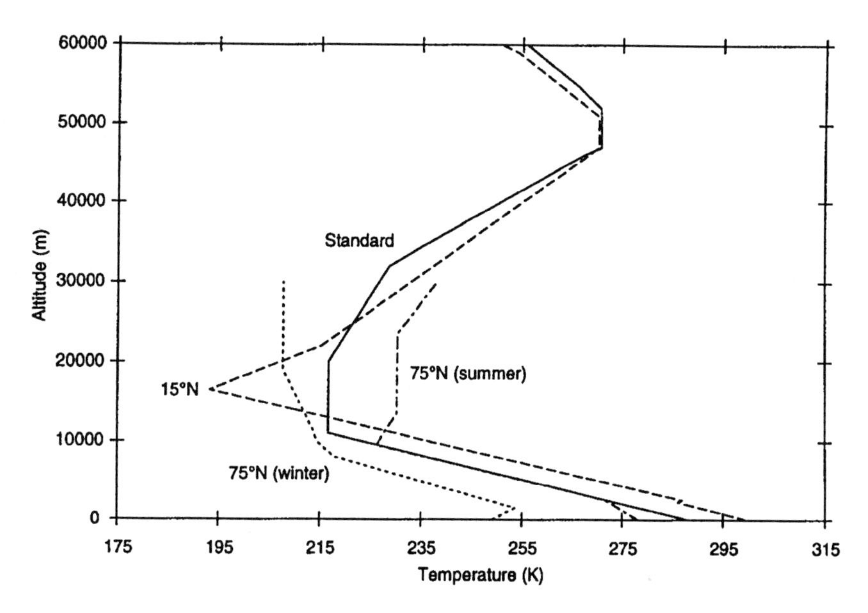

Figure 3.4. Temperature structure of the atmosphere as represented by example standard atmospheres.

The most commonly used sets of refractivity constants are those of Smith and Weintraub [1953] and Thayer [1974] (see Table 3.1).

Table 3.1. Experimentally-determined values for the refractivity constants (K_1 and K_2 are in K $mbar^{-1}$, K_3 is in K^2 $mbar^{-1}$).

	Smith and Weintraub [1953]	Thayer [1974]
K_1	77.61 ± 0.01	77.60 ± 0.014
K_2	72 ± 9	64.8 ± 0.08
K_3	$(3.75 \pm 0.03) \times 10^5$	$(3.776 \pm 0.004) \times 10^5$

For radio frequencies up to about 30 GHz, the troposphere is non-dispersive (except for the anomalous dispersion of the water vapour and oxygen spectral lines) and hence N is independent of frequency.

In radio meteorology, the equation for refractivity of air is most often written in the form

$$N = K_1 \frac{P}{T} + K_2^* \frac{e}{T^2} \tag{3.22}$$

where

$$K_2^* = [\,(K_2 - K_1)\,T + K_3\,]. \tag{3.23}$$

Equation (3.22) may be approximated as

$$N = 77.6 \frac{P}{T} + 3.73 \times 10^5 \frac{e}{T^2} \tag{3.24}$$

and is referred to as the Smith-Weintraub equation, see Smith and Weintraub [1953]. This equation is accurate to about 0.5% (roughly 1.5 N-units at the earth's surface under normal conditions) at frequencies below 30 GHz. However, the formulation of equation (3.21) when used with Thayer's values for the refractivity constants yields accuracies of from about 0.05 N-units for dry air to 0.2 N-units for extremely moist air.

The first and second terms of equation (3.22) are commonly referred to as the *dry* and *wet* components of refractivity. Alternatively, refractivity may be expressed as

$$N = K_1 \frac{M}{M_d} \frac{P}{T} - (K_1 \frac{M}{M_d} - K_2) \frac{e}{T} + K_3 \frac{e}{T^2} \tag{3.25}$$

where

$$\frac{M}{M_d} = \frac{T}{T'} = (1 + 0.3780 \, \frac{e}{P})^{-1} \tag{3.26}$$

in which T' is virtual temperature, and M and M_d denote the molar mass of moist and dry air respectively. The first term in equation (3.25) is referred to as the *hydrostatic* component of refractivity, Davis [1986] as it is a function of the density of moist air which may be assumed to be in hydrostatic equilibrium:

$$\nabla P = \rho \mathbf{g} \tag{3.27}$$

where P is the total pressure, ρ is the total density of moist air, and **g** is the acceleration of gravity. If we integrate equation (3.27) in the zenith direction, we get

$$P_s = \int_{z_s}^{z_a} \rho(z) g(z) dz \tag{3.28}$$

where P_s is the total pressure at the base of the vertical column and where the integration is performed from the earth's surface (z_s) to the top of the neutral atmosphere (z_a).

Alternatively, equation (3.25) may be written as

$$N = K_1 R_d \rho + K_2' R_w \rho_w + K_3 R_w \frac{\rho_w}{T} \tag{3.29}$$

where R_d and R_w are the gas constants for dry air and water vapour respectively, ρ_w is the density of water vapour, and

$$K_2' = \left(K_2 - \frac{M_w}{M_d} K_1 \right) \tag{3.30}$$

with M_w the molar mass of water vapour.

The formulation of equations (3.25) and (3.30) is useful as the zenith delay (see below) based on the hydrostatic component is not influenced by water vapour content unlike the dry component formalism.

Modeling the Delay. The range bias experienced by a signal propagating from a GPS satellite to the ground may be expressed in first approximation by the integral equation

$$d_{trop} = \int_{r_s}^{r_a} [n(r) - 1] \csc\theta(r) dr + \left[\int_{r_s}^{r_a} \csc\theta(r) dr - \int_{r_s}^{r_a} \csc\varepsilon(r) dr \right] \tag{3.31}$$

where n is the refractive index, r is the geocentric radius with r_s the radius of the earth's surface and r_a the radius of the top of the neutral atmosphere, and θ and ε respectively denote the refracted (apparent) and non-refracted (geometric or true) satellite elevation angle. This equation holds for a spherically symmetric atmosphere for which n varies only as a function of geocentric radius. The first integral accounts for the difference in the electromagnetic and geometric lengths of the refracted transmission path. The bracketed integrals account for path curvature; i.e., the difference in the refracted and rectilinear path lengths. Note that in this chapter we use d_{trop} as the symbol for tropospheric propagation delay, rather than T as used in other chapters, since the latter symbol is used here for temperature.

The integral equation can be evaluated given knowledge of the actual refractive index profile or it may be approximated by an analytical function. In applications in satellite ranging, the latter approach is most common with the use of a closed-form or truncated-series approximation based upon a simplified atmospheric model. In most cases, water vapour and the hydrostatic component are considered separately. Each component is usually written as the product of a zenith delay term, approximating the integral of the refractive index profile in the vertical direction, and a mapping function which maps the increase in delay with decreasing elevation angle. In general form

$$d_{trop} = d_h^z m_h(\varepsilon_s) + d_{wv}^z m_{wv}(\varepsilon_s) \tag{3.32}$$

where

d_h^z	is the zenith delay due to the hydrostatic component
d_{wv}^z	is the zenith delay due to water vapour
m_h	is the hydrostatic mapping function
m_{wv}	is the water vapour mapping function
ε_s	is the non-refracted elevation angle at the ground station

Zenith Delays. For a signal coming from the direction of the zenith, equation (3.31) becomes

$$\begin{aligned} d_{trop}^z &= \int_{r_s}^{r_a} [n(r) - 1] dr \\ &= 10^{-6} \int_{r_s}^{r_a} N(r) dr. \end{aligned} \tag{3.33}$$

This is the (total) tropospheric zenith delay. Sea level values of the total tropospheric delay in the zenith direction are of the order of 2.3 to 2.6 metres. The zenith delay can be expressed as the sum of a hydrostatic and wet component using the formalism of equation (3.29):

$$d_{trop}^{z} = 10^{-6} K_1 R_d \int_{z_s}^{z_a} \rho(z) dz + 10^{-6} R_w \int_{z_s}^{z_a} \left[K_2' + \frac{K_3}{T(z)} \right] \rho_w(z) dz. \tag{3.34}$$

The hydrostatic term accounts for roughly accounts for 90% of the total delay and can be obtained from the total surface pressure with an accuracy of a few millimetres by assuming the atmosphere to be in a state of hydrostatic equilibrium. The frequently used Saastamoinen [1973] hydrostatic zenith delay model is given by

$$d_{dry}^{z} = 10^{-6} K_1 R_d \frac{P_s}{g_m} \tag{3.35}$$

where g_m, the magnitude of gravity at the centroid of the atmospheric column is given by

$$g_m = 9.784(1 - 0.0026 \cos 2\phi - 0.00028 H) \tag{3.36}$$

where ϕ is the (geocentric) latitude of the station and H is the station orthometric height in kilometres.

The wet component is a function of the water vapour along the signal path. Unlike the hydrostatic delay, the wet delay is highly variable both spatially and temporally and a model prediction using surface meteorology yields an accuracy no better than 1 to 2 cm, depending on the atmospheric conditions.

It should be noted that there is a very small propagation delay due to liquid water in the form of clouds and rain along the signal path. The size of this delay is typically well below one centimetre and is generally ignored.

Mapping Functions. Over the past 20 years or so, geodesists and radio meteorologists have developed a variety of model profiles and mapping functions for the evaluation of the delay experienced by signals propagating through the troposphere at arbitrary elevation angles.

The simplest mapping function is the cosecant of the elevation angle which assumes that spherical constant-height surfaces can be approximated as plane surfaces. This is a reasonably accurate approximation only for high elevation angles and with a small degree of bending.

Marini [1972] showed that the elevation angle dependence of the tropospheric delay could be expressed as a continued fraction form in terms of the sine of the elevation angle:

$$m(\theta) = \cfrac{1}{\sin\theta + \cfrac{a}{\sin\theta + \cfrac{b}{\sin\theta + \cfrac{c}{\sin\theta + \dots}}}} \tag{3.37}$$

where the coefficients a, b, c, ..., are constants or linear functions. Most of the mapping functions that have been developed are based on a truncation of the continued fraction form. Note that $m(\theta=90°) \neq 1$. Some mapping functions accordingly use a normalised form of equation (3.37).

Among the large number of mapping functions that have been developed are those by Baby et al. [1988], Black [1978], Black and Eisner [1984], Chao [1972], Davis et al. [1985], Goad and Goodman [1974], Herring [1992], Hopfield [1969], Ifadis [1986], Lanyi [1984], Marini and Murray [1973], Moffett [1973], Niell [1993, 1995], Rahnemoon [1988], Saastamoinen [1973], Santerre [1987], and Yionoulis [1970]. The performance of these models has been assessed by Janes et al. [1989, 1991], Mendes and Langley [1994], and Estefan and Sovers [1994].

Janes et al. [1989, 1991] benchmarked delay predictions of the models and mapping functions against values obtained by ray-tracing the U.S. Standard Atmosphere, 1976, see NOAA/NASA/USAF [1976] and the associated U.S. Standard Atmosphere Supplements, 1966, ESSA/NASA/USAF [1966] which, as noted earlier, incorporate latitudinal and seasonal departures from the Standard. The authors concluded from their analysis that, of the models tested, the explicit form of the Saastamoinen zenith delay expressions, see Saastamoinen [1973] in combination with the Davis (also called CfA-2.2) hydrostatic, see Davis [1986] and Goad and Goodman water [1974] vapour mapping functions would provide superior performance to the other models under most conditions.

Mendes and Langley [1994] assessed the accuracy of most of the available mapping functions using ray-tracing through an extensive radiosonde data set covering different climatic regions as "ground truth." Ray-tracing was performed for different elevation angles starting at 3°. Virtually all of the tested mapping functions provided sub-centimetre accuracy for elevation angles above 15°. The precision of the Niell, Herring, and Ifadis mapping functions stood out from the rest even at high elevation angles. Their performance at low elevation angles (less than about 10°) was found to be quite remarkable. Lanyi's mapping function was also found to be a good performer although it does not appear to be quite as accurate as the other three and is not as efficient in terms of ease of implementation and computational speed.

Estefan and Sovers [1994] contrasted the Lanyi, Davis et al., Ifadis, Herring, and Niell mapping functions against the seasonal Chao model which had been implemented through look-up tables in the Jet Propulsion Laboratory's operational Orbit Determination Program (ODP). They reported that all of the tested mapping functions demonstrated superior accuracies compared to the old Chao model. They concluded that "no one 'best' tropospheric mapping function exists for every application and all ranges of elevation angles; however, based on the comparative survey presented, the authors recommend that the Lanyi and Niell mapping functions be incorporated into the ODP ..."

Other interesting observations on the performance of mapping functions are those by Herring [1992] who reported that the typical r.m.s. difference between ray-tracing at a 5° elevation angle and his mapping functions is 30 mm for the hydrostatic delay and 10 mm for the wet delay. Davis et al. [1991] examined errors in the Davis mapping function using data from a series of special very long baseline interferometry (VLBI) experiments. They found that mapping function errors do not exhibit a coherent annual signature but rather appear to be random over the long term.

The Water Vapour Problem. As previously mentioned, whereas the hydrostatic component of the vertical delay can generally be well modelled using accurate surface values of total pressure, the same is not true for the wet component. The water vapour in the troposphere is not well mixed and its distribution is therefore usually spatially and temporally inhomogeneous. The variability is highest in the atmospheric boundary layer, which extends from ground level up to a height of about 1.5 kilometres, and in the cloud layers that do not usually extend much beyond a height of about 4 kilometres.

Water vapour radiometers (WVRs) have been developed in an effort to remotely sense the amount of water vapour along a ray path. A WVR measures atmospheric black body radiation which is affected by the presence of water vapour. Although the technology has evolved considerably (e.g., Elgered et al. [1991]; Rocken et al. [1991]), the use of WVRs, at least in VLBI, is reported to provide only a marginal improvement in accuracy, Kuehn et al. [1991]. However, Tralli et al. [1988], based on the analysis of GPS data collected on baselines across the Gulf of California, suggest that the use of WVR data for tropospheric path delay calibration in humid regions appears to be important for achieving highest possible baseline accuracies. Experiments to further reduce and evaluate WVR instrumental errors are continuing [Kuehn et al., 1993].

The Residual Delay. The residual tropospheric range bias remaining after the application of one of the zenith delay models and associated mapping function can, in most instances, be estimated using the range data itself. Such estimation can take the form of a single scale bias or residual zenith delay estimated for observations spanning many hours, hourly estimates, or stochastic estimation using the Kalman filter approach, see Lindqwister et al. [1990]. Kalman filtering is an attractive alternative to water vapour radiometry both from the point of view of cost and accuracy. In fact, Tralli and Lichten [1990] have shown that stochastic estimation of total zenith path delays yields baseline repeatabilities of a few parts in 10^8, results which are comparable to or better than those obtained after path delay calibration using WVR and or surface meteorological measurements.

Tropospheric Error and Vertical Position. Very often an elevation cut-off angle of 15° or 20° is used in processing GPS data. Such a cut-off angle minimizes problems with noisy data and cycle slips (see Chapter 4) and minimizes the effects of errors in mapping a fixed zenith delay to low elevation angles. However, Yunck [1993] has pointed out that because the functions of tropospheric delay vs. elevation angle and change in signal propagation time due to a change in vertical position of the receiver's antenna vs. elevation angle are similar down to elevation angles of 20°, a fixed zenith delay error will cause an error in the estimated vertical position of the antenna which will increase as lower elevation data are included in the solution. An attempt to solve simultaneously for the zenith delay and the position of the antenna will be aided by the inclusion of low elevation angle data — provided that the mapping function is valid.

Special Problems: Small Networks and Valleys. Most tropospheric delay models and mapping functions for predicting tropospheric delay assume a laterally homogeneous atmosphere. In the actual atmosphere, the decorrelation of signal paths for a network of stations is governed by lateral gradients in atmospheric pressure, temperature, and humidity, and by differences in station elevation. Beutler et al. [1988] have shown that the effect of the differential troposphere on local GPS networks leads to a relative height error that can be written in a first approximation as

$$\Delta h_e = \Delta d^z_{trop} \sec \psi_{max} \tag{3.38}$$

where Δd^z_{trop} denotes the difference in zenith delay between co-observing stations and ψ_{max} is the maximum zenith angle observed. Neglect of the differential troposphere leads to approximately 3 to 5 mm of relative height error for every millimetre change in zenith delay between stations for ψ_{max} = 70-80°.

Janes et al. [1991] intimate that users of GPS data processing software that incorporates a tropospheric delay model driven by surface measurements of temperature, pressure, and relative humidity should be aware of potential pitfalls when processing data collected on baselines of local scale. They suggest that modelling of the differential troposphere between co-observing stations is only advisable when the meteorological gradients clearly exceed the accuracy to which surface meteorological parameters can be measured. Where horizontal gradients or station height differences are significant, careful measurement of surface meteorology is essential for proper modelling of the differential tropospheric delay. Temperature inversions, anomalous humidity profiles, and the use of inappropriate upper air profile lapse rate parameters can significantly reduce the accuracy of the delay model. Where gradients and height differences for a small network are slight, it may be more prudent to assume a laterally homogeneous atmosphere based either upon standard conditions (scaled to height) or upon averaged local meteorological measurements. Beutler et al. [1990] have also pointed out the difficulties associated with modelling the troposphere for small GPS networks.

3.5.2 Ionosphere

The ionosphere is that region of the earth's atmosphere in which ionizing radiation (principally from solar ultraviolet and x-ray emissions) causes electrons to exist in sufficient quantities to affect the propagation of radio waves. This definition does not impose specific limits on the height of the ionosphere. Nevertheless, it is useful to delineate some sort of boundary to the region. The height at which the ionosphere starts to become sensible is about 50 km and it stretches to heights of

1,000 km or more. Indeed, some would argue for an upper limit of 2,000 km. The upper boundary depends on what particular plasma density one uses in the definition since the ionosphere can be interpreted as thinning into the interplanetary plasma. Although the interplanetary plasma affects the propagation of the signals from space probes and the quasar signals observed in VLBI, it may be considered to lie beyond the orbits of GPS satellites and therefore will be ignored here.

The ionosphere is a dispersive medium for radio waves; that is, its refractive index is a function of the frequency. The refractive index is given by the Appleton-Hartree theory of electromagnetic wave propagation in an ionized medium in which there are an equal number of positive ions and free electrons. It is assumed that a uniform magnetic field is present and that the ions (being relatively massive) have negligible effect on radio waves. The complex refractive index, n, at angular frequency, ω, is given by (e.g., Bradley [1989])

$$\begin{aligned} n^2 &= (\mu - i\chi)^2 \\ &= 1 - \frac{X}{1 - iZ - \dfrac{Y_T^2}{2(1-X-iZ)} \pm \left(\dfrac{Y_T^4}{4(1-X-iZ)^2} + Y_L^2\right)^{\frac{1}{2}}} \end{aligned} \qquad (3.39)$$

where

$$X = \frac{N_e e^2}{\varepsilon_0 m_e \omega^2}, \quad Y_L = \frac{eB_L}{m_e \omega}, \quad Y_T = \frac{eB_T}{m_e \omega}, \quad \text{and} \quad Z = \frac{\nu}{\omega}.$$

In equation (3.39), N_e is the electron density, e and m_e are the electron charge and mass respectively, ε_0 is the permittivity of free space, and ν is the electron collision frequency. The subscripts T and L refer to the transverse and longitudinal components of the earth's magnetic field, B. The quantity

$$f_0 = \frac{\omega_0}{2\pi} = \frac{1}{2\pi} \frac{N_e e^2}{\varepsilon_0 m_e} \qquad (3.40)$$

is known as the electron plasma resonance frequency. Typically, $f_0 < 30$ MHz. For frequencies $f >> f_0$, such as those used by GPS, typical values for the components of the earth's magnetic field, and ignoring electron collisions, the refractive index of the ionosphere can be well approximated by

$$n = 1 - \frac{X}{2(1 \pm Y_L)}. \qquad (3.41)$$

The refractive index may be further approximated by ignoring the effect of the longitudinal components of the earth's magnetic field. If this is done, to first order, the phase refractive index of the ionosphere, appropriate for carrier phase observations, is given by

$$n_\phi = 1 - \frac{\alpha N_e}{f^2} \tag{3.42}$$

where α is a constant. Since group refractive index is defined as (see also equation (3.16))

$$n_g = n + f\frac{dn}{df} \tag{3.43}$$

we have for the ionospheric group refractive index, appropriate for pseudorange observations,

$$n_p = 1 + \frac{\alpha N_e}{f^2}\,. \tag{3.44}$$

If N_e has units of reciprocal metres cubed and f is given in Hz, then a has the value 40.28.

Ionospheric Phase Advance and Group Delay of GPS Signals. The integration of the expressions for n_ϕ and n_p along the path followed by a radio signal yields the electromagnetic path lengths

$$\rho_\phi = \int_S \left(1 - \frac{\alpha N_e}{f^2}\right) dS = \rho - d_{ion} \tag{3.45}$$

and

$$\rho_p = \int_S \left(1 + \frac{\alpha N_e}{f^2}\right) dS = \rho + d_{ion} \tag{3.46}$$

where ρ is the true geometric range and d_{ion} is the ionospheric range error which (ignoring path bending) is given by

$$d_{ion} = \frac{\alpha\,TEC}{f^2} \tag{3.47}$$

with TEC (total electron content) being the integrated electron density along the signal path. Carrier phase measurements of the range between a satellite and the ground are reduced by the presence of the ionosphere (the phase is advanced) whereas pseudorange measurements are increased (the signal is delayed) — by the same amount. Note that in concert with our use of the symbol d_{trop} for the tropospheric propagation delay, we have used d_{ion} to represent the ionospheric propagation delay. This contrasts with the use of I to represent this delay in other chapters.

TEC is highly variable both temporally and spatially. The dominant variability is diurnal. There are also solar-cycle and seasonal periodicities as well as short-term variations with commonly-noted periods of 20 to over 100 minutes. Typical daytime values of vertical TEC for mid-latitude sites are of the order of 10^{18} m^{-2} with corresponding night-time values of the order of 10^{17} m^{-2}. However, such typical day-time value can be exceeded by a factor of two or more, especially in near-equatorial regions. For a discussion of the variability of TEC values, see Jursa [1985].

Values for d_{ion} at the GPS L1 frequency of 1575.42 MHz in the zenith direction can reach 30 metres or more and near the horizon this effect is amplified by a factor of about three.

Corrections and Models. Dual frequency positioning systems take advantage of the dispersive nature of the ionosphere for correcting for its effect. In the case of GPS, for example, a linear combination of the L1 and L2 pseudorange measurements may be formed to estimate and subsequently remove the ionospheric bias from the L1 measurements:

$$d_{ion,1} = \frac{f_2^2}{f_2^2 - f_1^2}\left[P_1 - P_2\right] + e \tag{3.48}$$

where f_1 and f_2 are the L1 and L2 carrier frequencies respectively, P_1 and P_2 are the L1 and L2 pseudorange measurements, and e represents random measurement errors and unmodelled biases. A similar approach is used to correct carrier phase measurements with

$$d_{ion,1} = \frac{f_2^2}{f_2^2 - f_1^2}\left[\left(\lambda_1 N_1 - \lambda_2 N_2\right) - \left(\Phi_1 - \Phi_2\right)\right] + \varepsilon \tag{3.49}$$

where Φ_1 and Φ_2 are the L1 and L2 carrier phase measurements (in units of length) respectively, λ_1 and λ_2 are the L1 and L2 carrier wavelengths respectively, N_1 and N_2 are the L1 and L2 integer cycle ambiguities respectively, and ε represents random measurement errors and unmodelled biases. In practice, N_1 and N_2 cannot be determined but as long as the phase measurements are continuous (no cycle slips — see Chapter 4) they remain constant. Hence the carrier phase measurements can be used to determine the variation in the ionospheric delay — the so-called differential delay — but not the absolute delay at any one epoch. The estimation of the differential delay is this way (having ignored third and higher order effects) is good to a few centimetres. Brunner and Gu [1991] have proposed an improved model which accounts for the higher order terms neglected in the first order approximation, the geomagnetic field effect, and ray path bending. Numerical simulations showed that the residual range errors associated with the new model are less than two millimetres.

Note that in the dual-frequency correction approach, it is assumed that the L1 and L2 signals follow the same path through the ionosphere. While this is not quite true (at an elevation angle of 15°, for example, the maximum separation of the ray paths for a high TEC value of 1.38×10^{18} m^{-2} is about 35 metres, Brunner and Gu [1991]), the error induced is generally negligible except under conditions of severe ionospheric turbulence.

If measurements are made at only one carrier frequency, then an alternative procedure for correcting for ionospheric bias must be used. The simplest approach, of course, is to ignore the effect. This approach is often followed by surveyors carrying out relative positioning using single frequency GPS receivers. Differencing between the observations made by simultaneously observing receivers removes that part of the ionospheric range error that is common to the measurements at both stations. The remaining residual ionospheric range error results from the fact that the signals received at the two stations have passed through the ionosphere at slightly different elevation angles. Therefore, the TEC along the two signal paths is slightly different, even if the vertical ionospheric profile is identical at the two stations. It has been shown, see e.g. Georgiadou and Kleusberg [1988] that the main result of this effect in differential positioning is a baseline shortening proportional to the TEC and proportional to the baseline length. Beutler et al. [1988] and Santerre [1989, 1991] have examined the effect of the GPS satellite sky distribution on the propagation of residual ionospheric errors into estimated receiver positions. Such errors can introduce significant scale and orientation biases in relative coordinates. For example, at a typical mid-latitude site using an elevation cut-off angle of 20°, a horizontal scale bias of –0.63 parts per million is incurred for each 1×10^{17} m^{-2} of TEC not accounted for, see Santerre [1991].

It is also possible to use an empirical model to correct for ionospheric bias. The GPS broadcast message, for example, includes the parameters of a simple prediction model, Klobuchar [1986]; ARINC [1991]. Recent tests of this model against a limited set of dual-frequency GPS data showed that this broadcast model can perform very well. Newby and Langley [1992] showed that the model accounted for approximately 70 to 90% of the daytime ionospheric delay and 60 to 70% of the night-time delay at a mid-latitude site during a time of high solar activity. These results indicate that the broadcast model can, at times, remove more than the 50 to 60% r.m.s. of the ionosphere's effect generally acknowledged as the performance level of the model, see e.g., Klobuchar [1986]; Feess and Stephens [1986]. This same study showed that more sophisticated ionospheric models (the Bent Ionospheric Model, Bent and Llewellyn [1973], the 1986 International Reference Ionosphere (IRI), Rawer et al. [1981], and the Ionospheric Conductivity and Electron Density Profile (ICED), Tascoine et al. [1988] — see also Bilitza [1990]) did not appear to perform significantly better, on average, than the broadcast model. In fact, the performance of the ICED model was markedly poorer. Brown et al. [1991] have also evaluated the usefulness of ionospheric models as predictors of TEC. They concluded that none of the six models tested do a very good job probably because the top part of the ionosphere is inaccurately represented. Leitinger and Putz [1988] have looked at the use of the Bent and IRI models in providing information for higher order corrections of the ionospheric range bias.

Georgiadou and Kleusberg [1988] developed a model for the correction of carrier phase GPS observations from a network of single frequency receivers using estimated vertical ionospheric biases derived from the observations of a dual frequency receiver in the vicinity of the network. Webster and Kleusberg [1992] have recently extended this technique to correct the observations from an airborne single frequency receiver moving in the vicinity of three ground-based dual frequency receivers. A similar approach has been followed by Wild et al. [1989] and Wild [1994].

Ionospheric Scintillation and Magnetic Storms. If the number of electrons along a signal path from a satellite to a receiver changes rapidly, the resulting rapid change in the phase of the carrier may present difficulties for the carrier tracking loop in the receiver (see Chapter 4). For a GPS receiver tracking the L1 signal, a change of only 1 radian of phase (corresponding to 0.19×10^{16} m^{-2} change in TEC, or only 0.2% of a typical 10^{18} m^{-2} TEC) in a time interval equal to the inverse of the receiver bandwidth is enough to cause problems for the receiver's tracking loop. If the receiver bandwidth is only 1 Hz (which is just wide enough to accommodate the geometric Doppler shift) then when the second derivative of the phase exceeds 1 Hz per second, loss of lock will result. During such occurrences, the amplitude of the signal is generally fading also. These short-term (1 to 15 seconds) variations in the amplitude and phase of signals are known as ionospheric scintillations.

The loss of lock results in a phase discontinuity or cycle slip. A cycle slip must be repaired before the data following the slip can be used. Large variations in ionospheric range bias over short intervals of time can make the determination of the correct integer number of cycles associated with these phase discontinuities difficult. If the variations of the ionospheric range bias exceed one half of a carrier cycle, they may be wrongly interpreted in the data processing as a cycle slip.

There are two regions where irregularities in the earth's ionosphere often occur causing short term signal fading which can severely test the tracking capabilities of a GPS receiver: the region extending ±30° either side of the geomagnetic equator and the auroral and polar cap regions (see, e.g., Héroux and Kleusberg [1989] and Wanninger [1993]). The fading can be so severe that the signal level drops completely below the signal lock threshold of the receiver. When this occurs, data is lost until the receiver reacquires the signal. The process of loss and re-acquisition of signals may go on for several hours.

Such signal fading is also associated with geomagnetic storms. Magnetic storms (and the associated ionospheric storms) occur when high-energy charged particles from solar flares, eruptive prominences, or coronal holes arrive at the earth causing perturbations in the earth's magnetic field. The charged particles interact with the earth's neutral atmosphere producing excited ions and additional electrons. The strong electric fields that are generated cause significant changes to the morphology of the ionosphere, greatly changing the propagation delay of GPS pseudoranges and the advance in the carrier phases within time intervals as short as one minute. Such changes in the polar and auroral ionospheres can last for several hours.

Occasionally magnetic storm effects extend to the mid-latitudes. During the magnetic storm that occurred in March 1989, range-rate changes produced by rapid variations in TEC exceeded 1 Hz in one second, Klobuchar [1991]. As a result, GPS receivers with a narrow 1 Hz bandwidth were continuously losing lock during the worst part of the storm because of their inability to follow the changes.

GPS as a Tool for Studying the Atmosphere. Ionospheric scientists have used the satellites of the U.S. Navy Navigation Satellite System (Transit) as satellites of opportunity for studying the ionosphere for more than 30 years (e.g. de Mendonca [1963]; Leitinger et al. [1975; 1984]). By recording the Doppler shift on the two Transit frequencies, the change in TEC during a satellite pass may be determined. If data from a satellite pass can be acquired at several stations, it is possible to obtain a two-dimensional image of ionospheric electron density by applying the techniques of computerized tomography (e.g. Austen et al. [1987]). The signals from the constellation of GPS satellites are also being used to study the ionosphere (e.g. Lanyi and Roth [1988]; Clynch et al. [1989]; Melbourne [1989]; Coco [1991]). Monaldo [1991] used dual frequency GPS data to assess spatial variability of the ionosphere and estimate its potential impact on the monitoring of mesoscale ocean circulation using data from altimetric satellites. The troposphere is also being studied using GPS; see, for example, Kursinski [1994].

3.6 SIGNAL MULTIPATH AND SCATTERING

The environment surrounding the antenna of a GPS receiver can, at times, significantly affect the propagation of GPS signals and, as a result, the measured values of pseudorange and carrier phase. The chief effects caused by the environment are multipath and scattering.[6]

3.6.1 Multipath

Multipath is the phenomenon whereby a signal arrives at a receiver's antenna via two or more different paths. The difference in path lengths causes the signals to interfere at the antenna. This phenomenon was quite familiar to television viewers before cable became so pervasive. In dense urban areas, television signals could arrive from the transmitter by the direct, line-of-sight, route and possibly reflected off one or more nearby buildings. The reflected signal, usually weaker than the

---

[6] GPS signals are also susceptible to interference from certain kinds of signals emitted by nearby radio transmitters.

direct signal, produced a "ghost" image. For GPS, multipath is usually noted when operating near large reflecting obstacles such as buildings. In GPS usage, we consider multipath reflections to include all reflected signals from objects external to the antenna. A groundplane is considered to be an intrinsic part of the antenna and so reflection of signals from such a groundplane would not be treated as multipath.

A related phenomenon, somewhat similar to multipath, is imaging, which also involves large nearby reflecting obstacles. The reflecting object produces an "image" of the antenna and the resulting amplitude and phase characteristics are no longer those of the isolated antenna but of the combination of the antenna and its image. Of particular concern is the effect this has on the phase characteristics of the antenna (see Chapter 4).

When a circularly polarised wave is reflected from a surface such as a wall or the ground, the sense of polarization is changed. An antenna designed for RHCP signals will, in theory, infinitely attenuate a LHCP signal although, in practice, attenuations greater than 30 dB are rare and may be much less.

Multipath propagation affects both pseudorange and carrier phases measurements. According to work done for the GPS Joint Program Office by General Dynamics [1979] and reported by Bishop et al. [1985]:

- Multipath can cause both increases and decreases in measured pseudoranges.
- The theoretical maximum pseudorange error for P-code measurements is about 15 metres when the reflected/direct signal amplitude ratio is 1 (and by inference, 150 metres for C/A-code measurements).
- Because of the coded pulse nature of the signal, GPS P-code receivers can discriminate against multipath signals delayed by more than 150 ns (45 metres).
- Typical pseudorange errors show sinusoidal oscillations of periods of 6 to 10 minutes.

Evans and Hermann [1990] reported measured multipath on P-code pseudoranges of between 1.3 metres in a benign environment and 4 to 5 metres in a highly reflective environment. Martin [1978, 1980] assumes an error budget allocation for multipath with an r.m.s. value of 1 to 3 metres for P-code measurements and values an order of magnitude larger for C/A-code measurements.

As described by Seeber [1993], multipath effects on carrier phase observations can amount to a maximum of about 5 cm. If the direct and reflected signals are represented by

$$\begin{aligned} A_D &= A\cos\Phi_D \\ A_R &= \alpha A\cos(\Phi_D + \Phi) \end{aligned} \qquad (3.50)$$

where

A_D is the amplitude of the direct signal

A_R is the amplitude of the reflected signal

α is an attenuation factor ($0 \leq a \leq 1$) (0 = no reflection; 1 = reflected signal at same strength as direct signal)

Φ_D is the phase of the direct signal
Φ is the phase shift of the reflected signal with respect to the direct signal.
The superposition of both signals gives

$$A_\Sigma = A_D + A_R = A\cos\Phi_D + \alpha A\cos(\Phi_D + \Phi) = \beta A\cos(\Phi_D + \Theta) \tag{3.51}$$

With $A_{D,max} = A$ and $A_{R,max} = \alpha A$, then the resultant multipath error in the carrier phase measurement is

$$\Theta = \arctan\left(\frac{\sin\Phi}{\alpha^{-1} + \cos\Phi}\right). \tag{3.52}$$

The amplitude of the signal is

$$B = \beta A = A\sqrt{1 + \alpha^2 + 2\alpha\cos\Phi}\,. \tag{3.53}$$

The above equations indicate that for $\alpha = 1$, the maximum value of Θ is $\Theta = 90°$. Therefore the maximum error on an L1 carrier phase measurement is 0.25 x 19.05 cm or about 5 cm.

Multipath and imaging effects in a highly reflective environment are likely to be limiting factors for single epoch static pseudorange applications at the 10 m level, for static carrier applications at the few centimetre level, and for kinematic applications due to the higher noise level as well as to multipath induced loss of lock.

Multipath effects, when averaged over a long enough time for the relative phase of the direct and reflected signals to have changed by at least one cycle, will be considerably reduced. This is true only for static applications. Imaging effects, on the other hand, cannot be averaged out and may leave biases in the measurements. Multipath and imaging effects are closely repeatable from day to day for the same satellite/antenna site pair, hence monitoring of changes of the antenna coordinates at the centimetre and sub-centimetre level (as required for geodetic applications) may well be possible even in the presence of significant multipath.

The multipath and imaging errors in pseudorange and carrier phase measurements will map into computed receiver positions. It is therefore important to avoid these effects if at all possible. Possible mitigating measures are (see also Chapter 4)

- Careful selection of antenna locations.
- Carefully designed antennas (microstrip; choke ring); use of extended antenna ground planes.
- Use of radio frequency absorbing material near the antenna.
- Receiver design to discriminate against multipath (narrow correlators, multipath-estimating multiple-correlator channels).

3.6.2 Scattering

Another related phenomenon to multipath is signal scattering. Elósegui et al. [1994] have reported that a GPS signal scattered from the surface of a pillar on which a GPS antenna is mounted interferes with the direct signal. The error depends on the elevation angle of the satellite, varies slowly with elevation angle and time, does not necessarily cancel out for different antenna setups and/or long baselines, and introduces systematic errors at the centimetre-level in the estimates of all parameters including site coordinates and residual tropospheric propagation delays.

3.7 SUMMARY

In this chapter, we have examined the generation of the GPS signals and their propagation from the satellites to the antenna of a GPS receiver. After reviewing the fundamentals of electromagnetic wave propagation, we looked at the structure of the GPS signals, and then looked in some detail at the effects that the troposphere and the ionosphere have on the signals. Finally, we looked at propagation effects in the immediate vicinity of the GPS receiver's antenna with an examination of multipath and scattering.

Acknowledgements

Thanks are extended to Virgílio B. Mendes who drafted part of the section on tropospheric propagation delay and for comments on a draft of this chapter.

References

AGU (1994), *1994 Fall Meeting. EOS*, Transactions of the American Geophysical Union, Vol. 75, No. 44, Supplement.

ARINC (1991), Interface Control Document. Navstar GPS Space Segment / Navigation User Interfaces, *ICD-GPS-200*, ARINC Research Corp., Fountain Valley, CA, 3 July, 115 pp.

Ashjaee, J. and R. Lorenz (1992), "Precision GPS Surveying after Y-code." *Proceedings of ION GPS-92*, the Fifth International Technical Meeting of the Satellite Division of The Institute of Navigation, Albuquerque, NM, 16-18 September, pp. 657-659.

Austen, J.R., S.J. Franke, and C.H. Liu (1987), "Ionospheric imaging using computerized tomography." In The Effect of the Ionosphere on Communication, Navigation, and Surveillance Systems, *Proceedings of the 5th Ionospheric Effects Symposium*, Springfield, VA, 5-7 May, pp. 101-106.

Baby, H. B., P. Golé, and J. Lavergnat (1988), "A model for the tropospheric excess path length of radio waves from surface meteorological measurements." *Radio Science*, November-December, Vol. 23, No. 6, pp. 1023-1038.

Bent, R.B. and S.K. Llewellyn (1973), *Documentation and Description of the Bent Ionospheric Model. Space and Missiles Organization*, Los Angles, CA. AFCRL-TR-73-0657.

Beutler, G., I. Bauersima, W. Gurtner, M. Rothacher, T. Schildknecht, and A. Geiger (1988), "Atmospheric refraction and other important biases in GPS carrier phase observations." In *Atmospheric Effects on Geodetic Space Measurements*, Monograph 12, School of Surveying, University of New South Wales, Kensington, N.S.W., Australia, pp. 15-43.

Beutler, G., W. Gurtner, M. Rothacher, U. Wild, and E. Frei (1990), "Relative static positioning with the Global Positioning System: Basic technical considerations." In Global Positioning System: An Overview, *proceedings of International Association of Geodesy,* Symposium No. 102, Edinburgh, Scotland, 7-8 August, 1991, Springer-Verlag, New York; pp. 1-23.

Bilitza, D. (1990), Solar-terrestrial Models and Application Software. *NSSDC/WDC-A-R&S 90-19,* National Space Science Data Center/World Data Center A for Rockets and Satellites, Goddard Space Flight Center, Greenbelt, MD, July, 98 pp.

Bishop, G.J., J.A. Klobuchar, and P.H. Doherty (1985). "Multipath effects on the determination of absolute ionospheric time delay from GPS signals." *Radio Science*, Vol. 20, No. 3, pp. 388-396.

Black, H. D. (1978). "An easily implemented algorithm for the tropospheric range correction." *Journal of Geophysical Research*, 10. April, Vol. 83, No. B4, pp. 1825-1828.

Black, H. D., and A. Eisner (1984). "Correcting satellite Doppler data for tropospheric effects." *Journal of Geophysical Research*, 20. April, Vol. 89, No. D2, pp. 2616-2626.

Bradley, P.A. (1989). "Propagation of radiowaves in the ionosphere." *In Radiowave Propagation*, Eds. M.P.M. Hall and L.W. Barclay, Peter Peregrinus Ltd. (on behalf of the Institution of Electrical Engineers), London, England, U.K.

Brown, L.D., R.E. Daniell, Jr., M.W. Fox, J.A. Klobuchar, and P.H. Doherty (1991). "Evaluation of six ionospheric models as predictors of total electron content." *Radio Science*, Vol. 26, No. 4, pp. 1007-1015.

Brunner, F.K. (ed.) (1988). Atmospheric Effects on Geodetic Space Measurements. *Monograph 12*, School of Surveying, University of New South Wales, Kensington, N.S.W., Australia, 110 pp.

Brunner, F.K. (1991). "Wave propagation in refractive media: A progress report." *Report of International Association of Geodesy Special Study Group 4.93* (1987 - 1991).

Brunner, F.K. and M. Gu (1991). "An improved model for the dual frequency ionospheric correction of GPS observations." *Manuscripta Geodaetica*, Vol. 16, pp. 205-214.

Brunner, F.K. and W.M. Welsch (1993). "Effect of the troposphere on GPS measurements." *GPS World*, Vol. 4, No. 1, pp. 42-51.

Chao, C. C. (1972). A model for tropospheric calibration from daily surface and radiosonde balloon measurement. Jet Propulsion Laboratory, Pasadena, Calif., 8. August, *Technical Memorandum 391-350*, 16 pp.

Clynch, J.R., D.S. Coco, and C.E. Coker (1989). "A versatile GPS ionospheric monitor: High latitude measurements of TEC and scintillation." *In Proceedings of the Institute of Navigation Satellite Division Conference*, Colorado Springs, CO, pp. 445-450.

Coco, D. (1991). "GPS – Satellites of opportunity for ionospheric monitoring." *GPS World*, Vol. 2, No. 9, pp. 47-50.

Davis, J.L. (1986). Atmospheric Propagation Effects on Radio Interferometry. Ph.D. Dissertation. Air Force Geophysics Laboratory *Technical Report AFGL-TR-86-0243*, Hanscom AFB, MA, 276 pp.

Davis, J.L., T.A. Herring, and I.I. Shapiro (1991). "Effects of atmospheric modeling errors on determinations of baseline vectors from very long baseline interferometry." *Journal of Geophysical Research*, Vol. 96, pp. 643-650.

Davis, J. L., T. A. Herring, I. I. Shapiro, A. E. E. Rogers, and G. Elgered (1985). "Geodesy by radio interferometry: Effects of atmospheric modeling errors on estimates of baseline length." *Radio Science*, November-December, Vol. 20, No. 6, pp. 1593-1607.

de Mendonca, F. (1963), "Ionospheric electron content and variations measured by Doppler shifts in satellite transmissions." *Journal of Geophysical Research*, Vol. 67, No. 6, pp. 2315-2337.

de Munck, J.C. and T.A.Th. Spoelstra (eds.) (1992), *Proceedings of the Symposium on Refraction of Transatmospheric Signals in Geodesy*, The Hague, The Netherlands, 19-22 May, Netherlands Geodetic Commission, Publications on Geodesy, Delft, The Netherlands, No. 36, New Series.

Elgered, G., J.L. Davis, T.A. Herring, and I.I. Shapiro (1991), "Geodesy by radio interferometry: Water vapor radiometry for estimation of the wet delay." *Journal of Geophysical Research*, Vol. 96, pp. 6541-6555.

Elósegui, P., J.L. Davis, R.T.K. Jaldehag, J.M. Johansson, A.E. Niell, and I.I. Shapiro (1994), "Effects of signal scattering on GPS estimates of the atmospheric propagation delay." Presented at the *1994 Fall Meeting of the American Geophysical Union*, San Francisco, CA, 5-9 December. Abstract: EOS, Vol. 75, No. 44, Supplement, p. 173.

Environmental Science Services Administration, National Aeronautics and Space Administration, and United States Air Force (1966), *U.S. Standard Atmosphere Supplements*, 1966. U.S. Government Printing Office, Washington, D.C., 290 pp.

Estefan, J.A. and O.J. Sovers (1994), A Comparative Survey of Current and Proposed Tropospheric Refraction-delay Models for DSN Radio Metric Data Calibration. *JPL Publication 94-24*, Jet Propulsion Laboratory, Pasadena, CA, October, 53 pp.

Evans, A.G. and B.R. Hermann (1990), "A comparison of several techniques to reduce signal multipath from the Global Positioning System" In: Eds. Y. Bock and N. Leppard, Global Positioning System: An Overview; *Proceedings of International Association of Geodesy Symposium* No. 102; 7-8 August 1989; Edinburgh, Scotland; New York, Berlin; Springer-Verlag; 1990; pp. 74-81.

Feess, W.A. and S.G. Stephens (1986), "Evaluation of GPS ionospheric time delay algorithm for single-frequency users." *Proceedings of the PLANS-86 conference*, Las Vegas, NV, pp. 280-286.

Feynman, R.P., R.B. Leighton, and M. Sands (1964), *The Feynman Lectures on Physics, Vol. II* — Mainly Electromagnetism and Matter. Addison-Wesley Publishing Company, Reading, MA,

General Dynamics (1979), "Final user field test report for the NAVSTAR global positioning system phase I, major field test objective no. 17: Environmental effects, multipath rejection." *Rep. GPS-GD-025-C-US-7008*, sect. II, pp. 1-7. Electronics Division, General Dynamics, San Diego, California, 28 March.

Georgiadou, Y. and A. Kleusberg (1988), "On the effect of ionospheric delay on geodetic relative GPS positioning." *Manuscripta Geodaetica* , Vol. 13, pp. 1-8.

Goad, C.C. and L. Goodman (1974), "A modified Hopfield tropospheric correction model." *Presented at the American Geophysical Union Fall Annual Meeting*, San Francisco, CA, 12-17 December, 28 pp.

Héroux, P. and A. Kleusberg (1989), "GPS precise relative positioning and the ionosphere in auroral regions." *Proceedings of the 5th International Geodetic Symposium on Satellite Positioning*, Las Cruces, NM, pp. 475-486.

Herring, T.A. (1992), "Modeling atmospheric delays in the analysis of space geodetic data." Proceedings of the Symposium on Refraction of Transatmospheric Signals in Geodesy, Eds. J. C. de Munck, T. A. Th. Spoelstra, The Hague, The Netherlands, 19-22 May, *Netherlands Geodetic Commission, Publications on Geodesy*, Delft, The Netherlands, No. 36, New Series, pp. 157-164.

Hopfield, H. S. (1969), "Two-quartic tropospheric refractivity profile for correcting satellite data." *Journal of Geophysical Research*, 20. August, Vol. 74, No. 18, pp. 4487-4499.

Ifadis, I.I. (1986), The Atmospheric Delay of Radio Waves: Modelling the Elevation Dependence on a Global Scale. *Technical Report #38L*, Chalmers University of Technology, Göteborg, Sweden.

Janes, H.W., R.B. Langley, and S.P. Newby (1989), "A comparison of several models for the prediction of tropospheric propagation delay." *Proceedings of the 5th International Geodetic Symposium on Satellite Positioning*, Las Cruces, NM, pp. 777-788.

Janes, H.W., R.B. Langley, and S.P. Newby (1991), "Analysis of tropospheric delay prediction models: comparisons with ray-tracing and implications for GPS relative positioning." *Bulletin Géodésique*, Vol. 65, pp. 151-161.

Jursa, A.S., Ed. (1985), "Ionospheric Radio Wave Propagation." *Chapter 10 of Handbook of Geophysics and the Space Environment. Air Force Geophysics Laboratory*, Air Force Systems Command, United States Air Force. Available as Document ADA 167000 from the National Technical Information Service, Springfield, VA, U.S.A.

Klobuchar, J.A. (1986), "Design and characteristics of the GPS ionospheric time delay algorithm for single frequency users." *Proceedings of the PLANS-86 conference*, Las Vegas, NV, pp. 280-286.

Klobuchar, J.A. (1991), "Ionospheric effects on GPS." *GPS World*, Vol. 2, No. 4, pp. 48-51.

Kraus, J.D. (1950), *Antennas*. McGraw-Hill Book Company, New York.

Kuehn, C.E., W.E. Himwich, T.A. Clark, and C. Ma (1991), "An evaluation of water vapor radiometer data for calibration of the wet path delay in very long baseline interferometry experiments." *Radio Science*, Vol. 26, No. 6, pp. 1381-1391.

Kuehn, C.E., G. Elgered, J.M. Johansson, T.A. Clark, and B.O. Rönnäng (1993), "A microwave radiometer comparison and its implication for the accuracy of wet delays." *Contributions of Space Geodesy to Geodynamics*: Technology, Eds. D.E. Smith and D.L. Turcotte, American Geophysical Union Geodynamics Series, Vol. 25, pp. 99-114.

Kursinski, R. (1994), "Monitoring the earth's atmosphere with GPS." *GPS World*, Vol. 5, No. 3, pp. 50-54.

Langley, R.B. (1990), "Why is the GPS signal so complex?" *GPS World*, Vol. 1, No. 3, pp. 56-59.

Langley, R.B. (1992), "The effect of the ionosphere and troposphere on satellite positioning systems." *Proceedings of the Symposium on Refraction of Transatmospheric Signals in Geodesy*. Eds. J. C. de Munck, T. A. Th. Spoelstra, The Hague, The Netherlands, 19-22

May, Netherlands Geodetic Commission, Publications on Geodesy, Delft, The Netherlands, No. 36, New Series, p. 97 (abstract only).

Langley, R.B., Wells, W. and Mendes, V.B. (1995), *Tropospheric Propagation Delay*: A Bibliography. 2nd edition. March (unpublished).

Lanyi, G. (1984), "Tropospheric delay affecting radio interferometry." Jet Propulsion Laboratory, Pasadena, CA, *TDA Progress Report 42-78*, April-June, pp. 152-159.

Lanyi, G.E. and T. Roth (1988), "A comparison of mapped and measured total ionospheric electron content using global positioning system and beacon satellite observations." *Radio Science*, Vol. 23, pp. 483-492.

Leitinger, R. and E. Putz (1988), "Ionospheric refraction errors and observables." In *Atmospheric Effects on Geodetic Space Measurements*, Monograph 12, School of Surveying, University of New South Wales, Kensington, N.S.W., Australia, pp. 81-102.

Leitinger, R., G. Schmidt, and A. Tauriainen (1975), "An evaluation method combining the differential Doppler measurements from two stations that enables the calculation of the electron content of the ionosphere." *Journal of Geophysics*, Vol. 41, pp. 201-213.

Leitinger, R., G.K. Hartmann, F.J. Lohmar, and E. Putz (1984), "Electron content measurements with geodetic Doppler receivers." *Radio Science*, Vol. 19, pp. 789-797.

Lindqwister, U. J., J. F. Zumberge, G. Blewitt, and F. Webb (1990), "Application of stochastic troposphere modeling to the California permanent GPS geodetic array." Presented at the *American Geophysical Union Fall Meeting*, San Francisco, CA, 6 December, 14 pp.

Lorrain, P. and D.R. Corson (1970), *Electromagnetic Fields and Waves. 2nd. edition*. W.H. Freeman and Company, San Francisco, CA, 706 pp.

Lutgens, F.K. and E.J. Tarbuck (1989*), The Atmosphere: An Introduction to Meteorology. 4th edition*. Prentice Hall, Englewood Cliffs, NJ, 491 pp.

Marini, J.W. (1972), "Correction of satellite tracking data for an arbitrary atmospheric profile." *Radio Science*, Vol. 7, No. 2, pp. 223-231.

Marini, J.W. and C.W. Murray (1973), Correction of Laser Range Tracking Data for Atmospheric Refraction at Elevations above 10 Degrees. Goddard Space Flight Center *Report X-591-73-351*, NASA GSFC, Greenbelt, MD.

Martin, E.H. (1978, 1980), "GPS user equipment error models." Navigation, *Journal of the (U.S.) Institute of Navigation*, Vol. 25, No. 2, pp. 201-210 and reprinted in Global Positioning System — Papers Published in *Navigation* (Vol. I of "The Red Books"), Institute of Navigation, Alexandria, VA, pp. 109-118.

Melbourne, W.G. (1989), "The Global Positioning System for study of the ionosphere: An overview" Presented at the 1989 Fall Meeting of the American Geophysical Union, San Francisco, CA, 4-8 December. Abstract: *EOS*, Vol. 70, No. 43, p. 1048.

Mendes, V.B. and R.B. Langley (1994), "A comprehensive analysis of mapping functions used in modeling tropospheric propagation delay in space geodetic data." *KIS94, Proceedings of the International Symposium on Kinematic Systems in Geodesy, Geomatics and Navigation*, Banff, Alberta, 30 August - 2 September, The University of Calgary, Calgary, Alberta, Canada, pp. 87-98.

Moffett, J.B. (1973), Program requirements for two-minute integrated Doppler satellite navigation solution. *Technical Memorandum TG 819-1*, Applied Physics Laboratory, The Johns Hopkins University, Laurel, MD.

Monaldo, F. (1991), "Ionospheric variability and the measurement of ocean mesoscale circulation with a spaceborne radar altimeter." *Journal of Geophysical Research*, Vol. 96, pp. 4925-4937.

National Oceanic and Atmospheric Administration, National Aeronautics and Space Administration, and United States Air Force (1976), U.S. Standard Atmosphere, 1976. U.S. Government Printing Office, Washington, D.C., *NOAA-S/T 76-1562*, 227 pp.

Newby, S.P. and R.B. Langley (1992), "Three alternative empirical ionospheric models -- Are they better than the GPS Broadcast Model?" *Proceedings of the 6th International Geodetic Symposium on Satellite Positioning*, Columbus, OH, 17-20 March, pp. 240-244.

Niell, A. E. (1993), "A new approach for the hydrostatic mapping function." *Proceedings of the International Workshop for Reference Frame Establishment and Technical Development in Space Geodesy*, Communications Research Laboratory, Koganei, Tokyo, Japan, 18-21 January, pp. 61-68.

Niell, A.E. (1995), "Global mapping functions for the atmospheric delay at radio wavelengths." VLBI Geodetic Technical Memo No. 13, Haystack Observatory, Massachussetts Institute of Technology, Westford, MA. Submitted to *Journal of Geophysical Research*.

Nieuwejaar, P.W. (1988), "GPS signal structure." The NAVSTAR GPS System, *AGARD Lecture Series No. 161*, Advisory Group for Aerospace Research and Development, North Atlantic Treaty Organization, Neuilly sur Seine, France.

Owens, J.C. (1967), "Optical refractive index of air: Dependence on pressure, temperature and composition." *Applied Optics*, Vol. 6, No. 1, pp. 51-59.

Rahnemoon, M. (1988), Ein neues Korrekturmodell für Mikrowellen — Entfernungsmessungen zu Satelliten. Dr. -Ing. *dissertation Bayerischen Akademie der Wissenschaften, Deutsche Geodätische Kommission*, Munich, F. R. G., 188 pp.

Rawer, K., J.V. Lincoln, R.O. Conkright, D. Bilitza, B.S.N. Prasad, S. Mohanty, and F. Arnold (1981), International Reference Ionosphere. World Data Center A for Solar-Terrestrial Physics, NOAA, Boulder, *CO. Report UAG-82*.

Rocken, C., J.M. Johnson, R.E. Nielan, M. Cerezo, J.R. Jordan, M.J. Falls, L.D. Nelson, R.H. Ware, and M. Hayes (1991), "The measurement of atmospheric water vapor: Radiometer comparison and spatial variations." *IEEE Transactions on Geoscience and Remote Sensing*, GE-29, p. 3-8.

Roddy, D. and J. Coolen (1984), *Electronic Communications*. 3rd edition. Reston Publishing Company, Inc., Reston, VA.

Saastamoinen, J. (1973), "Contributions to the theory of atmospheric refraction." In three parts. *Bulletin Géodésique*, No. 105, pp. 279-298; No. 106, pp. 383-397; No. 107, pp. 13-34.

Santerre, R. (1987), Tropospheric refraction effects in GPS positioning. *SE 6910 graduate seminar Department of Surveying Engineering*, University of New Brunswick, Fredericton, N. B., December, 22 pp.

Santerre, R. (1989), GPS Satellite Sky Distribution: Impact of the Propagation of Some Important Errors in Precise Relative Positioning. Ph.D. Dissertation. Department of Surveying Engineering *Technical Report No. 145*, University of New Brunswick, Fredericton, N.B., Canada, 204 pp.

Santerre, R. (1991), "Impact of GPS satellite sky distribution." *Manuscripta Geodaetica*, Vol. 16, pp. 28-53.

Seeber, Günter (1993), *Satellite Geodesy: Foundations, Methods, and Applications*. Walter de Gruyter, Berlin and New York. 531 pp.

Smith, E.K. and S. Weintraub (1953), "The constants in the equation of atmospheric refractive index at radio frequencies." *Proceedings of the Institute of Radio Engineers*, Vol. 41, No. 8, pp. 1035-1037.

Spilker, J.J., Jr. (1978, 1980), "GPS Signal Structure and Performance Characteristics." *Navigation*, Journal of the (U.S.) Institute of Navigation, Vol. 25, No. 2, pp. 121-146 and reprinted in Global Positioning System — Papers Published in *Navigation* (Vol. I of "The Red Books"), Institute of Navigation, Alexandria, VA, pp. 29-54.

Tascoine, T.F., H.W. Kroehl, R. Creiger, J.W. Freeman, R.A. Wolf, R.W. Spiro, R.V. Hilmer, J.W. Shade, and B.A. Hausman (1988), "New ionospheric and magnetospheric specification models." *Radio Science*, Vol. 23, No. 3, pp. 211-222.

Thayer, G.D. (1974), "An improved equation for the radio refractive index of air." *Radio Science*, Vol. 9, No. 10, pp. 803-807.

Tralli, D.M., T.H. Dixon, and S.A. Stephens (1988), "Effect of wet tropospheric path delays on estimation of geodetic baselines in the Gulf of California using the Global Positioning System." *Journal of Geophysical Research*, Vol. 93, pp. 6545-6557.

Tralli, D.M. and S.M. Lichten (1990), "Stochastic estimation of tropospheric path delays in Global Positioning System geodetic measurements." *Bulletin Géodésique*, Vol. 64, pp. 127-159.

Van Dierendonck, A.J., S.S. Russell, E.R. Kopitzke, and M. Birnbaum (1978, 1980), "The GPS Navigation Message." *Navigation*, Journal of the (U.S.) Institute of Navigation, Vol. 25, No. 2, pp. 147-165 and reprinted in Global Positioning System — Papers Published in *Navigation* (Vol. I of "The Red Books"), Institute of Navigation, Alexandria, VA, pp. 55-73.

Wanninger, L. (1993), "Effects of the equatorial ionosphere on GPS." *GPS World*, Vol. 4, No. 7, pp. 48-54.

Webster, I. and A. Kleusberg (1992), "Regional modelling of the ionosphere for single frequency users of the Global Positioning System." *Proceedings of the 6th International Geodetic Symposium on Satellite Positioning*, Columbus, OH, 17-20 March, pp. 230-239.

Wild, U. (1994), Ionosphere and Geodetic Satellite Systems: Permanent GPS Tracking Data for Modelling and Monitoring. Ph.D. Thesis, Astronomical Institute, University of Bern. *Geodätisch-geophysikalische Arbeiten in der Schweiz*, Bern, Switzerland, Vol. 48, 155 pp.

Wild, U., G. Beutler, W. Gurtner, and M. Rothacher (1989), "Estimating the ionosphere using one or more dual frequency GPS receivers." *Proceedings of the 5th International Geodetic Symposium on Satellite Positioning*, Las Cruces, NM, pp. 724-736.

Yionoulis, S. M. (1970), "Algorithm to compute tropospheric refraction effects on range measurements." *Journal of Geophysical Research*, 20. December, Vol. 75, No. 36, pp. 7636-7637.

Yunck, T.P. (1993), "Coping with the atmosphere and ionosphere in precise satellite and ground positioning." Environmental Effects on Spacecraft Positioning and Trajectories, Ed. A Vallance Jones, based on papers presented at a Union Symposium held at the XXth General Assembly of the International Union of Geodesy and Geophysics, Vienna, August, 1991. *Geophysical Monograph No. 73*, American Geophysical Union, Washington, D.C., pp. 1-16.

4. GPS RECEIVERS AND THE OBSERVABLES

Richard B. Langley
Geodetic Research Laboratory, Department of Geodesy and Geomatics Engineering, University of New Brunswick, P.O. Box 4400, Fredericton, N.B. Canada E3B 5A3

4.1 INTRODUCTION

We saw in Chapter 3 that at a sufficiently large distance from a transmitter, the electromagnetic waves that it emits can be considered to be spherical. We can represent the electric field intensity of a spherical electromagnetic wave of frequency ω and wave number k at some distance r from the transmitter as

$$\mathbf{E} = \frac{\mathbf{E_0}}{r} e^{i(\omega t - kr)}. \tag{4.1}$$

The signal from a GPS satellite when it arrives at a receiver can be taken to be such a wave and if we replace r by ρ, we can represent the signal in simplified form as

$$y = A \cos(\omega t - k\rho + \phi') \tag{4.2}$$

where A is the signal amplitude, t is the elapsed time measured from the start of transmission from the satellite, ρ is the distance travelled from the satellite to the receiver, and ϕ' is a phase bias term which is the phase of the wave at the satellite at $t = 0$.

The distance travelled by the wave between the satellite and the receiver may be determined by one of two methods. At a fixed position in space, the phase of a received wave is ωt plus some unchanging constant. Let us set the constant to zero. Now, if we could identify the beginning of a particular cycle in the wave and if we knew that it was transmitted by the satellite at a certain time $t = 0$, say, then when we receive that particular cycle, the phase of the wave will be ωT, where T is the elapsed time between transmission of the particular cycle and its reception. The distance to the satellite could then be computed by multiplying the elapsed time by the speed of propagation. The beginning of a particular cycle could be identified by superimposing a modulation on the wave which we could then refer to as a carrier wave. Such modulated carrier waves are used in the technique of pseudoranging. Accurate timing information is required for this technique.

An alternative approach would be to count the number of full and fractional cycles in the carrier wave between the satellite and the receiver at a given instant in time — the carrier phase. This number equals the phase of the wave at the receiver assuming zero phase at the satellite. The distance to the satellite could then be determined by dividing the phase at the receiver by the propagation wave number. Unfortunately, no way exists to count directly the number of cycles between a satellite and a receiver at a given instant in time. How this problem is resolved in practice will be explained shortly.

We will describe these two basic GPS observables, the pseudorange and the carrier phase, in this chapter but before we do, we need to examine the instrument that provides us with measurements of these observables: the GPS receiver.

4.2 GPS RECEIVERS

A GPS receiver consists of a number of basic building blocks (see Figure 4.1): an antenna and associated preamplifier, a radio frequency or RF front end section, a signal tracker block, a command entry and display unit, and a power supply. The overall operation of the receiver is controlled by a microprocessor which also computes the receiver's coordinates. Some receivers also include a data storage device and/or an output to interface the receiver to a computer. We'll examine each of these components in turn, starting with the antenna. This discussion of the basics of how GPS receivers work is based on Langley [1991]. Further details on the operation of GPS receivers can be found in Spilker [1978, 1980] and Van Dierendonck [1995].

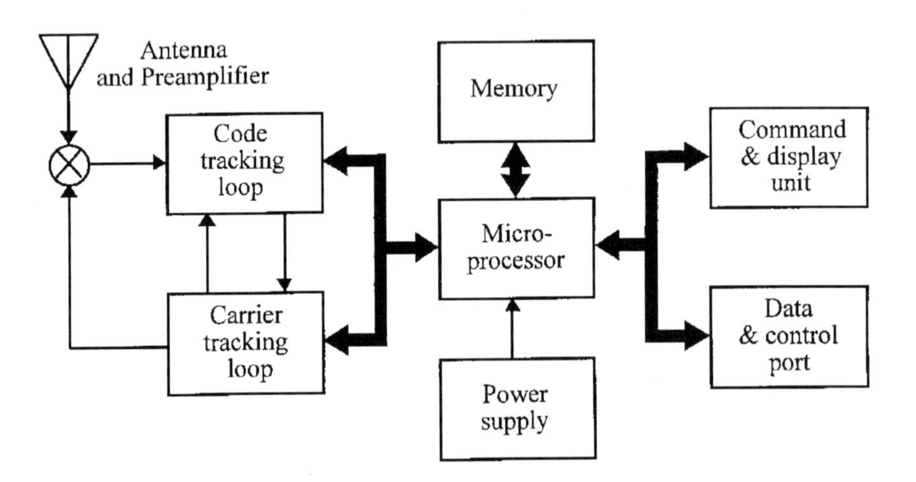

Figure 4.1. The major components of a generic one-channel GPS receiver.

4.2.1 The Building Blocks

Antennas. The job of the antenna is to convert the energy in the electromagnetic waves arriving from the satellites into an electric current which can be handled by the electronics in the receiver. The size and shape of the antenna are very important as these characteristics govern, in part, the ability of the antenna to pick up and pass on to the receiver the very weak GPS signals. The antenna may be required to operate at just the L1 frequency or, more typically for receivers used for geodetic work, at both the L1 and L2 frequencies. Also because the GPS signals are right-hand circularly polarized (RHCP), all GPS antennas must be RHCP as well. Despite these restrictions, there are several different types of antennas that are presently available for GPS receivers. These include monopole or dipole configurations, quadrifilar helices (also known as volutes), spiral helices, and microstrips.

Perhaps the most common antenna is the microstrip because of its raggedness and relative ease of construction. It can be circular or rectangular in shape and is similar in appearance to a small piece of copper-clad printed circuit board. Made up of one or more patches of metal, microstrips are often referred to as patch antennas. They may have either single or dual frequency capability and their exceptionally low profile makes them ideal for airborne and some hand-held applications.

Other important characteristics of a GPS antenna are its gain pattern which describes its sensitivity over some range of elevation and azimuth angles; its ability to discriminate against multipath signals, that is, signals arriving at the antenna after being reflected off nearby objects; and for antennas used in very precise positioning applications, the stability of its phase centre, the electrical centre of the antenna to which the position given by a GPS receiver actually refers.

A GPS antenna is typically omnidirectional. Such an antenna has an essentially nondirectional pattern in azimuth and a directional pattern in elevation angle. At the zenith, the antenna typically has a few dB of gain with respect to a circularly polarized isotropic radiator (dBic), a hypothetical ideal reference antenna. The gain gradually drops down to a few dB below that of a circularly polarized isotropic radiator at an elevation angle of 5° or so.

Some antennas, such as the microstrip, require a ground plane to make them work properly. This is usually a flat or shaped piece of metal on which the actual microstrip element sits. In geodetic surveying, the ground plane of the antenna is often extended with a metal plate or plates to enhance its performance in the presence of multipath. This is done through beam shaping (reducing the gain of the antenna at low elevation angles) and enhancing the attenuation of LHCP (reflected) signals. One form of ground plane is the choke ring, Tranquilla et al. [1989]; Yunck et al. [1989]. A choke ring consists of several concentric hoops, or thin-walled hollow cylinders, of metal mounted on a circular base at the centre of which is placed a microstrip patch antenna. Choke rings are particularly effective in reducing the effects of multipath.

Usually, GPS antennas are protected from possible damage by the elements or other means by the use of a plastic housing (radome) which is designed to minimally attenuate the signals. The signals are very weak; they have roughly the same strength as those from geostationary TV satellites (the strength of the received GPS signals is further discussed in section 4.4.1). The reason a GPS receiver does not need an antenna the size of those in some people's backyards has to do with the structure of the GPS signal and the ability of the GPS receiver to de-spread it (see Chapter 3). The power to extract a GPS signal out of the general background noise of the ether is concentrated in the receiver rather than the antenna. Nevertheless, a GPS antenna must generally be combined with a low noise preamplifier that boosts the level of the signal before it is fed to the receiver itself. In systems where the antenna is a separate unit, the preamplifier is housed in the base of the antenna and receives power from the same coaxial cable along which the signal travels to the receiver.

GPS signals suffer attenuation when they pass through most structures. Some antenna/receiver combinations are sensitive enough to work with signals received inside wooden frame houses and on the dashboards of automobiles and in the window recesses of aircraft, for example, but it is generally recommended that antennas be mounted with a clear view of the satellites. Even outdoors, dense foliage, particularly when it is wet, can attenuate the GPS signals sufficiently that many antenna/receiver combinations have difficulty tracking them.

Two or more GPS receivers can share the same antenna if an antenna splitter is used. The splitter must block the preamplifier DC voltage supplied by all but one of the receivers. The splitter should provide a degree of isolation between the receiver ports so that there is no mutual interference between receivers. Unless the splitter contains an active preamplifier, there will be at least a 3 dB loss each time the signal from the antenna is split.

Assessments of the performance of different antennas used with geodetic-quality GPS receivers have been presented by Schupler and Clark [1991] and Schupler et al. [1994]. Interest in modelling and improving the performance of GPS antennas was shown by the convening of a special session entitled "GPS Antennas" at the American Geophysical Union Fall Meeting in December, 1994. AGU [1994].

An excellent general reference on antennas, including microstrips, is the *Antenna Engineering Handbook,* Johnson [1993].

Mixing Antennas. Ideally, the phase centre of a GPS antenna is independent of the direction of arrival of the signals. However, in practice, there may be small (sub-centimetre in the case of well-designed, geodetic-quality antennas) displacements of the phase centre with changing azimuth and elevation angle. Antennas of the same make and model will typically show similar variations so that their effects can be minimized by orienting antennas on regional baselines to the same direction, say magnetic north. For a well designed antenna, the average horizontal position of the phase centre is usually coincident with the physical centre of the antenna. The vertical position of the phase centre with respect to an accessible physical plane through the antenna has to be established by anechoic chamber measurements. Note that the L1 and L2 phase centres of dual frequency antennas may be different. Now, as long as one is using the same make and model of antenna at both ends of a baseline, the actual position of the phase centre is not usually important; only the vertical heights of a specific point on the exterior of the antennas (say on the base of the preamplifier housing) above the geodetic makers needs to be measured. However, if a mixture of antennas of different make and/or model is used on a baseline or in a network, then the data processing software must know the heights of the phase centres of the antennas with respect to the physical reference points on the antennas so that the appropriate corrections can be made.

Bourassa [1994] carried out of study of the effects of the variation in phase centre position. He found that the observation site, length of observation session, use of ground planes, choice of elevation cut-off angle, orientation of the antenna, and frequency all could have an effect on the estimated coordinates of the antenna. The maximum sizes of the effects ranged from a few millimetres to over a centimetre.

Some success has been reported recently in the application of azimuth and elevation angle-dependent phase centre corrections in processing GPS data using a mixture of antennas, see Gurtner et al. [1994]; Braun et al. [1994].

Transmission Lines. The signals received by the antenna are passed to the receiver along a coaxial transmission line. The signals are attenuated with the degree of attenuation, referred to as insertion loss, dependent on the type and length of coaxial cable used. RG-58C has an insertion loss of about 0.8 dB/m at a frequency of 1575 MHz. The thicker Belden 9913, on the other hand, has an insertion loss of only 0.2 dB/m. For long cable runs, low loss cable is required or an additional low noise preamplifier may be placed between the antenna and the cable.

There is a small delay experienced by the signals travelling from the antenna to the receiver. However, this delay is the same for the signals simultaneously received from different satellites and so acts like a receiver clock offset.

The RF Section. The job of the RF section of a GPS receiver is to translate the frequency of the signals arriving at the antenna to a lower one, called an intermediate frequency or IF which is more easily managed by the rest of the receiver. This is done by combining the incoming signal with a pure sinusoidal signal generated by a component in the receiver known as a local oscillator. Most GPS receivers use precision quartz crystal oscillators, enhanced versions of the regulators commonly found in wristwatches. Some geodetic quality receivers have the provision for supplying the local oscillator signal from an external source such as an atomic frequency standard (rubidium vapour, cesium beam, or hydrogen maser). The IF signal contains all of the modulation that is present in the transmitted signal; only the carrier has been shifted in frequency. The frequency of the shifted carrier is simply the difference between the original received carrier frequency and that of the local oscillator. It is often called a beat frequency in analogy to the beat note that is heard when two musical tones very close together are played simultaneously. Most receivers employ multiple IF stages, reducing the carrier frequency in steps. The final IF signal passes to the work horse of the receiver, the signal tracker.

The Signal Trackers. The omnidirectional antenna of a GPS receiver simultaneously intercepts signals from all satellites above the antenna's horizon. The receiver must be able to isolate the signals from each particular satellite in order to measure the code pseudorange and the phase of the carrier. The isolation is achieved through the use of a number of signal channels in the receiver. The signals from different satellites may be easily discriminated by the unique C/A-code or portion of the P-code they transmit and are assigned to a particular channel.

The channels in a GPS receiver may be implemented in one of two basic ways. A receiver may have dedicated channels with which particular satellites are continuously tracked. A minimum of four such channels tracking the L1 signals of four satellites would be required to determine three coordinates of position and the receiver clock offset. Additional channels permit tracking of more satellites or the L2 signals for ionospheric delay correction or both.

The other channelization concept uses one or more sequencing channels. A sequencing channel "listens" to a particular satellite for a period of time, making measurements on that satellite's signal and then switches to another satellite. A single channel receiver must sequence through four satellites to obtain a three-dimensional position "fix". Before a first fix can be obtained, however, the receiver has to dwell on each satellite's signal for at least 30 seconds to acquire sufficient data from the satellite's broadcast message. The time to first fix and the time between position updates can be reduced by having a pair of sequencing channels.

A variation of the sequencing channel is the multiplexing channel. With a multiplexing channel, a receiver sequences through the satellites at a fast rate so that all of the broadcast messages from the individual satellites are acquired essentially simultaneously. For a multiplexing receiver, the time to first fix is 30 seconds or less, the same as for a receiver with dedicated multiple channels.

Receivers with single channels are cheaper but because of their slowness are restricted to low speed applications. Receivers with dedicated channels have greater sensitivity because they can make measurements on the signals more often but they have inter-channel biases which must be carefully calibrated. This calibration is usually done by the receiver's microprocessor. Most geodetic-quality GPS receivers have 8 to 12 dedicated channels for each frequency and can track the signals from all satellites in view.

The GPS receiver uses its tracking channels to make pseudorange measurements and to extract the broadcast message. This is done through the use of *tracking loops*. A tracking loop is a mechanism which permits a receiver to "tune into" or track a signal which is changing either in frequency or in time. It is a feedback device which basically compares an incoming (external) signal against a locally-produced (internal) signal, generates an error signal which is the difference between the two, and uses this signal to adjust the internal signal to match the external one in such a way that the error is reduced to zero or minimized. A GPS receiver contains two kinds of tracking loops: the delay lock, or code tracking, loop and the phase lock, or carrier tracking, loop.

The delay lock loop is used to align a pseudorandom noise (PRN) code sequence (from either the C/A or P-code) that is present in the signal coming from a satellite with an identical one which is generated within the receiver using the same algorithm that is employed in the satellite. Alignment is achieved by appropriately shifting the receiver-generated code chips in time so that a particular chip in the sequence is generated at the same instant its twin arrives from the satellite.

A correlation comparator in the delay lock loop continuously cross-correlates the two code streams. This device essentially performs a multiply and add process that produces a relatively large output only when the code streams are aligned. If the output is low, an error signal is generated and the code generator adjusted. In this way, the replicated code sequence is locked to the sequence in the incoming signal. The signals from other GPS satellites will have essentially no effect on the tracking process because the PRN codes of all the satellites were chosen to be orthogonal to each other. This orthogonality property means that a very low output is always produced by the correlator whenever the code sequences used by two different satellites are compared.

Because the P-code sequence is so long, a P-code tracking loop needs some help in setting its code generator close to the right spot for obtaining lock with the satellite signal. Its gets this help from information included in the HOW of the broadcast message which is available to the receiver by first tracking the C/A-code.

The time shift required to align the code sequences is, in principle, the time required for a signal to propagate from the satellite to the receiver. Multiplying this time interval by the speed of light gives us the distance or range to the satellite. But because the clocks in a receiver and in a satellite are, in general, not synchronized and run at slightly different rates, the range measurements are biased. These biased ranges are called *pseudoranges*. Since the chips in the satellite code sequences are generated at precisely known instants of time, the alignment of the receiver and satellite code sequences also gives us a reading of the satellite clock at the time of signal generation.

Once the code tracking loop is locked, the PRN code can be removed from the satellite signal by mixing it with the locally generated one and filtering the resultant signal. This procedure de-spreads the signal, shrinking its bandwidth down to about 100 Hz. It is through this process that the GPS receiver achieves the necessary signal to noise ratio to offset the gain limitation of a physically small antenna (see section 4.4.1).

The de-spread IF signal then passes to the phase lock loop which demodulates or extracts the satellite message by aligning the phase of the receiver's local oscillator signal with the phase of the IF or beat frequency signal. If the phase of the oscillator signal is not correct, this is detected by the demodulator in the phase lock loop and a correction signal is then applied to the oscillator. Once the oscillator is locked to the satellite signal, it will continue to follow the variations in the phase of the carrier as the range to the satellite changes.

Most implementations of carrier tracking use the Costas Loop, a variation of the phase lock loop designed for binary biphase modulated signals such as those transmitted by the GPS satellites.

The carrier beat phase observable is obtained in principle simply by counting the elapsed cycles and by measuring the fractional phase of the locked local oscillator signal. The phase measurement when converted to units of distance is then an ambiguous measurement of the range to the satellite. It is ambiguous because a GPS receiver cannot distinguish one particular cycle of the carrier from another and hence assumes an arbitrary number of full cycles of initial phase when it first locks onto a signal. This initial ambiguity must be solved for mathematically along with the coordinates of the receiver if phase observations are used for positioning. Because this ambiguity is constant as long as the receiver maintains lock on the received signal, the time rate of change of the carrier phase is freed from this ambiguity. This quantity is related to the Doppler shift of the satellite signal and is used, for example, to determine the velocity of a moving GPS receiver such as that in an aircraft.

After the carrier tracking loop locks onto a satellite signal, the bits in the broadcast message are subsequently decoded using standard techniques of bit synchronization and a data detection filter.

Codeless Phase Tracking. There are other ways to measure the carrier beat phase other than the standard code tracking / Costas Loop combination and one of these methods must be used to measure L2 carrier phases under AS. The simplest approach is the so-called signal squaring technique. The GPS signal is simply a constant carrier who's phase is shifted by exactly 180° more than a million times each second as a result of modulation by the PRN codes and the broadcast message. These phase reversals can be considered as a change in the amplitude of the signal from +1 to –1 or from –1 to +1 and the instantaneous amplitude is therefore either plus or minus one. Electronically squaring the signal results in a signal with a constant amplitude of unity, although with a frequency equal to twice the original. However, the phase of this signal is easily related to the phase of the original carrier. Of course, in the squaring process both the codes and the broadcast message are lost so code-derived pseudorange measurements are not possible and the information describing the orbits of the satellites as well as their health and the other details in the message must come from another source. There is an inherent signal to noise loss of 30 dB or more in the squaring process compared to code tracking which may result in noisier phase measurements. The codeless squaring technique is illustrated in Figure 4.2 (a) (this and the following three figures are after Van Dierendonck [1995]). In the figure, A represents the amplitude of the incoming signal, D(t) represents the navigation message data, C(t) represents the P-code, and E(t) represents the encryption W-code. The original frequency is f_0 and after squaring it is $2f_0$.

One of the first commercially available GPS receivers, the Macrometer™, used the squaring technique and a number of circa 1990 dual frequency receivers use this approach for measurements on the L2 frequency. A variation of this technique had been used in receivers which measured the *phase* of the code modulations without having to know the actual code sequences.

A serious limitation of the codeless squaring technique is that we end up with a carrier at twice the frequency of the original modulated carrier. As a result, carrier-phase ambiguities can be resolved to only half of the original carrier wavelength which significantly increases the multidimensional search for the correct integer ambiguities. To circumvent this problem, the codeless cross-correlation technique was developed (see Figure 4.2 (b)). With this technique, the L1 signal ($S_1(t)$) is delayed and mixed with the L2 signal ($S_2(t)$). In the mixed signal, with the appropriate delay, Δ, to compensate for the dispersive effect of the ionosphere (see Chapter 3), the codes and the message data will again cancel as in the squaring technique. The resulting signal has a frequency equal to the difference of the L1 and L2 frequencies. The corresponding wavelength of this frequency is about 86 centimetres or 4.52 times that of the L1 frequency which is of considerable help in resolving the ambiguities. Note also that the amplitude of the mixed signal is proportional to the autocorrelation function of the P-code evaluated at the delay Δ: $R(\Delta)$. By maximizing the amplitude, an estimate of the ionospheric delay is obtained.

a)

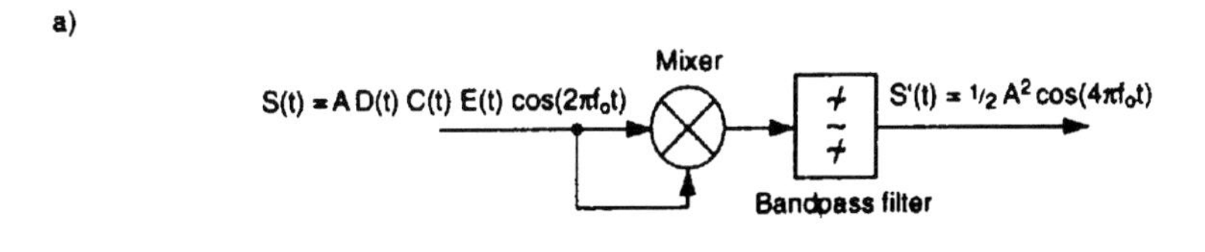

b)

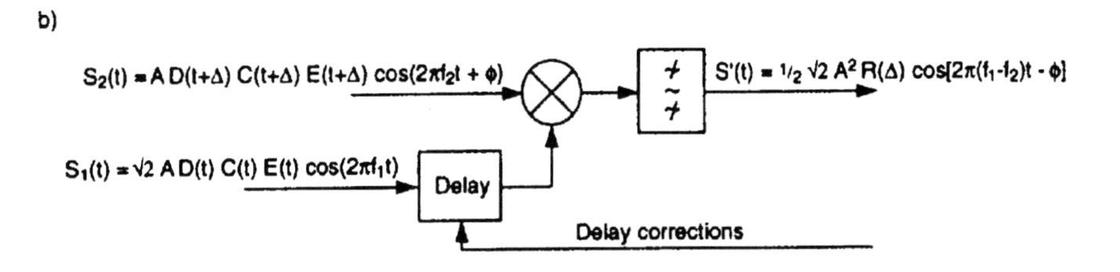

Figure 4.2. (a) The codeless squaring technique; (b) codeless cross-correlation technique.

Significant gains in signal to noise ratio are obtained for a technique that make use of the knowledge of the approximate chipping rate of the W-code. This allows the processing predetection filter bandwidth to be reduced from about 10 MHz to 500 kHz — still not as narrow as in a true P-code correlating receiver, but a significant improvement nevertheless. The gain in signal to noise ratio is about 13 dB. There are two versions of the technique (termed semicodeless): one that uses squaring and one that uses cross-correlation. The semicodeless squaring technique (see Figure 4.3 (a)) removes the encryption code and doubles the carrier frequency. The semicodeless cross-correlation technique (see Figure 4.3 (b)) removes the encryption code, detects the L1-L2 delay, and differences the carrier frequency.

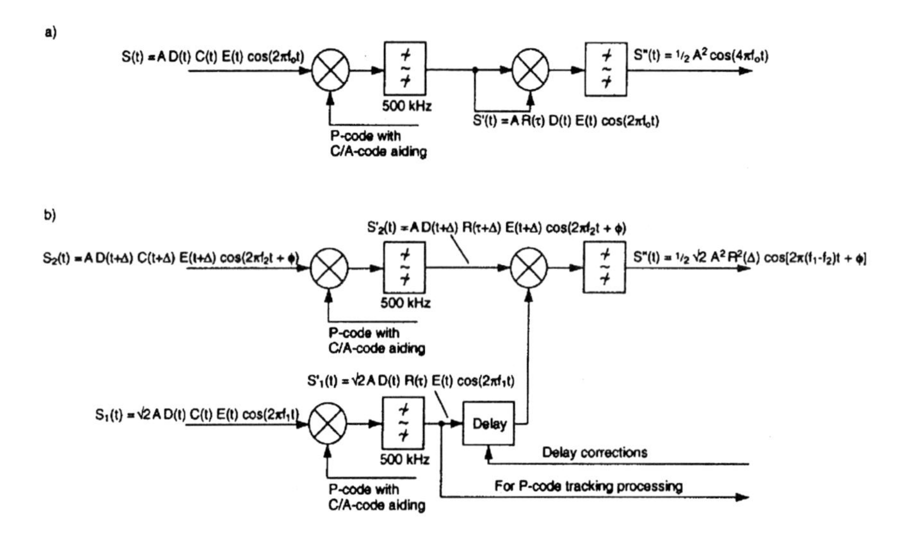

Figure 4.3. (a) Semicodeless squaring technique; (b) semicodeless cross-correlation technique.

Microprocessor, Interfaces, and Power Supply. In this section, we'll take a look at the role of the microprocessor embedded in a GPS receiver; the interfaces which allow us or an external device such as a computer to interact with the receiver; and the receiver's power requirements.

The Microprocessor. Although the bulk of a GPS receiver could be built using analogue techniques, the trend in receiver development has been to make as much of the receiver as possible digital, resulting in smaller, cheaper units. In fact, it is possible for the IF signal to be digitized and to perform the code and carrier tracking with software inside the microprocessor. So in some respects, a GPS receiver may have more in common with your compact disc player than it does with your AM radio. Because it has to perform many different operations such as initially acquiring the satellite signals as quickly as possible once the receiver is turned on, tracking the codes and carriers of the signals, decoding the broadcast message, determining the user's coordinates, and keeping tabs on the other satellites in the constellation, a GPS receiver's operation (even an analogue one) is controlled by a microprocessor. The microprocessor's software, that is the instructions for running the receiver, is imbedded in memory chips within the receiver.

The microprocessor works with digital samples of pseudorange and carrier phase. These are acquired as a result of analogue to digital conversion at some point in the signal flow through the receiver. It is these data samples that the receiver uses to establish its position and which may be recorded for future processing. The microprocessor may run routines which do some filtering of this raw data to reduce the effect of noise or to get more reliable positions and velocities when the receiver is in motion.

The microprocessor may also be required to carry out the computations for waypoint navigation or convert coordinates from the standard WGS 84 geodetic datum to a regional one. It also manages the input of commands from the user, the display of information, and the flow of data through its communication port if it has one.

The Command Entry and Display Unit. The majority of self-contained GPS receivers have a keypad and display of some sort to interface with the user. The keypad can be used to enter commands for selecting different options for acquiring data, for monitoring what the receiver is doing, or for displaying the computed coordinates, time or other details. Auxiliary information such as that required for waypoint navigation or weather data and antenna height for geodetic surveying may also be entered. Most receivers have well-integrated command and display capabilities with menus, prompting instructions, and even "on line" help. It should be mentioned that some receivers have a basic default mode of operation which requires no user input and can be activated simply by turning the receiver on.

Some GPS receivers are designed as sensors to be integrated into navigation systems and therefore don't have their own keypads and displays; input and output is only via data ports.

Data Storage and Output. In addition to a visual display, many GPS receivers including even some hand-held units provide a means of saving the carrier phase and/or pseudorange measurements and the broadcast messages. This feature is a necessity for receivers used for geodetic surveying and for differential navigation.

In geodetic surveying applications, the pseudorange and phase observations must be stored for combination with like observations from other simultaneously observing receivers and subsequent analysis. Usually the data is stored internally in the receiver using semiconductor memory. Some receivers can store data directly to hard or floppy disk using an external microcomputer.

Some receivers, including those which store their data internally for subsequent analysis and those used for real-time differential positioning, have an RS-232-C or some other kind of communications port for transferring data to and from a computer, modem or data radio. Some receivers can be remotely controlled through such a port.

The Power Supply. Most GPS receivers have internal DC power supplies, usually in the form of rechargeable nickel-cadmium (NiCd) batteries. The latest receivers have been designed to draw as little current as possible to extend the operating time between battery charges. Most receivers also make a provision for external power in the form of a battery pack or AC to DC converter.

4.3 GPS OBSERVABLES

Let us now turn our attention to the GPS observables. This discussion draws, in large measure, on a previously published article , Langley [1993].

The basic observables of the Global Positioning System — at least those which permit us to determine position, velocity, and time — are the pseudorange and the carrier phase. Additional observables that have certain advantages can be generated by combining the basic observables in various ways.

4.3.1 The Pseudorange

Before discussing the pseudorange, let's quickly review the structure of the signals transmitted by the GPS satellites (see Chapter 3). Each GPS satellite transmits two signals for positioning purposes: the L1 signal, centred on a carrier frequency of 1575.42 MHz, and the L2 signal, centred on 1227.60 MHz. Modulated onto the L1 carrier are two pseudorandom noise (PRN) ranging codes: the 1 millisecond-long C/A-code with a chipping rate of about 1 MHz and a week-long segment of the encrypted P-code with a chipping rate of about 10 MHz. Also superimposed on the carrier is the navigation message, which among other items, includes the ephemeris data describing the position of the satellite and predicted satellite clock correction terms. The L2 carrier is modulated by the encrypted P-code and the navigation message — the C/A-code is not present.

As we have seen, the PRN codes used by each GPS satellite are unique and have the property that the correlation between any pair of codes is very low. This characteristic allows all of the satellites to share the same carrier frequencies.

The PRN codes transmitted by a satellite are used to determine the pseudorange — a measure of the range, or distance, between the satellite antenna and the antenna feeding a GPS receiver. The receiver makes this measurement by replicating the code being generated by the satellite and determining the time offset between the arrival of a particular transition in the code and that same transition in the code replica. The time offset is simply the time the signal takes to propagate from the satellite to the receiver (see Figure 4.4). The pseudorange is this time offset multiplied by the speed of light. The reason the observable is called a pseudorange is that it is biased by the lack of time synchronization between the clock in the GPS satellite governing the generation of the satellite signal and the clock in the GPS receiver governing the generation of the code replica. This synchronization error is determined by the receiver along with its position coordinates from the pseudorange measurements. The pseudorange is also biased by several other effects including ionospheric and tropospheric delay, multipath, and receiver noise. As shown in Chapter 5 (note the use of a slightly different symbol notation in this section), we can write an equation for the pseudorange observable that relates the measurement and the various biases:

$$P = \rho + c \cdot (dt - dT) + d_{ion} + d_{trop} + e \tag{4.3}$$

where p is the measured pseudorange, ρ is the geometric range to the satellite, c is the speed of light, dt and dT are the offsets of the satellite and receiver clocks from GPS Time, d_{ion} and d_{trop} are the delays imparted by the ionosphere and troposphere respectively, and e represents the effect of multipath and receiver noise. The receiver coordinates are hidden in the geometric range along with the coordinates of the satellite. The objective in GPS positioning is to mathematically describe all of the terms on the right-hand side of the equation — including the initially unknown receiver coordinates in the geometric range term — so that the sum of the terms equals the measurement value on the left-hand side. Any error in the description of the terms will result in errors in the derived receiver coordinates. For example, both the geometric range term and the satellite clock term may include the effects of SA which, if uncompensated, introduce errors into the computed position of the receiver.

Figure 4.5 illustrates the variation in the pseudorange of a particular satellite as measured by a stationary GPS receiver. The large variation is of course dominated by the change in the geometric range due to the satellite's orbital motion and the rotation of the earth.

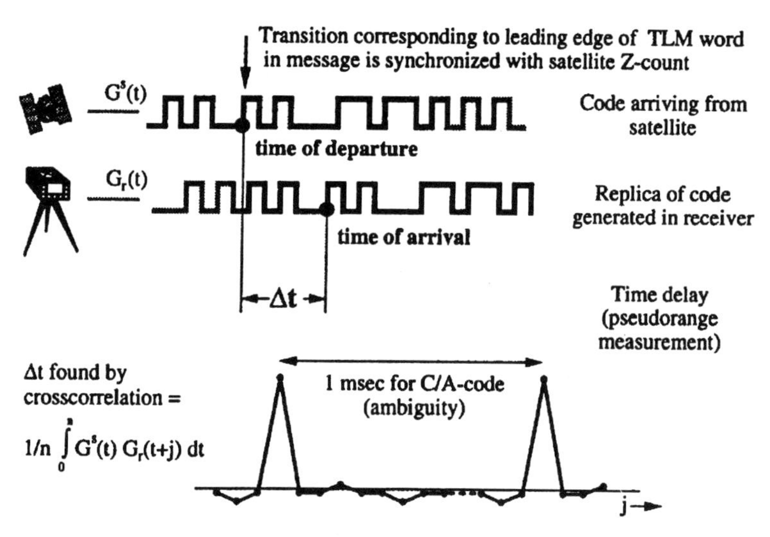

Figure 4.4. How the pseudorange is measured.

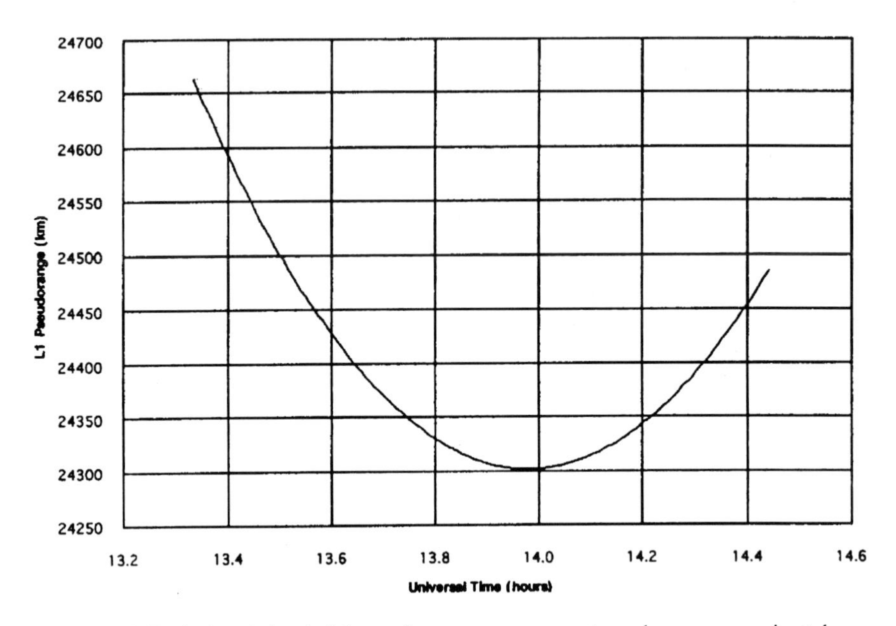

Figure 4.5. Typical variation in L1 pseudorange measurements made over approximately a one-hour period.

Pseudoranges can be measured using either the C/A-code or the P-code. Figure 4.6 shows typical C/A-code pseudorange noise for circa 1990 geodetic-quality GPS receivers.

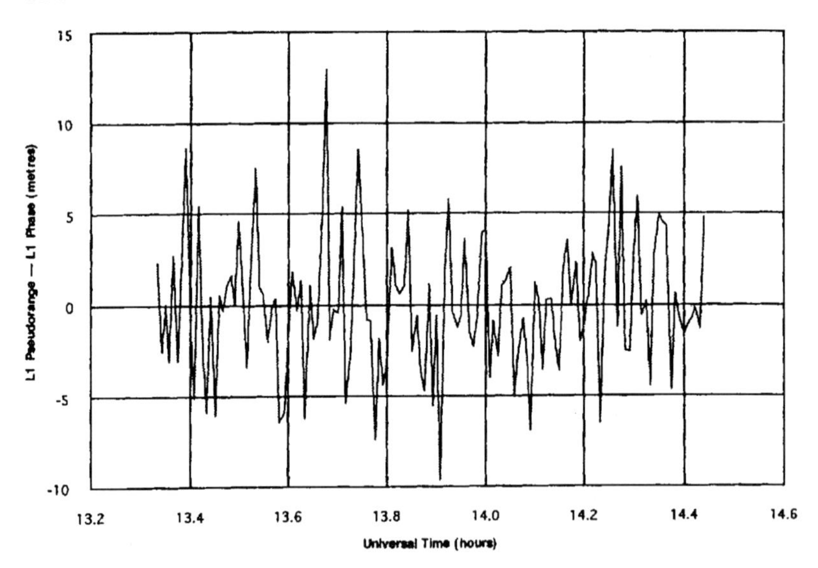

Figure 4.6. The difference between the L1 pseudorange measurements shown in Figure 4.5 and the corresponding phase measurements.

This "noise record" was obtained by subtracting the geometric range, clock, and atmospheric contributions from the pseudorange measurements illustrated in Figure 4.5. What remains is chiefly pseudorange multipath and receiver measurement noise. Because of its higher chipping rate, the P-code generally provides higher precision observations. However recent improvements in receiver technologies have resulted in higher precision C/A-code measurements than were previously achievable (see section 4.4.1).

4.3.2 The Carrier Phase

Even with the advances in code measurement technology, a far more precise observable than the pseudorange is the phase of the received carrier with respect to the phase of a carrier generated by an oscillator in the GPS receiver. The carrier generated by the receiver has a nominally constant frequency whereas the received carrier is changing in frequency due to the Doppler shift induced by the relative motion of the satellite and the receiver. The phase of the received carrier is related to the phase of the carrier at the satellite through the time interval required for the signal to propagate from the satellite to the receiver.

So, ideally, the carrier phase observable would be the total number of full carrier cycles and fractional cycles between the antennas of a satellite and a receiver at any instant. As we have seen earlier, the problem is that a GPS receiver has no way of distinguishing one cycle of a carrier from another. The best it can do, therefore, is to measure the fractional phase and then keep track of changes to the phase; the initial phase is undetermined, or ambiguous, by an integer number of cycles. In order to use the carrier phase as an observable for positioning, this unknown number of cycles or *ambiguity*, N, must be estimated along with the other unknowns — the coordinates of the receiver.

If we convert the measured carrier phase in cycles to equivalent distance units by multiplying by the wavelength, λ, of the carrier, we can express the carrier phase observation equation (see Chapter 5) as

$$\Phi = \rho + c \cdot (dt - dT) + \lambda \cdot N - d_{ion} + d_{trop} + \varepsilon \qquad (4.4)$$

which is very similar to the observation equation for the pseudorange — the major difference being the presence of the ambiguity term. In fact, the carrier phase can be thought of as a biased range measurement just like the pseudorange. Note also that the sign of the ionospheric term in the carrier phase equation is negative whereas in the pseudorange equation it is positive. As we have seen in Chapter 3, this comes about because the ionosphere, as a dispersive medium, slows down the speed of propagation of signal modulations (the PRN codes and the navigation message) to below the vacuum speed of light whereas the speed of propagation of the carrier is actually increased beyond the speed of light. Don't worry, Einstein's pronouncement on the sanctity of the speed of light has not been contradicted. The speed of light limit only applies to the transmission of information and a pure continuous carrier contains no information.

Although all GPS receivers must lock onto and track the carrier of the signal in order to measure pseudoranges, they may not measure or record carrier phase observations for use in navigation or positioning. Some however, may internally use carrier phase measurements to smooth — reduce the high frequency noise — the pseudorange measurements.

Incidentally, in comparison with the carrier phase, pseudoranges when measured in units of the wavelengths of the codes (about 300 meters for the C/A-code and 30 meters for the P-code) are sometimes referred to as code phase measurements.

4.3.3 Data Recording

The rate at which a GPS receiver collects and stores pseudorange and carrier phase measurements is usually user-selectable. Recording intervals of 15-30 seconds might be used for static surveys and up to 2 minutes for permanently operating GPS networks. In kinematic surveying, typical recording intervals are 0.5 to 5 seconds. Generally, for kinematic positioning applications using carrier phase observations, the higher the data rate the better. A high data rate helps in the detection and correction of cycle slips. Sometimes there may be a trade-off between the desired data rate and the amount of memory available in the receiver for data storage.

The data collected by a GPS receiver (time-tagged pseudorange and carrier phase measurements on one or both carrier frequencies and signal to noise ratios for all satellites simultaneously tracked referenced to a common epoch, the broadcast satellite ephemerides and clock coefficients, and (optionally) any meteorological data entered into the receiver) is usually stored in the receiver in proprietary binary-formatted files. These files are downloaded to a computer for post-processing either using manufacturer-supplied software or, which is usually the case for geodetic surveys, one of the software packages developed by university or government research groups.

RINEX. The babel of proprietary data formats could have been a problem for geodesists and others doing postprocessed GPS surveying, especially when combining data from receivers made by different manufacturers. Luckily, a small group of such users had the foresight about 1989 to propose a receiver-independent format for GPS data — RINEX, the Receiver-Independent Exchange format, Gurtner [1994]. This format has been adopted as the *lingua franca* of GPS postprocessing software, and most receiver manufacturers now offer a facility for providing data in this format. It replaced several earlier formats that had been in limited use for data exchange: FICA (Floating Integer Character ASCII) developed by the Applied Research Laboratory of the University of Texas; ARGO (Automatic Reformatting (of) GPS Observations) developed by the U.S. National Geodetic Survey; and an ASCII exchange format developed at the Geodetic Survey of Canada for internal use.

RINEX uses ASCII (plain text) files to ensure easy portability between different computer operating systems and easy readability by software and users alike. The current version of RINEX (Version 2) defines three file types: observation files, broadcast navigation message files, and meteorological data files. Each file consists of one or more header record sections describing the contents of the file and a section (or sections) containing the actual data. Each RINEX observation file usually contains the data collected by one receiver at one station during one session but can also contain all the data collected in sequence by a roving receiver during rapid static or kinematic surveys.

4.4 OBSERVATION MEASUREMENT ERRORS

In this section, we will examine the errors in the measurement of the GPS observables. In so doing, we will use an example of real data collected by a pair of Ashtech Z-12 geodetic-quality receivers. This data was collected in conjunction with receiver acceptance tests on behalf of Public Works and Government Services Canada, Wells et al. [1995].

In an effort to determine the C/A-code pseudorange noise of the Z-12 receivers, receiver-satellite pseudorange double differences (see Chapter 5) were formed using the data collected during a zero baseline test (an antenna splitter was used to supply the same antenna signal to two receivers). One hour of data was collected with a once per second recording rate. The Ashtech raw ASCII data was used to carry out our investigation because we had found that the RINEX data file generated by Ashtech's GPPS 5.1 software package contains smoothed pseudoranges instead of the original raw observations.

In this section, one receiver is referred to as the "base" receiver, the other as the "rover" receiver.

The pseudorange measurements were corrected for exact one millisecond jumps due to the resetting of the clocks in the receivers. Two such jumps occurred in the base station data (about every 28 minutes) and five in the rover receiver (about every 12 minutes). The time tags of the pseudorange and carrier phase data were not corrected for these jumps.

The C/A-code pseudorange noise was examined by first forming an observable which only contains receiver noise and multipath effects. Such an observable can be created by differencing the raw pseudorange measurement with its ionospheric delay removed and the raw carrier phase measurement with its ionospheric delay removed. The C/A-code pseudorange measurement on L1, measured in distance units, can be represented by

$$P_1 = \rho + c(dt - dT) + d_{ion_1} + d_{trop} + mp_{P_1} + noise_{P_1} \tag{4.5}$$

and the carrier phase measurement on L1 and L2, measured in distance units, by

$$\Phi_1 = \rho + c(dt - dT) + \lambda_1 N_1 - d_{ion_1} + d_{trop} + mp_{\Phi_1} + noise_{\Phi_1} \tag{4.6}$$

and

$$\Phi_2 = \rho + c(dt - dT) + \lambda_2 N_2 - d_{ion_2} + d_{trop} + mp_{\Phi_2} + noise_{\Phi_2} \tag{4.7}$$

respectively where ρ is the geometric distance between the satellite antenna and receiver antenna phase centres, c is the speed of light in a vacuum, dt is offset of the satellite clock from GPS time, dT is the offset of the receiver clock from GPS Time, d_{ion} is the ionospheric phase delay, d_{trop} is the tropospheric delay, and mp is the effect of multipath. The equations are essentially the same as equations (4.3) and (4.4) except that we have explicitly indicated the multipath components.

Since to an excellent approximation (see Chapter 3)

$$d_{ion_2} = d_{ion_1} \frac{f_1^2}{f_2^2} \tag{4.8}$$

the ionospheric delay on L1 (within an additive constant and with multipath and noise contributions) can be computed by forming the difference of the L1 and L2 carrier phase measurements:

$$\Phi_2 - \Phi_1 = d_{ion_1} - d_{ion_2} + \lambda_2 N_2 - \lambda_1 N_1 + mp_{\Phi_2} - mp_{\Phi_1} + noise_{\Phi_2} - noise_{\Phi_1} \tag{4.9}$$

and rearranging:

$$d_{ion_2} - d_{ion_1} = \Phi_1 - \Phi_2 + \lambda_2 N_2 - \lambda_1 N_1 + mp_{\Phi_2} - mp_{\Phi_1} + noise_{\Phi_2} - noise_{\Phi_1} \tag{4.10}$$

or

$$d_{ion_1} \frac{f_1^2}{f_2^2} - d_{ion_1} = \Phi_1 - \Phi_2 + \lambda_2 N_2 - \lambda_1 N_1 + mp_{\Phi_2} - mp_{\Phi_1} + noise_{\Phi_2} - noise_{\Phi_1} \,. \tag{4.11}$$

Solving for d_{ion_1} gives

$$d_{ion_1} = \left(\frac{f_2^2}{f_1^2 - f_2^2} \right) \cdot (\Phi_1 - \Phi_2 + \lambda_2 N_2 - \lambda_1 N_1 + mp_{\Phi_2} - mp_{\Phi_1} + noise_{\Phi_2} - noise_{\Phi_1}) \tag{4.12}$$

or

$$d_{ion_1} = 1.5457 \cdot (\Phi_1 - \Phi_2) + 1.5457 \cdot (\lambda_2 N_2 - \lambda_1 N_1 + mp_{\Phi_2} - mp_{\Phi_1} + noise_{\Phi_2} - noise_{\Phi_1}). \tag{4.13}$$

This measure of the L1 ionospheric delay could theoretically be used to correct the C/A-code pseudorange measurement as well as the L1 carrier phase measurement.

Then differencing these corrected measurements would give

$$\begin{aligned}(P_1 - d_{ion_1}) - (\Phi_1 + d_{ion_1}) &= \rho + c(dt - dT) + d_{trop} + mp_{P_1} + noise_{P_1} \\ &\quad -[\rho + c(dt - dT) + \lambda_1 N_1 + d_{trop} + mp_{\Phi_1} + noise_{\Phi_1}] \\ &= mp_{P_1} + noise_{P_1} - \lambda_1 N_1 - mp_{\Phi_1} - noise_{\Phi_1}.\end{aligned} \tag{4.14}$$

Actually, we cannot compute d_{ion_1} exactly as we don't know the values of the integer carrier phase ambiguities (nor the carrier phase multipath and noise). At best, we can compute a relative ionospheric delay which includes a (constant) contribution from the integer carrier phase ambiguities and the multipath and noise terms:

$$\begin{aligned}d^*_{ion_1} &= \left(\frac{f_2^2}{f_1^2 - f_2^2}\right) \cdot (\Phi_1 - \Phi_2) \\ &= 1.5457 \cdot (\Phi_1 - \Phi_2).\end{aligned} \tag{4.15}$$

The relative ionospheric delay, $d^*_{ion_1}$, computed from the PRN 1 carrier phase observations recorded at the base receiver are shown in Figure 4.7 (PRN 1 was selected arbitrarily). Also shown in Figure 4.7 is the ionospheric delay obtained by scaling the difference of the synthetic (obtained from semicodeless cross-correlation of the Y-codes on L1 and L2) P-code pseudoranges. This too is a relative measure of the ionospheric delay as satellite and receiver differential delays (between the L1 and L2 signals) are not taken into account. The noise and multipath on the pseudorange estimate of the ionospheric delay is clearly evident.

Although the estimate of the ionospheric delay from the carrier phase measurements is biased by the integer ambiguities, when we use it to correct both the L1 pseudorange and carrier phase observations and difference the results, we get

$$\begin{aligned}&\left[P_1 - \left(\frac{f_2^2}{f_1^2 - f_2^2}\right) \cdot (\Phi_1 - \Phi_2)\right] - \left[\Phi_1 + \left(\frac{f_2^2}{f_1^2 - f_2^2}\right) \cdot (\Phi_1 - \Phi_2)\right] \\ &= [P_1 - 1.5457 \cdot (\Phi_1 - \Phi_2)] - [\Phi_1 + 1.5457 \cdot (\Phi_1 - \Phi_2)]\end{aligned} \tag{4.16}$$

or

$$P_1 - \left(\frac{f_1^2 + f_2^2}{f_1^2 - f_2^2}\right) \cdot \Phi_1 + \left(\frac{2f_2^2}{f_1^2 - f_2^2}\right) \cdot \Phi_2 \tag{4.17}$$

$$= P_1 - 4.0914\Phi_1 + 3.0914\Phi_2 .$$

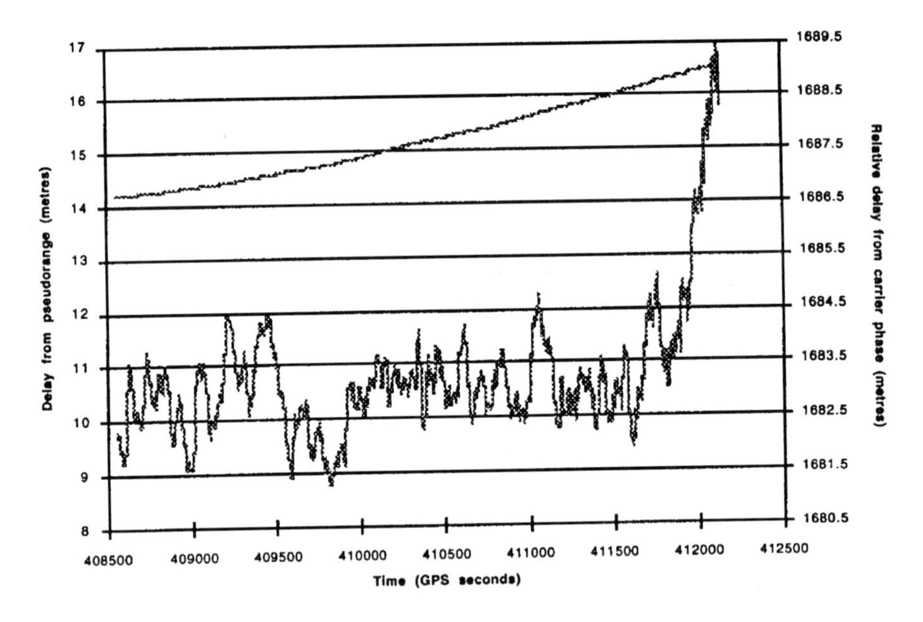

Figure 4.7. L1 ionospheric delay computed from the PRN 1 pseudorange and carrier phase measurements of the base receiver.

What effects remain in this linear combination? Using the basic equations for the pseudorange and carrier phase observations, we get

$$\begin{aligned} &P_1 - 4.0914\Phi_1 + 3.0914\Phi_2 \\ &= \rho + c(dt - dT) + d_{ion_1} + d_{trop} + mp_{P_1} + noise_{P_1} \\ &-4.0914[\rho + c(dt - dT) + \lambda_1 N_1 - d_{ion_1} + d_{trop} + mp_{\Phi_1} + noise_{\Phi_1}] \\ &+3.0914[\rho + c(dt - dT) + \lambda_2 N_2 - d_{ion_2} + d_{trop} + mp_{\Phi_2} + noise_{\Phi_2}] \end{aligned} \tag{4.18}$$

or

$$P_1 - 4.0914\Phi_1 + 3.0914\Phi_2 = mp_{P_1} + noise_{P_1}$$
$$-4.0914(\lambda_1 N_1 + mp_{\Phi_1} + noise_{\Phi_1}) \tag{4.19}$$
$$+3.0914(\lambda_2 N_2 + mp_{\Phi_2} + noise_{\Phi_2}).$$

In arriving at this result, we have assumed that the geometric distance, ρ, between the satellite antenna and receiver antenna phase centres; the receiver clock offset, dT; and the satellite clock offset, dt, are the same for L1 and L2 carrier phase and pseudorange measurements. This assumption was implicit in our use of the same symbols for these parameters in the first three equations of this section.

With the understanding that the multipath and noise in the carrier phase measurements is insignificant in comparison to C/A-code multipath and noise, this linear combination of the code and phase measurements essentially gives the C/A-code multipath and noise — offset by a constant DC-component due to the carrier phase ambiguities. The C/A-code multipath plus noise on the PRN 1 measurements recorded by the base receiver computed in this fashion are shown in Figure 4.8. The large offset due to the phase ambiguities has been removed by subtracting the computed value for the first epoch from all the data values. Figure 4.9 shows the similar results obtained using the data from the rover receiver. Figures 4.10 and 4.11 show the results for PRN 25 using the data from the base and rover receivers respectively. These plots are dominated by multipath which, since the two receivers shared the same antenna, should be identical. The plots bear this out with very similar variations noted for both receivers. The peak-to-peak variation for PRN 1 is about 4.5 metres and for PRN 25 about 3 metres.

To examine the noise component of the data series in Figures 4.8 to 4.11, we differenced the data between receivers (between receiver single differences — see Chapter 5). The time spans over which data was collected by the two receivers are slightly offset from each other. After matching time tags, we ended up with 3,549 differences. Figure 4.12 shows the C/A-code noise for PRN 1 computed in this fashion. Figure 4.12 has the same scale as the previous four figures and the change in the nature of the plots is quite apparent. Figure 4.13 is the same series plotted using an enlarged scale. Figures 4.14 and 4.15 show the results for PRN 25.

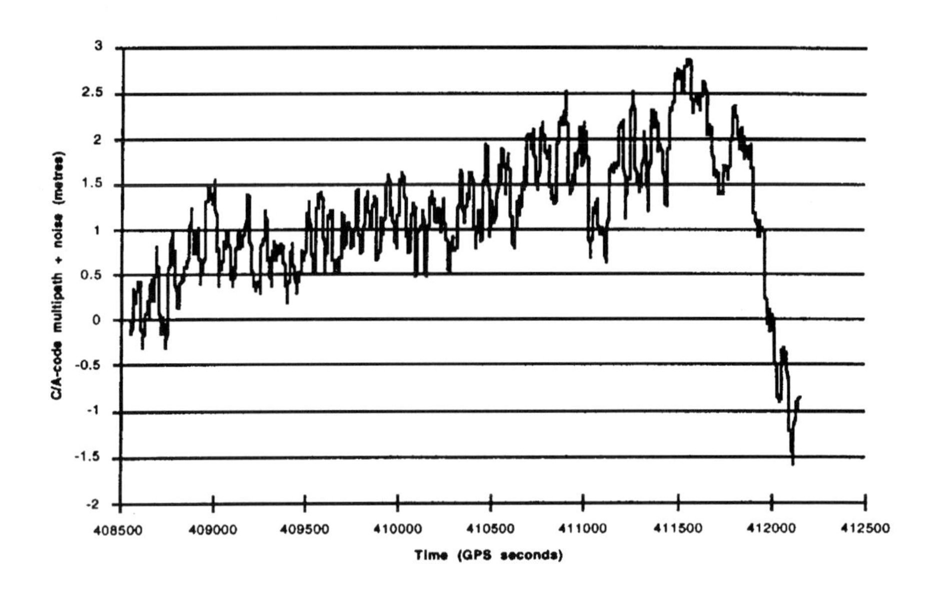

Figure 4.8. C/A-code multipath plus noise from base receiver measurements on PRN 1.

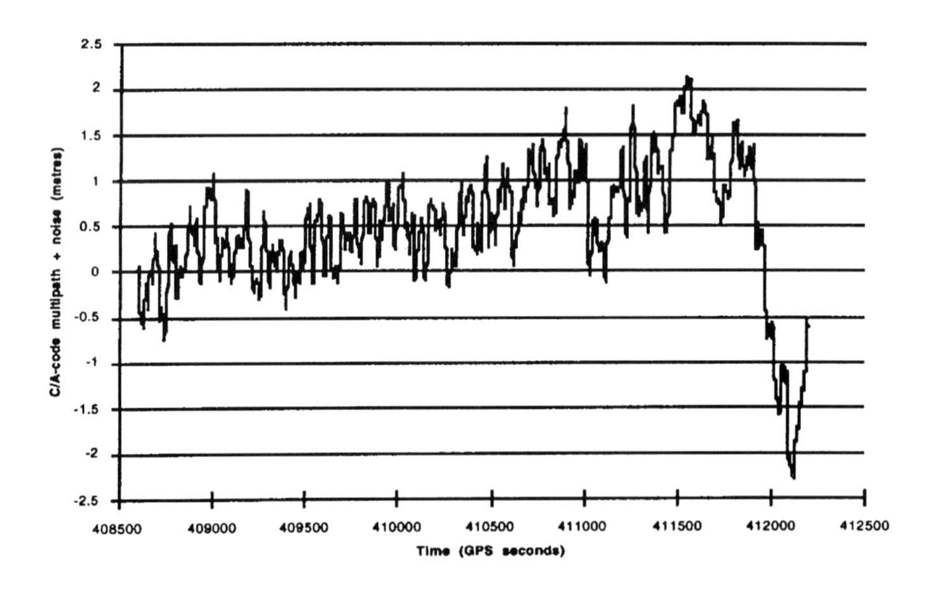

Figure 4.9. C/A-code multipath plus noise from rover receiver measurements on PRN 1.

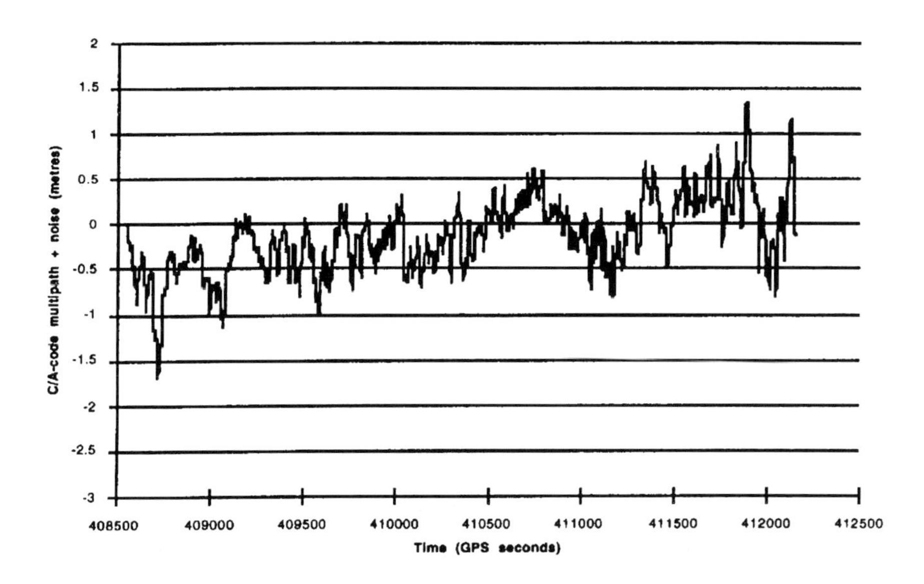

Figure 4.10. C/A-code multipath plus noise from base receiver measurements on PRN 25.

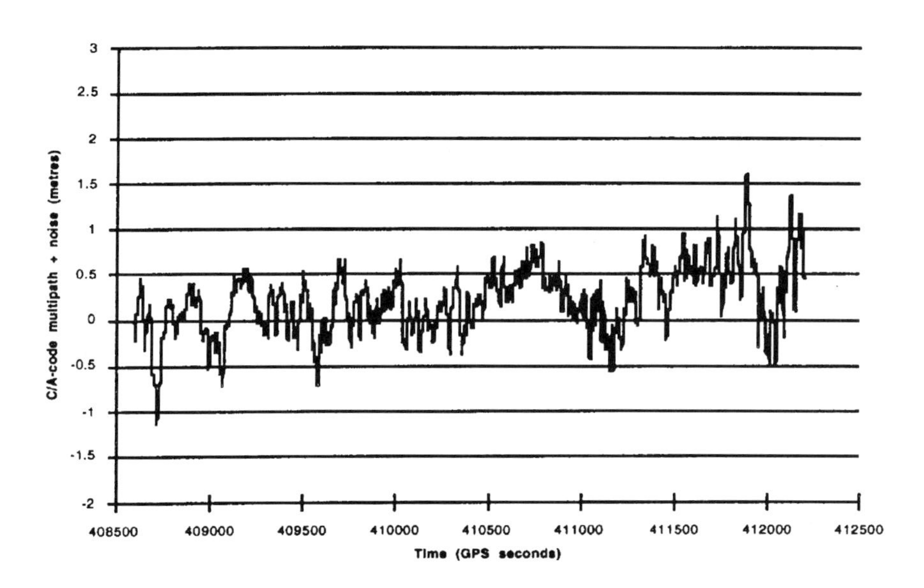

Figure 4.11. C/A-code multipath plus noise from rover receiver measurements on PRN 25.

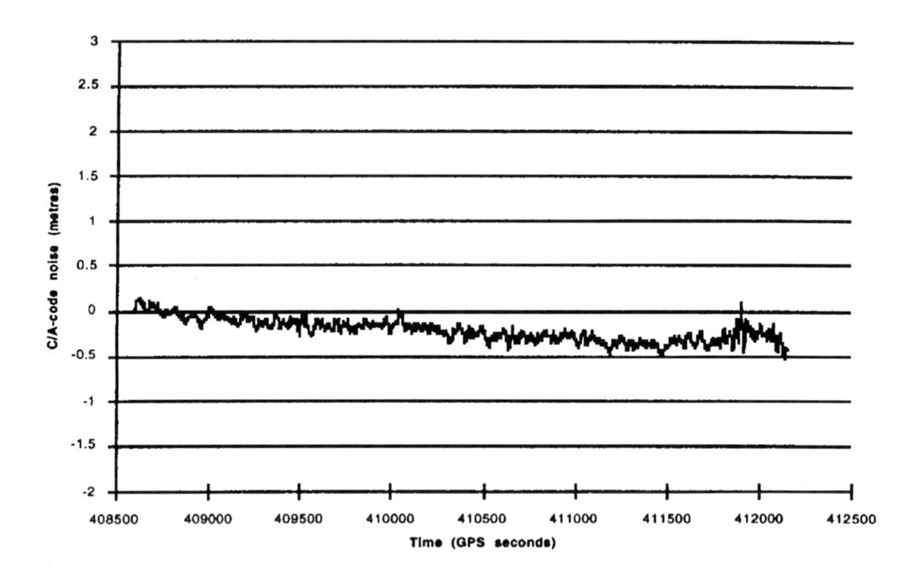

Figure 4.12. C/A-code noise from differencing rover and base receiver measurements on PRN 1.

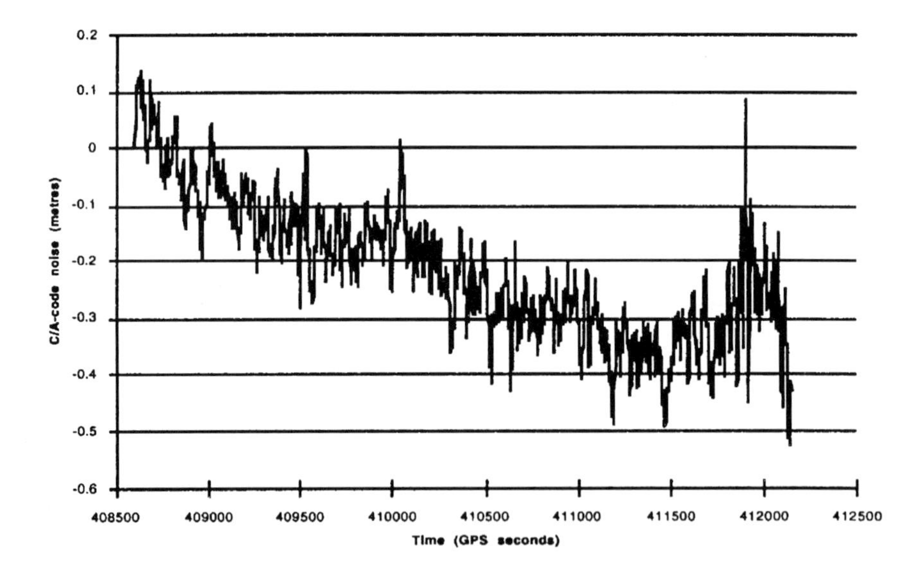

Figure 4.13. C/A-code noise from differencing rover and base receiver measurements on PRN 1 (enlarged scale).

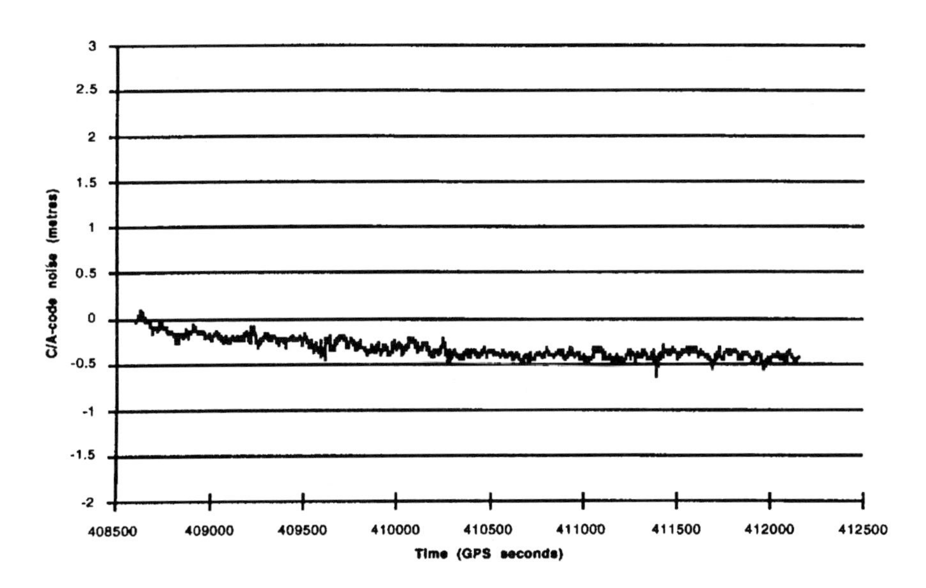

Figure 4.14. C/A-code noise from differencing rover and base receiver measurements on PRN 25.

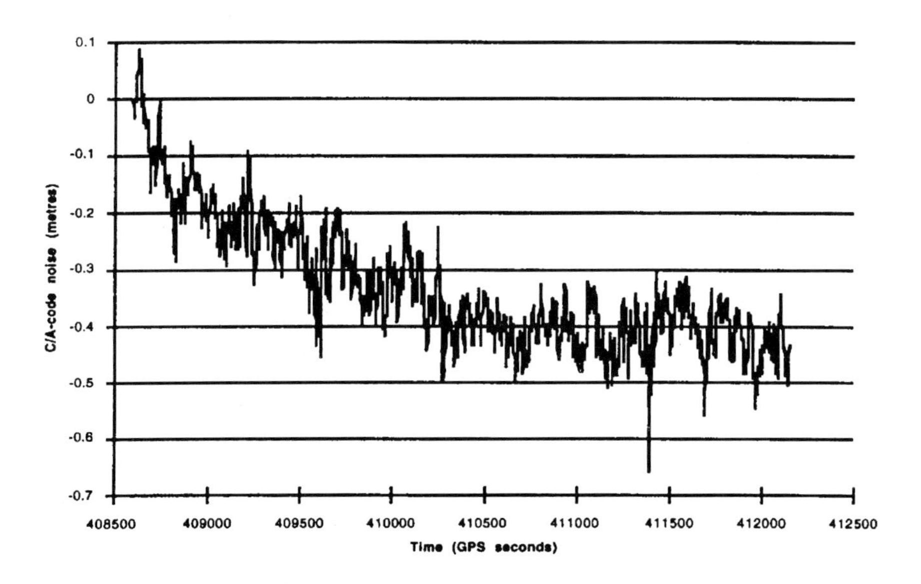

Figure 4.15. C/A-code noise from differencing rover and base receiver measurements on PRN 25 (enlarged scale).

The peak-to-peak variations in the plots for both PRN 1 and PRN 25 are about 60 to 70 centimetres and are dominated by a slow drift in the computed values. This drift — which appears to be quadratic — is more or less the same for both satellites and so seems to be receiver related. In fact, such a drift can be seen in the C/A-code multipath plus noise plots of Figures 4.8 to 4.13 with the base receiver showing slightly more drift than the rover receiver. This phenomenon may be due to the difference in the way the pseudorange and carrier phase measurements are made inside the receiver. Remember, we had assumed the same receiver clock effect on pseudorange and carrier phase, dT, when we differenced the pseudorange and carrier phase measurements in order to isolate the C/A-code multipath and noise. If, in fact, the clock behaviour in the two observables is slightly different, then this difference would show up in the plots. The fact that the effect appears to level off with time may indicate a temperature-related cause — possibly heating effects on frequency synthesizer components in the receiver. Another possible explanation is that the clock effects in the L1 and L2 carrier phase observations are different. Once again, such a difference would be present in the computed C/A-code pseudorange multipath plus noise observable. Campbell [1993] noted a similar phenomenon when computing between receiver differences of L1 minus L2 carrier phase observations for various receiver combinations on a 50 metre baseline.

To remove the drifts and any other non-noise receiver-related effects, we differenced the computed between-receiver data between satellites (double differences — see Chapter 5). The resulting values are shown in Figure 4.16. The arithmetic mean of the values has been subtracted from the data. The peak-to-peak variation of the values is 69.5 centimetres with the spike around 411900 seconds making a significant contribution. The r.m.s. of the values is only 7.8 cm. Since this value represents the noise coming from double-difference observations — two receivers and two satellites — we should divide it by 2 to get an estimate of the noise associated with C/A-code observations by a single receiver. For this value we get 3.9 centimetres. Although our analysis was performed for only one satellite pair, we have no reason to expect a significantly different value for the receiver C/A-code noise level.

In principle, we could have performed our analyses using the pseudorange data alone and simply performed satellite-receiver double differencing without invoking the carrier phase data for the ionospheric correction (the effect of the ionosphere is identical for the two receivers on a zero baseline). However, associated with the receiver 1-millisecond clock jumps are changes in the time tags of the pseudorange and carrier phase measurements equal to the accumulated clock jumps. Since the jumps occur at different times in the two receivers, this leads to slightly mismatched time tags for the data collected by two receivers. If such slight offsets are ignored and the time tags simply rounded off to the nearest second, one is left with anomalous jumps and drifts in the double-difference data which mask the receiver noise one is trying to assess.

Although we estimated the C/A-code pseudorange noise in this example to be at about the 4 centimetre level, this noise will be heavily dominated by the effects of multipath in most practical situations.

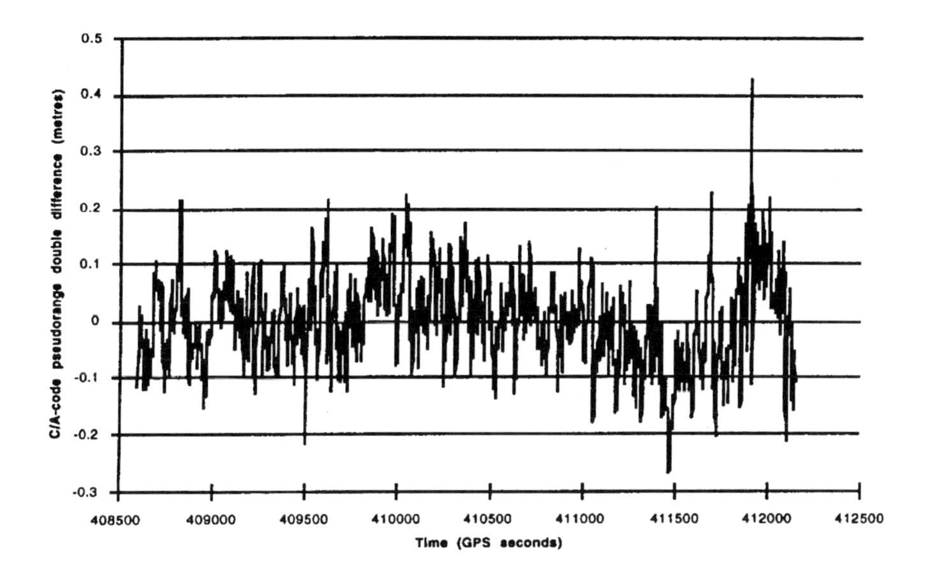

Figure 4.16. Zero-baseline C/A-code pseudorange double difference noise.

4.4.1 Thermal Noise

GPS receivers are not perfect devices: the measurement of the GPS observables cannot be made with infinite precision. There is always some level of noise contaminating the observations as we have seen from the case study in the previous section of this chapter. The most basic kind of noise is that produced by the movement of electrons in any substance (including electronic components such as resistors and semiconductors) that has a temperature above absolute zero (0 K). The electrical current generated by the random motion of the electrons is known as thermal noise (also called thermal agitation noise, resistor noise, or Johnson noise after J.B. Johnson who analyzed the effect in 1928). The noise occupies a broad frequency spectrum with the power in a given passband independent of the passband's centre frequency. The noise power is also proportional to the absolute temperature of the device in which the noise current flows. We can express these relationships as

$$p = kTB \tag{4.20}$$

where p is the thermal noise power, k is Boltzmann's constant (1.380 662 x 10^{-23} J K^{-1}), T is the temperature in kelvins, and B is the bandwidth in hertz.

In the absence of any GPS signal, the receiver and its associated antenna and preamplifier will detect a certain noise power, N. The ratio of the power of a received signal, S, and the noise power, N, measured at the same time and place in a circuit is used as a measure of signal strength. Obviously, the larger the S/N value, the stronger the signal.

Signal-to-noise measurements are usually made on signals at baseband (the band occupied by a signal after demodulation). At RF and IF, it is common to describe the signal level with respect to the noise level using the carrier-to-noise-power-density ratio, C/N_0. This is the ratio of the power level of the signal carrier to the noise power in a 1 Hz bandwidth. It is a key parameter in the analysis of GPS receiver performance and has a direct bearing on the precision of the receiver's pseudorange and carrier phase observations.

In Chapter 3, we saw that the expected minimum received C/A-code signal level is -160 dBW. This is the signal level referenced to a 0 dBic gain antenna. An actual GPS receiver omnidirectional antenna may have a few dB of gain near the zenith and negative gain at very low elevation angles. Also, there will be 1 or 2 dB of cable and circuit losses. So we may take -160 dBW as a strawman minimum carrier power level. To determine the noise power density, we need to determine the effective noise temperature of the receiving system. This is not simply the ambient temperature. The noise temperature of the system, or simply the system temperature, is a figure of merit of the whole receiving system. It is composed of the antenna temperature, a correction for losses in the antenna cable, and the equivalent noise temperature of the receiver itself. All of the temperatures are referred to the receiver input.

The antenna temperature is the equivalent noise temperature of the antenna. If the antenna is replaced with a resistance equal to the impedance of the antenna and heated up until the thermal noise it produces is the same as that with the antenna connected, then the temperature of the resistance is the noise temperature of the antenna. So the antenna temperature, T_a, is a measure of the noise power produced by the antenna; it is not the actual physical temperature of the antenna material. The noise in the antenna output includes the contributions from anything the antenna "sees" including radiation from the ground, the atmosphere, and the cosmos. T_a must be corrected for the contribution by the cable between the antenna and the receiver input. The cable is a "lossy" device: a signal travelling through it is attenuated (see section 4.2.1). But not only does a lossy component reduce the signal level, it also adds to the noise. It can be shown in Stelzried [1968], that if L is the total loss in the cable (power in divided by power out; L>1), then the total antenna temperature is given by

$$T_{ant} = \frac{T_a}{L} + \frac{L-1}{L} T_0 \tag{4.21}$$

where T_0 is the ambient temperature of the cable. Alternatively, we may write this as

$$T_{ant} = \alpha T_a + (1-\alpha) T_0 \tag{4.22}$$

where α, the fractional attenuation (0-1), is just the inverse of L. This equation is of the same form as the equation of radiative transfer found in physics. In fact, the physics of emission and absorption of electromagnetic radiation by a cloud of matter is similar to the emission and absorption taking place in an antenna cable.

The noise temperature of a receiver, T_r, is the noise temperature of a noise source at the input of an ideal noiseless receiver which would produce the same level of receiver output noise as the internal noise of the actual receiver. A typical home radio or television receiver might have a noise temperature of 1500 K whereas a receiver used in radio astronomy might have a noise temperature of less than 10 K. Instead of specifying the noise temperature of the receiver, it is common to use the noise factor, F, where

$$F = \frac{N_{out}}{GkT_0B} \tag{4.23}$$

and where N_{out} is the output noise power of the receiver, G its gain, and B its effective bandwidth. If a noise source connected to the input of a receiver has a noise temperature of T_0, the noise power at the output is given by

$$N_{out} = GkT_0B + GkT_rB = Gk(T_0 + T_r)B. \tag{4.24}$$

So that

$$F = 1 + \frac{T_r}{T_0}. \tag{4.25}$$

Typically, T_0 is taken to be the standard reference temperature of 290 K. If T_r is also 290 K, for example, then F = 2. It is convenient to express the noise factor in dB. It is then correctly referred to as a noise figure. For our example, the noise figure is 3.01 dB. Note that the term "noise figure" is often used arbitrarily for both the noise factor and its logarithm.

Since $T_r = (F - 1)T_0$, we have then for the system temperature:

$$T_{sys} = \frac{T_a}{L} + \frac{L-1}{L}T_0 + (F-1)T_0. \tag{4.26}$$

A typical value for T_{sys} of a GPS receiving system is 630 K, Spilker [1978; 1980]. The corresponding noise power density is 8.7 x 10^{-21} watts per hertz. Or, in logarithmic measure, -200.6 dBW-Hz. Using the value of -160 dBW for the received C/A-code carrier power and ignoring signal gains and losses in the antenna, cable, and receiver, we have a value for the carrier-to-noise-density ratio of about 40 dBW-Hz. Actually, C/N_0 values experienced in practice will vary a bit from this value depending on the actual power output of the satellite transmitter and variations in the space loss with changing distance between satellite and receiver, variations in antenna gain with elevation angle and azimuth of arriving signals, and signal losses in the preamplifier, antenna cable, and receiver. Ward [1994] gives a value of 38.4 dB-Hz for a minimum L1 C/A-code C/N_0 value assuming a unit gain antenna and taking into account typical losses. It turns out that all GPS satellites launched thus far have transmitted at levels where the received power levels have exceeded the minimum specified levels by 5 to 6 dB, Nagle et al. [1992]. Nominal C/N_0 values are therefore usually above 45 dB-Hz, see Van Dierendonck et al. [1992] and values of 50 dB-Hz are typically experienced in modern high-performance GPS receivers, Nagle et al. [1992].

Note that even a strong C/A-code signal with a level of -150 dBW is buried in the ambient noise which, in the approximately 0.9 MHz C/A-code 3 dB bandwidth, has a power of -141 dBW, some 9 dB stronger than the signal. Of course, the signal is raised out of the noise through the code correlation process. The process or spreading gain of a GPS receiver is the ratio of the bandwidth of the transmitted signal to the navigation message data rate, Dixon [1984]. For the C/A-code signal, using the 3 dB bandwidth, this works out to be about 43 dB. So, after despreading, the S/N for a strong signal is about 34 dB.

The C/N_0 value determines, in part, how well the tracking loops in the receiver can track the signals and hence the precision of the pseudorange and carrier phase observations. In the following discussion, we will only consider the effect of noise on code and carrier tracking loops in a standard code-correlating receiver. S/N losses experienced with codeless techniques have been discussed in section 4.2.1.

Code Tracking Loop. The code tracking loop — or delay lock loop, DLL — jitter, for an early/late one-chip-spacing correlator is given by Spilker [1977]; Ward [1994]

$$\sigma_{DLL} = \sqrt{\frac{\alpha B_L}{c/n_0}\left[1+\frac{2}{T\,c/n_0}\right]}\,\lambda_c \tag{4.27}$$

where α is the dimensionless DLL discriminator correlator factor (1 for a time-shared tau-dither early/late correlator, 0.5 for dedicated early and late correlators); B_L is the equivalent code loop noise bandwidth (Hz); c/n_0 is the carrier-to-noise density expressed as a ratio ($=10^{(C/N_0)/10}$ for C/N_0 expressed in dB-Hz); T is the predetection integration time (in seconds; T is the inverse of the predetection bandwidth or, in older receivers, the post-correlator IF bandwidth); and λ_c is the wavelength of the PRN code (29.305 m for the P-code; 293.05 m for the C/A-code). The second term in the brackets in equation (4.27) represents the so-called squaring loss.

Typical values for B_L for modern receivers range from less than 1 Hz to several Hz. If the code loop operates independently of the carrier tracking loop, then the code loop bandwidth needs to be wide enough to accommodate the dynamics of the receiver. However, if the code loop is aided through the use of an estimate of the dynamics from the carrier tracking loop, then the code loop can maintain lock without the need for a wide bandwidth. The code loop bandwidth need only be wide enough to track the ionospheric divergence between pseudorange and carrier phase so that it is not uncommon for carrier-aided receivers to have a code loop bandwidth on the order of 0.1 Hz, Braasch [1994]. Note that B_L in equation (4.27) is not necessarily the code loop bandwidth. If post-measurement smoothing (or filtering) of the pseudoranges using the much lower noise carrier phase observations is performed, then

$$B_L = \frac{1}{2T_s} \tag{4.28}$$

where T_s is the smoothing interval, RTCM [1994].

The predetection integration time, T, is typically 0.02 seconds (the navigation message data bit length). Increasing T reduces the squaring loss and this can be advantageous in weak signal situations.

For moderate to strong signals ($C/N_0 \gtrsim 35$ dB-Hz), equation (4.27) is well approximated by

$$\sigma_{DLL} \approx \sqrt{\frac{\alpha B_L}{c/n_0}}\,\lambda_c . \tag{4.29}$$

Using this approximation with $\alpha = 0.5$, $C/N_0 = 45$ dB-Hz, and $B_L = 0.8$ Hz, σ_{DLL} for the C/A-code is 1.04 m.

The most recently developed high-performance GPS receivers use narrow correlators in which the spacing between the early and late versions of the receiver-generated reference code is less than one chip, see Van Dierendonck et al. [1992]. For such receivers, equation (4.29) for signals of nominal strength can be rewritten as

$$\sigma_{DLL} \approx \sqrt{\frac{\alpha B_L d}{c/n_0}}\,\lambda_c \tag{4.30}$$

where d is the correlator spacing in chips. For a spacing of 0.1 chips, and with the same values for the other parameters as used for the evaluation of the one chip correlator, σ_{DLL} for the C/A-code is 0.39 cm. With post-measurement smoothing, this jitter can be made even smaller.

Carrier Tracking Loop. The analysis of the jitter in the carrier tracking loop of a GPS receiver, proceeds in a similar manner as that for the code tracking loop. In fact, the expression for the jitter in a Costas-type phase lock loop has the same form as that for the code tracking loop, Ward [1994]:

$$\sigma_{PLL} = \sqrt{\frac{B_P}{c/n_0}\left[1 + \frac{1}{2T\,c/n_0}\right]}\,\frac{\lambda}{2\pi} \tag{4.31}$$

where B_P is the carrier loop noise bandwidth (Hz), λ is the wavelength of the carrier, and the other symbols are the same as before. Bp must be wide enough for the tracking loop to follow the dynamics of the receiver. For most geodetic applications, the receiver is stationary and so bandwidths of 2 Hz or less can be used. However, a tracking loop with such a narrow bandwidth might have problems follow rapid variations in phase caused by the ionosphere. Some receivers adjust the loop bandwidth dynamically or allow the operator to set the bandwidth manually.

For signals of nominal strength, equation (4.31) is well approximated by

$$\sigma_{PLL} = \sqrt{\frac{B_P}{c/n_0}} \frac{\lambda}{2\pi} . \tag{4.32}$$

Using this approximation with $C/N_0 = 45$ dB-Hz and $B_P = 2$ Hz, σ_{PLL} for the L1 carrier phase is 0.2 mm.

4.4.2 Other Measurement Errors

Other errors affecting the receiver's measurement precision include local oscillator instability, crosstalk, inter-channel biases, drifts, and quantization noise.

Local oscillator instability can contribute errors to both pseudorange and carrier phase measurements. If a single local oscillator is used in the receiver and all measurements on all visible satellites are made at the same instant, the measurements will share the same oscillator instability which can be solved for or differenced out in the usual fashion. However, in a sequencing receiver where signals share channels, the measurements on different satellites and perhaps on different frequencies for dual frequency receivers, are not simultaneous. Therefore, oscillator instability over the internal sampling period will contribute uncorrelated noise to the measurements. At sampling periods of 20 milliseconds (the navigation message bit length), the instability of a typical quartz oscillator is much better than 1 part in 10^8 and so the contributed error is conservatively estimated to be on the order of a centimetre or less, Cohen [1992].

Crosstalk is the interference between RF paths. Signal energy from one path is coupled into another. This phenomenon can be particularly troublesome in receivers where there is high gain on the paths. A high level of isolation between paths is required to keep crosstalk within acceptable levels. Careful attention in receiver design can keep the effect of crosstalk to a level of 0.5 millimetres or lower, Cohen [1992].

As discussed earlier in this chapter, inter-channel bias is an error encountered in multiple channel receivers. The signal path lengths through the channels may be slightly different and therefore there will be an unequal error in measurements made on the signals in different channels at the same instant. However, in modern receivers, these biases can be calibrated out at the level of 0.1 millimetres or better, Hofmann-Wellenhof et al. [1994].

In a similar fashion to inter-channel biases, drifts in effective path lengths shared by the channels in the receiver can contribute errors to the measurements.

Such drifting is often associated with temperature changes inside the receiver. We saw what was believed to be an example of such drifting earlier in this chapter. Since such drifts are the same for simultaneous observations, they are removed in the standard double-differencing data processing approach.

Quantization noise results from the imprecision in analogue to digital conversion in the receiver. Unlike the situation with analogue GPS receivers, however, quantization noise can usually be neglected in a digital receiver, Ward [1981].

4.5 SUMMARY

In this chapter, we have looked at the basic operation of a GPS receiver, how the primary observables are measured, and the precision with which the measurements can be made. The discussion was purposely kept at a fairly introductory level. The reader interested in obtaining a deeper understanding of the operation of a GPS receiver should consult the appropriate references listed below.

References

AGU (1994), 1994 Fall Meeting. *EOS*, Vol. 75, No. 44, Supplement.

Bourassa, M. (1994), *Etude des Effets de la Variation des Centres de Phase des Antennes GPS. M.Sc.* thesis, Département des Science Géodésiques et de Télédétection, Université Laval, Ste-Foy, PQ, Canada, 109 pp.

Braasch, M.S. (1994), "Isolation of GPS multipath and receiver tracking errors." *Navigation*, Journal of the (U.S.) Institute of Navigation, Vol. 41, No. 4, pp. 415-434.

Braun, J., C. Rocken, and J. Johnson (1994), "Consistency of high precision GPS antennas." Presented at the *1994 Fall Meeting of the American Geophysical Union*, San Francisco, CA, 5-9 December. Abstract: EOS, Vol. 75, No. 44, Supplement, p. 172.

Campbell, J. (1993), "Instrumental effects on the ionospheric rate of change observed by dual L-band GPS carrier phases." *Proceedings of the GPS/Ionosphere Workshop "Modelling the Ionosphere for GPS Applications*," Neustrelitz, Germany, 29-30 September, p. 78.

Cohen, C.E. (1992), *Attitude Determination Using GPS: Development of an All Solid-state Guidance, Navigation, and Control Sensor for Air and Space Vehicles Based on the Global Positioning System*. Ph.D. thesis, Department of Aeronautics and Astronautics, Stanford University, Stanford, CA, 184 pp.

Gurtner, W. (1994), "RINEX: The receiver-independent exchange format." *GPS World*, Vol. 5, No. 7, pp. 48-52.

Gurtner, W., M. Rothacher, S. Schaer, L. Mervart, and G. Beutler (1994), "Azimuth- and elevation-dependent phase corrections for geodetic GPS antennas." Presented at the *1994 Fall Meeting of the American Geophysical Union*, San Francisco, CA, 5-9 December. Abstract: EOS, Vol. 75, No. 44, Supplement, p. 172.

Hofmann-Wellenhof, B., H. Lichtenegger, and J. Collins (1994), *Global Positioning System*: Theory and Practice. 3rd edition. Springer-Verlag, Vienna, 355 pp.

Johnson, R.C. (Ed.) (1993), *Antenna Engineering Handbook*. 3rd edition. Mc-Graw Hill, Inc., New York, NY.

Langley, R.B. (1991), "The GPS receiver: An introduction." *GPS World*, Vol. 2, No. 1, pp. 50-53.

Langley, R.B. (1993), "The GPS observables." *GPS World*, Vol. 4, No. 4, pp. 52-59.

Nagle, J.R., A.J. Van Dierendonck, and Q.D. Hua (1992), "Inmarsat-3 navigation signal C/A-code selection and interference analysis." *Navigation*, Journal of the (U.S.) Institute of Navigation, Vol. 39, No. 4, pp. 445-461.

RTCM (1994), *RTCM Recommended Standards for Differential Navstar GPS Service*. Version 2.1. RTCM Special Committee No. 104, Radio Technical Commission for Maritime Services, Washington, D.C., January.

Schupler, B.R. and T.A. Clark (1991), "How different antennas affect the GPS observable." *GPS World*, Vol. 2, No. 10, pp. 32-36.

Schupler, B.R., R.L. Allshouse, and T.A. Clark (1994), "Signal characteristics of GPS user antennas." *Navigation*, Journal of the (U.S.) Institute of Navigation, Vol. 41, No. 3, pp. 277-295.

Spilker, J.J., Jr. (1977), *Digital Communications by Satellite*. Prentice-Hall, Inc., Englewood Cliffs, NJ, 672 pp.

Spilker, J.J., Jr. (1978, 1980), "GPS signal structure and performance characteristics." *Navigation*, Journal of the (U.S.) Institute of Navigation, Vol. 25, No. 2, pp. 121-146 and reprinted in Global Positioning System — Papers Published in *Navigation* (Vol. I of "The Red Books"), Institute of Navigation, Alexandria, VA, pp. 29-54.

Stelzried, C.T. (1968), "Microwave thermal noise standards." *IEEE Transactions on Microwave Theory and Techniques*, Vol. MTT-16, No. 9, pp. 646-655.

Tranquilla, J.M., B.G. Colpitts, and J.P. Carr (1989), "Measurement of low-multipath antennas for TOPEX." *Proceedings of the Fifth International Geodetic Symposium on Satellite Positioning*, Las Cruces, NM, 13-17 March, Vol. I, pp. 356-361.

Van Dierendonck, A.J., P. Fenton, and T. Ford (1992), "Theory and performance of narrow correlator spacing in a GPS receiver." *Navigation*, Journal of the (U.S.) Institute of Navigation, Vol. 39, No. 3, pp. 265-283.

Van Dierendonck, A.J. (1995), "Understanding GPS receiver terminology: A tutorial." *GPS World*, Vol. 6, No. 1, pp. 34-44.

Ward, P. (1981), "An inside view of pseudorange and delta pseudorange measurements in a digital NAVSTAR GPS receiver." Paper presented at the *ITC/USA/'81 International Telemetering Conference on GPS Military and Civil Applications*, San Diego, CA, October.

Ward, P.W. (1994), "GPS Receiver RF Interference Monitoring, Mitigation, and Analysis Techniques." *Navigation*, Journal of the (U.S.) Institute of Navigation, Vol. 41, No. 4, pp. 367-391.

Wells, D., R. Langley, A. Komjathy, and D. Dodd (1995), Acceptance Tests on Ashtech Z-12 Receivers. *Final report for Public Works and Government Services Canada*, February, 149 pp.

Yunck, T.P., S.C. Wu, S.M. Lichten, W.I. Bertiger, U.J. Lindqwister, and G. Blewitt, (1989), "Toward centimeter orbit determination and millimeter geodesy with GPS." *Proceedings of the Fifth International Geodetic Symposium on Satellite Positioning*, Las Cruces, NM, 13-17 March, Vol. I, pp. 272-281.

5. GPS OBSERVATION EQUATIONS AND POSITIONING CONCEPTS

Peter J.G. Teunissen [(1)] and Alfred Kleusberg[(2)]
(1) Faculty of Civil Engineering and Geosciences, Section MGP, University of Technology, Thijsseweg 11, 2629 JA Delft, The Netherlands
(2) Institut für Navigation, Universität Stuttgart, Geschwister-Scholl-Str. 24 D, 70174 Stuttgart, Germany

5.1 INTRODUCTION

The purpose of this chapter is four-fold: First, it provides the connection between chapters 1 through 4 outlining the individual components of the Global Positioning System, and chapters 6 through 16 describing the models used in different geodetic applications of GPS. This connection is introduced through the observation equations for pseudorange and carrier phase measurements in section 5.2 *GPS Observables*, relating the measured quantities described in chapter 4 to geometrical and physical parameters of interest in a geodetic context. Typically, these equations will be non-linear with respect to some of the parameters, most notably with respect to the coordinates of the satellite and the receiver.

Second, the section 5.3 *Linear Combinations* explores the elimination and/or isolation of some of these geometrical and physical parameters through linear combinations of carrier phases and pseudoranges measured simultaneously with a single receiver, through linear combinations over time of carrier phases, and through linear combinations of the same type of observable measured simultaneously with different receivers and/or different satellite signals.

Third, section 5.4 *Single - Receiver Non-Positioning Models* takes simplified versions of the non-linear observation equations and some of their linear combinations for an analysis of the estimability of parameters and linear combination of parameters. Also introduced in this section is the redundancy accumulating if time series of measurements are analyzed.

Fourth, in sections 5.5 *The Linearized Observation Equations for Positioning* and 5.6 *Relative Positioning Models*, the simplified observation equations linearized with respect to receiver and satellite coordinates are introduced to assess the simultaneous estimability of receiver coordinates and other parameters not related to positions. Again, the accumulation of redundancy both through more-than-required measurements and through repetition of measurements over time are investigated. The analysis is split into section 5.5 for the single receiver case ("absolute positioning") and section 5.6 for the case of at least two simultaneously operating receivers ("relative positioning").

5.2 GPS OBSERVABLES

In this section we derive the non linear observation equations for GPS measurements. The purpose of the section is to provide the connection between the output of a GPS receiver and some physically meaningful quantities, the parameters of the observation equations.

We begin by relating the basic quantities *time*, *frequency*, and *phase* of a sinusoidal signal generated by an oscillator. The frequency f of a signal is the time derivative of the phase ϕ of the signal, and conversely, the phase of the signal is the time integral of the signal frequency:

$$f(t) = \frac{d\phi(t)}{dt} \tag{5.1}$$

$$\phi(t) = \int_{t_0}^{t} f(\tau)d\tau + \phi(t_0) \tag{5.2}$$

where $\phi(t_0)$ is the initial phase of the signal for zero time. The signal phase is measured in units of cycles, the frequency in units of Hertz (Hz).
Alternatively, the phase of the signal can be represented in units of radians through multiplication by 2π. Then the frequency $f(t)$ is simultaneously changed into *angular frequency* $\omega(t)$.

The phase of a signal can be converted to *time* t_i through subtraction of the initial phase $\phi(t_0)$ and subsequent division by the *nominal oscillator frequency* f_0. We use subscript i in t_i to indicate that it is different from time t.

$$\begin{aligned} t_i(t) &= \frac{\phi(t) - \phi(t_0)}{f_0} \\ &= \frac{\phi(t)}{f_0} - t_i(t_0) \end{aligned} \tag{5.3}$$

If the frequency of the oscillator is constant and equal to its nominal frequency, then equations (5.2) and (5.3) yield

$$t_i(t) = t - t_i(t_0) \tag{5.4}$$

i.e., the signal phase is a true measure of time, up to a constant term resulting from non-zero initial phase. Otherwise we obtain

$$t_i(t) = \frac{1}{f_0}\int_{t_0}^{t} f(\tau)d\tau - t_i(t_0)\,. \tag{5.5}$$

Separating the actual frequency into the nominal frequency and the frequency deviation or *frequency error* δf, we obtain

$$t_i(t) = \frac{1}{f_0}\int_{t_0}^{t} \{f_0 + \delta f(\tau)\}d\tau - t_i(t_0)\,. \tag{5.6}$$

Further introducing the time deviation δt_i as the integral of the relative frequency deviation

$$\delta t_i(t) = \int_{t_0}^{t} \frac{\delta f(\tau)}{f_0}d\tau \tag{5.7}$$

we obtain the final relation between the time measured from the phase of an oscillator generated signal, and true *time t*.

$$t_i(t) = t + \delta t_i(t) - t_i(t_0) \tag{5.8}$$

The time displayed by an oscillator output is equal to the sum of true time, a term accounting for the deviation of the actual oscillator frequency from its nominal frequency, and the effect of non zero initial phase of the oscillator. Often, it is appropriate to lump together the last two terms on the right hand side of equation (5.8) into the *clock error* dt_i, leading to

$$t_i(t) = t + dt_i(t)\,. \tag{5.9}$$

For the case of time displayed by oscillators in GPS satellites (referred to in the following as satellite time) and GPS receivers (referred to as receiver time), the time as maintained by the GPS control segment (referred to as GPS time) is a realization of the true time, t.

5.2.1 The Pseudorange

The pseudorange measurement, P_i^k, is equal to the difference between receiver time t_i at signal reception and satellite time t^k at signal transmission, scaled by the nominal speed of light in a vacuum, c. Pseudoranges are measured through *P*-code correlation (*Y*-code correlation) on signal frequencies f_1 and f_2, and/or through C/A-code correlation on the signal frequency f_1.

$$P_i^k(t) = c[t_i(t) - t^k(t - \tau_i^k)] + e_i^k \tag{5.10}$$

τ_i^k is the signal travel time from the signal generator in the satellite to the signal correlator in the GPS receiver; e_i^k is the pseudorange measurement error. Receiver time and satellite time are equal to GPS time plus the respective clock errors (offsets) as described by equation (5.9).

$$t_i(t) = t + dt_i(t) \tag{5.11}$$

$$t^k(t - \tau_i^k) = t - \tau_i^k + dt^k(t - \tau_i^k) \,. \tag{5.12}$$

Inserting equations (5.11) and (5.12) into (5.10) yields

$$P_i^k(t) = c\tau_i^k + c[dt_i(t) - dt^k(t - \tau_i^k)] + e_i^k \,. \tag{5.13}$$

The signal travel time τ_i^k can be split into three separate terms: the signal delay d^k occurring between the signal generation in the satellite and the transmission from the satellite antenna, the signal travel time $\delta\tau_i^k$ from the transmitting antenna to the receiver antenna, and the signal delay d_i between the receiving antenna and the signal correlator in the receiver.

$$\tau_i^k = d^k + \delta\tau_i^k + d_i \tag{5.14}$$

The signal travel time between the antennas is a function of the signal propagation speed ν along the signal path.

$$\nu = \frac{ds}{dt} \qquad \text{(5.15}$$

The signal propagation speed is related to the propagation speed in a vacuum, c, through the refractive index of the medium, n, by

$$\nu = \frac{c}{n} \,. \tag{5.16}$$

Combining these two equations for the signal propagation speed gives the differential relation between travel time and travelled distance

$$c\,dt = n\,ds \tag{5.17}$$

and integration along the signal path finally yields

$$c\,\delta\tau_i^k = \int_{\text{path}} n\,ds \,. \tag{5.18}$$

This integral is conveniently split into three separate terms according to

$$c\,\delta\tau_i^k = \int_{\text{geom}} ds + \int_{\text{geom}} (n-1)\,ds + \{\int_{\text{path}} n\,ds - \int_{\text{geom}} n\,ds\}\,. \tag{5.19}$$

The first term is the line integral along a straight line geometric connection between the transmitting and receiving antennas. In an ideal environment, this term is equal to the geometric distance $\rho_i^k(t, t-\tau_i^k)$ between the satellite antenna at signal transmission time, and the receiver antenna at signal reception time. However, if the straight line signal is interfering with other copies of the signal, which have propagated along different paths, the first term will be the sum of the geometric distance ρ_i^k and the multipath error, dm_i^k. Multipath can be caused by signal reflection at conducting surfaces of the satellite (satellite multipath) or in the vicinity of the receiver (receiver multipath).

$$\int_{\text{geom}} ds = \rho_i^k + dm_i^k \tag{5.20}$$

The second and third term in equation (5.19) describe the effect of atmospheric refraction, i.e., the effect resulting from the deviation from unity of the refractive index of the propagation medium. The second term describes the bulk of the delay introduced by the change of the signal propagation speed through atmospheric refraction. The third term describes the delay resulting from signal propagation along an actual signal path different from the straight line connection. This term is caused by ray bending through atmospheric refraction. It is much smaller than the second term, and often neglected.

For reasons discussed in previous chapters, the effect of atmospheric refraction is usually split into the *ionospheric refraction effect* I resulting from non-unity of the ionospheric refraction index n_I

$$I_i^k = \int_{\text{geom}} (n_I - 1)\,ds + \{\int_{\text{path}} n_I\,ds - \int_{\text{geom}} n_I\,ds\} \tag{5.21}$$

and the *tropospheric refraction effect* T resulting from non-unity of the tropospheric refraction index n_T

$$T_i^k = \int_{\text{geom}} (n_T - 1)\,ds + \{\int_{\text{path}} n_T\,ds - \int_{\text{geom}} n_T\,ds\}\,. \tag{5.22}$$

We can now insert equations (5.14) and (5.18) through (5.22) into (5.13) to obtain a more familiar variation of the observation equation for GPS pseudorange measurements.

$$\begin{aligned} P_i^k(t) &= \rho_i^k(t, t-\tau_i^k) + I_i^k + T_i^k + dm_i^k + \\ &\quad c[dt_i(t) - dt^k(t-\tau_i^k)] + c[d_i(t) + d^k(t-\tau_i^k)] + e_i^k . \end{aligned} \tag{5.23}$$

The final step in the derivation of the pseudorange observation equation is the introduction of the eccentricities between the centre of mass of the satellite and the satellite antenna, and between the receiver antenna and the point of interest (e.g. for positioning).

Denoting the position vector of the centre of mass by $\boldsymbol{r}^k$, the position vector of the terrestrial point of interest by $\boldsymbol{r}_i$, the eccentricity vector of the receiver antenna by $d\boldsymbol{r}_i$, and the eccentricity vector of the transmitting antenna by $d\boldsymbol{r}^k$, we obtain the following relation

$$\rho_i^k = \| (\boldsymbol{r}^k + d\boldsymbol{r}^k) - (\boldsymbol{r}_i + d\boldsymbol{r}_i) \| \tag{5.24}$$

with double bars indicating the length of a vector. Inserting equation (5.24) into (5.23) we obtain the final pseudorange observation equation:

$$\begin{aligned} P_i^k(t) &= \| (\boldsymbol{r}^k(t-\tau_i^k) + d\boldsymbol{r}^k(t-\tau_i^k)) - (\boldsymbol{r}_i(t) + d\boldsymbol{r}_i(t)) \| + \\ &\quad I_i^k + T_i^k + c[dt_i(t) - dt^k(t-\tau_i^k)] + \\ &\quad c[d_i(t) + d^k(t-\tau_i^k)] + dm_i^k + e_i^k . \end{aligned} \tag{5.25}$$

The right hand side contains in sequence the geometric distance between the transmitting and receiving antennas expressed by the positions of the terrestrial reference point (survey marker or similar) and the centre of mass of the GPS satellite and the corresponding eccentricities, the ionospheric delay effect, the tropospheric delay effect, the effect of satellite and receiver clock errors, the effect of satellite and receiver equipment delays, the effect of signal multipath, and the measurement error.

The receiver and satellite positions and satellite clock errors as well as the tropospheric delay effect are independent of the signal frequency. All other terms, including the eccentricity vectors, will in general be different for different signal frequencies.

5.2.2 The Carrier Phase

The carrier phase ϕ_i^k is equal to the difference between the phase ϕ_i of the receiver generated carrier signal at signal reception time, and the phase ϕ^k of the satellite generated carrier signal at signal transmission time. Only the fractional carrier phase can be measured when a satellite signal is acquired, i.e. an integer number N of full cycles is unknown. N is called the carrier phase ambiguity.

$$\phi_i^k(t) = \phi_i(t) - \phi^k(t-\tau_i^k) + N_i^k + \varepsilon_i^k \tag{5.26}$$

Applying equations (5.3), (5.8) and (5.9) to the phases on the right hand side according to

$$\begin{aligned} \phi_i(t) &= f_0 t_i(t) + \phi_i(t_0) \\ &= f_0(t + dt_i(t)) + \phi_i(t_0) \end{aligned} \tag{5.27}$$

$$\phi^k(t-\tau_i^k) = f_0(t - \tau_i^k + dt^k(t-\tau_i^k)) + \phi^k(t_0) \tag{5.28}$$

we obtain for the carrier phase observation equation

$$\phi_i^k(t) = f_0[\tau_i^k + dt_i(t) - dt^k(t-\tau_i^k)] + [\phi_i(t_0) - \phi^k(t_0)] + N_i^k + \varepsilon_i^k\,. \tag{5.29}$$

In order to transform this equation into units of distance, it is multiplied by the nominal wave length of the carrier signal

$$\lambda = \frac{c}{f_0} \tag{5.30}$$

to yield

$$\lambda\phi_i^k(t) = c\tau_i^k + c[dt_i(t) - dt^k(t-\tau_i^k)] + \lambda[\phi_i(t_0) - \phi^k(t_0)] + \lambda N_i^k + \lambda\varepsilon_i^k\,. \tag{5.31}$$

The first two terms on the right hand side represent the carrier signal travel time and the satellite/receiver clock errors similarly to the observation equation (5.13) for pseudoranges. The third term is constant and represents the non-zero initial phases of the satellite and receiver generated signals and the fourth term represents the integer carrier phase ambiguity.

The carrier signal travel time can be expanded similarly to the derivations for the pseudorange in equations (5.14) through (5.22). This results in:

$$\begin{aligned} \Phi_i^k(t) = {} & \rho_i^k(t, t-\tau_i^k) - I_i^k + T_i^k + \delta m_i^k + c[dt_i(t) - dt^k(t-\tau_i^k)] + \\ & c[\delta_i(t) + \delta^k(t-\tau_i^k)] + \lambda[\phi_i(t_0) - \phi^k(t_0)] + \lambda N_i^k + \varepsilon_i^k \end{aligned} \tag{5.32}$$

We have replaced on the left hand side the product of the carrier phase measurement and nominal wave length by Φ, the carrier phase measurement in units of distance. We have also omitted the wave length in front of the measurement error.

Comparing equation (5.32) to the corresponding pseudorange observation equation (5.23), we note:

- both contain the geometric distance $\rho_i^k(t, t-\tau_i^k)$
- both contain the clock error terms $c[dt_i(t) - dt^k(t-\tau_i^k)]$
- both contain the tropospheric refraction effect T_i^k
- the sign of ionospheric refraction effect I_i^k is reversed (c.f. chapter 3)
- the pseudorange multipath error dm_i^k has been replaced by the carrier phase multipath error δm_i^k
- the pseudorange equipment delay terms $c[d_i(t) + d^k(t-\tau_i^k)]$ have been replaced by the carrier phase equipment delay terms $c[\delta_i(t) + \delta^k(t-\tau_i^k)]$
- the carrier phase observation equation contains the additional terms $\lambda[\phi_i(t_0) - \phi^k(t_0)]$ resulting from the non-zero initial phases, and the carrier phase ambiguity term λN_i^k.

In the last step, the geometric distance is expanded into satellite and receiver coordinates and the related eccentricities, as described for the pseudorange observation in equation (5.24). Such expansion yields finally

$$
\begin{aligned}
\Phi_i^k(t) = {} & \| (\boldsymbol{r}^k(t-\tau_i^k) + \delta\boldsymbol{r}^k(t-\tau_i^k)) - (\boldsymbol{r}_i(t) + \delta\boldsymbol{r}_i(t)) \| + \\
& -I_i^k + T_i^k + \delta m_i^k + c[dt_i(t) - dt^k(t-\tau_i^k)] + \\
& c[\delta_i(t) + \delta^k(t-\tau_i^k)] + \lambda[\phi_i(t_0) - \phi^k(t_0)] + \lambda N_i^k + \varepsilon_i^k .
\end{aligned}
\tag{5.33}
$$

We have used $\delta\boldsymbol{r}^k$ and $\delta\boldsymbol{r}_i$ to denote the eccentricities pertaining to the carrier phase measurements. In general, these will be different from the eccentricities pertaining to the pseudorange measurements, since the effective antenna centres are different.

There is one additional and more hidden difference between the equations: The total signal travel time is slightly different for pseudorange and carrier phase measurements because of differences in the ionospheric effect and the equipment delays. As a result, the time argument for the evaluation of the satellite coordinates is also slightly different.

5.3 LINEAR COMBINATIONS

The purpose of this section is to derive the equations for certain linear combinations of GPS measurements. Such linear combinations are often used in the analysis of GPS observations, and some of the linear combinations are directly measurable with appropriately equipped GPS receivers.

5.3.1 Single Receiver, Single Satellite, Single Epoch Linear Combinations

In this sub-section we will examine two particular linear combinations of measurements. The first of these combines pseudoranges or carrier phases measured at different signal frequencies. The second one combines pseudoranges and carrier phases measured at the same signal frequency.

Inter Frequency Linear Combination. Some GPS receivers provide simultaneous pseudorange and carrier phase measurements on both GPS frequencies f_1 and f_2. We identify quantities affected by signal frequency by the subscripts 1 and 2. From equation (5.25) we obtain for the inter frequency difference of pseudoranges

$$\begin{aligned} P_{i,2}^k(t) - P_{i,1}^k(t) = {} & \{ \| (\boldsymbol{r}^k(t-\tau_{i,2}^k) + d\boldsymbol{r}_{,2}^k(t-\tau_{i,2}^k)) - (\boldsymbol{r}_i(t)) + d\boldsymbol{r}_{i,2}(t)) \| - \\ & \| (\boldsymbol{r}^k(t-\tau_{i,1}^k) + d\boldsymbol{r}_{,1}^k(t-\tau_{i,1}^k)) - (\boldsymbol{r}_i(t) + d\boldsymbol{r}_{i,1}(t)) \| \} + \\ & \{ I_{i,2}^k - I_{i,1}^k \} + \{ c[dt_i(t) - dt^k(t-\tau_{i,2}^k)] - c[dt_i(t) - dt^k(t-\tau_{i,1}^k)] \} + \\ & \{ c[d_{i,2}(t) + d_{i,2}^k(t-\tau_{i,2}^k)] - c[d_{i,1}(t) + d_{i,1}^k(t-\tau_{i,1}^k)] \} + \\ & \{ dm_{i,2}^k - dm_{i,1}^k \} + \{ e_{i,2}^k - e_{i,1}^k \} . \end{aligned} \qquad (5.34)$$

The tropospheric refraction term has been omitted, since it affects the pseudoranges on both frequencies identically, and therefore cancels in the difference. Some of the remaining terms are also very small and can be neglected for most purposes.

The first term on the right is the difference in geometric ranges as measured on the two frequencies. It is caused by slightly different signal travel times, resulting in slightly different time arguments for the satellite position vector. This time difference is usually less than 0.1 micro second, with corresponding negligible sub-millimetre satellite position differences. This term also contains the eccentricities which can be considerably different for the two GPS frequencies.

The second term contains the difference of the ionospheric refraction effect at the two frequencies, a major constituent of this particular linear combination.

The third term contains the clock errors. Since the measurements are taken simultaneously at the receiver, the receiver clock errors cancel completely. The satellite clock error does not cancel exactly, as it is appearing with slightly different time arguments. This remaining difference, however, is negligible small.

The fourth term contains the differences in equipment delays for the two signal frequencies. This term is significant and needs to be retained. The same is true for the difference in the multipath error. These considerations lead to the final representation for the difference between pseudorange measurements at two frequencies.

$$\begin{aligned} P^k_{i,2}(t) - P^k_{i,1}(t) \approx\ & \{ \| (\boldsymbol{r}^k(t-\tau^k_{i,2}) + d\boldsymbol{r}^k_{,2}(t-\tau^k_{i,2})) - (\boldsymbol{r}_i(t) + d\boldsymbol{r}_{i,2}(t)) \| - \\ & \| (\boldsymbol{r}^k(t-\tau^k_{i,1}) + d\boldsymbol{r}^k_{,1}(t-\tau^k_{i,1})) - (\boldsymbol{r}_i(t) + d\boldsymbol{r}_{i,1}(t)) \| \} + \\ & c\{[d_{i,2}(t) + d^k_{i,2}(t-\tau^k_i)] - [d_{i,1}(t) + d^k_{i,1}(t-\tau^k_i)]\} + \\ & \{I^k_{i,2} - I^k_{i,1}\} + \{dm^k_{i,2} - dm^k_{i,1}\} + e^k_{i,2} - e^k_{i,1} \end{aligned} \tag{5.35}$$

We have used the 'approximately equal' sign to indicate, that certain negligible small quantities have been removed from the equation. The right hand side contains in sequence the effect of inter frequency difference in eccentricities, the inter frequency difference of equipment delays, the inter frequency difference of ionospheric refraction effects, the inter frequency difference of multipath errors, and the measurement noise term.

It should be noted, that this inter frequency difference of pseudoranges is a directly measurable quantity in some receivers.

An equation similar to (5.35) can be derived for the inter frequency difference of carrier phase measurements as described by their observation equation (5.33). Using again subscripts to identify the frequency, we obtain with a similar level of approximation

$$\begin{aligned} \Phi^k_{i,2}(t) - \Phi^k_{i,1}(t) \approx\ & \{ \| (\boldsymbol{r}^k(t-\tau^k_{i,2}) + \delta\boldsymbol{r}^k_{,2}(t-\tau^k_{i,2})) - (\boldsymbol{r}_i(t) + \delta\boldsymbol{r}_{i,2}(t)) \| - \\ & \| (\boldsymbol{r}^k(t-\tau^k_{i,1}) + \delta\boldsymbol{r}^k_{,1}(t-\tau^k_{i,1})) - (\boldsymbol{r}_i(t) + \delta\boldsymbol{r}_{i,1}(t)) \| \} + \\ & c\{[\delta_{i,2}(t) + \delta^k_{i,2}(t-\tau^k_i)] - [\delta_{i,1}(t) + \delta^k_{i,1}(t-\tau^k_i)]\} - \\ & \{I^k_{i,2} - I^k_{i,1}\} + \{\delta m^k_{i,2} - \delta m^k_{i,1}\} + \\ & \lambda_{,2}[\phi_{i,2}(t_0) - \phi^k_{,2}(t_0)] - \lambda_{,1}[\phi_{i,1}(t_0) - \phi^k_{,1}(t_0)] + \\ & \lambda_{,2}N^k_{i,2} - \lambda_{,1}N^k_{i,1} + \varepsilon^k_{i,2} - \varepsilon^k_{i,1} \,. \end{aligned} \tag{5.36}$$

Comparing equation (5.36) to the corresponding pseudorange difference observation equation (5.35), we note:

- both are free of the tropospheric refraction effect and of clock error terms
- both equations retain the respective term for inter frequency differences of the eccentricities, the equipment delays, and the multipath errors
- the sign of the ionospheric refraction effect is reversed
- the carrier phase related equation contains additional terms for the initial phases and the phase ambiguity terms.

It should also be noted, that the measurement noise was amplified when forming the linear combinations. Assuming equal noise level for both frequencies and absence of correlation, this noise amplification factor is $\sqrt{2}$.

It is illustrating to look at equations (5.35) and (5.36) under some simplifying assumptions. If we assume, that

- all eccentricities are properly calibrated and can be removed from the right hand side of the equations, and
- all equipment delays are constant over time, and
- there is no change over time in the carrier phase ambiguity,

then the equations simplify according to

$$P_{i,2}^k(t) - P_{i,1}^k(t) \approx C_p + \{I_{i,2}^k - I_{i,1}^k\} + \{dm_{i,2}^k - dm_{i,1}^k\} + e_{i,2}^k - e_{i,1}^k \tag{5.37}$$

and

$$\Phi_{i,2}^k(t) - \Phi_{i,1}^k(t) \approx C_\phi - \{I_{i,2}^k - I_{i,1}^k\} + \{\delta m_{i,2}^k - \delta m_{i,1}^k\} + \varepsilon_{i,2}^k - \varepsilon_{i,1}^k \,. \tag{5.38}$$

Both equations provide a measurement for the sum of a constant term, the inter frequency ionospheric delay difference, the inter frequency multipath error difference, and some measurement noise.

Difference Between Pseudorange and Carrier Phase, Same Frequency. Another single station, single satellite linear combination can be formed by subtracting the pseudorange measurement from the simultaneously measured carrier phase. These measurements are represented by the observation equations (5.25) and (5.33). To the same degree of approximation as in equations (5.35) and (5.36), this difference can be expressed by:

$$\begin{aligned} \Phi_i^k(t) - P_i^k(t) \approx\ & \{ \| (\boldsymbol{r}^k(t-\tau_i^k) + \delta\boldsymbol{r}^k(t-\tau_i^k)) - (\boldsymbol{r}_i(t) + \delta\boldsymbol{r}_i(t)) \| - \\ & \| (\boldsymbol{r}^k(t-\tau_i^k) + d\boldsymbol{r}^k(t-\tau_i^k)) - (\boldsymbol{r}_i(t) + d\boldsymbol{r}_i(t)) \| \} + \\ & -2I_i^k + \{\delta m_i^k - dm_i^k\} + \lambda[\phi_i(t_0) - \phi^k(t_0)] + \lambda N_i^k + \\ & c\{[\delta_i(t) + \delta^k(t-\tau_i^k)] - [d_i(t) + d^k(t-\tau_i^k)]\} + \{\varepsilon_i^k - e_i^k\} \,. \end{aligned} \tag{5.39}$$

The first term on the right hand side contains the effect of the differences between the eccentricities related to carrier phase measurements and those related to pseudorange measurements, i.e., differences between the effective antenna centres. The second term contains twice the ionospheric refraction effect.

The third term is the difference between the carrier phase multipath and the pseudorange multipath errors. In general, the pseudorange multipath error is dominant in this term. The fourth and fifth term are related to the non-zero initial phases, and the carrier phase ambiguity. The sixth term results from differences in equipment delay experienced by pseudorange and carrier phase measurements.

As in the previous sub-section, we again take a look at a more simplified version of this equation. The simplifying assumptions stated just before equation (5.37) lead to

$$\Phi_i^k(t) - P_i^k(t) \approx C_{\phi p} - 2I_i^k + \{\delta m_i^k - dm_i^k\} + \{\varepsilon_i^k - e_i^k\} . \tag{5.40}$$

This simplified equation shows, that the difference between carrier phase and pseudorange measurements is equal to a constant term, a term representing twice the negative ionospheric refraction effect, the difference between the multipath errors, and a receiver noise term.

5.3.2 Phase Difference Over Time

Typically, a GPS receiver will provide measurements at regular pre-selected intervals of time, Δt. The difference between two subsequent phase measurements can be written (cf. equation (5.33)):

$$\begin{aligned}\Phi_i^k(t+\Delta t) - \Phi_i^k(t) \approx\ & \| (\boldsymbol{r}^k(t+\Delta t-\tau_i^k) + d\boldsymbol{r}^k(t+\Delta t-\tau_i^k)) - (\boldsymbol{r}_i(t+\Delta t) + d\boldsymbol{r}_i(t+\Delta t)) \| - \\ & \| (\boldsymbol{r}^k(t-\tau_i^k) + d\boldsymbol{r}^k(t-\tau_i^k)) - (\boldsymbol{r}_i(t) + d\boldsymbol{r}_i(t)) \| + \\ & [I_i^k(t+\Delta t) - I_i^k(t)] + [T_i^k(t+\Delta t) - T_i^k(t)] + \\ & [\delta m_i^k(t+\Delta t) - \delta m_i^k(t)] + \\ & c\{[dt_i(t+\Delta t) - dt^k(t+\Delta t-\tau_i^k)] - [dt_i(t) - dt^k(t-\tau_i^k)]\} + \\ & c\{[\delta_i(t+\Delta t) - \delta^k(t+\Delta t-\tau_i^k)] - [\delta_i(t) - \delta^k(t-\tau_i^k)]\} + \varepsilon_i^k .\end{aligned} \tag{5.41}$$

The terms related to initial phases and to phase ambiguity are constant over time, and accordingly have disappeared from the right hand side in the differencing process. It is again illustrating to look at this equation under some simplifying assumptions. In the present case, these assumptions are, that the time interval is so short that changes in eccentricities, atmospheric refraction, multipath, and equipment delays can be neglected.

Under these assumptions, equation (5.41) changes to

$$\begin{aligned}\Phi_i^k(t+\Delta t) - \Phi_i^k(t) \approx\ & \| (\boldsymbol{r}^k(t+\Delta t-\tau_i^k) + d\boldsymbol{r}^k(t+\Delta t-\tau_i^k)) - (\boldsymbol{r}_i(t+\Delta t) + d\boldsymbol{r}_i(t+\Delta t)) \| \\ & - \| (\boldsymbol{r}^k(t-\tau_i^k) + d\boldsymbol{r}^k(t-\tau_i^k)) - (\boldsymbol{r}_i(t) + d\boldsymbol{r}_i(t)) \| + \\ & c\{[dt_i(t+\Delta t) - dt^k(t+\Delta t-\tau_i^k)] - [dt_i(t) + dt^k(t-\tau_i^k)]\} + \varepsilon_i^k ,\end{aligned} \tag{5.42}$$

that is, the change in carrier phase measurement is primarily related to changes in satellite and receiver position, and to changes in the satellite and receiver clock errors.

If further on the change in satellite and receiver position can be represented through a linear velocity term according to

$$\begin{aligned}
&\| (\boldsymbol{r}^k(t+\Delta t-\tau_i^k) + d\boldsymbol{r}^k(t+\Delta t-\tau_i^k)) - (\boldsymbol{r}_i(t+\Delta t) + d\boldsymbol{r}_i(t+\Delta t)) \| \\
&- \| (\boldsymbol{r}^k(t-\tau_i^k) + d\boldsymbol{r}^k(t-\tau_i^k)) - (\boldsymbol{r}_i(t) + d\boldsymbol{r}_i(t)) \| \\
&= \Delta t \frac{d}{dt} \| (\boldsymbol{r}^k(t-\tau_i^k) + d\boldsymbol{r}^k(t-\tau_i^k)) - (\boldsymbol{r}_i(t) + d\boldsymbol{r}_i(t)) \|
\end{aligned} \tag{5.43}$$

and the change in the clock errors can be represented through a linear frequency deviation term

$$dt(t+\Delta t) - dt(t) = (\delta f / f_0)\Delta t \,, \tag{5.44}$$

then equation (5.42) further simplifies to

$$\frac{\Phi_i^k(t+\Delta t) - \Phi_i^k(t)}{\Delta t} \approx \frac{d}{dt} \| (\boldsymbol{r}^k(t-\tau_i^k) + d\boldsymbol{r}^k(t-\tau_i^k)) - (\boldsymbol{r}_i(t) + d\boldsymbol{r}_i(t)) \| + \lambda[\delta f_i - \delta f^k] + \varepsilon_i^k \,. \tag{5.45}$$

This equation simply states, that the rate of change of the carrier phase is approximately equal to the rate of change of the satellite-to-receiver distance, plus the frequency deviations of the satellite and receiver oscillators.

Some receivers provide a measurement of the left hand side of equation (5.45). This measurement is often labelled "Doppler frequency shift".

5.3.3 Measurement Difference Between Receivers

The difference between the phase measurements of two receivers (subscripts j and i) of the same satellite signal (superscript k) can be written (c.f. equation (5.33))

$$\begin{aligned}
\Phi_j^k(t_j) - \Phi_i^k(t_i) = \; & \| (\boldsymbol{r}^k(t_j-\tau_j^k) + \delta\boldsymbol{r}^k(t_j-\tau_j^k)) - (\boldsymbol{r}_j(t_j) + \delta\boldsymbol{r}_j(t_j)) \| \\
& - \| (\boldsymbol{r}^k(t_i-\tau_i^k) + \delta\boldsymbol{r}^k(t_i-\tau_i^k)) - (\boldsymbol{r}_i(t_i) + \delta\boldsymbol{r}_i(t_i)) \| \\
& - I_j^k + I_i^k + T_j^k - T_i^k + \delta m_j^k - \delta m_i^k + \\
& c[dt_j(t_j) - dt^k(t_j-\tau_j^k)] - c[dt_i(t_i) - dt^k(t_i-\tau_i^k)] + \\
& c[\delta_j(t_j) + \delta^k(t_j-\tau_j^k)] - c[\delta_i(t_i) + \delta^k(t_i-\tau_i^k)] + \\
& \lambda[\phi_j(t_0) - \phi^k(t_0)] - \lambda[\phi_i(t_0) - \phi^k(t_0)] + \lambda N_j^k - \lambda N_i^k + \varepsilon_j^k - \varepsilon_i^k \,.
\end{aligned} \tag{5.46}$$

Inspection of the right hand side of equation (5.46) reveals immediately, that the effect of the non-zero initial phase, $\phi^k(t_0)$, of the satellite's oscillator completely cancels. Several other terms cancel approximately, and the remaining difference can be neglected.

Typical GPS receivers perform measurements at regularly spaced intervals, timed with the individual receiver clock. This means, that measurements can be performed in a truly simultaneous manner only, if the receiver clock errors are the same at both receivers. In general, this will not be the case, and this is reflected by the use of different time arguments on the left hand side of equation (5.46).

In modern GPS receivers, however, the receiver clock is continuously updated, and the remaining clock errors are small. For such "simultaneous" measurements, a number of additional terms on the right hand side of equation (5.46) will cancel. The difference between the time arguments for the evaluation of the satellite's position $\boldsymbol{r}^k$, clock error dt^k, and equipment delay δ^k pertaining to the two original phase measurements is caused by the difference in receiver clock errors, $dt_j - dt_i$, and by the difference between the signal travel times, $\tau_j^k - \tau_i^k$. This travel time difference is always smaller than 0.05 seconds. For this time scale, the satellite clock error, and equipment delay can be considered approximately constant.

$$dt^k(t_j - \tau_j^k) \approx dt^k(t_i - \tau_i^k) \tag{5.47}$$

$$\delta^k(t_j - \tau_j^k) \approx \delta^k(t_i - \tau_i^k) \tag{5.48}$$

With these approximations we obtain from equation (5.46)

$$\begin{aligned} \Phi_j^k(t_j) - \Phi_i^k(t_i) \approx\ & \| (\boldsymbol{r}^k(t_j - \tau_j^k) + \delta\boldsymbol{r}^k(t_j - \tau_j^k)) - (\boldsymbol{r}_j(t_j) + \delta\boldsymbol{r}_j(t_j)) \| \\ & - \| (\boldsymbol{r}^k(t_i - \tau_i^k) + \delta\boldsymbol{r}^k(t_i - \tau_i^k)) - (\boldsymbol{r}_i(t_i) + \delta\boldsymbol{r}_i(t_i)) \| \\ & - I_j^k + I_i^k + T_j^k - T_i^k + \delta m_j^k - \delta m_i^k + \\ & c[dt_j(t_j) - dt_i(t_i)] + c[\delta_j(t_j) - \delta_i(t_i)] + \\ & \lambda[\phi_j(t_0) - \phi_i(t_0)] + \lambda N_j^k - \lambda N_i^k + \varepsilon_j^k - \varepsilon_i^k \ . \end{aligned} \tag{5.49}$$

A further substantial simplification of this equation can be achieved by omitting the explicit time variables, and by using the abbreviations

$$(\cdot)_j - (\cdot)_i = (\cdot)_{ij} \ , \qquad (\cdot)^j - (\cdot)^i = (\cdot)^{ij} \tag{5.50}$$

leading to

$$\begin{aligned}\Phi_{ij}^{k} \approx\ & \| (\boldsymbol{r}^{k} + \delta\boldsymbol{r}^{k}) - (\boldsymbol{r}_{j} + \delta\boldsymbol{r}_{j}) \| - \| (\boldsymbol{r}^{k} + \delta\boldsymbol{r}^{k}) - (\boldsymbol{r}_{i} + \delta\boldsymbol{r}_{i}) \| \\ & - I_{ij}^{k} + T_{ij}^{k} + \delta m_{ij}^{k} + cdt_{ij} + c\delta_{ij} + \lambda\phi_{ij}(t_0) + \lambda N_{ij}^{k} + \varepsilon_{ij}^{k} \ .\end{aligned} \tag{5.51}$$

When using equation (5.51) instead of (5.49) one should be aware, that the satellite positions appearing in the first two terms on the right hand side of equation (5.51) are significantly different.

Going through the same procedure for the pseudorange measurements yields the following equation for the difference between receivers.

$$\begin{aligned}P_{ij}^{k} \approx\ & \| (\boldsymbol{r}^{k} + d\boldsymbol{r}^{k}) - (\boldsymbol{r}_{j} + d\boldsymbol{r}_{j}) \| - \| (\boldsymbol{r}^{k} + d\boldsymbol{r}^{k}) - (\boldsymbol{r}_{i} + d\boldsymbol{r}_{i}) \| \\ & + I_{ij}^{k} + T_{ij}^{k} + dm_{ij}^{k} + cdt_{ij} + cd_{ij} + e_{ij}^{k} \ .\end{aligned} \tag{5.52}$$

Obviously, the phase ambiguity and initial phase related terms do not appear in this equation, the ionospheric refraction term shows a reversed sign, and the phase related multipath and eccentricity terms have been replaced by their pseudorange related counterparts.

5.3.4 Measurement Difference Between Satellites

The difference between the simultaneous phase measurements of a receiver of the signals transmitted by two different satellites can be written

$$\begin{aligned}\Phi_i^{l}(t_i) - \Phi_i^{k}(t_i) =\ & \| (\boldsymbol{r}^{l}(t_i - \tau_i^{l}) + \delta\boldsymbol{r}^{l}(t_i - \tau_i^{l})) - (\boldsymbol{r}_i(t_i) + \delta\boldsymbol{r}_i(t_i)) \| \\ & - \| (\boldsymbol{r}^{k}(t_i - \tau_i^{k}) + \delta\boldsymbol{r}^{k}(t_i - \tau_i^{k})) - (\boldsymbol{r}_i(t_i) + \delta\boldsymbol{r}_i(t_i)) \| \\ & - I_i^{l} + I_i^{k} - T_i^{l} - T_i^{k} + \delta m_i^{l} - \delta m_i^{k} + \\ & c[dt_i(t_i) - dt^{l}(t_i - \tau_i^{l})] - c[dt_i(t_i) - dt^{k}(t_i - \tau_i^{k})] + \\ & c[\delta_i(t_i) + \delta^{l}(t_i - \tau_i^{l})] - c[\delta_i(t_i) + \delta^{k}(t_i - \tau_i^{k})] + \\ & \lambda[\phi_i(t_0) - \phi^{l}(t_0)] - \lambda[\phi_i(t_0) - \phi^{k}(t_0)] + \lambda N_i^{l} - \lambda N_i^{k} + \varepsilon_i^{l} - \varepsilon_i^{k} \ .\end{aligned} \tag{5.53}$$

We note through inspection of the right hand side of equation (5.53), that in addition to the effect of the non-zero initial phase $\phi_i(t_0)$ of the receiver's oscillator also the receiver clock and delay terms are exactly cancelled for simultaneous measurements. Using again the notation (5.50), and omitting the time arguments we obtain for the phase difference between satellites

$$\begin{aligned} \Phi_i^{kl} \approx\ & \| (\boldsymbol{r}^l + \delta \boldsymbol{r}^l) - (\boldsymbol{r}_i + \delta \boldsymbol{r}_i) \| - \| (\boldsymbol{r}^k + \delta \boldsymbol{r}^k) - (\boldsymbol{r}_i + \delta \boldsymbol{r}_i) \| \\ & - I_i^{kl} + T_i^{kl} + \delta m_i^{kl} + cdt^{kl} + c\delta^{kl} + \lambda \phi^{kl}(t_0) + \lambda N_i^{kl} + \varepsilon_i^{kl} \ . \end{aligned} \tag{5.54}$$

We note that this equation follows from the previous one without any approximations. Also, the receiver position vectors appearing in the first two terms on the right hand side are exactly the same, even if the receiver is in motion during the measurement.

Going through the same procedure for the pseudorange measurements yields the following equation for the difference between satellites

$$\begin{aligned} P_i^{kl} \approx\ & \| (\boldsymbol{r}^l + d\boldsymbol{r}^l) - (\boldsymbol{r}_i + d\boldsymbol{r}_i) \| - \| (\boldsymbol{r}^k + d\boldsymbol{r}^k) - (\boldsymbol{r}_i + d\boldsymbol{r}_i) \| \\ & + I_i^{kl} + T_i^{kl} + dm_i^{kl} + cdt^{kl} + cd^{kl} + e_i^{kl} \ . \end{aligned} \tag{5.55}$$

with differences compared to the corresponding phase related equation as explained at the end of section (5.3.3).

5.3.5 Measurement Difference Between Satellites and Receivers

The difference between the simultaneous phase measurements of a receiver of the signals transmitted by two different satellites, and the simultaneous measurements at the same nominal time of a second receiver of the same signals follows from equation (5.54) as:

$$\begin{aligned} \Phi_j^{kl} - \Phi_i^{kl} =\ & \| (\boldsymbol{r}^l + \delta \boldsymbol{r}^l) - (\boldsymbol{r}_j + \delta \boldsymbol{r}_j) \| - \| (\boldsymbol{r}^k + \delta \boldsymbol{r}^k) - (\boldsymbol{r}_j + \delta \boldsymbol{r}_j) \| \\ & - \| (\boldsymbol{r}^l + \delta \boldsymbol{r}^l) - (\boldsymbol{r}_i + \delta \boldsymbol{r}_i) \| + \| (\boldsymbol{r}^k + \delta \boldsymbol{r}^k) - (\boldsymbol{r}_i + \delta \boldsymbol{r}_i) \| \\ & - I_j^{kl} + I_i^{kl} + T_j^{kl} - T_i^{kl} + \delta m_j^{kl} - \delta m_i^{kl} + cdt^{kl} - cdt^{kl} \\ & + c\delta^{kl} - c\delta^{kl} + \lambda \phi^{kl}(t_0) - \lambda \phi^{kl}(t_0) + \lambda N_j^{kl} - \lambda N_i^{kl} + \varepsilon_j^{kl} - \varepsilon_i^{kl} \ . \end{aligned} \tag{5.56}$$

Obviously, the satellite initial phase cancels, and within the approximations outlined in section 5.3.3 the satellite clock error and the satellite equipment delays will cancel as well, resulting with the notation (5.50) in

$$\begin{aligned} \Phi_{ij}^{kl} \approx\ & \| (\boldsymbol{r}^l + \delta \boldsymbol{r}^l) - (\boldsymbol{r}_j + \delta \boldsymbol{r}_j) \| - \| (\boldsymbol{r}^k + \delta \boldsymbol{r}^k) - (\boldsymbol{r}_j + \delta \boldsymbol{r}_j) \| \\ & - \| (\boldsymbol{r}^l + \delta \boldsymbol{r}^l) - (\boldsymbol{r}_i + \delta \boldsymbol{r}_i) \| + \| (\boldsymbol{r}^k + \delta \boldsymbol{r}^k) - (\boldsymbol{r}_i + \delta \boldsymbol{r}_i) \| \\ & - I_{ij}^{kl} + T_{ij}^{kl} + \delta m_{ij}^{kl} + \lambda N_{ij}^{kl} + \varepsilon_{ij}^{kl} \ . \end{aligned} \tag{5.57}$$

The equation contains in sequence the linear combination of the four geometric distances between the two receiver antennas and the two satellite antennas, the linear combinations of four ionospheric and tropospheric delay terms, the combined multipath error term, the integer phase ambiguity term, and the combined measurement noise term. The corresponding equation for pseudoranges can be derived from equation (5.55).

$$\begin{aligned} P_{ij}^{kl} \approx\ & \|(\mathbf{r}^l + d\mathbf{r}^l) - (\mathbf{r}_j + d\mathbf{r}_j)\| - \|(\mathbf{r}^k + d\mathbf{r}^k) - (\mathbf{r}_j + d\mathbf{r}_j)\| \\ & - \|(\mathbf{r}^l + d\mathbf{r}^l) - (\mathbf{r}_i + d\mathbf{r}_i)\| + \|(\mathbf{r}^k + d\mathbf{r}^k) - (\mathbf{r}_i + d\mathbf{r}_i)\| \\ & + I_{ij}^{kl} + T_{ij}^{kl} + dm_{ij}^{kl} + e_{ij}^{kl} \end{aligned} \tag{5.58}$$

5.4 SINGLE-RECEIVER NONPOSITIONING MODELS

In this section we assume to have available one single GPS-receiver observing pseudoranges and/or carrier phases. The pseudoranges and carrier phases may be observed on L_1 only or on both of the two frequencies L_1 and L_2. The purpose of this section is to discuss the possibilities one has for parameter estimation and quality control of the data, when single-channel time series of pseudoranges and/or carrier phases are available. Estimability and the possible presence of redundancy will therefore be emphasized.

5.4.1 The Simplified Observation Equations

As our point of departure we start from the pseudorange and carrier phase observation equations (5.25) and (5.33). For the sake of convenience they are repeated once again:

$$\begin{aligned} P_i^k &= \|(\mathbf{r}^k + d\mathbf{r}^k) - (\mathbf{r}_i + d\mathbf{r}_i)\| + c(dt_i - dt^k) + T_i^k + I_i^k + c(d_i + d^k) + dm_i^k + e_i^k \\ \Phi_i^k &= \|(\mathbf{r}^k + \delta\mathbf{r}^k) - (\mathbf{r}_i + \delta\mathbf{r}_i)\| + c(dt_i - dt^k) + T_i^k - I_i^k + c(\delta_i + \delta^k) + \delta m_i^k + \\ &\quad + \lambda[\phi_i(t_0) - \phi^k(t_0)] + \lambda N_i^k + \varepsilon_i^k \end{aligned}$$

In this section we will work with a simplified form of the above observation equations. For the purpose of this simplification, the following assumptions are made:

(1) The difference in total signal travel time between pseudoranges and carrier phases will be neglected. Hence, the pseudorange clock terms will

be assumed to be identical to the corresponding carrier phase clock terms.

(2) The differences between the frequency dependent pseudorange and carrier phase *receiver* eccentricities will be neglected. Similarly, the differences between the frequency dependent pseudorange and carrier phase *satellite* eccentricities will be neglected. Hence, the geometric range from receiver i to satellite k is assumed to be independent of the frequency used and the same for both pseudoranges and carrier phases. This geometric range will be denoted as ρ_i^k.

(3) It follows from the structure of the above observation equations and the fact that we only consider the single-channel case, that not all parameters on the right-hand side of these equations are separably estimable. A number of these parameters will therefore be lumped together into one single parameter. For each channel, the receiver-satellite range ρ_i^k, the clock terms dt_i and dt^k, and the tropospheric delay T_i^k will be lumped together in one single parameter s_i^k:

$$s_i^k = \rho_i^k + c(dt_i - dt^k) + T_i^k .$$

For each channel, also the instrumental delays of both receiver and satellite, and the multipath delay will be lumped together. For the pseudoranges and carrier phases this gives:

$$d_i^k = c(d_i + d^k) + dm_i^k \text{ and } \delta_i^k = c(\delta_i + \delta^k) + \delta m_i^k .$$

Finally, for the carrier phase observation equation the non-zero initial phases will be lumped together with the carrier phase ambiguity term:

$$M_i^k = \phi_i(t_0) - \phi^k(t_0) + N_i^k .$$

With the above lumping of the parameters, the L_1 and L_2 pseudorange and carrier phase observation equations become

$$P_{i,1}^k = s_i^k + I_{i,1}^k + d_{i,1}^k + e_{i,1}^k \; ; \; \Phi_{i,1}^k = s_i^k - I_{i,1}^k + \delta_{i,1}^k + \lambda M_{i,1}^k + \varepsilon_{i,1}^k$$

$$P_{i,2}^k = s_i^k + I_{i,2}^k + d_{i,2}^k + e_{i,2}^k \; ; \; \Phi_{i,2}^k = s_i^k - I_{i,2}^k + \delta_{i,2}^k + \lambda M_{i,2}^k + \varepsilon_{i,2}^k$$

(4) Based on the first-order expression for the ionospheric range delay, I = 40.3 TEC/f^2, the ionospheric range delay on L_2 will be expressed in terms of the ionospheric range delay on L_1 as $I_{i,2}^k = \alpha\, I_{i,1}^k$ with $\alpha = f_1^2/f_2^2$.

(5) It will be assumed, unless stated otherwise, that during the observational time span, a continuous, uninterrupted tracking of the satellites takes place. Hence the carrier phase ambiguities $M_{i,1}^k$ and $M_{i,2}^k$ are assumed to be constant during the entire observational time span.

(6) For the moment, the delays $d^k_{i,1}$, $d^k_{i,2}$, $\delta^k_{i,1}$ and $\delta^k_{i,2}$ are assumed to be either known or to be so small that they can be neglected. In case the delays are known, it is assumed that the pseudoranges and carrier phases are already corrected for them.

(7) All remaining parameters on the right-hand side of the above four equations, except the carrier phase ambiguities, are assumed to change with time. However, the functional dependency on time will be assumed unknown.

(8) The unmodelled errors $e^k_{i,1}$, $e^k_{i,2}$, $\varepsilon^k_{i,1}$ and $\varepsilon^k_{i,2}$ will be treated as noise. We will therefore in the following, when linear combinations are taken of the observation equations, refrain from carrying them through explicitly. The corresponding error propagation, needed to obtain variances and co-variances of the derived observables, is left to the reader.

(9) Since we will be restricting our attention to the single-channel case only, we will from now on, in order to simplify our notation, skip the lower index "i" denoting the receiver and the upper index "k" denoting the satellite. We will also use the abbreviation $a_1 = \lambda_1 M_1$ and $a_2 = \lambda_2 M_2$.

Based on the above assumptions, the four observation equations for the pseudoranges and carrier phases may now be written in the more concise form

$$\begin{aligned} P_1 &= s + I_1 \quad ; \quad \Phi_1 = s - I_1 + a_1 \\ P_2 &= s + \alpha I_1 \quad ; \quad \Phi_2 = s - \alpha I_1 + a_2 \ . \end{aligned}$$

In these equations we recognize three different type of unknown parameters: the parameter s, containing the geometric range, the clock terms and the tropospheric delay; the parameter I_1, being the ionospheric delay for the L_1 frequency; and, a_1 and a_2, which are the unknown carrier phase ambiguities on L_1 and L_2 respectively. In the following we will consider different subsets of the above set of four observables and discuss the possibilities for the determination of the three type of parameters.

5.4.2 On Single and Dual Frequency Pseudoranges and Carrier Phases

In this section we will consider six different subsets of the set of four type of GPS-observables. They are:

$$\begin{aligned} &1)\ P_1 \text{ and } \Phi_1 \quad ; \quad 2)\ P_1 \text{ and } P_2 \quad ; \quad 3)\ \Phi_1 \text{ and } \Phi_2 \\ &4)\ P_1, \Phi_1 \text{ and } \Phi_2 \quad ; \quad 5)\ \Phi_1, P_1 \text{ and } P_2 \quad ; \quad 6)\ P_1, P_2, \Phi_1 \text{ and } \Phi_2 \ . \end{aligned}$$

For each subset it will be shown how one can separate the parameters s and I_1 by taking particular linear combinations of the observables. In the results so obtained, one will recognize three different classes of linear combinations.

The ionosphere-free linear combinations: This first class of linear combinations consists of those linear combinations that are independent of the ionospheric delay I_1. Since ionosphere-free linear combinations solely depend on s (and possibly on an additional time-invariant term), they are particularly useful for monitoring *SA*-effects. Remember that Selective Availability (*SA*) degrades the stability of the on-board atomic clocks and therefore affects s.

The geometry-free linear combinations: The second class of linear combinations consists of those linear combinations that are independent of s. Since geometry-free linear combinations solely depend on I_1 (and possibly on an additional time-invariant term), they are particularly useful for monitoring the ionosphere.

The time-invariant linear combinations: The third class of linear combinations consists of those linear combinations that are independent of both s and I_1. They are constant in time and are therefore referred to as the time-invariant linear combinations. It is with these linear combinations that redundancy enters, thus allowing one to adjust (smooth) the data and/or to test for deviations from time-invariance. Such deviations may be caused by, e.g., outliers, cycleslips or the presence of multipath.

Pseudorange and Carrier Phase on L_1. In this first case we assume to have pseudoranges and carrier phases available on frequency L_1 only. Then

$$\begin{aligned} P_1(t_i) &= s(t_i) + I_1(t_i) \\ \Phi_1(t_i) &= s(t_i) - I_1(t_i) + a_1 \ . \end{aligned} \tag{5.59}$$

We have explicitly included the time argument "t_i" so as to emphasize that we are working with a discrete time series of pseudoranges and carrier phases.

We can separate the two type of parameters s and I_1, if we premultiply (5.59) with the one-to-one transformation matrix

$$\begin{bmatrix} \frac{1}{2} & \frac{1}{2} \\ \frac{1}{2} & -\frac{1}{2} \end{bmatrix} .$$

Note that a one-to-one transformation always preserves the information content of the system of equations. Application of the above transformation to (5.59) gives

$$\begin{aligned} \tfrac{1}{2}[P_1(t_i) + \Phi_1(t_i)] &= s(t_i) + \tfrac{1}{2}a_1 \\ \tfrac{1}{2}[P_1(t_i) - \Phi_1(t_i)] &= I_1(t_i) - \tfrac{1}{2}a_1 \ . \end{aligned} \tag{5.60}$$

This system of equations is clearly underdetermined. We have two equations with three unknown parameters. The parameters are $s(t_i)$, $I_1(t_i)$ and a_1. The system also remains underdetermined when the time series is considered as a whole. When the number of epochs equals k $(i = 1,...,k)$, there are $2k$ number of equations and $2k + 1$ number of unknowns, leaving an underdeterminancy of one. Hence, the information content of the L_1 pseudoranges and L_1 carrier phases is not sufficient to determine the three types of parameters $s(t_i)$, $I_1(t_i)$ and a_1 separately. Due to the constancy in time of a_1 however, it is possible to determine the time increments of $s(t)$ and $I_1(t)$.

If the delays d_1 and δ_1 were to be included as unknown parameters, we would get instead of (5.60), the two equations

$$\begin{aligned} \tfrac{1}{2}[P_1(t_i) + \Phi_1(t_i)] &= s(t_i) + \tfrac{1}{2}[a_1 + d_1 + \delta_1] \\ \tfrac{1}{2}[P_1(t_i) - \Phi_1(t_i)] &= I_1(t_i) - \tfrac{1}{2}[a_1 + d_1 - \delta_1] \;. \end{aligned} \tag{5.61}$$

This shows that the time increments of $s(t)$ and $I_1(t)$ can still be determined, provided that the delays are constant in time.

Dual Frequency Pseudorange. In this case only dual-frequency pseudorange data are available. Then

$$\begin{aligned} P_1(t_i) &= s(t_i) + I_1(t_i) \\ P_2(t_i) &= s(t_i) + \alpha I_1(t_i) \;. \end{aligned} \tag{5.62}$$

We can separate the two type of parameters s and I_1, if we premultiply (5.62) with the one-to-one transformation matrix

$$\begin{bmatrix} \frac{\alpha}{(\alpha - 1)} & \frac{-1}{(\alpha - 1)} \\ \frac{-1}{(\alpha - 1)} & \frac{1}{(\alpha - 1)} \end{bmatrix} .$$

Using the notation $P_{12}(t_i) = P_2(t_i) - P_1(t_i)$ for the inter-frequency difference of the pseudoranges, this gives

$$\begin{aligned} P_1(t_i) - P_{12}(t_i)/(\alpha - 1) &= s(t_i) \\ P_{12}(t_i)/(\alpha - 1) &= I_1(t_i) \end{aligned} . \tag{5.63}$$

This shows that the information content of the dual-frequency pseudoranges is just enough to determine $s(t_i)$ and $I_1(t_i)$ uniquely. Hence, when compared to (5.60), we are now able to determine the absolute time behaviour of $s(t)$ and $I_1(t)$ instead of just only their time increments. This result is spoiled however when the delays d_1 and d_2 are included as unknown parameters. One can then again at the most determine the time increments of $s(t)$ and $I_1(t)$, provided the delays are constant in time.

Dual Frequency Carrier Phase. In this case only dual-frequency carrier phase data are available. Then

$$\begin{aligned} \Phi_1(t_i) &= s(t_i) - I_1(t_i) + a_1 \\ \Phi_2(t_i) &= s(t_i) - \alpha I_1(t_i) + a_2 \ . \end{aligned} \tag{5.64}$$

We can separate the two type of parameters s and I_1, if we premultiply (5.64) with the one-to-one transformation matrix

$$\begin{bmatrix} \frac{\alpha}{(\alpha-1)} & \frac{-1}{(\alpha-1)} \\ \frac{1}{(\alpha-1)} & \frac{-1}{(\alpha-1)} \end{bmatrix} .$$

Using the notation $\Phi_{12}(t_i) = \Phi_2(t_i) - \Phi_1(t_i)$ for the inter-frequency difference of the carrier phases, this gives

$$\begin{aligned} \Phi_1(t_i) - \Phi_{12}(t_i)/(\alpha-1) &= s(t_i) + b \\ -\Phi_{12}(t_i)/(\alpha-1) &= I_1(t_i) + c \ , \end{aligned} \tag{5.65}$$

with the time-invariant parameters $b = [\alpha a_1 - a_2]/(\alpha-1)$ and $c = [a_1 - a_2]/(\alpha-1)$. This system of equations is clearly underdetermined. We have two equations with four unknown parameters. The parameters are $s(t_i)$, $I_1(t_i)$, b and c. The underdeterminancy of two also remains when the time series is considered as a whole. As with (5.59), the linear combinations of (5.65) can be used to determine the time increments of $s(t)$ and $I_1(t)$. This also holds true when the delays δ_1 and δ_2 are included, provided that the delays are constant in time.

Pseudorange on L_1 and Dual Frequency Carrier Phase. In this case the pseudoranges on L_1 and the carrier phases on both L_1 and L_2 are assumed to be available. Then

$$\begin{aligned} P_1(t_i) &= s(t_i) + I_1(t_i) \\ \Phi_1(t_i) - \Phi_{12}(t_i)/(\alpha-1) &= s(t_i) + b \\ -\Phi_{12}(t_i)/(\alpha-1) &= I_1(t_i) + c \end{aligned} \ . \tag{5.66}$$

where the last two equations follow from (5.65). We can separate the two types of parameters s and I_i, if we premultiply (5.66) with the one-to-one transformation matrix

$$\begin{bmatrix} 1 & -1 & -1 \\ 0 & 1 & 0 \\ 0 & 0 & 1 \end{bmatrix} .$$

This gives

$$\begin{aligned} P_1(t_i) - \Phi_1(t_i) + 2\Phi_{12}(t_i)/(\alpha - 1) &= -b - c \\ \Phi_1(t_i) - \Phi_{12}(t_i)/(\alpha - 1) &= s(t_i) + b \; . \\ -\Phi_{12}(t_i)/(\alpha - 1) &= I_1(t_i) + c \end{aligned} \tag{5.67}$$

The first linear combination is time-invariant, the second is free of the ionosphere, and the third is free of the geometry. Because of the time-invariance of the first linear combination, the system of equations becomes redundant when the time series is considered as a whole. Since one of the parameters needs to be known before the other parameters can be estimated, however, the system of equations also has a rank defect of one. This implies that the parameters $s(t_i)$, $I_1(t_i)$, b and c cannot be estimated independently. The redundancy of the system of equations equals $(k-1)$, with k being the number of epochs in the observational time span.

The above conclusions are not affected when the delays d_1, δ_1 and δ_2 are included, provided that the delays are constant in time.

The redundancy in the above system of equations (5.67) stems from the time-invariance of the first linear combination. This time-invariance, therefore, can be used to adjust the data, and in particular to smooth the pseudorange data. When we apply the recursive least-squares algorithm to the first equation of (5.67) and approximate the statistics of the data by setting the variances of the carrier phases to zero, we obtain the *carrier phase smoothed pseudorange algorithm* as a result:

$$\begin{aligned} P_{1,k|k-1} &= P_{1,k-1|k-1} + \tfrac{1}{k}[P_{1,k} - P_{1,k-1|k-1}] && \text{for } k > 1 \\ P_{1,k|k} &= P_{1,k|k-1} + \tfrac{(k-1)}{k}[(\Phi_{1,k} - \Phi_{1,k-1} - 2(\Phi_{12,k} - \Phi_{12,k-1})/(\alpha - 1)] && \text{for } k \geq 1 \end{aligned} \tag{5.68}$$

The index k is used to denote the epoch. The first equation of (5.68) can be considered the 'predictor' and the second equation the 'filter'. The algorithm is initialized with $P_{1,0|0} := P_1(t_1)$.

Carrier Phase on L_1 and Dual Frequency Pseudorange. Complementary to the previous case, we now assume the carrier phases to be available on L_1 only, but the pseudoranges on both frequencies. Then

$$\begin{aligned} \Phi_1(t_i) &= s(t_i) - I_1(t_i) + a_1 \\ P_1(t_i) - P_{12}(t_i)/(\alpha - 1) &= s(t_i) \\ P_{12}(t_i)/(\alpha - 1) &= I_1(t_i) \ . \end{aligned} \tag{5.69}$$

where the last two equations follow from (5.63). If we premultiply (5.69) with the one-to-one transformation matrix

$$\begin{bmatrix} 1 & -1 & 1 \\ 0 & 1 & 0 \\ 0 & 0 & 1 \end{bmatrix}$$

we obtain

$$\begin{aligned} \Phi_1(t_i) - P_1(t_i) + 2P_{12}(t_i)/(\alpha - 1) &= a_1 \\ P_1(t_i) - P_{12}(t_i)/(\alpha - 1) &= s(t_i) \\ P_{12}(t_i)/(\alpha - 1) &= I_1(t_i) \ . \end{aligned} \tag{5.70}$$

Compare this result with that of (5.67). Again redundancy is present due to the time-invariance of the first equation. The redundancy equals $(k-1)$.

Dual Frequency Pseudorange and Dual Frequency Carrier Phase. In this case we assume the pseudoranges and carrier phases to be available on both frequencies L_1 and L_2. From (5.63) and (5.65) follows then

$$\begin{aligned} P_1(t_i) - P_{12}(t_i)/(\alpha - 1) &= s(t_i) \\ P_{12}(t_i)/(\alpha - 1) &= I_1(t_i) \\ \Phi_1(t_i) - \Phi_{12}(t_i)/(\alpha - 1) &= s(t_i) + b \\ -\Phi_{12}(t_i)/(\alpha - 1) &= I_1(t_i) + c \ . \end{aligned} \tag{5.71}$$

Premultiplication with the one-to-one transformation matrix

$$\begin{bmatrix} \boldsymbol{I}_2 & \boldsymbol{O}_2 \\ -\boldsymbol{I}_2 & \boldsymbol{I}_2 \end{bmatrix} ,$$

gives then

$$\begin{aligned} P_1(t_i) - P_{12}(t_i)/(\alpha - 1) &= s(t_i) \\ P_{12}(t_i)/(\alpha - 1) &= I_1(t_i) \\ \Phi_1(t_i) - P_1(t_i) - [\Phi_{12}(t_i) - P_{12}(t_i)]/(\alpha - 1) &= b \\ -[\Phi_{12}(t_i) + P_{12}(t_i)]/(\alpha - 1) &= c \ . \end{aligned} \tag{5.72}$$

This system of equations is uniquely solvable for one single epoch and it becomes redundant when more than one epoch is considered. The redundancy stems from the time-invariance of the last two equations. The first two equations are not redundant. For k number of epochs the redundancy of the system of equations equals $2(k-1)$.

All the four type of parameters s, I_1, b and c are estimable. Hence, this is for the first time where, through the estimation of b and c, also the L_1 and L_2 carrier phase ambiguities a_1 and a_2 can be estimated.

When one considers the structure of the four equations of (5.72), one may be inclined to conclude, since the carrier phases do not appear in the first two equations, that they fail to contribute to the determination of both s and I_1. This however is not true. It should be realized that the four derived observables of (5.72) are correlated. This implies therefore if a proper least-squares adjustment is carried out on the basis of the redundant equations, that one also obtains least-squares corrections for the first two observables of (5.72). And it is through these least-squares corrections that the carrier phases contribute to the determination of both s and I_1.

When the delays d_1, d_2, δ_1, δ_2 are included in (5.72), the redundancy remains the same, provided that they are constant in time. But then only the time increments of s and I_1 are estimable.

5.5 THE LINEARIZED OBSERVATION EQUATIONS FOR POSITIONING

In the previous section all observation equations were linear. The observation equations will become nonlinear however, if they are to be used for positioning. This is due to the fact that ρ_i^k, the distance between receiver i and satellite k, is a nonlinear function of the receiver coordinates. Since it is assumed that one will be working with standard linear least-squares adjustment algorithms, one will need to linearize the nonlinear observation equations.

In this section we will first discuss the linearization of the nonlinear observation equations. Then it is briefly discussed how these linearized observation equations can be used for both single-point positioning as well as for relative positioning. The advantages of relative positioning from a qualitative point of view over that of single-point positioning are highlighted.

5.5.1 The Linearization

Our discussion of the linearization will be kept as simple as possible. Although this compels us to ignore some important subtleties of the linearization process, the linear equations that we obtain will be satisfactory for our purposes. A more elaborate discussion of the linearization process will be given in the following chapters. It will also include a discussion on the computation of the approximate values and on the iteration process.

Our linearization will be illustrated using the pseudorange observation equation for L_1 as an example. The linearization of the other observation equations goes along similar lines. The observation equation for the pseudorange on L_1 is given as

$$P_{i,1}^k = \rho_i^k + c[dt_i - dt^k] + T_i^k + I_{i,1}^k + d_{i,1}^k + e_{i,1}^k \ . \tag{5.73}$$

In order to linearize we need approximate values for the parameters. They are denoted as

$(\rho_i^k)^o$: the approximate distance between receiver and satellite,
$(dt_i)^o$: the approximate receiver time error,
$(dt^k)^o$: the approximate satellite time error,
$(T_i^k)^o$: the approximate tropospheric range error,
$(I_{i,1}^k)^o$: the approximate ionospheric range error,
$(d_{i,1}^k)^o$: the approximate receiver/satellite equipment and multipath delays.

Based on the approximate values, an approximate pseudorange can be computed

$$(P_{i,1}^k)^o = (\rho_i^k)^o + c[(dt_i)^o - (dt^k)^o] + (T_i^k)^o + (I_{i,1}^k)^o + (d_{i,1}^k)^o \ . \tag{5.74}$$

The difference between the observed pseudorange $P_{i,1}^k$ and computed pseudorange $(P_{i,1}^k)^o$ follows from subtracting (5.74) from (5.73). Expressed in terms of the parameter increments, it reads

$$\Delta P_{i,1}^k = \Delta\rho_i^k + c[\Delta dt_i - \Delta dt^k] + \Delta T_i^k + \Delta I_{i,1}^k + \Delta d_{i,1}^k + e_{i,1}^k \ , \tag{5.75}$$

where the Δ-symbol is used as notation for both the "observed" minus "computed" pseudorange, as well as for the parameter increments.

Note that so far, no actual linearization has been carried out. Equation (5.75) is simply the result of the difference of two equations. The purpose of taking the difference between the two equations (5.73) and (5.74) is, however, to obtain increments of a sufficiently small magnitude. This allows one then to replace the increments that are nonlinearly related to the parameters of interest, by their first order approximation.

In this section we will only consider the linearization of the increment $\Delta\rho_i^k$. For some applications, however, it might be opportune to also apply a linearization to some of the other increments appearing in (5.75). For instance, if a tropospheric model is available, one might use this model to linearize ΔT_i^k with respect to some of the parameters that appear nonlinearly in the tropospheric model.

In order to linearize $\Delta\rho_i^k$, recall that $\rho_i^k = \|(\boldsymbol{r}_i + d\boldsymbol{r}_i) - (\boldsymbol{r}^k + d\boldsymbol{r}^k)\|$. The receiver- and satellite eccentricities will be assumed known. Hence $\Delta\rho_i^k$ will only be linearized with respect to $\boldsymbol{r}_i$ and $\boldsymbol{r}^k$. Linearization of $\Delta\rho_i^k$ with respect to both the receiver coordinates as well as satellite coordinates gives therefore

$$\Delta\rho_i^k = -(\boldsymbol{u}_i^k)^T \Delta\boldsymbol{r}_i + (\boldsymbol{u}_i^k)^T \Delta\boldsymbol{r}^k \ . \tag{5.76}$$

with $\boldsymbol{u}_i^k$ the unit vector from receiver to satellite, $\Delta\boldsymbol{r}_i$ the increment of the receiver position vector and $\Delta\boldsymbol{r}^k$ the increment of the satellite position vector. The unit vector $\boldsymbol{u}_i^k$ is computed from the approximate coordinates of the receiver and satellite. In the following it is therefore assumed known.

The linearized L_1-pseudorange observation equation follows now from substituting (5.76) into (5.75). Since the linearization of the other three observation equations goes along the same lines, the linearized versions of the two pseudorange observation equations are given as

$$\begin{aligned}
\Delta P_{i,1}^k &= -(\boldsymbol{u}_i^k)^T \Delta\boldsymbol{r}_i + c\Delta dt_i + (\boldsymbol{u}_i^k)^T \Delta\boldsymbol{r}^k - c\Delta dt^k + \Delta T_i^k + \Delta I_{i,1}^k + \Delta d_{i,1}^k + e_{i,1}^k \\
\Delta P_{i,2}^k &= -(\boldsymbol{u}_i^k)^T \Delta\boldsymbol{r}_i + c\Delta dt_i + (\boldsymbol{u}_i^k)^T \Delta\boldsymbol{r}^k - c\Delta dt^k + \Delta T_i^k + \alpha\Delta I_{i,1}^k + \Delta d_{i,2}^k + e_{i,2}^k
\end{aligned} \tag{5.77}$$

and those of the two carrier phase observation equations are given as

$$\begin{aligned}
\Delta\Phi_{i,1}^k &= -(\boldsymbol{u}_i^k)^T \Delta\boldsymbol{r}_i + c\Delta dt_i + (\boldsymbol{u}_i^k)^T \Delta\boldsymbol{r}^k - c\Delta dt^k + \Delta T_i^k - \Delta I_{i,1}^k + \Delta\delta_{i,1}^k + \lambda_1 M_{i,1}^k + \varepsilon_{i,1}^k \\
\Delta\Phi_{i,2}^k &= -(\boldsymbol{u}_i^k)^T \Delta\boldsymbol{r}_i + c\Delta dt_i + (\boldsymbol{u}_i^k)^T \Delta\boldsymbol{r}^k - c\Delta dt^k + \Delta T_i^k - \alpha\Delta I_{i,1}^k + \Delta\delta_{i,2}^k + \lambda_2 M_{i,2}^k + \varepsilon_{i,2}^k
\end{aligned} \tag{5.78}$$

Note that we have neglected the eccentricity-driven differences in the unit vector for pseudoranges and carrier phases.

In the following two subsections, it will be discussed how these observation equations can be used for positioning purposes. In section 5.5.2 the single-point positioning concept is briefly discussed and in section 5.5.3 the relative positioning concept.

In the following we will refrain from carrying the noise terms $e_{i,1}^k$, $e_{i,2}^k$, $\varepsilon_{i,1}^k$ and $\varepsilon_{i,2}^k$ through explicitly.

5.5.2 Single-Point Positioning

In section 5.4 we restricted ourselves to the data of a single channel, observed by a single receiver. In this subsection we will continue to assume that we have only one single GPS-receiver available, observing pseudoranges or carrier phases. Contrary to the single-channel assumption however, we will now consider the multi-channel situation. This allows us then to include the geometry of the receiver-satellite configuration and to explore the possibilities one has for determining the position of the single receiver i.

When we consider the linearized observation equations for the pseudoranges, (5.77), three groups of parameters can be recognized. One group of parameters that depends on the satellite k being tracked. A second group that depends on the propagation medium between satellite k and receiver i. And a third group that depends on the receiver i. These parameters appear in the observation equation as

$$\begin{aligned} \Delta P_i^k = & -(\boldsymbol{u}_i^k)^T \Delta \boldsymbol{r}_i + c\Delta dt_i + c\Delta d_i \quad (\textit{receiver dependent}) \\ & +(\boldsymbol{u}_i^k)^T \Delta \boldsymbol{r}^k - c\Delta dt^k + c\Delta d^k \quad (\textit{satellite dependent}) \\ & +\Delta T_i^k + \Delta I_i^k + \Delta dm_i^k \quad (\textit{atmosphere and multipath dependent}) \; . \end{aligned} \tag{5.79}$$

For positioning purposes the primary parameter of interest is of course $\Delta \boldsymbol{r}_i$, the unknown increment to the receiver's position. But it will be clear, considering the above observation equation, that the information content of the observables will not be sufficient to determine the receiver's position if in addition to $\Delta \boldsymbol{r}_i$ also the other parameters are unknown. One could think of reducing the number of parameters by lumping some of them together. For instance, with the lumped parameters

$$\Delta dt_i^{\prime} = \Delta dt_i + \Delta d_i \quad \text{and} \quad \nabla_i^k = (\boldsymbol{u}_i^k)^T \Delta \boldsymbol{r}^k - c\Delta dt^k + c\Delta d^k + \Delta T_i^k + \Delta I_i^k + \Delta dm_i^k$$

the pseudorange observation equation becomes

$$\Delta P_i^k = -(\boldsymbol{u}_i^k)^T \Delta \boldsymbol{r}_i + c\Delta dt_i^{\prime} + \nabla_i^k \; . \tag{5.80}$$

Unfortunately, however, this lumping of the parameters does not solve our problem completely. With (5.80) we are still faced with the parameter ∇_i^k, which introduces an unknown for each one of the observed pseudoranges. The only approach that therefore can be taken in the present context is to simply assume ∇_i^k

to be zero.

If the receiver tracks m satellites simultaneously (k = 1,...,m) and if ∇_i^k is assumed to be zero, the pseudorange observation equations can be written in the compact vector-matrix form as

$$\Delta \boldsymbol{p}_i(t_i) = \left(\boldsymbol{A}(t_i)\ \boldsymbol{l}_m\right)\begin{bmatrix}\Delta \boldsymbol{r}_i(t_i)\\ c\Delta dt_i^l(t_i)\end{bmatrix}, \tag{5.81}$$

where: $\Delta \boldsymbol{p}_i(t_i) = \left(\Delta P_i^1(t_i), \Delta P_i^2(t_i), ..., \Delta P_i^m(t_i)\right)^T$, $\boldsymbol{A}(t_i) = \left(-\boldsymbol{u}_i^1(t_i), -\boldsymbol{u}_i^2(t_i), ..., -\boldsymbol{u}_i^m(t_i)\right)^T$ and $\boldsymbol{l}_m = (1,1,...,1)^T$. This system of observation equations consists of m equations in 4 unknown parameters. This implies, assuming the satellite configuration at epoch t_i to be such that the design matrix $\left(\boldsymbol{A}(t_i), \boldsymbol{l}_m\right)$ is of full rank, that the redundancy of the system of observation equations equals $(m-4)$. Hence, a minimum of four satellites are needed to determine the parameters $\Delta \boldsymbol{r}(t_i)$ and $c\Delta dt_i^l(t_i)$ uniquely.

The above shows that single-point positioning is feasible when one is willing to neglect ∇_i^k. The same can be shown to hold true when carrier phases are observed. In that case, however, one will need a minimum of two epochs of data, due to the presence of the unknown carrier phase ambiguities. But it will be clear that the quality of single-point positioning largely depends on the validity of $\nabla_i^k = 0$. For some applications this assumption may be acceptable. Unfortunately, however, this assumption is not acceptable for those applications where high positioning precision is required. We will therefore refrain from a further elaboration on the single-point positioning concept.

5.5.3 Relative Positioning

Relative positioning involves the simultaneous observation of m satellites by a minimum of two GPS-receivers. The advantage of relative positioning over single-point positioning lies in the fact that in case of relative positioning, the parameters of interest are much less sensitive to interfering uncertainties such as ephemeris-, clock- and atmospheric effects. This will be shown using the pseudorange observable as an example. But the same reasoning applies equally well to the carrier phase observable.

The principle advantage of relative positioning becomes apparent when we consider the so-called single-difference observables. Let P_i^k be the pseudorange observable of receiver i observing satellite k, and let P_j^k be the pseudorange observable of receiver j on the same frequency as P_i^k, observing the same satellite k at the same instant. The two corresponding linearized observation equations read then

$$\Delta P_i^k = -(\boldsymbol{u}_i^k)^T \Delta \boldsymbol{r}_i + c\Delta dt_i + (\boldsymbol{u}_i^k)^T \Delta \boldsymbol{r}^k - c\Delta dt^k + \Delta T_i^k + \Delta I_i^k + \Delta d_i^k$$

$$\Delta P_j^k = -(\boldsymbol{u}_j^k)^T \Delta \boldsymbol{r}_j + c\Delta dt_j + (\boldsymbol{u}_j^k)^T \Delta \boldsymbol{r}^k - c\Delta dt^k + \Delta T_j^k + \Delta I_j^k + \Delta d_j^k \ . \tag{5.82}$$

In the single-point positioning concept of the previous subsection we were forced to neglect the group of parameters that were dependent on the satellite being tracked. In the present context however, these parameters now appear in two observation equations instead of in one. This therefore gives us the opportunity to either eliminate these parameters or to considerably reduce for their effect.

If we take the difference of the two equations of (5.82) and introduce the notation as outlined in equation (5.50), we obtain

$$\begin{aligned} \Delta P_{ij}^k = \; & -(\boldsymbol{u}_j^k)^T \Delta \boldsymbol{r}_{ij} - (\boldsymbol{u}_{ij}^k)^T \Delta \boldsymbol{r}_i + c\Delta dt_{ij} + c\Delta d_{ij} && \textit{(receiver dependent)} \\ & + (\boldsymbol{u}_{ij}^k)^T \Delta \boldsymbol{r}^k && \textit{(satellite dependent)} \\ & + \Delta T_{ij}^k + \Delta I_{ij}^k + \Delta dm_{ij}^k && \textit{(atmosphere and multipath dependent)} \ . \end{aligned} \tag{5.83}$$

This observable is referred to as the single-difference pseudorange. We speak of relative positioning since the baseline vector $\boldsymbol{r}_{ij}$ is the primary quantity to be solved for. If we read the single-difference observation equation from right to left, the following remarks are in order:

Atmospheric and multipath delays. For two receivers located close together, the atmospheric delays are (almost) the same because the radio signals travel through the same portion of the atmosphere and thus experience the same changes in velocity and ray bending. Hence, the atmospheric delays, ΔT_{ij}^k and ΔI_{ij}^k, almost cancel in the difference for small interstation distances. In the following the multipath delay dm_{ij}^k will be assumed absent. Hence it will be assumed that provisions are made (e.g., in the location of the receivers) that prevent multipath.

Orbital uncertainty. The position increment of satellite k, $\Delta \boldsymbol{r}^k$, appears in the inner product $(\boldsymbol{u}_{ij}^k)^T \Delta \boldsymbol{r}^k$. The following upperbound can be given to this inner product:

$$|(\boldsymbol{u}_{ij}^k)^T \Delta \boldsymbol{r}^k| \leq \|\boldsymbol{u}_{ij}^k\| \; \|\Delta \boldsymbol{r}^k\| \ .$$

Furthermore we have: $\|\boldsymbol{u}_{ij}^k\|^2 = 2[1 - (\boldsymbol{u}_i^k)^T (\boldsymbol{u}_j^k)] = 2(1 - \cos\alpha)$, with α being the angle between the two unit vectors $\boldsymbol{u}_i^k$ and $\boldsymbol{u}_j^k$. We also have by means of the cosine-rule: $\|\boldsymbol{r}_{ij}\|^2 = \|\boldsymbol{r}_i^k\|^2 + \|\boldsymbol{r}_j^k\|^2 - 2\|\boldsymbol{r}_i^k\| \; \|\boldsymbol{r}_j^k\| \cos\alpha \cong 2\|\boldsymbol{r}_j^k\|^2 (1 - \cos\alpha)$, since $\|\boldsymbol{r}_i^k\| \cong \|\boldsymbol{r}_j^k\|$. It follows, therefore, that $\|\boldsymbol{u}_{ij}^k\| \cong \|\boldsymbol{r}_{ij}\| / \|\boldsymbol{r}_j^k\|$. Hence, the above inequality may be approximated as

$$|(\boldsymbol{u}_{ij}^k)^T \Delta \boldsymbol{r}^k| \leq \left\{ \frac{\|\boldsymbol{r}_{ij}\|}{\|\boldsymbol{r}_j^k\|} \right\} \|\Delta \boldsymbol{r}^k\| \ . \tag{5.84}$$

But this shows, when the baseline length $\| \boldsymbol{r}_{ij} \|$ is small compared to the high altitude orbit of the GPS-satellite, $\| \boldsymbol{r}_j^k \|$, that the effect of the orbital uncertainty, $\Delta \boldsymbol{r}^k$, gets drastically reduced.

Instrumental delays and clock errors. Note that both the instrumental delay of the satellite, d^k, as well as the satellite clock error, dt^k, have been eliminated from the single-difference equation. The relative receiver clock error dt_{ij} is the only clock error remaining. It will be lumped together with the relative instrumental delay of the two receivers: $dt_{ij}^{/} = dt_{ij} + d_{ij}$.

Receiver positioning error. In the single-difference equation (5.83), the position vectors of the two receivers have been parametrized into a baseline vector $\boldsymbol{r}_{ij}$ and a position vector of receiver i, $\boldsymbol{r}_i$. It will be clear that if $\boldsymbol{r}_i$ is known and $\boldsymbol{r}_{ij}$ is solved for, that also $\boldsymbol{r}_j$ is known. It will be assumed in the following that $\boldsymbol{r}_i$ is known. This allows us to set the increment $\Delta \boldsymbol{r}_i$ equal to zero. Note, however, that analogously to (5.84), we have

$$|(\boldsymbol{u}_{ij}^k)^T \Delta \boldsymbol{r}_i| \leq \left\{ \frac{\| \boldsymbol{r}_{ij} \|}{\| \boldsymbol{r}_j^k \|} \right\} \| \Delta \boldsymbol{r}_i \| \ . \tag{5.85}$$

This shows that also the effect of the uncertainty in the position of receiver i, $\Delta \boldsymbol{r}_i$, gets drastically reduced in the single-difference equation.

It follows from the above discussion that we are now - in contrast to the situation of the previous subsection - in a much better position to neglect the satellite dependent parameters. The concept of relative positioning, therefore, will be further explored in the next section.

5.6 RELATIVE POSITIONING MODELS

In this section we consider relative positioning based on pseudoranges and based on carrier phases. Both the single-frequency case as well as dual-frequency case will be considered. Particular attention will be given to the estimability of the parameters. First the pseudoranges will be considered.

5.6.1 Relative Positioning Using Pseudoranges

First we will consider the single-frequency case, then the dual-frequency case. In the single-frequency case the ionospheric delay will be assumed absent. In the dual-frequency case, however, the ionospheric delay will be included in the observation equations.

The Single Frequency Case. It is assumed that two GPS-receivers, i and j, simultaneously observe the L_1-pseudoranges to m satellites. Hence, at the time of observation the following pseudoranges become available: $P_{i,1}^k$ and $P_{j,1}^k$ for $k = 1,...,m$. Instead of working with these two types of undifferenced pseudorange observables, $P_{i,1}^k$ and $P_{j,1}^k$, we may as well work with the single single-differenced pseudorange observable $P_{ij,1}^k$. This data compression is permitted since - as we have seen earlier - the unknown satellite clock error dt^k plus satellite delay d^k gets eliminated when taking the difference.

The m single-differenced linearized pseudorange observation equations read

$$\Delta P_{ij,1}^k(t_i) = -\left(\boldsymbol{u}_j^k(t_i)\right)^T \Delta \boldsymbol{r}_{ij}(t_i) + c\Delta dt_{ij}^{/}(t_i), \quad k = 1,...,m \; . \tag{5.86}$$

Note that we have assumed the increments $\Delta \boldsymbol{r}_i$, $\Delta \boldsymbol{r}^k$, Δdm_{ij}^k, ΔT_{ij}^k and ΔI_{ij}^k to be zero. In vector-matrix form the above observation equations read

$$\Delta \boldsymbol{p}_{ij,1}(t_i) = \left(A(t_i) \;\; \boldsymbol{l}_m\right) \begin{bmatrix} \Delta \boldsymbol{r}_{ij}(t_i) \\ c\Delta dt_{ij}^{/}(t_i) \end{bmatrix} \; . \tag{5.87}$$

Compare this result with that of (5.81) and note that the single receiver quantities have now been replaced by the single-differences $\Delta \boldsymbol{p}_{ij,1}(t_i)$, $\Delta \boldsymbol{r}_{ij}(t_i)$ and $c\Delta dt_{ij}^{/}(t_i)$. In the present context, one is thus solving for the baseline vector $\boldsymbol{r}_{ij}$ and the relative receiver clock error $dt_{ij}^{/}$ instead of the single position vector $\boldsymbol{r}_i$ and single receiver clock error $dt_i^{/}$. And since the approximations involved in the relative positioning concept are less crude than those that were made in the single-point positioning concept, the accuracy with which $\boldsymbol{r}_{ij}$ and $dt_{ij}^{/}$ can be determined is higher than the accuracy with which $\boldsymbol{r}_i$ and $dt_i^{/}$ can be determined in the single-point positioning concept.

The Dual Frequency Case. We will now consider the dual frequency case and include the ionospheric delays in the observation equations. In the single-frequency case the two undifferenced pseudoranges, $P_{i,1}^k$ and $P_{j,1}^k$, were replaced by one single single-differenced pseudorange $P_{ij,1}^k$. This reduction from two observables to one single observable was allowed, since the unknown satellite clock error got eliminated in the single-differenced observable. In the dual-frequency case, however, we are not allowed - if we want to retain the same level of redundancy - to simply replace the four undifferenced pseudorange observables, $P_{i,1}^k$, $P_{j,1}^k$, $P_{i,2}^k$ and $P_{j,2}^k$, by the two single-differenced pseudoranges $P_{ij,1}^k$ and $P_{ij,2}^k$. In that case the satellite clock error would be eliminated twice. Thus in order to preserve the information content, we should go from four undifferenced pseudoranges to three instead of two differenced pseudoranges. The three differenced pseudoranges that will be taken as our starting point, are

$$P_{ij,1}^{k} = P_{j,1}^{k} - P_{i,1}^{k}$$
$$P_{ij,2}^{k} = P_{j,2}^{k} - P_{i,2}^{k}$$
$$P_{j,12}^{k} = P_{j,2}^{k} - P_{j,1}^{k} \; .$$

Note that the first two are between-receiver differences, one for each of the two frequencies, whereas the last one is a between-frequency difference. In order to deal with the ionospheric delays separately, we now transform - in analogy of section 5.4.2.2 - these differenced pseudoranges by pre-multiplication with

$$\begin{bmatrix} \frac{\alpha}{(\alpha-1)} & \frac{-1}{(\alpha-1)} & 0 \\ \frac{-1}{(\alpha-1)} & \frac{1}{(\alpha-1)} & 0 \\ 0 & 0 & \frac{1}{(\alpha-1)} \end{bmatrix} .$$

As a result this gives us the following observation equations

$$\begin{aligned} \Delta P_{ij,1}^{k}(t_i) - \Delta P_{ij,12}^{k}(t_i)/(\alpha-1) &= -\left(\boldsymbol{u}_j^k(t_i)\right)^T \Delta \boldsymbol{r}_{ij}(t_i) + c\Delta dt_{ij}^{//}(t_i) \\ \Delta P_{ij,12}^{k}(t_i)/(\alpha-1) &= \Delta I_{ij,1}^{k}(t_i) + c d_{ij,12}/(\alpha-1) \\ \Delta P_{j,12}^{k}(t_i)/(\alpha-1) &= \Delta I_{j,1}^{k}(t_i) + c(d_{j,12} + d_{,12}^{k})/(\alpha-1) , \end{aligned} \tag{5.88}$$

with $dt_{ij}^{//} = dt_{ij} + (\alpha d_{ij,1} - d_{ij,2})/(\alpha-1)$. Note that the last two equations do not contribute to the solution of the baseline $\Delta \boldsymbol{r}_{ij}$. Hence, if one is only interested in positioning, the first of the above three observation equations can be used to obtain in vector-matrix form the system

$$\Delta \boldsymbol{p}_{ij,1}(t_i) - \Delta \boldsymbol{p}_{ij,12}(t_i)/(\alpha-1) = \left(\boldsymbol{A}(t_i) \;\; \boldsymbol{l}_m\right) \begin{bmatrix} \Delta \boldsymbol{r}_{ij}(t_i) \\ c\Delta dt_{ij}^{//}(t_i) \end{bmatrix} . \tag{5.89}$$

Compare to (5.87). If we assume the instrumental delays to be known or absent, then the last two type of observation equations of (5.88) can be used for both absolute and relative ionospheric monitoring purposes. In vector-matrix form they read

$$\begin{aligned} \Delta \boldsymbol{p}_{ij,12}(t_i)/(\alpha-1) &= \Delta \boldsymbol{I}_{ij,1}(t_i) \\ \Delta \boldsymbol{p}_{j,12}/(\alpha-1) &= \Delta \boldsymbol{I}_{j,1}(t_i) \; . \end{aligned} \tag{5.90}$$

Note that this system of equations can either be used on its own for ionospheric monitoring purposes, or in combination with (5.89). Due to the existing correlation between the observables of (5.89) and (5.90), the best precision for the ionospheric delay estimates will be obtained with the latter approach.

5.6.2 Relative Positioning Using Carrier Phases

In this subsection we will restrict ourselves to the carrier phases. First we will consider the single frequency case, then the dual frequency case. In the dual frequency case, the ionospheric delay will be included again.

The Single-Frequency Case. Instead of using the undifferenced carrier phases, $\Phi^k_{i,1}$ and $\Phi^k_{j,1}$, we will make use of the single-differenced carrier phase $\Phi^k_{ij,1}$. Since the structure of the carrier phase observation equation is, apart from the carrier phase ambiguity, quite similar to that of the pseudorange observation equation, the system of L_1 carrier phase observation equations for observation epoch t_i follows in analogy to (5.87) as

$$\Delta\boldsymbol{\phi}_{ij,1}(t_i) = \left(\boldsymbol{A}(t_i)\ \boldsymbol{l}_m\ \boldsymbol{I}_m\right)\begin{bmatrix}\Delta\boldsymbol{r}_{ij}(t_i)\\ c\Delta dt^{l}_{ij}(t_i)\\ \boldsymbol{a}_{ij,1}\end{bmatrix}, \tag{5.91}$$

with $\Delta\boldsymbol{\phi}_{ij,1}(t_i) = \left(\Delta\Phi^1_{ij,1}(t_i), \Delta\Phi^2_{ij,1}(t_i), \ldots, \Delta\Phi^m_{ij,1}(t_i)\right)^T$ and $\boldsymbol{a}_{ij,1} = (\lambda_1 M^1_{ij,1}, \lambda_1 M^2_{ij,1}, \ldots, \lambda_1 M^m_{ij,1})^T$. Note that $(\boldsymbol{A}(t_i),\ \boldsymbol{l}_m,\ \boldsymbol{I}_m)(\boldsymbol{I}_3,\ 0,\ -\boldsymbol{A}(t_i)^T)^T = \boldsymbol{0}$ and $(\boldsymbol{A}(t_i),\ \boldsymbol{l}_m,\ \boldsymbol{I}_m)\,(\boldsymbol{0}_3,\ 1,\ -\boldsymbol{l}^T_m)^T = \boldsymbol{0}$. This shows that the design matrix of (5.91) has a rank defect of 4 and that the linear dependent combinations of the column vectors of the design matrix are given by $\left(\boldsymbol{I}_3, 0,\ -\boldsymbol{A}(t_i)^T\right)^T$ and $\left(\boldsymbol{0}_3, 1, -\boldsymbol{l}^T_m\right)^T$. This shows that the rank defect is due to the presence of the unknown ambiguity vector $\boldsymbol{a}_{ij,1}$. The conclusion reads therefore that - in contrast to the pseudorange case - the relative position of the two receivers cannot be determined from carrier phase data of a single epoch only. Accordingly, we need a minimum of two epochs of carrier phase data. In that case the system of observation equations becomes

$$\begin{bmatrix}\Delta\boldsymbol{\phi}_{ij,1}(t_i)\\ \Delta\boldsymbol{\phi}_{ij,1}(t_j)\end{bmatrix} = \begin{bmatrix}\boldsymbol{A}(t_i) & \boldsymbol{l}_m & \boldsymbol{0} & \boldsymbol{0} & \boldsymbol{I}_m\\ \boldsymbol{0} & \boldsymbol{0} & \boldsymbol{A}(t_j) & \boldsymbol{l}_m & \boldsymbol{I}_m\end{bmatrix}\begin{bmatrix}\Delta\boldsymbol{r}_{ij}(t_i)\\ c\Delta dt^{l}_{ij}(t_i)\\ \Delta\boldsymbol{r}_{ij}(t_j)\\ c\Delta dt^{l}_{ij}(t_j)\\ \boldsymbol{a}_{ij,1}\end{bmatrix}. \tag{5.92}$$

Note that we have assumed an uninterrupted tracking of the satellites over the time span between t_i and t_j. Hence, the ambiguity vector $\boldsymbol{a}_{ij,1}$ appears in both sets of observation equations, namely in those of epoch t_i and in those of epoch t_j. Also note that we have assumed $\Delta\boldsymbol{r}_{ij}(t_j) \neq \Delta\boldsymbol{r}_{ij}(t_i)$. Hence, the position of at least one of the two receivers, i and j, is assumed to have changed between the epochs t_i and t_j.

Considering the design matrix of the system (5.92) we observe a rank deficiency of 1. The linear dependent combination of the column vectors of the design matrix is given by $\left(\mathbf{0},1,\mathbf{0},1,-\boldsymbol{l}_m^T\right)^T$. This shows that the information content of the carrier phase data of the two epochs is not sufficient to separately determine the two receiver clock parameters $c\Delta dt_{ij}^{\prime}(t_i)$ and $c\Delta dt_{ij}^{\prime}(t_j)$, and the full vector of ambiguities. One can eliminate this rank defect by lumping one of the two receiver clock parameters with the ambiguity vector. This gives

$$\begin{bmatrix} \Delta\phi_{ij,1}(t_i) \\ \Delta\phi_{ij,1}(t_j) \end{bmatrix} = \begin{bmatrix} A(t_i) & \mathbf{0} & \mathbf{0} & \boldsymbol{I}_m \\ \mathbf{0} & A(t_j) & \boldsymbol{l}_m & \boldsymbol{I}_m \end{bmatrix} \begin{bmatrix} \Delta\boldsymbol{r}_{ij}(t_i) \\ \Delta\boldsymbol{r}_{ij}(t_j) \\ c\Delta dt_{ij}^{\prime}(t_i,t_j) \\ \boldsymbol{a}_{ij,1} + \boldsymbol{l}_m c\Delta dt_{ij}^{\prime}(t_i) \end{bmatrix} \tag{5.93}$$

with $c\Delta dt_{ij}^{\prime}(t_i,t_j) = c\Delta dt_{ij}^{\prime}(t_j) - c\Delta dt_{ij}^{\prime}(t_i)$. This shows that one can only determine the difference in time of the receiver clock parameters. Note that $c\Delta dt_{ij}^{\prime}(t_i,t_j)$ becomes independent of the instrumental receiver delays, if these delays are constant in time. Also note that the redundancy of the above system equals m-7. This shows that 7 satellites are minimally needed for a unique solution.

The above system (5.93) can be reduced to a form that closely resembles that of the system of pseudorange observation equations (5.87). To see this, first note that the $2m$ observation equations of (5.93) can be reduced by m if one is only interested in determining the two position differences $\boldsymbol{r}_{ij}(t_i)$ and $\boldsymbol{r}_{ij}(t_j)$. By taking the difference in time of the carrier phases one eliminates m observation equations as well as m ambiguities. As a result one obtains from (5.93):

$$\Delta\phi_{ij,1}(t_i,t_j) = \left(A(t_i,t_j) \; A(t_j) \; \boldsymbol{l}_m\right) \begin{bmatrix} \Delta\boldsymbol{r}_{ij}(t_i) \\ \Delta\boldsymbol{r}_{ij}(t_i,t_j) \\ \Delta dt_{ij}^{\prime}(t_i,t_j) \end{bmatrix} , \tag{5.94}$$

with $\Delta\phi_{ij,1}(t_i,t_j) = \Delta\phi_{ij,1}(t_j) - \Delta\phi_{ij,1}(t_i)$, $A(t_i,t_j) = A(t_j) - A(t_i)$ and $\Delta\boldsymbol{r}_{ij}(t_i,t_j) = \Delta\boldsymbol{r}_{ij}(t_j) - \Delta\boldsymbol{r}_{ij}(t_i)$.

So far the position of at least one of the two receivers was allowed to change in time. If we assume the two receivers to be stationary, however, we have $\Delta \boldsymbol{r}_{ij}(t_i) = \Delta \boldsymbol{r}_{ij}(t_j) = \Delta \boldsymbol{r}_{ij}$, in which case (5.94) reduces to

$$\Delta\phi_{ij,1}(t_i,t_j) = \begin{pmatrix} \boldsymbol{A}(t_i,t_j) & \boldsymbol{l}_m \end{pmatrix} \begin{bmatrix} \Delta \boldsymbol{r}_{ij} \\ c\Delta dt^{/}_{ij}(t_i,t_j) \end{bmatrix} . \tag{5.95}$$

Hence, through the elimination of the ambiguity vector we obtained a system of carrier phase observation equations that resembles that of the system of pseudorange observation equations (5.87). Due to the additional condition that $\Delta \boldsymbol{r}_{ij}(t_i) = \Delta \boldsymbol{r}_{ij}(t_i) = \Delta \boldsymbol{r}_{ij}$, the redundancy of (5.95) has increased by 3 when compared to the redundancy of (5.93). Hence, 4 satellites are now minimally needed for a unique solution.

In subsection 5.6.3 we will continue our discussion of the system of carrier phase observation equations and in particular consider the complicating factor that in case of GPS, the receiver-satellite configurations only change slowly with time. First, however, we will consider the dual-frequency case.

The Dual-Frequency Case. In analogy of the dual-frequency pseudorange case, we will take as our starting point the following three differenced carrier phases

$$\begin{aligned} \Phi^k_{ij,1} &= \Phi^k_{j,1} - \Phi^k_{i,1} \\ \Phi^k_{ij,2} &= \Phi^k_{j,2} - \Phi^k_{i,2} \\ \Phi^k_{j,12} &= \Phi^k_{j,2} - \Phi^k_{j,1} \, . \end{aligned}$$

Again, the first two are between-receiver differences, one for each of the two frequencies, whereas the last one is a between-frequency difference. In order to deal with the ionospheric delays separately, we now transform - in analogy of section 5.4.2.3 - these differenced carrier phases by

$$\begin{bmatrix} \frac{\alpha}{(\alpha-1)} & \frac{-1}{(\alpha-1)} & 0 \\ \frac{1}{(\alpha-1)} & \frac{-1}{(\alpha-1)} & 0 \\ 0 & 0 & \frac{-1}{(\alpha-1)} \end{bmatrix} .$$

As a result this gives us the following observation equations

$$\begin{aligned} \Delta\Phi^k_{ij,1}(t_i) - \Delta\Phi^k_{ij,12}(t_i)/(\alpha-1) &= -\left(\boldsymbol{u}^k_j(t_i)\right)^T \Delta \boldsymbol{r}_{ij}(t_i) + c\Delta dt^{//}_{ij}(t_i) + b^k_{ij} \\ -\Delta\Phi^k_{j,12}(t_i)/(\alpha-1) &= \Delta I^k_{ij,1}(t_i) - c\delta_{ij,12}/(\alpha-1) + c^k_{ij} \\ -\Delta\Phi^k_{j,12}(t_i)/(\alpha-1) &= \Delta I^k_{j,1}(t_i) - c(\delta_{j,12} + \delta^k_{,12})/(\alpha-1) + c^k_j \end{aligned} \tag{5.96}$$

with $dt_{ij}^{//} = dt_{ij} + (\alpha\delta_{ij,1} - \delta_{ij,2})/(\alpha - 1)$ and $b_{ij}^{k} = [\alpha a_{ij,1}^{k} - a_{ij,2}^{k}]/(\alpha - 1)$, $c_{ij}^{k} = [a_{ij,1}^{k} - a_{ij,2}^{k}]\ /(\alpha - 1)$ and $c_{j}^{k} = [a_{j,1}^{k} - a_{j,2}^{k}]/(\alpha - 1)$.

Compare with (5.88). If we are only interested in positioning, the first of the above three observation equations can be used to obtain in vector-matrix form the system

$$\Delta\phi_{ij,1}(t_i) - \Delta\phi_{ij,12}(t_i)/(\alpha - 1) = \left(A(t_i)\ \boldsymbol{l}_m\ \boldsymbol{I}_m\right) \begin{bmatrix} \Delta\boldsymbol{r}_{ij}(t_i) \\ c\Delta dt_{ij}^{//}(t_i) \\ \boldsymbol{b}_{ij} \end{bmatrix} . \tag{5.97}$$

Compare with (5.91) and note the correspondence in structure of the system of observation equations. Hence, the remarks made with respect to the system (5.91) also apply to the above system of equations.

5.6.3 On The Slowly Changing Receiver-Satellite Geometry.

We will now continue our discussion of the system of carrier phase observation equations (5.93). We will only consider the single-frequency case. The dual-frequency case is left to the reader. For the purpose of this subsection we extend the system (5.93) by including data from a third observational epoch t_k. The corresponding system of observation equations becomes then

$$\begin{bmatrix} \Delta\phi_{ij,1}(t_i) \\ \Delta\phi_{ij,1}(t_j) \\ \Delta\phi_{ij,1}(t_k) \end{bmatrix} = \begin{bmatrix} A(t_i) & 0 & 0 & 0 & 0 & \boldsymbol{I}_m \\ 0 & A(t_j) & \boldsymbol{l}_m & 0 & 0 & \boldsymbol{I}_m \\ 0 & 0 & 0 & A(t_k) & \boldsymbol{l}_m & \boldsymbol{I}_m \end{bmatrix} \begin{bmatrix} \Delta\boldsymbol{r}_{ij}(t_i) \\ \Delta\boldsymbol{r}_{ij}(t_j) \\ c\Delta dt_{ij}^{/}(t_i,t_j) \\ \Delta\boldsymbol{r}_{ij}(t_k) \\ c\Delta dt_{ij}^{/}(t_i,t_k) \\ \boldsymbol{a}_{ij,1} + \boldsymbol{l}_m c\Delta dt_{ij}^{/}(t_i) \end{bmatrix} . \tag{5.98}$$

This system has been formulated for three epochs of data, t_i, t_j and t_k. But it will be clear, that it can be easily extended to cover more epochs of data. When the number of epochs equals T, the redundancy of the system becomes $(T-1)(m-4)-3$. This shows that a minimum of 7 satellites needs to be tracked, when only two epochs of data are used. And when three epochs of data are used, the minimum of satellites to be tracked equals 6. The redundancy increases of course when two or more of the unknown baselines are identical. For instance, when both receivers are assumed to be stationary over the whole observational time span, then all the baselines are identical and the redundancy becomes $(T-1)(m-1)-3$. In that case a minimum of 4 satellites needs to be tracked, when only two epochs of data are used.

When we consider the design matrix of the above system, we note that it is still rank defect when the three matrices $A(t_i)$, $A(t_j)$ and $A(t_k)$ are identical. The linear dependent combinations of the column vectors of the design matrix that define the rank defect are given by $\left(\boldsymbol{I}_3, \boldsymbol{I}_3, 0, \boldsymbol{I}_3, 0, -A(t_i)^T\right)^T$. We also note that this rank defect is absent when at least two out of the three matrices differ, i.e., when either $A(t_i) \neq A(t_j)$, $A(t_j) \neq A(t_k)$ or $A(t_i) \neq A(t_k)$.

In our discussion of the pseudorange case the above type of rank defect did not occur, simply because one single epoch of pseudorange data is in principle sufficient for solving $\Delta \boldsymbol{r}_{ij}$. In the carrier phase case, however, we need - due to the presence of the unknown ambiguities - a minimum of two epochs of data. This is why in the carrier phase case, somewhat closer attention needs to be paid to the time dependency of the receiver-satellite geometry.

In this subsection four strategies will be discussed that can be used to overcome the above mentioned rank deficiency problem. These four strategies can either be used on a stand alone basis or in combination with one another. They can be characterized as follows:

(i) use of long observational time spans
(ii) using the antenna swap technique
(iii) starting from a known baseline
(iv) using integer ambiguity fixing

(i) The use of long observational time spans. Strictly speaking we of course have $A(t_i) \neq A(t_j)$, when $t_i \neq t_j$. But it will also be clear, although a strict rank defect is absent when $A(t_i) \neq A(t_j)$ holds true, that near rank deficiencies will be present when $A(t_i) \cong A(t_j) \cong A(t_k)$. And if this so happens to be the case, the parameters of (5.98) will be very poorly estimable indeed. One way to avoid near rank deficiencies of the above type, is to ensure that at least two out of the three receiver-satellite geometries, which are captured in the three matrices $A(t_i)$, $A(t_j)$, $A(t_k)$, are sufficiently different. Assuming that $t_i < t_j < t_k$, this implies, since the receiver-satellite configuration changes only slowly with time due to the high altitude orbits of the GPS satellites, that the time span between the two epochs t_i and t_k should be sufficiently large.

(ii) The use of the 'antenna swap' technique. Instead of using a long observational time span so as to ensure that the receiver-satellite geometry has changed sufficiently, the 'antenna swap' technique solves the problem of near rank deficiency by the artifice of moving what is normally the stationary antenna i to the initial position of the moving antenna j while, at the same time, moving the mobile antenna from its initial position to the position of the stationary antenna. The implications of this 'antenna swap' technique are best explained by

referring to the system of carrier phase observation equations (5.98). Before the 'antenna swap', the carrier phases of epoch t_i refer to the baseline $\boldsymbol{r}_{ij}(t_i)$ and after the 'antenna swap' the carrier phases of epoch t_j refer to the baseline $\boldsymbol{r}_{ij}(t_j)$. The swapping of the two antennas implies now that these two baselines are identical apart from a change of sign. Hence, $\boldsymbol{r}_{ij}(t_j) = -\boldsymbol{r}_{ij}(t_i)$. Note that the other parameters in the observation equations remain unchanged, since an uninterrupted tracking of the satellites is still assumed. The with the 'antenna swap' technique corresponding system of observation equations follows therefore from substituting $\boldsymbol{r}_{ij}(t_j) = -\boldsymbol{r}_{ij}(t_i)$ into (5.98) as

$$\begin{bmatrix} \Delta\phi_{ij,1}(t_i) \\ \Delta\phi_{ij,1}(t_j) \\ \Delta\phi_{ij,1}(t_k) \end{bmatrix} = \begin{bmatrix} A(t_i) & 0 & 0 & 0 & I_m \\ -A(t_j) & l_m & 0 & 0 & I_m \\ 0 & 0 & A(t_k) & l_m & I_m \end{bmatrix} \begin{bmatrix} \Delta\boldsymbol{r}_{ij}(t_i) \\ c\Delta dt_{ij}^{/}(t_i,t_j) \\ \Delta\boldsymbol{r}_{ij}(t_k) \\ c\Delta dt_{ij}^{/}(t_i,t_k) \\ \boldsymbol{a}_{ij,1} + \boldsymbol{l}_m c\Delta dt_{ij}^{/}(t_i) \end{bmatrix} . \tag{5.99}$$

This system is now still of full rank even when $A(t_i) = A(t_j) = A(t_k)$. Note that the system can be solved recursively if so desired. First, the first two sets of equations are used to solve for $\Delta\boldsymbol{r}_{ij}(t_i)$, $c\Delta dt_{ij}^{/}(t_i,t_j)$ and $\left(\boldsymbol{a}_{ij,1} + \boldsymbol{l}_m c\Delta dt_{ij}^{/}(t_i)\right)$. One could call this the initialization step. Then, the estimate of $\left(\boldsymbol{a}_{ij,1} + \boldsymbol{l}_m c\Delta dt_{ij}^{/}(t_i)\right)$ together with the carrier phase data of the following epoch, $\Delta\phi_{ij,1}(t_k)$, are used to solve for $\Delta\boldsymbol{r}_{ij}(t_k)$ and $c\Delta dt_{ij}^{/}(t_i,t_k)$. In this way one can continue for the next and following epochs as well. The advantage of the 'antenna swap' technique over the first approach is thus clearly that it allows for a reduction in the total observation time.

(iii) The use of a known baseline. Still another approach to deal with the near rank deficiency is to make use of a known baseline. This method therefore requires that in the vicinity of the survey area at least two stations with accurately known coordinates are available. With the baseline $\boldsymbol{r}_{ij}(t_i)$ known, we have $\Delta\boldsymbol{r}_{ij}(t_i) = \boldsymbol{0}$, from which it follows that (5.98) reduces to

$$\begin{bmatrix} \Delta\phi_{ij,1}(t_i) \\ \Delta\phi_{ij,1}(t_j) \\ \Delta\phi_{ij,1}(t_k) \end{bmatrix} = \begin{bmatrix} 0 & 0 & 0 & 0 & I_m \\ A(t_j) & l_m & 0 & 0 & I_m \\ 0 & 0 & A(t_k) & l_m & I_m \end{bmatrix} \begin{bmatrix} \Delta\boldsymbol{r}_{ij}(t_j) \\ c\Delta dt_{ij}^{/}(t_i,t_j) \\ \Delta\boldsymbol{r}_{ij}(t_k) \\ c\Delta dt_{ij}^{/}(t_i,t_k) \\ \boldsymbol{a}_{ij,1} + \boldsymbol{l}_m c\Delta dt_{ij}^{/}(t_i) \end{bmatrix} . \tag{5.100}$$

This system is clearly of full rank even when $A(t_i) = A(t_j) = A(t_k)$. Also this system can be solved recursively.

(iv) The use of integer ambiguity fixing. As was mentioned earlier the pseudorange case is not affected by the rank deficiencies caused by the slowly changing receiver-satellite geometry. This is simply due to the absence of the ambiguities in the pseudorange observation equations. The idea behind the present approach is therefore to find a way of removing the unknown ambiguities from the system of carrier phase observation equations. This turns out to be possible if one makes use of the fact that the so-called double-difference ambiguities are integer-valued.

The lumped parameter vector $\left(\boldsymbol{a}_{ij,1} + \boldsymbol{l}_m c\Delta dt_{ij}^{\prime}(t_i)\right)$ in (5.98), has entries which all are real-valued. It is possible, however, to reparametrize this m-vector such that a new vector is obtained of which only one entry is real-valued. The remaining $(m-1)$-number of entries of this transformed parameter vector will then be integer-valued. The transformed parameter vector is defined as

$$\begin{bmatrix} a \\ \boldsymbol{N} \end{bmatrix} = (\boldsymbol{l}_m \;\; \boldsymbol{D})^{*}\left(\boldsymbol{a}_{ij,1} + \boldsymbol{l}_m c\Delta dt_{ij}^{\prime}(t_i)\right)/\lambda_1 \tag{5.101}$$

in which $(\boldsymbol{l}_m, \boldsymbol{D})$ is an m-by-m matrix of full rank and matrix $\boldsymbol{D}$ is of the order m-by-$(m-1)$ with the structure

$$\boldsymbol{D}^{T} = \begin{bmatrix} 1 & & & -1 & & & \\ & \ddots & & \vdots & & & \\ & & 1 & -1 & & & \\ & & & -1 & 1 & & \\ & & & \vdots & & \ddots & \\ & & & -1 & & & 1 \end{bmatrix}.$$

The scalar parameter a in (5.101) is real-valued, but all the entries of the $(m-1)$-vector $\boldsymbol{N}$ are integer-valued. The integerness of the entries of $\boldsymbol{N}$ can be verified as follows. Since $\boldsymbol{D}^{*}\boldsymbol{l}_m = 0$, it follows from (5.101) that $\boldsymbol{N} = \boldsymbol{D}^{*}\boldsymbol{a}_{ij,1}/\lambda_1$, which shows that the entries of $\boldsymbol{N}$ are simply differences of the integer single-difference ambiguities $N_{ij,1}^{k}$. The entries of $\boldsymbol{N}$ are therefore referred to as double-difference ambiguities and they are integer-valued.

The inverse of (5.101) reads

$$\boldsymbol{a}_{ij,1} + \boldsymbol{l}_m c\Delta dt_{ij}^{\prime}(t_i) = \lambda_1 \boldsymbol{l}_m a/m + \lambda_1 \boldsymbol{D}(\boldsymbol{D}^{*}\boldsymbol{D})^{-1}\boldsymbol{N}. \tag{5.102}$$

If we substitute (5.101) into (5.98) we obtain

$$\begin{bmatrix} \Delta\phi_{ij,1}(t_i) \\ \Delta\phi_{ij,1}(t_j) \\ \Delta\phi_{ij,1}(t_k) \end{bmatrix} = \begin{bmatrix} A(t_i) & \mathbf{0} & \mathbf{0} & \mathbf{0} & \mathbf{0} & \boldsymbol{l}_m & \lambda_1 \boldsymbol{D}(\boldsymbol{D}^*\boldsymbol{D})^{-1} \\ \mathbf{0} & A(t_j) & \boldsymbol{l}_m & \mathbf{0} & \mathbf{0} & \boldsymbol{l}_m & \lambda_1 \boldsymbol{D}(\boldsymbol{D}^*\boldsymbol{D})^{-1} \\ \mathbf{0} & \mathbf{0} & \mathbf{0} & A(t_k) & \boldsymbol{l}_m & \boldsymbol{l}_m & \lambda_1 \boldsymbol{D}(\boldsymbol{D}^*\boldsymbol{D})^{-1} \end{bmatrix} \begin{bmatrix} \Delta \boldsymbol{r}_{ij}(t_i) \\ \Delta \boldsymbol{r}_{ij}(t_j) \\ c\Delta dt_{ij}^{l}(t_i,t_j) \\ \Delta \boldsymbol{r}_{ij}(t_k) \\ c\Delta dt_{ij}^{l}(t_i,t_k) \\ (\lambda_1 a/m) \\ \boldsymbol{N} \end{bmatrix} . \tag{5.103}$$

With this reparametrized system of carrier phase observation equations we are now in the position to make explicit use of the fact that all entries of $\boldsymbol{N}$ are integer-valued. It will be clear that this additional information strengthens the above system of observation equations in the sense that it puts additional constraints on the admissible solution space of the parameters.

Very sophisticated and successful methods have been developed for determining the integer-values of the double-difference ambiguities (the theory and concepts of integer ambiguity fixing is treated in chapter 8). Once these integer ambiguities are fixed, the above system of carrier phase observation equations becomes of full rank and reads

$$\begin{bmatrix} \Delta\phi_{ij,1}(t_i) - \lambda_1 \boldsymbol{D}(\boldsymbol{D}^*\boldsymbol{D})^{-1}\boldsymbol{N} \\ \Delta\phi_{ij,1}(t_j) - \lambda_1 \boldsymbol{D}(\boldsymbol{D}^*\boldsymbol{D})^{-1}\boldsymbol{N} \\ \Delta\phi_{ij,1}(t_k) - \lambda_1 \boldsymbol{D}(\boldsymbol{D}^*\boldsymbol{D})^{-1}\boldsymbol{N} \end{bmatrix} = \begin{bmatrix} A(t_i) & \mathbf{0} & \mathbf{0} & \mathbf{0} & \mathbf{0} & \boldsymbol{l}_m \\ \mathbf{0} & A(t_j) & \boldsymbol{l}_m & \mathbf{0} & \mathbf{0} & \boldsymbol{l}_m \\ \mathbf{0} & \mathbf{0} & \mathbf{0} & A(t_k) & \boldsymbol{l}_m & \boldsymbol{l}_m \end{bmatrix} \begin{bmatrix} \Delta \boldsymbol{r}_{ij}(t_i) \\ \Delta \boldsymbol{r}_{ij}(t_j) \\ c\Delta dt_{ij}^{l}(t_i,t_j) \\ \Delta \boldsymbol{r}_{ij}(t_k) \\ c\Delta dt_{ij}^{l}(t_i,t_k) \\ (\lambda_1 a/m) \end{bmatrix} . \tag{5.104}$$

For positioning purposes the primary parameters of interest are of course the baseline vectors $\Delta \boldsymbol{r}_{ij}(t_i)$, $\Delta \boldsymbol{r}_{ij}(t_j)$, $\Delta \boldsymbol{r}_{ij}(t_k)$. It is possible to reduce the above system of observation equations to one in which as parameters only the baseline vectors appear. If we premultiply each of the three m-vectors of observables in (5.104) with the $(m-1)$-by-m matrix $\boldsymbol{D}^*$, the above system reduces, because of $\boldsymbol{D}^*\boldsymbol{l}_m = \mathbf{0}$, to

$$\begin{bmatrix} \Delta \boldsymbol{D}^{*}\boldsymbol{\phi}_{ij,1}(t_i) - \lambda_1 \ \boldsymbol{N} \\ \Delta \boldsymbol{D}^{*}\boldsymbol{\phi}_{ij,1}(t_j) - \lambda_1 \ \boldsymbol{N} \\ \Delta \boldsymbol{D}^{*}\boldsymbol{\phi}_{ij,1}(t_k) - \lambda_1 \ \boldsymbol{N} \end{bmatrix} = \begin{bmatrix} \boldsymbol{D}^{*}A(t_i) & 0 & 0 \\ 0 & \boldsymbol{D}^{*}A(t_j) & 0 \\ 0 & 0 & \boldsymbol{D}^{*}A(t_k) \end{bmatrix} \begin{bmatrix} \Delta \boldsymbol{r}_{ij}(t_i) \\ \Delta \boldsymbol{r}_{ij}(t_j) \\ \Delta \boldsymbol{r}_{ij}(t_k) \end{bmatrix} . \qquad (5.105)$$

In (5.104) the elements of $\boldsymbol{\phi}_{ij}$ are referred to as single-differenced carrier phases, whereas in (5.105) the elements of $\boldsymbol{D}^{*}\boldsymbol{\phi}_{ij}$ are referred to as double-differenced carrier phases. When we compare (5.105) with (5.104) we note that the number of observables of (5.104) equals $3m$, whereas the number of observables of (5.105) equals $3m$-3. Hence, 3 observation equations have been eliminated in our transformation from (5.104) to (5.105). At the same time however, also 3 unknown parameters have been eliminated. They are the two clock parameters $\Delta dt_{ij}^{l}(t_i,\ t_j)$ and $\Delta dt_{ij}^{l}(t_i,\ t_k)$, and the real-valued scalar ambiguity a. The two systems of observations equations, (5.104) and (5.105), are therefore equivalent in the sense that the redundancy has been retained under the transformation. Both systems will therefore give identical estimates for the unknown baseline vectors.

5.7 SUMMARY

In this introductory chapter, the GPS observation equations were derived and a conceptual overview of their use for positioning was given.

In section 5.2 the basic GPS observables, being the pseudorange observable and the carrier phase observable, were introduced. It was shown how these observables can be parametrized in geometrically and physically meaningful quantities. In this parametrization, leading to the nonlinear observation equations, it has been attempted to include all significant terms.

Certain linear combinations of the GPS observables were studied in section 5.3. Some of these were single-receiver linear combinations, while others were dual-receiver linear combinations. The former are particularly useful for single-receiver nonpositioning GPS data analysis (cf. section 5.4), while the latter are used for relative positioning applications (cf. section 5.7).

Based on different subsets of the GPS observables, estimability and redundancy aspects of the single-receiver linear combinations were discussed in section 5.4. Time series of these linear combinations, possibly expanded with an additional modelling for time dependency, usually form the basis for single-receiver (e.g. GPS reference station) quality control and integrity monitoring.

In section 5.5, a linearization with respect to the relevant geometric unknown parameters of the nonlinear observation equations was carried out. Based on this linearization, both the single-point positioning and relative positioning concept was discussed. The advantages of relative positioning over the single-point positioning concept were shown.

The relative positioning concept was further explored in section 5.6. Based on single- or dual-frequency, pseudoranges or carrier phases, this final section presented a conceptual overview of the corresponding relative positioning models. In order to obtain a better understanding of the implications of the different structures of these models, particular attention was given to aspects of estimability, redundancy and rank deficiency.

References

ARINC (1991), *Interface Control Document ICD-GPS-200 Rev. B-PR. ARINC* Research Corporation, 11770 Warner Avenue Suite 210, Fountain Valley, CA 92708.

Brunner, F.K. (1988), *Atmospheric Effects on Geodetic Space Measurements*. Monograph 12, School of Surveying, University of New South Wales, Kensington, Australia.

Hofmann-Wellenhof, B., H. Lichtenegger, J. Collins (1997), *GPS: Theory and Practice*, 4th edition, Springer Verlag, New York.

Leick, A (1990), *GPS Satellite Surveying*. John Wiley & Sons, New York.

Seeber, G. (1993), *Satellite Geodesy*. Walter de Gruyter, Berlin.

Wells, D., N. Beck, D. Delikaraoglou, A. Kleusberg, E. Krakiwsky, G. Lachapelle, R. Langley, M. Nakiboglou, K.-P. Schwarz, J. Tranquilla and P. Vanicek (1987), *Guide to GPS Positioning*. Canadian GPS Associates, Fredericton, N.B. Canada, E3B 5A3.

6. GPS DATA PROCESSING METHODOLOGY: FROM THEORY TO APPLICATIONS

Geoffrey Blewitt
Department of Geomatics, University of Newcastle,
Newcastle upon Tyne, NE1 7RU, United Kingdom

6.1 INTRODUCTION

The idea behind this chapter is to use a few fundamental concepts to help develop a way of thinking about GPS data processing that is intuitive, yet has a firm theoretical foundation. Intuition is based on distilling an alarming array of information into a few core concepts that are basically simple. The fundamental concepts I have chosen to explore and develop here are generally based on equivalence principles and symmetry in problems. This involves looking at the same thing from different ways, or looking at apparently different things in the same way. Using symmetry and equivalence, we can often discover elegant explanations to problems.

The ultimate goal is that, the reader will be able to see answers to apparently complicated questions from first principles. An immediate goal, is to use this theoretical-intuitive approach as a vehicle to introduce a broad variety of algorithms and their application to high precision geodesy.

6.1.1 Background

It is useful to begin by placing this work into context briefly, by listing some of the common features of GPS data processing for high precision geodesy:

User Input Processing

- operating system interface
- interactive user control
- automation (batch control, defaults, contingency rules, etc.)
- help (user manual, on-line syntax help, on-line module guide, etc.)

Data Preprocessing

- GPS observation files, site database, Earth rotation data, satellite ephemerides, surface meteorological data, water vapor radiometer data formatting
- tools (satellite removal, data windowing, concatenation, etc.)
- editing (detecting and removing outliers and cycle slips)
- thinning (data decimation, data smoothing)
- data transformation (double differencing, ionosphere-free combination, etc.)
- ambiguity initialization (and possible resolution)

Observation Models
- nominal values for model parameters
- (observed - computed) observations and partial derivatives
- orbit dynamics and satellite orientation
- Earth rotation and surface kinematics
- media propagation (troposphere, ionosphere)
- clocks
- relativistic corrections (clocks, space-time curvature)
- antenna reference point and phase centre offset
- antenna kinematics
- phase modeling (phase centre variation, polarization, cycle ambiguity)

Parameter Estimation
- parameter selection
- stochastic model and a priori constraints
- inversion (specific algorithms, filtering, blocking techniques, etc.)
- residual analysis (outliers, cycle slips) and re-processing
- sensitivity analysis (to unestimated parameters)

Solution Processing
- a priori constraints
- ambiguity resolution
- solution combination and kinematic modelling
- frame projection and transformation tools
- statistics (formal errors, repeatability, in various coordinate systems, etc.)

Output Processing
- archive solution files
- information for the user
- export formatting (RINEX, SINEX, IONEX, etc.)
- presentation formatting (e.g., graphics)

6.1.2 Scope and Content

This chapter introduces some theoretical ideas behind GPS data processing, leading to discussions on how this theory relates to applications. It is certainly not intended to review specific software, but rather to point to concepts underlying the software.

Obviously, it would be beyond the scope of this text to go into each of the above items in detail. Observation models have already been covered in depth in previous chapters. I have therefore chosen to focus on three topics that generally lie within the areas of data preprocessing, parameter estimation, and solution processing.

I'll start with a very practical equivalence, the *equivalence of pseudorange and carrier phase*, which can be used to develop data processing algorithms. Then I explain what I mean by the *equivalence of the stochastic and functional model*, and show how this leads to different (but equivalent) methods of estimating parameters. Finally, I discuss *frame invariance and estimability* to (1) introduce geometry from a relativistic perspective, and (2) help the reader to distinguish between what can and what cannot be inferred from GPS data. In each case, I begin with a theoretical development of the concept, followed by a discussion, and then the ideas are used for a variety of applications.

6.2 EQUIVALENCE OF PSEUDORANGE AND CARRIER PHASE

To enhance intuition on the development of data processing algorithms, it can be useful to forget that carrier phase has anything to do with cycles, and instead think of it as a precise pseudorange with an additional bias. If we multiply the carrier phases from a RINEX file, which are in units of cycles, by their nominal wavelengths, the result is a set of data in distance units ($\Phi_1 \equiv \lambda_1 \varphi_1$ and $\Phi_2 \equiv \lambda_2 \varphi_2$). The advantage of expressing both pseudorange and carrier phase observables in the same units is that the symmetry in the observation equations is emphasized, thus assisting in our ability to visualize possible solutions to problems.

6.2.1 Theoretical Development

We start with an equation that will serve as a useful working model of the GPS observables, which can be manipulated to develop suitable data processing algorithms. In chapter 5, we see carrier phase developed in units of distance. Simplifying equations (5.23) and (5.32), the dual-frequency carrier phase and pseudorange data can be expressed in a concise and elegant form (where we purposely space the equations to emphasize the symmetry):

$$
\begin{aligned}
\Phi_1 &= \rho \qquad\qquad\quad - I + \lambda_1 N_1 + \delta m_1 \\
\Phi_2 &= \rho - (f_1/f_2)^2 I + \lambda_2 N_2 + \delta m_2 \\
P_1 &= \rho \qquad\qquad\quad + I \qquad\qquad + dm_1 \\
P_2 &= \rho + (f_1/f_2)^2 I \qquad\qquad + dm_2
\end{aligned}
\tag{6.1a}
$$

The reader should be aware that all of parameters in this equation are generally biased, so should not be interpreted literally except in a few special cases which will be discussed. The term ρ is the satellite-receiver range; but it is biased by clock errors, S/A, and tropospheric delay. It is often called the *non-dispersive delay* as it is identical for all four data types. The term I is the ionospheric group delay at the L1 frequency, which has the opposite sign as phase delay. It is a biased parameter, as the L1 and L2 signals are transmitted at slightly different times for different satellites. The terms N_1 and N_2 are the ambiguity parameters which, it should be remembered, are biased by initialization constants, and are therefore generally not integers; however they can change by integer amounts due to cycle slips. We call $\lambda_1 N_1$ and $\lambda_2 N_2$ the *carrier phase biases* (which have distance units). Finally, the last column of parameters are multipath terms, where it has been assumed that most of the error is due to multipath rather than receiver noise.

There are a few terms missing from equation (6.1a) which will be referred to below in a discussion on systematic errors. These errors will negligibly affect most algorithms developed from this equation, however, any limitations should be kept in mind.

Equation (6.1a) can be conveniently arranged into matrix form. Since this is really the same equation but in a matrix form, we denote it as equation (6.1b):

$$\begin{bmatrix} \Phi_1 \\ \Phi_2 \\ P_1 \\ P_2 \end{bmatrix} = \begin{bmatrix} 1 & -1 & 1 & 0 \\ 1 & -(f_1/f_2)^2 & 0 & 1 \\ 1 & +1 & 0 & 0 \\ 1 & +(f_1/f_2)^2 & 0 & 0 \end{bmatrix} \begin{bmatrix} \rho \\ I \\ \lambda_1 N_1 \\ \lambda_2 N_2 \end{bmatrix} + \begin{bmatrix} \delta m_1 \\ \delta m_2 \\ dm_1 \\ dm_2 \end{bmatrix} \tag{6.1b}$$

We note that the above equation has been arranged so that the coefficient matrix has no units. This proves to be convenient when analyzing the derived covariance matrix. It is worth commenting that, when performing numerical calculations, the coefficient for the L2 ionospheric delay should always be computed exactly using $f_1/f_2 \equiv 154/120$.

6.2.2 Discussion

Interpreting the Terms. As will now be explained, not only can we apply equation (6.1) to raw, undifferenced observation data, but also to single and double difference data, and to observation residuals. Depending on the application, the terms have different interpretations. In some cases, a particular term might have very predictable behavior; in others, it might be very unpredictable, and require stochastic estimation.

For example, in the case of the double difference observation equations, the ambiguity parameters N_1 and N_2 are not biased, but are truly integers. Moreover, the ionosphere parameter I is truly an unbiased (but differential) ionospheric parameter. For short enough distances and depending on various factors that affect the ionosphere, it might be adequate to ignore I when using double differences. Chapter 11 goes in this in more detail

Equation (6.1) might also be interpreted as a residual equation, where a model for the observations have been subtracted from the left hand side. In this case, the parameter terms are to be interpreted as residual offsets to nominal values. For example, if the equation is applied to double difference residuals, and if the differential tropospheric delay can be adequately modeled, then the range term ρ can be interpreted as a double difference range residual due to errors in the nominal station coordinates.

All parameters generally vary from one epoch to another, often unpredictably. Whether using undifferenced, single differenced, or double differenced data, any cycle slip or loss of lock that occurs will induce a change in the value of the ambiguity parameters, by exactly an integer. For undifferenced data, the range term ρ is typically extremely unpredictable due to the combined effects of S/A and receiver clock variation. The ionospheric delay I can often be predicted several minutes ahead using polynomials, but it can also exhibit wild fluctuations. Double differencing will lead to smooth, predictable behavior of ρ (for static surveying).

It is typical for carrier phase multipath, δm_1 and δm_2, to be at the level of a few millimetres, sometimes deviating as high as a few cm, level; whereas the level of pseudorange multipath dm_1 and dm_2 is generally greater by two orders of magnitude (decimetres to metres). It is extremely difficult to produce a functional model for multipath from first principles, and it is more typical to model it either empirically (from its daily repeating signature), or stochastically (which in its simplest form amounts to adjusting the data weight from stations with high multipath).

Using Equation (6.1). Can we use equation (6.1) to form a least-squares solution for the unknown parameters? Even if we interpret the multipath terms as residuals to be minimized, we would have 4 parameters at each epoch, and only 4 observational data. We can therefore construct an exact solution for each epoch, if we ignore the multipath terms. (Once again, we caution that any numerical computations should use exact values for $f_1/f_2 \equiv 154/120$).

$$
\begin{aligned}
\begin{bmatrix} \rho \\ I \\ \lambda_1 N_1 \\ \lambda_2 N_2 \end{bmatrix}
&= \begin{bmatrix} 1 & -1 & 1 & 0 \\ 1 & -\left(f_1/f_2\right)^2 & 0 & 1 \\ 1 & +1 & 0 & 0 \\ 1 & +\left(f_1/f_2\right)^2 & 0 & 0 \end{bmatrix}^{-1}
\begin{bmatrix} \Phi_1 \\ \Phi_2 \\ P_1 \\ P_2 \end{bmatrix} \\
&= \begin{bmatrix} 0 & 0 & +f_1^2/\left(f_1^2-f_2^2\right) & -f_2^2/\left(f_1^2-f_2^2\right) \\ 0 & 0 & -f_2^2/\left(f_1^2-f_2^2\right) & +f_2^2/\left(f_1^2-f_2^2\right) \\ 1 & 0 & -\left(f_1^2+f_2^2\right)/\left(f_1^2-f_2^2\right) & +2f_2^2/\left(f_1^2-f_2^2\right) \\ 0 & 1 & -2f_1^2/\left(f_1^2-f_2^2\right) & +\left(f_1^2+f_2^2\right)/\left(f_1^2-f_2^2\right) \end{bmatrix}
\begin{bmatrix} \Phi_1 \\ \Phi_2 \\ P_1 \\ P_2 \end{bmatrix} \qquad (6.2) \\
&\cong \begin{bmatrix} 0 & 0 & +2.546 & -1.546 \\ 0 & 0 & -1.546 & +1.546 \\ 1 & 0 & -4.091 & +3.091 \\ 0 & 1 & -5.091 & +4.091 \end{bmatrix}
\begin{bmatrix} \Phi_1 \\ \Phi_2 \\ P_1 \\ P_2 \end{bmatrix}
\end{aligned}
$$

Note that the carrier phase biases are constant until lock is lost on the satellite, or until a cycle slip occurs. We can therefore use these equations to construct algorithms that (1) resolve ambiguities, and (2) detect and solve for cycle slips. The second point to notice, is that between cycle slips, we know that the ambiguities are constant. If we are interested in only the variation in the other parameters (rather than the absolute values), then we are free to ignore any constant terms due to the ambiguity parameters.

We can rearrange equation (6.1) to reflect this idea, by attaching the ambiguity parameters to the carrier phase observations. Of course, we might not know these parameters perfectly, but that will have no effect on the estimated *variation* in the other parameters. Furthermore, we can explicitly introduce the pseudorange multipath terms into the parameter vector:

$$\begin{aligned}\begin{bmatrix}\tilde{\Phi}_1\\ \tilde{\Phi}_2\\ P_1\\ P_2\end{bmatrix} &\equiv \begin{bmatrix}\Phi_1-\lambda_1 N_1\\ \Phi_2-\lambda_2 N_2\\ P_1\\ P_2\end{bmatrix}\\ &= \begin{bmatrix}1 & -1 & 0 & 0\\ 1 & -(f_1/f_2)^2 & 0 & 0\\ 1 & +1 & 1 & 0\\ 1 & +(f_1/f_2)^2 & 0 & 1\end{bmatrix}\begin{bmatrix}\rho\\ I\\ dm_1\\ dm_2\end{bmatrix}+\begin{bmatrix}\delta m_1\\ \delta m_2\\ e_1\\ e_2\end{bmatrix}\end{aligned} \tag{6.3}$$

where we have explicitly included pseudorange measurement noises e_1 and e_2. As we did for equation (6.1), equation (6.3) can be inverted:

$$\begin{aligned}\begin{bmatrix}\rho\\ I\\ dm_1\\ dm_2\end{bmatrix} &= \begin{bmatrix}1 & -1 & 0 & 0\\ 1 & -(f_1/f_2)^2 & 0 & 0\\ 1 & +1 & 1 & 0\\ 1 & +(f_1/f_2)^2 & 0 & 1\end{bmatrix}^{-1}\begin{bmatrix}\tilde{\Phi}_1\\ \tilde{\Phi}_2\\ P_1\\ P_2\end{bmatrix}\\ &= \begin{bmatrix}+f_1^2/(f_1^2-f_2^2) & -f_2^2/(f_1^2-f_2^2) & 0 & 0\\ +f_2^2/(f_1^2-f_2^2) & -f_2^2/(f_1^2-f_2^2) & 0 & 0\\ -(f_1^2+f_2^2)/(f_1^2-f_2^2) & +2f_2^2/(f_1^2-f_2^2) & 1 & 0\\ -2f_1^2/(f_1^2-f_2^2) & +(f_1^2+f_2^2)/(f_1^2-f_2^2) & 0 & 1\end{bmatrix}\begin{bmatrix}\tilde{\Phi}_1\\ \tilde{\Phi}_2\\ P_1\\ P_2\end{bmatrix}\\ &\cong \begin{bmatrix}+2.546 & -1.546 & 0 & 0\\ +1.546 & -1.546 & 0 & 0\\ -4.091 & +3.091 & 1 & 0\\ -5.091 & +4.091 & 0 & 1\end{bmatrix}\begin{bmatrix}\tilde{\Phi}_1\\ \tilde{\Phi}_2\\ P_1\\ P_2\end{bmatrix}\end{aligned} \tag{6.4}$$

Notice the striking similarity in equations (6.2) and (6.4), and reversal of roles between carrier phase and pseudorange. One can see the familiar *ionosphere-free linear combination* of data as solutions for the range term; whereas in equation (6.2) it applies to pseudorange, in equation (6.4) it applies to carrier phase. Similarly, the ionospheric delay term is equal the familiar *ionospheric* or *geometry-free linear combination* of pseudorange in equation (6.2), and of carrier phase in equation (6.4).

The coefficients for the ambiguity and multipath estimates are symmetrical between equations (6.2) and (6.4). We can interpret this as follows. In equation (6.2), the pseudorange is being effectively used as a model for the carrier phase due in order to infer the carrier phase bias parameters. On the other hand, in equation (6.4) the carrier phase is effectively being used to model time variations in the pseudorange in order to infer pseudorange multipath variations. The symmetry of the coefficients in the two equations is therefore not surprising given this explanation.

Statistical Errors. Since the level of errors are strikingly higher for pseudorange as compared with carrier phase, we should look at the propagation of errors into the above parameters. The method used here is similar to the familiar computation of *dilution of precision* for point positioning. The covariance matrix for the parameter estimates given by equation (6.2) can be computed by the usual procedure as follows:

$$\mathbf{C} = \left(\mathbf{A}^{\mathrm{T}}\mathbf{C}_{\mathrm{data}}^{-1}\,\mathbf{A}\right)^{-1} \tag{6.5}$$

where $\mathbf{A}$ is the coefficient matrix in equations (6.1) or (6.3), and $\mathbf{C}_{\mathrm{data}} = \mathbf{W}^{-1}$ is the data covariance matrix. If we assume that the data covariance is diagonal, that there is no difference between the level of errors on L1 and L2, and that the variance for carrier phase is negligible compared to the pseudorange then we write the data covariance:

$$\mathbf{C}_{\mathrm{data}} = \lim_{\varepsilon \to 0} \begin{pmatrix} \varepsilon\sigma^2 & 0 & 0 & 0 \\ 0 & \varepsilon\sigma^2 & 0 & 0 \\ 0 & 0 & \sigma^2 & 0 \\ 0 & 0 & 0 & \sigma^2 \end{pmatrix} \tag{6.6}$$

Recall that in a real situation, a typical value might be $\varepsilon \approx 10^{-4}$, which justifies our simplification by taking the limit $\varepsilon \to 0$. Applying equations (6.5) and (6.6) to equation (6.2), and substituting values for the frequencies, we find the parameter covariance:

$$\mathbf{C}_{(6.2)} = \sigma^2 \begin{pmatrix} 8.870 & -6.324 & -15.194 & -19.286 \\ -6.324 & 4.779 & 11.103 & 14.194 \\ -15.194 & 11.103 & 26.297 & 33.480 \\ -19.286 & 14.194 & 33.480 & 42.663 \end{pmatrix} \tag{6.7}$$

The formal standard deviations for the parameters are the square root of the diagonal elements:

Table 6.1: Formal errors of parameter derived at a single epoch using dual frequency code and carrier phase data

Parameter	Standard Deviation
ρ	2.978σ
I	2.186σ
N_1	$5.128\,\sigma/\lambda_1$
N_2	$6.532\,\sigma/\lambda_2$

The formal error for the ionosphere-free range term is approximately 3 times the level of the measurement errors. This result also applies to the estimates of range variation in equation (6.4) which uses the ionosphere-free carrier phase data. It illustrates the problem for short baselines, where there is a trade-off between raising the effective measurement error, versus reducing systematic error from the ionosphere (see chapter 13). The formal error for the ionospheric delay (at the L1 frequency) is approximately 2 times the level of the measurement errors, which shows that the L1 and L2 signals are sufficiently well separated in frequency to resolve ionospheric delay. The large scaling factors of 5.128 to 6.532 for the carrier phase ambiguities shows that pseudorange multipath must be adequately controlled if there is any hope to resolve ambiguities (or detect cycle slips) using pseudorange data. For example, if we aim for an N_1 standard deviation of 0.25 cycles, then the pseudorange precision must be approximately 5 times smaller than this, which is less than 1 cm!

Systematic Errors. At this point, it is worth recalling that we have not used any functional model for the range term or the ionospheric term, other than that they satisfy the following assumptions:

- The range term (which includes range, tropospheric delay, and clock offsets) are identical for all observables.
- Ionospheric delay varies as the inverse square of the frequency, with the phase delay having the same magnitude but opposite sign to the group delay

Equations (6.2) and (6.4) tells us that we can form an estimators for the carrier phase ambiguities and pseudorange multipath variation, even in the extreme situation when we have no functional model for range, tropospheric delay, clocks, and ionospheric delay (other than the above simple assumptions). For example, no assumptions have been made concerning motion of the GPS antenna, and we can therefore derive algorithms to fix cycle slips and resolve carrier phase ambiguities that are suitable for kinematic applications. Similarly, pseudorange multipath can be assessed as an antenna is moved through the environment.

In the next section, we derive algorithms that can be considered application independent. This is only strictly true as far as the above assumptions are valid. For completeness, we list here reasons why the above assumptions might not be valid:

- The carrier signal is circularly polarized, and hence the model should really include the *phase wind up effect* caused by relative rotation between the GPS satellite's antenna and the receiver's antenna, Wu et al. [1993]. This is particularly important for moving antennas with data editing algorithms operating on undifferenced data. It has also happened in the past that one of the GPS satellites began to spin due to some malfunction, thus causing a dramatic phase wind up effect in the carrier phase, which was of course not observed in the pseudorange. One way around such problems is to use the widelane phase combination, which is rotationally invariant $\varphi_W \equiv (\varphi_1 - \varphi_2) = (\Phi_1/\lambda_1 - \Phi_2/\lambda_2)$. The phase wind up effect can be almost eliminated by double differencing, or by using an iterative procedure to account for antenna orientation (which can often be modelled adequately using a preliminary solution for the direction of motion). An interesting twist on this is to try to use the observed phase wind up to help with models of vehicle motion. For this purpose, the single differenced ionospheric phase could be used between a nearby reference antenna, and an antenna on a moving vehicle. Over short distances, the ionospheric delay would almost cancel, leaving a clear signature due to antenna rotation.
- The model should really include antenna phase centre offsets and antenna phase centre variation. We are free to define any such errors under the umbrella term *multipath*, but it is advisable to correct the data for such effects. Double differencing over short baselines almost eliminates phase centre effects, provided the same antenna types and method of mounting are used.
- There is generally a slight difference in the time of transmission for the L1 and L2 signals, which is different for each satellite (typically a few metres). Moreover, the effective time of reception might be slightly different in receivers which make no attempt to self-calibrate this interchannel delay. Certainly, for precise geodetic applications, such a bias is irrelevant, as it would either cancel in double differencing, or be harmlessly absorbed as a clock parameter; however, for ionospheric applications, these biases must be modelled.
- There might be a slight variable bias in the receiver between the different observable types due to any internal oscillators and electronics specific to the L1 and L2 frequencies. Hence the assumption that the range term is the same for all observables becomes invalid. Receivers are not supposed to do this, but

hardware problems have been known to cause this effect, which is often temperature dependent. The effect is mitigated by double differencing, but can be problematic for undifferenced processing, for data processing algorithms, and for ionospheric estimation software.

- There might be slight variable biases due to systematic error in the receiver, such as tracking loop errors that are correlated with the Doppler shift. For well designed, geodetic-class receivers, these biases should be down at the millimetre level.

6.2.3 Applications

Multipath Observables. This is the simplest and most obvious application from the previous discussion. Equation (6.4) shows how pseudorange multipath can be estimated epoch by epoch. It relies on the assumption that the carrier phase biases are constant. If not, then the data should first be edited to correct for any cycle slips. It should also be remembered that such multipath estimates are biased; therefore, only multipath variation and not the absolute multipath can be inferred by this method. This method is particularly useful for assessing the quality of the environment at a GPS station. This might be used for site selection, for example. Another application is to look at the multipath statistics. These could then be used to compute pseudorange data weights in least squares estimation, or for other algorithms that use the pseudorange.

Data Editing. Data editing includes the process of outlier detection, cycle slip detection, and cycle slip correction. Equations (6.1-6.4) point to a possible method for data editing, as it shows that the parameters are correlated, and therefore perhaps at each epoch, the set of four observations can be assessed for self-consistency. But outlier detection requires data redundancy, which we do not have for individual epochs.

However, we can monitor the solution for the parameters, equation (6.2), and ask whether they are behaving as expected. This line of thought leads naturally to a sophisticated approach involving a Kalman filter to predict the solution at the next epoch, and then compare this prediction with the new data. If the data rate is sufficiently high that prediction becomes meaningful, then this approach might be useful.

However, experience by the author and those developing the GIPSY software at the Jet Propulsion Laboratory showed this approach to be problematic, at least for undifferenced data. The presence of selective availability, the possible occurrence of high variability in ionospheric total electron content, and the poor frequency stability of receiver's internal oscillators limit the usefulness of Kalman filtering for typical geodetic data taken at a rate of 1 per 30 seconds. Even going to higher rates does not significantly improve the situation if the receiver clock is unpredictable. Moreover, results were difficult to reproduce if the analyst were allowed to tune the filter.

As a result, a simpler, fully automatic algorithm was developed known as

TurboEdit, see Blewitt [1990], which uses the positive aspects of filtering (i.e., noise reduction through averaging, and using prediction as a means of testing new data). The new algorithm attempted to minimize sensitivity to unusual, but acceptable circumstances, by automatically adapting its procedures to the detected level of noise in the data. The specific TurboEdit algorithm will not be described in detail here, but rather we shall next focus on some principles upon which data editing algorithms can be founded.

Firstly, we shall look at the use of pseudorange to help detect and correct for cycle slips. (In this context, by cycle slip we mean a discontinuity in the integer ambiguity parameter, which can be caused by the receiver incorrectly counting the cycles, or if the receiver loses lock on the signal). From Table 6.1, we see that a pseudorange error of 1 cm would result in an instantaneous estimate for the carrier phase biases at the level of 5 to 6 cm, which corresponds to approximately one quarter of a wavelength. Therefore, it would seem that pseudorange multipath would have to be controlled at this level if we were to simply use these estimates to infer whether the integer ambiguity had changed by one cycle. This seems like a dire situation.

This situation can be improved, if we realize three useful observations of experience: (1) Most often, cycle slips are actually caused by loss of lock, in which case the slip is much greater than one cycle; therefore, detection is simpler than correction. (2) Unless we have conditions so adverse that any result would be questionable, most often we have many epochs of data until we reach a cycle slip; therefore, we can average the results from these epochs to estimate the current value of the carrier phase bias. (3) If we have an initial algorithm that flags epochs where it suspects a cycle slip may have occurred, we can separately average the carrier phase bias solutions either side of the suspected cycle slip, and test the hypothesis that the carrier phase bias has changed by at least an integer number of wavelengths.

Following this logic, we can estimate how many epochs of data are required either side of the cycle slip so that the pseudorange errors can be averaged down sufficiently for us to test the hypothesis that a cycle slip has occurred. Using the results in Table 6.1, and assuming the errors average down as the square root of the number of epochs, we can, for example, write the error in the estimated cycle slip on L1 as:

$$\sigma_{\text{slip}}(n) = \sqrt{2}\,\frac{5.128\sigma}{\sqrt{n}} \tag{6.8}$$

where n is the number of epochs either side of the hypothesized cycle slip. For example, if we insist that this computed error be less than one quarter of a cycle, i.e., approximately 5 cm, and if we assume that the pseudorange error s is approximately 50 cm, then we see that the number of epochs must be greater than approximately 5000. This is clearly an unrealistic approach as it stands.

We can use the above approach, if instead we use the well known widelane carrier phase ambiguity. From equation (6.7) and using the law of propagation of errors, we can compute the formal variance in the widelane ambiguity,

$$N_W \equiv N_1 - N_2$$

$$\begin{aligned}
\mathbf{C}_W &= \sigma^2 \begin{pmatrix} 1/\lambda_1 & -1/\lambda_2 \end{pmatrix} \begin{pmatrix} 26.297 & 33.480 \\ 33.480 & 42.663 \end{pmatrix} \begin{pmatrix} 1/\lambda_1 \\ -1/\lambda_2 \end{pmatrix} \\
&= \sigma^2 \left(26.297/\lambda_1^2 - 2 \times 33.480/\lambda_1 \lambda_2 + 42.663/\lambda_2^2 \right) \\
&= \left(\sigma/\lambda_1 \right)^2 \left(26.297 - 66.960 \times \left(120/154 \right) + 42.663 \times \left(120/154 \right)^2 \right) \\
&= \left(0.15720 \sigma/\lambda_1 \right)^2
\end{aligned} \tag{6.9}$$

This derivation uses the exact relation $(\lambda_1/\lambda_2) = (f_2/f_1) = (120/154)$. (As a rule for numerical stability, it is always wise to substitute explicitly for the L1 carrier wavelength only at the very last step).

Remarkably, the standard error in the widelane wavelength does not reach 1 cycle until the pseudorange errors approach $\sigma = \lambda_1/0.15720 = 6.3613\lambda_1 \approx 120\text{cm}$. We can therefore use such widelane estimates on an epoch by epoch basis as an algorithm to flag possible cycle slips. The hypothesis can then be tested by averaging down the pseudorange noise either side of the proposed slip, as discussed previously.

Widelaning data editing methods are generally very successful for modern geodetic receivers, which have well behaved pseudorange. However, they do not distinguish as to whether the slip occurred on L1, L2, or both.

This problem can be resolved by looking at either the biased range parameter ρ or biased ionospheric parameter I in equation (6.4). Note that the carrier phase ambiguity parameters appear to the right side of this equation. Were there to be a cycle slip, it would manifest itself as a discontinuity in both parameters I and ρ. Using this method requires that either one of these parameters be predictable, to effectively bridge the time period during which the receiver lost lock on the signal. For double differenced data, this should be rather straightforward for both parameters, particularly for short baselines. For undifferenced data, parameter ρ tends to be too unpredictable due to S/A and receiver clock variation; however, I is usually sufficiently well behaved that time periods of up to several minutes can be bridged. A low order polynomial predictor could be used for this purpose.

Data editing algorithms can be designed to be adaptive to the changing quality of the data and the predictability of the parameters. The level of pseudorange noise can be easily monitored, as discussed, using equation (6.4) to estimate the multipath terms (taking care to correct for cycle slips detected so far).

The predictability of the parameters can be tested by applying the prediction algorithm backwards in time to *previous* data which are known to be clean or corrected for cycle slips. For example, if we have a loss of lock and subsequent data outage of 5 minutes, we might want to test a simple algorithm which predicts the I parameter using a second order polynomial on 15 minutes of data prior to the loss of lock. The test could be conducted by extrapolating the polynomial backwards, and comparing it with existing data.

Data Thinning. For static GPS where data are collected over a period of several hours, a carrier phase data rate of 1 epoch every 5 minutes should be more than sufficient to achieve high precision results. In fact, using higher rate data is unlikely to improve the result significantly. The reason for this is that if we continue to increase the data rate, we may well be able to reduce the contribution of measurement error to errors in the parameter estimates; however, we will do little to reduce the effect of systematic error, for example, low frequency components of multipath.

Therefore, if we are presented with a file with carrier phase data every 30 seconds, a simple and effective way to speed up the processing is to decimate the data, only accepting one point every 5 minutes.

For pseudorange data, however, a higher data rate often leads to improved results, presumably because measurement error and high frequency components of multipath continue to be significant error sources. A better approach than decimation would be to interpolate the high rate pseudorange data to every 5 minute data epoch, because the interpolation process would help average down the high frequency noise. For subsequent least squares analysis to be valid, the interpolator should strictly only independent 5 minute segments of high rate data, so that no artificial correlation's are introduced (which could, for example, confound other algorithms in your software).

A convenient method of interpolation is to use the carrier phase as a model for the pseudorange. The multipath expressions in Equation (6.4) provides us the solution to this problem. For example, we can rearrange (6.4) to express a model of the pseudorange data in terms of the carrier phase data and the pseudorange multipath:

$$\begin{aligned} P_1 &= 4.091\tilde{\Phi}_1 - 3.091\tilde{\Phi}_2 + dm_1 \\ &= 4.091\Phi_1 - 3.091\Phi_2 + dm_1 + B \end{aligned} \qquad (6.10)$$

The carrier phase here is effectively being used to mimic the time variation in the pseudoranges, correctly accounting for variation in range and ionospheric delay. The constant B is due to the (unknown) carrier phase biases. We can proceed to express an estimator for P_1 as the expected value:

$$\begin{aligned} \hat{P}_1 &\equiv E(P_1) \\ &= E(4.091\Phi_1 - 3.091\Phi_2 + dm_1 + B) \\ &= 4.091\Phi_1 - 3.091\Phi_2 + E(B) \end{aligned} \qquad (6.11)$$

where, for the carrier phase data, the expected values are simply the actual data recorded at the desired 5-minute epoch, and we have assumed that the expected value for multipath is zero. If we wish our resulting estimate $\hat{P}_1$ to be truly independent from one 5 minute epoch to the next, then $E(B)$ can only be based on data found in a 5 minute window surrounding this epoch, giving us the following expression:

$$\hat{P}_1 = 4.091\Phi_1 - 3.091\Phi_2 + \langle P_1 - 4.091\Phi_1 + 3.091\Phi_2 \rangle \tag{6.12}$$

(where the angled brackets denote the time average operator). The result is a smoothed estimate for the pseudoranges. Remember, only the smoothed pseudorange that falls on the 5 minute epoch should to be saved.

Parameter Estimation. It is common in geodetic software to first use the pseudorange to produce a receiver clock solution, and then use the double differenced carrier phase to produce a precise geodetic solution. The reason we need to know the receiver clock time precisely is to determine the time the data were collected, and hence fix the geometry of the model at that time. Once this has been done, the clock parameters are then effectively removed from the problem by double differencing. The disadvantage to this scheme, is that we might be interested in producing a high precision clock solution.

One way of approaching this is to estimate clock parameters explicitly along with the geodetic parameters, using undifferenced carrier phase data. The problem with this is that there is an extremely high correlation with the (undifferenced) carrier phase bias parameters.

An alternative, elegant one-step procedure is to process undifferenced carrier phase and pseudorange data simultaneously, effectively using the pseudorange to break the correlation. For example, suppose we process dual frequency ionosphere-free carrier phase and pseudorange together. The models for both types of observables are identical, apart from the carrier phase bias (which can, in any case, assume a nominal value of zero), and apart from the lack of a phase wind-up effect in the pseudorange data. Similarly, the parameters we estimate would be identical; that is, no extra parameters are required. Apart from expanding the possible applications (where clocks are involved), this method provides an extra degree of reliability, especially for kinematic applications, where the pseudorange effectively provides a consistency check on the carrier phase solution.

One very interesting new application of this idea is called *precise point positioning*. Developed by researchers at JPL, see Zumberge et al. [1996], this technique is identical to conventional pseudorange point positioning, except that (1) both pseudorange and carrier phase data are processed simultaneously, and (2) precise satellite ephemerides are used. Precise point positioning allows a single receiver to be positioned with 1 cm accuracy in the global frame (ITRF). We return to this exciting new tool later in this chapter.

Ambiguity Resolution. The application of equation (6.2) to ambiguity resolution is basically very similar to the application to data editing and the correction of cycle slips. It must be remembered, however, that only the double differenced carrier phase biases are an integer number of wavelengths. Therefore, equation (6.2) should be interpreted as for double differenced data and parameters. Alternatively, undifferenced parameters can be estimated, and subsequently the estimates can be double differenced.

As discussed for cycle slip correction, the pseudorange multipath too large for reliable ambiguity resolution using equation (6.2) directly. On the other hand, the widelane carrier phase ambiguity $N_W \equiv N_1 - N_2$ can be fixed very reliably using pseudoranges, even of mediocre quality. The advantage to this method is that it is independent of baseline length.

As with correcting cycle slips, we need to address how we resolve the ambiguities for N_1 and N_2 separately. Assuming we know the widelane, the problem reduces to finding the correct value for N_1. Once again, one possible answer lies in the solutions for the (double differenced) ionospheric term I. Using equation (6.4), and assuming we have a very good model for I, we can find the best fitting values of the ambiguities for $\tilde{\Phi}_1 \equiv \Phi_1 - \lambda_1 N_1$ and $\tilde{\Phi}_2 \equiv \Phi_2 - \lambda_2 N_2$, subject to the constraint $N_1 - N_2 = \hat{N}_W$, where $\hat{N}_W$ is the widelane ambiguity, previously resolved using equation (6.2)

$$
\begin{aligned}
I &= 1.546\tilde{\Phi}_1 - 1.546\tilde{\Phi}_2 \\
I/1.546 &= \left(\Phi_1 - \lambda_1 N_1\right) - \left(\Phi_2 - \lambda_2 N_2\right) \\
&= \Phi_1 - \Phi_2 - \lambda_1 N_1 + \lambda_2 N_2 \\
&= \Phi_1 - \Phi_2 - \lambda_2 \hat{N}_w + \left(\lambda_2 - \lambda_1\right) N_1
\end{aligned}
\tag{6.13}
$$

This situation is relatively easy over baselines of a few km, where it can be assumed that, to a good approximation, $I = 0$. However, the coefficient $\left(\lambda_2 - \lambda_1\right) \approx 5.4\text{cm}$ is very small, so we can easily run into problems over 5 km during daylight hours, and over 30 km at night. However, it is an almost an instantaneous technique, and was used successfully for rapid static surveying of post-seismic motion following the Loma Prieta earthquake of 1989, Blewitt et al. [1990].

Over longer baselines, Melbourne [1985] suggested an approach that uses the ionosphere-free phase combination of equation (6.4) and a good model for the range term. Later experience has shown that the range model must be based on a preliminary bias-free solution (since our a priori knowledge of the troposphere is generally inadequate). From equation (6.4), we can find the value of N_1 that best fits the range model, subject to the usual widelane constraint:

$$
\begin{aligned}
\rho &= 2.546\tilde{\Phi}_1 - 1.546\tilde{\Phi}_2 \\
&= 2.546\left(\Phi_1 - \lambda_1 N_1\right) - 1.546\left(\Phi_2 - \lambda_2 N_2\right) \\
&= 2.546\Phi_1 - 1.546\Phi_2 - 2.546\lambda_1 N_1 + 1.546\lambda_2 N_2 \\
&= 2.546\Phi_1 - 1.546\Phi_2 - 1.546\lambda_2 \hat{N}_w + \left(1.546\lambda_2 - 2.546\lambda_1\right) N_1
\end{aligned}
\tag{6.14}
$$

The coefficient $\left(1.546\lambda_2 - 2.546\lambda_1\right) \approx 10.7\text{cm}$ shows that this method will work provided we can control our estimated (double differenced) range errors to within a few centimetres. Using precise orbit determination and stochastic tropospheric estimation, this method has proved successful over thousands of km, see Blewitt [1989], and even over global scales, Blewitt and Lichten [1992].

6.3 EQUIVALENCE OF STOCHASTIC AND FUNCTIONAL MODELS

We are familiar with standard least squares theory, where the observations have both a functional model, which tells us how to compute the observation, and a stochastic model, which tells us the expected statistics of the errors. If we decide to augment the functional model with extra parameters, an equivalent result can be obtained if instead we modify the stochastic model. As we shall see, this equivalence introduces great flexibility into estimation algorithms, with a wide variety of geodetic applications.

6.3.1 Theoretical Development

Terminology. Consider the linearized observation equations:

$$z = Ax + v \tag{6.15}$$

where z is the column vector of observed minus computed observations, A is the design matrix, x is the column vector of corrections to functional model parameters, and v is a column vector of errors. Let us assume the stochastic model

$$\begin{aligned} E(v) &= 0 \\ E(vv^T) &= C \equiv W^{-1} \end{aligned} \tag{6.16}$$

Assuming a well conditioned problem, the *best linear unbiased estimator* of x is:

$$\hat{x} = \left(A^T W A\right)^{-1} A^T W z \tag{6.17}$$

which has the following statistical properties:

$$\begin{aligned} E(\hat{x}) &= E(x) = x \\ E(\hat{x}\hat{x}^T) &= \left(A^T W A\right)^{-1} \equiv C_{\hat{x}} \end{aligned} \tag{6.18}$$

If we use a Bayesian approach to estimation, we may make the a priori assumption $E(x) = 0$ where we implicitly introduce pseudo-data $x = 0$ with an a priori covariance C_0. In this case, the estimator becomes:

$$\hat{x} = \left(A^T W A + C_0^{-1}\right)^{-1} A^T W z \tag{6.17b}$$

We see that (6.17b) approaches (6.17) in the limit $C_0 \rightarrow \infty$, hence we can consider (6.17) the special case of (6.17b) where we have no a priori information.

Augmented Functional Model. Suppose we aim to improve our solution by estimating corrections to an extra set of functional model parameters y. We therefore consider the augmented observation equations:

$$z = Ax + By + v \tag{6.19}$$

We can write this in terms of partitioned matrices:

$$z = \begin{pmatrix} A & B \end{pmatrix} \begin{pmatrix} x \\ y \end{pmatrix} + v \tag{6.20}$$

We can therefore see by analogy with (6.17) that the solution for the augmented set of parameters will be

$$\begin{aligned} \begin{pmatrix} \hat{x} \\ \hat{y} \end{pmatrix} &= \left(\begin{pmatrix} A^T \\ B^T \end{pmatrix} W \begin{pmatrix} A & B \end{pmatrix} \right)^{-1} \begin{pmatrix} A^T \\ B^T \end{pmatrix} Wz \\ &= \begin{pmatrix} A^T WA & A^T WB \\ B^T WA & B^T WB \end{pmatrix}^{-1} \begin{pmatrix} A^T Wz \\ B^T Wz \end{pmatrix} \end{aligned} \tag{6.21}$$

We now use the following lemma on matrix inversion for symmetric matrices, which can easily be verified:

$$\begin{pmatrix} \Lambda_1 & \Lambda_{12} \\ \Lambda_{21} & \Lambda_2 \end{pmatrix}^{-1} = \begin{pmatrix} \left(\Lambda_1 - \Lambda_{12}\Lambda_2^{-1}\Lambda_{21}\right)^{-1} & \left(\Lambda_{12}\Lambda_2^{-1}\Lambda_{21} - \Lambda_1\right)^{-1} \Lambda_{12}\Lambda_2^{-1} \\ \left(\Lambda_{21}\Lambda_1^{-1}\Lambda_{12} - \Lambda_2\right)^{-1} \Lambda_{21}\Lambda_1^{-1} & \left(\Lambda_2 - \Lambda_{21}\Lambda_1^{-1}\Lambda_{12}\right)^{-1} \end{pmatrix} \tag{6.22}$$

Applying this lemma, we can derive the following elegant result for the estimates of x, the parameters of interest (defining the projection operator P):

$$\hat{x} = \left(A^T WPA\right)^{-1} A^T WPz \quad \text{where} \quad P \equiv I - B\left(B^T WB\right)^{-1} B^T W \tag{6.23}$$

That is, we have derived a method of estimating parameters x, without having to go to the trouble of estimating y. The result (6.23) is remarkable, in that it has exactly the same form as equation (6.17), where we substitute the original weight matrix for *the reduced weight matrix*:

$$\begin{aligned} W' &\equiv WP \\ &= W - WB\left(B^T WB\right)^{-1} B^T W \end{aligned} \tag{6.24}$$

If we are in fact interested in obtaining estimates for y at each batch, we can backsubstitute $\hat{x}$ into (6.21) (for each batch) to obtain:

$$\hat{y} = \left(B^T WB\right)^{-1} B^T W(z - A\hat{x}) \tag{6.25}$$

Augmented Stochastic Model. We need to find a stochastic model that gives rise to the reduced weight matrix (6.24). A stochastic model is correctly stated in terms of the expectation values (6.16), but unfortunately, the reduced weight matrix is singular (because P is an idempotent matrix: $PP=P$). However, an interesting interpretation arises if we derive the stochastic model from first principles. If we treat the augmented part of the model as a source of noise (called *process noise*) rather than as part of the functional model, we can write the *augmented stochastic model* as follows:

$$\begin{aligned} C' &= E\left(v'v'^T\right) \\ &= E\left((By+v)(By+v)^T\right) \\ &= E\left(vv^T\right) + BE\left(yy^T\right)B^T \\ &= C + BC_y B^T \end{aligned} \tag{6.26}$$

Where C_y is, by definition, an *a priori covariance matrix* for the parameter, y. Note that we can choose C_y to be arbitrarily large, if we wish the data to completely influence the result. If we now invert this expression, it ought to correspond to the reduced weight matrix of (6.24). But first, we need to know another very useful matrix inversion lemma (worth remembering!):

$$\left(\Lambda_1 \pm \Lambda_{12}\Lambda_2^{-1}\Lambda_{21}\right)^{-1} = \Lambda_1^{-1} \mp \Lambda_1^{-1}\Lambda_{12}\left(\Lambda_2 \pm \Lambda_{21}\Lambda_1^{-1}\Lambda_{12}\right)^{-1}\Lambda_{21}\Lambda_1^{-1} \tag{6.27}$$

where $\Lambda_{12} \equiv \Lambda_{21}^T$. Applying this lemma to equation (6.26), we find:

$$
\begin{aligned}
W' &= \left(C + BC_y B^T\right)^{-1} \\
&= C^{-1} - C^{-1}B\left(BC^{-1}B^T + C_y{}^{-1}\right)^{-1} B^T C^{-1} \\
&= W - WB\left(BWB^T + C_y{}^{-1}\right)^{-1} B^T W
\end{aligned}
\tag{6.28}
$$

Comparing this expression with (6.24), we find the only difference is the presence of the a priori covariance matrix for parameters y. The functional and stochastic approach are equivalent in the limit that the a priori stochastic parameter covariance is made sufficiently large. (See the discussion following (6.17b)). For the two models to be equivalent, the augmented stochastic model should only account for correlations introduced by data's functional dependence on the process noise (as defined by the matrix B), with no a priori information on the actual variance of the process noise.

Stochastic Estimation. In the context of least squares analysis, a parameter in general is defined in terms of its linear relationship to the observable (i.e., through the design matrix). A *stochastic parameter* has the same property, but is allowed to vary in a way that can be specified statistically. In computational terms, a *constant parameter* is estimated as a constant over the entire data span, whereas a stochastic parameter is estimated as a constant over a specified batch interval, and is allowed to vary from one interval to the next. For example, a special case of this is where the stochastic parameter is allowed to change at every data epoch.

The least squares estimator includes specific a priori information on the parameter, in terms of (1) how its value propagates in time from one batch to the next, and (2) how its variance propagates to provide an a priori constraint on the next batch's estimate. Here, we introduce two of the most important models used in stochastic estimation for precise GPS geodesy:

(1) The simplest is the *white noise* model, which can be specified by:

$$
\begin{aligned}
E\left(y_i\right) &= 0 \\
E\left(y_i y_j\right) &= \sigma^2 \delta_{ij}
\end{aligned}
\tag{6.29}
$$

(2) The random walk model can be specified by

$$
\begin{aligned}
E\left(y_i - y_j\right) &= 0 \\
E\left(\left(y_i - y_j\right)^2\right) &= \vartheta\left(t_i - t_j\right)
\end{aligned}
\tag{6.30}
$$

In the absence of data, the white noise parameters become zero with a constant assumed variance, whereas the random walk parameters retain the last estimated value, with a variance that increases linearly in time. The white noise model is useful where we wish to impose no preconceived ideas as to how a parameter might vary, other than (perhaps) its expected average value. As (6.30) does not require us to specify $E(y_i)$, the random walk model is particularly useful for cases where we do expect small variations in time, but we might have little idea on what to expect for the overall bias of the solution.

The white noise and random walk models are actually special cases of the first order Gauss-Markov model of process noise, see Bierman [1977], however, this general model is rarely used in GPS geodesy.

6.3.2 Discussion

Model Equivalence. The equivalence of (6.23) with (6.21), and (6.24) with (6.26) proves the correspondence between modifying the functional model and modifying the stochastic model. Instead of estimating extra parameters, we can instead choose to modify the stochastic model so as to produce the reduced weight matrix, or equivalently, an augmented covariance. Note that, as we would expect, the weight matrix is reduced in magnitude, which is why it is said that estimating extra parameters *weakens the data strength.* It follows that the corresponding covariance matrices for the data and for the estimated parameters will increase.

We can summarize the above theoretical development by the maxim:

(covariance augmentation) $\equiv$ *(weight matrix reduction)* $\equiv$ *(parameter estimation)*

That is, augmenting the stochastic model can be considered implicit estimation of additional parameters, with the advantage of that there is a saving in computation. The only disadvantage is that the full covariance matrix between all x and y parameters is not computed. Fortunately, there are many applications where the full covariance matrix is of little interest, particularly for problems that are naturally localised in space and time.

Stochastic Parameters. The above theory indicates possible ways to deal with *stochastic parameters*, that are allowed to vary in time according to some stochastic model. Equations (6.23), (6.26) and (6.28) provides a simple mechanism for us to estimate a special class of stochastic parameters called *white noise parameters*, that are allowed to vary from one (specified) batch of data to the next, with no a priori correlation between batches. The a priori covariance matrix in (6.28) can be ignored if we wish, but it can be useful if we believe we know the parameter variance a priori (from some other source), and we do not wish to weaken the data strength unnecessarily.

For example, if we have some parameters which are stochastic in time, we could group the data into batches covering a set time interval, and apply equation (6.23) to estimate the x parameters at each batch interval. The final x parameter estimates could then be derived by accumulating the normal equations from every batch, and then inverting.

The formalism presented above also suggests a method for implementing random walk parameter estimation. Specifically, (6.28) allows for the introduction of an a priori covariance, which could come from the previous batch interval solution, augmented by the model (6.30). Several convenient formulisms have been developed for the step-by-step (batch sequential) approach to estimation, including algorithms such as the Kalman Filter. It is beyond the scope of this chapter to go into specific algorithms, but we shall describe filtering in general terms.

Filtering. In *filtering* algorithms, the a priori estimate for each batch is a function of the current running estimate mapped from the previous batch. The current estimate is then specified as a weighted linear combination of the a priori estimate, and the data from the current batch. The relative weights are determined by the *gain matrix*, which can also account for the a priori correlations between stochastic parameters in accordance with the user-specified stochastic model (6.29) or (6.30). The principles of separating stochastic from global parameters are the same as described earlier. The process of backsubstitution in this context is called *smoothing*, which is essentially achieved by running the filter backwards to allow earlier data to be influenced by the later data in a symmetric way.

Algorithm Equivalence. Whatever algorithm is used, we should always remember that it is the underlying stochastic and functional models that determine the solution. That is, it is possible to construct a conventional weighted least-squares estimator to produce the same answer as, say, a Kalman filter. The choice of algorithm is largely one of computational efficiency, numerical stability, and convenience in being able to control the stochastic model.

Although we have not shown it here, there is a similar equivalence between stochastic estimation and applying a transformation to remove so-called *nuisance parameters*. A simple example of this is the ionospheric linear combination of data, which removes ionospheric delay. This is equivalent to estimating ionospheric delay as *white noise* for each observation. Likewise, the double differencing transformation is equivalent to estimating white noise clock parameters (assuming all available data are effectively transformed). There are parameter estimation algorithms that make use of this kind of equivalence, for example, the use of the Householder transformations in the square root information filter (SRIF) to produce a set of statistically uncorrelated linear combinations of parameters as a function of linear combinations of data. Hence, in the SRIF, there is no longer a distinction between stochastic and functional model, and algorithm development becomes extremely easy (for example, as used by Blewitt [1989] to facilitate *ambiguity bootstrapping*).

In summary, one can effectively implement the same functional and stochastic model in data estimation using the following methods:

(1) explicit estimation by augmenting the functional model;

(2) implicit estimation by augmenting the stochastic model;

(3) parameter elimination by transforming the data and stochastic model.

This presents a rich variety of possible techniques to deal with parameters, which partly explains the very different approaches that software packages might take. Specific algorithms, such as the square root information filter may effectively embody approaches at once, which illustrates the point that the algorithm itself is not fundamental, but rather the underlying functional and stochastic model.

6.3.3 Applications

Global and Arc Parameters. We sometimes call x *global parameters* and y *local parameters* (if they are localized in space, e.g., for a local network connected to a global network through a subset of stations) or *arc parameters* (if they are localized in time, e.g., coordinates for the Earth's pole estimated for each day). More generally y can be called *stochastic parameters*, since it allows us to estimate a parameter that varies (in some statistical way) in either space or time. As we have seen, we don't actually have to explicitly estimate y, if all we are interested in are the global parameters, x.

Earth Rotation Parameters. A typical daily solution for a global GPS network might contain coordinates for all the stations, plus parameters to model the orientation of the Earth's spin axis in the conventional terrestrial frame and its rate of rotation (for example, X and Y pole coordinates, and length of day). We can then combine several day's solutions for the station coordinates, in which case the station coordinates can be considered global parameters. It is also possible to estimate station velocity at this stage, to account for tectonic motion. Next, we can orient this station coordinate (and velocity) solution to a conventional frame, such as the ITRF (IERS Terrestrial Reference Frame). If we then wished to produce improved estimates for daily Earth rotation parameters in this frame, we could then apply equation (6.25) to compute the corrections:

$$\Delta\hat{y} = -\left(B^T WB\right)^{-1}\left(B^T WA\right)\Delta\hat{x} \tag{6.31}$$

This can easily be done if the coefficient matrix relating $\Delta\hat{y}$ to $\Delta\hat{x}$ is stored along with each daily solution. This is an example of smoothing, without having to resort to the full Kalman filter formulism. Effectively, the Earth rotation parameter have been estimated as white noise parameters. The length of day estimates can then be integrated to form an estimate of variation in the Earth's hour angle (UT1-UTC), which would effectively have been modelled as random walk (which can be defined as integrated white noise).

Helmert Wolf Method. The spatial analogy to the above is sometimes called the *Helmert-Wolf Method* or *Helmert Blocking.*. The data are instead batched according to geographic location, where the stochastic y parameters are the station coordinates of a local network. The x parameters comprise station coordinates at the overlap (or *nodes*) between local networks. The x parameters are first estimated for each network according to (6.23); then these estimates are combined; finally the y parameters can obtained using (6.25). Helmert Blocking seen in this context is therefore simply a specific application of a more general concept.

Troposphere Estimation. The random walk model is commonly used for tropospheric zenith bias estimation, because (1) this closely relates to the expected physics of atmospheric turbulence, see Truehaft and Lanyi [1987], and (2) surface meteorological measurements don't provide sufficient information for us to constrain the overall expected bias.

Filtering algorithms can easily allow the tropospheric zenith bias to vary from one data epoch to the next. Traditional least-squares algorithms can also estimate the troposphere stochastically by either explicit augmentation of the set of parameters, or by using the reduced weight matrix (6.24). However, the traditional method is too cumbersome for dealing with a separate parameter at every data epoch, which is why it is common to estimate tropospheric biases which are constant over time periods of an hour or so. Results indicate that this works well for geodetic estimation, but of course, it might be unsatisfactory for tropospheric research.

Clock Estimation. The white noise model is commonly used for clock estimation when processing undifferenced data. This is partly because one does not have to worry about any type of glitch because the a prior correlation is assumed to be zero. As already mentioned, white noise clock estimation is an alternative to the double differencing approach. The advantage, of course, is that we obtain clock estimates, which leads us naturally to the application *precise point positioning.*

Precise Point Positioning. We can consider receiver coordinates as local parameters, connected to each other only through the global parameters (that affect all spatially separated observations), which include orbit, satellite clock, and Earth rotation parameters. The global network of permanent GPS stations is now reaching the point that the addition of an extra station would do very little to change the estimated orbit and satellite clock parameters. We can therefore take the global solution to be one using the current global network, and consider a user's receiver coordinates as the local parameters. Application of (6.25) to a single receivers carrier phase and pseudorange data using the global parameter solution for x would therefore give us a precise point position solution for y.

This can actually be simplified further. The term $\left(z - A\hat{x}\right)$ in (6.25) is simply the user's receivers data, minus a model computed using the global parameters (orbits, clocks, and Earth rotation parameters). Therefore, we can solve for y by only storing the global parameters and the single receiver's data. Further still, the orbit and Earth rotation parameters can be processed to produce a table of orbit positions in the Earth fixed frame.

Putting all of this together, we can therefore see that (1) producing a single receiver point position solution using a precise ephemerides in the Earth fixed frame is essentially equivalent to (2) processing the station's data as double differences together in a simultaneous solutions with the global network's data. The only difference is that the user's receiver cannot influence the orbit solution. This is astonishingly simple, and has revolutionized high precision geodetic research due to the very short time it takes to produce high precision results, which is typically a few minutes for a 24 hour data set, Zumberge et al. [1996].

6.4 FRAME INVARIANCE AND ESTIMABILITY

Strange as it may seem, station coordinates are generally not estimable parameters. This statement may appear ludicrous, given that GPS is supposedly designed to allow us to position ourselves. But position relative to what? In the simple case of point positioning, we are positioning ourselves relative to the given positions of the GPS satellites, in the reference frame known as WGS-84. How are the orbits known? They are determined using the Control Segment's tracking stations at known coordinates. How are these tracking station coordinates known? By using GPS. And so the questions continue, in a circular fashion.

To view this problem clearly, we consider the general case of the one step procedure, estimating all the satellite orbits and all the station coordinates at the same time. In this section, we consider the nature of these coordinates, and consider exactly what is estimable when attempting to position a global network of GPS stations.

6.4.1 Theoretical Development

Insight from Physics. "A view advanced by Einstein, there is a widespread belief among modern physicists that the fundamental equations of physics should possess the same form in all coordinates systems contemplated within the physical context of a given theory", Butkov [1968]. Although coordinates are essential for the computation of observable models, our intuition is better served if we think of *geometrical objects*, which we can define as *frame invariant* objects. In general terms, such geometrical objects are called *tensors*. Tensors are classified according to their rank. Formally, a tensor of rank r is defined as an *invariant linear function of r directions,* Butkov [1968]. The rank of a tensor (not to be confused with the rank of a matrix) tells you the number of indices required to specify the tensor's coordinates. Here, we stick to familiar objects; *vectors* which are tensors of rank 1 (which have a single direction in space), and *scalars* are tensors of rank zero (which have no directionality).

Equations can be explicitly expressed in terms of tensors without reference to coordinates. One must therefore be careful not to confuse the true vector, which is a geometrical object, with the column vector, which is a representation of the vector in a specific coordinate frame. For example, although the coordinates represented in a column vector change under a frame transformation, the true vector does not change.

Vectors and Transformations. A vector can be defined as an *invariant linear function of direction,* Butkov [1968]. We should really think of the vector as an axiomatic geometrical object, which represents something physical, and is therefore unaffected by frame transformations. We can write the vector **x** in frame F as, Mathews and Walker [1964]:

$$\mathbf{x} = \sum_{i} x_i \mathbf{e}_i \tag{6.32}$$

in terms of coordinates x_i and base vectors $\mathbf{e}_i$ (vectors which define the direction of the coordinate axes). The same vector can be written in frame F' as:

$$\mathbf{x} = \sum_{i} x'_i \mathbf{e}'_i \tag{6.33}$$

We can, for analytical convenience, write this equivalence in matrix form, where we define the row matrix **e** and column matrix x as follows:

$$\begin{aligned} \mathbf{x} &= \begin{pmatrix} \mathbf{e}_1 & \mathbf{e}_2 & \mathbf{e}_3 \end{pmatrix} \begin{pmatrix} x_1 \\ x_2 \\ x_3 \end{pmatrix} \equiv \mathbf{e}x \\ &= \begin{pmatrix} \mathbf{e}'_1 & \mathbf{e}'_2 & \mathbf{e}'_3 \end{pmatrix} \begin{pmatrix} x'_1 \\ x'_2 \\ x'_3 \end{pmatrix} \equiv \mathbf{e}'x' \end{aligned} \tag{6.34}$$

Notice that both coordinates and the base vectors must change such that the vector itself remains unchanged. This axiomatic invariance of a vector requires that the transformation for the base vectors is accompanied by a related (but generally different) transformation of the coordinates. We can start by following the convenient matrix form of (6.34) to define each base vector of the new frame $\mathbf{e}'_i$ as a vector in the old frame with coordinates γ_{ji}. Coordinates γ_{ji} are elements of the transformation matrix *G:*

$$\begin{aligned} \mathbf{e}'_i &= \sum_{j} \mathbf{e}_j \gamma_{ji} \\ \mathbf{e}' &= \mathbf{e}\Gamma \end{aligned} \tag{6.35}$$

Using the equivalence relation (6.34), we find the corresponding transformation for coordinates:

$$\begin{aligned}
\mathbf{x} &= \mathbf{e}x \\
&= \mathbf{e}\left(\Gamma\Gamma^{-1}\right)x \\
&= \left(\mathbf{e}\Gamma\right)\left(\Gamma^{-1}x\right) \\
&= \mathbf{e}'x' \\
\therefore x' &= \Gamma^{-1}x
\end{aligned} \tag{6.36}$$

Objects such as the coordinates are said to transform *contragradiently* to the base vectors. Objects which transform in the same way are said to transform *cogradiently*.

Scalar Functions and Transformations. The frame transformation, represented by G, is called a *vector function*, since it transforms vectors into vectors. In contrast, geodetic measurements can be generally called *scalar functions* of the vectors. The dot product between vectors is an example of a scalar function. Simply take a look at typical functional models used in geodesy, and you will find objects such as the dot product between vectors. We therefore need to look at the theory of scalar functions of vectors and how they transform.

A *linear scalar function* of a vector can be defined in terms of its effect on the basis vectors. For example, in 3-dimensional space, we can define the scalar function as a list of 3 numbers known as the *components* of the scalar function:

$$\begin{aligned}
\phi(\mathbf{e}_1) &= \alpha_1 \\
\phi(\mathbf{e}_2) &= \alpha_2 \\
\phi(\mathbf{e}_3) &= \alpha_3
\end{aligned} \tag{6.37}$$

When the scalar function is applied to a general vector **x**, the result can be written

$$\begin{aligned}
\phi(\mathbf{x}) &= \phi\left(x_1\mathbf{e}_1 + x_2\mathbf{e}_2 + x_3\mathbf{e}_3\right) \\
&= \alpha_1 x_1 + \alpha_2 x_2 + \alpha_3 x_3 \\
&= \begin{pmatrix} \alpha_1 & \alpha_2 & \alpha_3 \end{pmatrix} \begin{pmatrix} x_1 \\ x_2 \\ x_3 \end{pmatrix} \\
&\equiv \alpha x
\end{aligned} \tag{6.38}$$

The result must be independent of reference frame, because the geometrical vector **x** is frame invariant. Therefore we can derive the law of transformation for the scalar components a:

$$
\begin{aligned}
\phi(\mathbf{x}) &= \alpha x \\
&= \alpha\left(\Gamma\Gamma^{-1}\right)x \\
&= (\alpha\Gamma)\left(\Gamma^{-1}x\right) \\
&= \alpha' x' \\
\therefore \alpha' &= \alpha\Gamma
\end{aligned}
\tag{6.39}
$$

This proves that the scalar components transform cogradiently with the base vectors, and contragradiently with the coordinates.

It would appear that the scalar functions have very similar properties to vectors, but with slightly different rules about how to transform their components. The scalar function is said to form a *dual space*, with the same dimensionality as the original vectors.

Supposing we have a vector **y** in our geodetic system, we can define a special the scalar function that always forms the dot product with **y**. The result can be expressed in matrix form:

$$
\begin{aligned}
\phi_{\mathbf{y}}(\mathbf{x}) &= \mathbf{y}.\mathbf{x} \\
&= (\mathbf{e}y)^T.(\mathbf{e}x) \\
&= y^T\left(\mathbf{e}^T.\mathbf{e}\right)x \\
&= y^T g x
\end{aligned}
\tag{6.40}
$$

where g is the matrix representation of the *metric tensor*, which can be thought of as describing the unit of length for possible directions in space (here represented in 3 dimensions), Misner, Thorne, and Wheeler [1973]:

$$
g \equiv \begin{pmatrix} \mathbf{e}_1.\mathbf{e}_1 & \mathbf{e}_1.\mathbf{e}_2 & \mathbf{e}_1.\mathbf{e}_3 \\ \mathbf{e}_2.\mathbf{e}_1 & \mathbf{e}_2.\mathbf{e}_2 & \mathbf{e}_2.\mathbf{e}_3 \\ \mathbf{e}_3.\mathbf{e}_1 & \mathbf{e}_3.\mathbf{e}_2 & \mathbf{e}_3.\mathbf{e}_3 \end{pmatrix}
\tag{6.41}
$$

Comparing (6.38) and (6.40), we see that the components of the dot product scalar function are given in matrix form by

$$
\alpha_y = y^T g
\tag{6.42}
$$

Proper length. One can therefore easily construct such a scalar function for every vector, simply using the vector's coordinates, and the metric properties of the space.

$$
\begin{aligned}
l_{\mathbf{x}} &= \left|\phi_{\mathbf{x}}(\mathbf{x})\right|^{\frac{1}{2}} \\
&= \left(x^T g x\right)^{\frac{1}{2}}
\end{aligned}
\tag{6.43}
$$

It is easy to prove using all the above definitions, that the length of a vector, defined by (6.43) is completely frame invariant, no matter what kind of transformation is performed. For example, if the frame were scaled up so that a different *unit of length* were being used, the metric tensor would be scaled down to compensate.

In the language of relativity, such a length defined using a 4-dimensional space-time metric, is called a *proper length.* Proper length which is said to be a *scalar invariant* (i.e., a tensor of rank 0). The geometry expressed by (6.43) is known as a *Riemann geometry.* In a *Riemannian space* (e.g., the surface of a sphere), length is calculated along geodesics, which are in turn defined by the metric tensor. It reduces to Euclidean geometry in the special case that the metric is the identity matrix, in which case we have Cartesian coordinates.

In physics, the metric tensor (6.41) is a property of space-time, to be inferred by experiment. According to special relativity, a natural consequence of the universality of the speed of light is that space-time according to an *inertial observer* has the metric

$$
g_0 = \begin{pmatrix} 1 & 0 & 0 & 0 \\ 0 & 1 & 0 & 0 \\ 0 & 0 & 1 & 0 \\ 0 & 0 & 0 & -c^2 \end{pmatrix}
\tag{6.44}
$$

It might seem odd that a dot product $\mathbf{e}_t . \mathbf{e}_t$ has a negative value, but if we accept that any reasonable definition of "length" must be frame invariant, that's what experiment tells us!, Schutz [1990]. The proper length between two points of relative coordinates in the rest frame (Dx, Dy, Dz, Dt) is therefore defined as:

$$
l_0 \equiv \left(\Delta x^2 + \Delta y^2 + \Delta z^2 - c^2 \Delta t^2\right)^{\frac{1}{2}}
\tag{6.45}
$$

which reduces to our normal concept of *spatial length* between two points at the same time coordinate (Pythagoras Theorem):

$$
s_0 \equiv l_0(\Delta t = 0) = \left(\Delta x^2 + \Delta y^2 + \Delta z^2\right)^{\frac{1}{2}}
\tag{6.46}
$$

Proper length is known experimentally to be frame invariant, as is evidenced by the independence of the speed of light on the motion of the source; so in two different frames moving at constant velocity with respect to each other, we can write:

$$l' \equiv \left(\Delta x'^2 + \Delta y'^2 + \Delta z'^2 - c^2 \Delta t'^2\right)^{\frac{1}{2}} = l_0 \equiv \left(\Delta x^2 + \Delta y^2 + \Delta z^2 - c^2 \Delta t^2\right)^{\frac{1}{2}} \quad (6.47)$$

But, in general, our normal concept of the *spatial length* would be different!

$$s' \equiv \left(\Delta x'^2 + \Delta y'^2 + \Delta z'^2\right)^{\frac{1}{2}} \neq s_0 \equiv \left(\Delta x^2 + \Delta y^2 + \Delta z^2\right)^{\frac{1}{2}} \quad (6.48)$$

In general relativity, spatial length is affected not only by relative motion, but also by the gravitational potential. The geometrical error in assuming a 3-dimensional Euclidean space amounts to about 2 cm in the distance between satellites and receivers. The relative geometrical error amounts to 1 part per billion, which is not insignificant for today's high precision capabilities. For convenience, three dimensional Euclidean space underlies GPS spatial models, with relativity applied to geometrical delay as corrections to Pythagoras Theorem (and with relativistic corrections applied to compute the coordinate time of signal transmission).

Scalar Function Equivalence in Euclidean Space. We therefore proceed assuming 3-dimensional Euclidean geometry is adequate, assuming relativistic corrections are applied. In this case, the metric is represented by a 3×3 identity matrix. By inspection of (6.41), we see that the basis vectors would be orthonormal, hence defining a Euclidean space where points are represented by Cartesian coordinates.

In Euclidean space, the components of the scalar function are simply the Cartesian components of the vector. For each and every Cartesian vector, there is a corresponding scalar function with identical components as the vector. One can think of this scalar function as the operator that projects any vector onto itself. In a sense, the vector has been redefined in terms of its own functionality. The set of scalar functions defining the dot product operator with respect to each vector in the geodetic system completely describes the geometry as far as scalar observations are concerned. We can therefore conclude that the dot product operator has an equal footing in representing geometry as the vectors themselves. (Length *is* geometry!, MTW [1973]).

Measurements and Geometry. From a modern physical point of view, measurement models should be independent of the selected coordinate frame. Therefore, the measurement model must be a function of objects which are frame invariant. In geodesy, we are seeking to estimate spatial vectors in a Euclidean space. This seems at first to be problematic, since measurements are scalars, not spatial vectors. This situation can be theoretically resolved, once we realize that vectors have a one to one correspondence with scalar functions defined as the dot products between the vectors. These dot products define the length of a vector, and the angles between them. Fundamentally, the above theoretical development proves that the dot products contain all the geometrical information than did the original vectors. A pertinent example, is the differential delay of a quasar signal between two VLBI telescopes. It represents a projection of the baseline vector connecting the two telescopes in the quasar direction. Many observations would reveal the vector as it projects onto different directions.

We must therefore be careful when it comes to estimating coordinates. From what has been said, only the dot products between vectors are estimable, not the coordinates themselves. Since we have freedom to select different frames it is clear that the set of coordinates in a given frame must be redundant, which is to say, there are more coordinates than is necessary to define the vectors, and therefore scalar functions and modelled observations.

Assuming the above Euclidean model, we can show explicitly the source of this redundancy. Following again the VLBI analogy, consider a baseline vector **x**, and a quasar direction **y**, which is represented in two different solutions, one in frame F, the other in frame F'. Projection is formally represented by the dot product between the two vectors. The equivalence relation (6.34) tells us that the dot product between any two vectors must give the same answer no matter which frame is used. Since we assume we can construct orthonormal base vectors (of unit length, and at right angles to each other, as in a Cartesian frame), we can write the dot product for both frames as:

$$s \equiv \mathbf{x}.\mathbf{y} = x^T y = x'^T y' \tag{6.49}$$

What types of coordinate transformation are allowed that satisfy (6.41)? It can easily be shown that the transformation matrix must be orthogonal; that is its transpose equals its inverse. In matrix notation, let us consider equation (6.49), where we apply a transformation R to go from frame F coordinates to frame F' coordinates:

$$\begin{aligned} x^T y &= x'^T y' \\ &= (Rx)^T (Ry) \\ &= x^T \left(R^T R\right) y \\ \therefore R^T &= R^{-1} \end{aligned} \tag{6.50}$$

Such transformations are called *rotations,* and (6.50) shows the property of rotation matrices. We therefore deduce that global rotations have no effect on dot products computed according to (6.49) (which assumed Euclidean frames).

The analogy in special relativity is the *Lorentz transformation*, which can be considered as a rotation in 4-dimensional space-time (3 rotations + 3 velocity boosts). Relativistic transformations preserve proper length, but can change spatial length. The only change in scale which is physically acceptable is that due to the relativistic choice of reference frame, which depends on relative speed (special relativity) and the gravitational potential (general relativity). For example, VLBI solutions computed in the barycentric frame (origin at the centre of mass of the solar system) produce baselines with a different scale to SLR solutions computed in the geocentric frame.

6.4.2 Discussion

Space Geodetic Consequences. The corollary of equation (6.50) and the correspondence between vectors and scalar functions, is that space geodetic data cannot provide any information on global rotations of the *entire* system. Since this arbitrariness has 3 degrees of freedom (a rotation), it results in a 3-rank deficient problem. This justifies our original statement that coordinates are not estimable.

In the case of VLBI, the entire system includes quasars, so the rotation of the Earth is still accessible. However, in GPS, the system has satellites that can be only be considered approximately fixed (through dynamical laws which have systematic errors), and therefore we can conclude the data only weakly constrains long period components of Earth rotation. As for relative rotation of the Earth's plates, we can conclude that, on purely geometrical grounds, there is no advantage of VLBI over GPS (this is not to exclude other classes of arguments to favor one or the other).

Geometrical Paradigm. The paradigm for this section is geometrical. In the spirit of Einstein, although conventions and coordinates provide a convenient representation of reality for computational purposes, our intuition is often better served by a geometrical model that is independent of these conventions. The relationship between the geometrical paradigm and the conventional model is discussed below, where we refer the reader to Chapter 1 for a more complete description of the conventional terms.

Consider a network of GPS stations, tracking all the GPS satellites. Using the GPS data, we can estimate the *geometrical* figure defined by the stations and the satellite orbits. That is, GPS provides information on internal geometry, including the distances between stations, and the angles between baselines, and how these parameters vary in time. The geometrical figure defined by the stations is sometimes called *the polyhedron*, particularly in IGS jargon. This is to remind us that, fundamentally, the data can tell us precisely the shape of the figure described by the network of points. For permanent tracking networks, the data can also tell us how the polyhedron's internal geometry changes over time. The elegant aspect of this geometrical picture, is that it more closely relates to quantities that can actually be measured in principle, such as the time it takes for light to travel from one station to another. This is in contrast to coordinates which are frame dependent.

Since GPS orbits can be well modelled over an arc length of a day (2 complete orbits), we have access to an instantaneous inertial frame, which by definition, is the frame in which Newton's laws appear to be obeyed. In historical terminology, GPS data together with dynamical orbit models give us access to an *inertial frame of date*. A frame determined this way, cannot rotate significantly relative to inertial space, otherwise the orbits would not appear to obey Newton's laws.

The system can therefore determine the direction of the instantaneous spin axis of the Earth with respect to the polyhedron. Although the spin axis is not tangible like stations and satellites, it is an example of an estimable vector. For example, GPS can tell us unambiguously the angles between any baseline and the instantaneous spin axis (called the Celestial Ephemeris Pole, CEP). We can therefore determine a station's latitude relative to the CEP without any problem. However, the direction of the CEP as viewed by the polyhedron does wander from one day to the next, a phenomenon known as *polar motion*. It would therefore be impractical to define a station's latitude this way, so instead, a conventional reference pole direction is defined (called the conventional terrestrial pole, or CTP).

The problem is, the choice of CTP is arbitrary, and fundamentally has nothing to do with GPS data. Therefore, conventional station latitudes (relative to the CTP) strictly cannot be estimated, but only true latitudes can (relative to the CEP). This state of affairs is not hopeless; for example, the CTP can be defined by constraining at 2 station latitudes. If we allow for time evolution of the polyhedron (which we must), then we must also specify the time evolution of the CTP with respect to the polyhedron, which again goes beyond GPS, and into the domain of conventions.

GPS is also sensitive to the rate of rotation of the Earth about the CEP. Again, this is because the satellites are forced to obey Newton's laws in our model. Since the spin rate can be estimated, our model can map the time series of station positions in the instantaneous inertial frame back to an arbitrary reference time. We can therefore determine the relative longitude between stations, as angles subtended around the CEP. However, just as for latitudes, the longitudes determined this way would wander from one day to the next due to polar motion (an effect that is maximum near the poles, and negligible at the equator). Longitudes are therefore also dependent on the choice of CTP. Moreover, only relative longitude can be inferred, since GPS data has no way to tell us exactly the location of the Prime Meridian. Once again, we would have to go beyond GPS, and arbitrarily fix some station's longitude to a conventional value (preferably, near the equator), thus effectively defining the Prime Meridian.

We note in passing that the CEP also varies in inertial space (by nutation and precession). We only need to model this variation over the period for which we are modelling the satellite dynamics, which is typically over a day or so. GPS is therefore insensitive to nutation and precession errors longer than this period, because, in effect, we are defining a brand new inertial frame of date every time we reset the orbit model. The reason for having relatively short orbit arcs (as compared to SLR) is not because of fears about nutation, but rather because of inadequacies in the orbit model. But an orbit arc of a day is sufficient for the purpose of precisely determining the polyhedron, which implicitly requires a sufficiently precise determination of polar motion and Earth spin rate. (The Earth's spin rate is often parameterized as the excess length of day, or variation in UT1-UTC, that is the variation in the Earth's hour angle of rotation relative to atomic time).

Finally, there is another geometrical object to which GPS is sensitive, and that is the location of the Earth's centre of mass within the geometrical figure of the polyhedron. In Keplerian terms, the Earth centre of mass is at the focus for each and every GPS elliptical orbit. Of course, Kepler's laws are only approximate. More precisely, the Earth's centre of mass is the dynamical origin of the force models used to compute the GPS satellite orbits.

If we arbitrarily displaced the polyhedron relative to this origin, we would find the satellite orbits appearing to violate Newton's laws. We therefore can say that GPS can locate *the geocentre*, which is to say that it can determine a displacement of the centre of figure with respect to the centre of mass, Vigue et al. [1992]. Effectively, GPS therefore allows us to estimate geocentric station height, which is the radial distance from the Earth centre of mass. However, it should be kept in mind, that the geocentre estimate is very sensitive to the accuracy of orbit force models, and is not determined as precisely as the geometry of the figure. In fact, vary rarely is true geocentric height variation shown from GPS analyses, but rather height relative to the average figure, which is an order of magnitude more precise, with the (apparent) geocentre variation often displayed separately as a global parameter.

6.4.3 Applications

Free Network Solutions. If we estimated all station coordinates and satellite positions, the 3-rank deficiency in the problem would imply that a solution could not be obtained. However, suppose we apply very loose a priori constraints to the station coordinates. The above theory predicts that our resulting coordinates would still be ill-defined, however the geometry of the figure would be estimable, Heflin et al. [1992]. That is, if we were to compute the dot product between any two vectors in the system, we would find it to be well constrained by the data. Such a solution has been called a *free network solution*, a *fiducial-free solution*, or a *loose solution*.

We discuss below several applications of free network solutions, which for example can be used directly to estimate geophysical parameters of interest, since geophysical parameters depend on scalar functions of the vectors, not the coordinates. For some applications, though, a frame definition may be necessary.

Frame Definition. Although *conventional reference systems* may have ideal notions of the basis vectors and origin, *conventional terrestrial frames* today are *defined* through the coordinates of a set of points co-located with space geodetic instruments. These points serve to define implicitly the directions of the coordinate axes (i.e., the basis vectors). Such frames contain an extremely redundant number of points, and so might also serve as a source of a priori geometrical information.

One could choose to fix all these points, which might be advantageous for a weak data set, where a priori information may improve the parameters of interest.

If the data set is strong, finite errors in the frame's geometry will conflict with the data, producing systematic errors. To avoid this problem, we should only impose *minimal constraints*. The large redundancy allows one to define the frame statistically, so that errors on the redundant set of definitions for the X,Y, and Z axis directions are averaged down. This can be achieved using either 3 equations of constraint for orientation, or by using the fiducial free, or free network approach, where all station coordinates are estimated, with the final indeterminate solution being rotated to agree on average with the frame.

Quality Assessment. Internal quality assessment involves looking at the residuals to the observations, after performing the least-squares solution, and assessing the significance of deviations. Residuals are estimable even in the absence of frame definition, and so it is recommended to assess internal quality of the data using free network solutions, prior to applying any constraints, otherwise it would be impossible to distinguish data errors from systematic errors arising from the constraints.

External quality assessment involves comparing geodetic solutions. How can we tell if the geometry of the solutions (i.e., the vectors) are the same, if the solution is only represented by coordinates?

If we do happen to know how the base vectors are related between the two systems, then we can simply transform one set of coordinates and do a direct comparison. This is rarely the case, unless the base vectors are implicitly defined through constrained coordinates in the system.

Alternatively, we can use the fact that, fundamentally, a vector reveals itself through its scalar functions, and check all of the dot products in the system. Obvious candidates for this include baseline length, and angles between baselines.

Thirdly, one can solve for a rotation transformation between the two frames, apply the transformation, and compare all the vector coordinates, which it must be stressed, are to be constructed between *physical* points to which the observations are sensitive.

This last point requires clarification. The position coordinates of a point, for example, do not constitute the coordinates of a physical vector, unless the origin has some physical significance in the model. For VLBI it does not, for SLR it does, and for GPS, there is a degree of sensitivity which depends on global coverage, and other issues. We are allowed to use the Earth centre of mass as a physical point for satellite systems, so "position vector" does become physically meaningful in the special case that the origin is chosen to be at the Earth centre of mass. So, strictly, the theory does not permit a direct comparison of VLBI station coordinates with, say, GPS; however, it does permit a comparison of the vectors. However, if one were to insist on comparing single station positions, one could remove an estimated translational bias between the frames, but the resulting station coordinates would logically then be some linear combination of all estimated station coordinates, making interpretation potentially difficult.

Finally, as already discussed, a change in scale is not considered an acceptable transformation between frames assuming Euclidean space. Apart from relativistic considerations, scaling between solutions must be considered as systematic error rather than a valid frame transformation.

Coordinate Precision. We should now be able to see that coordinates could not be estimated unless we have observational access to any physical objects that might have been used to define the unit vectors. (For example, a physical inscription marking the Prime Meridian). Coordinate precision therefore not only reflects the precision to which we have determined the true geometry of the figure, but also the precision to which we have attached ourselves to a particular frame. Coordinate precision (e.g., as formally given by a covariance matrix computation) can therefore be a very misleading measure of the geometrical precision.

Geophysical Interpretation. Our ability to attach ourselves to a particular frame has absolutely no consequence to the fundamental physics to be investigated (say, of the Earth, or of satellite orbit dynamics). However particular frames may be easier to express the dynamic models. For example, the inertial frame is better for describing the satellite equations of motion. A terrestrial (co-rotating) frame is easier for describing motion of the Earth's crust. Nevertheless, the estimable quantities will be frame independent.

One pertinent example here is the relative Euler pole of rotation between two plates, with the estimable quantities being, for example, the relative velocity along the direction of a baseline crossing a plate boundary. Another example is crustal deformation due to strain accumulation. Here, the invariant geometrical quantity is the symmetric strain tensor, with the invariant estimable quantities being scalar functions of the strain tensor. However, space geodesy cannot unambiguously state, for example, the velocity coordinates of a point, since that requires arbitrarily defined axes.

When comparing geophysically interesting parameters, one must take care to ensure frame invariance, or at least, approximate frame invariance. For example, comparing station velocity components between solutions, or Euler poles of individual plates will generally show discrepancies that relate to frame definition.

Ambiguity Resolution. This section could more generally refer to all inherently scalar parameters, such as tropospheric or clock parameters. Like these parameters, the carrier phase ambiguities are manifestly frame independent quantities. As a consequence, no frame constraints are necessary at all to estimate ambiguity parameters. In fact, there are good reasons for not including frame constraints. Frame constraints, if not minimal, can distort solutions due to systematic error in the a priori geometry. This can be very undesirable where the a priori information is suspect.

As a test of this concept, Blewitt and Lichten [1992] solved for ambiguities on a global scale using a network solution free of frame constraints, and found they could resolve ambiguities over even the longest baselines (up to 12,000 km).

Covariance Projection. One might wish to compare coordinates after applying a rotation between solutions. Or perhaps one wishes to assess the geometrical strength of the free network solution. In both cases, it is useful to consider the coordinate error as having a component due to internal geometry, and external frame definition. A free network solution is ill-defined externally, but well defined internally. How can we compute a covariance matrix that represents the internal errors?

We can apply a *projection operator*, which is defined as the estimator for coordinate residuals following a least squares solution to rotation. Consider the linearized observation equation which rotates the coordinates into another frame, accounting for possible measurement error:

$$x = Rx' + v \tag{6.51}$$

This can be rearranged so that the 3 unknown angles contained in R are put into a column vector θ, and defining a rectangular matrix A as a linear function of θ such that:

$$A\theta \equiv Rx' \tag{6.52}$$

Therefore, after substitution in to (6.23), we find the least squares estimator for the errors:

$$\begin{aligned} \hat{v} &= x - A\hat{\theta} \\ &= x - A\left(A^T W A\right)^{-1} A^T W x \\ &= \left(I - A\left(A^T W A\right)^{-1} A^T W\right) x \end{aligned} \tag{6.53}$$

The covariance matrix for the estimated errors is therefore:

$$\begin{aligned} C_{\hat{v}} &= \left(I - A\left(A^T W A\right)^{-1} A^T W\right) C_x \left(I - A\left(A^T W A\right)^{-1} A^T W\right)^T \\ &= C_x - A\left(A^T W A\right)^{-1} A^T \\ &= C_x - A C_{\hat{\theta}} A^T \end{aligned} \tag{6.54}$$

This is called *projecting* the covariance matrix onto the space of errors, Blewitt et al. [1992]. Since these errors are scalar quantities (independent of frame), they represent the geometrical errors of the system. Therefore, the projected covariance matrix is a formal computation of the precision to which the geometry has been estimated, without us having to define a frame.

Note from (6.54) that we can write the original covariance matrix for coordinates as:

$$
\begin{aligned}
C_x &= C_{\hat{v}} + A\left(A^T W A\right)^{-1} A^T \\
&= C_{\hat{v}} + A C_{\hat{\theta}} A^T \\
&= C_{\text{internal}} + C_{\text{external}}
\end{aligned} \tag{6.55}
$$

This shows explicitly that the coordinate covariance can be decomposed into a covariance due to internal geometrical errors, and an external term which depends on the level of frame attachment.

Loosening Transformation. In the case of *loose solutions*, in effect the orientation parameters have loose a priori constraints. If this constraint can be represented by the (large) a priori covariance matrix E (*external*), equation (6.54) would more correctly read (see 6.17b):

$$
\begin{aligned}
C_x &= C_{\hat{v}} + A\left(A^T W A + E^{-1}\right)^{-1} A^T \\
&\approx C_{\hat{v}} + A E A^T
\end{aligned} \tag{6.56}
$$

where we use the fact that the data themselves provide no information on global orientation, hence the components of $A^T W A = C_{\hat{\theta}}^{-1}$ can be considered negligibly small.

We call (6.56) a *loosening transformation*, or a *covariance augmentation.* The resulting covariance is often called a *loosened covariance*, or *loosened solution* (even though we have not changed the estimates themselves). It can be applied, for example, to network solutions that have a well defined constraint in orientation, for applications where we wish to effectively remove the frame definition. Once augmented in this way, the coordinate covariance can then be projected onto another frame, applying the projection operator.

Equation (6.55) should look familiar. We have actually seen it before in equation (6.26), in the context of augmenting the stochastic model as an alternative to estimating extra parameters. Effectively, this is telling us that a combination of loosened solutions is equivalent to estimating and removing a relative rotation between constrained networks and combining them. It also tells us that it is unnecessary to estimate and remove relative rotations between loose solutions prior to combination.

This has very practical applications when combining network solutions from various analysis groups, who might apply different minimal coordinate constraints. Upon receiving a coordinate solution with full covariance matrix, one can proceed to loosen the covariance matrix prior to combination with other solutions. Therefore, one does not have to estimate and apply transformation parameters every time the coordinate solution is processed. Moreover, covariance loosening has the elegant aspect in that the fundamental rank-3 deficiency is represented in an obvious way to the user, as the diagonal elements of the covariance matrix will be large, with the off-diagonal elements containing the geometrical information to which the data are truly sensitive.

As an example, the IGS Densification Program (IDP) currently uses the above concept of combining loose covariance matrices from a variety of analysis centres. Algorithm development for manipulating such solutions becomes very straightforward, when one does not have to worry about solutions being constrained to different frames. The IDP also, once again, illustrates the concept of global and local parameters, with each regional network being connected to the global network using 3 common *anchor stations*. Using 3 anchor stations allows for the implicit estimation of relative rotation when combining regional and global network solutions, Blewitt et al [1993 and 1995].

6.5 SUMMARY AND CONCLUSIONS

We are now in a position to summarize some of the most important conclusions in terms of a few maxims, which are purposely expressed in an informal way to appeal to an intuitive mode of thinking.

6.5.1 Equivalence of Pseudorange and Carrier Phase

Models for the pseudorange can be constructed using carrier phase, and visa versa.

This allows us to develop algorithms that use both data types to:

- Estimate pseudorange multipath
- Edit carrier phase and pseudorange data for outliers
- Edit carrier phase data for cycle slips
- Smooth the pseudorange using the carrier phase
- Process undifferenced data without a preliminary point position solution
- Resolve carrier phase ambiguities in a model independent way

6.5.2 Equivalence of the Stochastic and Functional Models

(covariance augmentation)≡(weight reduction)≡(estimation)≡(data combination)

This allows us to develop algorithms to:

- separately estimate global and local parameters
- partition problems in time
- partition problems in space
- remove nuisance parameters
- implicity estimate parameters
- estimate stochastic parameters
- estimate precise point positions using single receivers

6.5.3 Frame Invariance and Estimability

(invariant geometry)≡(tensors, vectors, scalars)≡(scalar functions)≡(observations)
This allows us to understand:

- a geometrical paradigm for space geodesy
- which parameters are estimable
- problems with coordinate estimability
- frame definition
- the importance and utility of free network solutions
- internal and external components of coordinate error
- covariance projection as a means to quantify geometrical error
- loosening transformation to remove rotational information
- network combination analysis

6.5.4 Concluding Remark

What I hope to have achieved in this chapter is (1) specifically, to impart an intuitive understanding of certain aspects of GPS data processing and estimation, and (2) more generally, to have inspired a change in the way we might think about GPS data processing problems in general, by looking for patterns, symmetries, and equivalence's, and exploiting these so that answers to questions become more obvious.

Acknowledgement

I would like to thank the late Professor Richard P. Feynman as a teacher and researcher during my years at the California Institute of Technology, for his inspiration.

References

Bierman, G. (1977), *Factorization methods for discrete sequential estimation,* Academic Press, New York, NY.

Blewitt, G. (1989), Carrier phase ambiguity resolution for the Global Positioning System applied to geodetic baselines up to 2000 km, *Journ. Geophys. Res., Vol. 94*, No. B8, pp. 10187-10283.

Blewitt, G. (1990), An automatic editing algorithm for GPS data, *Geophys. Res. Lett., Vol. 17*, No. 3, pp. 199-202.

Blewitt, G. (1990), A new tool for dense monitoring of crustal strain: GPS rapid static surveying, *Eos. Trans. Am. Geophys. U., Vol. 71*, No. 17, p. 483.

Blewitt, G., and S.M. Lichten (1992), Carrier phase ambiguity resolution up to 12000 km: Results from the GIG'91 experiment, in *Proc. of the Sixth Int. Symp. on Satellite Positioning*, Columbus, Ohio State University.

Blewitt, G. (1992), M.B. Heflin, F.H. Webb, U.J. Lindqwister, and R. P. Malla, Global coordinates with centimeter accuracy in the International Terrestrial Reference Frame using the Global Positioning System, *Geophy. Res. Lett., 19*, pp. 853-856.

Blewitt, G., Y. Bock, and G. Gendt (1993), Regional clusters and distributed processing, in *Proc. of the IGS Analysis Center Workshop*, Ed. by J. Kouba, pp. 62-91, Ottawa, Canada.

Blewitt, G., Y. Bock, and J. Kouba (1995), "Constructing the IGS polyhedron by distributed processing," in Proceedings of the IGS Workshop, ed. by J. Zumberge, IGS Central Bureau, Pasadena, Calif., USA, p. 21-36.

Butkov, E. (1968), *Mathematical physics*, Addison-Wesley, Amsterdam.

Heflin, M.B., W.I. Bertiger, G. Blewitt, A.P. Freedman, K.J. Hurst, S.M. Lichten, U.J. Lindqwister, Y. Vigue, F.H. Webb, T.P. Yunck, and J.F. Zumberge (1992), Global geodesy using GPS without fiducial sites, *Geophysical Research Letters, Vol. 19*, pp. 131-134.

Mathews (1964), J., and R.L. Walker, *Mathematical methods of physics*, Benjamin Cummings, Amsterdam.

Melbourne, W.G. (1985), The case for ranging in GPS based geodetic systems, in Proceedings of 1st Int. Symp. on Precise Positioning with the Global Positioning System, U.S. Dept. of Commerce, Rockville, MD.

Misner, C. W., K.S. Thorne, and J.A. Wheeler (1973), *Gravitation*, Freeman, San Francisco.

Wu, J.T., S.C. Wu, G.A. Hajj, W.I. Bertiger, and S.M. Lichten (1993), Effects of antenna orientation on GPS carrier phase, *Manuscripta Geodaetica, 18,* pp. 91-93.

Schutz, B.F. (1990), A first course in general relativity, Cambridge Univ. Press, Cambridge.

Truehaft, R.N. and G.E. Lanyi (1987), The effects of the dynamic wet troposphere on radio interferometric measurements, *Radio Science, 22*, pp. 251-265.

Vigue, Y., S.M. Lichten, G. Blewitt, M.B. Heflin, and R.P. Malla (1992), Precise determination of the Earth's center of mass using measurements from the Global Positioning System," *Geophys. Res. Lett., 19,* pp. 1487-1490.

Zumberge, J.F., M.B. Heflin, D.C. Jefferson, M.M. Watkins, and F.H. Webb (1997), Precise point positioning for the efficient and robust analysis of GPS data from large networks, *Journ. Geophys. Res., 102*, No. B3, pp. 5005-5018.

7. QUALITY CONTROL AND GPS

Peter J.G. Teunissen
Faculty of Civil Engineering and Geosciences, section MGP,
Thijsseweg 11, 2629 JA Delft, The Netherlands

7.1 INTRODUCTION

As in many other sciences, empirical data are used in geodesy to make inferences about physical reality. In order to get a workable hypothesis, we try to capture the conglomerate of physical reality by means of a mathematical model. This model consists of two parts: the functional model and the stochastic model. By means of the functional model, we try to describe the relations that are believed to exist between the observables themselves or between the observables and the unknown parameters of the model. The stochastic model is used to capture the expected uncertainty or variability of the empirical data. Philosophically speaking, every model is misspecified. The question of the model's validity is therefore not one of proving its correctness, but rather one of showing whether the mathematical model, given its intended purposes, is sufficiently consistent with the data observed.

The present chapter develops tools for model validation and qualification. As such, the material presents an introduction into the theory of quality control, together with its application in GPS positioning. Though the notion of 'quality' is much abused and subject to inflation, its importance in any consumer-producer relationship cannot be denied. Based on a quantification of the client's demands, the quality of a product or service can be said to express the extent in which these demands are met. Although quality comprises many different factors (e.g. completeness, timeliness, costs, etc.), we will restrict attention to a more narrower notion of quality, namely that a (GPS) survey is considered to be qualified for its purpose when it is delivered with sufficient precision and reliability. Precision as expressed by the a posteriori covariance matrix of the coordinates, measures the survey's characteristics in propagating random errors, while reliability describes the survey's ability to check for the presence of modelling errors.

When linking the mathematical model to the data, the utmost care has to be exercized in formulating the observation equations (functional model) and the covariance matrix of the observables (stochastic model). In case of GPS, the set of observables that can be used, consists of carrier phases and pseudoranges (code) on L_1 and L_2. These observables can be linked to the baseline coordinates and the carrier phase ambiguities. Apart from these unknown parameters, the observation equations may also include additional parameters, such as those needed for ionospheric and tropospheric refraction, or for the (receiver and/or satellite) clock errors, or for the signal delays in the hardware or for multipath.

Whether or not all these parameters need to be included as well, depends very much on the hardware used, the circumstances of measurement and the particular application at hand. For instance, the ionospheric delays could be negligible in case of sufficiently short baselines, or the multipath may be avoided when sufficient precautions in the measurement set-up are taken.

In the stochastic model we try to capture the intrinsic uncertainties of the measurements. For this we need a proper understanding of the instrumentation and measurement procedures used. The stochastic model is not restricted to the precision of the individual measurements. In case of GPS one should also consider the possibility of cross-correlation and/or time-correlation. Depending on how the measurement process is implemented in the GPS-receivers, the observables may or may not be cross-correlated. In the presence of anti-spoofing for instance, some receivers use a hybrid technique to provide dual frequency code measurements. As a result the code measurements become cross-correlated. Also the presence of time-correlation should be considered, in particular when use is made of high sampling rates.

The mathematical model used, depends on the application at hand and thus also on the GPS mode of operation. There is a whole suite of GPS models one can think of. A GPS model for relative positioning may be based on the simultaneous use of two receivers (single baseline) or more than two receivers (multi baseline or network). It may have the relative receiver-satellite geometry included (geometry based) or excluded (geometry free). When it is excluded, not the baseline components are involved as unknowns in the model, but instead, the receiver-satellite ranges themselves. GPS models may also be discriminated as to whether the slave receiver(s) are in motion (non-stationary) or not (stationary). When in motion, one solves for one or more trajectories, since with the receiver-satellite geometry included, one will have a new baseline for each new epoch.

The fact that a whole suite of different GPS models exists, implies that there are different stages at which quality control can be excercized. Roughly speaking, one can discriminate between the following four levels:

1. Receiver level: in principle it is already possible to validate the time series of undifferenced data of a single receiver. By lumping some of the parameters (e.g. range, clock errors, tropospheric delay, orbital uncertainty) and by introducing a smoothness constraint on the time behaviour of the ionospheric delays, the necessary redundancy enters which is needed for validation. Single receiver quality control is very useful for reference receivers that are used in active GPS control networks or in DGPS.

2. Baseline level: in this case the observation equations are parametrized in terms of the baseline vector of the two receivers. Here the redundancy primarily stems from the presence in the design matrix of the receiver-satellite geometry and from the assumed constancy over time of the ambiguities. Additional redundancy enters when the baseline is considered stationary instead of moving.

3. Network level: when sufficient (independent) baselines are used to form a network, redundancy enters by enforcing the closure of 'baseline loops'. The redundancy characteristics of a baseline network are very similar to that of a classical levelling network.

4. Connection level: additional redundancy enters again, when a free GPS network is connected to points of an existing geodetic control. In that case it is the shape of the free network which is compared with that of the existing control network.

If the way in which the input dataset presents itself in these four cases is compared, we note that the 'time factor' is usually absent in the last two cases, but that it is present in the first two cases. In the first two cases, the dataset consists of time series of carrier phases and pseudoranges, while some of the unknown parameters may change from epoch to epoch as well. This is not true for the last two cases. In fact, these last two cases are not unlike the traditional (terrestrial) networks, which usually are first adjusted and validated as a free-network, before they are connected to an existing (higher order) control. They can therefore be treated accordingly and thus adjusted in batch mode. Although a batch mode approach is applicable in principle to the first two cases as well, it is seldomly used for the data processing of a single receiver. Here one usually opts for a recursive approach. A recursive approach is also natural for the case of a moving baseline, in particular if real-time or near real-time results are required. For the case of a stationary baseline the situation is not that clear-cut. Here both approaches, batch and recursive, are viable alternatives.

Since batch and recursive data processing are both used with GPS, the quality control of both approaches will be treated in this chapter. We will start with the batch approach, since this is probably the one with which one is the most familiar. This is done in section 2 *Validation of batch solutions*. It is shown how modelling errors, such as for instance outliers, cycle slips, baseline errors or antenna height errors, can be detected and identified. It is also shown how the reliability of the various tests can be expressed in terms of the minimal detectable biases. Section 3 *Validation of recursive solutions* parallels section 2. There are however also a number of marked differences, two of which are the recursion itself and the timeframe used for testing. First, since the estimation procedure is recursive, the quality control procedure is formulated in recursive form as well. This applies to all three steps of detection, identification and adaptation. Second, we make a distinction between local tests and global tests. Local tests are based on information currently available, whereas global tests also make use of information from the past. Local tests are attractive since they make immediate corrective action possible. Global tests on the other hand are more powerful than their local counterparts. In section 4 *Application to some GPS models* we apply the theory to some of the more common single baseline models. Particular attention is given to the minimal detectable biases.

7.2 VALIDATION OF BATCH SOLUTIONS

In this section we discuss ways of checking the quality of batch solutions. Reliability and precision of a least-squares estimator refer respectively to the first two moments of the estimator. Since the least-squares estimator is unbiased under the null hypothesis, its quality under the null hypothesis is captured by its precision. The precision can be tested by comparing the covariance matrix with a criterion matrix. The estimator becomes biased when the functional model is misspecified. The risk of having a biased solution can be controlled by testing the null hypothesis against alternative hypotheses. Likely model errors are specified by means of the alternative hypotheses. The testing procedure consists of the detection and identification of model errors and if needed, of an adaptation of the null hypothesis. The strength of the testing procedure is coupled to the power of the tests and described by the reliability of the final solution.

7.2.1 Least-Squares Estimation

The need for an adjustment of observations or the estimation of parameters arises usually when one has to solve an inconsistent system of equations. In geodesy this is most often the case, when one has to solve a redundant system of observation equations. The adjustment principle often used is that of least-squares. This principle is not only simple to apply, but it also has important optimality properties. Least-squares estimation is equivalent to maximum likelihood estimation in case the observables are normally distributed, and in the linear case, it is also equivalent to best linear unbiased estimation. A prerequisite for applying this principle in a proper way, is that a number of basic assumptions need to be made about the input data, the measurements. Since measurements are always uncertain to a certain degree, they are modelled as sample values of a random vector, the m-vector of observables $\underline{y}$ (*note:* the underscore will be used to denote random variables). In case the vector of observables is normally distributed, its distribution is uniquely characterized by the first two (central) moments: the expectation (or mean) $E\{\underline{y}\}$ and the dispersion (or variance) $D\{\underline{y}\}$. Information on both the expectation and dispersion needs to be provided, before any adjustment can be carried out.

Functional model: Knowing the information content of the observables allows one to link them to the unknown parameters of interest. These parameters are then to be estimated. The link between the m-vector of observables $\underline{y}$ and the n-vector of unknown parameters x is established by means of the system of m observation equations $\underline{y} = Ax + \underline{e}$, where $\underline{e}$ describes the discrepancy between $\underline{y}$ and Ax. If we assume that $\underline{e}$ models the random nature of the variability in the measurements, it seems acceptable to assume that this variability is zero 'on the average' and therefore that its mean is zero, $E\{\underline{e}\} = 0$. Combined with the previous equation, this gives

$$E\{\underline{y}\} = Ax \tag{7.1}$$

This system of equations is referred to as the functional model. It is given once the design matrix A of order $m \times n$ is specified.

The system as it is given here is linear in x. Quite often however, the observation equations are nonlinear. In that case a linearization needs to be carried to make the system linear again. The parameter vector x usually consists of coordinates and possibly additional nuisance parameters, such as for instance receiver clock errors in case of GPS. The coordinates could be of any type. For instance, they could be Cartesian coordinates or geographic coordinates. The choicc of the type of coordinates is not essential for the adjustment, but is more a matter of convenience and depends on what is required for the particular application at hand.

Stochastic model: Measurements are intrinsically uncertain. Remeasurement of the same phenomenon under similar circumstances, will usually give slightly different results. This variability in the outcomes of measurements is modelled through the probability density function of $\underline{y}$. In case of the normal distribution, it is completely captured by its dispersion. In order to properly weigh the observables in the adjustment process, the dispersion needs to be specified beforehand. It is given as

$$D\{\underline{y}\} = Q_y \tag{7.2}$$

This is the stochastic model, with Q_y being the $m \times m$ variance matrix of the observables. In these lecture notes, we will assume that Q_y is known. Hence, unknown variance components and their estimation are not treated.

Since the variance matrix describes the variability one can expect of the measurements when they are repeated under similar circumstances, it is said to describe the precision of the observables. In order to be able to specify Q_y correctly, a good understanding of the measurement equipment and the measurement procedures used, is needed. Quite often the variance matrix Q_y can be taken as a diagonal matrix. This is the case, when the measurements have been obtained independently from one another. The variance matrix becomes full (nondiagonal) however, when for instance, the measurements themselves are the result of a previous adjustment. In case of GPS, the variance matrix is nondiagonal when use is made of vectorial double-differenced code and/or phase observables or when time correlation is present.

Least-squares: Once the measurements have been collected and the functional model and the stochastic model have been specified, the actual adjustment or estimation can be carried out. The least-squares estimator of the unknown parameter vector x is given as

$$\hat{\underline{x}} = \left(A^T Q_y^{-1} A\right)^{-1} A^T Q_y^{-1} \underline{y} \tag{7.3}$$

It depends on the design matrix A, the variance matrix Q_y and the vector of observables $\underline{y}$. With $\hat{\underline{x}}$, one can compute the adjusted observables as $\hat{\underline{y}} = A\hat{\underline{x}}$ and the least-squares residuals as $\hat{\underline{e}} = \underline{y} - \hat{\underline{y}}$.

The above expression for the least-squares estimator is based on a functional model which is linear. In the nonlinear case, one will first have to apply a linearization before the above expression can be applied. For the linearization one will need approximate values for the unknown parameters. It may happen that one already has a fair idea about the values the parameters will take. These values could then be taken as a first approximation. If this is not the case, then a minimum set of the observations themselves will have to be used for computing approximate coordinates. In case the approximate values of the unknown parameters are rather poor, one often will have to iterate the least-squares solution.

Quality: Every function of a random vector is itself a random variable as well. Thus $\hat{\underline{x}}$ is a random vector, just like the vector of observables $\underline{y}$ is. And when $\hat{\underline{x}}$ is linearly related to $\underline{y}$, it will have a normal distribution whenever $\underline{y}$ has one. In that case also the distribution of $\hat{\underline{x}}$ can be uniquely characterized by means of its expectation and dispersion. Its expectation reads

$$E\{\hat{\underline{x}}\} = x \tag{7.4}$$

Thus the expectation of the least-squares estimator equals the unknown, but sought for parameter vector x. This property is known as unbiasedness. From an empirical point of view, the equation implies, that if the adjustment would be repeated, each time with measurements collected under similar circumstances, then the different outcomes of the adjustment would on the average coincide with x. It will be clear, that this is a desirable property indeed.

The dispersion of $\hat{\underline{x}}$, describing its precision, is given as

$$D\{\hat{\underline{x}}\} = Q_{\hat{x}} = \left(A^T Q_y^{-1} A\right)^{-1} \tag{7.5}$$

This variance matrix is independent of $\underline{y}$. This is a very useful property, since it implies that one can compute the precision of the least-squares estimator without having the actual measurements available. Only the two matrices A and Q_y need to be known. Thus once the functional model and stochastic model have been specified, one is already in a position to know the precision of the adjustment result. It also implies, that if one is not satisfied with this precision, one can change it by changing A and/or Q_y. This is typically done at the design stage of a geodetic project, prior to the actual measurement stage. Changing the geometry of the network and/or adding/deleting observables, will change A. Using different measurement equipment and/or different measurement procedures, changes Q_y.

Precision testing: Although we know how to change $Q_{\hat{x}}$ for the better or the worse, we of course still need a way of deciding when the precision, as expressed by the variance matrix, is good enough. It will be clear that this depends very much on the particular application at hand. What is important though, is that one has a *precision criterion* available by which the precision of the least-squares solution can be judged. The following are some approaches that can be used for testing the precision of the solution.

It may happen in a particular application, that one is only interested in one particular function of the parameters, say $\theta = f^T x$ In that case it becomes very easy to judge its quality. One then simply has to compare its variance with the given criterium

$$\sigma_{\hat{\theta}}^2 = f^T Q_{\hat{x}} f < \text{criterion} \tag{7.6}$$

The situation becomes more difficult though, when the application at hand requires that the precision of more than one function needs to be judged upon. Let us assume that the functions of interest are given as $\theta = F^T x$, where F is a matrix. The corresponding variance matrix is then given as $Q_{\hat{\theta}} = F^T Q_{\hat{x}} F$. One way to judge the precision in this case, is by inspecting the variances and covariances of the matrix $Q_{\hat{\theta}}$. An alternative way would be to use the average precision as precision measure. In that case one relies on the trace of the variance matrix

$$\tfrac{1}{p}\,\text{trace}\left(F^T Q_{\hat{x}} F\right) < \text{criterion} \tag{7.7}$$

where p is the dimension of the vector θ. When using the trace one has to be aware of the fact that one is not taking all the information of the variance matrix $Q_{\hat{\theta}}$ into account. It depends only on the variances and not on the covariances. Hence, the correlation between the entries of $\hat{\theta}$ are then not taken into account. When using the trace one should also make sure that it makes sense to speak of an average variance. In other words the entries of θ should contain the same type of variables. It would not make sense to use the trace, when θ contains completely different variables, each having their own physical dimension.

The trace of $Q_{\hat{\theta}}$ equals the sum of its eigenvalues. Instead of the sum of eigenvalues, one might decide that it suffices to consider the largest eigenvalue $\lambda_{\max}$ only,

$$\lambda_{\max}\left(F^T Q_{\hat{x}} F\right) < \text{criterion} \tag{7.8}$$

In that case one is thus testing whether the function of θ which has the poorest precision, still passes the precision criterion. When this test is passed successfully, one knows that all other functions of θ will also have a precision which is better than the criterion. For some applications this may be an advantage, but it could be a disadvantage as well. It could be a disadvantage in the sense that the above test could be overly conservative. That is, when the function having the poorest precision passes the above test, all other functions, which by definition have a better precision, may turn out to be unnecessarily precise.

So far we assumed that the precision criterion was given in scalar form. But this need not be the case. The precision criterion could also be given in matrix form. In that case one is working with a *criterion matrix*, which we will denote as C_x The precision test amounts then to testing whether the precision as expressed by the actual variance matrix $Q_{\hat{\theta}} = F^T Q_{\hat{x}} F$ is better than or as good as the precision expressed by the criterium matrix $F^T C_x F$. Also this test can be executed by means of solving an eigenvalue problem, but now it will be a *generalized* eigenvalue problem

$$\left| F^T Q_{\hat{x}} F - \lambda F^T C_x F \right| = 0 \tag{7.9}$$

Note that when the matrix $F^T C_x F$ is taken as the identity matrix, the largest eigenvalue of (7.9) reduces to that of (7.8). Using the generalized eigenvalue problem is thus indeed more general than the previous discussed approaches. It is characterized by the fact that it allows one to compare the actual variance matrix $Q_{\hat{\theta}}$ directly with its criterion. The two matrices $F^T Q_{\hat{x}} F$ and $F^T C_x F$ are identical, when all eigenvalues equal one, and all functions of $\hat{\theta}$ have a better precision than the criterion when the largest generalized eigenvalue is less than one.

So far we assumed the matrix A to be of full rank. The variance matrix $Q_{\hat{x}}$ will not be unique however, when A has a rank defect. In that case the variance matrix depends on the chosen set of minimal constraints. Since these constraints should not affect our conclusion when evaluating the precision, we have to make sure that our procedure of precision testing is invariant for the minimal constraints. This is possible when we make use of S-transformations, see Baarda [1973] and Teunissen [1985]. Let C_x be the criterium matrix and $Q_{\hat{x}_b}$ the variance matrix of a minimally constrained solution. The eigenvalues of the generalized eigenvalue problem

$$\left| Q_{\hat{x}_b} - \lambda S_b C_x S_b^T \right| = 0 \tag{7.10}$$

where S_b is the S-transformation that corresponds to the minimal constraints of $Q_{\hat{x}_b}$, are then invariant to the chosen set of minimal constraints. Thus when the largest eigenvalue is less than one, it is guaranteed that all functions of $\underline{\hat{x}}_b$ have a precision which is better than what they would have were $Q_{\hat{x}_b}$ be replaced by the criterion matrix.

7.2.2 Model Validation

Applying only an adjustment to the observed data is not enough. The result of an adjustment and its quality rely heavily on the validity of the functional and stochastic model. Errors in one of the two, or in both, will invalidate the adjustment results. One therefore needs, in addition to the methods of adjustment theory, also methods that allow one to check the validity of the assumptions underlying the functional and stochastic model. These methods are provided for by the theory of statistical testing.

Null and alternative hypothesis: Before any start can be made with the application of the theory of statistical testing, one needs to have a clear idea of the null hypothesis H_0 and of the alternative hypothesis H_a. The null hypothesis consists of the model that one believes to be valid. But in order to have enough confidence in this belief, the null hypothesis is compared with or tested against an alternative hypothesis. As null hypothesis we take

$$H_0: \; E\{\underline{y}\} = Ax \,, \;\; D\{\underline{y}\} = Q_y \tag{7.11}$$

We have seen, that in order to be able to apply the least-squares principle, only the first two moments of the random vector of observables $\underline{y}$ need to be specified. The first moment (mean) $E\{\underline{y}\} = Ax$ and the second moment (variance matrix) $D\{\underline{y}\} = Q_y$. In the case of statistical testing however, this is not sufficient. In addition, one will have to specify the type of probability function of $\underline{y}$ as well. Since most observational data in geodesy can be modelled as samples drawn from a normal distribution, we will include in the null hypothesis the assumption that $\underline{y}$ has the normal (or Gaussian) probability density function.

In order to test the null hypothesis against an alternative hypothesis, we need to know what type of misspecifications one can expect in the null hypothesis. Of course, every part of the null hypothesis could have been specified incorrectly. The mean Ax can be wrong, the variance matrix Q_y can be wrong and/or $\underline{y}$ could have a distribution other than the normal distribution. In these lecture notes we will restrict ourselves to misspecifications in the mean. Experience has shown, that these are by large the most common errors that occur when formulating the model. As alternative hypothesis we will therefore use

$$H_a: \; E\{\underline{y}\} = Ax + C\nabla \,, \;\; D\{\underline{y}\} = Q_y \tag{7.12}$$

where C is a known matrix of order $m \times q$ and ∇ is an unknown vector of order $q \times 1$. With H_a we thus oppose the null hypothesis H_0 to a more relaxed alternative hypothesis, in which more explanatory parameters, namely ∇, are introduced. These additional parameters are then supposed to model those effects which were assumed absent in H_0. For instance, through $C\nabla$ one may model the presence of one or more blunders (outliers) in the data, or the presence of refraction, or any other systematic effect which was not taken into account in H_0. The relation between $E\{\underline{y}\}$ and ∇ is specified through the matrix C.

Examples of matrix *C*: The model errors could be one dimensional ($q = 1$) or higher dimensional ($1 < q \leq \text{redundancy}$). In the one dimensional case, the matrix C reduces to a vector c and ∇ becomes a scalar. Some examples of c-vectors are:

1. An outlier in the ith code observable: Assuming that $\underline{y}$ consists of code observables (pseudoranges), the c-vector takes the form $(0,\ldots,0,1,0.\ldots,0)^T$, with the 1 as the ith entry of c.

2. A cycle slip in a phase observable that starts at epoch $l \leq k$: Assuming that $\underline{y}$ consists of a phase observable that spans k epochs, the c-vector takes the form $(0,\ldots,0,1,1,\ldots,1,1)^T$, where the 1's occupy the last $k - l + 1$ entries of c.

3. An antenna height error at one receiver location: Assuming that $\underline{y}$ consists of GPS-baselines, the c-vector takes the form $(0,\ldots,0,h_x,h_y,h_z,0,\ldots,0)^T$, with the non-zero entries being the elements of the linearized transformation matrix from Cartesian to ellipsoidal coordinates.

As an example of a higher dimensional model error we have the unspecified error in the ith baseline of a GPS-baseline network: Assuming that $\underline{y}$ consists of GPS-baselines, the C-matrix takes the form $(0_{3\times3},\ldots,0_{3\times3},I_3,0_{3\times3},\ldots,0_{3\times3})^T$.

The test statistic: The purpose of testing H_0 against H_a is to infer whether the data supports H_0 or whether the data rejects H_0 on the basis of H_a. In order to make such a decision, we first need a function of $\underline{y}$, the test statistic, that 'measures' to what extent H_0 or H_a is supported by the data. It reads

$$\underline{T}_q = \frac{1}{q}\underline{\hat{e}}^T Q_y^{-1} C \left[C^T Q_y^{-1} Q_{\hat{e}} Q_y^{-1} C \right]^{-1} C^T Q_y^{-1} \underline{\hat{e}} \tag{7.13}$$

The test statistic $\underline{T}_q$ has a central F-distribution with q and ∞ degrees of freedom when H_0 is true, but a non-central F-distribution with non-centrality parameter λ when the alternative hypothesis H_a is true. Thus

$$H_0 : \underline{T}_q \sim F(q,\infty,0) \text{ and } H_a : \underline{T}_q \sim F(q,\infty,\lambda) \tag{7.14}$$

One decides to reject the null hypothesis when the outcome of the test statistic is too large. That is, when $T_q \geq F_{\alpha_q}(q,\infty,0)$, where α_q is the chosen level of significance.

7.2.3 Detection, Identification and Adaptation

Above we gave the test statistic for testing the null hypothesis H_0 against a particular alternative hypothesis H_a. In most practical applications however, it is usually not only one model error one is concerned about, but quite often many more than one. This implies that one needs a *testing procedure* for handling the various alternative hypotheses. In this subsection we will discuss a way of structuring such a testing procedure. The DIA-procedure consists of the following three steps:

1. *Detection*: An overall model test is performed to diagnose whether an unspecified model error occurred.

2. *Identification*: After detection of a model error, identification of the potential source of model error is needed.

3. *Adaptation*: After identification of a model error, adaptation of the null hypothesis is needed to eliminate the presence of biases in the solution.

Detection

Since one usually first wants to know whether one can have any confidence in the assumed null hypothesis without the need to specify any particular alternative hypothesis, the first step consists of a check on the *overall* validity of H_0. This implies that one opposes the null hypothesis to the most relaxed alternative hypothesis possible. The most relaxed alternative hypothesis is the one that leaves the observables completely free. Hence, under this alternative hypothesis no restrictions at all are imposed on the observables. We therefore have the situation

$$H_0 : E\{\underline{y}\} = Ax \text{ versus } H_a : E\{\underline{y}\} \in R^m \tag{7.15}$$

Since $E\{y\} \in R^m$ implies that matrix (A, C) is square and regular and that $q = m - n$, it follows from (7.13) that the appropriate test statistic for detection reads as

$$\underline{T}_{m-n} = \frac{\hat{\underline{e}}^T Q_y^{-1} \hat{\underline{e}}}{m-n} \tag{7.16}$$

The appropriate test statistic for testing the null hypothesis against the most relaxed alternative hypothesis is thus equal to the weighted sum-of-squares of the least-squares residuals, divided by the redundancy. The null hypothesis will be rejected when $T_{m-n} > F_{\alpha_{m-n}}(m-n, \infty, 0)$.

Identification

In the detection phase, one tests the overall validity of the null hypothesis. If this leads to a rejection of the null hypothesis, one has to search for possible model misspecifications. That is, one will then have try to identify the model error which caused the rejection of the null hypothesis. This implies that one will have to specify through the matrix C, the type of likely model errors. This specification of possible alternative hypotheses is application dependent and is one of the more difficult tasks in hypothesis testing. It depends namely very much on ones experience, which type of model errors one considers to be likely.

The 1-dimensional case: In case the model error can be represented by a scalar, $q=1$ and matrix C reduces to a vector which will be denoted by the lowercase character c. This implies that the alternative hypothesis takes the form

$$H_a : E\{\underline{y}\} = Ax + c\nabla \tag{7.17}$$

The alternative hypothesis is specified, once the vector c is specified. The appropriate test statistic for testing the null hypothesis against the above alternative hypothesis H_a follows when the vector c is substituted for the C-matrix in (7.13). It gives $\underline{T}_{q=1} = (\underline{w})^2$, with

$$\underline{w} = \frac{c^T Q_y^{-1} \underline{\hat{e}}}{\sqrt{c^T Q_y^{-1} Q_{\hat{e}} Q_y^{-1} c}} \tag{7.18}$$

This test statistic has a standard normal distribution $N(0,1)$ under H_0.

A typical example where one has to deal with more than one one-dimensional alternative hypotheses, occurs when one wants to check all individual observations for potential outliers. In that case there are as many alternative hypotheses as there are observations. Hence, there are also m different c-vectors and m different test statistics $\underline{w}$. In case of outliers, the c-vectors take the form $c_i = (0,\ldots,0,1,0,\ldots,0)^T$. Thus c_i is a unit vector having the 1 as its ith entry. The corresponding test statistic is denoted as $\underline{w}_i$. By letting i run from 1 up to and including m, one can screen the whole data set on the presence of potential blunders in the individual observations. The test statistic $\underline{w}_i$ which returns the in absolute value largest value, then pinpoints the observation which is most likely corrupted with a blunder. Its significance is measured by comparing the value of the test statistic with the critical value. Thus the jth observation is suspected to have a blunder, when

$$|w_j| \geq |w_i| \;\; \forall i \;\text{ and }\; |w_j| > N_{\alpha_1/2}(0,1) \tag{7.19}$$

This procedure of screening each individual observation for the presence of an outlier, is known as *data snooping,* Baarda [1968].

In many applications in practice, the variance matrix Q_y is diagonal. If that is the case, the expression of the above test statistic simplifies considerably. With a diagonal Q_y-matrix, we have

$$\underline{w}_i = \frac{\hat{\underline{e}}_i}{\sigma_{\hat{e}_i}}$$

The appropriate test statistic is then thus equal to the least-squares residual of the ith observation divided by the standard deviation of the residual.

The higher dimensional case: It may happen that a particular model error can not be represented by a single scalar. In that case $q > 1$ and ∇ becomes a vector. The appropriate test statistic is then the one we met earlier, namely (7.13). It could happen that one has alternative hypotheses with different degrees of freedom q_i. The most likely alternative hypothesis is then the one for which

$$\frac{T_{q_i}}{F_{\alpha_{q_i}}(q_i, \infty, 0)}$$

is maximum. The levels of significance α_{q_i} of these tests are coupled as $\lambda(\alpha_{q_i}, q_i, \gamma)$=constant, see the section on reliability.

Adaptation

Once one or more likely model errors have been identified, a corrective action needs to be undertaken in order to get the null hypothesis accepted. Here, one of the two following approaches can be used in principle. Either one replaces the data or part of the data with new data such that the null hypothesis does get accepted, or, one replaces the original null hypothesis with a new hypothesis that does take the identified model errors into account. The first approach amounts to a remeasurement of (part of) the data. This approach is feasible for instance, when in case of datasnooping some individual observations are identified as being potentially corrupted by blunders. These are then the observations which get remeasured. In the second approach no remeasurement is undertaken. Instead the model of the null hypothesis is enlarged by adding additional parameters such that all identified model errors are taken care off. Thus with this approach, the identified alternative hypothesis becomes the new null hypothesis.

Once the adaptation step is completed, one of course still has to make sure whether the newly created situation is acceptable or not. This at least implies a repetition of the detection step. When adaptation is applied, one also has to be aware of the fact that since the model may have changed, also the 'strength of the model' may have changed. In fact, when the model is adapted through the addition of more explanatory parameters, the model has become weaker in the sense that the test statistics will now have less detection and identification power. That is, the reliability has become poorer. It depends on the particular application at hand whether this is considered acceptable or not.

7.2.4 Reliability

It is common to measure the quality of the results of estimation by means of the precision of the estimators. However, precision measures only one aspect of the estimator, namely the amount of expected variability in samples of the estimator around its mean. Such a measure may suffice in case the estimator is unbiased. But the unbiasedness of the estimator can not be guaranteed, since it depends on the validity of the model used. The purpose of model testing, as discussed above, is to minimize the risk of having a biased solution. One should realize however, that like any result of estimation, also the outcomes of the statistical tests are not exact and therefore prone to errors. It depends on the 'strength' of the model how much confidence one can have in the outcomes of these statistical tests. A measure for this confidence is provided by the concept of reliability.

Power of the tests

The test statistics are random variables, which have a probability density function that depends on which of the hypotheses is true. For our test statistic $\underline{T}_q$ we have

$$H_0 : \underline{T}_q \sim F(q,\infty,0) \text{ and } H_a : \underline{T}_q \sim F(q,\infty,\lambda) \tag{7.20}$$

with the *noncentrality parameter*

$$\lambda = \nabla^T C^T Q_y^{-1} Q_{\hat{e}} Q_y^{-1} C \nabla \tag{7.21}$$

When testing, two type of errors can be made. A type I error is made, when one decides to reject the null hypothesis while in fact it is true. The probability of such an error is given by the level of significance α and reads

$$\alpha = P\left[\underline{T}_q > F_\alpha(q,\infty,0) \middle| H_0\right] = \int_{F_\alpha(q,\infty,0)}^{\infty} p_F(x|q,0)dx \tag{7.22}$$

where $p_F(x|q,0)$ is the probability density function of the central F-distribution, having q and ∞ degrees of freedom.

A type II error is made, when one decides to accept the null hypothesis while in fact it is false. The probability of such an error is given as

$$\beta = P\left[\underline{T}_q \leq F_\alpha(q,\infty,0) \mid H_a\right].$$

Instead of working with β, one can also work with its complement $\gamma = 1 - \beta$, which is known as the *power of the test*. It is the probability of correctly rejecting the null hypothesis. Hence, it is the probability

$$\gamma = P\left[\underline{T}_q > F_\alpha(q,\infty,0) \middle| H_a\right] = \int_{F_\alpha(q,\infty,0)}^{\infty} p_F(x|q,\lambda)dx \tag{7.23}$$

Note that the power γ depends on the three parameters α, q and λ. Using a shorthand notation, we write

$$\gamma = \gamma(\alpha, q, \lambda) \tag{7.24}$$

When testing, we of course would like to have a sufficiently high probability of correctly detecting a model error when it occurs. One can make γ larger by increasing α, or, by decreasing q, or, by increasing λ. This can be seen as follows. A larger α implies a smaller critical value $F_\alpha(q,\infty,0)$ and via the integral (7.23) thus a larger power γ. Thus if we want a smaller probability for the error of the first kind (α smaller), this will go at the cost of a smaller γ as well. That is, one can not simultaneously decrease α and increase γ.

The power γ also gets larger, when q gets smaller. This is also understandable. When q gets smaller, the less additional parameters are used in H_a and therefore the more "information" is used in formulating H_a. For such an alternative hypothesis one would expect that if it is true, the probability of accepting it will be higher than for an alternative hypothesis that contains more additional parameters. Finally, the power γ also gets larger, when λ gets larger. This is understandable when one considers (7.21). For instance, one would expect to have a higher probability of correctly rejecting the null hypothesis, when the model error gets larger. And when ∇ gets larger, also the non-centrality parameter λ gets larger.

Using α and/or q as tuning parameters to increase γ, does not make much sense however. The parameter q can not be changed at will, since it depends on the type of model error one is considering. And increasing α, also does not make sense, since it would lead to an increased probability of an error of the first kind. Hence, this leaves us with λ. According to (7.21), the non-centrality parameter depends on

1. the model error $C\nabla$
2. the variance matrix Q_y
3. the design matrix A

Since one can not change the model error at will, it is through changes in the variance matrix Q_y and/or in the design matrix A that one can increase λ, thereby improving the detection power of the test. Thus the detection power γ of the tests can, if needed, be improved by using more precise measurement equipment, by adding more observations to the model and/or by changing the structure of the design matrix A.

The minimal detectable bias

The power γ is the probability with which a model error $C\nabla$ can be found with the appropriate test. Thus given the model error $C\nabla$, we can compute the noncentrality parameter as

$$\lambda = \nabla^T C^T Q_y^{-1} Q_{\hat{e}} Q_y^{-1} C\nabla \tag{7.25}$$

and from it, together with α and q, the power as $\gamma = \gamma(\alpha, q, \lambda)$. We can however also follow the inverse route. That is, given the power γ, the level of significance α and the dimension q, the non-centrality parameter can be computed as $\lambda = \lambda(\alpha, q, \gamma)$. If we combine this function with (7.25), we get a *quadratic equation* in ∇

$$\lambda(\alpha, q, \gamma) = \nabla^T C^T Q_y^{-1} Q_{\hat{e}} Q_y^{-1} C\nabla \tag{7.26}$$

This equation is said to describe the *internal reliability* of the null hypothesis $E\{\underline{y}\} = Ax$ with respect to the alternative hypothesis $E\{\underline{y}\} = Ax + C\nabla$. Each model error $C\nabla$ that satisfies the quadratic equation, can be found with a probability γ. Since it is often more practical to know the size of the model error that can be found with a certain probability, than knowing the probability with which a certain model error can be found, we will use the inverse of (7.26).

In the one dimensional case $q = 1$ the matrix C reduces to the vector c. Inverting (7.26) becomes then rather straightforward. As a result we get for the size of the model error

$$|\nabla| = \sqrt{\frac{\lambda(\alpha_1, 1, \gamma)}{c^T Q_y^{-1} Q_{\hat{e}} Q_y^{-1} c}} \tag{7.27}$$

The variate $|\nabla|$ is known as the *Minimal Detectable Bias (MDB)*. It is the size of the model error that can just be detected with a probability γ, using the appropriate $\underline{w}$-test statistic. Larger errors will be detected with a larger probability and smaller errors with a smaller probability.

In order to guarantee that the model error $c\nabla$ is detected with the same probability by both the w-test and the overall model test, one will have to relate the critical values of the two tests. This can be done by equalizing both the powers of the two tests and their non-centrality parameters. Thus if $\lambda(\alpha_{m-n}, m-n, \gamma_{m-n})$ is the non-centrality parameter of the $\underline{T}_{m-n}$-test statistic and $\lambda(\alpha_1, 1, \gamma_1)$ the non-centrality parameter of the $\underline{T}_1$-test statistic, we have

$$\lambda(\alpha_{m-n}, m-n, \gamma) = \lambda(\alpha_1, 1, \gamma) \tag{7.28}$$

From this relation, α_{m-n} and thus the critical value of the overall model test, can be computed, once the power γ and the level of significance α_1 is chosen. Common values in case of geodetic networks, are $\alpha_1 = 0.001$ and $\gamma_0 = 0.80$, from which it follows that $\lambda_0 = \lambda(\alpha_1, 1, \gamma_0) = 17.075$.

Data snooping: Note that the MDB can be computed for each alternative hypothesis, once the vector c of that particular alternative hypothesis has been specified. Thus in case of data snooping one can compute m MDB's. As with the w_i-test statistic, also these MDB's simplify considerably when the variance matrix Q_y is diagonal. In that case we have

$$|\nabla_i| = \sigma_{y_i} \sqrt{\frac{\lambda_0}{1 - \sigma^2_{\hat{y}_i} / \sigma^2_{y_i}}} \tag{7.29}$$

where $\sigma^2_{y_i}$ is the *a priori* variance of the *i*th observation and $\sigma^2_{\hat{y}_i}$ is the *a posteriori* variance of this observation. We thus clearly see that a better precision of the observation as well as a larger amount in which its precision gets improved by the adjustment, will improve the internal reliability, that is, will result in a smaller MDB.

The higher dimensional case: When $q > 1$, the inversion is a bit more involved. In this case ∇ is a vector, which implies that its direction needs to be taken into account as well. We use the factorization $\nabla = \|\nabla\| d$, where d is a unit vector ($d^T d = 1$). If we substitute the factorization into (7.26) and then invert the result, we get

$$\|\nabla\| = \sqrt{\frac{\lambda(\alpha_q, q, \gamma)}{d^T C^T Q_y^{-1} Q_{\hat{e}} Q_y^{-1} C d}} \quad (d = \text{unit vector}) \tag{7.30}$$

The size of the model error now depends on the chosen direction vector d. But by letting d vary over the unit sphere in R^q, one can obtain the whole range of MDB's that can be detected with a probability γ.

With the MDB's one is said to describe the internal reliability. In practical applications however, one often not only wants to know the size of the model errors that can be detected with a certain probability, but also what their impact would be on the estimated parameters. After all, it are the estimated parameters, or functions thereof, which constitute the final result of an adjustment. The *external reliability* describes the influence of model errors of the size of the MDB's on the final result of the adjustment.

An example: MDB's for GPS code data

This example, taken from Teunissen [1991], shows how the receiver-satellite geometry affects the MDB's. Figure 7.1 shows the distribution of the six satellites and Table 7.1 shows the MDB's of a single-epoch, code-only solution using these satellites. In this case the average horizontal precision in terms of the horizontal dilution of precision was good, *HDOP*=1.23. The results of Table 7.1 show however that a sufficient precision in the position solution need not necessarily correspond with a sufficiently small MDB.

When the skyplot of Figure 7.1 is compared with Table 7.1, the following remarks can be made. First note that satellites 1 and 5 are close together. The two pseudoranges to these two satellites therefore check each other well. This explains the relatively small and almost equal MDB's for the first and fifth pseudorange. Also note that in the absence of satellite 3, one would expect because of symmetry, the MDB's of the pseudoranges 2 and 6 to be of the same order. With satellite 3 however, additional redundancy for checking pseudorange 2 enters, which explains why its MDB is smaller than that of pseudorange 6. Finally note the large value for the MDB of the fourth pseudorange. This is due to the fact that the line-of-sight vector to satellite 4 is not too far from the symmetry axis of the remaining line-of-sight vectors. See also Teunissen [1990].

The external reliability is given in Table 7.1. Shown are the influence of the six MDB's on the least-squares position solution. Note the large impact of the MDB of the fourth pseudorange. Also note that the MDB of the sixth pseudorange mainly affects the north component of position. Compare this with the skyplot of Figure 7.1.

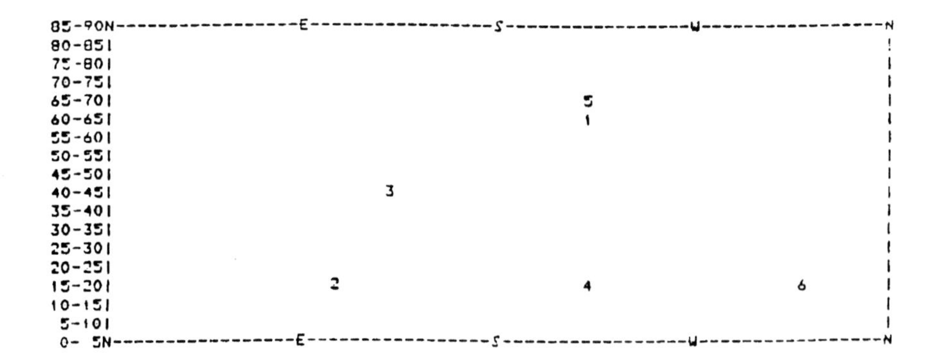

Figure 7.1. Skyplot of six GPS satellites

Table 7.1. MDB's of pseudorange divided by σ_p

No.	Satellite PRN	$\lvert\nabla\rvert / \sigma_p$
1	PRN 16	5.41
2	PRN 18	8.81
3	PRN 2	5.25
4	PRN 9	69.55
5	PRN 6	5.62
6	PRN 17	22.63

Table 7.2. GPS external reliability

Satellite	$\nabla x / \sigma_p$ (East)	$\nabla y / \sigma_p$ (North)	$\nabla z / \sigma_p$ (Up)
1 PRN 16	0.50	-0.19	-3.76
2 PRN 18	-4.13	-0.11	4.60
3 PRN 2	-1.41	0.52	-1.25
4 PRN 9	32.37	50.09	65.67
5 PRN 6	0.26	-0.57	-4.50
6 PRN 17	2.99	-15.14	6.01

7.3 VALIDATION OF RECURSIVE SOLUTIONS

Since recursive solutions enable *sequential*, rather than *batch*, processing of the measurement data, they are particularly suited for the processing of long time series of data and/or the estimation of time-varying parameters. In this section we develop the necessary tools for validating these recursive solutions. As a consequence, the quality control procedure becomes now recursive as well.

7.3.1 Recursive Least-Squares Estimation

In the previous section the parameter vector x was assumed to be time-invariant and completely unknown. In the present section we will relax these assumptions. First, the parameter vector x (sometimes also referred to as the state vector) will be allowed to vary with time. Thus instead of the notation x, we will now use the notation x_k to indicate the (discrete) time epoch k to which x refers. Secondly, we will assume that the parameter vector, instead of being deterministic, is now a random vector as well. This will be indicated by means of the underscore, $\underline{x}$.

Functional model: The functional model consists of two parts. As before we have an observational model that links the observables to the parameter vector:

$$\underline{y}_k = A_k \underline{x}_k + \underline{e}_k \tag{7.31}$$

where $\underline{y}_k$ is an $m_k \times 1$ vector of observables of epoch k; A_k is the corresponding design matrix; $\underline{x}_k$ is the $n \times 1$ random state vector of epoch k and $\underline{e}_k$ is the vector of measurement noise, for which it is again assumed that $E\{\underline{e}_k\} = 0$.

In addition to the observational model, we also have a so-called dynamic model. It describes the variability of the state vector with time

$$\underline{x}_k = \Phi_{k,k-1} \underline{x}_{k-1} + \underline{d}_k \tag{7.32}$$

where $\Phi_{k,k-1}$ is the transition matrix and $\underline{d}_k$ is the dynamic noise (sometimes also referred to as process noise or system noise), for which it is assumed that $E\{\underline{d}_k\} = 0$. Note that the transition matrix describes the link in time between the means of the state vectors of different epochs.

Stochastic model: There are three type of random vectors for which the dispersion needs to be specified. They are $\underline{e}_k, \underline{d}_k$ and $\underline{x}_0$. Once their dispersion is known, the dispersion of the other random vectors can be derived using the above functional model. It will be assumed that these three type of random vectors are not correlated and that

$$E\left\{\underline{e}_k \underline{e}_l^T\right\} = Q_{e_k} \delta_{k,l}\ ,\ E\left\{\underline{d}_k \underline{d}_l^T\right\} = Q_{d_k} \delta_{k,l}\ ,\ D\{x_0\} = Q_{x_0} \tag{7.33}$$

where $\delta_{k,l} = 1$ when $k = l$ and $\delta_{k,l} = 0$ when $k \neq l$.

Recursive least-squares: Depending on the application at hand, one might wish to obtain an estimate of the state vector at a certain time k, which depends on all observations taken up to and including time $k+l$. If $l<0$ the estimation process is called *prediction*. The estimated state vector depends then on the observations taken prior to the desired time of estimation. If $l=0$ the process is called *filtering* and in this case the state estimate depends on all the observations prior to and at time k. Finally, if $l>0$ the process is called *smoothing* and the state vector estimate depends on the observations taken prior to, at and after time k.

In these lecture notes we will restrict ourselves to recursive prediction and filtering. This is a 'forward' estimation process which uses past and present data to estimate the state vector. Again the least-squares principle is used to obtain the estimators, but now in recursive form. Due to the recursive nature, it is especially suited for real-time applications or applications where long time series of data need to be processed. Both situations occur in GPS data processing.

For each epoch, the recursive procedure consist of two steps: a prediction step (the time-update) and a filter step (the measurement-update). They are

$$\begin{aligned} \hat{x}_{k|k-1} &= \Phi_{k,k-1}\hat{x}_{k-1|k-1} \\ \hat{x}_{k|k} &= \hat{x}_{k|k-1} + K_k\left[y_k - A_k\hat{x}_{k|k-1}\right] \end{aligned} \tag{7.34}$$

where

$$K_k = Q_{\hat{x}_{k|k}} A_k^T Q_{e_k}^{-1} = Q_{\hat{x}_{k|k-1}} A_k^T \left[Q_{e_k} + A_k Q_{\hat{x}_{k|k-1}} A_k^T\right]^{-1}$$

The above pair of equations is often referred to as the Kalman-filter, Kalman [1960] and the matrix K_k as the Kalman gain-matrix. The first equation of the above pair, is the *time-update*. Using the dynamic model it propagates the state vector estimate of epoch $k-1$ to the next epoch k, thus giving the predicted state vector of epoch k. The second equation of the above pair is *the measurement-update*. It combines in a least-squares sense, the predicted state vector with the observations of epoch k, to produce the filtered state vector of epoch k.

The variance matrices of the predicted and filtered state vector are given as

$$\begin{aligned} Q_{\hat{x}_{k|k-1}} &= \Phi_{k,k-1} Q_{\hat{x}_{k-1|k-1}} \Phi_{k,k-1}^T + Q_{d_k} \\ Q_{\hat{x}_{k|k}} &= \left[I_n - K_k A_k\right] Q_{\hat{x}_{k|k-1}} \end{aligned} \tag{7.35}$$

These variance matrices are related as $Q_{\hat{x}_{k|k-1}} \geq Q_{\hat{x}_{k-1|k-1}}$ and $Q_{\hat{x}_{k|k}} \leq Q_{\hat{x}_{k|k-1}}$. Due to the presence of the system's noise, the predicted state is usually less precise than the previous filtered state. Due to the additional measurements however, the filtered state is usually of a better precision than its predicted counterpart.

More information on the above recursive filter, including extensions such as state vector augmentation for time-correlation or iterations due to non-linearities, can be found in Jazwinsky [1970], Sorenson [1970], Gelb [1974], Brammer and Siffling [1989] and Teunissen [1990c]. See also chapter *The GPS as a Tool in Global Geodynamics*, where the above recursive filter is applied for tropospheric zenith corrections. Factorization methods of the filter are discussed in Bierman [1973].

7.3.2 Model Validation

The above given recursive estimation procedure produces optimal estimators of the state vector with well defined statistical properties. The state estimators are unbiased and have minimum variance within the class of linear unbiased estimators. This optimality is only guaranteed however as long as the assumptions underlying the model hold. Misspecifications in the model will invalidate the results of estimation and thus also any conclusions based on them.

The null and alternative hypothesis: To have some safeguard against misspecifications, it is important to validate the model. As null hypothesis H_0 we consider the functional and stochastic model which was specified above. With respect to the model errors that might occur, we again restrict attention to misspecifications in the mean of $\underline{y}_k$, although the theory is also applicable to situations where one has to deal with misspecifications in the mean of $\underline{x}_k$, see Teunissen [1990a]. The null hypothesis and alternative hypothesis read then

$$H_0 : \underline{y}_k = A_k \underline{x}_k + \underline{e}_k \text{ versus } H_a : \underline{y}_k = A_k \underline{x}_k + C_k \nabla + \underline{e}_k \tag{7.36}$$

where the $m_k \times q$ matrices C_k specify the type of model error that occured. These matrices are zero for those epochs where the model error is absent and they are non-zero for the epochs where the model error is assumed present.

An important role in the process of model validation is played by the so-called *predicted residual*. They play a role which is similar to the role played by the least-squares residual of the previous section. The predicted residual is defined as the difference between the actual vector of observables and the vector of predicted observables, which is based on the predicted state vector: $\underline{v}_k = \underline{y}_k - A_k \hat{\underline{x}}_{k|k-1}$. Under the working hypothesis H_0, the predicted residual is Gaussian distributed with zero mean and dispersion $E\{\underline{v}_k \underline{v}_l^T\} = Q_{v_k} \delta_{kl}$ where $Q_{v_k} = [Q_{e_k} + A_k Q_{\hat{x}_{k|k-1}} A_k^T]$. Note that the predicted residuals do not correlate in time. It is this property, together with the fact that the predicted residual are readily available at each measurement update, which makes it possible to formulate a recursive quality control procedure that can be executed in real-time.

If we collect all k vectors of predicted residuals in one vector $\underline{v} = (\underline{v}_1^T, \ldots, \underline{v}_k^T)^T$ the null and alternative hypothesis, when expressed in terms of the predicted residuals, read

$$H_0 : E\{\underline{v}\} = 0 \text{ and } H_a : E\{\underline{v}\} = C_v \nabla \tag{7.37}$$

with dispersion $Q_v = \text{blockdiag}(Q_{v_1}, \ldots, Q_{v_k})$. The $\sum_{i=1}^{k} m_i \times q$ matrix $C_v = (C_{v_1}^T, \ldots, C_{v_k}^T)^T$ of the alternative hypothesis follows from the propagation of the C_k matrices through the time-update and measurement-update equations.

The test statistic: Due to the above formulation of the null hypothesis and alternative hypothesis in terms of the predicted residuals, the appropriate test statistic can be formulated in terms of these predicted residuals as well. This is an advantage over the ordinary least-squares residuals, since the predicted residuals are readily available from each measurement update, see (7.34). The test statistic for testing H_0 against H_a is given as

$$\begin{aligned} \underline{T}_q &= \frac{1}{q}\underline{v}^T Q_v^{-1} C_v \left[C_v^T Q_v^{-1} C_v\right]^{-1} C_v^T Q_v^{-1} \underline{v} \\ &= \frac{1}{q}\left[\sum_{i=1}^k C_{v_i}^T Q_{v_i}^{-1} \underline{v}_i\right]^T \left[\sum_{i=1}^k C_{v_i}^T Q_{v_i}^{-1} C_{v_i}\right]^{-1} \left[\sum_{i=1}^k C_{v_i}^T Q_{v_i}^{-1} \underline{v}_i\right] \end{aligned} \quad (7.38)$$

The second expression follows from the first, since Q_v is block diagonal. The test statistic has a central F-distribution with q and ∞ degrees of freedom, when H_0 is true, but a non-central F-distribution with non-centrality parameter λ, when the alternative hypothesis H_a is true.
Thus

$$H_0: \underline{T}_q \sim F(q,\infty,0) \text{ and } H_a: \underline{T}_q \sim F(q,\infty,\lambda) \quad (7.39)$$

with non-centrality $\lambda = \nabla^T C_v^T Q_v^{-1} C_v \nabla$. The null hypothesis is rejected when $T_q \geq F_{\alpha_q}(q,\infty,0)$.

7.3.3 Detection, Identification and Adaptation

In order to handle the various alternative hypotheses that might occur, we again make use of the three steps: detection, identification and adaptation. However, when compared to the DIA-procedure which is applicable for batch solutions, there are now some important additional requirements that need to be taken into account. First, since the estimation procedure is recursive, the DIA-procedure itself has to be recursive as well. In many real-time applications, it would not be of much use to be forced to perform the quality control in batch-mode while the estimation is done in recursive mode. That is, one would then like to execute the DIA-procedure in real-time as well, parallel to the estimation procedure. A second difference is that we are now dealing with a time-series of measurements. For the identification step this implies, that not only the most likely type of model error needs to be identified, but also its most likely time of occurrence. And finally as a third difference, we usually do not have the freedom in the adaptation step to remeasure the data. In case of a geodetic network for instance, one can usually remeasure those measurements that were identified as being corrupted with outliers. In case of a dynamic system for which real-time results are requested, this is rather difficult or even impossible. In that case one is forced to explicitly model the identified errors.

The recursive detection and identification steps of the DIA-procedure were first discussed in Teunissen and Salzmann [1988, 1989] and subsequently extended with the adaptation step in Teunissen [1990a,b]. Other applications of the theory of this section can be found in Salzmann [1991, 1994, 1995], Tiberius [1991], Abousalem et al. [1994], Cross [1994], De Jong [1994, 1996], Jin et al. [1995], Jin [1996, 1997], Gillissen and Elema [1996].

In the following we make a distinction between *local* model testing and *global* model testing. We speak of local model testing when the tests performed at time k only depend on the predicted state vector at time k and the observations of time k. If the test takes more than one epoch into account we speak of global testing. This distinction is introduced to have the possibility to discriminate between model errors that have either a local or a more global character. By definition, the local DIA-procedure can be executed in real-time. That is, the corrective action of the adaptation step is designed to coincide with the moment the model error occurred. Global testing is of course more powerful than local testing. Hence, for certain model errors local testing may be too insensitive. Model errors that slowly built up as time proceeds may have a high probability of passing the local tests unnoticed. For such type of errors global testing is needed, due to their built in memory capabilities. The price paid for the higher power of the global tests is the delay between the time the model error started to occur and the time of detection. But such a delay will be acceptable if it is considered more important to detect the model error then not detect it at all.

Local DIA-procedure

For the local case we will assume that no invalidation of the model has taken place prior to the present time of testing k. This implies that attention can be restricted to the following *local* hypotheses

$$H_0^k : \; E\{\underline{v}_k\} = 0 \text{ and } H_a^k : \; E\{\underline{v}_k\} = C_{v_k}\nabla \tag{7.40}$$

Local detection: In order to test the overall validity of the local hypothesis, no restrictions are imposed on the mean of the predicted residual under H_a. Hence $H_0^k : E\{\underline{v}_k\} = 0$ and $H_a^k : \; E\{\underline{v}_k\} \in R^{m_k}$ .This implies that matrix C_{v_k} is chosen as a square and regular matrix and thus that $q = m_k$. Since $C_{v_{k_i}} = 0$ for $i < k$, the corresponding test statistic for the *local overall model* (LOM) test follows from (7.38) as

$$\underline{T}_{\text{LOM}}^k = \frac{\underline{v}_k^T Q_{v_k}^{-1} \underline{v}_k}{m_k} \tag{7.41}$$

Its expectation under H_0 equals one. The LOM-test for the detection of unspecified model errors in the above local null hypothesis reads therefore as follows: An unspecified model error is considered present at time k when $T^k_{\mathrm{LOM}} \geq F_{\alpha_{m_k}}(m_k, \infty, 0)$.

Local identification: For identification, candidate alternative hypotheses need to be specified explicitly. This is done through the matrix C_{v_k}. In the local case we have $C_{v_k} = C_k$. We will restrict attention to one dimensional model errors. Thus $q = 1$ and C_k reduces to a vector, which will be denoted with the lower case letter c_k. The null and alternative hypothesis are then given as: $H^k_0 : E\{\underline{v}_k\} = 0$ and $H^k_a : E\{\underline{v}_k\} = c_k \nabla$.
Since $C_{v_i} = 0$ for $i < k$ and $C_{v_k} = c_k$, it follows from (7.38) that we can write $\underline{T}_{q-1} = (\underline{t}^k)^2$ with

$$\underline{t}^k = \frac{c_k^T Q_{v_k}^{-1} \underline{v}_k}{\sqrt{c_k^T Q_{v_k}^{-1} c_k}} \tag{7.42}$$

This is the test statistic which is used for identification. Note that there are as many of these test statistics as there are candidate alternative hypotheses. Each one of them is specified with the appropriate c_k-vector. The test statistic has a standard normal distribution $N(0,1)$ under H^k_0. Identification consists now of the following steps. First a search is performed for the largest value of $|t^k|$ among all specified alternative hypotheses. This will give us then the most likely model error. The test statistic of the most likely model error is then compared with the critical value $N_{\alpha_1/2}(0,1)$. If it exceeds this value, the model error is considered sufficiently likely.

The power of the above test statistic is measured by its MDB, which reads

$$\left|\nabla^k\right| = \sqrt{\frac{\lambda_0}{c_k^T Q_{v_k}^{-1} c_k}} \tag{7.43}$$

Local adaptation: After identification of the most likely alternative hypothesis, adaptation of the model and thus of the estimation equations is needed to eliminate the presence of biases in the filtered state vector. In the local case, immediate corrective action is possible by resetting the filter so that the adapted state and its variance matrix read at time k as

$$\begin{aligned} \underline{\hat{x}}^a_{k|k} &= \underline{\hat{x}}^0_{k|k} - K_k c_k \underline{\hat{\nabla}}^k \\ Q^a_{\hat{x}_{k|k}} &= Q^0_{\hat{x}_{k|k}} + K_k c_k \sigma^2_{\hat{\nabla}^k} c_k^T K_k^T \end{aligned} \tag{7.44}$$

where $\underline{\hat{\nabla}}^k$ is the least-squares estimator of the identified model error and where $\sigma^2_{\hat{\nabla}^k}$ is its variance. Both are easily constructed from the identification test statistic as

$$\underline{\hat{\nabla}}^k = \underline{t}^k / \sqrt{c_k^T Q_{v_k}^{-1} c_k} \quad \text{and} \quad \sigma^2_{\hat{\nabla}^k} = \frac{1}{c_k^T Q_{v_k}^{-1} c_k}$$

In case of GPS, the model errors that are likely to occur are outliers in the code data and/or cycle slips in the phase data. For both these type of model errors the c_k-vector will be of the form $(0,\ldots,0,1,0,\ldots,0)^T$ where the 1 corresponds with the corrupted entry of $\underline{y}_k$. For both these type of model errors the above adaptation only needs to be performed once. After adaptation at time k, one can switch back to the standard filter under H_0.

An example: outliers in GPS code data

By means of an example taken from Tiberius [1996] it will be shown how the DIA-procedure performs. The model on which this example is based is one of instantaneous relative positioning using single frequency code data. Both receivers tracked the same 7 satellites every epoch using a sampling interval of 10 seconds. Atmospheric delays were neglected because of the short baseline length ($\approx$ 2.3 km). The standard deviation of the undifferenced code observable was set at $\sigma_p = 30$ cm. The skyplot over the observation period is shown in Figure 7.2

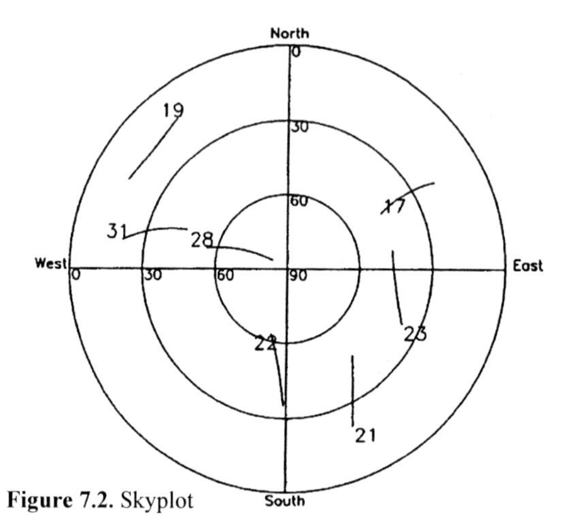

Figure 7.2. Skyplot

At six different epochs over the observation period, one of the seven pseudoranges was artificially corrupted by an outlier. The epoch number (k), the corrupted pseudorange (PRN) and the size of the outlier (∇) are shown in the first three columns of Table (7.3). The last column shows the least-squares estimate $\hat{\nabla}^k$ of the outlier. Over the observation period the MDB's were about 2-3 metre.

Table 7.3. Pseudorange outliers and their estimated value

Epoch	PRN	∇ (m)	$\hat{\nabla}^k$ (m)
k=50	22	10.0	9.5
k=100	31	-10.0	-9.3
k=150	19	10.0	9.4
k=200	17	5.0	5.1
k=250	23	2.0	2.1
k=300	31	2.0	2.2

The results of the detection step are depicted in Figure 7.3. Along the vertical axis $T^k_{\mathrm{LOM}} / F_\alpha(3,\infty,0)$ is shown. The null hypothesis is rejected when this value exceeds one. The figure clearly shows that model errors are indeed detected at the above given six different epochs. Identification of the model errors is therefore needed for these six epochs. For each epoch this is done with the identification test statistic $\underline{t}^k$. For each epoch, there are seven of such test statistics, one for each (single differenced) pseudorange.

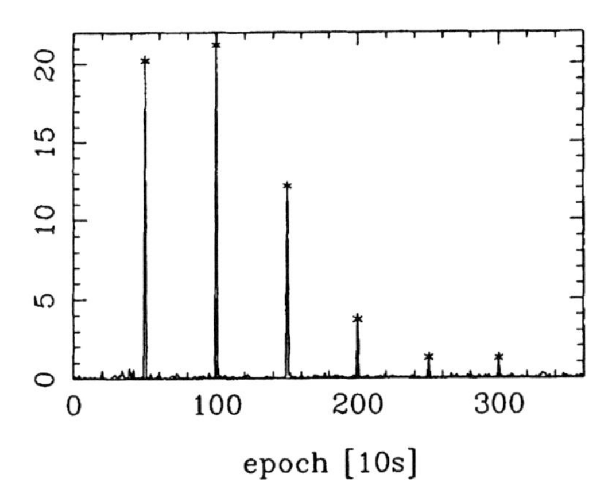

Figure 7.3. Time series of $T^k_{\mathrm{LOM}} / F_\alpha(3,\infty,0)$

The values of these test statistics are shown in Table 7.4. The in absolute value largest test statistics are underlined. They indicate the pseudoranges which are most likely corrupted by outliers. The values underlined all exceed the critical value $N_{\alpha_1/2}(0,1) = 3.29$. All six outliers are correctly detected and identified.

Table 7.4. Sample values per epoch of $\underline{t}^k$ for the seven pseudoranges.

PRN	k=50	k=100	k=150	k=200	k=250	k=300
17	1.9	-4.6	-6.4	<u>6.8</u>	-3.4	0.2
19	7.5	14.8	<u>12.4</u>	-3.9	0.6	-3.1
21	-7.5	5.5	-1.8	-0.1	-0.5	0.3
22	<u>15.9</u>	3.5	3.7	2.5	-1.0	-0.7
23	-2.7	-4.0	2.5	-6.0	<u>3.9</u>	0.7
31	-2.2	<u>-16.3</u>	-11.4	1.7	1.1	<u>3.9</u>
28	-13.9	6.8	3.5	-0.4	-2.7	-3.6

Table 7.5. Internal and external reliability $(\alpha = 0.001,\ \gamma = 0.80)$

PRN	MDB (cycle)	position bias (m)
17	0.16	0.12
19	0.16	0.07
21	0.19	0.10
22	0.13	0.06
23	0.12	0.06
31	0.13	0.07
28	0.13	0.05

An example: cycle slip MDB and external reliability

The same set-up is used as in the previous example. Instead of the single frequency code data however, we now use single frequency phase data $(\sigma = 0.03 \text{ m})$. The integer ambiguities are assumed to be resolved. This implies that the carrier phase data will act as very precise code data. For the theory of integer ambiguity fixing, we refer to the chapter *GPS Carrier Phase Ambiguity Fixing Concepts.*

Based on local testing, the seven cycle slip MDB's together with their potential impact on the baseline, are shown in Table 7.5. Figure 7.4 shows the size and the orientation of the seven potential baseline biases.

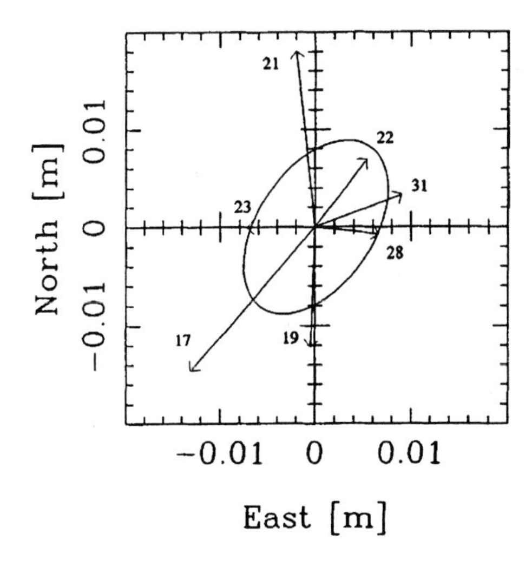

Figure 7.4. External reliability: potential bias in horizontal position.

Global DIA-procedure

The LOM-test discussed above may turn out to be too insensitive to detect global unmodelled trends. For those cases we consider the following *global* hypotheses

$$H_0^{l,k}:\ E\{\underline{v}^{l,k}\}=0 \text{ and } H_a^{l,k}:\ E\{\underline{v}^{l,k}\}=C_v^{l,k}\nabla \tag{7.45}$$

with the $\sum_{i=l}^{k} m_i \times 1$ vector of predicted residuals $\underline{v}^{l,k}=\left(\underline{v}_l^T,\ldots,\underline{v}_k^T\right)^T$.

Global detection: In order to test the overall validity of the global hypothesis, no restrictions are imposed on the mean of the vector of predicted residuals $\underline{v}^{l,k}$ under $H_a^{l,k}$. Hence: $H_0^{l,k}$: $E\{\underline{v}_k\}=0$ and $H_a^{l,k}$: $E\{\underline{v}_k\}\in R^{\sum_{i=l}^{k} m_i}$. This implies that matrix $C_v^{l,k}$ is chosen as a square and regular matrix and thus that $q=\sum_{i=l}^{k} m_i$. Since $C_{v_i}=0$ for $i<l$, the corresponding test statistic follows from (7.38) as

$$\underline{T}_{\text{GOM}}^{l,k}=\frac{\sum_{i=l}^{k}\underline{v}_i^T Q_{v_i}^{-1}\underline{v}_i}{\sum_{i=l}^{k} m_i} \tag{7.46}$$

Note that this test statistic reduces to its local counterpart when $l = k$. As with the LOM test statistic, its expectation equals one under the null hypothesis. The global test statistic can be written in recursive form as

$$\underline{T}_{\text{GOM}}^{l,k} = \underline{T}_{\text{GOM}}^{l,k-1} + G_{l,k}\left[\underline{T}_{\text{LOM}}^{k} - \underline{T}_{\text{GOM}}^{l,k-1}\right] \tag{7.47}$$

with the scalar gain: $G_{l,k} = m_k / \sum_{i=l}^{k} m_i$. The *global overall model* (GOM) test reads now as follows: An unspecified global model error is considered present in the time interval $[l,k]$ if and only if $\underline{T}_{\text{GOM}}^{l,k} \geq F_\alpha(\sum_{i=l}^{k} m_i, \infty, 0)$.

A practical problem is the choice of l, the time that the model error is assumed to be starting to occur. Since the starting time of the model error is unknown a priori, one has to start in principle with $l = 1$. A fixed value of l however, implies a growing memory recursion, with the practical problem of a possible long delay in time of detection. Rejection of $H_0^{l,k}$ at time k with the GOM-test may imply namely that a global model error started to occur as early as time $l = 1$. In order to reduce the time of delay, it is worthwhile to consider a moving *window* of length N by constraining l to $k - N + 1 \leq l \leq k$. When choosing N one of course has to make sure that the detection power of the test is still sufficient. This is typical a problem one should take into consideration when designing the filter. Instead of using a finite window, one could also consider using a fading window. This implies that an alternative gain $G_{l,k}$ is used, see Teunissen [1990a].

Global identification: In order to identify global model errors that have been detected at time k for the interval $[l,k]$, alternative hypotheses need to be specified explicitly. This is done through the matrix $C_v^{l,k}$. As before we restrict attention to one dimensional model errors in the observational model. Thus $C_v^{l,k}$ reduces to a vector , $c_v^{l,k} = (c_{v_l}^T, \ldots, c_{v_k}^T)^T$ and the hypotheses read: $H_0^{l,k}$: $E\{\underline{v}^{l,k}\} = 0$ and $H_a^{l,k}$: $E\{\underline{v}^{l,k}\} = c_v^{l,k}\nabla$. With $C_{v_i} = 0$ for $i < l$, it follows from (7.38) that we can write $\underline{T}_{q=1} = (\underline{t}^{l,k})^2$ with

$$\underline{t}^{l,k} = \frac{\sum_{i=l}^{k} c_{v_i}^T Q_{v_i}^{-1} \underline{v}_i}{\sqrt{\sum_{i=l}^{k} c_{v_i}^T Q_{v_i}^{-1} c_{v_i}}} \tag{7.48}$$

This test statistic has a standard normal distribution under $H_0^{l,k}$. Just like in the case of detection, we can write the global identification test statistic in recursive form, so that the local identification test statistic is used as input for each new epoch:

$$\left(\underline{t}^{l,k}\right)^2 = \left(\underline{t}^{l,k-1}\right)^2 + g_{l,k}\left[\left(\underline{t}^{k}\right)^2 - \left(\underline{t}^{l,k-1}\right)^2\right] \tag{7.49}$$

with the scalar gain: $g_{l,k} = c_{v_k}^T Q_{v_k}^{-1} c_{v_k} / \sum_{i=l}^{k} c_{v_i}^T Q_{v_i}^{-1} c_{v_i}$. In order to perform the recursion, we still need the vectors c_{v_i} for $i = l,\ldots,k$. They can be computed recursively as

$$\begin{aligned} c_{v_i} &= c_i - A_i x_{i,l} \ , i = l,\ldots,k \\ x_{i+1,l} &= \Phi_{i+1,i}\left[x_{i,l} + K_i c_{v_i}\right], \ x_{l,l} = 0 \end{aligned}$$

where c_i specifies the one dimensional model error for each epoch *i.* Note that the above recursion corresponds with the propagation of the recursive estimation process itself.

The power of the above test statistic is measured by its MDB, which reads

$$\left|\nabla^{l,k}\right| = \sqrt{\frac{\lambda_0}{\sum_{i=l}^{k} c_{v_i}^T Q_{v_i}^{-1} c_{v_i}}} \tag{7.50}$$

Strictly speaking the above test statistic (7.48) has to be computed for each alternative hypothesis considered and for each epoch $k \geq l$. However, since l is unknown a priori, one has to start in principle with $l = 1$. This implies that one has to compute k number of test statistics per alternative hypothesis at the time of testing k. As a result one obtains a *test matrix* of increasing order with $|t^{l,k}|$ as its entries. An example is shown in Figure 7.5a.

$$\begin{bmatrix} t^{1,1} & t^{1,2} & t^{1,3} & \cdots \\ & t^{2,2} & t^{2,3} & \cdots \\ & & t^{3,3} & \cdots \\ & & & \ddots \end{bmatrix}_{H_a} \quad \begin{bmatrix} t^{1,1} & t^{1,2} & \bullet & \bullet \\ & t^{2,2} & t^{2,3} & \bullet \\ & & t^{3,3} & \ddots \\ & & & \ddots \end{bmatrix}_{H_a} \quad \begin{bmatrix} \bullet & t^{1,2} & \bullet & \bullet \\ & \bullet & t^{2,3} & \bullet \\ & & \bullet & \ddots \\ & & & \bullet \end{bmatrix}_{H_a}$$

(a) (b) (c)

Figure 7.5. Test matrix for identification with (a) no window, (b) a moving window with $k-2 < l \leq k$, and (c) a moving window with $k-2 < l \leq k-1$

This is clearly unpractical, both from a computational point of view as well as because of the possible increase of the delay in time of identification. Fortunately not all entries of the test matrix may be necessary if one studies the power of the test statistics. Although the power will increase theoretically for an increasing size of the interval $[l,k]$, the gain in power could be negligible for all practical purposes. This motivates, in accordance with our discussion of detection, the use of a moving window. This is shown in Figure 7.5b for the case $k-N+1 \leq l \leq k$ and in Figure 7.5c for the case $k-N+1 \leq l \leq k-M$. The rational behind this last constraint is that in some applications the test statistic may be too insensitive for global identification if $l > k - M$

With the windows introduced, we are now in a position to describe the identification procedure. At the time of testing k, one first determines per alternative hypothesis the value of l in the window for which $|t^{l,k}|$ is at its maximum. In other words, the kth column of the test matrix is searched for the in absolute value largest entry. The corresponding row number of the test matrix then identifies l as the most likely time of occurrence of the model error if the corresponding alternative hypothesis would be true. In order to find both the most likely alternative hypothesis and most likely value of l, the values of $\max_{k-N+1 \leq l \leq k-M} |t^{l,k}|$ for the different alternative hypotheses are compared. The maximum of this set finally identifies the most likely time of occurrence l and the most likely alternative hypothesis. Its likelihood is then tested against the critical value $N_{\alpha_1/2}(0,1)$.

Global adaptation: In the global case one is confronted with a delay in time of identification. This implies that in principle one has to correct the filtered states from time l to the present time k. This as a consequence would involve smoothing, which also can be executed recursively. However, adaptation based on smoothing is a corrective action after the fact, which may not be needed for real-time applications. It could also be that the recursive smoothing presents a too heavy computational burden. If recursive smoothing is not considered a viable option, the following approach to adaptation presents itself. At time k, when the most likely alternative hypothesis has been identified, one resets the filter by correcting the filtered state vector and its variance matrix as

$$\begin{aligned} \hat{\underline{x}}^a_{k|k} &= \hat{\underline{x}}^0_{k|k} - \Phi_{k,k+1}\, x_{k+1,l}\, \hat{\underline{\nabla}}^{l,k} \\ Q^a_{\hat{x}_{k|k}} &= Q^0_{\hat{x}_{k|k}} + \Phi_{k,k-1} x_{k+1,l} \sigma^2_{\hat{\nabla}^{l,k}} x^T_{k+1,l} \Phi^T_{k,k-1} \end{aligned} \tag{7.51}$$

where

$$\hat{\underline{\nabla}}^{l,k} = \underline{t}^{l,k} \Big/ \sqrt{\Sigma_{i=l}^{k} c^T_{v_i} Q^{-1}_{v_i} c_{v_i}}$$

is the least-squares estimator of the identified model error and where

$$\sigma^2_{\hat{\nabla}^{l,k}} = 1 \Big/ \sqrt{\Sigma_{i=l}^{k} c^T_{v_i} Q^{-1}_{v_i} c_{v_i}}$$

is its variance. Both can be computed in recursive form, Teunissen [1990a].

For outliers in GPS code data or slips in GPS phase data, the above adaptation only needs to be performed at time k. After adaptation at time k, one can switch back to the standard filter under the null hypothesis.

The above approach of adaptation implies of course that the filtered states remain biased between times l and k. This bias may be considered negligible however, if the built up of the global model error is still too small to be detected with the GOM-test. This again points out that when designing the filter one should carefully consider the choice of window length and the detection power of the test in relation to the biases one is willing to accept.

7.4 APPLICATION TO SOME GPS MODELS

In this section we will apply the theory of the previous two sections to some common single baseline models. Special attention is given to the test statistics for the identification of outliers in the code data and cycle slips in the phase data. The with the test statistics corresponding MDB's are also presented.

7.4.1 The Geometry-Free Model

The relative receiver-satellite geometry is absent from the model when the GPS observation equations are parametrized in terms of the receiver-satellite ranges instead of in the baseline components. This model is therefore referred to as the geometry-free model. The single-differenced (SD) phase and code observation equations of the geometry-free model are given for a single epoch i as

$$\begin{aligned} \phi_1(i) &= \rho(i) - \mu_1 I(i) + \lambda_1 a_1 \\ \phi_2(i) &= \rho(i) - \mu_2 I(i) + \lambda_2 a_1 \\ p_1(i) &= \rho(i) + \mu_1 I(i) \\ p_2(i) &= \rho(i) + \mu_2 I(i) \end{aligned} \tag{7.52}$$

where ϕ_1 and ϕ_2 are the SD phase observables on L_1 and L_2, expressed in units of range rather than in units of cycles; p_1 and p_2 are the SD code observables on L_1 and L_2; ρ is the unknown parameter that contains in lumped form the SD range from the two receivers to the same satellite, plus the receiver clock errors and the SD-form of the tropospheric delay; I is the SD-form of the unknown ionospheric delay; and a_1 and a_2 are the unknown SD carrier phase ambiguities; they contain the non-zero initial phases plus the integer ambiguities. The known wavelengths are denoted as λ_1 and λ_2. Since the ionospheric delay is to a first order inversely proportional to the square of the frequency, we have to the same degree of approximation

$$\mu_1 = \frac{\lambda_1}{\lambda_2} = \frac{60}{77}, \; \mu_2 = \frac{\lambda_2}{\lambda_1} = \frac{77}{60}$$

It will be assumed that the GPS observables are uncorrelated between channels nor correlated in time. For k epochs $(i = 1, \ldots, k)$, their variance matrix is assumed to be given as

$$I_k \otimes \text{diag}\left(\sigma_\phi^2, \sigma_\phi^2, \sigma_p^2, \sigma_p^2\right)$$

where I_k is the unit matrix of order k and $\otimes$ denotes the Kronecker product.

Single frequency, short baseline case

The ionospheric delays may be neglected if the interstation distance between the two receivers is short enough. If we also assume to have only single frequency data available, the above set of SD observation equations reduces to

$$\begin{aligned} \phi(i) &= \rho(i) + \lambda a \\ p(i) &= \rho(i) \end{aligned} \tag{7.53}$$

where the index denoting the frequency has been omitted. Elimination of $\rho(i)$ gives $\phi(i) - p(i) = \lambda a$, from which it follows that the single-epoch ambiguity variance is given as

$$\sigma_{a(i)}^2 = \frac{\sigma_p^2}{\lambda^2}(1 + \varepsilon),$$

with the phase-code variance ratio $\varepsilon = \sigma_\phi^2 / \sigma_p^2$. Note that ε is very small indeed (e.g. 10^{-4}), due to the relatively poor precision of the code data. For k epochs, the redundancy equals $(k-1)$ provided that the ambiguity a remains constant.

The above model is one of the simplest models one will meet in GPS data processing. In the following we will present the test statistics for the identification of respectively a spike (outlier) in the code observable and a slip (cycleslip) in the phase observable. We also give the corresponding MDB's. They show how well a spike or a slip can be found with the test statistics.

Spike in code p at epoch l: The test statistic for the identification of an outlier in the code observable $p(l)$ using k epochs of data, reads

$$t^{l,k} = \frac{\lambda\left[a(l) - \hat{a}_k\right]}{\sigma_p\sqrt{(1+\varepsilon)\left(1-\frac{1}{k}\right)}} \tag{7.54}$$

with $\lambda a(l) = \phi(l) - p(l)$ and $\lambda \hat{a}_k = \frac{1}{k}\sum_{i=1}^{k}\left(\phi(i) - p(i)\right)$. Note that the test statistic is based on the difference between the single-epoch ambiguity solution and the least-squares ambiguity based on k epochs. The test statistic is normalized to have a standard normal distribution under H_0.

The corresponding MDB for an outlier in the code observable is given as

$$|\nabla| = \sigma_p\sqrt{\frac{(1+\varepsilon)\lambda_0}{1-\frac{1}{k}}} \tag{7.55}$$

Note that the MDB becomes infinite when $k = 1$. This is understandable, since redundancy is absent in case of only one epoch. With $\lambda_0 \approx 17$ and $\varepsilon \approx 0$, it follows for k large enough that

$$|\nabla| \approx 4.13\,\sigma_p = 0.88 \text{ mtr. for } \sigma_p^2 = 2(15\text{cm})^2$$

The MDB is thus at best more than four times the SD standard deviation of code.

Slip in carrier ϕ at epoch *l*: The test statistic for the identification of a cycle slip in the phase observable $\phi(l)$ using k epochs of data, reads

$$t^{l,k} = \frac{\lambda\left[\hat{a}_{l-1} - \hat{a}_k\right]}{\sigma_p\sqrt{(1+\varepsilon)\left(\frac{1}{l-1}-\frac{1}{k}\right)}} \tag{7.56}$$

Note that the test statistic is based on the difference between the least-squares ambiguity using the first $l-1$ epochs and the one which uses all k epochs. If the cycle slip occurred at epoch l, the first estimate is *unbiased* by the slip, whereas the second is *biased* by it.

The corresponding MDB for a slip in the phase observable reads in units of cycles, as

$$|\nabla| = \frac{\sigma_p}{\lambda}\sqrt{\frac{(1+\varepsilon)\lambda_0}{(l-1)\left(1-\frac{l-1}{k}\right)}} \tag{7.57}$$

Note that the MDB becomes infinite in case $k = 1$ and in case $l = 1$. In the first case there is no redundancy and in the second case the slip in the ambiguity cannot be separated from the ambiguity itself.

In case the slip occurred at the epoch of testing itself, we have $l = k$ and the MDB reduces to

$$|\nabla| = \frac{\sigma_p}{\lambda}\sqrt{\frac{(1+\varepsilon)\lambda_0}{1-\frac{1}{k}}}$$

Compare this with the outlier MDB of (7.55). It shows that one can only expect to find slips in the phase data, if these slips are a multiple of one cycle. The model is simply to weak to be able to find slips of one cycle, when $l = k$. In order to find smaller slips, k needs to be much larger than l. For k large enough, we get

$$|\nabla| = \frac{\sigma_p}{\lambda}\sqrt{\frac{(1+\varepsilon)\lambda_0}{(l-1)}}$$

which shows that the MDB gets smaller for larger l. The overall conclusion for the present model reads thus that one will never be able to find slips of one cycle if they occur at the beginning or at the end of the time series. In order to find sufficiently small slips, k must be much larger than l and l itself must also be large enough.

Dual frequency phase, short baseline case

The above single frequency, geometry-free model turned out to be too weak to identify sufficiently small slips in the phase data. Let us now investigate what the inclusion of phase on a second frequency brings. Instead of model (7.53), we now have the SD observation equations

$$\begin{aligned} \phi_1(i) &= \rho(i)+\lambda_1 a_1 \\ \phi_2(i) &= \rho(i)+\lambda_2 a_2 \\ p_1(i) &= \rho(i) \end{aligned} \tag{7.58}$$

Elimination of $\rho(i)$ gives the pair of equations

$$\begin{aligned} \phi_1(i)-p(i) &= \lambda_1 a_1 \\ \phi_2(i)-p(i) &= \lambda_2 a_2 \end{aligned}$$

from which the single-epoch ambiguity variance matrix follows as

$$Q_{a(i)} = \frac{\sigma_p^2}{\lambda_1\lambda_2}\begin{bmatrix} \mu_2(1+\varepsilon) & 1 \\ 1 & \mu_1(1+\varepsilon) \end{bmatrix}$$

Note that the two ambiguities are highly correlated since $\sigma_\phi << \sigma_p$. The least-squares solutions of the two ambiguities, $\hat{a}_{1,k}$ and $\hat{a}_{2,k}$, are given as the time averages of $\phi_1(i) - p(i)$ and $\phi_2(i) - p(i)$ times $1/\lambda$.

Spike in code *p* at epoch *l*: The test statistic for the identification of an outlier in the code observable $p(l)$ using *k* epochs of data, reads

$$t^{l,k} = \frac{\lambda_1\left[a_1(l) - \hat{a}_{1,k}\right] + \lambda_2\left[a_2(l) - \hat{a}_{2,k}\right]}{\sigma_p\sqrt{2(1+2\varepsilon)\left(1-\frac{1}{k}\right)}} \tag{7.59}$$

Note that the test statistic is based on the L_1 and L_2 *sum* of the difference between the single-epoch ambiguity solution and the least-squares ambiguity based on *k* epochs.

The corresponding MDB for an outlier in the code observable is given as

$$|\nabla| = \sigma_p\sqrt{\frac{\left(1+\frac{1}{2}\varepsilon\right)\lambda_0}{1-\frac{1}{k}}} \tag{7.60}$$

Note, since $\varepsilon \approx 0$, that this MDB is practically identical to the one of (7.55). Thus the inclusion of phase on a second frequency did not really improve the situation for outlier detection in the code observables.

Slip in phase ϕ_1 at epoch *l*: The test statistic for the identification of a cycle slip in the phase observable $\phi_1(l)$ using *k* epochs of data, reads

$$t^{l,k} = \frac{(1+\varepsilon)\lambda_1\left[\hat{a}_{1,l-1} - \hat{a}_{1,k}\right] - \lambda_2\left[\hat{a}_{2,l-1} - \hat{a}_{2,k}\right]}{\sigma_\phi\sqrt{(2+\varepsilon)(1+\varepsilon)\left(\frac{1}{l-1}-\frac{1}{k}\right)}} \tag{7.61}$$

Note, since $\varepsilon \approx 0$, that the test statistic is essentially based on the L_1 and L_2 *difference* of the difference between the least-squares ambiguity using the first $l-1$ epochs and the one which uses all *k* epochs.

The corresponding MDB for a cycle slip in the L_1 phase observable reads in units of cycles, as

$$|\nabla| = \frac{\sigma_\phi}{\lambda_1}\sqrt{\frac{(2+\varepsilon)\lambda_0}{(1+\varepsilon)(l-1)\left(1-\frac{l-1}{k}\right)}} \tag{7.62}$$

If we neglect the very small phase-code variance ratio ε and set $l = k$, we get

$$|\nabla| \approx \frac{\sigma_\phi}{\lambda_1}\sqrt{\frac{2\lambda_0}{1-\frac{1}{k}}}$$

For $k>1$, this MDB is very small indeed. This shows that the inclusion of phase on a second frequency has a very favourable impact on ones ability to find small slips in the phase data.

Dual frequency, long baseline case

So far, the SD ionospheric delays have been neglected. This is allowed when the interstation distance between the two receivers is sufficiently small. For longer baselines however, the SD ionospheric delays cannot be neglected anymore. Inclusion of the ionospheric delays as unknown parameters into the model will of course weaken the model. It will therefore also have its deteriorating effect on ones ability to detect model errors in the model.

Dual frequency data are needed perse when the ionospheric delays are included in the model. We therefore assume to be working with the four SD observation equations of (7.52). Again we can apply the theory of the previous two chapters to derive the appropriate test statistics with their corresponding MDB's. In order to infer how well slips in the phase data can be found, we need the cycle slip MDB. For a slip in ϕ_1 that starts at epoch l, using $k \geq l$ epochs of data, the MDB reads

$$|\nabla| = \frac{\sigma_p}{\lambda_1}\sqrt{\frac{\left[(1+\varepsilon)^2+4\upsilon_1^2\varepsilon\right]\lambda_0}{\left[\upsilon_1^2\left(1+\upsilon_2^2\right)+\varepsilon\right](l-1)\left(1-\frac{l-1}{k}\right)}} \approx \frac{0.6\,\sigma_p}{\lambda_1\sqrt{(l-1)\left(1-\frac{l-1}{k}\right)}} \tag{7.63}$$

with

$$\upsilon_1 = \frac{\mu_2+\mu_1}{\mu_2-\mu_1} \text{ and } \upsilon_2 = \frac{2\mu_2}{\mu_1+\mu_2}$$

This MDB is somewhat smaller than the slip MDB (7.57) of the single frequency, short baseline case, but it is clearly much larger than the slip MDB (7.62). This shows the deteriorating effect of being forced to include the ionospheric delays as unknown parameters into the model. For $l=k$, the above MDB (7.63) equals about 1 cycle in case $\sigma_p^2 = 2(30cm)^2$ and about 0.5 cycle in case $\sigma_p^2 = 2(15cm)^2$. Smaller slips can be found when $l<k$. But this of course has the practical drawback that no immediate identification of the slip is possible. That is, one has to wait $(k-l+1)$ epochs before the test statistic is powerful enough to identify smaller slips.

An alternative way to obtain more powerful test statistics is to strengthen the model by making use of a priori information on the ionospheric delays. For instance, the ionosphere is known to decorrelate as function of the interstation distance between the two receivers. This suggests an a priori weighting of the ionospheric delays as function of the baseline length. In this approach, an ionospheric observation equation is added to the original model (7.52) for each epoch. The sample value for the ionospheric delay can then be taken from an externally provided ionospheric model, see e.g. Georgiadou [1994], Wild [1994], Wanninger [1994]. In some applications it may even suffice to take zero as sample value. The a priori uncertainty in the ionospheric delays is modelled through an appropriate variance covariance matrix, see e.g. Schaffrin and Bock [1988], Wild and Beutler [1991], Schaer [1994], Teunissen [1996]. See also the chapter on *Medium Distance GPS Measurements*.

Single receiver quality control

Quality control of GPS data is also possible in case data of only a single GPS receiver is processed. In this case however, the spatial decorrelation property of the ionosphere is not of much use. One is namely forced by definition of the single receiver set-up, to work with undifferenced data. Despite this drawback though, there is still some a priori information on the characteristics of the ionospheric delays left, namely that in general these delays behave as more or less smooth functions of time. This suggests that the undifferenced ionospheric delays can be described by a dynamic model based on a polynomial of order s. Such an approach can then be easily combined with a standard Kalman filter set-up. The $(s+1)\times(s+1)$ transition matrix of the ionospheric delays and its time derivatives reads then

$$\Phi_{k,k-1} = \begin{bmatrix} 1 & \Delta t & . & . & . & \frac{\Delta t^{s}}{s!} \\ & 1 & \Delta t & . & . & \frac{\Delta t^{(s-1)}}{(s-1)!} \\ & & . & . & . & . \\ & & & . & . & . \\ & & & & . & . \\ & & & & & 1 \end{bmatrix}$$

If we assume the input of the dynamic system to consist of white noise with spectral density q^{s+1}, the (i,j)th entry of the variance matrix Q_{d_k} of the process noise reads

$$Q_{d_k}(i,j) = q^{s+1} \frac{\Delta t^{2s-i-j+1}}{(s-i)!(s-j)!(2s-i-j+1)}$$

For the single receiver case, the observation equations are identical in form as the ones given in (7.52). The only difference is that they should now be read as *undifferenced* observation equations. Thus also the unknown parameters are now of an undifferenced nature. This implies for instance, that ρ is now made up of the receiver-satellite range, the receiver and satellite clock error and the tropospheric delay.

Based on the theory of the previous two sections, single receiver quality control algorithms have been developed in De Jong [1994], Jin et al. [1995] and Jin [1996]. For dual frequency phase and code data $\left(\sigma_\phi \approx 2\,\text{mm.}, \sigma_p \approx 30\,\text{cm.}\right)$, using a polynomial model with $s=1$ and a spectral density of $q^2 = 10^{-8}$, De Jong [1994] reports MDB's for outliers in the code data of about 1 metre and MDB's for slips in the phase data of only a few centimetres, when $l = k$. See also the chapter *Active GPS Control Stations.*

7.4.2 The Geometry-Based Model

In the previous section we considered the geometry-free model. This model dispenses with the receiver-satellite geometry and it thus allows *single-channel* data processing. In the present section, we will replace the parametrization in terms of the receiver-satellite ranges by a parametrization in terms of the three baseline components. This implies that the data processing will now be *a multi-channel* based one. It also implies that the receiver-satellite geometry enters the model, thus giving a way to strengthen the model. In the following, we will consider two cases. In the first case we allow the two receivers to change position from epoch to epoch. In the second case, the two receivers are assumed to be stationary over the complete observation time span.

The short and moving baseline case

The baseline is assumed to be short enough to permit the negligence of the ionospheric delays. We also restrict our attention to the single frequency case. The development of the dual frequency case goes along similar lines. Furthermore, we assume that the same m satellites are tracked during the complete observation time span. The vectorial SD observation equations for epoch i, are then given in linearized form as

$$\begin{aligned} \Phi(i) &= A(i)x(i) + \lambda a \\ P(i) &= A(i)x(i) \end{aligned} \tag{7.64}$$

where the m-vector $\Phi(i)$ contains the 'observed minus computed' SD phases of epoch i; the m-vector $P(i)$ contains the 'observed minus computed' SD pseudoranges of epoch i; the unknown parameter vector $x(i)$ contains the three baselines components and the SD receiver clock error of epoch i; and the m-vector a contains the SD ambiguities. The receiver-satellite geometry at epoch i is captured in the $m \times 3$ SD design matrix $A'(i)$. This matrix together with the m-

vector e_m of 1's, equals $A(i) = [A'(i), e_m]$. The coefficients of e_m correspond with the SD receiver clock error. The two variance matrices of $\Phi(i)$ and $P(i)$ are assumed to be given as $Q_\Phi = \sigma_\phi^2 I_m$ and $Q_P = \sigma_p^2 I_m$. Thus cross-correlation between the observables is assumed to be absent. We also assume time-correlation to be absent.

Apart from the ionospheric delays, also the tropospheric delays are assumed absent. Note that this is a marked difference with the geometry-free model. In the geometry-free model the inclusion of the tropospheric delays was easy, since they were lumped together with the receiver-satellite ranges. For more details on the above model, we refer to chapter *GPS Observation Equations and Positioning Concepts.*

As before, we will consider a spike (outlier) in the code data at epoch l and a slip (cycle slip) in the phase data that starts at epoch l. Since we are now working with vectorial observation equations, we also need to indicate which entry of Φ or P is affected by the model error.

Spike in the ith entry of P at epoch l: The appropriate test statistic for the identification of an outlier in the ith entry of $P(l)$, using k epochs of data, reads

$$t_i^{l,k} = \frac{c_i^T\left[P(l) - A(l)\hat{x}_{l|k}\right]}{\sqrt{\sigma_p^2 - c_i^T A(l) Q_{\hat{x}_{l|k}} A(l)^T c_i}} \tag{7.65}$$

where $\hat{x}_{l|k}$ is the least-squares estimate of $x(l)$ based on k epochs and $Q_{\hat{x}_{l|k}}$ is its variance matrix. The ith entry of the m-vector c_i equals one, while all its other entries are zero.

Due to the high altitude orbits of the GPS satellites, the relative receiver-satellite geometry is known to change slowly with time. This implies $A'(i) \approx A$ for not too long observation time spans. In the following, we will use this approximation when presenting the MDB's. It will result in tractable expressions for the MDB's, without sacrificing too much rigor.

The MDB that corresponds with the above test statistic, is given as

$$|\nabla| = \sigma_p \sqrt{\frac{\lambda_0}{1 - \left[\frac{1}{k} + \frac{\varepsilon}{1+\varepsilon}\left(1 - \frac{1}{k}\right)\right]\left[\frac{1}{m} + c_i^T P_{\bar{A}} c_i\right]}} \tag{7.66}$$

with the projectors

$$P_{\bar{A}} = P_D A\left[A^T P_D A\right]^{-1} A^T P_D \text{ and } P_D = D\left(D^T D\right)^{-1} D^T = I_m - \frac{1}{m} e_m e_m^T$$

where the $(m-1)\times m$ matrix D^T equals the double difference (DD) matrix operator. Thus if A captures the SD receiver-satellite geometry, the DD geometry is captured by $D^T A$.

Note that the contribution of the number of epochs used (k), the number of satellites tracked (m), the measurement precision (ε) and the receiver-satellite geometry (A), is clearly visible in the above expression. The contribution of the receiver-satellite geometry is absent in the absence of satellite redundancy. In that case $m=4$ and $\frac{1}{m}+c_i^T P_{\bar{A}} c_i = 1$. The MDB reduces then to the one we met earlier when discussing the geometry-free model, see (7.55).

In the absence of phase data, the first set of observation equations of (7.64) are missing. As a consequence all parameters become disconnected in time, implying that the least-squares solution of $x(i)$ will be based on single epoch data only. The corresponding MDB can be obtained from (7.66) by setting $\sigma_\phi = \infty$. As a result the MDB simplifies to

$$|\nabla| = \sigma_p \sqrt{\frac{\lambda_0}{1-\frac{1}{m}-c_i^T P_{\bar{A}} c_i}}$$

Additional redundancy enters when the phase data are included in the model as well. As a consequence one may expect smaller MDB's. Since the phase data are much more precise than the code data, we have $\sigma_\phi \ll \sigma_p$. If we neglect the phase variance in (7.66), we get the approximation

$$|\nabla| \approx \sigma_p \sqrt{\frac{\lambda_0}{1-\frac{1}{k}\left[\frac{1}{m}+c_i^T P_{\bar{A}} c_i\right]}}$$

This shows that apart from the contribution of m and of the receiver-satellite geometry through $P_{\bar{A}}$, an additional contribution enters depending on the number of epochs used. A rough approximation of the MDB can be obtained if we make use of the projector property of $P_{\bar{A}}$. Since the trace of this matrix equals its rank $r(P_{\bar{A}})=3$, it follows that the average value of the diagonal entries of the projector equals $\frac{3}{m}$. We therefore have as an approximation $c_i^T P_{\bar{A}} c_i \approx \frac{3}{m}$.

Slip in the *i*th entry of Φ at epoch *l*: The appropriate test statistic for the identification of a slip that starts at epoch l in the ith entry of Φ, using k epochs of data, reads

$$t_i^{l,k} = \frac{c_i^T Q_{\hat{a}_{l-1}}^{-1}\left[\hat{a}_{l-1}-\hat{a}_k\right]}{\sqrt{c_i^T Q_{\hat{a}_{l-1}}^{-1}\left[Q_{\hat{a}_{l-1}}-Q_{\hat{a}_k}\right]Q_{\hat{a}_{l-1}}^{-1} c_i}} \tag{7.67}$$

Note that the ambiguity variance matrix $Q_{\hat{a}_{l-1}}$ will generally be a full matrix and thus not a diagonal matrix. This implies that *all m* entries of both $\hat{a}_{l-1}$ and $\hat{a}_k$ contribute to the test statistic. This shows that for the purpose of cycle slip detection, a simple monitoring of the individual SD ambiguities is not optimal in terms of the detection power of the test. Monitoring of the individual ambiguities is only optimal when satellite redundancy is absent. In that case the ambiguity variance matrices reduce to scaled unit matrices.

The MDB of the above test statistic (7.67) equals

$$|\nabla| = \sqrt{\frac{\lambda_0}{c_i^T Q_{\hat{a}_{l-1}}^{-1} \left[Q_{\hat{a}_{l-1}} - Q_{\hat{a}_k} \right] Q_{\hat{a}_{l-1}}^{-1} c_i}} \tag{7.68}$$

If we assume $A'(i) \approx A$, then $Q_{\hat{a}_k} = \frac{\sigma_\phi^2}{k}\left[I_m - \frac{1}{1+\varepsilon}\left(I_m - P_D + P_{\bar{A}} \right) \right]^{-1}$ and the MDB, when expressed in units of cycles, becomes

$$|\nabla| = \frac{\sigma_\phi}{\lambda} \sqrt{\frac{\lambda_0}{\left[1 - \frac{1}{1+\varepsilon}\left(\frac{1}{m} + c_i^T P_{\bar{A}} c_i \right) \right] (l-1)\left(1 - \frac{l-1}{k} \right)}} \tag{7.69}$$

The contribution of the receiver-satellite geometry is again clearly visible. This contribution is absent in the absence of satellite redundancy. When $m = 4$, then $(\frac{1}{m} + c_i^T P_{\bar{A}} c_i) = 1$ and the MDB reduces to the one we met earlier when discussing the geometry-free model, see (7.57). In that case the MDB was too large to find sufficiently small slips when $l = k$. In the present case however, we get for $l = k$, even in the absence of code data $(\sigma_p = \infty)$,

$$|\nabla| = \frac{\sigma_\phi}{\lambda} \sqrt{\frac{\lambda_0}{\left(1 - \frac{1}{m} - c_i^T P_{\bar{A}} c_i \right)\left(1 - \frac{1}{k} \right)}}$$

This shows that very small slips can be found, even when $l = k > 1$.

The short but stationary baseline case

So far the baseline was allowed to change from epoch to epoch. Now we will assume that the baseline remains constant. The SD receiver clock error however, is still allowed to change from epoch to epoch. Thus only the first three entries of the parameter vector $x(i)$ are assumed to be time-invariant. Since this will introduce additional redundancy, one can expect that the MDB's will become smaller. That is, smaller model errors can be found with the same probability. In order to infer how much smaller the MDB's become, we again consider a spike in the code data and a slip in the phase data.

Spike in the *i*th entry of *P* at epoch *l*: The MDB for an outlier in one of the code observables equals

$$|\nabla| = \sigma_p \sqrt{\frac{\lambda_0}{1 - \frac{1}{mk} - \frac{\varepsilon}{1+\varepsilon}\frac{1}{m}\left(1 - \frac{1}{k}\right) - \frac{1}{k} c_i^T P_{\bar{A}} c_i}} \tag{7.70}$$

Compare this result with (7.66). Although the above MDB is smaller than the one of (7.66), the difference is minor due to the small value of ε. This shows that the identification of outliers in the code data does not really improve when the baseline is considered to be stationary instead of moving.

Slip in the *i*th entry of Φ at epoch *l*: The MDB for a slip that starts at epoch l in one of the phase observables, reads

$$|\nabla| = \frac{\sigma_\phi}{\lambda} \sqrt{\frac{\lambda_0}{\left(1 - \frac{1}{1+\varepsilon}\frac{1}{m}\right)(l-1)\left(1 - \frac{l-1}{k}\right)}} \tag{7.71}$$

Compare this result with (7.69). Note that the projector $P_{\bar{A}}$ is absent from this expression. This is due to our approximation $A'(i) \approx \bar{A}$. It shows that the receiver-satellite geometry itself has no influence on the MDB. Only the number of satellites that are tracked, counts. Also note that the above expression reduces to that of the geometry-free model (7.57), when $m = 1$.

Summary

For easy reference, the single-frequency MDB's of the geometry-free model and of the geometry-based model, for the moving and stationary baseline case, are summarized in Table 7.6. From these results we can conclude that for *outliers* in the code data:

- The MDB's are (practically) independent of the precision of the phase data.
- The MDB's of the moving baseline case are (practically) identical to those of the stationary baseline case.
- In all three cases, the minimum value of the MDB's equals $\sigma_p\sqrt{\lambda_0}$

- In the geometry-free model $k>1$ is needed per se, whereas, due to the presence of the relative receiver-satellite geometry, $k=1$ is possible for the geometry-based models. In these last two cases we have for $k=1, m=6, \delta=0.5$, an MDB of $\sigma_p\sqrt{3\lambda_0}$

In a similar way, we may conclude for *cycle slips* in the phase data:

- The MDB of the geometry-free model is governed by the (poor) precision of the code data, whereas the two MDB's of the geometry-based model are governed by the (high) precision of the phase data.

- The MDB's are infinite when either $k=1$ or $l=1$.

- For $l=k$, the minimum values of the MDB's (in cycles) equal $\frac{\sigma_p}{\lambda}\sqrt{\lambda_0}$ for the geometry-free model and $\frac{\sigma_\phi}{\lambda}\sqrt{\lambda_0}$ for the geometry-based model.

- For $l=k=2, m=6,$ and $\delta=0.5$ the MDB equals $\frac{\sigma_\phi}{\lambda}\sqrt{6\lambda_0}$ for the moving baseline and a factor 0.6 larger, for the stationary baseline.

Table 7.6. Single frequency MDB's for a spike/slip at epoch l in code/phase based on k epochs, using phase and code data, with the approximations $\varepsilon \approx 0$ and $A'(i) \approx A$, with $\delta = c_i^T P_{\bar{A}} c_i$ (the average value of δ equals $\frac{3}{m}$ and $\lambda_0 \approx 17$ for $\alpha = 0.001$ and $\gamma = 0.80$).

	geometry-free	moving baseline	stationary baseline
Code spike at l	$\sigma_p\sqrt{\frac{\lambda_0}{1-\frac{1}{k}}}$	$\sigma_p\sqrt{\frac{\lambda_0}{1-\frac{1}{k}\left(\frac{1}{m}+\delta\right)}}$	$\sigma_p\sqrt{\frac{\lambda_0}{1-\frac{1}{k}\left(\frac{1}{m}+\delta\right)}}$
Phase slip at l	$\frac{\sigma_p}{\lambda}\sqrt{\frac{\lambda_0}{(l-1)\left(1-\frac{l-1}{k}\right)}}$	$\frac{\sigma_\phi}{\lambda}\sqrt{\frac{\lambda_0}{\left(1-\frac{1}{m}-\delta\right)(l-1)\left(1-\frac{l-1}{k}\right)}}$	$\frac{\sigma_\phi}{\lambda}\sqrt{\frac{\lambda_0}{\left(1-\frac{1}{m}\right)(l-1)\left(1-\frac{l-1}{k}\right)}}$

7.5 SUMMARY AND CONCLUSIONS

In this chapter an introduction was given to the theory of GPS quality control. Statistical tests are performed in order to validate the assumptions underlying the GPS models used. We discussed the validation of batch solutions as well as of recursive solutions. Validation consists of the three steps: detection, identification and adaptation. In the detection step the overall model is checked for possible misspecifications. A specification of the likely model error is not needed in this step. In case of a batch solution, the test statistic for detection equals a quadratic form in the least-squares residuals. In case of a recursive solution, it equals a quadratic form in the predicted residuals, which can be computed recursively parallel to the recursive (Kalman) filter.

Identification of the most likely model error is needed in case detection leads to a rejection of the null hypothesis. For identification one needs to decide upon the class of alternative hypotheses. Each alternative hypothesis is specified through the use of the C-matrices or c-vectors. Examples of c-vectors are those that are needed for outliers and/or cycle slips, and examples of C-matrices are those that are needed for the identification of baseline errors in a GPS baseline network. Also the test statistics for identification can be computed recursively parallel to the recursive (Kalman) filter.

Once the most likely model error has been identified, model adaptation is needed so as to eliminate the presence of biases in the computed solutions. This implies either a remeasurement of part of the data, or a replacement of the null hypothesis with the identified alternative hypothesis. The switch from the null hypothesis to the alternative hypothesis can also be performed recursively.

The expected performance of the various tests depends in a large part on the 'strength' of the mathematical model used. The size of the model errors that one can expect to find with a certain probability, is described by the minimal detectable biases (MDB's). For different GPS single-baseline models, examples were given of the outlier and cycle slip tests statistics together with their corresponding MDB's.

References

Abousalem, M.A., M.A. Forkheim, J.F. McLellan (1994), Quality GPS post processing. *Proceedings KIS94*, Banff, Canada, Aug. 30 - Sept. 2, pp. 197-206.

Baarda, W. (1968), A testing procedure for use in geodetic networks. *Neth. geod. Comm.,* Vol. 2, No. 5, Delft.

Baarda, W. (1973), S-transformations and criterion matrices. *Neth. Geod. Comm.*, Vol. 5, No. 1, Delft.

Bierman, G.J. (1977), *Factorization methods for discrete sequential estimation*, Academic Press, Orlando.

Brammer, K., G. Siffling (1989), *Kalman-Bucy Filters*, Artech House.

Cross, P. (1994), Quality measures for differential GPS positioning. *The Hydrographical Journal*, No. 72, pp. 17-22.

De Jong, C.D. (1994), Real-time integrity monitoring of single and dual frequency GPS observations. In: *GPS newsletter*, Vol. 9, No. 1, pp. 33-49.

De Jong, C.D. (1996), Real-time integrity monitoring of dual-frequency GPS observations form a single receiver. *Acta Geod. Geoph. Hung.*, Vol. 31 (1-2), pp. 37-46.

Gelb, A. (Ed.) (1974), *Applied optimal estimation*. MIT-Press, Cambridge, M.A.

Georgiadou, Y. (1994), Modelling the ionosphere for an active control network of GPS stations. *LGR-Series*, No. 7, Delft Geodetic Computing Centre.

Gillissen, I., I.A. Elema (1996), Test results of DIA: A real-time adaptive integrity monitoring procedure, used in an integrated navigation system. *International Hydrographic Review*, LXXIII(1):75-103.

Jazwinski, A.H. (1970), *Stochastic Processes and Filtering Theory*, Academic Press, New York.

Jin, X., H. van der Marel, C.D. de Jong (1995), Computation and quality control of differential GPS corrections. *Proceedings ION GPS-95*, Palm Springs, California, USA, pp. 1071-1079.

Jin, X. (1996), *Theory of carrier adjusted DGPS positioning approach and some experimental results*. PhD-thesis, Delft.

Jin, X. (1997), Algorithm for carrier adjusted DGPS positioning and some numerical results. Journal of Geodesy, Vol. 71, No. 7, June 1997, pp. 411-422.

Kalman, R.E., (1960), A new approach to linear filtering and prediction problems. *J. Basic Engng.*, 82, 34-45.

Kleusberg, A., P.J.G. Teunissen (Eds.) (1996), GPS for Geodesy, *Lecture Notes in Earth Sciences*, Vol. 60, Springer Verlag.

Leick, A. (1995), *GPS Satellite Surveying*. 2nd edition, John Wiley, New York.

Parkinson, B. et al. (Eds.) (1996), *GPS: Theory and Applications*. Volumes 1 and 2. AIAA.

Salzmann, M. (1991), MDB: A design tool for integrated navigation systems. *Bulletin Geodesique*, 65:109-115.

Salzmann, M. (1994), A real-time quality control procedure for use in integrated navigation systems. *The Hydrographic Journal*, 72:25-30.

Salzmann, M. (1995), Real-time adaptation for model errors in dynamic systems. *Bulletin Geodesique*, 69:81-91.

Schaer, S. (1994), *Stochastische Ionosphärenmodellierung beim 'Rapid Static Positioning' mit GPS*. Astronomisches Institut, Universität Bern.

Schaffrin, B. and Y. Bock (1988), A unified scheme for processing GPS dual-band observations. *Bulletin Geodesique*, 62:142-160.

Seeber, G. (1993), *Satellite Geodesy*. Walter de Gruyter, Berlin/New York.

Sorenson, H.W. (1970), Least-squares estimation: from Gauss to Kalman. *IEEE Spectrum*, Vol.7, pp. 63-68.

Teunissen, P.J.G. (1985), Generalized inverses, adjustment, the datum problem and S-transformations. In: *Optimization and design of geodetic networks*, E. Grafarend and F. Sanso (Eds.), Springer Verlag.

Teunissen, P.J.G. (1990), GPS op afstand bekeken. In: *Snellius Anniversary* Volume 1985-1990, pp. 215-233.

Teunissen, P.J.G. (1990a), An integrity and quality control procedure for use in multi sensor integration. In: *Proceedings ION GPS-90*, pp. 513-522.

Teunissen, P.J.G. (1990b), Quality control in integrated navigation systems. *IEEE Aerospace and Electronics Systems Magazine*, Vol. 5(7), pp. 35-41.

Teunissen, P.J.G. (1990c), *Dynamic Data Processing*, Lecture Notes in Mathematical Geodesy, Delft, 342 p.

Teunissen, P.J.G. (1991), Differential GPS: Concepts and Quality Control. In: *Proceedings NIN Global Positioning System*. 43 p.

Teunissen, P.J.G. (1996), The geometry-free GPS ambiguity search space in case the ionosphere is weighted. *Journal of Geodesy*, Vol. 71, No. 6, May 1997, pp. 370-383.

Teunissen, P.J.G., M. Salzmann (1988), *Performance analysis of Kalman filters*. Report 88.2, Section Mathematical and Physical Geodesy, Delft.

Teunissen, P.J.G., M. Salzmann (1989), A recursive slippage test for use in state space filtering. *Manuscripta Geodaetica*, Vol. 14(6), pp. 383-390.

Tiberius, C.C.J.M. (1991), Quality control and integration aspects of vehicle location systems. *LGR-Series* No. 1.

Tiberius, C.C.J.M. (1996), *Kinematic GPS surveying*, Internal LGR report.

Wanninger, L. (1995), Improved ambiguity resolution by regional differential modelling of the ionosphere. In: *Proceedings ION GPS-95*, September 12-15, 1995, Palm Springs.

Wild, U. (1994), Ionosphere and Geodetic Satellite Systems: Permanent GPS Tracking Data for Modelling and Monitoring. *Geodätisch-geophysikalische Arbeiten in der Schweiz*, Band 48.

Wild, U., G. Beutler (1991), Deterministische und stochastische Ionosphärenmodelle. Zeitschrift für Vermessung, Photogrammetrie und Kulturtechnik, Heft 6, pp. 298-302.

8. GPS CARRIER PHASE AMBIGUITY FIXING CONCEPTS

Peter J.G. Teunissen
Department of Geodetic Engineering, Delft University of Technology,
Thijsseweg 11, 2629 JA Delft, The Netherlands

8.1 INTRODUCTION

High precision relative GPS positioning is based on the very precise carrier phase measurements. A prerequisite for obtaining high precision relative positioning results is that the double-differenced carrier phase ambiguities become sufficiently separable from the baseline coordinates. Different approaches are in use and have been proposed to ensure a sufficient separability between these two groups of parameters. In particular the approaches that explicitly aim at resolving the integer-values of the double-differenced ambiguities have been very successful. Once the integer ambiguities are successfully fixed, the carrier phase measurements will start to act as if they were high-precision pseudorange measurements, thus allowing for a baseline solution with a comparable high precision. The fixing of the ambiguities on integer values is however a non-trivial problem, in particular if one aims at numerical efficiency. This topic has therefore been a rich source of GPS-research over the last decade or so. Starting from rather simple but timeconsuming integer rounding schemes, the methods have evolved into complex and effective algorithms.

Among the different approaches that have been proposed for carrier phase ambiguity fixing are those documented in Counselman an Gourcvitch [1981], Remondi [1984; 1986; 1991], Hatch [1986; 1989; 1991], Hofmann-Wellenhof and Remondi [1988]. Seeber and Wübbena [1989], Blewitt [1989], Abott et al. [1989], Frei and Beutler [1990], Euler and Goad [1990], Kleusberg [1990], Frei [1991], Wübbena [1991], Euler and Landau [1992], Erickson [1992], Goad [1992], Teunissen [1993a; 1994a,b], Hatch and Euler [1994], Mervart et al. [1994], De Jonge and Tiberius [1994], Goad and Yang [1994].

The purpose of the present lecture notes is to present the theoretical concepts of the GPS ambiguity fixing problem, to formulate procedures of solving it and to outline some of the intricacies involved. Several examples are included in the lecture notes for both quantitative as well as qualitative purposes. To gain a firm footing with the GPS ambiguity fixing problem, it is cast in the familiar framework of least-squares adjustment and testing theory. Starting from the double-differenced carrier phase observation equations, the section 8.2 *Integer Least-Squares Adjustment and Testing* presents an overview of both the ambiguity estimation part as well as the ambiguity validation part of the GPS ambiguity fixing problem. It shows how the fixed solution can be arrived at via the float solution and it shows how both these solutions can be validated.

In section 8.3 *Search for the Integer Least-Squares Ambiguities* two concepts for numerically solving the integer least-squares problem, are discussed. The first concept is based on using the ellipsoidal planes of support and the other is based on using a sequential conditional least-squares adjustment of the ambiguities. In case of short observational time span based carrier phase data, both concepts - when applied to the traditional double-differenced ambiguities - suffer from the fact that the least-squares ambiguities are highly correlated. In order to corroborate this, quantitative indications are given of the elongation of the ambiguity search space, the precision and correlation of the least-squares ambiguities and of the signature of the spectrum of conditional variances. The poor performance of the search is also exemplified by means of both an analytical example as well as an illustrative numerical example.

In section 8.4 *The Invertible Ambiguity Transformations* the concept of integer ambiguity reparametrization is introduced. Starting from the nonuniqueness of the double-differenced ambiguities and the idea of considering linear combinations of the double-differenced carrier phase observables, the class of invertible single-channel ambiguity transformations is identified and then generalized to the multi-channel case. The importance of this class is that it provides significant leeway to influence the dependence of the double-differenced ambiguity variance-covariance matrix on the design matrix containing the receiver-satellite geometry. Members from this class allow one to replace the original integer least-squares problem with an equivalent formulation that is much easier and hence much faster to solve.

In section 8.5 *The LSQ Ambiguity Decorrelation Adjustment* it is shown how the original integer least-squares problem can be reparametrized so as to obtain a formulation which is easier to solve. The basic idea that lies at the root of the method - both in the construction of the ambiguity transformation as in the formulation of the search bounds - is that integer least-squares ambiguity estimation becomes trivial once all least-squares ambiguities are fully decorrelated. Although the integer nature of the ambiguities generally prohibits a full decorrelation of the ambiguities, the presence of the discontinuity in the spectrum of conditional variances still enables one to decorrelate the ambiguities to a large extent. It is shown how the decorrelation can be achieved by means of using integer approximations of the fully decorrelating conditional least-squares transformations. Results of the decorrelation are shown in terms of the elongation of the transformed ambiguity search space, the precision and correlation of the transformed least-squares ambiguities, and the flattened and lowered spectrum of transformed conditional variances. The section is concluded with both a qualitative and quantitative based discussion on the characteristics of the GPS spectrum.

8.2 INTEGER LEAST-SQUARES ADJUSTMENT AND TESTING

In this section we will give an overview of the least-squares based concepts of GPS ambiguity fixing. The GPS ambiguity fixing problem consists of two distinct parts:

1. The ambiguity *estimation* problem, and
2. The ambiguity *validation* problem.

Given a model of observation equations, the estimation part addresses the problem of finding optimal estimators for the unknown parameters. Since optimality will be based on the principle of least-squares, the task is to find the least-squares solution for the unknown integer ambiguities. The second part of the GPS ambiguity fixing problem is concerned with the validation of the estimated integer ambiguities. The validation part is of importance in its own right and quite distinct from the estimation part. One will namely always be able to compute an integer least-squares solution, whether it is of poor quality or not. The question addressed by the validation part is therefore, whether the quality of the computed integer least-squares solution is such that one is also willing to accept this solution.

8.2.1 The Double-Differenced Carrier Phase Observation Equations

The GPS observables are code-derived pseudorange measurements and carrier phase measurements. The GPS observables relate the measured quantities described in chapter 4 to geometrical and physical parameters of interest in a geodetic context. As we have seen in section 5.2, linear combinations of the GPS observables can be taken so as to eliminate and/or isolate these geometrical and physical parameters. In this section we will start from the so-called double-differenced (DD) carrier phase observables. They follow from phase measurement differences between satellites and receivers (cf. section 5.2.5).

The non-linear observation equation for the difference between the simultaneous phase measurements of a receiver j of the signals transmitted by two different satellites, k and l, and the simultaneous measurements at the same nominal time t of a second receiver i of the same signals, reads (cf. equation (5.57))

$$\Phi_{ij}^{kl}(t) = \rho_{ij}^{kl} - I_{ij}^{kl} + T_{ij}^{kl} + \delta m_{ij}^{kl} + \lambda N_{ij}^{kl} + \varepsilon_{ij}^{kl} \ . \tag{8.1}$$

This linear combination Φ_{ij}^{kl} will be referred to as the *double differenced* (DD) phase measurement. If we assume the positions of the satellites k and l, and of receiver i to be known, the unknown parameters in equation (8.1) are: (*i*) the linear combination of the four geometric distances between the two receivers, i and j, and the two satellites, k and l; it depends in a nonlinearly way on the unknown position of receiver j; (*ii*) the two linear combinations, I_{ij}^{kl} and T_{ij}^{kl}, of four ionospheric and tropospheric delay terms; (*iii*) the combined multipath term δm_{ij}^{kl}; and (*iv*) the DD phase ambiguity N_{ij}^{kl}.

An interesting feature of the above observation equation is that not all parameters are real-valued. We know a priori, that the DD phase ambiguity N_{ij}^{kl} can only take on *integer* values. Within the context of classical (least-squares) adjustment theory this is a rather unusual situation. Classical adjustment theory has been developed on the basis of the premises that all parameters are real-valued. This implies that the well-known methods of classical adjustment theory are not really applicable here. Of course, we could still try to apply classical adjustment theory. The space of integers is namely a subset of the space of reals. Hence, one could decide to disregard the integer nature of the DD ambiguities and simply treat them as reals. The consequence of such a decision is however that not all information is taken into account, information which in principle can have a very beneficial impact on the estimability of the unknown parameters. The goal of this chapter is therefore to show how one can incorporate the integerness of the DD ambiguities in the parameter estimation process and to give an outline of how to proceed when one wants to compute estimates of these parameters.

In order to keep things as simple as possible, we will simplify the above observation equation by stripping it from its atmospheric and multipath delay terms, I_{ij}^{kl}, T_{ij}^{kl} and δm_{ij}^{kl}. Whether this is allowed and under what circumstances it is allowed, will not be of our concern in this chapter (refer to the chapters on short, medium and global distances). It is remarked however, that this simplification is not a prerequisite for the theory that will be developed in this chapter. The stripped version of equation (8.1) reads

$$\Phi_{ij}^{kl}(t) = \rho_{ij}^{kl} + \lambda N_{ij}^{kl} + \varepsilon_{ij}^{kl} . \tag{8.2}$$

The parameters that remain are therefore the unknown real-valued baseline components of receiver j with respect to receiver i and the unknown, but integer-valued DD ambiguity N_{ij}^{kl}.

In the following it will be assumed that both receivers, i and j, are stationary, and that at each observational time epoch t a sufficient number of satellites, say $(m+1)$, are simultaneously tracked. This implies, if satellite k is taken as reference satellite, that we have the following DD carrier phase measurements at our disposal at time t: $\Phi_{ij}^{kl}(t), \Phi_{ij}^{k2}(t), \ldots, \Phi_{ij}^{k(k-1)}(t), \Phi_{ij}^{k(k+1)}(t), \ldots, \Phi_{ij}^{k(m+1)}(t)$. This implies, if the total number of observational time epochs equals T, that the total number of DD carrier phase measurements equals mT. The corresponding total number of unknown parameters equals $(m+3)$. There are 3 unknown baseline components and m unknown DD ambiguities. The tracking of the satellites is assumed to be uninterrupted during the observational time span. Hence, cycle slips are assumed to be absent.

With the above mT DD carrier phases we can form a system of observation equations, which after linearization with respect to the unknown parameters, gives the linear system of equations

$$y = Aa + Bb + e\,, \tag{8.3}$$

where y is the vector of mT observed minus computed DD carrier phases, a is the unknown vector of m DD ambiguities, b is the unknown vector of 3 baseline components, A and B are the corresponding design matrices for the ambiguities and baseline components, and e is the vector that contains the mT measurement noise terms.

The above system (8.3) will be taken as our point of departure for computing estimates of the unknown parameters a and b. The system (8.3) has been constructed from carrier phases on a single frequency only. It will be clear however, when dual frequency data are available, that the carrier phases on the second frequency can be incorporated in the system in a similar way as it has been done for the carrier phases on the first frequency. Also pseudorange data, when available, can be incorporated in the system. When forming the system (8.3), it was also assumed that only two receivers, i and j, were tracking the satellites. It will be clear however, that a similar system of linear equations can be constructed when more than two receivers track the same satellites simultaneously. For instance, if 3 stationary receivers h, i and j are tracking, the vector b will consist of 6 baseline components, e.g., the baseline components of receivers h and j with respect to receiver i.

Our estimation criterion for solving the above system (8.3) will be based on the principle of least-squares. From a statistical viewpoint this choice is motivated by the fact that in the absence of modelling errors, properly weighted linear least-squares estimators are identical to unbiased minimum variance estimators. Furthermore, these estimators are also maximum likelihood estimators if the assumption of normality holds for the GPS observables.

8.2.2 The Float and Fixed Least-Squares Solution

The least-squares criterion for solving the linear system of observation equations (8.3) reads

$$\min_{a,b} \| y - Aa - Bb \|^2_{Q_y}\,, \tag{8.4}$$

where $\| \cdot \|^2_{Q_y} = (\cdot)^T Q_y^{-1} (\cdot)$ and Q_y is the variance-covariance matrix of the DD observables. The minimization problem (8.4) would be an ordinary unconstrained least-squares problem if all the parameters were allowed to range through the space of reals, i.e. if

$$a \in R^m \text{ and } b \in R^3, \tag{8.5}$$

would hold. In our case however, we do have the additional information that all the DD ambiguities are integer-valued. Instead of (8.5), we therefore have

$$a \in Z^m \text{ and } b \in R^3, \tag{8.6}$$

with Z^m being the m-dimensional space of integers. The minimization problem (8.4) together with (8.6) will be referred to as an integer least-squares problem. It is a constrained least-squares problem due to the integer-constraint $a \in Z^m$. The solution of the integer least-squares problem will be denoted as $\check{a}$ and $\check{b}$, and the solution of the corresponding unconstrained least-squares problem will be denoted as $\hat{a}$ and $\hat{b}$. The estimates $\check{a}$ and $\check{b}$ will be referred to as the *fixed least-squares* solution and the estimates $\hat{a}$ and $\hat{b}$ as the *float least-squares* solution.

It is of interest to consider the relationship that exists between the float and fixed solution. For that purpose, we decompose the quadratic objective function of (8.4) into the following sum of three squares

$$\| y - Aa - Bb \|^2_{Q_y} = \| \hat{e} \|^2_{Q_y} + \| \hat{b}(a) - b \|^2_{Q_{\hat{b}(a)}} + \| \hat{a} - a \|^2_{Q_{\hat{a}}}, \tag{8.7}$$

where $\hat{e}$ is the unconstrained least-squares residual vector; $\hat{b}(a)$ is the least-squares estimate of b, but *conditioned* on a; $Q_{\hat{b}(a)}$ is the variance-covariance matrix of $\hat{b}(a)$; and $Q_{\hat{a}}$ is the variance-covariance matrix of $\hat{a}$. The geometry of the above *orthogonal* decomposition is shown in Figure 8.1.

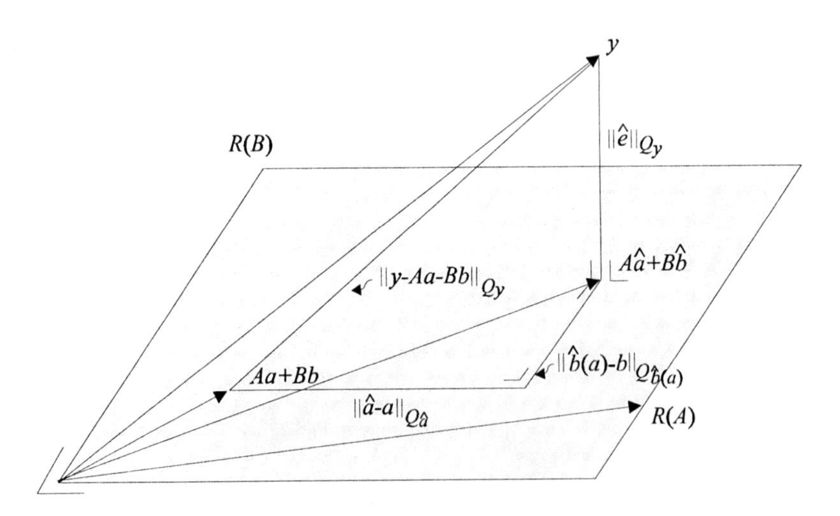

Figure 8.1. Orthogonal decomposition of $\| y - Aa - Bb \|^2_{Q_y}$.

From the above decomposition follows, that the last two squares on the right-hand side of (8.7) vanish identically if the objective function would be minimized as function of $a \in R^m$ and $b \in R^3$. Hence, the minimizers would then be given by $\hat{a} \in R^m$ and $\hat{b} \in R^3$, and the minimum of the objective function would be given by the squared norm of the least-squares residual vector $\hat{e}$,

$$\| y - A\hat{a} - B\hat{b} \|^2_{Q_y} = \| \hat{e} \|^2_{Q_y} . \tag{8.8}$$

In our case, the objective function needs to be minimized as function of $a \in Z^m$ and $b \in R^3$. In that case, only the second square on the right-hand side of (8.7) vanishes identically and the minimizers are given as $\check{a} \in R^m$ and $\check{b} = \hat{b}(\check{a}) \in R^3$. The corresponding minimum of the objective function reads then,

$$\| y - A\check{a} - B\check{b} \|^2_{Q_y} = \| \check{e} \|^2_{Q_y} = \| \hat{e} \|^2_{Q_y} + \| \hat{a} - \check{a} \|^2_{Q_{\hat{a}}} . \tag{8.9}$$

The minimum (8.9) is clearly larger than, or at the most, equally large as the floated value of the objective function (8.8). This difference is of course due to the fact that (8.9) is based on the additional constraint of restricting the ambiguity vector a to the space of integers.

The above shows that we may follow a two-step procedure for solving the integer least-squares problem. The first step consists then of solving the *unconstrained* least-squares problem. As a result of this first step, real-valued estimates for both the ambiguities and baseline components are obtained, together with their corresponding variance-covariance matrices:

$$\begin{bmatrix} \hat{a} \\ \hat{b} \end{bmatrix} , \begin{bmatrix} Q_{\hat{a}} & Q_{\hat{a}\hat{b}} \\ Q_{\hat{b}\hat{a}} & Q_{\hat{b}} \end{bmatrix} . \tag{8.10}$$

This result forms then the input for the second step. In the second step one first solves for the vector of integer least-squares estimates of the ambiguities, $\check{a}$. It follows from solving

$$\min_a \| \hat{a} - a \|^2_{Q_{\hat{a}}}, \text{ with } a \in Z^m . \tag{8.11}$$

Once the solution $\check{a}$ has been obtained, the residual $(\hat{a} - \check{a})$ is used to adjust the *float* solution $\hat{b}$ of the first step, to get $\check{b} = \hat{b}(\check{a})$. As a result, the final *fixed* baseline solution is obtained as

$$\check{b} = \hat{b}(\check{a}) = \hat{b} - Q_{\hat{b}\hat{a}} Q_{\hat{a}}^{-1} (\hat{a} - \check{a}) . \tag{8.12}$$

This equation shows the relation that exists between the fixed and float solution, $\check{b}$ and $\hat{b}$. It shows how the difference of these two baseline estimates depends on the difference between the real-valued least-squares ambiguity estimate $\hat{a}$ and integer least-squares ambiguity estimate $\check{a}$.

If we apply the error propagation law to (8.12) and assume the integer estimate $\check{a}$ to be nonstochastic, the variance-covariance matrix of $\check{b}$ follows as

$$Q_{\check{b}} = Q_{\hat{b}} - Q_{\hat{b}\hat{a}} Q_{\hat{a}}^{-1} Q_{\hat{a}\hat{b}} \,. \tag{8.13}$$

This result shows that $Q_{\check{b}} < Q_{\hat{b}}$. Hence, the fixed baseline solution is of a better precision than the corresponding float solution. This is of course understandable if we think of the additional information which has been used in computing $\check{b}$. It can be shown, when short observational time spans are used, that we in fact have $Q_{\check{b}} << Q_{\hat{b}}$. This can be explained as follows. Since GPS satellites are in very high altitude orbits, their relative positions with respect to the receivers change slowly, which implies in case of short observational time spans, that the ambiguities - when treated as being real-valued - become very poorly separable from the baseline coordinates. As a result, the precision with which the baseline can be estimated will be rather poor. However, when one explicitly aims at resolving for the integer-values of the ambiguities, the high precision carrier phase observables will start to act as if they were high precision pseudorange observables. As a result, the baseline coordinates become estimable with a comparable high precision and $Q_{\check{b}} << Q_{\hat{b}}$ holds true. The sole purpose of *ambiguity-fixing* is thus, to be able, via the inclusion of the integer-constraint $a \in Z^m$, to obtain a drastic improvement in the precision of the baseline solution. When successful, ambiguity fixing is thus a way to avoid long observational time spans, which otherwise would have been needed if the ambiguities were treated as being real-valued.

8.2.3 Validating the Float and Fixed Solution

In the previous section it was shown how the integer least-squares estimation problem can be solved in two steps. The first step provides the float solution and in the second step, after the integer least-squares ambiguities have been found, the fixed solution is obtained. It is of importance to realize however, that computing integer least-squares estimates is one thing, validating them is quite another. That is, one will always be able to compute an integer least-squares solution, whether it is of poor quality or not. We therefore still need to consider the question, whether we are willing to accept the computed integer least-squares solution. In this section the means for validating both the float and fixed solution will be discussed. For that purpose use will be made of concepts from the standard theory of statistical hypothesis testing, see e.g., Baarda [1968], Koch [1987], Teunissen [1994c] and chapter *Quality Control and GPS*. At the end of this section also some theoretical shortcomings of the current approaches of integer ambiguity validation will be discussed.

We start defining three classes of hypotheses. They are

$$
\begin{aligned}
H_1&: y = Aa + Bb + e, \; Q_y = \sigma^2 G_y, \; a \in R^m, \; b \in R^n, \; e \in R^k \\
H_2&: y \in R^k \qquad\qquad\quad , \; Q_y = \sigma^2 G_y \\
H_3&: y = A\check{a} + Bb + e, \; Q_y = \sigma^2 G_y, \; b \in R^n, \; e \in R^k .
\end{aligned}
\tag{8.14}
$$

The first hypothesis, H_1, considers the model of observation equations without the constraint that the ambiguities are to be integer-valued. Hence, the least-squares solution under H_1 will give the float solution, i.e. $\hat{a}$, $\hat{b}$ and $\hat{e}$. Under the third hypothesis, H_3, we assume to know a priori what the correct integer values of the ambiguities are. In practice, the value $\check{a}$ in H_3 is chosen to be equal to the integer least-squares solution for the ambiguities. Hence, the least-squares solution under H_3 will give the fixed solution, i.e. $\check{a}$, $\check{b}$, $\check{e}$. Note that the first hypothesis H_1 is more relaxed than the third hypothesis H_3 and that $H_3 \subset H_1$. The second hypothesis, H_2, is the most relaxed hypothesis. That is, under H_2 no restrictions at all are placed on $y \in R^k$. In terms of subsets, the three hypotheses can therefore be ordered as: $H_3 \subset H_1 \subset H_2$.

In the following we will assume that the k-vector of observables y is normally distributed with a zero-mean residual vector e. The variance-covariance matrix Q_y has been factored as $Q_y = \sigma^2 G_y$, with the variance-factor of unit weight σ^2 and the cofactor matrix G_y. Both the cases where σ^2 is assumed known and where it is assumed unknown, will be considered. The unbiased estimates of the variance-factor of unit weight σ^2 under respectively H_1 and H_3 are given as

$$
H_1: \hat{\sigma}^2 = \frac{\hat{e}' G_y^{-1} \hat{e}}{k-m-n} \quad \text{and} \quad H_3: \check{\sigma}^2 = \frac{\check{e}' G_y^{-1} \check{e}}{k-n} .
\tag{8.15}
$$

In the denominators of these two expressions, we recognize the redundancy under H_1, $k-m-n$, and the redundancy under H_3, $k-n$.

The first question we would like to answer is whether the model on the basis of which the float solution is computed, H_1, can be considered valid or not. This is an important question in its own right, since the data can still be contaminated with undetected errors (e.g., outliers or cycle slips) and/or the chosen system of observation equations can still fail to capture some geometrical and physical effects (e.g., atmospheric delays or multipath). The test statistic which allows us to test H_1 against the most relaxed alternative hypothesis, i.e. which tests H_1 against H_2, is given by the ratio $\hat{\sigma}^2/\sigma^2$. This test statistic is distributed under H_1 as $1/(k-m-n)$ times the χ^2-distribution, $\chi^2(k-m-n)$, or, as the F-distribution, $F(k-m-n,\infty)$. Its mean and variance under H_1 are equal to respectively 1 and $2/(k-m-n)$. The decision to accept H_1 is made, when the value of the test statistic is less than the critical value $F_\alpha(k-m-n,\infty)$, with α being the chosen level of significance. Thus H_1 is accepted when

$$\hat{\sigma}^2/\sigma^2 < F_\alpha(k-m-n,\infty) \,. \tag{8.16}$$

If the value of the test statistic fails to pass this test, it is likely that either the data are still contaminated with errors and/or that the observation equations fail to capture all relevant geometrical and physical effects. As a consequence, the corresponding float solution is contaminated with these unmodelled effects as well. When the above test fails, one will therefore have to try to identify the cause for the failure. This can be done by applying in succession, tests for the identification of these model errors (e.g., datasnooping for outliers, cycle slip testing, etc.), see e.g., Baarda [1968], Van der Marel [1990], Teunissen [1994c]. In case the processing is based on recursive least-squares algorithms like the Kalman filter, then the detection, identification and adaption procedure of Teunissen [1990a], can be used, see also, e.g., Teunissen and Salzmann [1989], De Jong [1994], Jin [1995], De Jong [1996], Gillissen and Elema [1996].

The question as to whether the model under H_3 can be considered valid or not, can be handled along similar lines as discussed above. That is, the appropriate test statistic for testing H_3 against the most relaxed alternative H_2, is given by the ratio $\check{\sigma}^2/\sigma^2$. This test statistic is distributed under H_3 as $F(k-n,\infty)$. Its mean and variance under H_3 are equal to respectively 1 and $2/(k-n)$. The decision to accept H_3 is made, when the value of the test statistic is less than the critical value $F_\alpha(k-n,\infty)$, with α the chosen level of significance. Thus H_3 is accepted when

$$\check{\sigma}^2/\sigma^2 < F_\alpha(k-n,\infty) \,. \tag{8.17}$$

Note, that this test tests H_3 against the most relaxed alternative hypothesis H_2. An alternative test for the validation of H_3 and one which is more powerful, can be constructed if we are willing to accept that the first hypothesis H_1 is true. In that case we can test H_3 against H_1, instead of against H_2. This test is therefore focused on answering the question whether the fixing of the ambiguity vector a on the value $\check{a}$ is valid or not. The appropriate test statistic for testing H_3 against H_1 is given as $(\hat{a}-\check{a})^T G_{\hat{a}}^{-1}(\hat{a}-\check{a})/m\sigma^2$, with $G_{\hat{a}}$ being the cofactor of $Q_{\hat{a}}$. It is distributed under H_3 as $F(m,\infty)$. Its mean and variance under H_3 are equal to respectively 1 and $2/m$. The decision to consider the value $\check{a}$ valid, is then made when

$$(\hat{a}-\check{a})^T G_{\hat{a}}^{-1}(\hat{a}-\check{a})/m\sigma^2 < F_\alpha(m,\infty) \,. \tag{8.18}$$

This shows not surprisingly, that this test is based on the distance - as measured in the metric defined by $Q_{\hat{a}}$ - between the integer ambiguity vector $\check{a}$ and the centre $\hat{a}$ of the ambiguity search space. If the value of the test statistic fails to pass the test (8.18), the conclusion reads that the value $\check{a}$ for a is rejected. In that case, the confidence in the value $\check{a}$ is low, implying that one should refrain from using the fixed solution. Instead, one should then either base the results on the hypothesis H_1 and thus be content with the float solution, or alternatively, gather more data (e.g., make use of longer observational time spans, or, include dual frequency data if applicable, or, include pseudorange data if applicable) and then repeat the whole estimation and validation process.

Instead of using the expression of (8.18) for the test statistic, we may also use an expression in which both of the test statistics of (8.16) and (8.17) occur. This follows from the identity

$$(\hat{a}-\check{a})^T G_{\hat{a}}^{-1}(\hat{a}-\check{a})/m\sigma^2 = \frac{k-n}{m}(\check{\sigma}^2/\sigma^2) - \frac{k-m-n}{m}(\hat{\sigma}^2/\sigma^2)\,, \tag{8.19}$$

which is easily verified when using (8.9). Hence, the test statistic of (8.18) is a weighted difference of the two test statistics of (8.16) and (8.17).

In case the value of the test statistic passes the test (8.18), the conclusion reads that there is no evidence to reject the value $\check{a}$ for a. Still however, one should be careful to conclude from this that one can safely fix the ambiguities and provide the fixed solution to the user. The fact that there is no evidence to reject the value $\check{a}$, does not mean that $\check{a}$ is the one and only integer ambiguity vector for which such an evidence is lacking. There still could exist integer ambiguity vectors other than $\check{a}$, that pass the test (8.18). In that case, the likelihood that $\check{a}$ is the correct integer ambiguity vector would not differ too much from the likelihood that some other integer vector, say $\check{a}'$, would be the correct ambiguity vector. Fixing the ambiguities on $\check{a}$ should therefore be avoided in this case, because of the existing high likelihood of fixing the ambiguities to a wrong value. And fixing the ambiguity vector to a wrong value, can have dramatic consequences for the fixed baseline solution.

To summarize: (*i*) we know by definition, if $\check{a}$ is chosen as the integer least-squares ambiguity vector, that $\check{a}$ is the most likely integer candidate for a; (*ii*) we also know, when the test (8.18) passes, that the most likely candidate $\check{a}$ is indeed a likely candidate; but, (*iii*) we do not know yet, how the likelihood of the candidate $\check{a}$ compares to the likelihood of other integer vectors. We therefore need an additional test, in order to be able to compare the likelihoods of integer candidates, see e.g., Abbot et al. [1989], Wübbena [1991], Frei [1991], Euler and Schaffrin [1991], Erickson [1992], Rothacher [1993], Betti et al. [1993], Tiberius and De Jonge [1995]. Next to the most likely candidate $\check{a}$, one therefore usually also makes use of the second most likely integer candidate, which will be denoted as $\check{a}'$. The idea is now, that we should try to find a test statistic which in some way measures the likelihood of $\check{a}'$ relative to the likelihood of $\check{a}$. An intuitively appealing test statistic for that purpose is given by the ratio $\check{\sigma}'^2/\check{\sigma}^2$. By the definition of $\check{a}$ and $\check{a}'$ this ratio is always larger than one. The second most likely value $\check{a}'$is then considered to be far less likely than the most likely value $\check{a}$, if the ratio $\check{\sigma}'^2/\check{\sigma}^2$ is significantly larger than one. Thus if $\check{a}$ passes the test (8.18) and

$$\check{\sigma}'^2 / \check{\sigma}^2 > c\,, \tag{8.20}$$

in which $c > 1$ is a to be chosen critical value, the decision reads that the most likely value $\check{a}$ is not only likely enough, but also far more likely than the second most likely value $\check{a}'$. Hence, in that case one decides to make use of the fixed solution.

Within the context of GPS ambiguity fixing, the acceptance test (8.20), or variations thereof, has been in use for quite some time now and it appears to work satisfactorily. There is however one pitfall that should be avoided. In the GPS-literature it is sometimes claimed that the test statistic of (8.20) has an F-distribution. Unfortunately, this is not true. The two quadratic forms in the nominator and denominator of the test statistic are namely not independent. This implies for instance, that once a value for c is chosen, one is not allowed to make use of the F-distribution for the computation of the corresponding level of significance.

As an alternative to test (8.20), one may also consider to make use of a test similar to that of (8.18). That is, one may decide to make use of the fixed solution if both the test (8.18) and the test

$$(\hat{a} - \check{a}')^T G_{\hat{a}}^{-1} (\hat{a} - \check{a}') / m\sigma^2 > F_{\alpha'}(m,\infty) \geq F_{\alpha}(m,\infty) \tag{8.21}$$

are passed. The rationale behind using the combined test is, when (8.18) and (8.21) are satisfied, that the value $\check{a}$ may be considered validated and the value $\check{a}'$ invalidated. In order to make sure that $\check{a}'$ is sufficiently less likely than $\check{a}$, one will have to choose $F_{\alpha'}(m,\infty)$ sufficiently larger than $F_{\alpha}(m,\infty)$. The acceptance region for the combined test, (8.18) and (8.21), is shown in Figure 8.2.

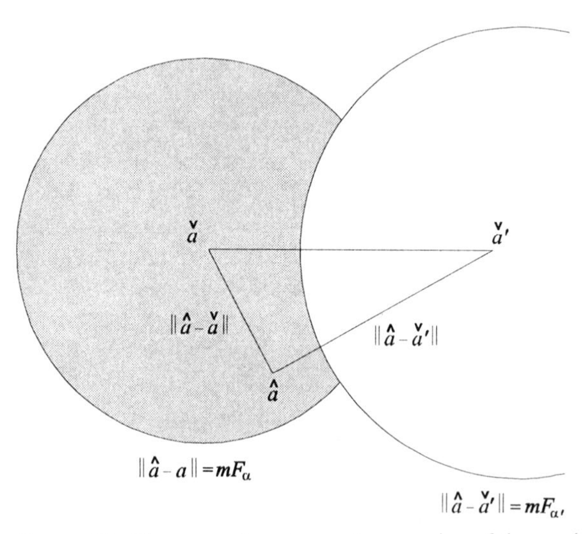

Figure 8.2. The moon-shaped acceptance region of the combined test (8.18) and (8.21)

Up to this point, the variance-factor of unit weight σ^2 was assumed known. If σ^2 is unknown however, both the tests (8.16) and (8.17) cannot be executed. That is, when σ^2 is unknown a priori, one will not be able to test the hypotheses H_1 and H_3 against their most relaxed alternative H_2. In this case it will however still be possible to test H_3 against H_1. The appropriate test statistic for testing H_3 against H_1, when σ^2 is unknown, follows when we replace σ^2 in expression (8.18) by $\hat{\sigma}^2$. The resulting test statistic will then have the distribution $F(m,k-m-n)$ under H_3.
Hence, instead of (8.18) the test becomes then

$$(\hat{a}-\check{a})^T G_{\hat{a}}^{-1}(\hat{a}-\check{a})/m\hat{\sigma}^2 < F_{\alpha}(m,k-m-n)\,. \tag{8.22}$$

The means of the test statistics of (8.16), (8.17) and (8.18) were all equal to 1. The mean of the test statistic of (8.22) under H_3 is however not equal to 1. It equals $(k-m-n)/(k-m-n-2)$. Hence, it is larger than 1, but very close to it when $(k-m-n)$ is large. In fact the distribution of the test statistic of (8.22) tends to that of (8.18) when $(k-m-n)$ increases. As it was the case with the test statistic of (8.18), also the test statistic of (8.22) can be expressed in terms of the two test statistics of (8.16) and (8.17). Instead of a weighted difference however, it now becomes dependent on the ratio of the two test statistics of (8.16) and (8.17). This follows from the identity

$$[(\hat{a}-\check{a})^T G_{\hat{a}}^{-1}(\hat{a}-\check{a})/m\hat{\sigma}^2-1] = \frac{k-n}{m}[\check{\sigma}^2/\hat{\sigma}^2-1]\,. \tag{8.23}$$

Above, we have discussed the validation of both the float and the fixed solution using standard concepts from the theory of statistical hypothesis testing. This is also the approach which is still often used. Although these concepts appear to work satisfactorily in practice, it is not without importance to point out that there are some theoretical shortcomings associated with these concepts, see e.g. Teunissen [1997]. These shortcomings are not related to the way the validation of the float solution is handled. The validation of the float solution as it has been discussed above is theoretically sound, of course provided that the underlying assumptions are valid. The theoretical shortcomings, as the author sees it, are directed towards the way the validation of the fixed solution is handled. It already starts with the formulation of the third hypothesis H_3 of (8.14). In the formulation of this hypothesis, the integer ambiguity vector $\check{a}$ is treated as a deterministic vector of which the values of its entries are independently set. This however is not true. First of all, the entries of $\check{a}$ are not independently chosen. Instead, they depend on the same vector of observables y as it is used in the formulation of the hypothesis. Hence, when the values of the entries of y change, also the integer values of the entries of $\check{a}$ might change. Secondly, since the vector of observables y is assumed to be a random vector, also the integer least-squares ambiguity vector $\check{a}$ is stochastic and not deterministic. The conclusion reads therefore, that instead of H_3 of (8.14), the correct hypothesis should read

$$H_3: y = Aa + Bb + e, \; Q_y = \sigma^2 G_y, \; a \in Z^m, \; b \in R^n. \tag{8.24}$$

That is, H_3 should read as H_1 with the additional integer constraint $a \in Z^m$ and not read as H_1 with the additional constraint of $a = \check{a}$. The consequence of formulation (8.24), as opposed to the formulation of H_3 in (8.14) is, that in order to test H_3 of (8.24) against H_1, one will have to take the stochasticity of the integer estimator of the ambiguity vector into account. This is a nontrivial problem, since the probability density function of $\check{a}$ is of the discrete type. For a discussion see Blewitt [1989], Teunissen [1990b, 1997], Betti et al. [1993]. Fortunately, the practical relevance of the above pitfall may be minor, in particular when it can be assured that $\check{a}$ is the only integer candidate of sufficient likelihood. One of the features of a proper validation procedure should namely be to verify whether or not sufficient probability mass is located at a single grid point of Z^m. This can be done without the need to know the complete discrete distribution of $\check{a}$. Only the probability of correct integer estimation is needed. And when this probability is sufficiently close to one, the influence of the stochastity of $\check{a}$ may be negelected.

8.3 SEARCH FOR THE INTEGER LEAST-SQUARES AMBIGUITIES

In this section it will be shown how the integer least-squares problem (8.11) can be solved numerically. The solution will be found by means of a search process. Two different concepts will be discussed, one which is based on the idea of using planes of support and one which is based on the idea of using a sequential conditional least-squares adjustment of the ambiguities. The concept which is based on using the ellipsoidal planes of support parallels the use of simultaneous confidence intervals in statistics for multiple comparisons Scheffé [1956]. Within the context of GPS ambiguity fixing the method of Frei and Beutler [1990, 1991] is based on it. The method of Teunissen [1993a, 1994a] is based on the second concept. When interpreted algebraically instead of statistically, it parallels the use of a triangular decomposition. Within the context of GPS ambiguity fixing, alternative approaches that make use of a triangular decomposition are proposed in Blewitt [1989], Wübbena [1991], Euler and Landau [1992]. In this section, we will also discuss the dependency of the search performance on the statistical characteristics of the least-squares ambiguities. In particular, it will be explained why the search for the integer least-squares ambiguities performs so poorly, when only short observational time span carrier phase data are used.

8.3.1 The Ambiguity Search Space and its Planes of Support

Up to this point we did not show how the integer least-squares problem (8.11) can actually be solved. As it turns out, the computation of the integer minimizer of (8.11) is a far from trivial problem. There are namely in general no standard techniques available for solving (8.11) as they are available for solving ordinary least-squares problems. It is therefore that with the minimization problem (8.11),

$$\min_{a} \; (\hat{a}-a)^T Q_{\hat{a}}^{-1}(\hat{a}-a), \text{ with } a \in Z^m, \tag{8.25}$$

the intricacy of the integer ambiguity estimation problem manifests itself. To solve it we will resort to methods that in one way or another make use of a discrete search strategy. The idea is to replace the space of all integers, Z^m, with a smaller subset that can be enumerated and that still contains the integer least-squares solution. This smaller subset will be chosen as a region bounded by an hyper-ellipsoid, where the hyper-ellipsoid is based on the objective function of (8.25). This ellipsoidal region is given by

$$(\hat{a}-a)^T Q_{\hat{a}}^{-1}(\hat{a}-a) \leq \chi^2, \tag{8.26}$$

and it will be referred to as the *ambiguity search space* (see Figure 8.3). It is centred at the real-valued least-squares estimate $\hat{a}$, its shape is governed by the variance-covariance matrix $Q_{\hat{a}}$ and its size can be controlled through the selection of the positive constant χ^2.

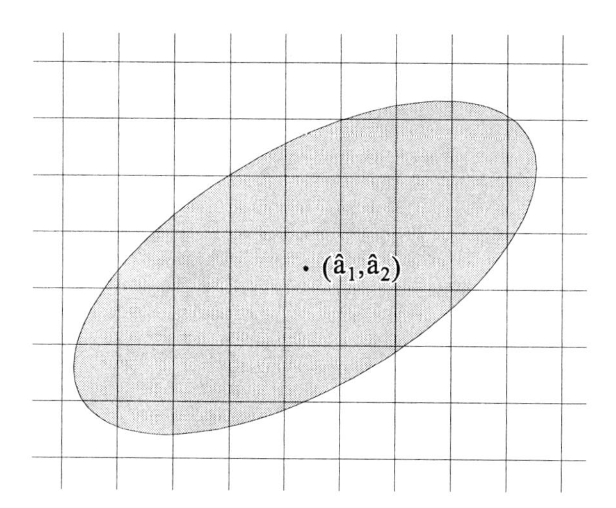

Figure 8.3. The ambiguity search space and integer grid.

One way of finding the minimizer of (8.25) is to identify first the set of integer ambiguity vectors a that satisfy the inequality (8.26), i.e. to identify the set of gridpoints that lie within the ambiguity search space, and then to select from this set that gridpoint that gives the smallest value for the objective function of (8.25). The quadratic form of (8.26) however, can not be used as such to identify the set of candidate gridpoints. The first idea that comes in mind is therefore to replace inequality (8.26) with an equivalent description that is based on using the *planes of support* of the ellipsoid. This equivalence can be constructed as follows. Let d be an arbitrary vector of R^m and let $(\hat{a}-a)$ be orthogonally projected onto d.

The orthogonal projection of $(\hat{a}-a)$ onto d, where orthogonality is measured with respect to the metric of $Q_{\hat{a}}$, is then given as $d(d^TQ_{\hat{a}}^{-1}d)^{-1}d^TQ_{\hat{a}}^{-1}(\hat{a}-a)$. The square of the length of this vector equals $[d^TQ_{\hat{a}}^{-1}(\hat{a}-a)]^2/(d^TQ_{\hat{a}}^{-1}d)$. Since the length of the orthogonal projection of a vector onto an arbitrary direction is always less than or equal to the length of the vector itself, we have

$$(\hat{a}-a)^TQ_{\hat{a}}^{-1}(\hat{a}-a) = \max_{d\in R^m}[d^TQ_{\hat{a}}^{-1}(\hat{a}-a)]^2/(d^TQ_{\hat{a}}^{-1}d). \tag{8.27}$$

Hence, it follows from this equality that, when d is replaced by $Q_{\hat{a}}c$, we obtain the equivalence

$$(\hat{a}-a)^TQ_{\hat{a}}^{-1}(\hat{a}-a) \leq \chi^2 \iff [c^T(\hat{a}-a)]^2/(c^TQ_{\hat{a}}c) \leq \chi^2, \forall c\in R^m. \tag{8.28}$$

Both inequalities describe the ambiguity search space. In the second inequality we recognize $c^T(\hat{a}-a) = \pm(c^TQ_{\hat{a}}c)^{1/2}\chi$, which is the pair of parallel planes of support of the ambiguity search space having vector c as normal. The above equivalence therefore states that the ambiguity search space coincides with the region that follows from taking all intersections of the areas between each pair of planes of support. Hence, in order to find the set of candidate gridpoints that satisfy (8.26), we may as well make use of the planes of support (see Figure 8.4). That is, instead of working with the single quadratic inequality (8.26), we may work with the family of scalar inequalities

$$[c^T(\hat{a}-a)]^2 \leq (c^TQ_{\hat{a}}c)\chi^2, \ \forall c\in R^m. \tag{8.29}$$

When working with the above inequalities, there are however two restrictions that need to be appreciated. First of all, the above equivalence (8.28) only holds for the *infinite* set of planes of support. But for all practical purposes one can only work with a finite set. Working with a finite set implies however, that the region bounded by the planes of support will be larger in size than the original ambiguity search space. Of course, one could think of minimizing the increase in size by choosing an appropriate set of normal vectors c. For instance, if the normal vectors c are chosen to lie in the direction of the major and minor axes of the ambiguity search space, then the resulting region will fit the ambiguity search space best. But here is where the second restriction comes into play. One simply has no complete freedom in choosing the planes of support. Their normals c should namely be chosen such that the resulting interval (8.29) can indeed be used for selecting candidate gridpoints. Hence, the normal vectors c cannot be chosen arbitrarily.

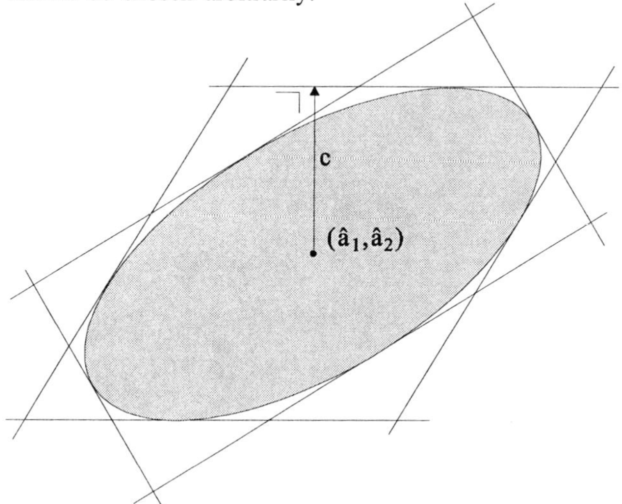

Figure 8.4. Ambiguity search space and planes of support.

The above two restrictions imply that one has to be aware of the following serious pitfall when using the planes of support approach, Teunissen et al. [1997]. When bounding the search space with the planes of support one has to make sure that the size of the resulting space is large enough to contain the integer least-squares solution and in case of validation, the second best solution as well. When the search space is set, one will always be able to find a best and second best solution *within* this search space. But this does not necessarily imply that they are the best and second best solution for the *entire* space of integers as well. This can only be guaranteed when one knows beforehand that these two solutions are contained in the search space. In the following we will assume that this is the case.

The most rudimentary approach would be to circumscribe the ambiguity search space with an m-dimensional rectangular box of which the sides are perpendicular to the coordinate axes in R^m. This is achieved when the normals c are chosen as $c_i = (0,...,1,0,...0)^T$, with the 1 as the ith-coordinate. The region bounded by the corresponding planes of support is then described by the following finite set of m scalar inequalities:

$$(\hat{a}_i - a_i)^2 \leq \sigma^2_{\hat{a}_i}\chi^2, \text{ for } i = 1,...,m \,, \quad (8.30)$$

where $\sigma^2_{\hat{a}_i}$ is the variance of the ith ambiguity. The intervals of (8.30) can now be used to select candidate ambiguity integers from which then the minimizer of (8.25) can be chosen. It will be clear that the m-dimensional rectangular box described by the inequalities of (8.30), fits the ambiguity search space best if this search space would be spheroidal or at least would have its principal axes parallel to the coordinate axes. The fit will be rather poor however, when the ambiguity search space is both elongated and rotated with respect to the coordinate axes. An improvement of the region circumscribing the ambiguity search space can be achieved by introducing additional planes of support. As an example, consider the case that one is working with dual-frequency carrier phase data instead of with single-frequency carrier phase data. With (8.2), the difference between the L_2 and L_1 DD carrier phases follows then as

$$\Phi^{kl}_{ij,12} = \lambda_2 N^{kl}_{ij,2} - \lambda_1 N^{kl}_{ij,1} + \varepsilon^{kl}_{ij,12} .$$

This shows that the linear combination $\lambda_2 N^{kl}_{ij,2} - \lambda_1 N^{kl}_{ij,1}$ can be estimated with a high precision. Thus, if the ambiguity vector a is partitioned as $a = (a_1, a_2)^T$ with a_1 having as its entries all L_1 DD ambiguities and a_2 all L_2 DD ambiguities, it follows with the choice

$$c = (-\lambda_1, 0, \ldots, \lambda_2, 0, \ldots, 0)^T$$

that $c^T Q_{\hat{a}} c$ will be very small indeed. Hence with this type of choice for the normal c, the bound $[c^T(\hat{a}-a)]^2 \leq c^T Q_{\hat{a}} c \chi^2$ introduces in addition to (8.30) a tight constraint on the candidate L_1 and L_2 ambiguity integer pairs.

8.3.2 Sequential Conditional Least-Squares Ambiguities

In the previous section we have seen that the planes of support of the ambiguity search space allow us to formulate a set of scalar inequalities on the basis of which the search for the minimizer of the integer least-squares problem (8.25) can be performed. It was noted however, that the region bounded by the chosen finite set of planes of support may not necessarily follow the shape and orientation of the ambiguity search space. This observation suggests, since the shape and orientation of the ambiguity search space is governed by the ambiguity variance-covariance matrix $Q_{\hat{a}}$, that we consider in somewhat more detail the impact of the structure of the ambiguity variance-covariance matrix.

To start, it helps if we ask ourselves the question what the structure of (8.25) must be in order to be able to apply the simplest of all integer estimation methods. Clearly, the simplest integer estimation method is "rounding to the nearest" integer. In general this approach will not give us the correct answer to the integer least-squares problem (8.25). However, it does give the correct answer, when the ambiguity variance-covariance matrix $Q_{\hat{a}}$ is diagonal, i.e.

when all least-squares ambiguities are fully decorrelated. A diagonal $Q_{\hat{a}}$ implies namely that (8.25) reduces to a minimization of a sum of independent squares

$$\underset{a_1,\ldots,a_m \in Z}{\text{minimize}} \sum_{i=1}^{m} (\hat{a}_i - a_i)^2 / \sigma^2_{\hat{a}_i} . \tag{8.31}$$

Hence, in that case we can work with m separate scalar integer least-squares problems, and the integer minimizers of each of these individual squares are then simply given by the integer nearest to $\hat{a}_i$. The conclusion reads therefore, that the ambiguity integer least-squares problem becomes trivial when all least-squares ambiguities are fully decorrelated.

In reality, the least-squares ambiguities are usually highly correlated and the variance-covariance matrix $Q_{\hat{a}}$ is far from being diagonal. Still however, it is possible to recover a sum-of-squares structure of the objective function, similar to that of (8.31), if we diagonalize $Q_{\hat{a}}$. Not every diagonalization works however. What is needed in addition, is that the diagonalization realizes, like in (8.31), that the individual ambiguities can be assigned to the individual squares in the total sum-of-squares. This for instance, rules out a diagonalization based on the eigenvalue decomposition of the ambiguity variance-covariance matrix. In the same spirit of decomposition (8.7), we will therefore apply a conditional least-squares decomposition to the ambiguities. And this will be done on an ambiguity-by-ambiguity basis. Hence, we will introduce the *sequential conditional least-squares ambiguities* $\hat{a}_{i|I}$, $i = 1,\ldots,m$, see e.g., Teunissen [1993a]. The estimate $\hat{a}_{i|I}$ is the least-squares estimate of the *i*th ambiguity a_i, conditioned on a fixing of the previous (*i*-1) ambiguities. The shorthand notation $\hat{a}_{i|I}$ stands therefore for $\hat{a}_{i|(i-1),\ldots,1}$. The sequential conditional least-squares ambiguities follow from the ordinary least-squares ambiguities as

$$\hat{a}_{i|I} = \hat{a}_i - \sum_{j=1}^{i-1} \sigma_{\hat{a}_i \hat{a}_{j|J}} \sigma^{-2}_{\hat{a}_{j|J}} (\hat{a}_{j|J} - a_j) . \tag{8.32}$$

In this expression $\sigma_{\hat{a}_i \hat{a}_{j|J}}$ denotes the covariance between $\hat{a}_i$ and $\hat{a}_{j|J}$. An important property of the $\hat{a}_{i|I}$ is that they do not correlate. Hence their variance-covariance matrix is diagonal. It follows from (8.32) that the ambiguity difference $(\hat{a}_i - a_i)$ can be written in terms of the differences $(\hat{a}_{j|J} - a_j)$, $j = 1,\ldots,i$ as $(\hat{a}_i - a_i) = (\hat{a}_{i|I} - a_i) + \sum_j \sigma_{\hat{a}_i \hat{a}_{j|J}} \sigma^{-2}_{\hat{a}_{j|J}} (\hat{a}_{j|J} - a_j)$. Hence, when this is written out in vector-matrix form, using the notation $\hat{d} = (\hat{a}_1, \hat{a}_{2|1},\ldots,\hat{a}_{m|M})^T$, and the error propagation law is applied, it follows, because of the fact that the conditional least-squares ambiguities are mutually uncorrelated, that

$$(\hat{a} - a) = L(\hat{d} - a) \text{ and } Q_{\hat{a}} = LDL^T, \tag{8.33}$$

where: $D = diag(\ldots, \sigma^2_{\hat{a}_{i|I}},\ldots)$ and $(L)_{ij} = 0$ for $1 \leq i < j \leq m$ and $(L)_{ij} = 1$ for $i = j$ and $(L)_{ij} = \sigma_{\hat{a}_i \hat{a}_{j|J}} \sigma^{-2}_{\hat{a}_{j|J}}$ for $1 \leq j < i \leq m$. The above matrix decomposition is well-known and is usually referred to as the LDL^T-decomposition, see, e.g., Golub and Van Loan [1986]. With our "re-discovery" of the LDL^T-decomposition, we now can give a clear statistical interpretation to each of the entries of the lower triangular matrix L and to each of the entries of the diagonal matrix D. This interpretation will also be of help, when we discuss ways of improving the search for the integer least-squares ambiguities (cf. section 8.5.3).

Since the sequential conditional least-squares ambiguities are mutually uncorrelated, substitution of (8.32) into (8.25) gives the desired sum-of-squares structure and allows us to rewrite the integer least-squares problem as

$$\underset{a_1,\ldots,a_m \in Z}{\text{minimize}} \sum_{i=1}^{m} (\hat{a}_{i|I} - a_i)^2 / \sigma^2_{\hat{a}_{i|I}} . \tag{8.34}$$

Note the similarity between (8.31) and (8.34). In fact, the minimization problem (8.34) reduces to that of (8.31) when all least-squares ambiguities would be fully decorrelated. In that case the ordinary least-squares ambiguities become identical to their conditional counterparts.

Based on the sum-of-squares structure of (8.34), we may now formulate a search for the integer least-squares ambiguities. Using the above sum-of-squares structure, the ambiguity search space can be described as

$$\sum_{i=1}^{m} (\hat{a}_{i|I} - a_i)^2 / \sigma^2_{\hat{a}_{i|I}} \leq \chi^2 , \tag{8.35}$$

and the scalar bounds on the individual ambiguities become

$$\begin{cases} (\hat{a}_1 - a_1)^2 \leq \sigma^2_{\hat{a}_1} \chi^2 \\ (\hat{a}_{2|1} - a_2)^2 \leq \sigma^2_{\hat{a}_{2|1}} [\chi^2 - (\hat{a}_1 - a_1)^2 / \sigma^2_{\hat{a}_1}] \\ \quad \cdot \\ \quad \cdot \\ (\hat{a}_{m|M} - a_m)^2 \leq \sigma^2_{\hat{a}_{m|M}} [\chi^2 - \sum_{j=1}^{m-1} (\hat{a}_{j|J} - a_j)^2 / \sigma^2_{\hat{a}_{j|J}}] . \end{cases} \tag{8.36}$$

Note that the bounds of (8.36) are sharper than those of (8.30).

In order to discuss our search based on (8.36), the two-dimensional case will be used as an illustrative example. In the two-dimensional case, the ambiguity search space is given by the inequality

$$(\hat{a}_1 - a_1)^2 / \sigma^2_{\hat{a}_1} + (\hat{a}_{2|1} - a_2)^2 / \sigma^2_{\hat{a}_{2|1}} \leq \chi^2 . \tag{8.37}$$

This two-dimensional ambiguity search space is shown in Figure 8.5. In the figure we have also drawn the line passing through the centre of the ellipse, $(\hat{a}_1, \hat{a}_2)$, having $(1, \sigma_{\hat{a}_2\hat{a}_1}\sigma_{\hat{a}_1}^{-2})$ as direction vector. This line intersects the ellipse at two points where the normal of the ellipse is directed along the a_1-axis. Note that the point $(\hat{a}_1, \hat{a}_{2|1})$ moves along this line when $\hat{a}_1$ is varied.

Also shown in the figure is the rectangular box that encloses the ellipse. It is described by the two scalar inequalities

$$\begin{cases} (\hat{a}_1 - a_1)^2 \leq \sigma_{\hat{a}_1}^2 \chi^2 \\ (\hat{a}_2 - a_2)^2 \leq \sigma_{\hat{a}_2}^2 \chi^2 \end{cases} \tag{8.38}$$

Hence, these two inequalities are the two-dimensional counterparts of (8.30). But instead of using these two inequalities, the sum-of-squares structure of (8.37) allows us to formulate the following two bounds on the two ambiguities a_1 and a_2,

$$\begin{cases} (\hat{a}_1 - a_1)^2 \leq \sigma_{\hat{a}_1}^2 \chi^2 \\ (\hat{a}_{2|1} - a_2)^2 \leq \sigma_{\hat{a}_{2|1}}^2 \lambda(a_1)\, \chi^2, \end{cases} \tag{8.39}$$

with $\lambda(a_1) = 1 - (\hat{a}_1 - a_1)^2/\sigma_{\hat{a}_1}^2 \chi^2$. These two intervals and their lengths are also shown in Figure 8.5.

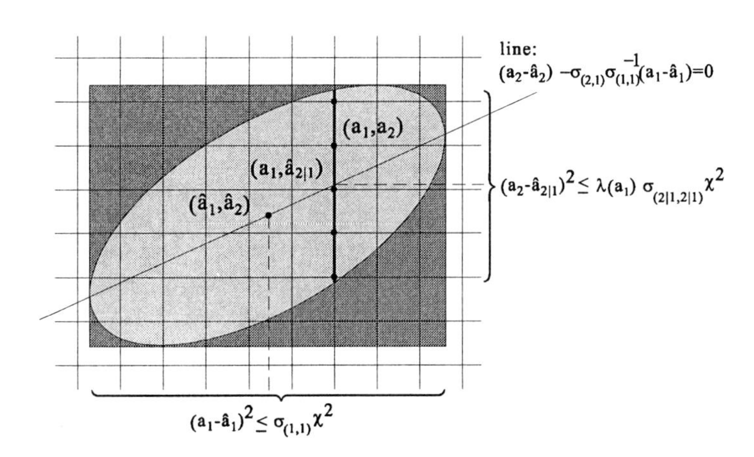

Figure 8.5. Ambiguity search space and search bounds.

Based on the two scalar inequalities of (8.39), our search for the integer least-squares ambiguities may now be described as follows. First one selects an integer ambiguity a_1 that satisfies the first bound of (8.39). Then based on this chosen integer ambiguity value a_1, the conditional least-squares estimate $\hat{a}_{2|1}$ and scalar $\lambda(a_1)$ are computed. These values are then used to select an integer ambiguity a_2 that satisfies the second bound of (8.39). Since we aim at finding the integer minimizer, it is natural to choose the integer candidates in such a way that the individual squares in the sum-of-squares (8.37) are made as small as possible. This implies that a_2 should always be chosen as the integer nearest to $\hat{a}_{2|1}$. But remember that $\hat{a}_{2|1}$ depends on a_1. If one then fails to find an integer a_2 that satisfies the second bound, one restarts and chooses for a_1 the second nearest integer to $\hat{a}_1$, and so on. Note that in this way, one is roughly following the direction of the line $(a_1, \hat{a}_{2|1})$, working with a_1 along the a_1-axis from the inside of the ellipse, in an alternating fashion, towards the bounds of the ellipse. This process is continued until an admissible integer-pair (a_1, a_2) is found, i.e. until a gridpoint is found that lies inside the ambiguity search space. Then a shrinking of the ellipse is applied, by applying an appropriate downscaling of χ^2, after which one continues with the next and following nearest integers to $\hat{a}_1$. This process is continued until one fails to find an admissible integer for a_1. The last found integer-pair is then the sought for integer least-squares solution.

Example 1:

This example illustrates the above described search procedure. The least-squares estimates of the two ambiguities and their variance-covariance matrix are given as

$$\hat{a} = \begin{bmatrix} 1.05 \\ 1.30 \end{bmatrix} \quad ; \quad Q_{\hat{a}} = \begin{bmatrix} 53.40 & 38.40 \\ 38.40 & 28.00 \end{bmatrix} .$$

The χ^2-value is given as $\chi^2 = 1.5$ and the corresponding ambiguity search space with integer grid is shown in Figure 8.6.

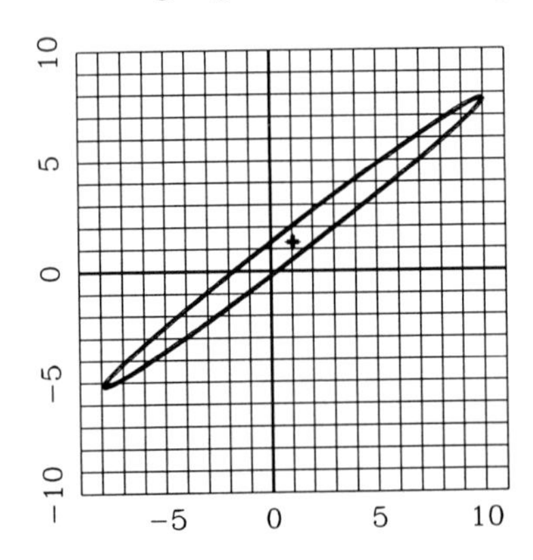

Figure 8.6. The 2D-ambiguity search space and integer grid.

The complete set of integer pairs (a_1, a_2) that lie inside the ambiguity search space is given in Table 8.1. This table also gives the corresponding values for the objective function $F(a_1, a_2) = (\hat{a}_1 - a_1)^2/\sigma^2_{\hat{a}_1} + (\hat{a}_{2|1} - a_2)^2/\sigma^2_{\hat{a}_{2|1}}$.

Let us now consider the actual results of the search based on the two sequential bounds of (8.39). See also Table 8.2. Since $\hat{a}_1 = 1.05$, we choose a_1 as its nearest integer, which is $a_1 = 1.00$. Based on this integer value for a_1, the conditional least-squares estimate for the second ambiguity reads $\hat{a}_{2|1} = 1.26$. Since its nearest integer reads 1.00, we choose a_2 as $a_2 = 1.00$. Hence, we now have an integer pair $(a_1, a_2) = (1.00, 1.00)$ which lies inside the ambiguity search space. Since the value of the objective function of this integer pair equals $F(1.00, 1.00) = 0.1804$, we may now shrink the ellipse and set the χ^2-value at $\chi^2 = 0.1804$. It will be clear that the second nearest integer of $\hat{a}_{2|1} = 1.26$, being 2.00, will give an integer pair (1.00, 2.00) that lies outside the ellipse. Hence, in order to continue we go back to the first ambiguity and consider the second nearest integer to $\hat{a}_1 = 1.05$, which is $a_1 = 2.00$. Based on this value for a_1, the conditional least-squares estimate for the second ambiguity reads $\hat{a}_{2|1} = 1.98$. Since its nearest integer reads 2.00, we now choose a_2 as $a_2 = 2.00$. It follows that this new integer pair $(a_1, a_2) = (2.00, 2.00)$ lies inside the shrunken ellipse and that the value of its objective function reads $F(2.00, 2.00) = 0.0176$. We may now again shrink the ellipse and set $\chi^2 = 0.0176$. It will be clear that the second nearest integer of $\hat{a}_{2|1} = 1.98$, being 1.00, will give an integer pair (2.00, 1.00) that lies outside the shrunken ellipse.

Table 8.1. The set of integer candidates and their function values.

No.	a_1	a_2	$F(a_1,a_2)$	No.	a_1	a_2	$F(a_1,a_2)$
1	-6	-4	1.0680	11	1	2	1.4014
2	-5	-3	0.6921	12	2	2	0.0176
3	-4	-2	0.7618	13	3	2	1.3471
4	-3	-2	0.6959	14	3	3	0.3006
5	-3	-1	1.2773	15	4	3	0.6223
6	-2	-1	0.2037	16	4	4	1.0293
7	-1	0	0.1572	17	5	4	0.3432
8	0	0	0.7890	18	6	5	0.5099
9	0	1	0.5564	19	7	6	1.1223
10	1	1	0.1804	20	8	6	1.1339
				21	9	7	1.1843

Table 8.2. The integer pairs that are encountered during the search.

$\chi^2 = 1.5$		$\chi^2 = 0.1804$		$\chi^2 = 0.0176$	
a_1	a_2	a_1	a_2	a_1	a_2
1.00	1.00	1.00	-		
		2.00	2.00	2.00	-
				-	-

Hence, we again go back to the first ambiguity and now consider the third nearest integer to $\hat{a}_1 = 1.05$, which is 0.00. It follows however that this value for a_1 does not satisfy the first bound of (8.39) for $\chi^2 = 0.0176$. As a result, the search stops and the last found integer pair is provided as the solution sought. That the integer pair $(a_1, a_2) = (2.00, 2.00)$ indeed equals the integer least-squares solution can be verified by means of Table 8.1.

8.3.3 On the DD Ambiguity Precision and Correlation

In the previous two sections we have introduced two concepts that can be used for solving the integer least-squares problem (8.25). First the use of the ellipsoidal planes of support was discussed. The search based on this concept is rather straightforward, but as it was pointed out, the bounds that follow from using the planes of support may be rather conservative, in particular when the ambiguity search space is elongated and rotated with respect to the grid axes. Moreover, these bounds are fixed from the outset. The second concept that we discussed made use of a sequential conditional least-squares adjustment of the ambiguities, thus achieving a sum-of-squares structure for the objective function that has to be minimized. As a consequence we obtained bounds for the individual ambiguities that are less conservative and that are also not fixed from the outset. These bounds adjust themselves depending on the stage of progress of the search process.

So far, no quantitative indications were given of how well the search for the integer least-squares ambiguities will perform. We stressed though, that the elongation and orientation of the ambiguity search space is an important factor for the performance of the search. If the ambiguity search space turns out to be spheroidal or an hyper-ellipsoid with its principal axes parallel to the grid axes, then a simple rounding to the nearest integer will suffice. A slight difference between the direction of the principal axes and the grid axes however, may already render the approach of rounding to the nearest integer useless. The search based on the use of the ellipsoidal planes of support may suffice in its most rudimentary form, where use is made of the enclosing m-dimensional rectangular box, if the ambiguity search space, although rotated with respect to the grid axes, is still quite close to a spheroid. The fit of the m-dimensional rectangular box will become poorer though, the more elongated the ambiguity search space gets. Overall, the search bounds that follow from the sequential conditional least-squares adjustment of the ambiguities, follow the shape of the ambiguity search space best. But then again, in order to get a better insight as to its performance, we still need to know more about the behaviour of these adjustable bounds.

The above motivates us to have a somewhat closer look at the structure of the ambiguity variance-covariance matrix $Q_{\hat{a}}$. In this section we will therefore give some quantitative indications of the ambiguity search space. More elaborate examples of the numerical characteristics of the ambiguity search spaces can be found in Teunissen [1994a,d], De Jonge and Tiberius [1994], Teunissen and Tiberius [1994e]. First however, we will consider an example of a synthetic 2×2 ambiguity variance-covariance matrix. The structure of this matrix has been chosen such that it resembles the structure of the actual m×m ambiguity variance-covariance matrices. The example will illustrate some of the main features of this variance-covariance matrix and show what the implications of the particular structure of this matrix are for the integer ambiguity search.

Example 2:

Let the variance-covariance matrix of the two least-squares ambiguities $\hat{a}_1$ and $\hat{a}_2$ be given as

$$\begin{bmatrix} \sigma^2_{\hat{a}_1} & \sigma_{\hat{a}_1,\hat{a}_2} \\ \sigma_{\hat{a}_2,\hat{a}_1} & \sigma^2_{\hat{a}_2} \end{bmatrix} = \sigma^2 \begin{bmatrix} 1 & 0 \\ 0 & 1 \end{bmatrix} + \begin{bmatrix} \beta_1 \\ \beta_2 \end{bmatrix} \begin{bmatrix} \beta_1 \\ \beta_2 \end{bmatrix}^T . \tag{8.40}$$

It will be assumed that

$$\sigma^2 << \beta_1^2,\ \beta_2^2 \ ;\ \beta_1^2 \cong \beta_2^2 . \tag{8.41}$$

Note that the above 2×2 matrix is given as the sum of a scaled rank-2 matrix and a rank-1 matrix. And because of (8.41), the entries of the rank-2 matrix are very much smaller than the entries of the rank-1 matrix. In order to give some qualitative indications as to how the two-dimensional ambiguity search space and the statistics of the two ambiguities are affected by the particular structure of the variance-covariance matrix, we will consider the elongation of the ambiguity search space, the correlation coefficient of the two ambiguities and their conditional variances.

First we consider the elongation of the ambiguity search space. Elongation will be denoted by e and it is given as the ratio of the largest and smallest lengths of the principal axes of the ambiguity search space. It follows from

(8.40) that the elongation squared is given as

$$e^2 = 1 + \beta_1^2/\sigma^2 + \beta_2^2/\sigma^2 . \tag{8.42}$$

This shows that $e = 1$ when $\beta_1 = \beta_2 = 0$. In that case, the ambiguity search space equals a perfect circle. In our case however, (8.41) holds true, which implies that $\beta_1^2/\sigma^2 >> 1$ and $\beta_2^2/\sigma^2 >> 1$. Hence, in our case the ambiguity search space is extremely elongated.

In order to measure the statistical dependency between the two ambiguities, we first consider their correlation. It follows from (8.40) that the square of the correlation coefficient is given as

$$\rho^2 = ((1+\sigma^2/\beta_1^2)(1+\sigma^2/\beta_2^2))^{-1} \tag{8.43}$$

Together with (8.41) this shows that $\rho^2 \cong 1$. Hence, the two ambiguities are very heavily correlated. As a consequence of this extreme correlation, one will observe a large discontinuity in the conditional variances. To show this, consider the variance $\sigma^2_{\hat{a}_1}$ and the conditional variance $\sigma^2_{\hat{a}_{2|1}}$. It follows from (8.40) that

$$\sigma^2_{\hat{a}_1} = \sigma^2 + \beta_1^2 \quad ; \quad \sigma^2_{\hat{a}_{2|1}} = \sigma^2 + \beta_2^2 \, \frac{\sigma^2/\beta_1^2}{1+\sigma^2/\beta_1^2} . \tag{8.44}$$

Together with (8.41) this shows that $\sigma^2_{\hat{a}_{2|1}} << \sigma^2_{\hat{a}_1}$. Hence, there is a tremendous drop in value when one goes from the variance of the first ambiguity to the conditional variance of the second ambiguity. With β_1^2 sufficiently large, we approximately have $\sigma^2_{\hat{a}_1} \cong \beta_1^2$ and $\sigma^2_{\hat{a}_{2|1}} \cong 2\sigma^2$. The very important implication of this result for the search of the integer least-squares ambiguities is the following. When $\sigma^2_{\hat{a}_1}$ is large and $\sigma^2_{\hat{a}_{2|1}}$ extremely small, the problem of search-halting will be significant. A large $\sigma^2_{\hat{a}_1}$ implies namely, that the first bound of (8.39) will be rather loose. Quite a number of integers will therefore satisfy this first bound. This on its turn implies, when we go to the second bound of (8.39), which is very tight due to $\sigma^2_{\hat{a}_{2|1}} << \sigma^2_{\hat{a}_1}$, that we have a high likelihood of not being able to find an integer that satisfies this second bound. The potential of halting is therefore very significant when one goes from the first to the second bound. As a consequence a large number of trials are required, before one is able to find a candidate integer-pair.

In order to corroborate the above given qualitative results, we will now consider an example based on actual GPS data. Quantitative results, which are thought to be representative, are given for the elongation of the ambiguity search space and for the precision, correlation and spectrum of the DD ambiguities. Our example is based on a 7 satellite configuration, using dual frequency carrier phase data only. Hence, pseudorange data has not been used. The results that will be shown are based on the mere use of two epochs of data separated by only one second. The reason for choosing for our example the short observational time span of one second using the minimum number of two epochs, is to illustrate the extreme values the statistics of the DD ambiguities can reach. The a priori standard deviation of both the L_1 and L_2 carrier phases was set at the value of $\sigma = 3$ mm. Correlation in time and correlation between the channels were assumed to be nonexistent. Also atmospheric delays and multipath were assumed to be absent.

Figure 8.7 shows the elongation of the ambiguity search space as function of the observational time span. Note the extremely large elongation for short observational time spans. The elongation improves when the spacing in time of the data increases, that is when the observational time span gets longer.

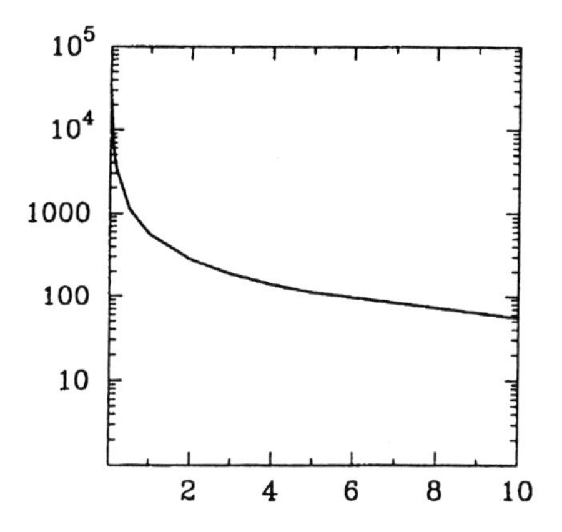

Figure 8.7. Elongation as function of the observational time span in minutes.

Figure 8.8 shows the precision of the twelve DD ambiguities and Figure 8.9 shows the histogram of the absolute values of the DD ambiguity correlation coefficients. Note that the precision of the least-squares DD ambiguities is extremely poor, since their standard deviations range from 60 cycles to 250 cycles. This is an indication that the size of the ambiguity search space will be rather large when compared to the unit grid spacing of one cycle. Hence, the 12-dimensional rectangular box enclosing the ambiguity search space is prone to have a very large amount of candidate grid points.

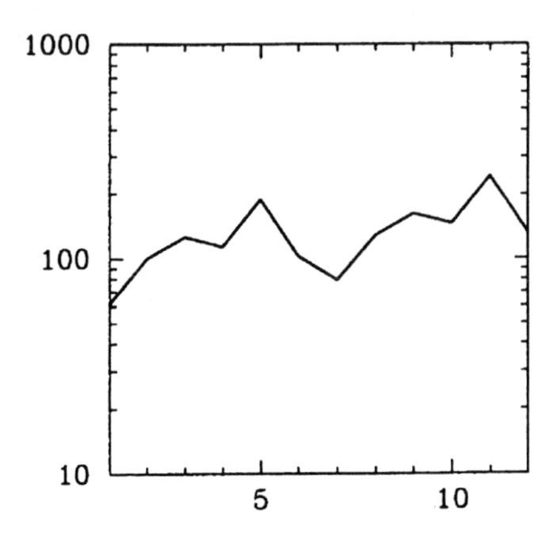

Figure 8.8. The standard deviations of the 12 DD ambiguities in cycles.

Figure 8.9 clearly shows that the majority of the sixty-six correlation coefficients are larger than a half in absolute value. Quite a few are even very close to one in absolute value. This shows that the DD ambiguities are highly correlated indeed. Hence, the ambiguity variance-covariance matrix $Q_{\hat{a}}$ can be considered to be far from diagonal. The presence of high correlation is an indication that the unconditional standard deviations of the ambiguities are likely to differ significantly from their conditional counterparts. The bounds in (8.30) are determined by the unconditional standard deviations. It are the conditional standard deviations however, that play a decisive role in the bounds of (8.36).

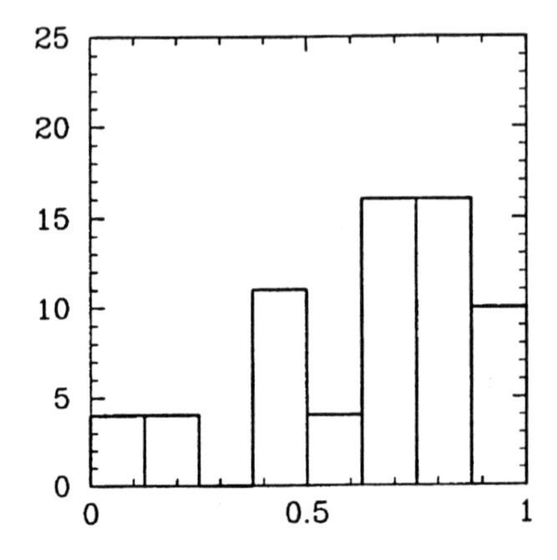

Figure 8.9. Histogram of the absolute values of the 66 DD ambiguity correlation coefficients.

The spectrum of the twelve conditional standard deviations is shown in Figure 8.10. Note the logarithmic scale along the vertical axis. Figure 8.10 clearly shows that quite a few of the conditional standard deviations are very small indeed. There are three large conditional standard deviations and nine extremely small ones. This shape of the spectrum is very typical for GPS single baseline positioning. A somewhat similar shape of the spectrum will be found in case one considers GPS multi baseline positioning, see Teunissen et al. [1994]. In that case however, the location of the discontinuity will be different. The discontinuity is a consequence of the intrinsic structure of the carrier phase model of observation equations and the chosen parametrization in terms of the DD ambiguities. Although it is possible to prove analytically that the spectrum of the DD ambiguities must have a large discontinuity of the size shown in Figure 8.10, it suffices for our purposes to give a more intuitive explanation for this discontinuity. The discontinuity in the spectrum is located when passing from the third to the fourth conditional standard deviation. This location is completely determined by the dimension of the parameter b, which equals three in our single baseline case. The fourth and following conditional standard deviations have to be small for the following reason. If we assume that three or more of the ambiguities are fixed, the corresponding highly precise carrier phases will allow us to determine the baseline with a comparable high precision. But with the baseline determined with such a high precision, the remaining carrier phases allow us to determine their ambiguities also with such a high precision. Hence, it follows indeed that the conditional standard deviations of these ambiguities have to be very small.

The shape of the spectrum shown in Figure 8.10 has an extremely important impact on the bounds of (8.36) and consequently on the performance of the search.

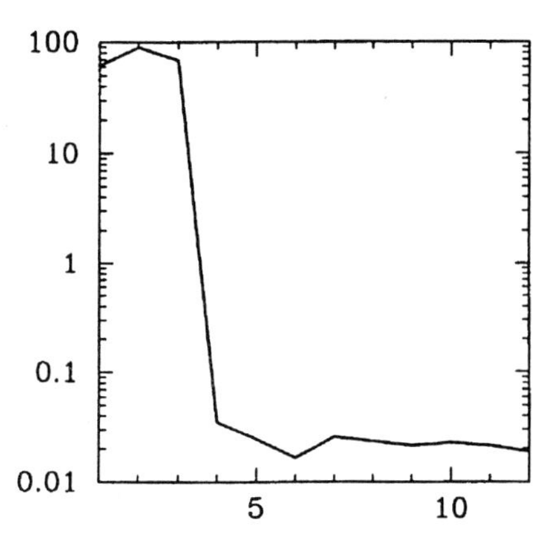

Figure 8.10. The spectrum of the conditional standard deviations of the DD ambiguities in cycles.

Since the first three conditional variances are rather large, the first three bounds (i=1,2,3) of (8.36) will be rather loose. Hence, quite a number of integer triples will satisfy these first three bounds. The remaining conditional variances however are very small. The corresponding bounds of (8.36) will therefore be very tight indeed. This implies, when we go from the third to the fourth ambiguity that we have a high likelihood of not being able to find an integer quartet that satisfies the first four bounds. Hence, the potential of halting is very significant when one goes from the third to the fourth ambiguity. As a consequence a large number of trials are required, before one is able to find an m-tuple that satisfies all m bounds. This is therefore the reason why in case of very short observational time spans based on carrier phase data only, the search for the integer least-squares DD ambiguities performs so poorly.

The above discussed phenomenon of halting can also be illustrated by showing the number of integer ambiguity vectors (or number of integer candidates) that progressively satisfy the bounds of (8.36). In Figure 8.11 the number of integer candidates is shown as function of the number of sequential bounds they satisfy. Note the logarithmic scaling of the vertical axis. Starting from the first bound (with about 1000 candidates), the figure shows that the number of integer candidates increases, that this number reaches its maximum for the first three bounds ($3\text{-}4.10^8$ number of candidate integer triples) and that from then on the number of integer candidates decreases again. Note that in this case the ambiguity search space contained only one integer vector. The behaviour shown is completely in agreement with the shape of the spectrum shown in Figure 8.10. The sharp decrease which sets in after the maximum has been reached, stipulates that the number of integer candidates that satisfy the first $j > 3$ bounds of (8.36) is significantly much smaller than the number of integer candidates that satisfy the first three bounds of (8.36).

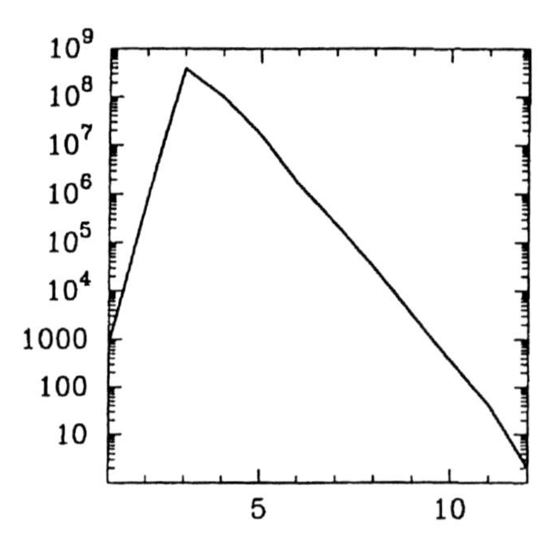

Figure 8.11. The number of integer candidates per number of sequential bounds.

8.4 THE INVERTIBLE AMBIGUITY TRANSFORMATIONS

In the processing of GPS data a prominent role is played by certain linear combinations of the GPS observables. Depending on the application, derived observables can be formed with certain desirable properties, such as for instance geometry-free and ionosphere-free linear combinations (cf. chapter 5). Within the context of ambiguity fixing, the DD linear combinations of the carrier phase observables play a prominent role, because of the integer nature of their ambiguities. Also in case of dual frequency data, linear combinations of the DD observables have been studied and are in use, such as for instance the narrow-lane, the wide-lane and extra wide-lane linear combinations, see, e.g., Wübbena [1989], Allison [1991], Goad [1992]. But also other wide-lane linear combinations have been studied Cocard and Geiger [1992]. It is the purpose of this section to introduce and discuss the class of linear combinations that can be of use in aiding the search for the integer least-squares ambiguities. As a result this will lead to the class of invertible ambiguity transformations as introduced by Teunissen [1993b].

8.4.1 The DD ambiguities are not Unique

In the previous section 8.3.3, we have seen - in case short observational time spans are used based on carrier phase data only - that the search for the integer least-squares ambiguities suffers from the fact that the DD ambiguity search space is highly elongated, that the least-squares DD ambiguities are highly correlated and that the spectrum of the conditional standard deviations of the DD ambiguities contains a large discontinuity. All these characteristics are completely determined by the structure of the variance-covariance matrix $Q_{\hat{a}}$ of the DD ambiguities. A change in the ambiguity variance-covariance matrix $Q_{\hat{a}}$ will therefore change the shape of the ambiguity search space and hence, will have its effect on the performance of the search. This observation suggests that it may be worthwhile to consider ways of changing the variance-covariance matrix $Q_{\hat{a}}$, so as to improve the performance of the search. One approach would be to include more data into the model, for instance by including precise pseudorange data into the model. Another approach would be to prolong the observational time span. Alternatively however, one could also think of ways of changing the ambiguity variance-covariance matrix, while basing the results on the same type and amount of data. That this is possible, can be made clear if we have a closer look at how the DD ambiguities are defined.

Recall that the DD ambiguities are defined as

$$N_{ij}^{kl} = N_{ij}^{l} - N_{ij}^{k} = N_j^l - N_i^l - N_j^k + N_i^k , \qquad (8.45)$$

with N_{ij}^{l} and N_{ij}^{k} being the so-called single-differenced ambiguities and with N_j^l, N_i^l, N_j^k and N_i^k being the so-called undifferenced ambiguities. Now let us assume that we have two receivers i and j available and that three satellites, numbered as 1, 2 and 3, are tracked. Then the number of undifferenced ambiguities equals 6, the number of independent single-differenced ambiguities equals 3 and the number of independent DD ambiguities equals 2. The fact that in this case only 2 independent DD ambiguities exists, does not imply however that this pair of DD ambiguities is unique. There are different ways of constructing an independent set of DD ambiguities. For instance, let us assume that $k = 1$ and $l = 2, 3$. The two corresponding independent DD ambiguities are then given as

$$N_{ij}^{12} = N_{ij}^{2} - N_{ij}^{1} \text{ and } N_{ij}^{13} = N_{ij}^{3} - N_{ij}^{1} . \tag{8.46}$$

In this case satellite $k = 1$ is taken as the reference satellite in the formation of the DD ambiguities. But instead of taking the first satellite as reference satellite, one may also choose the second or third satellite as reference satellite. For instance, if the second satellite is taken as reference satellite, we have $k = 2$ and $l = 1, 3$. The two corresponding independent DD ambiguities are then given as

$$N_{ij}^{21} = N_{ij}^{1} - N_{ij}^{2} \text{ and } N_{ij}^{23} = N_{ij}^{3} - N_{ij}^{2} . \tag{8.47}$$

It will be clear that these DD ambiguities differ from those of (8.46). Since

$$N_{ij}^{21} = -N_{ij}^{12} \text{ and } N_{ij}^{23} = N_{ij}^{13} - N_{ij}^{12} , \tag{8.48}$$

it follows that the two sets are related through the one-to-one transformation

$$\begin{pmatrix} N_{ij}^{21} \\ N_{ij}^{23} \end{pmatrix} = \begin{pmatrix} -1 & 0 \\ -1 & 1 \end{pmatrix} \begin{pmatrix} N_{ij}^{12} \\ N_{ij}^{13} \end{pmatrix} . \tag{8.49}$$

The above example was based on the situation of three satellites. It will be clear however, that the same can be said for any number of satellites being tracked. For instance, when 5 satellites are tracked, we have 4 independent DD ambiguities. Two sets of 4 independent DD ambiguities that can be defined are then: N_{ij}^{12}, N_{ij}^{13}, N_{ij}^{14}, N_{ij}^{15} and N_{ij}^{31}, N_{ij}^{32}, N_{ij}^{34}, N_{ij}^{35}. The first set has satellite 1 as reference and the second set has satellite 3 as reference. It is easily verified that these two sets are related through the one-to-one transformation

$$\begin{bmatrix} N_{ij}^{12} \\ N_{ij}^{13} \\ N_{ij}^{14} \\ N_{ij}^{15} \end{bmatrix} = \begin{bmatrix} -1 & 1 & 0 & 0 \\ -1 & 0 & 0 & 0 \\ -1 & 0 & 1 & 0 \\ -1 & 0 & 0 & 1 \end{bmatrix} \begin{bmatrix} N_{ij}^{31} \\ N_{ij}^{32} \\ N_{ij}^{34} \\ N_{ij}^{35} \end{bmatrix} . \tag{8.50}$$

The conclusion that we can draw from the above discussion is that the DD ambiguities are not unique. They depend on the choice made for the reference satellite. An important consequence of the nonuniqueness of the DD ambiguities is that we must conclude that their variance-covariance matrix is nonunique as well. That is, if we change our choice of reference satellite in the definition of the DD ambiguities, not only the DD ambiguities change, but their variance-covariance matrix $Q_{\hat{a}}$ changes as well. A change in the variance-covariance matrix however, will affect the ambiguity search space and thus the performance of the search. This shows that one set of independent DD ambiguities might have a somewhat better search-performance than another set of independent DD ambiguities.

8.4.2 Linear Combinations of the L1 and L2 DD Carrier Phases

We have seen in the previous section that DD ambiguities of a particular set can be generated by taking certain linear combinations of the DD ambiguities from another set. Up to now however, the only linear combinations considered are those that follow from a change of reference satellite. Furthermore, the linear combinations considered so far, are linear combinations of DD ambiguities belonging to carrier phases of one and the same frequency. It is therefore of interest to investigate whether it is possible to generalize the idea of taking linear combinations of the DD ambiguities. In this section we will consider linear combinations of the L_1 and L_2 DD carrier phases.

Let us assume that we have dual-frequency carrier phase data available. For each of the two frequencies, the stripped versions of the DD carrier phase observation equations read then (cf. equation (8.2))

$$\begin{cases} \Phi_{ij,1}^{kl}(t) = \rho_{ij}^{kl} + \lambda_1 N_{ij,1}^{kl} + \varepsilon_{ij,1}^{kl} \\ \Phi_{ij,2}^{kl}(t) = \rho_{ij}^{kl} + \lambda_2 N_{ij,2}^{kl} + \varepsilon_{ij,2}^{kl} \end{cases} \tag{8.51}$$

In order to simplify the notation somewhat, we will omit in this section the four indices k, l, i and j, and the time argument t. We will now consider linear combinations of the above two DD carrier phases, Φ_1 and Φ_2. By defining the linear combination as

$$\Phi_{\alpha\beta} = \frac{\alpha\lambda_2}{\alpha\lambda_2+\beta\lambda_1}\Phi_1 + \frac{\beta\lambda_1}{\alpha\lambda_2+\beta\lambda_1}\Phi_2 , \tag{8.52}$$

where α and β are two scalars, we obtain from (8.51) the derived carrier phase observation equation

$$\Phi_{\alpha\beta} = \rho + \lambda_{\alpha\beta}N_{\alpha\beta} + \varepsilon_{\alpha\beta} , \tag{8.53}$$

with:

$\lambda_{\alpha\beta} = \lambda_1\lambda_1/(\alpha\lambda_2+\beta\lambda_1)$: the wavelength of $\Phi_{\alpha\beta}$,
$N_{\alpha\beta} = (\alpha N_1 + \beta N_2)$: the ambiguity of $\Phi_{\alpha\beta}$, and
$\varepsilon_{\alpha\beta}$: the measurement noise of $\Phi_{\alpha\beta}$.

Note, that the structure of (8.53) is similar to that of the observation equations in (8.51). Again we recognize a geometric term, ρ, an ambiguity multiplied with the wavelength, $\lambda_{\alpha\beta}N_{\alpha\beta}$, and a measurement noise term $\varepsilon_{\alpha\beta}$. It will be clear, that in order for the ambiguity $N_{\alpha\beta}$ to become integer-valued, both α and β need to be chosen as integers.

Two well-known examples of linear combinations of the L_1 and L_2 DD ambiguities are the so-called narrow-lane and wide-lane ambiguities, see, e.g., Wübbena [1989], Allison [1991], Mervart et al. [1994]. The narrow-lane and wide-lane ambiguities are both integer-valued. The narrow-lane ambiguity is obtained by setting $\alpha = \beta = 1$. It is referred to as the 'narrow-lane', since the wavelength of the narrow-lane carrier phase is approximately 11 cm and therefore much smaller than the L_1 and L_2 wavelengths. The wide-lane ambiguity is obtained by setting $\alpha = -\beta = 1$. The wavelength of the wide-lane carrier phase is approximately 86 cm. Apart from the narrow-lane and wide-lane phases, there are of course an infinitely many other linear combinations that one might consider.

Above, *one* single derived carrier phase observation equation (cf. equation (8.52)) was obtained from *two* DD carrier phase observation equations (cf. equation (8.51)). It will be clear that the single derived observation equation (8.53) contains less information than the two original DD observation equations. In order to retain the information content of the two original DD carrier phase observation equations, we therefore need to work with two independent derived carrier phase observation equations instead of with one. If we define the two derived carrier phases as

$$\begin{bmatrix} \Phi_{\alpha\beta} \\ \Phi_{\gamma\delta} \end{bmatrix} = \begin{bmatrix} \frac{\alpha\lambda_2}{\alpha\lambda_2+\beta\lambda_1} & \frac{\beta\lambda_1}{\alpha\lambda_2+\beta\lambda_1} \\ \frac{\gamma\lambda_2}{\gamma\lambda_2+\delta\lambda_1} & \frac{\delta\lambda_1}{\gamma\lambda_2+\delta\lambda_1} \end{bmatrix} \begin{bmatrix} \Phi_1 \\ \Phi_2 \end{bmatrix}, \tag{8.54}$$

their observation equations become

$$\begin{cases} \Phi_{\alpha\beta} = \rho + \lambda_{\alpha\beta} N_{\alpha\beta} + \varepsilon_{\alpha\beta} \\ \Phi_{\gamma\delta} = \rho + \lambda_{\gamma\delta} N_{\gamma\delta} + \varepsilon_{\gamma\delta} \end{cases} \tag{8.55}$$

with the ambiguities

$$\begin{bmatrix} N_{\alpha\beta} \\ N_{\gamma\delta} \end{bmatrix} = \begin{bmatrix} \alpha & \beta \\ \gamma & \delta \end{bmatrix} \begin{bmatrix} N_1 \\ N_2 \end{bmatrix}. \tag{8.56}$$

A necessary and sufficient condition for transformation (8.54) to be one-to-one is that the determinant of the transformation matrix of (8.54) differs from zero. From this follows, that

$$\alpha\delta - \gamma\beta \neq 0 \tag{8.57}$$

must hold true. Note that this condition is also necessary and sufficient for the ambiguity transformation (8.56) to be invertible. It will be clear that the derived carrier phases, $\Phi_{\alpha\beta}$ and $\Phi_{\gamma\delta}$, contain the same information as the original two DD carrier phases, Φ_1 and Φ_2, when the condition (8.57) is satisfied. However, if the objective is to use the transformed phase observation equations (8.55) for *ambiguity fixing*, there are - apart from condition (8.57) - two additional conditions that need to be fulfilled. Firstly, in order for the transformed ambiguities, $N_{\alpha\beta}$ and $N_{\gamma\delta}$, to be integers, the four scalar entries of the ambiguity transformation matrix (8.56), α, β, γ and δ, need to be integers as well. Secondly, the entries of the inverse of the ambiguity transformation matrix of (8.56) also need to be integers. The reason for including this second condition can be made clear as follows. If the scalars α, β, γ and δ are integers, then so are the transformed ambiguities $N_{\alpha\beta}$ and $N_{\gamma\delta}$, when the original DD ambiguities N_1 and N_2 are integers. However, the converse of this statement is not necessarily true. That is, when the ambiguities $N_{\alpha\beta}$ and $N_{\gamma\delta}$ are integers, then the ambiguities N_1 and N_2 need not be integers, even when the scalars α, β, γ and δ are integers. But this situation is clearly not acceptable, since it could imply that an integer fixing of the transformed ambiguities, $N_{\alpha\beta}$ and $N_{\gamma\delta}$, corresponds to a fixing of the original ambiguities, N_1 and N_2, on *noninteger* values. We therefore need to ensure that integer values of $N_{\alpha\beta}$ and $N_{\gamma\delta}$ correspond to integer values of N_1 and N_2. And this is only possible by enforcing the condition that the entries of the inverse of the ambiguity transformation matrix (8.56) are integers as well. The important conclusion that is reached, reads therefore that both the transformation matrix (8.56) and its inverse must have entries that are integer.

With the above stated conditions, we are now in the position to infer which of the different integer ambiguities can be taken as pairs. This is illustrated in the following 4 examples.

Example 3:

The transformation from the L_1 and L_2 DD ambiguities, N_1 and N_2, to the narrow-lane and wide-lane ambiguities reads

$$\begin{bmatrix} N_{11} \\ N_{1,-1} \end{bmatrix} = \begin{bmatrix} 1 & 1 \\ 1 & -1 \end{bmatrix} \begin{bmatrix} N_1 \\ N_2 \end{bmatrix} . \tag{8.58}$$

The integer entries of this matrix ensure that the ambiguities N_{11} and $N_{1,-1}$ are integer, whenever the ambiguities N_1 and N_2 are integer. Note however, that with N_1 and N_2 being integer, the range of the above transformation is not sufficient to cover all integer-pairs N_{11} and $N_{1,-1}$. For instance, the above two linearly independent equations are inconsistent when $N_{11} = 1$ and $N_{1,-1} = 0$. That is, when $N_{11} = 1$ and $N_{1,-1} = 0$, no integer values for N_1 and N_2 can be found as a solution to the above equations. And this also happens, for instance, when $N_{11} = 0$ and $N_{1,-1} = 1$, or when $N_{11} = 2$ and $N_{1,-1} = 1$. The reason for this situation becomes clear when we consider the inverse of the above transformation. The inverse is given as

$$\begin{bmatrix} N_1 \\ N_2 \end{bmatrix} = \begin{bmatrix} 1/2 & 1/2 \\ 1/2 & -1/2 \end{bmatrix} \begin{bmatrix} N_{11} \\ N_{1,-1} \end{bmatrix} . \tag{8.59}$$

This result clearly shows that the noninteger entries of the inverse are causing the original two equations to be inconsistent for certain integer values of N_{11} and $N_{1,-1}$. The interesting conclusion is therefore reached, that one cannot pair the narrow-lane ambiguity to the wide-lane ambiguity. If one would namely use the narrow-lane phase together with the wide-lane phase, instead of the original DD phases Φ_1 and Φ_2, for ambiguity fixing, the outcome could be that by integer-fixing N_{11} and $N_{1,-1}$, one in fact is fixing N_1 and N_2 to noninteger values.

Example 4:

The previous example showed that the wide-lane ambiguity cannot be paired with the narrow-lane ambiguity. It is possible however, to pair the wide-lane ambiguity with one of the two original DD ambiguities. The ambiguity transformation from N_1 and N_2 to N_1 and $N_{1,-1}$ reads

$$\begin{bmatrix} N_1 \\ N_{1,-1} \end{bmatrix} = \begin{bmatrix} 1 & 0 \\ 1 & -1 \end{bmatrix} \begin{bmatrix} N_1 \\ N_2 \end{bmatrix} . \tag{8.60}$$

The inverse of this transformation is given as

$$\begin{bmatrix} N_1 \\ N_2 \end{bmatrix} = \begin{bmatrix} 1 & 0 \\ 1 & -1 \end{bmatrix} \begin{bmatrix} N_1 \\ N_{1,-1} \end{bmatrix} . \tag{8.61}$$

This shows, that whenever N_1 and N_2 are integer, so are N_1 and $N_{1,-1}$, and vice versa. Note that, apart from a change in sign, the matrix of (8.60) is identical to the matrix of (8.49). This illustrates by means of an example, that ambiguity transformations that realize a change in reference satellite indeed retain the integer nature of the ambiguities.

Example 5:

The ambiguity transformation from N_1 and N_2 to N_{11} and $N_{4,-5}$ reads

$$\begin{bmatrix} N_{11} \\ N_{4,-5} \end{bmatrix} = \begin{bmatrix} 1 & 1 \\ 4 & -5 \end{bmatrix} \begin{bmatrix} N_1 \\ N_2 \end{bmatrix} . \tag{8.62}$$

The inverse of this transformation is given as

$$\begin{bmatrix} N_1 \\ N_2 \end{bmatrix} = \begin{bmatrix} 5/9 & 1/9 \\ 4/9 & -1/9 \end{bmatrix} \begin{bmatrix} N_1 \\ N_{4,-5} \end{bmatrix} . \tag{8.63}$$

This shows that it is not allowed to pair the narrow-lane ambiguity to $N_{4,-5}$.

Example 6:

The ambiguity transformation from N_1 and N_2 to $N_{-60,77}$ and $N_{-7,9}$ reads

$$\begin{bmatrix} N_{-60,77} \\ N_{-7,9} \end{bmatrix} = \begin{bmatrix} -60 & 77 \\ -7 & 9 \end{bmatrix} \begin{bmatrix} N_1 \\ N_2 \end{bmatrix} . \tag{8.64}$$

The inverse of this transformation is given as

$$\begin{bmatrix} N_1 \\ N_2 \end{bmatrix} = \begin{bmatrix} -9 & 77 \\ -7 & 60 \end{bmatrix} \begin{bmatrix} N_{-60,77} \\ N_{-7,9} \end{bmatrix} . \tag{8.65}$$

This shows that the pair of ambiguities $N_{-60,77}$ and $N_{-7,9}$ are indeed admissible.

8.4.3 Single-Channel Ambiguity Transformations

In the previous section we have looked at the transformed carrier phase observation equations (8.55), having as ambiguities the integers $N_{\alpha\beta}$ and $N_{\gamma\delta}$. It is however not really necessary to work explicitly with the derived phase observables $\Phi_{\alpha\beta}$ and $\Phi_{\gamma\delta}$. Instead, one might as well work with the original DD carrier phases Φ_1 and Φ_2 and then use the inverse of the ambiguity transformation

$$\begin{bmatrix} N_{\alpha\beta} \\ N_{\gamma\delta} \end{bmatrix} = \begin{bmatrix} \alpha & \beta \\ \gamma & \delta \end{bmatrix} \begin{bmatrix} N_1 \\ N_2 \end{bmatrix} , \tag{8.66}$$

so as to *reparametrize* the ambiguities from N_1, N_2 to $N_{\alpha\beta}$, $N_{\gamma\delta}$. The inverse of (8.66) reads

$$\begin{bmatrix} N_1 \\ N_2 \end{bmatrix} = \frac{1}{\alpha\delta - \beta\gamma} \begin{bmatrix} \delta & -\beta \\ -\gamma & \alpha \end{bmatrix} \begin{bmatrix} N_{\alpha\beta} \\ N_{\gamma\delta} \end{bmatrix} . \tag{8.67}$$

As we know from the previous section, the ambiguity transformation (8.66) is admissible if and only if the matrix entries of both (8.66) and (8.67) are integer valued. Instead of explicitly checking the integerness of the entries of both the transformation matrix and its inverse, we may also infer the admissibility of the ambiguity transformation from the entries of one of the two matrices. This can be seen as follows. We start from the assumption that the four scalars α, β, γ and δ are all integer valued. From (8.67) follows then that the entries of the inverse are also integer, when $\alpha\delta\text{-}\gamma\beta = \pm 1$ holds true. This condition is therefore in addition to the condition that the scalars α, β, γ and δ must be integers, a sufficient condition for the admissibility of the ambiguity transformation (8.66). The question is now, whether it is also a necessary condition. The answer to this question is in the affirmative, as the following shows. Let us denote the entries of the inverse as $\bar{\alpha}$, $\bar{\beta}$, $\bar{\gamma}$ and $\bar{\delta}$. If the ambiguity transformation matrix and its inverse have integer entries, then both their determinants, $\alpha\delta$-$\beta\gamma$ and $\bar{\alpha}\bar{\delta}-\bar{\beta}\bar{\gamma}$, are integers as well and $(\alpha\delta-\beta\gamma)(\bar{\alpha}\bar{\delta}-\bar{\beta}\bar{\gamma}) = 1$. From this follows then that $\alpha\delta-\beta\gamma = \pm 1$ must hold.

Hence, with this result we are now in the position to conclude that the condition that the entries of both the ambiguity transformation matrix and its inverse must be integers, can be replaced by the condition that the entries of the transformation matrix need to be integer and that its determinant needs to equal ± 1. This shows that instead of considering the inverse explicitly, it suffices to check the value of the determinant of the ambiguity transformation.

We may now use the inverse (8.67), knowing that $\alpha\delta-\beta\gamma = \pm 1$, and replace the DD ambiguities N_1 and N_2 in the original L_1 and L_2 DD phase observation equations,

$$\begin{cases} \Phi_1 = \rho + \lambda_1 N_1 + \varepsilon_1 \\ \Phi_2 = \rho + \lambda_2 N_2 + \varepsilon_2 \end{cases} \tag{8.68}$$

by the new ambiguities $N_{\alpha\beta}$ and $N_{\gamma\delta}$. As a result the reparametrized observation equations become

$$\begin{cases} \Phi_1 = \rho + \pm\lambda_1\, \delta\, N_{\alpha\beta} - \pm\lambda_1\, \beta\, N_{\gamma\delta} + \varepsilon_1 \\ \Phi_2 = \rho - \pm\lambda_2\, \gamma\, N_{\alpha\beta} + \pm\lambda_2\, \alpha\, N_{\gamma\delta} + \varepsilon_2\,. \end{cases} \tag{8.69}$$

The ambiguity transformation (8.66) is referred to as a *single-channel* transformation, since it operates on the DD ambiguities of one single channel. Hence, if the ambiguity transformation (8.66) is applied to all channels, the DD ambiguities are transformed on a channel-by-channel basis.

If we base our least-squares adjustment on the observation equations (8.69), the variance-covariance matrix of the ambiguities becomes dependent on the scalars α, β, γ and δ. Hence, we may now think of choosing 'suitable' values for these scalars, so as to improve the performance of the ambiguity search process. It is generally believed that for the purpose of ambiguity fixing, only those integer linear combinations are of value that produce a phase observable which has a relatively long wavelength, a relatively low noise behaviour and a reasonable small ionospheric delay. Indeed, these properties are beneficial to the integer ambiguity fixing process. One should recognize however, that a more complete picture is obtained once one knows how for a particular case, the combination of carrier phase noise and chosen functional model, propagates into the variance-covariance matrix of the ambiguities. Hence, the choice for certain linear combinations should not so much be made on the basis of only phase noise and wavelength, but more on how the variance-covariance matrix is affected by the choice. As we have seen earlier, it is namely the ambiguity variance-covariance matrix that dictates the performance of the integer ambiguity search.

In order to show how one can influence the spectrum of conditional variances through the use of (8.66), the following three single-channel ambiguity transformations were chosen as an example

$$Z_1^T = \begin{bmatrix} 1 & -1 \\ 0 & 1 \end{bmatrix}, \; Z_2^T = \begin{bmatrix} 4 & -5 \\ 1 & -1 \end{bmatrix}, \; Z_3^T = \begin{bmatrix} -60 & 77 \\ -7 & 9 \end{bmatrix}.$$

It is easily verified that these transformations are indeed admissible. Based on our 7 satellite configuration using two epochs of dual frequency carrier phase data with an observational time span of only one second, Figure 8.12 shows the original and the three transformed spectra of conditional standard deviations.

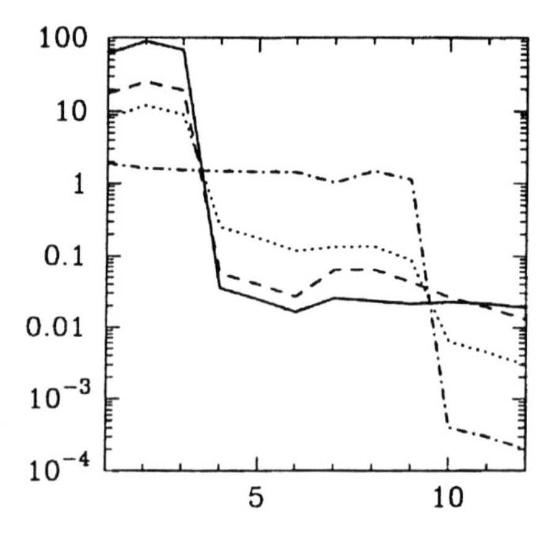

Figure 8.12. The original and the single-channel transformed spectra in cycles; the DD L_1/L_2-spectrum = full curve, Z_1-spectrum = dashed curve, Z_2-spectrum = dotted curve, Z_3-spectrum = dash-dotted curve.

The figure clearly shows that all of the first three large conditional standard deviations in the transformed spectra are smaller than the ones of the original DD spectrum. This implies, when use is made of the sequential bounds (8.36), that the number of candidate integers of the transformed ambiguities satisfying the first three bounds, is smaller than the number of candidate integers of the original DD ambiguities satisfying these first three bounds. We also observe that the large discontinuity which is present in the original DD spectrum gets reduced in the transformed spectra. This has as consequence that the search for the transformed integer least-squares ambiguities is less likely to halt between the third and fourth bound, than the search for the original DD ambiguities. We also observe from Figure 8.12, that the Z_3-spectrum is almost flat in the beginning, but that it drops when one passes the ninth conditional standard deviation. Hence, in this case one can expect that halting takes place between the ninth and tenth bound.

8.4.4 Multi-Channel Ambiguity Transformations.

In the previous section we identified the class of single-channel ambiguity transformations. It was shown how some of these a priori chosen ambiguity transformations affect the spectrum of conditional variances. Although some improvement could be seen, it is clear that the usage of these single-channel ambiguity transformations is limited and that one cannot expect to obtain results that are overall satisfactorily. Moreover, the transformations used were obtained on quite an ad hoc basis. In this section it will therefore be discussed how these single-channel ambiguity transformations can be generalized.

In the previous section the four entries of transformation (8.66) were assumed to be the same for all channels. This however, is not really necessary. One can in principle choose different sets of values for the different channels. In this way one can accommodate to the different entries in the variance-covariance matrix of the ambiguities. It also seemed in the previous section that the transformation was restricted to the dual frequency case. That is, N_1 was assumed to be an L_1 DD ambiguity and N_2 an L_2 DD ambiguity. But this is not necessary either. The transformation can namely also be used in case one is dealing with L_1 data only. In that case N_1 and N_2 are simply DD ambiguities of two different channels. This observation also suggests a generalization to more than two channels. And indeed, there is no reason for restricting the order of the transformation to two. This then brings us to the multi-channel ambiguity transformations.

Let a be our vector of m DD ambiguities. The entries of a may be ambiguities of the L_1-type only, of the L_2-type only or of both types. Since all the entries of a are integer valued, we have $a \in Z^m$, with Z^m being the m-space of integers. Let Z be an $m \times m$ matrix of full rank. Then, if we transform a with Z^T, we would like all the entries of the transformed ambiguity vector, $z = Z^T a$, to be integer as well. This implies, since the entries of a are integer valued, that all the entries of the matrix Z need to be integer as well. This condition however, is necessary but not yet sufficient. That is, we do not only want $z = Z^T a$ to be integer whenever a is integer, but also that $a = (Z^T)^{-1} z$ is integer whenever z is integer. From this follows that matrix Z is an admissible ambiguity transformation matrix if and only if both Z and its inverse Z^{-1} have entries which are integer. Note that this is in agreement with the results of the previous section. The integerness of the entries of Z^{-1} is needed to avoid that one would be fixing the DD ambiguities on noninteger values.

In order to give an illustration of admissible ambiguity transformation, some simple examples will be given. The identity-matrix is of course a trivial example. But also all permutation matrices belong to the class of admissible ambiguity transformations. The entries of permutation matrices and of their inverses are all integer valued. The permutation matrices are in fact implicitly used, when one reorders the entries of the ambiguity vector. Also all ambiguity transformations that change the choice of reference satellite in the DD ambiguities, are admissible. The three examples considered so far, were all matrices that have entries equal to 1 in absolute value. An example of an admissible ambiguity transformation matrix for which this does not necessarily hold is given by the integer triangular matrix having ones on its diagonal. It is easily verified that its inverse is again an integer triangular matrix with ones on its diagonal. It is also of importance to understand that once certain ambiguity transformations are identified, other ambiguity transformations can be derived from them by performing certain standard matrix operations, like: inversion, transposition and multiplication. For instance, when Z_1 and Z_2 are two given ambiguity transformations, then so are Z_1^{-1}, Z_1^T and Z_1Z_2.

Now that the class of admissible multi-channel ambiguity transformations has been identified, we can generalize the idea of section 8.4.3 to reparametrize the DD carrier phase observation equations (cf. equations (8.68) and (8.69)). The original system of DD carrier phase observation equations was given as (cf. equation (8.3))

$$y = Aa + Bb + e . \tag{8.71}$$

Let Z^T be an admissible ambiguity transformation and let $z = Z^T a$ be the vector of transformed ambiguities. Then $a = Z^{-T} z$ and the system of observation equations (8.71) can be reparametrized as

$$y = AZ^{-T} z + Bb + e . \tag{8.72}$$

We may now choose to base our least-squares adjustment either on the original system of observation equations, (8.71), or on the reparametrized system of observation equations, (8.72). The solution for the baseline vector will of course not be affected by the reparametrization. The least-squares estimates for the ambiguities will differ however. The DD and transformed ambiguity estimates and their variance-covariance matrices are related as

$$\hat{z} = Z^T \hat{a} \text{ and } Q_{\hat{z}} = Z^T Q_{\hat{a}} Z . \tag{8.73}$$

Since $Q_{\hat{z}}$ differs from $Q_{\hat{a}}$, also the performance of the search for the integer least-squares ambiguities is affected by the reparametrization. Hence, the reparametrization provides us now with the opportunity to consider ambiguity transformations that allow us to improve the performance of the integer ambiguity search. The question which of the ambiguity transformations suffices for that purpose, is taken up in the next section.

8.5 THE LSQ AMBIGUITY DECORRELATION ADJUSTMENT

This section is devoted to the elimination of the potential problem of search halting. The original integer least-squares problem is reparametrized such that an equivalent formulation is obtained, but one that is much easier to solve. Since the poor performance of the integer ambiguity search was shown to be due to the high correlation, the aim will be to decorrelate the least-squares ambiguities. The ambiguity transformation that will be constructed, removes the discontinuity from the spectrum of ambiguity conditional variances and provides ambiguities that show a dramatic improvement in both precision and correlation. As a result the search for the transformed ambiguities can be performed in a highly efficient manner. The method of the least-squares ambiguity decorrelation adjustment was introduced in Teunissen [1993a] and examples of its performance can be found in, e.g., de Jonge and Tiberius [1994], Teunissen [1994a], Goad and Yang [1994], Hein and Werner [1995], Tiberius and De Jonge [1995], Rizos and Han [1995], de Jonge and Tiberius [1996], de Jonge et al. [1996], Li and Gao [1997], Boon and Ambrosius [1997], Han et al. [1997], Teunissen et al. [1997].

8.5.1 The Reparametrized Integer Least-Squares Problem

In the previous section 8.4.4 the class of admissible ambiguity transformations was identified. Members from this class can now be used to aid the ambiguity fixing process. Let Z^T be an ambiguity transformation, which is used to transform the DD ambiguities as

$$z = Z^T a, \quad \hat{z} = Z^T \hat{a}, \quad Q_{\hat{z}} = Z^T Q_{\hat{a}} Z \; . \tag{8.74}$$

The ambiguity integer least-squares problem (8.25) would then transform accordingly into the equivalent minimization problem

$$\min_z \; (\hat{z}-z)^T Q_{\hat{z}}^{-1} (\hat{z}-z) \quad , \quad \text{with } z \in Z^m \; . \tag{8.75}$$

Similarly, the original ambiguity search space (8.35) would transform into the new ambiguity search space

$$\sum_{i=1}^{m} (\hat{z}_{i|I} - z_i)^2 / \sigma^2_{\hat{z}_{i|I}} \leq \chi^2 . \tag{8.76}$$

The shape and orientation of this ambiguity search space differs from that of the original ambiguity search space (8.35), except of course in case $Z = I_m$. Despite this difference however, the transformed ambiguity search space (8.76) has, as it should be, the same number of candidate gridpoints as the original ambiguity search space.

Based on the transformed ambiguity search space (8.76), the sequential bounds of the transformed ambiguities become

$$\begin{cases} (\hat{z}_1 - z_1)^2 & \leq \sigma^2_{\hat{z}_1} \chi^2 \\ (\hat{z}_{2|1} - z_2)^2 & \leq \sigma^2_{\hat{z}_{2|1}} [\chi^2 - (\hat{z}_1 - z_1)^2 / \sigma^2_{\hat{z}_1}] \\ & \cdot \\ & \cdot \\ (\hat{z}_{m|M} - z_m)^2 & \leq \sigma^2_{\hat{z}_{m|M}} [\chi^2 - \sum_{j=1}^{m-1} (\hat{z}_{j|J} - z_j)^2 / \sigma^2_{\hat{z}_{j|J}}] \end{cases} \tag{8.77}$$

These bounds can now be used in exactly the same way as it has been described earlier in section 8.3.2, for the computation of the integer least-squares solution. Once the integer least-squares solution $\check{z}$ has been found, the integer minimizer of (8.25) can be recovered from invoking the inverse relation $\check{a} = Z^{-T}\check{z}$. The fixed baseline solution follows then from (8.12). Alternatively, one could also use

$$\check{b} = \hat{b} - Q_{\hat{b}\hat{z}} Q_{\hat{z}}^{-1} (\hat{z} - \check{z}) \tag{8.78}$$

to obtain the fixed baseline.

In order to have any use for our ambiguity transformation Z, we should aim at finding a transformation that makes the transformed integer least-squares problem (8.75) easier to solve than the original problem (8.25). Note, that the ambiguity transformation has no effect - as it should be - on the validation part of the ambiguity fixing problem. The test statistics which are used for validation (cf. section 8.2.3), are invariant for the ambiguity transformation. With (8.74), we have the equality

$$(\hat{a} - \check{a})^T Q_{\hat{a}}^{-1} (\hat{a} - \check{a}) = (\hat{z} - \check{z})^T Q_{\hat{z}}^{-1} (\hat{z} - \check{z}) . \tag{8.79}$$

Hence, the only purpose of the ambiguity transformation is to lighten the computational burden. Clearly the ideal situation would be, to have a transformation Z that allows for a full decorrelation of the ambiguities. In that case, $Q_{\hat{z}}$ is diagonal and (8.75) can simply be solved by rounding the entries of $\hat{z}$ to their nearest integer. Unfortunately however, the restrictions on Z do generally not allow for a complete diagonalization of the ambiguity variance-covariance matrix. For instance, the choice where Z contains the eigenvectors of $Q_{\hat{a}}$ is generally not allowed, since the entries of the eigenvectors are usually noninteger. Also a diagonalization based on $Z^T = L^{-1}$, with L being the triangular factor of $Q_{\hat{a}}$, is not admissible. Again, the non-zero off-diagonal entries of L will generally be noninteger. These two examples show that in terms of diagonality, one will have to be content with a somewhat less perfect result. Nevertheless a decrease in correlation, although not complete, will already be very helpful, since it would improve the performance of the integer ambiguity search process. In the next section, we will consider the decorrelation of the ambiguities in the two-dimensional case.

8.5.2 A 2D-Decorrelating Ambiguity Transformation

In order to answer the question as to how to construct our ambiguity transformation Z^T, we first consider the problem in two dimensions. Let the ambiguities and their variance-covariance matrix be given as

$$a = \begin{bmatrix} \hat{a}_1 \\ \hat{a}_2 \end{bmatrix} \quad \text{and} \quad Q_{\hat{a}} = \begin{bmatrix} \sigma^2_{\hat{a}_1} & \sigma_{\hat{a}_1\hat{a}_2} \\ \sigma_{\hat{a}_2\hat{a}_1} & \sigma^2_{\hat{a}_2} \end{bmatrix} . \tag{8.80}$$

We know that the sequential conditional least-squares ambiguities are fully decorrelated. The idea is therefore to start from the conditional least-squares based transformation. When (8.32) is written in vector-matrix form, we obtain for the two-dimensional case, the transformation

$$\begin{bmatrix} \hat{a}_1 \\ \hat{a}_{2|1} \end{bmatrix} = \begin{bmatrix} 1 & 0 \\ -\sigma_{\hat{a}_2\hat{a}_1}\sigma^{-2}_{\hat{a}_1} & 1 \end{bmatrix} \begin{bmatrix} \hat{a}_1 \\ \hat{a}_2 \end{bmatrix} . \tag{8.81}$$

Since we are studying the effect of transformations on $Q_{\hat{a}}$, we have for reasons of convenience skipped the elements a_1 and a_2 in the above transformation. Note, that this transformation not only decorrelates, but in line with the correspondence between linear least-squares estimation and best linear unbiased estimation, also returns $\hat{a}_{2|1}$ as the element which has the best precision of all linear unbiased functions of $\hat{a}_1$ and $\hat{a}_2$. Also note, that both the transformation matrix of (8.81) as well as its inverse would have integer entries if the scalar $-\sigma_{\hat{a}_2\hat{a}_1}\sigma^{-2}_{\hat{a}_1}$ would be integer. Hence, the above transformation would be an admissible ambiguity transformation if the scalar $-\sigma_{\hat{a}_2\hat{a}_1}\sigma^{-2}_{\hat{a}_1}$ would be integer.

Unfortunately however, the scalar $-\sigma_{\hat{a}_2\hat{a}_1}\sigma^{-2}_{\hat{a}_1}$ generally fails to be an integer. This shortcoming however, is easily repaired. We simply approximate the above transformation by replacing $-\sigma_{\hat{a}_2\hat{a}_1}\sigma^{-2}_{\hat{a}_1}$ by $[-\sigma_{\hat{a}_2\hat{a}_1}\sigma^{-2}_{\hat{a}_1}]$, where [.] stands for 'rounding to the nearest integer'. This gives

$$\begin{bmatrix} \hat{a}_1 \\ \hat{a}_{2'} \end{bmatrix} = \begin{bmatrix} 1 & 0 \\ -[\sigma_{\hat{a}_2\hat{a}_1}\sigma^{-2}_{\hat{a}_1}] & 1 \end{bmatrix} \begin{bmatrix} \hat{a}_1 \\ \hat{a}_2 \end{bmatrix} . \tag{8.82}$$

It is easily verified that this transformation is admissible.

In the conditional least-squares transformation of (8.81), the choice was made to keep $\hat{a}_1$ unchanged and to replace $\hat{a}_2$ with $\hat{a}_{2|1}$. Instead of this choice however, we could also think of interchanging the role of the two ambiguities. In that case, we will get instead of the transformation (8.81), the conditional least-squares transformation

$$\begin{bmatrix} \hat{a}_{1|2} \\ \hat{a}_2 \end{bmatrix} = \begin{bmatrix} 1 & -\sigma_{\hat{a}_1\hat{a}_2}\sigma^{-2}_{\hat{a}_2} \\ 0 & 1 \end{bmatrix} \begin{bmatrix} \hat{a}_1 \\ \hat{a}_2 \end{bmatrix} . \tag{8.83}$$

Both transformations (8.81) and (8.83) fully decorrelate. Geometrically, these two transformations can be given the following useful interpretation (see Figure 8.13). Imagine the original two-dimensional ambiguity search space centred at $\hat{a} = (\hat{a}_1, \hat{a}_2)^T$. A *full* decorrelation between the two ambiguities can be realized, if we push the two horizontal tangents of the ellipse from the $\pm\chi\sigma_{\hat{a}_2}$ level towards the $\pm\chi\sigma_{\hat{a}_{2|1}}$ level, while at the same time keeping fixed the area of the ellipse and the location of the two vertical tangents (see Figure 8.13, left). This is precisely what transformation (8.81) does. Alternatively, one can also achieve a full decorrelation, if instead of the two horizontal tangents, the two vertical tangents are pushed from the $\pm\chi\sigma_{\hat{a}_1}$ level towards the $\pm\chi\sigma_{\hat{a}_{1|2}}$ level (see Figure 8.13, right). This is precisely what transformation (8.83) does.

The two transformations (8.81) and (8.83) fully decorrelate, but are unfortunately not admissible. Transformation (8.82) on the other hand is admissible, but will not achieve a full decorrelation. Still however it will achieve a decorrelation to some extent. And the same holds true for the integer approximation of transformation (8.83). The idea is therefore, instead of using (8.81) and (8.83), to make use of their integer approximations. And this will be done in an alternating fashion so as to interchange the role of the two ambiguities. That is, the admissible transformation (8.82) is applied first and then followed by an integer approximation of the type (8.83). The second admissible ambiguity transformation reads therefore

$$\begin{bmatrix} \hat{a}_{1'} \\ \hat{a}_{2'} \end{bmatrix} = \begin{bmatrix} 1 & -[\sigma_{\hat{a}_1\hat{a}_{2'}}\sigma_{\hat{a}_{2'}}^{-2}] \\ 0 & 1 \end{bmatrix} \begin{bmatrix} \hat{a}_1 \\ \hat{a}_{2'} \end{bmatrix} . \tag{8.84}$$

The first transformation (8.82) thus pushes the two *horizontal* tangents of the ambiguity search space from the $\pm\chi\sigma_{\hat{a}_2}$ level towards the $\pm\chi\sigma_{\hat{a}_{2'}}$ level, while at the same time keeping fixed the area of the search space and the location of the *vertical* tangents. The second transformation (8.84) then pushes the two vertical tangents from the $\pm\chi\sigma_{\hat{a}_1}$ level towards the $\pm\chi\sigma_{\hat{a}_{1'}}$ level, while at the same time keeping fixed the area of the search space and the location of the horizontal tangent. And this process is continued until the next transformation reduces to the trivial identity transformation, implying that no further decorrelation is achievable any more. The amount of decorrelation that can be achieved is discussed in Teunissen [1993a]. Note, since the area of the search space is kept constant at all times, whereas the area of the enclosing box is reduced in each step, that the ambiguity search space is forced to become more sphere-like (for a proof see Teunissen [1994a]) and that the transformed ambiguities are of a better precision than the original DD ambiguities.

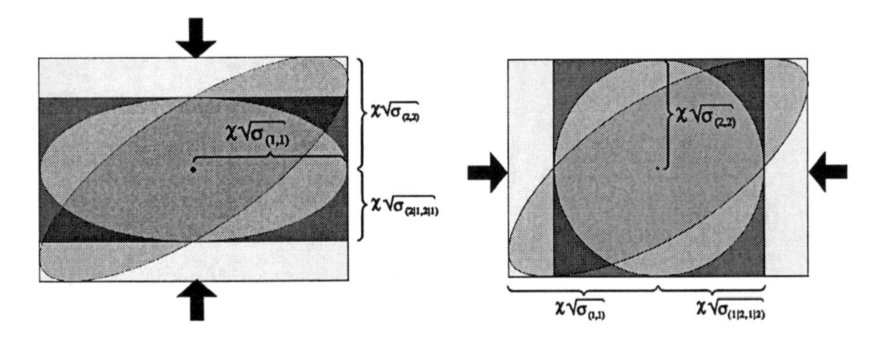

Figure 8.13. Decorrelating ambiguities by pushing tangents

Example 7:

This example is a continuation of example 1. First we will consider the construction of the decorrelating ambiguity transformation Z^T and the transformed ambiguity search space. After that, we will consider the search based on the transformed ambiguities.

The variance-covariance matrix of the two original ambiguities reads

$$Q_{\hat{a}} = \begin{bmatrix} 53.4 & 38.4 \\ 38.4 & 28.0 \end{bmatrix} .$$

Starting with the less precise ambiguity, the ambiguity transformation of the type of (8.84) gives for our present example

$$Z_1^T = \begin{bmatrix} 1 & -1 \\ 0 & 1 \end{bmatrix} .$$

Hence, after this first step the transformed variance-covariance matrix reads

$$Z_1^T Q_{\hat{a}} Z_1 = \begin{bmatrix} 4.6 & 10.4 \\ 10.4 & 28.0 \end{bmatrix} .$$

Now we will tackle the second ambiguity. Using the ambiguity transformation of the type of (8.82) gives then

$$Z_2^T = \begin{bmatrix} 1 & 0 \\ -2 & 1 \end{bmatrix} .$$

With this second step the transformed variance-covariance matrix becomes

$$Z_2^T Z_1^T Q_{\hat{a}} Z_1 Z_2 = \begin{bmatrix} 4.6 & 1.2 \\ 1.2 & 4.8 \end{bmatrix} .$$

In order to see whether a next step is required, we again go back to the first ambiguity and again consider an ambiguity transformation of the type of (8.84). For our present example however, this results in the identity transformation which shows that no further steps are required. Our decorrelating ambiguity transformation reads therefore

$$Z^T = Z_2^T Z_1^T = \begin{bmatrix} 1 & -1 \\ -2 & 3 \end{bmatrix} .$$

The with the above steps corresponding original, intermediate and transformed ambiguity search spaces are shown in Figure 8.14.

Diagnostics that give a quantitative indication of the performance of our decorrelating ambiguity transformation are given as

$$\sigma^2_{\hat{a}_1} = 53.4,\ \sigma^2_{\hat{a}_{2|1}} = 0.387 \quad , \quad \sigma^2_{\hat{z}_1} = 4.6,\ \sigma^2_{\hat{z}_{2|1}} = 4.487$$

and

$$\rho_{\hat{a}} = 0.993,\ \rho_{\hat{z}} = 0.255 \quad ; \quad e_{\hat{a}} = 17.861,\ e_{\hat{z}} = 1.300\ .$$

First note that the discontinuity which is present in the spectrum of the conditional variances of the original ambiguities, has largely been eliminated. In correspondence with this, we also observe that the new ambiguities are far less correlated than the original ambiguities. And finally we note that the elongation of the ambiguity search space has indeed been pushed close to its minimum value of one.

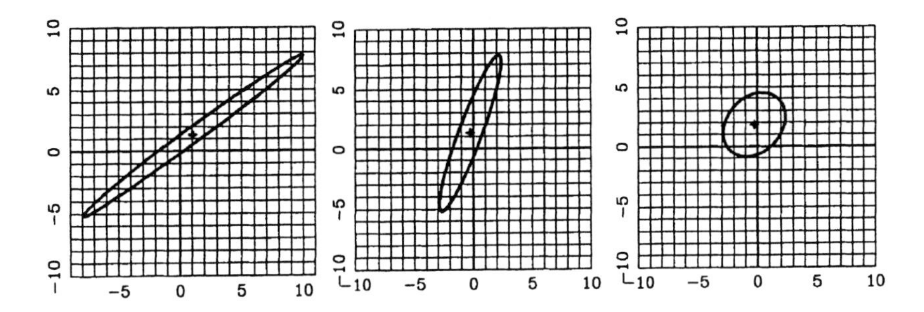

Figure 8.14. The original, the intermediate and the transformed ambiguity search spaces.

Let us now consider the search in term of our new ambiguities. With our decorrelating ambiguity transformation Z^T, the least-squares estimates $\hat{z}_1$ and $\hat{z}_2$ follow as

$$\hat{z} = \begin{bmatrix} -0.25 \\ 1.80 \end{bmatrix} = \begin{bmatrix} 1 & -1 \\ -2 & 3 \end{bmatrix} \begin{bmatrix} 1.05 \\ 1.30 \end{bmatrix} .$$

The results of the search are shown in Table 8.3. The steps that are followed are identical to the ones discussed in section 8.3.2. Since $\hat{z}_1 = -0.25$, we choose z_1 as its nearest integer, which is $z_1 = 0.00$. Based on this integer value for z_1, the conditional least-squares estimate for the second ambiguity reads $\hat{z}_{2|1} = 1.87$. Since its nearest integer reads 2.00, we choose z_2 as $z_2 = 2.00$. Hence, we now have an integer pair $(z_1, z_2) = (0.00, 2.00)$ which lies inside the transformed ambiguity search space. Since the value of the objective function $F(z_1, z_2) = (\hat{z}_1 - z_1)^2/\sigma^2_{\hat{z}_1} + (\hat{z}_{2|1} - z_2)^2/\sigma^2_{\hat{z}_{2|1}}$ of this integer pair equals $F(0.00, 2.00) = 0.0176$, we may now shrink the ellipse and set the χ^2-value at $\chi^2 = 0.0176$. It will be clear that the second nearest integer of $\hat{z}_{2|1} = 1.87$, being 1.00, will give an integer pair (0.00, 1.00) that lies outside the ellipse. Hence, in order to continue we go back to the first ambiguity and consider the second nearest integer to $\hat{z}_1 = -0.25$, which is -1.00. It follows however that this value does not satisfy its bound for $\chi^2 = 0.0176$. As a result, the search stops and the integer least-squares solution

Table 8.3 The integer pairs that are encountered during the search.

$\chi = 1.5$		$\chi^2 = 0.0176$	
z_1	z_2	z_1	z_2
0.00	2.00	0.00	-
		-	-

is provided as $(\check{z}_1, \check{z}_2) = (0.00, 2.00)$. In order to obtain the integer least-squares solution for the original ambiguities, we invoke $\check{a} = Z^{-T}\check{z}$ and get

$$\check{a} = \begin{bmatrix} 2.00 \\ 2.00 \end{bmatrix} = \begin{bmatrix} 3 & 1 \\ 2 & 1 \end{bmatrix} \begin{bmatrix} 0.00 \\ 2.00 \end{bmatrix},$$

which is of course identical to the solution found in example 1. With the present example we have illustrated that the performance of the search improves, when use is made of the less correlated and transformed ambiguities z_1 and z_2. In the present example, the gain is of course not spectacular. The gain will become spectacular however, when the original ambiguity search space is much more elongated (which is the case for short timespan carrier phase data) and when one treats the problem in higher dimensions (which is also the case with GPS).

8.5.3 The Decorrelated Least-Squares Ambiguities

In the previous section it was shown how to decorrelate the two ambiguities, thereby removing the gap between $\sigma^2_{\hat{a}_1}$ and $\sigma^2_{\hat{a}_{2|1}}$. The two-dimensional ambiguity transformation was constructed from a sequence of transformations of the following two types:

$$Z_1^T = \begin{bmatrix} 1 & 0 \\ z_{21} & 1 \end{bmatrix} \text{ and } Z_2^T = \begin{bmatrix} 1 & z_{12} \\ 0 & 1 \end{bmatrix}, \tag{8.85}$$

in which z_{21} and z_{12} are appropriately chosen integers. To generalize this to the m-dimensional case, we first need to generalize these type of transformations accordingly. Although one can think of different generalizations, we will follow the simplest approach and use the two-dimensional ambiguity transformation for the m-dimensional case as well.

Transformations of the type (8.85) are known as Gauss-transformations and they are considered to be the basic tools for zeroing entries in matrices Golub and Van Loan [1986]. In our case, due to the integer nature of z_{12} and z_{21}, they will be used to decrease the conditional correlations instead of zeroing them, thereby trying to flatten the spectrum of ambiguity conditional variances. As it was shown in section 8.3.3, it is the large discontinuity in the spectrum of conditional variances that forms a hindrance for the efficient search of the integer least-squares ambiguities. A flattened spectrum will therefore be very beneficial for our search. In case of a single baseline model, the discontinuity is located at the two neighbouring conditional variances $\sigma^2_{\hat{a}_{i|I}}$ and $\sigma^2_{\hat{a}_{i+1|I+1}}$ for i=3. Hence, if we let $\hat{a}_{i|I}$ and $\hat{a}_{i+1|I}$, for i=3, play the role of our two ambiguities $\hat{a}_1$ and $\hat{a}_2$ of the previous section, we should be able to remove this discontinuity from the spectrum by using the decorrelating two-dimensional ambiguity transformation of the previous section. The variance-covariance matrix of the conditional least-squares ambiguities $\hat{a}_{i|I}$ and $\hat{a}_{i+1|I}$, which is needed to construct the two-dimensional transformation, is easily found from the LDL^T-decomposition of $Q_{\hat{a}}$. With the diagonal matrix D partitioned as D = diag (D_{11}, D_{22}, D_{33}), where the 2×2 diagonal matrix D_{22} contains the conditional variances $\sigma^2_{\hat{a}_{i|I}}$ and $\sigma^2_{\hat{a}_{i+1|I+1}}$ for i=3, and with the lower triangular matrix L partitioned accordingly, it follows that $(L_{21}D_{11}L_{21}^T + L_{22}D_{22}L_{22}^T)$ is the variance-covariance matrix of the least-squares ambiguities $\hat{a}_i$ and $\hat{a}_{i+1}$ and that $L_{22}D_{22}L_{22}^T$ is the variance-covariance matrix of the conditional least-squares ambiguities $\hat{a}_{i|I}$ and $\hat{a}_{i+1|I}$. It is this last matrix that is now used for the construction of the two-dimensional ambiguity transformation. As a result of this transformation, we are able to close the large gap that exists between the third and fourth conditional variance in the spectrum. But of course, after this transformation has been applied, other, but smaller discontinuities emerge. They however, can also be removed by applying the two-dimensional transformation. The idea is therefore to continue applying the transformation to pairs of neighbouring ambiguities until the complete spectrum of conditional variances is flattened. Once this has been completed, the m-dimensional ambiguity transformation Z^T is known and the original least-squares ambiguity vector $\hat{a}$ can be transformed as $\hat{z} = Z^T\hat{a}$.

The following example shows how the above least-squares ambiguity decorrelation adjustment method works when applied to a synthetic 3×3 variance-covariance matrix.

Example 8:

In this example we will consider a synthetic variance-covariance matrix which has been chosen such that its structure is similar to the actual variance-covariance matrix of the DD-ambiguities. The synthetic variance-covariance matrix is given as the sum of a scaled unit matrix and a rank-2 matrix

$$Q_{\hat{a}} = \sigma^2 I_3 + (\beta_1 \beta_2)(\beta_1 \beta_2)^T \text{ with } \beta_i \in R^3,\ i = 1,2\,.$$

It is furthermore assumed that the diagonal entries of the rank-2 matrix are all of the same order and significantly larger than the scale factor σ^2 of the scaled unit matrix. For the present example the scale factor is chosen as $\sigma^2 = 0.04$ and the entries of the rank-2 matrix as

$$\beta_1 = \begin{bmatrix} 0.218 \\ -2.228 \\ -2.462 \end{bmatrix}, \quad \beta_2 = \begin{bmatrix} 2.490 \\ 1.135 \\ 0.434 \end{bmatrix}.$$

Based on these chosen values, the variances respectively the sequential conditional variances can be computed. They read

i	1	2	3
$\sigma^2_{\hat{a}_i}$	6.288	6.292	6.290
$\sigma^2_{\hat{a}_{i\vert I}}$	6.288	5.420	0.089

From these results we see that there is a relative large drop in value when going from the second to the third conditional variance. The location of this discontinuity is due to the fact that the second matrix in the sum of $Q_{\hat{a}}$ is of rank-2 and the size of the discontinuity is due to the differences in size between the entries of the two matrices in the sum.

First we will consider the search based in the original ambiguities. The least-squares estimates of the ambiguities are given as $\hat{a}_1 = 2.97$, $\hat{a}_2 = 3.10$ and $\hat{a}_3 = 5.45$. The sequential bounds of (8.36) are used for the search, with $m = 3$. The constant χ^2 is chosen to be equal to one. In Figure 8.15 the ellipsoid is depicted in which the search for the integer-triples is performed.

The grid points in the projected ellipse in the 1-2 plane are the integer pairs (a_1, a_2) that satisfy the first two bounds of (8.36). The with these integer pairs corresponding intervals for a_3 and the a_3-integers within them are depicted as respectively the little bars, and the dots. The large values of the conditional variances of a_1 and a_2 compared to the one for a_3 lead to intervals for a_1 and a_2 that are significantly larger than the intervals for a_3. Moreover, since the intervals for a_3 are small compared to the grid spacing, there is a high probability that there will be no a_3-integer within them.

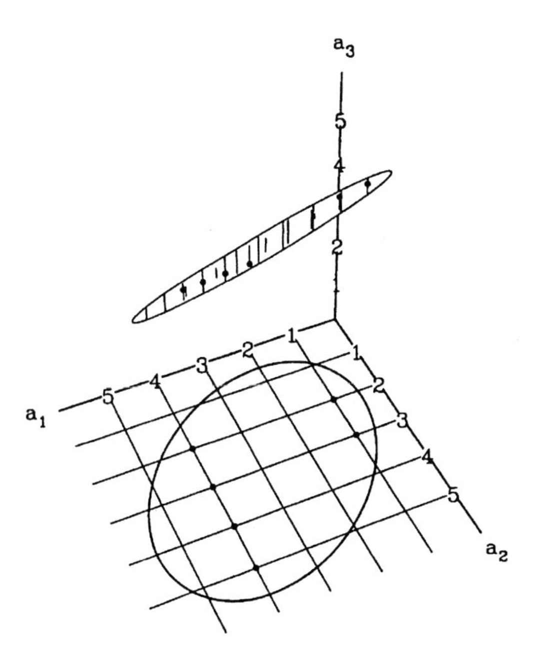

Figure 8.15. The 3D-ambiguity search space and its perpendicular projection onto the 1-2 plane.

In Table 8.4. the results of the search process are shown. In the search process the integers are kept as close as possible to their corresponding conditional estimates. The results show that quite some trials are required to find an integer triple satisfying all three bounds. This is of course due to the fact that the first two bounds are rather loose whereas the third bound is tight. The integer triple found reads (4, 3, 5), and it allows us to shrink the ellipsoid by setting the χ^2-value to $\chi^2 = 0.218$. Continuation of the search in the same manner shows that no other integer triples lie inside the shrunken ellipsoid. Hence, the search stops and it is concluded that the integer triple (4, 3, 5) equals the sought for integer least-squares solution.

Table 8.4. The integer triples that are encountered during the search.

$\chi^2 = 1.0$			$\chi^2 = 0.218$		
a_1	a_2	a_3	a_1	a_2	a_3
3	3	-			
3	4	-			
3	2	-			
3	5	-			
3	1	-			
2	3	-			
2	2	-			
2	4	-			
2	1	-			
4	3	5	4	3	-
			4	4	-
			-	-	-

We will now consider the search based on the transformed ambiguities. From the original variance-covariance matrix

$$Q_{\hat{a}} = \begin{bmatrix} 6.288 & 2.340 & 0.544 \\ 2.340 & 6.292 & 5.978 \\ 0.544 & 5.978 & 6.290 \end{bmatrix},$$

the 3D-decorrelating ambiguity transformation Z^T is constructed in four steps as

$$Z^T = \begin{bmatrix} 1 & 0 & 0 \\ -1 & 1 & 0 \\ 0 & 0 & 1 \end{bmatrix} \begin{bmatrix} 1 & 0 & 3 \\ 0 & 1 & 0 \\ 0 & 0 & 1 \end{bmatrix} \begin{bmatrix} 1 & 0 & 0 \\ 0 & 1 & 1 \\ 0 & 0 & 1 \end{bmatrix} \begin{bmatrix} 1 & 0 & 0 \\ 0 & 1 & 0 \\ 0 & -1 & 1 \end{bmatrix} = \begin{bmatrix} 1 & -3 & 3 \\ -1 & 3 & -2 \\ 0 & -1 & 1 \end{bmatrix}.$$

The transformed variance-covariance matrix reads therefore

$$Q_{\hat{z}} = Z^T Q_{\hat{a}} Z = \begin{bmatrix} 1.146 & 0.334 & 0.082 \\ 0.334 & 4.476 & 0.230 \\ 0.082 & 0.230 & 0.626 \end{bmatrix}.$$

The variances respectively the sequential conditional variances of the transformed ambiguities read

i	1	2	3
$\sigma^2_{\hat{z}_i}$	1.146	4.476	0.626
$\sigma^2_{\hat{z}_{i\|I}}$	1.146	4.376	0.610

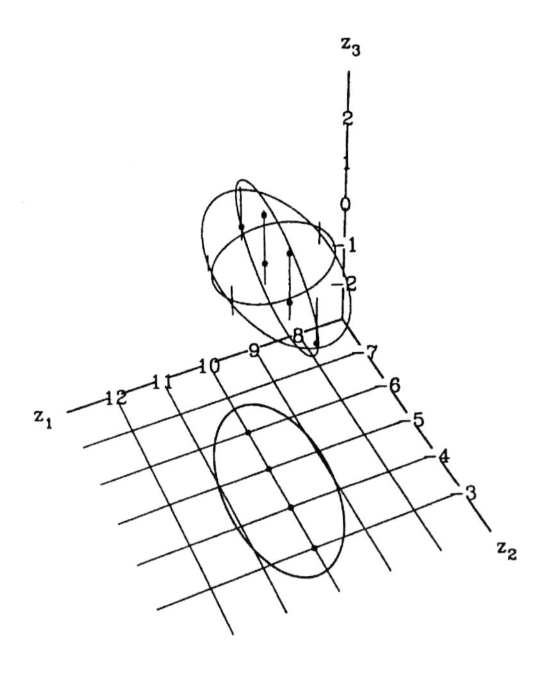

Figure 8.16. The transformed search space and its perpendicular projection onto the 1-2 plane.

Note that the relatively large drop in value which was present in the original spectrum has now been diminished in size in the transformed spectrum. Also note that the new ambiguities are more precise that the original ones. The transformed ambiguities are also less correlated and their search space is less elongated. The elongation has been pushed from its original value $e_{\hat{a}} = 17.965$ to the smaller value of $e_{\hat{z}} = 2.734$. In Figure 8.16 the transformation search space is depicted. It is centred at $\hat{z}_1 = 10.02$, $\hat{z}_2 = -4.57$, $\hat{z}_3 = 2.35$ which follows from $\hat{z} = Z^T\hat{a}$. The smaller elongation of the transformed search space can be clearly seen; the intervals for the third ambiguity z_3 are now in general larger than the grid spacing, and we see that two intervals contain more than one integer. In Table 8.5 the results for the search of the transformed ambiguities are shown. Comparing it with Table 8.4. learns us that the halting problem has indeed been eliminated. The integer least-squares solution found reads $(\check{z}_1, \check{z}_2, \check{z}_3) = (10, -5, 2)$. The corresponding solution values for the original ambiguities follow from $\check{a} = (Z^T)^{-1}\check{z}$ as $(\check{a}_1, \check{a}_2, \check{a}_3) = (4, 3, 5)$, which is of course identical to the solution found earlier.

Table 8.5. The integer triples that are encountered during the search in the transformed ellipsoid.

$\chi^2 = 1.0$			$\chi^2 = 0.218$		
z_1	z_2	z_3	z_1	z_2	z_3
10	-5	2	10	-5	-
			10	-4	-
			-	-	-

In order to illustrate the least-squares ambiguity decorrelation adjustment method with actual GPS data, the same 7 satellite configuration using dual frequency carrier phase data of section 8.3.3 is used. Application of the method to the data of this example resulted in the following multi-channel ambiguity transformation

$$Z^T = \begin{bmatrix} 1 & -5 & 3 & 4 & -1 & -1 & -2 & 4 & -4 & -1 & 3 & 3 \\ -2 & -4 & 1 & 0 & 1 & 5 & 5 & 4 & -1 & 2 & -2 & -8 \\ 0 & 2 & -4 & -4 & -3 & 8 & -5 & 3 & 2 & -1 & 2 & -4 \\ 5 & 1 & 1 & 1 & -1 & -7 & -2 & 1 & 0 & -6 & -2 & 3 \\ 0 & 1 & -3 & 1 & -4 & 1 & 5 & 4 & -3 & 0 & -5 & 5 \\ 2 & 0 & 0 & 1 & 2 & -4 & -1 & 3 & -3 & -2 & -7 & 8 \\ 0 & 5 & -5 & 3 & -5 & 5 & 0 & -1 & -1 & -3 & -1 & 4 \\ 1 & 2 & 0 & 2 & -3 & 1 & 0 & 4 & -4 & -8 & -3 & 2 \\ 5 & -3 & -4 & -1 & 3 & -2 & -6 & 1 & 6 & 5 & -2 & 2 \\ -5 & 2 & -2 & 1 & 0 & 1 & 2 & -4 & 5 & -3 & 5 & -2 \\ -4 & -3 & 1 & 3 & 4 & -5 & 1 & 6 & 0 & -10 & -3 & 6 \\ 1 & -1 & -2 & 6 & 0 & 3 & -1 & -1 & -1 & 1 & -1 & 3 \end{bmatrix} . \tag{8.86}$$

Note that Z^T is truly a multi-channel transformation. Every new ambiguity is formed as a linear combination of all original DD ambiguities. With (8.86), the original DD ambiguities can be transformed as $z = Z^T a$. In order to recover a from z, the inverse of Z^T is needed. It reads

$$Z^{-T} = \begin{bmatrix} 86 & 145 & -281 & -127 & -276 & -136 & 607 & -50 & 172 & -249 & 127 & -435 \\ -362 & -258 & 589 & 530 & 417 & -290 & -598 & -675 & -566 & -308 & 281 & 865 \\ -213 & -249 & 68 & -36 & 195 & -349 & -335 & 113 & -190 & -54 & 127 & 308 \\ -231 & -204 & 426 & 326 & 340 & -104 & -580 & -322 & -367 & -59 & 95 & 643 \\ 5 & 113 & -9 & 136 & -131 & -91 & 385 & -376 & -77 & -258 & 172 & -118 \\ 59 & 54 & -299 & -231 & -199 & -154 & 394 & 190 & 231 & -154 & 59 & -367 \\ 67 & 113 & -219 & -99 & -215 & -106 & 473 & -39 & 134 & -194 & 99 & -339 \\ -282 & -201 & 459 & 413 & 325 & -226 & -466 & -526 & -441 & -240 & 219 & 674 \\ -166 & -194 & 53 & -28 & 152 & -272 & -261 & 88 & -42 & -148 & 99 & 240 \\ -180 & -159 & 332 & 254 & 265 & -81 & -452 & -251 & -286 & -46 & 74 & 501 \\ 4 & 88 & -7 & 106 & -102 & -71 & 300 & -293 & -60 & -201 & 134 & -92 \\ 46 & 42 & -233 & -180 & -155 & -120 & 307 & 148 & 180 & -120 & 46 & -286 \end{bmatrix} . \tag{8.87}$$

Note that all entries of the inverse are indeed integer. Also note that the first six rows of the inverse are to a good approximation scaled versions of the last six rows. The scale factor equals 77/60, which is the ratio of the L_2-wavelength and the L_1-wavelength.

Using the ambiguity transformation (8.86), we obtain the new ambiguity variance-covariance matrix $Q_{\hat{z}}$ from the original DD ambiguity variance-covariance matrix $Q_{\hat{a}}$, as $Q_{\hat{z}} = Z^T Q_{\hat{a}} Z$. In order to illustrate the performance of transformation (8.86), we will compare the elongations of the original and transformed ambiguity search spaces and the correlation and precision of the original and transformed least-squares ambiguities.

Figure 8.17 shows the elongation of both the original and the transformed ambiguity search space as function of the observational time span. Note the dramatic decrease in elongation which is achieved. Even when the two observation epochs are separated by 10 minutes, an improvement by a factor of about ten is reached.

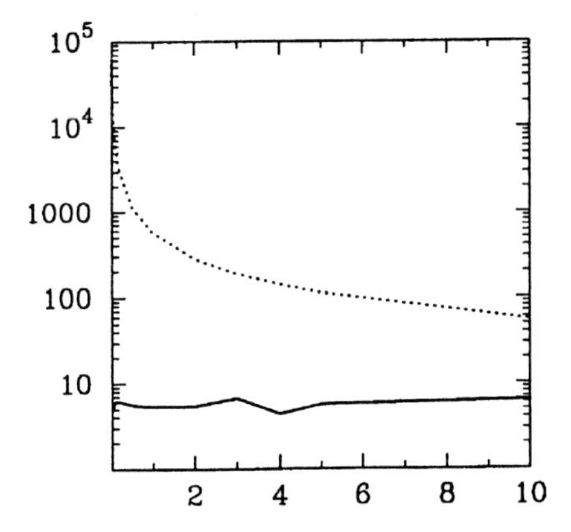

Figure 8.17. Elongation of both the original (dotted curve) and the transformed (full curve) ambiguity search space as function of the observational time span in minutes.

Figure 8.18 shows the two histograms of the absolute values of the correlation coefficients of the DD ambiguities $\hat{a}_i$ and the transformed ambiguities $\hat{z}_i$. It follows upon comparing the two histograms that the ambiguity transformation (8.86) has indeed achieved a large decrease in correlation between the ambiguities. None of the correlation coefficients $\rho_{z_i z_j}$ is close to ± 1 and the largest is even smaller than half in absolute value.

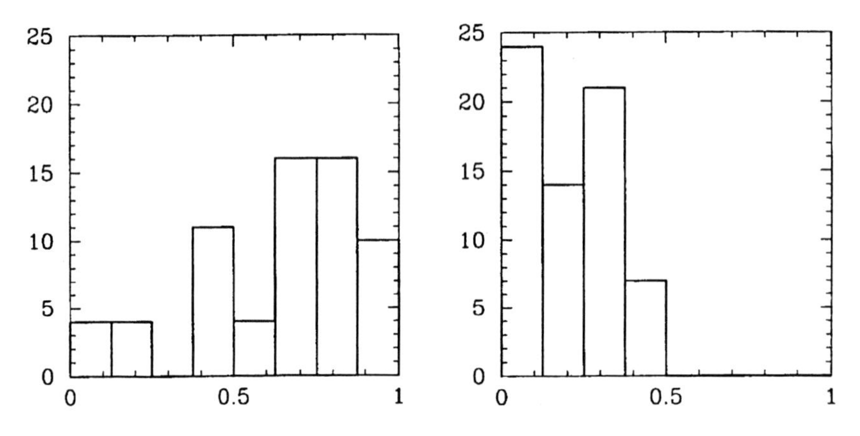

Figure 8.18. Histograms of $|\rho_{\hat{a}_i \hat{a}_j}|$ (left) and $|\rho_{\hat{z}_i \hat{z}_j}|$ (right).

As it was shown in section 8.3.3, the original spectrum contained a large discontinuity when passing from the third to the fourth conditional standard deviation. The first three conditional standard deviations were rather large, whereas the remaining nine conditional standard deviations were very small indeed. And it was due to this large drop in value of the conditional standard deviations that the search for the integer least-squares ambiguities was hindered by a high likelihood of halting. Figure 8.19 shows both the original and transformed spectrum of conditional standard deviations.

The improvement in the spectrum is clearly visible from Figure 8.19. The discontinuity has disappeared and all conditional standard deviations are now of the same small order. Due to this low level of the flattened spectrum, the search for the transformed integer least-squares ambiguities - based on the sequential bounds of (8.77) - can be executed in a highly efficient manner.

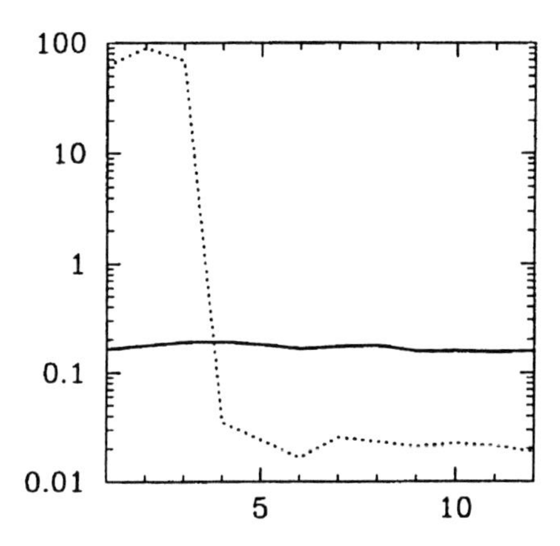

Figure 8.19. Original (dotted curve) and transformed (full curve) spectrum of conditional standard deviations.

In line with this, we observe from Figure 8.20 that the number of transformed integer candidates that progressively satisfy the sequential bounds, is indeed dramatically much smaller than the number of original integer candidates that satisfy their sequential bounds. The dotted and full curve of Figure 8.20 of course meet when all twelve bounds are considered; the original and transformed ambiguity search space both contain the same number of integer vectors.

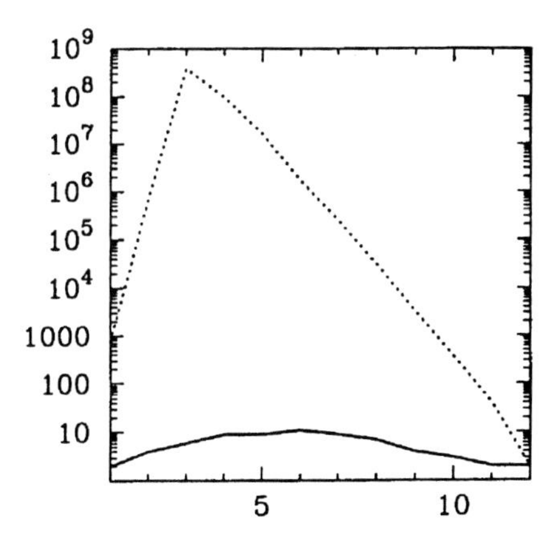

Figure 8.20. The number of original (dotted curve) and transformed (full curve) integer candidates per number of sequential bounds.

To accentuate the fact that the transformed ambiguities are indeed of a very high precision, Table 8.6 gives an overview of the least-squares ambiguity estimates themselves, all expressed in cycles. Shown are the ordinary noninteger least-squares estimates and their precision of both the original DD ambiguities, $\hat{a}_i$, as well as of the transformed ambiguities, $\hat{z}_i$. Also shown are the corresponding integer least-squares estimates, $\check{a}_i$ and $\check{z}_i$, and the differences between the noninteger and integer solution. The high precision of the transformed ambiguities can clearly be seen from the table. In fact, for this particular case a simple 'rounding to the nearest integer' of the least-squares estimates of the transformed ambiguities already would suffice for finding the correct integer least-squares solution. It should be remarked however, that a high precision of the ambiguities is generally no guarantee that the integer least-squares solution is found by means of a simple rounding to the nearest integer. Very precise ambiguities could namely still be highly correlated. Hence, even if the ambiguities are of a high precision, only a search as advocated in section 8.3.2 guarantees that the integer least-squares solution is found.

Table 8.6. The noninteger and integer least-squares estimates of the original and the transformed ambiguities.

$\hat{a}_i$	$\sigma_{\hat{a}_i}$	$\check{a}_i$	$\hat{a}_i - \check{a}_i$	$\hat{z}_i$	$\sigma_{\hat{z}_i}$	$\check{z}_i$	$\check{z}_i - \hat{z}_i$
-587.29	61.72	-593	5.71	-3336478.09	0.16	-3336478	-0.09
-8069.63	99.90	-8073	3.37	-3338914.10	0.18	-3338914	0.10
2827.63	126.12	2842	-14.37	-2526738.94	0.21	-2526739	0.06
7054.42	113.66	7066	-11.58	-841205.96	0.22	-841206	0.04
-839102.6	189.00	-839083	-19.96	-3354032.85	0.20	-3354033	0.15
-5384.42	102.93	-5393	8.58	-2514838.94	0.22	-2514839	0.06
-753.67	79.21	-761	7.33	822591.25	0.21	822591	0.25
-10354.68	128.21	10359	4.32	-3361667.92	0.22	-3361668	0.08
3629.56	161.86	3648	-18.44	-827184.97	0.22	-827185	0.03
9062.12	145.86	9077	-14.88	3344449.04	0.21	3344449	0.04
-4055.60	242.55	-4030	-25.60	-5025486.06	0.23	502586	-0.06
-6909.99	132.10	-6921	11.01	845106.07	0.21	845106	0.07

Up to this point we tacitly assumed that the search space was appropriately scaled. That is, that it contained the solution sought and that it at the same time was small enough, so as not to contain too many grid points, see discussion in Teunissen [1993a], Teunissen et al. [1996], de Jonge and Tiberius [1996]. An appropriate value for χ^2 can be obtained as follows. Starting from the real-valued least-squares estimate of the transformed ambiguity vector, each of its entries is rounded to its nearest integer, either unconditionally or conditionally. This will give an integer vector, which then is substituted for z in (8.76). The value of χ^2 is then taken to be equal to the value of the quadratic form. This approach guarantees that the search space will at least contain one grid point and thus certainly the sought for integer least-squares solution. Also, the number of grid points contained in it, will not be too large. This is due to the high precision and low correlation of the transformed ambiguities. A similar approach can be used when one wants to guarantee that at least two grid points are contained in the search space, e.g., for the purpose of validation.

8.5.4 On the GPS Spectra of Ambiguity Conditional variances

We have seen that for the single baseline case, the GPS spectrum of DD ambiguity conditional variances shows a distinct discontinuity when passing from the third to the fourth conditional variance. It is the presence of this discontinuity in the spectrum that prohibits an efficient search. We have also seen how this discontinuity can be removed from the spectrum. This is made possible through a decorrelation of the least-squares ambiguities. As a result a lowered and flattened spectrum is obtained, which allows for a very efficient search for the integer least-squares solution. Since the signature of the spectrum of ambiguity conditional variances is decisive for the performance of the search, it is of interest to consider the spectrum in somewhat closer detail. In this section we will therefore consider the spectrum in relation to the available observational data. A list of qualitative conclusions about the signature of the spectrum will conclude this section.

The spectra that will be considered in this section are based on respectively:(*i*) L_1 carrier phase data only (σ_{ϕ_1} = 3mm); (*ii*) dual-frequency carrier phase data ($\sigma_{\phi_1} = \sigma_{\phi_2}$ = 3mm) using the wide-lane ambiguities; (*iii*) dual-frequency carrier phase data using the L_1 and L_2 DD ambiguities; and (*iv*) dual-frequency carrier phase data aided with pseudorange data (σ_p = 60 cm). Both the original and transformed spectra will be shown. The examples that will be shown, are again based on the same 7 satellite configuration which has been used earlier in section 8.3.3. Also the same underlying model of observation equations (cf. section 8.2.1, equation (8.2)) has been used.

In Figure 8.21 the single baseline spectra of both the original and the transformed spectra of conditional standard deviations are shown. The corresponding elongations and variances before and after the transformation are given in Table 8.7.

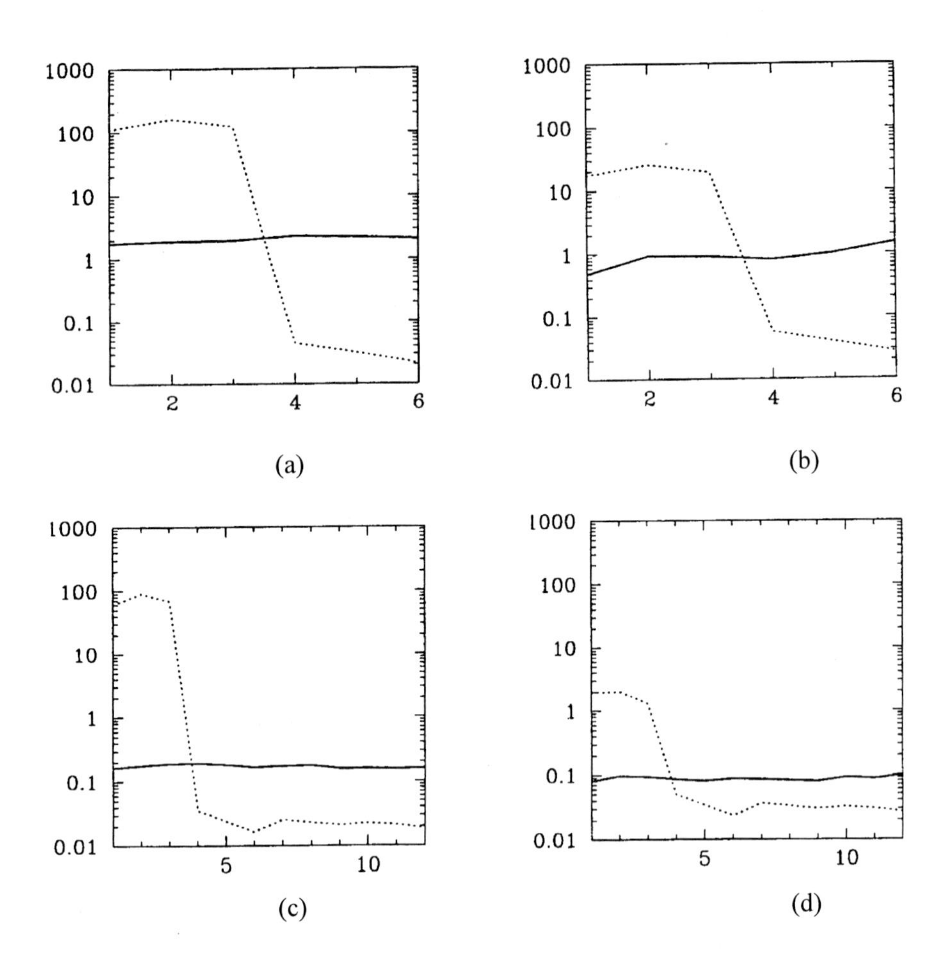

Figure 8.21. The original (dotted curve) and the transformed (full line) spectra of conditional standard deviations in cycles. (a) L_1-spectrum; (b) wide-lane (L_5) spectrum; (c) L_1/L_2-spectrum; (d) code-aided L_1/L_2 spectrum.

Figure 8.21(a) shows the L_1-spectrum of the original DD ambiguities and the transformed ambiguities. The discontinuity in the original spectrum is clearly visible. Figure 8.21(b) shows the spectrum of the wide-lane ambiguities. Again we observe a discontinuity when passing from the third to the fourth conditional variance. The size however, of the discontinuity in the wide-lane spectrum is

smaller than that of the L_1 spectrum (compare the dotted curves of Figures 8.21(a) and 8.21(b)). That is, the three large conditional standard deviations of the wide-lane ambiguities are smaller than those of the L_1 ambiguities, and the three small conditional standard deviations of the wide-lane ambiguities are larger than those of the L_1 ambiguities. In other words, the wide-lane spectrum is flatter than that of the L_1 spectrum. This shows, with reference to our earlier discussion on the search for the integer least-squares solution, that the search for the integer wide-lane ambiguities will be less hindered by the potential problem of halting than the search for the integer L_1 DD ambiguities.
Note however, that although the three large conditional standard deviations of the wide-lane ambiguities are very much smaller than those of the L_1 DD ambiguities, the difference between the third and fourth conditional standard deviations of the wide-lane ambiguities is still significant. Hence, the search for the integer least-squares wide-lane ambiguities still exhibits the potential problem of halting. This will not be the case however, with the decorrelated wide-lane ambiguities. The transformed spectrum of the wide-lane ambiguities is rather flat and its level is smaller than that of the transformed spectrum of the L_1 DD ambiguities. This shows, that the search for the integer least-squares solution of the transformed wide-lane ambiguities will have in this case a somewhat better performance than the search for the integer least-squares solution of the transformed L_1 DD ambiguities.

Wide-lane ambiguities can be constructed once dual-frequency carrier phase data are available. Instead of working with the wide-lane ambiguities however, one may also work with the original L_1 and L_2 DD ambiguities. In fact, if one is willing to believe that the simple and stripped version of the observation equations (cf. section 8.2.1, equation (8.2)) holds true, then Figure 8.21(c) shows that the L_1/L_2 ambiguities should be preferred over the wide-lane ambiguities. The level of the transformed L_1/L_2-spectrum is not only smaller than the level of the transformed L_1-spectrum, but also than that of the transformed wide-lane spectrum. The reason for this lower level is due to the presence of a larger number of very small conditional standard deviations in the original L_1/L_2-spectrum. The number of small conditional standard deviations equals 3 in both the original L_1-spectrum as in the original wide-lane spectrum. In the original L_1/L_2-spectrum however, this number equals 6. Since our decorrelating transformation leaves the volume of the ambiguity search space invariant, it also leaves the product of the conditional standard deviations invariant. This implies, if the number of small conditional standard deviations in the original spectrum increases, that the flattening of the spectrum due to the decorrelating transformation, will result in a lower level for the transformed spectrum.

Table 8.7. Elongation (e) and minimum and maximum standard deviations (σ(max), σ(min)) of the original and transformed ambiguities.

	$e_{\hat{a}}$	$\sigma_{\hat{a}}$ min.	max.	$e_{\hat{z}}$	$\sigma_{\hat{z}}$ min.	max.
L_1	29479.4	112.0	343.0	2.7	1.76	2.74
L_5	3630.2	17.5	53.6	3.7	0.49	1.58
L_1 and L_2	33913.8	61.7	242.6	4.8	0.16	0.23
L_1 and L_2 + code	469.0	1.4	4.1	6.1	0.08	0.13

Figure 8.21(d) shows the original and transformed code-aided L_1/L_2-spectrum. When the original L_1, L_1/L_2 and code-aided L_1/L_2 spectra are compared (the dotted curves in Figures 8.21(a), 8.21(c) and 8.21(d)), the following is observed. Inclusion of the L_2 carrier phases hardly affects the first three large conditional standard deviations. Instead, it results in an increase of the number of very small conditional standard deviations (compare dotted curves of Figures 8.21(a) and 8.21(c)). The inclusion of the pseudorange data however, hardly affects the small conditional standard deviations, but instead lowers the value of the first three large conditional standard deviations. As a result, the size of the discontinuity in the original code-aided L_1/L_2-spectrum is smaller than that of the original L_1/L_2-spectrum. This shows, as one might expect, that the performance of the search for the integer least-squares solution will be enhanced by including pseudorange data. Nevertheless, as Figure 8.21(d) shows, a further improvement is realized through the decorrelation of the ambiguities.

Based on the above findings and also on that what has been discussed in earlier sections, the following qualitative conclusions can be drawn concerning the signature of the GPS spectra of ambiguity conditional variances (remark: it is possible to verify these conclusions by means of an analytical proof):

(1) In case the single baseline model of observation equations is parametrized in no other unknown parameters than the 3 baseline coordinates and the *m* DD ambiguities, the spectrum of DD ambiguity conditional variances, when based on relatively short observational time spans, will always show a discontinuity when passing the third conditional variance. The location and/or size of the discontinuity will change, when the model of observation equations - apart from the *m* DD ambiguities - is based on more than 3 remaining unknown parameters. In the multi baseline case for instance, the number of large conditional variances will equal three times the number of baselines.

(2) When L_2 carrier phase data are included, the number of very small conditional variances increases by the number of additional DD ambiguities. The inclusion of the L_2 carrier phase data will not affect the

location of the discontinuity and will also have no large effect on the size of the discontinuity. It is the increase in the number of small conditional variances, which makes it possible to reach a lower level for the transformed spectrum.

(3) The number of very small conditional variances also increases, in case more satellites are tracked. They will increase by the number of additional satellites in case of L_1, and by twice that number in case of L_1/L_2. The inclusion of more satellites will not affect the location of the discontinuity and will also have no large effect on the size of the discontinuity. Again, it is the increase in the number of small conditional variances, which makes it possible to reach a lower level for the transformed spectrum. This conclusion and the previous one, make therefore quite clear what role is played by *satellite redundancy* and *dual frequency* data.

(4) The inclusion of pseudorange data, will hardly affect the very small conditional variances. Instead, it lowers the value of the large conditional variances and therefore achieves some flattening of the spectrum. As a result, the performance of the search for the integer least-squares DD ambiguities improves. Since the large conditional variances decrease, whereas the very small conditional variances remain largely unchanged, the inclusion of pseudorange data also results in a lower level for the transformed spectrum.

(5) A longer observational time span, i.e. a larger spacing between the observational epochs, has a similar effect on the spectrum as the inclusion of pseudorange data. In fact, it is possible in principle for a large enough observational time span, to obtain a completely flattened spectrum.

(6) The level of the transformed spectrum can be predicted, once the spectrum of the original ambiguities is given. This follows from the fact that the transformed spectrum is almost flat and that the product of conditional variances remains invariant under the transformation.

(7) The degree of success of our decorrelating transformation depends to a large extent on the presence of the discontinuity in the spectrum. In other words, a lessening of the correlation of the least-squares ambiguities is not possible, when their spectrum is flat already.

8.6 SUMMARY

In this contribution we presented the theoretical concepts of GPS carrier phase ambiguity fixing. The main purpose of ambiguity fixing is, to be able via the inclusion of the integer constraint $a \in Z^m$, to obtain a drastic improvement in the precision of the baseline solution. When successful, ambiguity fixing is thus a way to avoid long observational time spans, which otherwise would have been needed if the ambiguities were treated as being real-valued. GPS-ambiguity fixing consists of the following two distinct problems: (1) The ambiguity *estimation* problem, and (2) The ambiguity *validation* problem.

The ambiguity estimation problem can be formulated as the problem of finding the integer least-squares estimates of the carrier phase ambiguities. Although this problem is easily formulated mathematically, it is not so easy to solve. The integer least-squares estimates of the GPS carrier phase ambiguities, are the solution to the minimization problem

$$\min_{a} (\hat{a}-a)^T Q_{\hat{a}}^{-1} (\hat{a}-a) \quad , \quad a \in Z^m .$$

Due to the presence of the integer-constraint $a \in Z^m$ and the fact that the least-squares double-differenced ambiguities are usually highly correlated, the efficiency in computing the integer least-squares ambiguity vector $\check{a}$ is seriously hampered. In order to efficiently solve the above integer least-squares problem, an ambiguity reparametrization is carried out so as to obtain new ambiguities that are largely decorrelated. This method of the least-squares ambiguity decorrelation adjustment is based on using integer approximations of the conditional least-squares transformations. By introducing the reparametrization

$$z = Z^T a \ , \ \hat{z} = Z^T \hat{a} \ , \ Q_{\hat{z}} = Z^T Q_{\hat{a}} Z ,$$

in which Z is an admissible ambiguity transformation, we obtain the equivalent integer least-squares problem

$$\min_{z} (\hat{z}-z)^T Q_{\hat{z}}^{-1} (\hat{z}-z) \quad , \quad z \in Z^m .$$

The corresponding integer least-squares ambiguity vector $\check{z}$ is obtained from a search which is based on bounds that follow from a sequential conditional least-squares ambiguity adjustment. These bounds are given as

$$\begin{cases} (\hat{z}_1 - z_1)^2 & \leq \sigma_{\hat{z}_1}^2 \chi^2 \\ (\hat{z}_{2|1} - z_2)^2 & \leq \sigma_{\hat{z}_{2|1}}^2 \chi^2 [1 - (\hat{z}_1 - z_1)^2 / \sigma_{\hat{z}_1}^2 \chi^2] \\ & \vdots \\ (\hat{z}_{m|M} - z_m)^2 & \leq \sigma_{\hat{z}_{m|M}}^2 \chi^2 [1 - \sum_{j=1}^{m-1} (\hat{z}_{j|J} - z_j)^2 / \sigma_{\hat{z}_{j|J}}^2 \chi^2] . \end{cases}$$

Since the decorrelating ambiguity transformation Z achieves a flattening and lowering of the spectrum of ambiguity conditional variances, the potential problem of search halting has been largely eliminated. As a result an efficient search performance for the transformed integer least-squares ambiguities, $\check{z}_1$, $\check{z}_2,\ldots,\check{z}_m$, is obtained. Once the integer least-squares ambiguity vector $\check{z}$ has been computed, the corresponding integer least-squares ambiguity vector $\check{a}$ can be recovered from invoking $\check{a} = Z^{-T}\check{z}$.

Apart from solving the estimation step in the GPS ambiguity fixing problem, also the validation step needs to be considered. In the validation step the question is answered, whether we are willing to accept the computed integer least-squares solution. If so, then a computation of the corresponding fixed baseline solution $\check{b}$ makes sense. The fixed baseline solution $\check{b}$ follows then from the float solution $\hat{b}$ and the ambiguity residual $(\hat{a}-\check{a})$ as $\check{b} = \hat{b} - Q_{\hat{b}\hat{a}}Q_{\hat{a}}^{-1}(\hat{a}-\check{a})$.

Acknowledgement

The author acknowledges the assistance of ir. P.J. de Jonge and ir. C.C.J.M. Tiberius for preparing the numerical examples.

References

Abbot, R. I., C.C. Counselman III, S.A. Gourevitch (1989), GPS Orbit Determination: Bootstrapping to Resolve Carrier Phase Ambiguity. *Proceedings of the Fifth International Symposium on Satellite Positioning*, Las Cruces, New Mexico, pp. 224-233.

Allison, T. (1991), Multi-Observable Processing Techniques for Precise Relative Positioning. Proceedings *ION GPS-91*. Albuquerque, New Mexico, 11-13 September 1991, pp. 715-725.

Baarda, W. (1968), *A Testing Procedure for Use in Geodetic Networks*, Netherlands Geodetic Commission, Publications on Geodesy, New Series, Vol. 2, No. 5.

Betti, B., M Crespi, and F. Sanso (1993), A Geometric Illustration of Ambiguity Resolution in GPS Theory and a Bayesian Approach, *Manuscripta Geodaetica* 18: 317-330.

Blewitt, G. (1989), Carrier Phase Ambiguity Resolution for the Global Positioning System Applied to Geodetic Baselines up to 2000 km. *Journal of Geophysical Research*, Vol. 94, No. B8, pp. 10.187-10.203.

Boon, F., B. Ambrosius (1997), Results of a real-time application of the LAMBDA method in GPS based aircraft landings. *Proceedings KIS97*, Banff, Canada, pp. 339-344.

Cocard, C., A Geiger (1992), Systematic Search for all Possible Widelanes. Proceedings *6th Int. Geod. Symp. on Satellite Positioning*. Columbus, Ohio, 17-20 March 1992.

Counselman, C.C., S.A. Gourevitch (1981), Miniature Interferometer Terminal for Earth Surveying: Ambiguity and Multipath with Global Positioning System. *IEEE Transactions on Geoscience and Remote Sensing*, Vol. GE-19, No. 4, pp. 244-252.

Erickson, C. (1992), *Investigations of C/A code and carrier measurements and techniques for rapid static GPS surveys*. Report no. 20044, Department of Geomatics Engineering, Calgary, Alberta, Canada.

Euler, H.-J., C. Goad (1990), On Optimal Filtering of GPS Dual Frequency Observations without using Orbit Information. *Bulletin Geodesique*, Vol 65, pp. 130-143.

Euler, H,-J., B. Schaffrin (1991), On a Measure for the Discernability between Different Ambiguity Resolutions in the Static-Kinematic GPS-mode. Proceedings of *IAG International Symposium 107 on Kinematic Systems in Geodesy, Surveying and Remote Sensing*, Sept. 10-13 1990, Springer Verlag, New York, pp. 285-295.

Euler, H,-J., H. Landau (1992), Fast GPS Ambiguity Resolution On-The-Fly for Real-Time Applications. Proceedings *6th Int. Geod. Symp. on Satellite Positioning*. Columbus, Ohio, 17-20 March 1992, pp. 650-729.

Frei, E., G. Beutler (1990), Rapid Static Positioning Based on the Fast Ambiguity Resolution Approach FARA: Theory and First Results. *Manuscripta Geodaetica*, Vol. 15, No. 6, 1990.

Frei, E. (1991), *Rapid Differential Positioning with the Global Positioning System*. In: Geodetic and Geophysical Studies in Switzerland, Vol 44.

Gillesen, I., I.A. Elema (1996), Test results of DIA: a real-time adaptive integrity monitoring procedure used in an integrated navigation system. *International Hydrographic Review*, 73-1, pp. 75-103.

Goad, C. (1992), Robust Techniques for Determining GPS Phase Ambiguities. Proceedings *6th Int. Geod. Symp. on Satellite Positioning*. Columbus, Ohio, 17-20 March 1992, pp. 245-254.

Goad, C., M. Yang (1994), On Automatic Precision Airborne GPS Positioning. *Proceedings of the International Symposium on Kinematic Systems in Geodesy, Geomatics and Navigation KIS'94*. Banff, Alberta, Canada. August 30 - September 2, 1994, pp. 131-138.

Golub, G.H., C.F. Van Loan (1986), *Matrix Computations*. North Oxford Academic.

Han, S., K. Wong and C. Rizos (1997), Instantaneous resolution for real-time GPS attitude determination. In: *Proceedings KIS97*, Banff, Canada, pp. 409-416.

Hatch, R. (1986), Dynamic differential GPS at the Centimeter Level. *Proceedings 4th International Geod. Symp. in Satellite Positioning*, Austin, Texas, 28 April -2 May, pp. 1287-1298.

Hatch, R. (1989), Ambiguity Resolution in the Fast Lane. Proceedings *ION GPS-89*, Colorado Springs, CO, 27-29 September, pp. 45-50.

Hatch, R. (1991), Instantaneous Ambiguity Resolution. Proceedings of *IAG International Symposium 107 on Kinematic Systems in Geodesy, Surveying and Remote Sensing*, Sept. 10-13, 1990, Springer Verlag, New York, pp. 299-308.

Hatch, R., H.-J. Euler (1994), A Comparison of Several AROF Kinematic Techniques. *Proceedings of ION GPS-94*, Salt Lake City, Utah, USA, pp. 363-370.

Hein, G., W. Werner (1995), Comparison of different on-the-fly ambiguity resolution techniques. In: *Proceedings ION GPS95*, Part 2 of 2, Palm Springs, California, pp. 1137-1144.

Hofmann-Wellenhof, B., B.W. Remondi (1988), The Antenna Exchange: one Aspect of High-Precision Kinematic Survey. Presented at the International GPS Workshop, *GPS Techniques Applied to Geodesy and Surveying*, Darmstadt, FRG, 10-13 April.

Jin, X.X. (1995), A recursive procedure for computation and quality control of GPS differential corrections. Delft Geodetic Computing Centre, *LGR Series* No. 8.

Jong, C. de (1994), Real-Time integrity monitoring of single and dual frequency GPS observation. In: *GPS-nieuwsbrief*, 9e jaargang, no. 1, mei 1994.

Jong, C. de (1996), Real-time integrity monitoring of dual frequency GPS observations from a single receiver. *Acta Geod. Geoph.*, Hungary, Vol. 31 (1-2), pp. 37-46.

Jonge de P.J., C.C.J.M. Tiberius (1994), A new GPS ambiguity estimation method based on integer least-squares. *Proceedings Third International Symposium on Differential Satellite Navigation Systems DSNS'94*. London, England, April 18-22, 1994, paper no. 73.

Jonge de P.J., C.C.J.M. Tiberius (1996), The LAMBDA method for integer ambiguity estimation: implementation aspects. Delft Geodetic Computing Centre *LGR Series* No. 12.

Jonge de P.J., C.C.J.M. Tiberius and P.J.G. Teunissen (1996), Computational aspects of the LAMBDA method for GPS ambiguity resolution. In: *Proceedings ION GPS96*, Part 1 of 2, Kansas City, Missouri, pp. 935-944.

Kleusberg A. (1990), A Review of Kinematic and Static GPS Surveying Procedures. *Proceedings of the Second International Symposium on Precise Positioning with the Global Positioning system*, Ottawa, Canada, September 3-7 1990, pp. 1102-1113.

Koch, K.R. (1987), *Parameter Estimation and Hypothesis Testing in Linear Models*, Springer Verlag.

Li, Z., Y. Gao (1997), Construction of high dimensional ambiguity transformations for the LAMBDA method. In: *Proceedings KIS97*, Banff, Canada, pp. 305-309.

Marel, H. v.d. (1990), Statistical Testing and Quality Analysis of GPS Networks. In: *Proceedings Second International Symposium on Precise Positioning with the Global Positioning System*. Ottawa, 3-7 September 1990. pp. 935-949.

Mervart, L., G. Beutler, M. Rothacher and U. Wild (1994), Ambiguity Resolution Strategies using the Results of the International GPS Geodynamics Service (IGS) *Bulletin Geodesique*, 68: 29-38.

Remondi, B.W. (1984), *Using the Global Positioning System (GPS) Phase Observables for Relative Geodesy: Modelling, Processing, and Results*, Ph.D. Dissertation, NOAA, Rockville, 360 pp..

Remondi, B.W. (1986), Performing Centimeter-Level Surveys in Seconds with GPS Carrier Phase; Initial Results. *Journal of Navigation*, Volume III, the Institute of Navigation.

Remondi, B.W. (1991), Pseudo-Kinematic GPS Results Using the Ambiguity Function Method, *Journal of Navigation*, Vol. 38, No, 1, pp. 17-36.

Rizos, C., S. Han (1995), A new method for constructing multi-satellite ambiguity combinations for improved ambiguity resolution. In: *Proceedings ION GPS95*, part 2 of 2, Palm Springs, California, pp. 1145-1153.

Rothacher, M. (1993), *Bernese GPS Software Version 3.4: Documentation*. University of Berne, Switzerland.

Scheffé, H. (1956), *The Analysis of Variance*. John Wiley and Sons.

Seeber, G.G. Wübbena (1989), Kinematic Positioning With Carrier Phases and "On the Way" Ambiguity Solution. *Proceedings 5th Int. Geod. Symp. on Satellite Positioning*. Las Cruces, New Mexico, March 1989.

Teunissen, P.J.G., M.A. Salzmann (1989), A Recursive Slippage Test for Use in State-Space Filtering, *Manuscripta Geodaetica*, 1989, 14: 383-390.

Teunissen, P.J.G. (1990a), Quality Control in Integrated Navigation Systems. *IEEE Aerospace and Electronic Systems Magazine*, Vol. 5, No. 7, pp. 35-41.

Teunissen, P.J.G. (1990b), GPS op afstand bekeken (in Dutch). In: *Een halve eeuw in de goede richting*. Lustrumboek Snellius 1950-1990. pp. 215-233.

Teunissen, P.J.G. (1993a), *Least-Squares Estimation of the Integer GPS Ambiguities*. Delft Geodetic Computing Centre (LGR), 16p. Invited Lecture, Section IV Theory and Methodology. IAG General meeting, Beijing, China, August 1993. Also in LGR-Series No. 6.

Teunissen, P.J.G. (1993b), *The Invertible GPS Ambiguity Transformations*. Delft Geodetic Computing Centre (LGR), LGR-report No.9, 9 p.

Teunissen, P.J.G. (1994a), A New Method for Fast Carrier Phase Ambiguity Estimation. *IEEE Position Location and Navigation Symposium PLANS'94* Las Vegas, April 1994, pp. 562-573.

Teunissen, P.J.G. (1994b), *The Least-Squares Ambiguity Decorrelation Adjustment: A Method for Fast GPS Integer Ambiguity Estimation*. Delft Geodetic Computing Centre (LGR), LGR-report No.9, 18 p.

Teunissen, P.J.G. (1994c), *Testing Theory - An Introduction*. Lecture Notes Series Mathematical Geodesy. Department of Geodetic Engineering, Delft University of Technology.

Teunissen, P.J.G. (1994d), On the GPS Double-Difference Ambiguities and their Parial Search Spaces. *Hotine-Marussi Symposium on Mathematical Geodesy*, L'Aquila, Italy, May 29 - June 3, 1994, 10 p.

Teunissen, P.J.G., C.C.J.M. Tiberius (1994e), Integer Least-Squares Estimation of the GPS Phase Ambiguities. *Proceedings of the International Symposium on Kinematic Systems in Geodesy, Geomatics and Navigation KIS'94*. Banff, Alberta, Canada. August 30 - September 2, 1994, pp. 221-231.

Teunissen, P.J.G., P.J. de Jonge and C.C.J.M. Tiberius (1994), On the Spectrum of the GPS DD-ambiguities. *Proceedings of ION GPS-94, 7th International Technical Meeting of the Satellite Division of the Institute of Navigation*. Salt Lake City, Utah, USA. September 20-23, 1994, pp. 115-124.

Teunissen, P.J.G., P.J. de Jonge and C.C.J.M. Tiberius (1996), The volume of the GPS ambiguity search space and its relevance for integer ambiguity resolution. In: *Proceedings ION GPS96*, Part 1 of 2, Kansas City, Missouri, pp. 889-898.

Teunissen, P.J.G., P.J. de Jonge and C.C.J.M. Tiberius (1997), The performance of the LAMBDA method for fast GPS ambiguity resolution. *NAVIGATION*, Journal of the Institute of Navigation, Fall issue, 1997.

Teunissen, P.J.G. (1997), Some remarks on GPS ambiguity resolution. *Artificial Satellites*, Vol. 32, No. 3.

Tiberius, C.C.J.M., P.J. de Jonge (1995), Fast positioning using the LAMBDA method. In: *Proceedings DSNS 95*, Bergen, Norway, paper No. 30.

Wübbena, G. (1989), The GPS Adjustment Software Package - GEONAP - Concepts and Models. Proceedings *5th Int. Geod. Symp. on Satellite Positioning*. Las Cruces, New Mexico, 13-17 March 1989, pp. 452-461.

Wübbena, G. (1991), *Zur Modellierung von GPS Beobachtungen fur die hochgenaue Positionsbestimmung*. Hannover, 1991.

9. ACTIVE GPS CONTROL STATIONS Theory, Implementation and Application

Hans van der Marel
Faculty of Civil Engineering and Geosciences, Section MGP, University of Technology, Thijsseweg 11, 2629 JA Delft, The Netherlands

9.1 INTRODUCTION

An Active GPS Control Station consists essentially of a permanently operating GPS receiver, a computer and some supplementary components. The purpose of an active GPS control station is to provide raw phase (and code) data for geodynamics applications, reference frame maintenance, surveying or kinematic applications (post processing), Real Time Kinematic (RTK) data for surveying or Differential GPS (D-GPS) corrections for navigation applications, or a combination of these.

The main task of an active GPS control station is to collect GPS code and phase data, and distribute this data to users for various applications. Nevertheless, an active GPS control station is more than just a GPS receiver observing 24 hours a day, 7 days per week. Auxiliary data, like temperature, humidity and pressure, may have to be collected as well, but this is not crucial. A necessity for any active GPS control station is that the data must be collected, stored, processed and handled in various ways, and this makes a small computer with the appropriate software necessary. Functions that can be performed by this computer are: integrity monitoring of the GPS data, data archiving, compression and retrieval (bulletin board), broadcasting of D-GPS corrections and RTK data, remote control and operation.

Active GPS control stations may operate as an individual station or as part of a network. In a network there is usually a dedicated computing centre, which may be integrated with one of the control stations. Some functions of the active GPS control stations, such as the data archiving and retrieval functions (e.g. the bulletin board system) can be centralized in the computing centre. Further tasks for the computing centre are:

- regular check of the (other) active GPS control stations
- network integrity monitoring, which is more powerful than the integrity monitoring at the control stations
- extra processing, resulting in additional products (e.g. atmospheric models)
- operate an information system

The main advantages of a network of active GPS control stations are the redundancy, the improved availability and reliability of the control stations, and a central point of access for the user. A drawback of a network approach is the additional communication between the computing centre and reference stations.

Examples of active GPS control systems can be found on every continent of the world, ranging from rudimentary D-GPS reference stations, local and wide area differential GPS systems using national broadcasting networks or geostationary satellites to transmit corrections, national networks for surveying and geokinematics, regional networks of several hundreds of GPS receivers for earthquake monitoring like in Japan and California, up to the world wide IGS network and its subnetworks. It is outside the scope of this chapter to discuss particular systems, although some components of the Dutch Active GPS Reference System (AGRS-NL) will be taken as an example. More information can be found in the literature, see e.g., Abe & Tsusji [1994], Beutler et al. [1994], Bock & Shimada [1990], Bruyninx et al. [1996], Delikaraoglou et al. [1994], Frohlich [1994], Hedling & Jonsson [1995] Young et al. [1995], and on the World Wide Web. In fact, the explosive development of the Internet and World Wide Web makes it an attractive medium for the distribution of active GPS control station data.

In this chapter we take first a look at the components of an active GPS reference station. This includes software. Next we will have a close look at the GPS data which is collected by the reference stations. A regular and invertible transformation of the observations is used to separate as much as possible the underlying processes. This transformation is the basis for the real-time integrity monitoring, which is discussed in section four. Precise differential GPS corrections can be derived directly from the integrity monitoring. In the last section, some of the network concepts and data distribution aspects are discussed.

9.2 ACTIVE GPS CONTROL STATION COMPONENTS

In this section the hardware and software components of active GPS control stations are described. Due to the wide range of applications active GPS control stations have, which lead to differences in their design, it is not easy to give a single description which is valid for all stations. Therefore, most of the components and techniques discussed in this, and the following sections, should be considered multi functional building blocks, which may, or may not, be used. Variations in design should not be confused with lack of standardization. Standardization is very important because active GPS control stations are generally multi user systems. Data formats and user interfaces should be standardized, as is the case presently, but standardization of methods and control station design could also be very helpful.

9.2.1 Reference Station Components

The complete design of an active GPS control station must aim at stability and reliable operation. Requirements for an active GPS control station are:
- a well founded and stable antenna construction, on bedrock or stable sediments, clear from obstructions and reflecting surfaces which could cause multipath
- a room, or shelter, to house the GPS receiver and other equipment, with heating and/or cooling when required
- electrical power, communication facilities (phone and/or data lines)
- accessible by car for maintenance personal, but not for unauthorized visitors
- not susceptible to vandalism and protected from animals
- free from radio interference

Sometimes it may be possible to use existing facilities, otherwise, everything must be specially constructed for the active GPS control station. An other strong argument for the selection of a particular site is co-location with other techniques, such as laser, VLBI or a tide gauge station. For example, the five active GPS control stations in the Netherlands use existing facilities, like tide gauges at the North Sea coast and the river Maas, and facilities at astronomical and geodetic institutes for co-location with VLBI and laser ranging. Existing sites have also one drawback: obstructions. Therefore, in the Netherlands the antennas are mounted on a specially designed mast to keep clear from obstructions. The masts are founded on stable sediments by means of underground piles. As antenna a Dorne-Margolin antenna with chokerings is used. The antenna is protected from birds and weather by a radome. A good source of information on the planning of active GPS control stations is UNAVCO's World Wide Web site, which has some nice examples of active GPS control stations. Other nice examples of active GPS control stations can be found in Canada, Sweden, Austria, United States and Japan, see e.g., Quek et al. [1989], Pesec & Erker [1996], Hedling & Jonsson [1995], Young et al. [1995] and Abe & Tsuji [1994].

The hardware for an active GPS control station may consist of the following components:
- GPS receiver and antenna
- Meteorological equipment (optional)
- External clock (optional)
- Computer, with storage devices and extra serial ports
- Communication equipment for network access, modems and/or transmitter for D-GPS or RTK corrections
- Uninterruptable Power Supply (UPS)
- Remote Power On/Off switch

First of all, a Reference Station needs a geodetic dual frequency GPS receiver. Amongst the criteria for the selection of a receiver type are: number of channels (12), data availability, sampling rate, price and solidity of a receiver. Single frequency receivers are possible for simple D-GPS operation, but dual-frequency receivers offer much better performance of the integrity monitoring. Besides, the additional cost for a dual frequency receiver is negligible compared to the other infrastructural components.

The sample rate at which the receiver operates depends on the type of application. This may be one second for kinematic applications, five or ten seconds for rapid static surveying, or thirty seconds for precise surveying and geodynamics. If more applications, with different sample rates, have to be served at the same time, the receiver should operate with the highest sample rate. Software must then be used to reduce the data rate for other applications. High sample rates (1 second) are also an advantage for the integrity monitoring discussed in section 9.4.

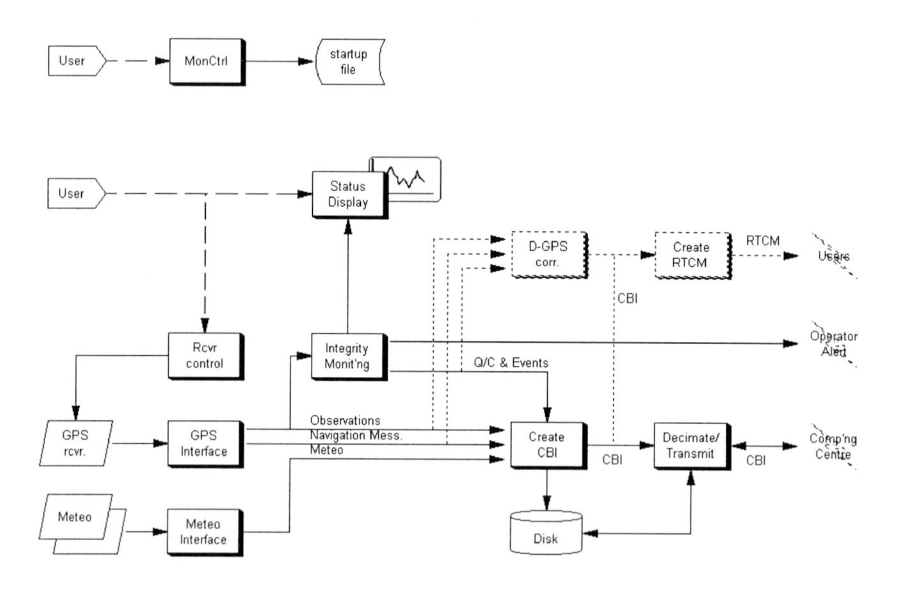

Figure 9.1. Active GPS control station software (AGRS-NL)

Reference stations are optionally equipped with an external clock and meteo equipment. The external clock will be connected directly to the receiver. The use of an external clock will enhance the integrity monitoring at the active control station. Meteorological measurements are important for atmospheric and environmental monitoring. Especially atmospheric pressure is needed when the stations are used to derive the precipital water vapour content of the atmosphere. Temperature and humidity provide additional information.

The GPS receiver (and meteorological equipment) is connected to a small computer to store the data locally, to send the data to users or a special computer centre, and to operate the receiver remotely (Figure 9.1). Modems, ISDN or internet can be used for the communication with users, the computing centre, and remote control by the operator. For example, in the Dutch AGRS-NL we use two high speed modems over ordinary telephone lines for the communication with the computing centre and the operator. This proved to be the most reliable form of communications so far, although we may upgrade to ISDN soon. ISDN (Integrated Services Digital Network) is a good alternative for the communication. ISDN operates over ordinary telephone lines as well, but bypasses the analog-digital conversion. ISDN is well suited for carrying large amounts of computer data, and can set up new connections much faster than by modem. Besides data collection and communication, the control station software is responsible for a first integrity check of the GPS data. In case of problems the software must alert the operator (using a pager) or the computing centre.

Furthermore, if this is not done by the computing centre, the computer at the active GPS control station must operate a bulletin board system from which the user can download the data. When the active GPS control station has to provide D-GPS or RTK corrections the receiver, or the computer, must be linked to a transmitter. The purchase, licensing and installation of a transmitter is a task which requires specialist help. Alternatively, the D-GPS corrections can be send by a direct modem line to a D-GPS provider (e.g. a national broadcasting company).

Finally, an uninterruptable power supply (UPS) is necessary to cover small outages and fluctuations in the AC power. In order to cover outages in the AC power over longer periods the station must be able to generate its own power. In some countries, like the Netherlands, outages over longer periods are very rare, so that it is not really necessary to invest in a very expensive autonomous power supply. A very useful tool to reset the station, in case it hangs, is a Remote Power On/Off switch. The remote power on/off switch can be activated to interrupt the power for 15-30 seconds, using a ordinary phone and a special dialling sequence. The result is that the computer and other equipment will reboot. The switch is installed between the telephone line and modem, but it does not depend on the modem for its operation (it only monitors the rings on the phone).

9.2.2 Reference Station Software

The software of an active GPS control station is, in the first place, charged with reading data from the GPS receiver and meteo sensors. Secondly, the software may have to do a quality check (integrity monitoring). The software has to store GPS and meteo data, events and quality information of the quality control, on disk, and if appropriate, transmit this data to the computing centre. Optionally, differential GPS corrections and RTK data may have to be generated in RTCM format. These are tasks for dedicated software, see e.g., De Jong [1994] or UNAVCO's world wide web pages. If the data is not transmitted to a computing centre the reference station may operate a bulletin board service, which is standard software. Also, for the remote control of the computer and file transfer standard software can be used.

The active GPS control station software must perform number of tasks at the same time. This is depicted in Figure 9.1 for the Dutch AGRS-NL system. Some of the tasks the control station software must take care of are:

GPS interface: (real-time) interface with the receiver. The interface is different for various type of receivers; each type of receiver has its own interface. This includes the control over receiver parameters like, elevation cutoff, sampling rate and communication options. The control function may run as a separate task using software from the manufacturer. In some cases the receiver may also be controlled directly from the real-time reference station software. For testing purposes and off-line processing it can also be useful if the software can read data from file (proprietary and/or RINEX format).

Meteo sensor interface: interface with the (optional) meteo sensors. The interface will be different for the various types of meteo sensors.

Integrity monitoring: first quality check of the data. The quality control can take place on separate channels (per GPS satellite) and is independent of broadcast ephemerides and station locations. In case of serious problems, or lack of data, this task may alert the operator (using for example a pager). The monitoring software developed for the Dutch AGRS-NL uses the DIA procedure (Detection, Identification and Adaption) of chapter 7, see e.g., De Jong [1994, 1997]. The integrity monitoring will be discussed in more detail in section 9.4. In the integrity monitoring an elevation dependent stochastic model is used.

Status display: the status display gives information about stations and satellites (skyplot), results of the quality control, meteo data and events.

Data storage: GPS observables, navigation messages, meteo data and results of the integrity monitoring are stored on disk. The data will expire after a selectable period and will then be deleted.

Communication with computing centre: after data rate has been adjusted to the user requirements, using the decimation procedure of section 9.3, it will be transmitted by modem, ISDN or other means, continuously or intermittently. For the Dutch AGRS-NL data is transmitted every hour, but other intervals, or continuous transmission can be selected. If the data transport fails, it will restart from the last transmitted observation. The data rate is at present 30 seconds (with plans to reduce this to ten seconds). The receiver itself will be operating with a 1 second sample rate. Batches of one second data may be transmitted on request to the computing centre. An efficient data format is needed to keep transmission times low.

Bulletin board: if data is not transmitted to the computing centre the control station may operate a bulletin board.

D-GPS and RTK corrections (RTCM data): when these are not computed by the receiver they have to be computed and formatted as RTCM data by the reference station software.

Remote Control software: in order to operate the active GPS control station from a distance, remote control software is needed. This is standard and widely available software (e.g., PC-AnyWhere, ReachOut, Carbon Copy, etc.).

Some of these tasks will be done in near real-time. For instance, D-GPS and RTK corrections must be computed in near real-time, which means that the receiver interface and integrity monitoring must also be done in near real-time. The software must then be in a real-time multitasking environment, and the various tasks have to be integrated in a single package. If there is no need for a near real-time operation some of the tasks may also be run in sequence. Other tasks, for instance the communication with the computing centre, may be started by external events (request for data).

9.2.3 Computing Centre Software

When there is more than one active control station it may be better to implement some tasks of a control station at a central location. This can be one of the active control stations, or a dedicated computing centre. The computing centre is then the place where the data is collected from the reference stations, processed, stored and distributed. Tasks of the computer centre are:

- Data reception;
- Processing and integrity monitoring;
- Data distribution and retrieval;

The lay-out of the Dutch AGRS-NL system, which we give as an example, is shown in Figure 9.2.

The main task of the data reception software is to collect data from the reference stations. Data will be saved on disk directly after reception. The data reception software used by the Dutch AGRS-NL has two modes, a continuously and a intermittent one. In case data from a reference station is not received within a certain time the operator is informed using a pager.

The purpose of the processing is twofold. First it can be used to check and control the quality of GPS data. This can now be done more rigorously than before, using single differences instead of raw observations. The stability of GPS antenna positions, quality of GPS satellites, broadcast ephemerides, code and phase observations (cycle slips and outliers) can be checked. In particular cycle slips can be corrected, and outliers, unhealthy satellites and GPS receivers with problems can be flagged. Secondly, from the processing additional information may be derived, for instance atmospheric parameters, which can be helpful for users, or may be used for non geodetic applications.

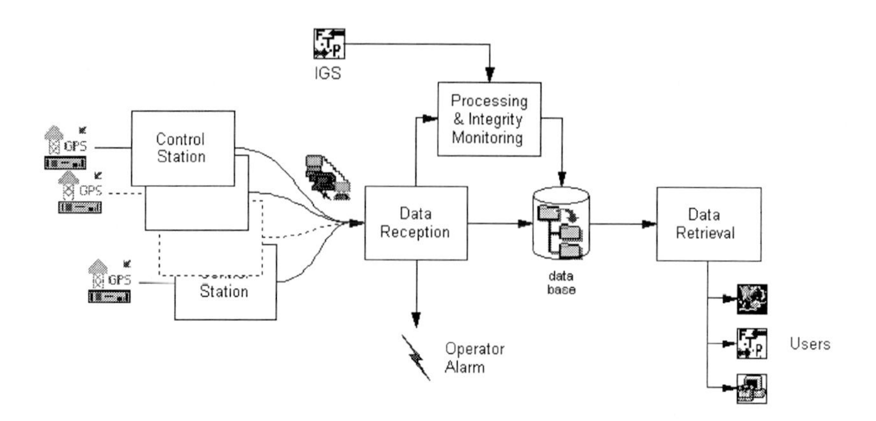

Figure 9.2. AGRS-NL system and computing centre

The data distribution and retrieval software makes the raw data and other products available to the users. It can be operated using standard bulletin board software, or using the World Wide Web system. With the World Wide Web system the number of users is not limited by number of modems at the computing centre, and charges for phone costs are generally smaller.

Standard data formats should be used as much as possible for the distribution of data, e.g., RINEX and compressed RINEX. For real time operations the RTCM format can be used. However, because those data formats are not designed for efficiency they are not suitable for the communication between reference stations and the computing centre. For the AGRS-NL a special data format has been developed, which is efficient, retains more information than in RINEX files, and is receiver independent. This data can be converted to RINEX, and in real time into RTCM as well.

9.3 SINGLE CHANNEL OBSERVATION EQUATIONS

In this section the GPS tracking modes and observation equations are introduced. A regular and invertible transformation for the dual frequency code and phase data is derived, which will break down the original observations into their 'principle components'. This transformation forms the basis for the integrity monitoring discussed in the next section, as well as some other important applications: computation of differential corrections, data decimation and compression. It is the same transformation as the transformation introduced in chapter 6.

9.3.1 GPS Tracking Modes

Dual frequency GPS receivers track the code delay and carrier beat phase of up to twelve satellites. Common tracking modes for dual frequency receivers are:

1 Code and carrier phase measurement of L1-C/A, L1-P and L2-P signals (six observation code correlation tracking)

2A Code and carrier phase measurements of L1-P and L2-P signals; L1-C/A is not observed in this case (four observation code correlation tracking)

2B Code and carrier phase measurement of L1-C/A and - cross correlated - L2⊗L1-Y signals (four observation cross correlation tracking)

3 Code and carrier phase measurement of L1-C/A signal and phase measurement of squared L2-Y (L2⊗L2-Y) signal; with L2 wavelength halved (three observation signal squaring tracking)

Although active GPS control stations will mostly be equipped with dual frequency GPS receivers, for completeness, also the tracking modes for single frequency GPS receivers will be given:

4 Code and carrier phase measurement of L1-C/A (or L1-P)

5 Code measurement of L1-C/A (or L1-P)

The actual tracking mode depends strongly on the type of receiver being used and whether Anti-Snooping (A-S) is on or off (see Chapter 4). In the above classification the notation for P and Y code is slightly synonymous. Receivers may be forced into cross correlation or signal squaring mode even when P code is available (case 2A and 3). On the other hand, some receivers, are capable of observing all these signals even in the case Anti-Spoofing is on, at the expense of a small loss of Signal to Noise ratio.

Different tracking modes may be used for different channels and satellites, for instance, depending on whether Anti-Snooping is on or off for a particular satellite. Occasionally receivers may loose one of the signals, therefore falling back into another - listed or unlisted - tracking mode (or single frequency mode). In some cases both L1-C/A and L1-P code measurements, but only one L1 phase measurement, are given, resulting in a "five observation" mode. This case, although very common, will be considered as a special case of 1. The measurements from the four observation cross correlation mode (2B) are easily converted into L2 measurements, giving full wavelength phase measurements on L2 (Chapter 4). The differences between the two tracking modes are thus mainly of a stochastic nature, see also section 9.3.6.

The most common tracking mode for Active GPS control stations is the four observation case. This mode will be the starting point in this chapter.

9.3.2 GPS Observation Equations

The observation equations for a code observation $P_{i,f}^{s}(t)$ and carrier phase observation $\phi_{i,f}^{s}(t)$ at time t are

$$
\begin{aligned}
P^s_{i,f}(t) &= \rho^s_i(t,t-\tau^s_{i,f}(t)) + c\delta t_i(t) - c\delta t^s(t-\tau^s_{i,f}(t)) + \gamma_f I^s_i + T^s_i \\
&\quad + m^s_{i,f} + d^s_{i,f} + e^s_{i,f} \\
\phi^s_{i,f}(t) &= \rho^s_i(t,t-\tau^s_{i,f}(t)) + c\delta t_i(t) - c\delta t^s(t-\tau^s_{i,f}(t)) - \gamma_f I^s_i + T^s_i + \lambda_f A^s_{i,f} \\
&\quad + \bar{m}^s_{i,f} + \bar{d}^s_{i,f} + \bar{e}^s_{i,f}
\end{aligned} \tag{9.1}
$$

with $\phi_{i,f}^s(t) = \lambda_f\, \varphi_{i,f}^s(t)$ the phase observation in units of meters, and $\varphi_{i,f}^s(t)$ the phase observation in units of L1 or L2 cycles. The index *i* is used to denote the receiver (only one in our case), *s* is used to identify the satellite. and *f* is used to indicate the frequency and signal type (C/A, P1, P2). The topocentric distance ρ is a function of the position of the receiver at time of reception, $\mathbf{r}_i(t)$, and satellite at time of transmission, $\mathbf{r}^s(t-\tau)$:

$$
\rho^s_i(t,t-\tau^s_{i,f}(t)) = \| \, r_i(t) - r^s(t-\tau^s_{i,f}(t)) \, \| \simeq \| \, r_i(t) - r^s(t-\tau^s_{i,0}(t)) \, \| \tag{9.2}
$$

with $\tau_{i,f}^s$ the travel time of the signal, including atmospheric delays, and $\tau_{i,0}^s$ the travel time in vacuum. The travel time is a function of the topocentric distance, ionospheric and tropospheric delays,

$$
\tau^s_{i,f}(t) = \frac{1}{c}\left(\rho^s_i(t,t-\tau^s_{i,f}(t)) \pm \gamma_f I^s_i + T^s_i \right) \simeq \frac{1}{c}\left(\rho^s_i(t,t-\tau^s_{i,0}(t)) \pm \gamma_f I^s_i + T^s_i \right) \tag{9.3}
$$

with c the speed of light in vacuum. The true travel time inside the expression for the topocentric distance can, but only there, conform chapter 5, safely be replaced by the travel time in vacuum. The travel time in vacuum is

$$
\tau^s_{i,0}(t) = \rho^s_i(t,t-\tau^s_{i,0}(t)) \,/c \tag{9.4}
$$

The travel time in vacuum appears on both sides of the equation, so this is an implicit equation. The equation can be solved by a simple iterate scheme or by approximating it further using a Taylor expansion (two iterations, or a second order Taylor expansion, are sufficient).

Other quantities in the observation equations are the receiver clock error $\delta t_i(t)$, the satellite clock error $\delta t^s(t-\tau)$, which may be approximated by $\delta t^s(t)$ conform chapter 5, the multipath error on the code and phase measurement, $m_{i,f}^s$ and $\bar{m}_{i,f}^s$, instrumental delays on code and phase, $d_{i,f}^s$ and $\bar{d}_{i,f}^s$, and code and phase measurement noise, $e_{i,f}^s$ and $\bar{e}_{i,f}^s$. The ionospheric delay on L1 is I_i^s, and the ionospheric delay on L2 is $f_1^2/f_2^2 I_i^s = \gamma_f I_i^s$. Therefore, $\gamma_f=1$ for L1, and $\gamma_f = f_1^2/f_2^2$ for L2. The tropospheric delay, T_i^s, is the same for L1 and L2, and code and phase measurements. Finally, $A_{i,f}^s$, is the - constant - phase ambiguity term.

The observation equations can be simplified considerably by lumping the topocentric distance, satellite and receiver clock error, and tropospheric delay, into a single parameter $S_i^s(t)$, with

$$S_i^s(t) \doteq \rho_i^s(t,t-\tau_{i,0}^s) + c\,\delta t_i(t) - c\,\delta t^s(t-\tau_{i,0}^s) + T_i^s \tag{9.5}$$

$S_i^s(t)$ is frequency independent, and is the same for code and phase measurements. Furthermore, when the multipath and instrumental parameters are combined, with

$$\begin{aligned} b_{i,f}^s &\doteq m_{i,f}^s + d_{i,f}^s \\ \bar{b}_{i,f}^s &\doteq \bar{m}_{i,f}^s + \bar{d}_{i,f}^s \end{aligned} \tag{9.6}$$

the observation equations are simply

$$\begin{aligned} P_{i,f}^s(t) &= S_i^s(t) + \gamma_f I_i^s(t) + b_{i,f}^s(t) + e_{i,f}^s(t) \\ \phi_{i,f}^s(t) &= S_i^s(t) - \gamma_f I_i^s(t) + \lambda_f A_{i,f}^s + \bar{b}_{i,f}^s(t) + \bar{e}_{i,f}^s(t) \end{aligned} \tag{9.7}$$

with $S_i^s(t)$ an - ionosphere free - measure for the satellite-receiver distance, biased by satellite and receiver clock errors, and with I_i^s the ionospheric delay on L1, $b_{i,f}^s$ and $\bar{b}_{i,f}^s$ small - time varying - bias terms due to multipath and instrumental effects and random noise terms, and $A_{i,f}^s$ the - constant (in time) - phase ambiguity term.

9.3.3 Transformation of the Observation Equations

When there is no confusion about the particular satellite or receiver being used, as is the case with single channel analysis, the subscript *i* for the receiver and superscript *s* for the satellite can be dropped. Further, when measurements are done at discrete epochs t_k, k=0,1,....,N, the time argument t_k can be replaced by a lower index *k*. The observation equations in the four observation case are then

$$\begin{aligned} \phi_{P1,k} &= S_k - I_k + \lambda_1 A_{P1} + \bar{b}_{P1,k} + \bar{e}_{P1,k} \\ \phi_{P2,k} &= S_k - \gamma I_k + \lambda_2 A_{P2} + \bar{b}_{P2,k} + \bar{e}_{P2,k} \\ P_{P1,k} &= S_k + I_k + b_{P1,k} + e_{P1,k} \\ P_{P2,k} &= S_k + \gamma I_k + b_{P2,k} + e_{P2,k} \end{aligned} \tag{9.8}$$

with $\gamma=f_1^2/f_2^2 \simeq 1.6469$. In matrix notation the observation equations are

$$\underbrace{\begin{pmatrix}\Phi_{P1}\\ \Phi_{P2}\\ P_{P1}\\ P_{P2}\end{pmatrix}_k}_{y_k} = \begin{pmatrix}1 & -1 & 0 & 0 & | & 1 & 0\\ 1 & -\gamma & 0 & 0 & | & 0 & 1\\ 1 & 1 & 1 & 0 & | & 0 & 0\\ 1 & \gamma & 0 & 1 & | & 0 & 0\end{pmatrix} \begin{pmatrix}S\\ I\\ b_{P1}\\ b_{P2}\\ \bar{b}_{P1}\\ \bar{b}_{P2}\end{pmatrix}_k + \begin{pmatrix}\lambda_1 & 0\\ 0 & \lambda_2\\ 0 & 0\\ 0 & 0\end{pmatrix}\begin{pmatrix}A_{P1}\\ A_{P2}\end{pmatrix} + \begin{pmatrix}\bar{e}_{P1}\\ \bar{e}_{P2}\\ e_{P1}\\ e_{P2}\end{pmatrix}_k$$

$$\qquad \cdots\cdots A_4 \cdots\cdots \tag{9.9}$$

with y_k the vector of observations at time t_k and $\mathbf{A}_4$, which will be needed later, a 4 by 4 subset of a larger matrix as indicated. The time dependent parameters are sorted in order of decreasing variability. S_k has the largest time variations; even after subtracting an a-priori range S_k^0, computed from a known station location and satellite ephemerides, S_k-S_k^0, can be as large as several hundreds of kilometres. The largest contribution is from receiver clock errors, especially for receivers without accurate clocks. When the receiver clock error is taken care of, periodic variations of 100-200 meters with periods of 5-7 minutes, caused by the S/A clock dither effect, remain. The second largest parameter is the ionospheric delay I_k; typical values are in the range of 1-50 meters, with a significant diurnal and satellite elevation dependent variation. The code multipath, which constitute the main part of $b_{P1,k}$ and $b_{P2,k}$, are the next largest effect. The variability is in the range of several decimeters up to a few meters.

The multipath and instrumental effects on the phase observations, $\bar{b}_{P1,k}$ and $\bar{b}_{P2,k}$ are the smallest systematic effects. Typical values are centimetres and subcentimetres (except for some effects such as antenna rotation). Finally, the phase ambiguities are real constants, except for an occasional cycle slip.

Transformation of the model. New observations may be formed which are "look alikes" of the first four parameters. If a regular and invertible transformation matrix **T**, such that $\mathbf{T}=\mathbf{A}_4^{-1}$, is chosen, with

$$T = \begin{bmatrix} \frac{\gamma}{\gamma-1} & \frac{-1}{\gamma-1} & 0 & 0 \\ \frac{1}{\gamma-1} & \frac{-1}{\gamma-1} & 0 & 0 \\ \frac{-(\gamma+1)}{\gamma-1} & \frac{2}{\gamma-1} & 1 & 0 \\ \frac{-2\gamma}{\gamma-1} & \frac{\gamma+1}{\gamma-1} & 0 & 1 \end{bmatrix} = \begin{bmatrix} T_{11} & 0 \\ T_{21} & I \end{bmatrix} \quad T_{11} = \begin{bmatrix} \frac{\gamma}{\gamma-1} & \frac{-1}{\gamma-1} \\ \frac{1}{\gamma-1} & \frac{-1}{\gamma-1} \end{bmatrix} \quad T_{21} = \begin{bmatrix} \frac{-(\gamma+1)}{\gamma-1} & \frac{2}{\gamma-1} \\ \frac{-2\gamma}{\gamma-1} & \frac{\gamma+1}{\gamma-1} \end{bmatrix} \tag{9.10}$$

where $\mathbf{T}_{21}\mathbf{T}_{21}=\mathbf{I}$, i.e. $\mathbf{T}_{21}^{-1}=\mathbf{T}_{21}$, and $\gamma=f_1^2/f_2^2 \simeq 1.6469$. Multiplying the system of equations in (9.9) by $\mathbf{T}$ gives the derived observations $\mathbf{y'}_k=\mathbf{T}\ \mathbf{y}_k$, with $\mathbf{y'}_k$

$$\begin{pmatrix} \tilde{S} \\ \tilde{I} \\ \tilde{b}_{P1} \\ \tilde{b}_{P2} \end{pmatrix}_k = \begin{pmatrix} S \\ I \\ b_{P1} \\ b_{P2} \end{pmatrix}_k + \begin{pmatrix} T_{11} \\ T_{21} \end{pmatrix} \begin{pmatrix} \lambda_1 A_{P1} + \bar{b}_{P1,k} \\ \lambda_2 A_{P2} + \bar{b}_{P2,k} \end{pmatrix} + \begin{pmatrix} T_{11} & 0 \\ T_{21} & I \end{pmatrix} \begin{pmatrix} \bar{e}_{P1} \\ \bar{e}_{P2} \\ e_{P1} \\ e_{P2} \end{pmatrix}_k \tag{9.11}$$

From the equations it is clear that the derived observations $\tilde{S}_k$ and $\tilde{I}_k$ are linear combinations of phase measurements only, with corresponding small observation noise. These are the well known ionosphere free and geometry free linear combinations. See also chapter 6. The observation equations for the derived observations $\mathbf{y'}_k$ are

$$\begin{pmatrix} \tilde{S} \\ \tilde{I} \\ \tilde{b}_{P1} \\ \tilde{b}_{P2} \end{pmatrix}_k \doteq \begin{pmatrix} \frac{\gamma}{\gamma-1} & \frac{-1}{\gamma-1} & 0 & 0 \\ \frac{1}{\gamma-1} & \frac{-1}{\gamma-1} & 0 & 0 \\ \frac{-(\gamma+1)}{\gamma-1} & \frac{2}{\gamma-1} & 1 & 0 \\ \frac{-2\gamma}{\gamma-1} & \frac{\gamma+1}{\gamma-1} & 0 & 1 \end{pmatrix} \begin{pmatrix} \Phi_{P1} \\ \Phi_{P2} \\ P_{P1} \\ P_{P2} \end{pmatrix}_k \simeq \begin{pmatrix} 2.546 & -1.546 & 0 & 0 \\ 1.546 & -1.546 & 0 & 0 \\ -4.092 & 3.092 & 1 & 0 \\ -5.092 & 4.092 & 0 & 1 \end{pmatrix} \begin{pmatrix} \Phi_{P1} \\ \Phi_{P2} \\ P_{P1} \\ P_{P2} \end{pmatrix}_k \tag{9.12}$$

Thus the new observations are all biased by functions of the constant phase ambiguities and the small - but time varying - phase biases.

Assuming the original observations are not correlated, their covariance Q_y matrix reads:

$$Q_{y_k} = \mathrm{Diag}\left(\lambda_1^2\sigma_{\varphi_{P1}}^2 \quad \lambda_2^2\sigma_{\varphi_{P2}}^2 \quad \sigma_{P_{P1}}^2 \quad \sigma_{P_{P2}}^2\right) \tag{9.13}$$

The standard deviations of the observations are a function of the Signal to Noise ratio of the signal. Since the Signal to Noise ratio decreases with the elevation, the standard deviation of the observations can be modelled as a function of the elevation. In Jin [1996a, 1996b] an exponential function of the elevation is used for the standard deviation of the code measurements:

$$\sigma = \sigma_0 + \sigma_1 \exp(-e/e_0) \tag{9.14}$$

with e the elevation of the satellite (using the same units as e_0). The shape of the function depends on σ_0, σ_1 and e_0. The standard deviation at 90^0 elevation is approximately σ_0 (if $e_0 \ll 90^0$), the standard deviation at zero degrees elevation is $\sigma_0+\sigma_1$. Values for σ_0, σ_1 and e_0 depend on the type of observation and type of receiver being used. Values have been computed in Jin [1996b], using the code predicted residuals from the integrity monitoring discussed in section 9.4. An example is for a TurboRogue SNR-8000 is given in Figure 9.3; the solid curves are $0.065+0.5\cdot\exp(-e/15)$ for P1 and $0.070+0.6\cdot\exp(-e/16)$ for P2.
The corresponding covariance matrix $Q_{y'}$ of the transformed observations follows from applying the covariance law to (9.11) and reads:

$$Q_{y_k'} = T Q_{y_k} T^* \tag{9.15}$$

The new observations are now, contrary to the original observations, correlated, hence $Q_{y'}$ is a full matrix.

(a)

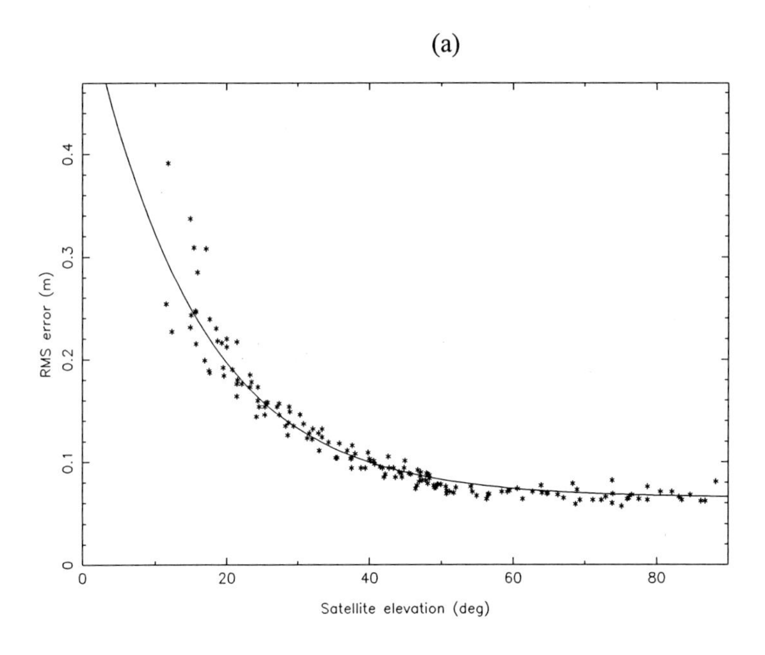

(b)

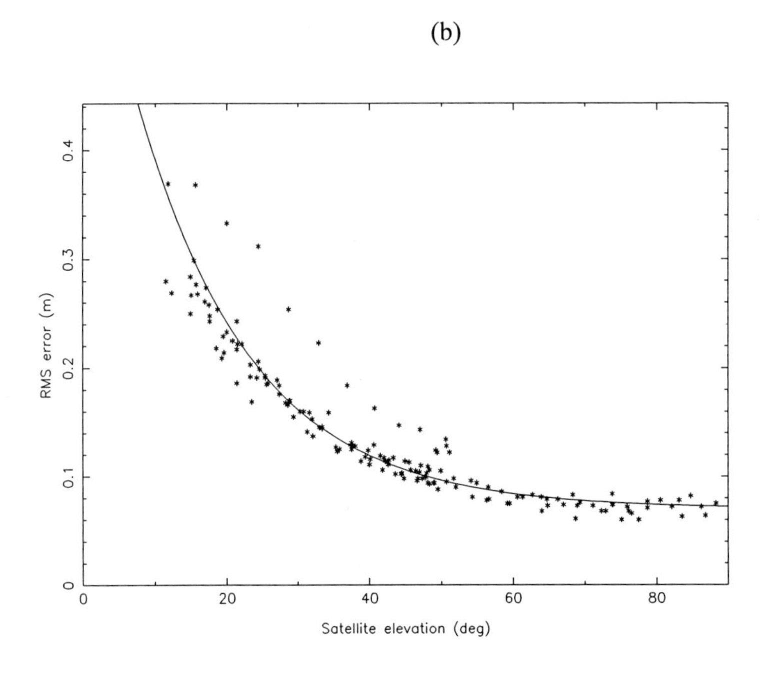

Figure 9.3. Standard deviation of TurboRogue SNR-8000 P1 code (a) and P2 code (b)

9.3.4 Elimination of the Phase Ambiguity

The new observations are all biased by the constant phase ambiguities. However, when is assumed that the average of the time varying bias parameters is zero, it is possible to estimate the phase ambiguities:

$$\begin{pmatrix} \lambda_1 \hat{A}_{P1} \\ \lambda_2 \hat{A}_{P2} \end{pmatrix} = \frac{1}{N+1} \sum_{k=0}^{N} T_{21} \begin{pmatrix} \tilde{b}_{P1} \\ \tilde{b}_{P2} \end{pmatrix}_k = \frac{1}{N+1} \sum_{k=0}^{N} \begin{pmatrix} I_2 & T_{21} \end{pmatrix} \begin{pmatrix} \phi_{P1} \\ \phi_{P2} \\ P_{P1} \\ P_{P2} \end{pmatrix}_k \tag{9.16}$$

with T_{21} a subset of the transformation matrix **T** and $\mathbf{I}_2$ the 2 by 2 identity matrix, such that

$$T_a \doteq \begin{pmatrix} I & T_{21} \end{pmatrix} = \frac{1}{\gamma - 1} \begin{pmatrix} \gamma-1 & 0 & -(\gamma+1) & 2 \\ 0 & \gamma-1 & -2\gamma & \gamma+1 \end{pmatrix} \tag{9.17}$$

is composed of parts of the transformation matrix **T**. The model of observation equations for the phase ambiguities is

$$\begin{pmatrix} \lambda_1 \hat{A}_{P1} \\ \lambda_2 \hat{A}_{P2} \end{pmatrix} = \begin{pmatrix} \lambda_1 A_{P1} \\ \lambda_2 A_{P2} \end{pmatrix} + \frac{1}{N+1} \sum_{k=0}^{N} T_a \begin{pmatrix} \bar{b}_{P1} \\ \bar{b}_{P2} \\ b_{P1} \\ b_{P2} \end{pmatrix}_k + \frac{1}{N+1} \sum_{k=0}^{N} T_a \begin{pmatrix} \bar{e}_{P1} \\ \bar{e}_{P2} \\ e_{P1} \\ e_{P2} \end{pmatrix}_k \tag{9.18}$$

The ambiguity estimates are biased by the mean multipath and instrumental delays. The covariance matrix of the ambiguity estimates (in units of meters) is

$$Q_a = \frac{1}{N+1} \sum_{k=0}^{N} \left(T_a \, Q_{y_k} \, T_a^* \right) \tag{9.19}$$

In other words, the longer the interval k=0,1,...,N is, the more precise the ambiguity estimates are.

In post processing the full uninterrupted satellite track can be used to estimate one set of ambiguities. This can be used to shift the transformed observations (9.11) by the same amount for all epochs, so that the bias due to unknown ambiguities is removed in a consistent manner, without changing the dynamical behaviour of the transformed observations. Multiplying the system of equations in (9.9) by **T**, and inserting the estimates for the phase ambiguities, gives the shifted observations $\mathbf{y}''_k$

$$\begin{pmatrix} \hat{S} \\ \hat{I} \\ \hat{b}_{P1} \\ \hat{b}_{P2} \end{pmatrix}_k \doteq \frac{1}{\gamma - 1} \begin{pmatrix} \gamma & -1 & 0 & 0 \\ 1 & -1 & 0 & 0 \\ -(\gamma+1) & 2 & \gamma - 1 & 0 \\ -2\gamma & \gamma+1 & 0 & \gamma - 1 \end{pmatrix} \begin{pmatrix} \phi_{P1_k} - \lambda_1 \hat{A}_{P1} \\ \phi_{P2_k} - \lambda_2 \hat{A}_{P2} \\ P_{P1_k} \\ P_{P2_k} \end{pmatrix} \tag{9.20}$$

The observation model for the derived observations $\mathbf{y}''_k$ is

$$\begin{aligned} \begin{pmatrix} \hat{S} \\ \hat{I} \\ \hat{b}_{P1} \\ \hat{b}_{P2} \end{pmatrix}_k = \begin{pmatrix} S \\ I \\ b_{P1} \\ b_{P2} \end{pmatrix}_k &+ \begin{pmatrix} \frac{\gamma}{\gamma-1} & \frac{-1}{\gamma-1} \\ \frac{1}{\gamma-1} & \frac{-1}{\gamma-1} \\ \frac{-(\gamma+1)}{\gamma-1} & \frac{2}{\gamma-1} \\ \frac{-2\gamma}{\gamma-1} & \frac{\gamma+1}{\gamma-1} \end{pmatrix} \begin{pmatrix} \bar{b}_{P1,k} - \frac{1}{N+1}\sum_{l=0}^{N} \bar{b}_{P1,l} \\ \bar{b}_{P2,k} - \frac{1}{N+1}\sum_{l=0}^{N} \bar{b}_{P2,l} \end{pmatrix} - \begin{pmatrix} \frac{-\gamma}{\gamma-1} & \frac{1}{\gamma-1} \\ \frac{1}{\gamma-1} & \frac{-1}{\gamma-1} \\ 1 & 0 \\ 0 & 1 \end{pmatrix} \begin{pmatrix} \frac{1}{N+1}\sum_{l=0}^{N} b_{P1,l} \\ \frac{1}{N+1}\sum_{l=0}^{N} b_{P2,l} \end{pmatrix} \\ &+ \begin{pmatrix} \frac{\gamma}{\gamma-1} & \frac{-1}{\gamma-1} & 0 & 0 \\ \frac{1}{\gamma-1} & \frac{-1}{\gamma-1} & 0 & 0 \\ \frac{-(\gamma+1)}{\gamma-1} & \frac{2}{\gamma-1} & 1 & 0 \\ \frac{-2\gamma}{\gamma-1} & \frac{\gamma+1}{\gamma-1} & 0 & 1 \end{pmatrix} \begin{pmatrix} \bar{e}_{P1} \\ \bar{e}_{P2} \\ e_{P1} \\ e_{P2} \end{pmatrix}_k - \begin{pmatrix} \frac{\gamma}{\gamma-1} & \frac{-1}{\gamma-1} & \frac{-\gamma}{\gamma-1} & \frac{1}{\gamma-1} \\ \frac{1}{\gamma-1} & \frac{-1}{\gamma-1} & \frac{1}{\gamma-1} & \frac{-1}{\gamma-1} \\ \frac{-(\gamma+1)}{\gamma-1} & \frac{2}{\gamma-1} & 1 & 0 \\ \frac{-2\gamma}{\gamma-1} & \frac{\gamma+1}{\gamma-1} & 0 & 1 \end{pmatrix} \begin{pmatrix} \frac{1}{N+1}\sum_{l=0}^{N} \bar{e}_{P1,l} \\ \frac{1}{N+1}\sum_{l=0}^{N} \bar{e}_{P2,l} \\ \frac{1}{N+1}\sum_{l=0}^{N} e_{P1,l} \\ \frac{1}{N+1}\sum_{l=0}^{N} e_{P2,l} \end{pmatrix} \end{aligned} \tag{9.21}$$

The new observations are biased by functions of the average phase and code biases (average multipath) and - time varying - phase biases. Although it is nice and easy to correct for the phase ambiguity term this way, there are complications. The first complication is that cycle slips should not occur in the interval over which the ambiguities are computed; either cycle slips must be fixed beforehand, or a new summation must be started. This implies that it is absolutely essential to have a means to detect cycle slips in the data. A procedure to detect cycle slips will be discussed in the next section. The second complication is that the standard deviation of the phase ambiguity estimate is proportional to the square root of the number of available epochs times the standard deviation of the code measurements. This effect has to be added to the standard deviation of the derived observations. Hence, the standard deviation of $\hat{S}_k$ and $\hat{I}_k$ is not any more a function of the standard deviation of the phase measurements only, but depends also on the - average - standard deviation of the code measurement and is proportional to the square root of the number of epochs. Fortunately, the relative precision between parameters of different type, and between two epochs, is not changed. The third complication is that the summation over all epochs can not be used in real-time applications.

In real time applications the ambiguity estimates can be computed recursively. The update for epoch k, based on equation (9.16), is:

$$\begin{pmatrix} \lambda_1 \hat{A}_{P1} \\ \lambda_2 \hat{A}_{P2} \end{pmatrix}_k = \frac{k}{k+1} \begin{pmatrix} \lambda_1 \hat{A}_{P1} \\ \lambda_2 \hat{A}_{P2} \end{pmatrix}_{k-1} + \frac{1}{k+1} \begin{pmatrix} \frac{-(\gamma+1)}{\gamma-1} & \frac{2}{\gamma-1} \\ \frac{-2\gamma}{\gamma-1} & \frac{\gamma+1}{\gamma-1} \end{pmatrix} \begin{pmatrix} \tilde{b}_{P1} \\ \tilde{b}_{P2} \end{pmatrix}_k \tag{9.22}$$

A similar equation may be derived for the original observations. Although the ambiguity itself is a constant, the estimates on different epochs are not the same. It is technically possible to use the estimate of the ambiguity parameters of epoch k to correct the observations of epoch k, and the estimate of epoch k+1 to correct the observations of epoch k+1, and so on, but this will change the dynamical properties of the resulting observations. The relative precision of $\hat{S}_k$ and $\hat{I}_k$ between two epochs is not any more a function of the standard deviation of the phase measurements only. Now, it also depends on the - average - standard deviation of the code measurement and is proportional to the square root of the number of epochs used in the computation of the phase ambiguity so far. In other words, the series will become smoother the longer a satellite is being tracked.

9.3.5 Computation of Differential Corrections

The differential correction for D-GPS applications is defined as

$$\Delta_k = S_k + I_k - S_k^0(r_i, r^s) \tag{9.23}$$

with S_k^0 the topocentric distance between the receiver and satellite, computed from a priori values. A simple way to compute differential corrections is to take the L1 code observations and subtract the topocentric distance. However, corrections computed this way suffer from all the problems associated to code observations, such as multipath and high observation noise. Using the definition of equation (9.23) it becomes as simple matter to compute the differential correction from the derived observations $\mathbf{y'}_k$ or $\mathbf{y''}_k$. This may result in more precise corrections. The observation model for the ambiguous differential correction, computed from $\mathbf{y'}_k$, is

$$
\begin{aligned}
\tilde{\Delta}_k &= \tilde{S}_k + \tilde{I}_k - S_k^0 = \begin{pmatrix} \frac{\gamma+1}{\gamma-1} & \frac{-2}{\gamma-1} \end{pmatrix} \begin{pmatrix} \Phi_{P1} \\ \Phi_{P2} \end{pmatrix} - S_k^0 = \\
& S_k - S_k^0 + I_k + \begin{pmatrix} \frac{\gamma+1}{\gamma-1} & \frac{-2}{\gamma-1} \end{pmatrix} \begin{pmatrix} \lambda_1 A_{P1} + \bar{b}_{P1,k} \\ \lambda_2 A_{P2} + \bar{b}_{P2,k} \end{pmatrix} + \begin{pmatrix} \frac{\gamma+1}{\gamma-1} & \frac{-2}{\gamma-1} \end{pmatrix} \begin{pmatrix} \bar{e}_{P1,k} \\ \bar{e}_{P2,k} \end{pmatrix}
\end{aligned}
\tag{9.24}
$$

Although the precision of this correction depends only on the precision of the phase observations, it is of little practical value because of the offset due to the -constant- phase ambiguities. Note that the differential correction cannot be computed by simply subtracting the topocentric range from the L1 phase measurement, because then the ionosphere term will have the wrong sign.

When the phase ambiguity estimate from the previous section is substituted, the observation model is

$$
\begin{aligned}
\hat{\Delta}_k &= \hat{S}_k + \hat{I}_k - S_k^0 = \begin{pmatrix} \frac{\gamma+1}{\gamma-1} & \frac{-2}{\gamma-1} \end{pmatrix} \begin{pmatrix} \Phi_{P1,k} \\ \Phi_{P2,k} \end{pmatrix} - \frac{1}{N+1} \sum_{l=0}^{N} \begin{pmatrix} \frac{\gamma+1}{\gamma-1} & \frac{-2}{\gamma-1} & -1 \end{pmatrix} \begin{pmatrix} \Phi_{P1,l} \\ \Phi_{P2,l} \\ P_{P1,l} \end{pmatrix} - S_k^0 = \\
& S_k - S_k^0 + I_k + \\
& \begin{pmatrix} \frac{\gamma+1}{\gamma-1} & \frac{-2}{\gamma-1} & -1 \end{pmatrix} \begin{pmatrix} \bar{b}_{P1,k} - \frac{1}{N+1} \sum_{l=0}^{N} \bar{b}_{P1,l} \\ \bar{b}_{P2,k} - \frac{1}{N+1} \sum_{l=0}^{N} \bar{b}_{P2,l} \\ - \frac{1}{N+1} \sum_{l=0}^{N} b_{P1,l} \end{pmatrix} + \begin{pmatrix} \frac{\gamma+1}{\gamma-1} & \frac{-2}{\gamma-1} & -1 \end{pmatrix} \begin{pmatrix} \bar{e}_{P1,k} - \frac{1}{N+1} \sum_{l=0}^{N} \bar{e}_{P1,l} \\ \bar{e}_{P2,k} - \frac{1}{N+1} \sum_{l=0}^{N} \bar{e}_{P2,l} \\ - \frac{1}{N+1} \sum_{l=0}^{N} e_{P1,l} \end{pmatrix}
\end{aligned}
\tag{9.25}
$$

Now the precision of the differential correction also depends on the quality of the average L1 code observation. The code observation on L2 does not play a role in the computation of the differential correction.

Unfortunately, the computation of the differential correction has the same complications as the transformed observations $\mathbf{y''}_k$. The first complication is that there may not be any cycle slips in the interval over which the ambiguities are computed. Cycle slips must be detected and corrected, or a new summation must be started. The second complication is that the standard deviation of the phase ambiguity estimate is proportional to the square root of the number of available epochs times the standard deviation of the code measurements. This effect has to be added to the standard deviation of the differential correction. Hence, the standard deviation of the differential corrections is not any more a function of the standard deviation of the phase measurements only, but depends also on the - average - standard deviation of the code measurement and is proportional to the square root of the number of epochs. Fortunately, the relative precision between two epochs is not changed.

The third complication is that for real-time applications it is impossible to sum over the future epochs of a satellite track. A useful equation may be derived from the original equation (9.16) if the sum over N+1 is replaced by k. Although the ambiguity itself is a constant, the estimates on different epochs are not the same. The relative precision of the differential correction between two epochs is not any more a function of the standard deviation of the phase measurements only. Now, it also depends on the - average - standard deviation of the code measurement and is proportional to the square root of the number of epochs used in the computation of the phase ambiguity so far. In other words, the series will become smoother the longer a satellite is being tracked without interruption or cycle slips.

A final complication is that not only the differential corrections is needed, but also the differential correction rate. This problem is solved together with the cycle slip problem in the next section.

9.3.6 Missing Observations and other Tracking Modes

Four observation code correlation tracking. The four observation model of equation (9.9) can still be used when the L1-code measurement, L2-code measurement, or both, are not available. For instance, if the L2-code measurement is missing one may skip the fourth row, with the derived observation $\tilde{b}_{P2}$, and the parameter b_{P2} from the equations. This does not affect the other derived observations. This is explained as follows: the inverse of the three by three submatrix consisting of the first three rows and columns of A_4, is equal to the three by three submatrix consisting of the first three rows and columns of $\mathbf{T}$. Therefore, removing the fourth row (observation) and fourth column (parameter), does not affect the other observables. The same is true for the L1-code measurement.

When some of the L1-code and/or L2-code measurement are missing, but not all, estimates for the phase ambiguities can be computed, as before, restricting the summation over the epochs with *both* L1-code and L2-code. However, if all L1-code, or L2-code, measurements are missing, which is the case in a three observation model, it is not possible to compute the ambiguities as before. In the presence of missing L1-code and L2-code observations it will be better to estimate a linear combination of the ambiguities,

$$\begin{pmatrix} \frac{-(\gamma+1)\lambda_1}{\gamma-1}\hat{A}_{P1} + \frac{2\lambda_2}{\gamma-1}\hat{A}_{P2} \\ \frac{-2\gamma\lambda_1}{\gamma-1}\hat{A}_{P1} + \frac{(\gamma+1)\lambda_2}{\gamma-1}\hat{A}_{P2} \end{pmatrix} = T_{21} \begin{pmatrix} \lambda_1 \hat{A}_{P1} \\ \lambda_2 \hat{A}_{P2} \end{pmatrix} = \frac{1}{N+1} \sum_{k=0}^{N} \left(T_{21} \quad I_2 \right) \begin{pmatrix} \phi_{P1} \\ \phi_{P2} \\ P_{P1} \\ P_{P2} \end{pmatrix}_k \tag{9.26}$$

with $\mathbf{T}_{21}$ a subset of the transformation matrix $\mathbf{T}$ and $\mathbf{I}_2$ the 2 by 2 identity matrix. Usually the two linear combinations will be computed separately, because the actual summation may be over different epochs because of missing code observations. Also it is possible to compute only one linear combination if all code observations of a particular frequency are missing. Multiplying the linear combination by $\mathbf{T}_{21}$ returns the original ambiguities (because $\mathbf{T}_{21}^{-1}=\mathbf{T}_{21}$). The first linear combination is identical to the linear combination of ambiguities used in the differential correction, hence, the differential correction can be computed without L2-code measurements.

So, when the L1-code and/or L2-code measurement are missing this does not change the computation much. The only difficulty is the computation of the ambiguities, but this can be solved partly by using a linear combination of ambiguities. However, if one of the phase measurements is missing, none of the derived observations can be computed using equation (9.9). If one of the phase measurements is missing a new model has to be formulated: the single frequency model.

Four observations cross correlation tracking. In this case the following observations are available:

- $\varphi_{C/A}$ - Carrier phase observation on L1, obtained from C/A-code tracking, in cycles
- $P_{C/A}$ - Code observation on L1, obtained from C/A-code tracking
- $\varphi_{P2\otimes P1}$ - Carrier phase observation on L2⊗L1, obtained from tracking the cross-correlated signal, in cycles of $c/(f_1\text{-}f_2)$ meters
- $P_{P2\otimes P1}$ - Direct measurement of differential group delay between L2 and L1

Using these four observations, the following L2 measurements are created:

$$\begin{aligned} \varphi_{L2\otimes} &= \varphi_{C/A} + \varphi_{L2\otimes L1} \\ P_{L2\otimes} &= P_{C/A} + P_{L2\otimes L1} \end{aligned} \tag{9.27}$$

with phase measurements in cycles. The phase measurement on L2, expressed in units of meters, is $\phi_{L2} = \lambda_2\, \varphi_{L2}$. Assuming the original observations are not correlated with covariance matrix

$$Q_y = \mathrm{Diag}\left(\sigma^2_{\varphi_{C/A}} \quad \sigma^2_{\varphi_{L1\otimes L2}} \quad \sigma^2_{P_{C/A}} \quad \sigma^2_{P_{L1\otimes L2}}\right) \tag{9.28}$$

the covariance matrix Q_y of the observations $y_k = (\, \phi_{C/A}\, \phi_{L2}\, P_{C/A}\; P_{L2}\,)^*$, with the phase in units of meters, is given by:

$$Q_y = \begin{pmatrix} \lambda_1^2\sigma^2_{\varphi_{C/A}} & \lambda_1\lambda_2\sigma^2_{\varphi_{C/A}} & & \\ \lambda_1\lambda_2\sigma^2_{\varphi_{C/A}} & \lambda_2^2\sigma^2_{\varphi_{C/A}} + \lambda_2^2\sigma^2_{\varphi_{L2\otimes L1}} & & \\ & & \sigma^2_{P_{C/A}} & \sigma^2_{P_{C/A}} \\ & & \sigma^2_{P_{C/A}} & \sigma^2_{P_{C/A}} + \sigma^2_{P_{L2\otimes l1}} \end{pmatrix} \tag{9.29}$$

When this matrix is compared to the covariance matrix of the four observations code correlation mode, we see that:

- the standard deviation for the L1 measurements is larger, because C/A code measurements are used instead of P code measurements
- the standard deviation for the L2 derived measurements is larger, because of
  - a) the summation of a L1-C/A and cross correlated measurement, instead of a direct L2 measurement
  - b) the larger standard deviation of the cross correlation measurement compared to a direct L2 measurement
- the L1 and L2 derived measurements are now correlated, because both are based on the same C/A code measurements.

Now, the transformation of equation (9.11) can be used to derive new observations S_k, I_k, $b_{C/A,k}$ and $b_{l2,k}$ as in the previous section. The corresponding covariance matrix $Q_{y'}$ of the transformed observations follows from (9.15), using Q_y from (9.29) instead of the diagonal matrix from (9.13). It may seem that the consequences of cross-correlation tracking are only minor, but this is not true. The correlation between the L1 and L2 phase measurement has some serious consequences for the cycle slip detection.

Six observation code correlation tracking. The observation equations in case of the six observation model are

$$\underbrace{\begin{pmatrix} \phi_{P1} \\ \phi_{P2} \\ P_{P1} \\ P_{P2} \\ P_{C/A} \\ \phi_{C/A} \end{pmatrix}_k}_{y_k} = \underbrace{\left(\begin{array}{cccccc|cc} 1 & -1 & 0 & 0 & 0 & 0 & 1 & 0 \\ 1 & -\gamma & 0 & 0 & 0 & 0 & 0 & 1 \\ 1 & 1 & 1 & 0 & 0 & 0 & 0 & 0 \\ 1 & \gamma & 0 & 1 & 0 & 0 & 0 & 0 \\ 1 & 1 & 0 & 0 & 1 & 0 & 0 & 0 \\ 1 & -1 & 0 & 0 & 0 & 1 & 0 & 0 \end{array}\right)}_{\cdots\cdots\cdots A_6 \cdots\cdots\cdots} \begin{pmatrix} S \\ I \\ b_{P1} \\ b_{P2} \\ b_{C/A} \\ \bar{b}_{C/A} \\ \bar{b}_{P1} \\ \bar{b}_{P2} \end{pmatrix}_k + \begin{pmatrix} \lambda_1 & 0 & 0 \\ 0 & \lambda_2 & 0 \\ 0 & 0 & 0 \\ 0 & 0 & 0 \\ 0 & 0 & 0 \\ 0 & 0 & \lambda_1 \end{pmatrix} \begin{pmatrix} A_{P1} \\ A_{P2} \\ A_{C/A} \end{pmatrix} + \begin{pmatrix} \bar{e}_{P1} \\ \bar{e}_{P2} \\ e_{P1} \\ e_{P2} \\ e_{C/A} \\ \bar{e}_{C/A} \end{pmatrix}_k \tag{9.30}$$

The time dependent parameters are again sorted in order of decreasing variabi-l-ity: first the term containing the geometry and clocks, S_k, or similarly S_k-$S_k^{\ 0}$, then the ionospheric delay I_k, followed by code multipath, with constitute the main part of $b_{P1,k}$, $b_{P2,k}$ and $b_{C/A,k}$, and finally multipath and instrumental errors on the phase observations, $\bar{b}_{P1,k}$, $\bar{b}_{P2,k}$ and $\bar{b}_{C/A,k}$, The phase ambiguities are real constants, except for an occasional cycle slip.

If a regular and invertible transformation matrix $\mathbf{T}$ is chosen, such that $\mathbf{T}=\mathbf{A}_6^{-1}$, with

$$\mathbf{T} = \frac{1}{\gamma-1}\begin{bmatrix} \gamma & -1 & 0 & 0 & 0 & 0 \\ 1 & -1 & 0 & 0 & 0 & 0 \\ -(\gamma+1) & 2 & \gamma-1 & 0 & 0 & 0 \\ -2\gamma & \gamma+1 & 0 & \gamma-1 & 0 & 0 \\ -(\gamma+1) & 2 & 0 & 0 & \gamma-1 & 0 \\ -(\gamma-1) & 0 & 0 & 0 & 0 & \gamma-1 \end{bmatrix} = \begin{bmatrix} T_{11} & 0 & 0 \\ T_{21} & I & 0 \\ T_{31} & 0 & I \end{bmatrix} \tag{9.31}$$

and $\mathbf{T}_{21}$ the same as in equation (9.10), new observations may be formed. Multi-plying the system of equations in (9.30) by $\mathbf{T}$ gives the derived observations $\mathbf{y'}_k=\mathbf{T}\ \mathbf{y}_k$, with $\mathbf{y'}_k$

$$\begin{pmatrix} \tilde{S} \\ \tilde{I} \\ \tilde{b}_{P1} \\ \tilde{b}_{P2} \\ \tilde{b}_{C/A} \\ \tilde{\Delta}_{L1} \end{pmatrix}_k \doteq \frac{1}{\gamma-1} \begin{pmatrix} \gamma & -1 & 0 & 0 & 0 & 0 \\ 1 & -1 & 0 & 0 & 0 & 0 \\ -(\gamma+1) & 2 & \gamma-1 & 0 & 0 & 0 \\ -2\gamma & \gamma+1 & 0 & \gamma-1 & 0 & 0 \\ -(\gamma+1) & 2 & 0 & 0 & \gamma-1 & 0 \\ -(\gamma-1) & 0 & 0 & 0 & 0 & \gamma-1 \end{pmatrix} \begin{pmatrix} \phi_{P1} \\ \phi_{P2} \\ P_{P1} \\ P_{P2} \\ P_{C/A} \\ \phi_{C/A} \end{pmatrix}_k \tag{9.32}$$

From the equations it is clear that the derived observations $\tilde{S}_k$ and $\tilde{I}_k$ are linear combinations of phase measurements only, with corresponding small observation noise. These are the well known ionosphere free and geometry free linear combinations. The observation equations for the derived observations $\mathbf{y}'_k$ are

$$\begin{pmatrix} \tilde{S} \\ \tilde{I} \\ \tilde{b}_{P1} \\ \tilde{b}_{P2} \\ \tilde{b}_{C/A} \\ \tilde{\Delta}_{L1} \end{pmatrix}_k = \begin{pmatrix} S \\ I \\ b_{P1} \\ b_{P2} \\ b_{C/A} \\ \bar{b}_{C/A}-\bar{b}_{P1} \end{pmatrix}_k + \frac{1}{\gamma-1} \begin{pmatrix} \gamma & -1 & 0 \\ 1 & -1 & 0 \\ -(\gamma+1) & 2 & 0 \\ -2\gamma & \gamma+1 & 0 \\ -(\gamma+1) & 2 & 0 \\ 0 & 0 & \gamma-1 \end{pmatrix} \begin{pmatrix} \lambda_1 A_{P1} + \bar{b}_{P1,k} \\ \lambda_2 A_{P2} + \bar{b}_{P2,k} \\ \lambda_1 A_{C/A} - \lambda_1 A_{P1} \end{pmatrix} + T \begin{pmatrix} \bar{e}_{P1} \\ \bar{e}_{P2} \\ e_{P1} \\ e_{P2} \\ e_{C/A} \\ \bar{e}_{C/A} \end{pmatrix}_k \tag{9.33}$$

The new observations are all biased by functions of the constant phase ambiguities and the small - but time varying - phase biases.

As can be seen from the four and six observation models the new observations are linear combinations of P1 and P2 phase measurements, sometimes in combination with one code measurement. Therefore, it will be difficult to adapt these models if either P1 or P2 phase measurements are missing: this will lead directly to a single frequency model. If one of the other observations is missing it is relatively easy to adapt the models.

If, in case of the six observation model, the phase measurement on the C/A-signal is missing, the models (9.30) and (9.33) can be adapted by removing the sixth row, and parameters $\bar{b}_{C/A}$, $A_{C/A}$ and $A_{C/A}$-A_{P1}. The matrix must be adjusted accordingly. If the C/A phase is missing for all epochs then the corresponding rows and parameters can be removed permanently: this gives the five observation model.

When there is no C/A code measurement in the six observation model, the model can be adapted simply by deleting the fifth row and parameters corresponding with it. When both C/A code and phase are missing the six observation model reduces to the four observation model discussed earlier.

Single frequency model. The observation equations in case of the single frequency model are

$$\begin{pmatrix} \Phi_{P1} \\ P_{P1} \end{pmatrix}_k = \left(\begin{array}{cc|cc} 1 & -1 & 0 & 1 \\ 1 & 1 & 1 & 0 \end{array}\right) \begin{pmatrix} S \\ I \\ b_{P1} \\ \bar{b}_{P1} \end{pmatrix}_k + \begin{pmatrix} \lambda_1 \\ 0 \end{pmatrix} A_{P1} + \begin{pmatrix} \bar{e}_{P1} \\ e_{P1} \end{pmatrix}_k \tag{9.34}$$

$$y_k \qquad \cdots A_2 \cdots$$

with y_k the vector of observations at time t_k and $\mathbf{A}_2$, which will be needed later, a 2 by 2 subset of a larger matrix as indicated. The time dependent parameters are again sorted in order of decreasing variability.

New observations may be formed, with the help of a regular and invertible transformation matrix $\mathbf{T}$, such that $\mathbf{T}=\mathbf{A}_2^{-1}$, with $\mathbf{y'}_k=\mathbf{T}\ y_k$:

$$\begin{pmatrix} \tilde{S} \\ \tilde{I} \end{pmatrix}_k \doteq \begin{pmatrix} \frac{1}{2} & \frac{1}{2} \\ -\frac{1}{2} & \frac{1}{2} \end{pmatrix} \begin{pmatrix} \Phi_{P1} \\ P_{P1} \end{pmatrix}_k \qquad \text{with} \qquad T = \begin{pmatrix} \frac{1}{2} & \frac{1}{2} \\ -\frac{1}{2} & \frac{1}{2} \end{pmatrix} \tag{9.35}$$

The observation equations are

$$\begin{pmatrix} \tilde{S} \\ \tilde{I} \end{pmatrix}_k = \begin{pmatrix} S \\ I \end{pmatrix}_k + \begin{pmatrix} \frac{1}{2} \\ -\frac{1}{2} \end{pmatrix} \lambda_1 A_{P1} + \begin{pmatrix} \frac{1}{2} & \frac{1}{2} \\ -\frac{1}{2} & \frac{1}{2} \end{pmatrix} \begin{pmatrix} \bar{b}_{P1} \\ b_{P1} \end{pmatrix}_k + \begin{pmatrix} \frac{1}{2} & \frac{1}{2} \\ -\frac{1}{2} & \frac{1}{2} \end{pmatrix} \begin{pmatrix} \bar{e}_{P1} \\ e_{P1} \end{pmatrix}_k \tag{9.36}$$

The new observations are biased by the constant phase ambiguity and the - time varying - phase and code biases. Compared to the dual frequency models, the derived observations $\tilde{S}_k$ and $\tilde{I}_k$ are not any more linear combinations of phase measurements only. This has consequences for the observation noise. Assuming the original observations are not correlated, their covariance Q_y matrix reads:

$$Q_{y_k} = \mathrm{Diag}\left(\lambda_1^2 \sigma_{\varphi_{P1}}^2 \quad \sigma_{P_{P1}}^2\right) \tag{9.37}$$

The covariance matrix for the transformed observations is then

$$Q_{y_k'} = \begin{pmatrix} \frac{1}{4}(\lambda_1^2\sigma^2_{\varphi_{P1}} + \sigma^2_{P_{P1}}) & \frac{1}{4}(\sigma^2_{P_{P1}} - \lambda_1^2\sigma^2_{\varphi_{P1}}) \\ \frac{1}{4}(\sigma^2_{P_{P1}} - \lambda_1^2\sigma^2_{\varphi_{P1}}) & \frac{1}{4}(\lambda_1^2\sigma^2_{\varphi_{P1}} + \sigma^2_{P_{P1}}) \end{pmatrix} \tag{9.38}$$

The new observations are now, contrary to the original observations, correlated, hence $Q_{y'}$ is a full matrix, with the largest contribution coming from the standard deviation of the code measurement.

The new observations are all biased by the constant phase ambiguity. Without the help of an ionosphere model it will not be possible to estimate them. Therefore, this model is of little practical use.

9.3.7 Applications

In the previous sections it is was shown that it is possible to transform the measurements from various tracking modes into - physically more meaningful - derived observations. Especially in dual frequency modes, the (relative) precision of the combined distance/clock term S_k and ionosphere term I_k is of the same order of magnitude as the precision of the phase measurements. There are many applications for this transformation. Some of them can be used at an active GPS control station. Others will have a broader application, and can be used on (between receiver) single differences as well.

Applications running at active GPS control stations

Integrity monitoring and cycle slip detection: For all applications of the transformation discussed in this section it is essential that cycle slips are detected and their effect is removed as much as possible. This can be done by introducing appropriate dynamical models for the transformed observations, and using the DIA procedure of chapter 7 to test for cycle slips and outliers in the data. Without this procedure most other applications given in this section become obsolete. This procedure will be discussed in detail in the next section.

D-GPS corrections: Differential GPS corrections with good internal precision can be computed directly from the transformed observations. However, for practical applications the computation must be combined with the integrity monitoring algorithm presented in the next section. The differential correction rate (and optionally the rate-rate) follow also directly from the integrity monitoring. A separate filter is not needed. More information on the D-GPS correction computation will be given in the next section.

Data thinning, decimation and adjustment: GPS applications may require different data rates (sample interval). For kinematic applications one second data may be needed, while for geodynamics monitoring thirty second data is the standard. Factors which also may limit the data rate are the bandwidth of the communication channel, telephone rates, and so on. For example in the Dutch AGRS-NL the sample rate of the receiver can be set to one second, but only 30 second data (or 10 second data) will be transmitted routinely to a computing centre. The one second data is transmitted only on a need to have basis.

One way to reduce the data rate is simple decimation: pick every N'th observation. However, the quality of the code information can be improved significantly using a data adjustment procedure, similar to the data thinning procedure of chapter 6:

1. use (9.11) to transform the observations over the decimation interval
2. compute the mean of b_{P1} and b_{P2} over the decimation interval
3. take the derived observations of a 'middle' epoch, replace $b_{P1,k}$ and $b_{P2,k}$ by the mean computed in step 2, and apply the inverse transformation

Of course, we have to check for cycle slips during the decimation period, hence, the integrity monitoring. There is no real limit to the decimation period. Care must be taken not to use the same measurement twice, otherwise time correlations are introduced, i.e. the decimation periods may not overlap.

The phase measurements are not changed by this approach. However, we may use a similar approach by smoothing the ionosphere term (using a mean, polynomial, or a Kalman smoother). Adding this to the second step will improve the phase measurements as well, especially the phase measurement on L2 in the presence of cross correlation. We can even smooth S_k in the same way, provided that the smoothing procedure is implemented such that clock errors and clock steering affect the smoothed observations to different satellites by the same amount. The duration over which we can apply this phase smoothing successfully is limited. According to Allen Osborne Associates (1995), The TurboRogue receivers use internally a similar approach for the code measurements. However, in the TurboRogue the phase measurements are adjusted differently, by using a polynomial fit of the L1 and L2 phase measurements over a maximum of 10 seconds.

Data compression: The amount of data produced by receivers operating with a sample rate of one second can be very large, especially when the data is stored as RINEX files (2-3 Mb/hour). Standard compression techniques, using the Lempel-Ziv or other algorithm, may reduce the size by a factor three. However, a much better compression is achieved when the data is stored as binary data, and not ascii, in the first place, using a minimum number of bits. In this case the transformed observations need less bits than the original observations. Additional savings can be obtained when only increments with respect to a reference epoch are stored.

The above described compression will yield much better results than a zipped

RINEX file. Further savings can be obtained when some loss of information is allowed. For example, b_{P1} and b_{P2} will not be stored every epoch, only their average over a period of time. This kind of compression only works for a stationary receiver: it is not allowed for a moving receiver.

Related (off-line) applications

Post processing analysis: Analysis in post processing mode has the advantage that the phase ambiguities can be removed without affecting the dynamical behaviour of the transformed observations. Straightforward plots of the bias and ionosphere observables will yield valuable information on the accuracy of the code measurements and the activity of the ionosphere. Before plotting the ambiguity parameters must be solved and cycle slips must be removed. Because of the high relative precision of $\tilde{I}_k$ the first and second divided difference may give important information of scintillations and short periodic variations in the ionosphere, and the precision of the phase measurements. Straightforward plots of S_k and its first divided difference are only useful once S_k^0 is subtracted. The second divided difference of S_k may however show short periodic clock effects, disclosing for some receivers, the clock steering they use.

For active GPS control stations these techniques may be used to study the code multipath at a reference station, and to select the best spot for the antenna. Later it can be used to monitor (regularly) the quality of the code data, compute elevation dependent models for the code noise, select and compare receivers.

Phase smoothed pseudo-ranges: The derived observations S_k (with ambiguities eliminated) can be considered as a special kind of phase smoothed pseudo-ranges for relative positioning applications with code observations. The phase smoothed pseudo-ranges can be used instead of code observations. Because of the S/A effect it makes no sence to use S_k for single point positioning. It is only after the S/A effect has been removed, by taking the single difference between two receivers, that positioning applications benefit from the smooth and ionosphere free nature of S_k. Single differences may be formed before or after the transformation.

Preprocessing: The transformation and integrity monitoring procedure may be used for preprocessing and cycle slip correction of (between receiver) single differences. A high sample rate is necessary. Usually this preprocessing step is followed by a final cycle slip correction procedure, e.g., based on triple differen-ces.

Ionospheric modelling: With the help of a mapping function a single layer model of the ionosphere may be estimated. If the period is long enough also the differential instrumental delays may be estimated, or in case of phase data, the L1-L2 phase ambiguity (see chapter 15).

9.4 REAL TIME INTEGRITY MONITORING

The transformation in the previous section was a regular and invertible transformation. No information is added, none is removed, and there is no redundancy. The transformation produced a series of S_k, I_k, $b_{P1,k}$, $b_{P2,k}$ for k=0,1..., each with distinct properties which can be described by a dynamical model. This model forms the basis of the integrity monitoring, using the procedures for the validation of recursive solutions, outlined in chapter 7.

9.4.1 Recursive (Kalman) Filtering

The real-time quality control of GPS code and carrier observations is implemented using discrete linear(ized) dynamic systems. The dynamics of $b_{P1,k}$, $b_{P2,k}$, I_k and sometimes S_k, will be modelled as a (low order) polynomial. The dynamical model for each reads:

$$x_{p;k+1} = \Phi_{p;k+1,k} x_{p;k} + d_{p;k} \tag{9.39}$$

where $\mathbf{x}_{p;k}$ is the m+1 dimensional state vector for S_k, I_k, or any of the other parameters, at time t_k, with

$$\mathrm{x}_p = \left(x_p \;\; x_p^{(1)} \;\; . \; . \; . \;\; x_p^{(m)} \right)^* \tag{9.40}$$

and $x_p^{(m)}$ the m'th derivative of x_p, and x_p one of the parameters of the previous section (S_k, I_k, etc.). $\Phi_{p;k+1,k}$ is the known m+1-by-m+1 transition matrix

$$\Phi_{p;k+1,k} = \begin{pmatrix} 1 & \Delta t & . & . & . & \frac{\Delta t^m}{m!} \\ & 1 & \Delta t & . & . & \frac{\Delta t^{m-1}}{(m-1)!} \\ & & . & & & \\ & & & . & & \\ & & & & . & \\ & & & & & 1 \end{pmatrix} \tag{9.41}$$

with $\Delta t = t_{k+1} - t_k$, and $d_{p;k}$ is the process noise, assumed to be Gaussian distributed with mean zero and known covariance matrix

$$E\{d_{p;k}d^{*}_{p;l}\} = Q_{p;k}\delta_{kl} \tag{9.42}$$

Assuming the input of this system consists of white noise with spectral density $q^{(m+1)}$, which has dimension $m^2/s^{(2m+1)}$, the process noise matrix $Q_{p;k}$ is given by:

$$Q_{p;k} = q^{(m+1)} \int_{t_{k-1}}^{t_k} r(\tau)r(\tau)^{*}d\tau \tag{9.43}$$

where:

$$r(\tau) = \begin{pmatrix} \frac{(t_k-\tau)^m}{m!} & . & . & . & t_k-\tau & 1 \end{pmatrix}^{*} \tag{9.44}$$

From the above follows that:

$$Q_{p;k}(i,j) = q^{(m+1)} \frac{(t_k-t_{k-1})^{2m-i-j+1}}{(m-i)!(m-j)!(2m-i-j+1)} \qquad i,j = 0,...,m \tag{9.45}$$

The individual state vectors can be combined into a single n-dimensional state vector x, with

$$x = \begin{pmatrix} x_S & x_I & . & . & . & x_{b_{P2}} \end{pmatrix}^{*} \tag{9.46}$$

The dynamics of the system are then modelled by the equation:

$$x_{k+1} = \Phi_{k+1,k}x_k + d_k \tag{9.47}$$

with $\Phi_{k+1,k}$ the known n-by-n transition matrix

$$\Phi_{k,k-1} = \begin{pmatrix} \Phi_{S;k,k-1} & & & \\ & \Phi_{I;k,k-1} & & \\ & & \ddots & \\ & & & \Phi_{b_{P2};k,k-1} \end{pmatrix} \tag{9.48}$$

and d_k is the process noise, assumed to be Gaussian distributed with mean zero and known covariance matrix:

$$E\{d_k d_l^*\} = Q_k \delta_{kl} \quad \text{with} \quad d = \begin{pmatrix} d_S & d_I & . & . & . & d_{b_{P2}} \end{pmatrix}^* \tag{9.49}$$

The process noise matrix is

$$Q_k = \begin{pmatrix} Q_{S;k} & & & \\ & Q_{I;k} & & \\ & & \ddots & \\ & & & Q_{b_{P2};k} \end{pmatrix} \tag{9.50}$$

The initial state is also Gaussian distributed with known mean $x_{0|0}$ and known variance matrix $Q_{x_{0|0}}$, independent of d_k.

The observables of the system are modelled by the equation:

$$y_k = A_k x_k + e_k \tag{9.51}$$

where A_k is a known m_k-by-n design matrix and the measurement noise e_k, independent of d_l and x_0, is Gaussian distributed with mean zero and known covariance matrix:

$$E\{e_k e_l^*\} = Q_{y_k} \delta_{kl} \tag{9.52}$$

The observables can be the original phase and code observations y_k, with A_k the design matrix from the previous section with additional columns composed of zeroes, corresponding to the derivatives in the state vector. The observations can also be the transformed observations y'_k. The design matrix A_k is then composed of zeroes and ones. In this case the covariance matrix Q_y is a full matrix. The constant bias in the observations is of no concern for the filter, since the dynamical model is insensitive for a constant offset.

Based on the above model, the optimal recursive prediction and filtering equations for the state estimate, under the null hypothesis H_0, read Teunissen [1990] and Salzmann [1993]:

$$\begin{aligned} x_{k|k-1} &= \Phi_{k.k-1} x_{k-1|k-1} \\ x_{k|k} &= x_{k|k-1} + K_k (y_k - A_k x_{k|k-1}) \end{aligned} \tag{9.53}$$

with corresponding covariance matrices:

$$Q_{x_{k|k-1}} = \Phi_{k,k-1} Q_{x_{k-1|k-1}} \Phi^{*}_{k,k-1} + Q_k$$
$$Q_{x_{k|k}} = (I - K_k A_k) Q_{x_{k|k-1}} \tag{9.54}$$

where:

$$K_k = Q_{x_{k|k-1}} A_k^{*} (Q_{y_k} + A_k Q_{x_{k|k-1}} A_k^{*})^{-1} \tag{9.55}$$

is the so-called Kalman gain matrix.

Initialization of the filter. Initialization of the recursive filter, i.e. computation of the state $x_{0|0}$ and its corresponding covariance matrix $Qx_{0|0}$ is accomplished through a batch least squares solution. The number of epochs to be used for this initialization depends on the order of the polynomial used to model the dynamics, and the available observations. With the model described above, at least m+1, epochs are required for the least squares solution.

The parameters in the observation equations must refer to a common reference epoch t_0. The observation equations are in this case

$$y_k = A_k x_k + e_k = A_k \Phi_{k,0} x_0 + e_k \tag{9.56}$$

The least squares solution for $x_{0|0}$ is given by:

$$x_{0|0} = \left(\sum_{k=0}^{m} (\Phi^{*}_{k,0} A_k^{*} R^{-1} A_k \Phi_{k,0}) \right)^{-1} \left(\sum_{k=0}^{m} (\Phi^{*}_{k,0} A_k^{*} R^{-1} y_k) \right) \tag{9.57}$$

and $Qx_{0|0}$ by:

$$Q_{x_{0|0}} = \left(\sum_{k=0}^{m} (\Phi^{*}_{k,0} A_k^{*} R^{-1} A_k \Phi_{k,0}) \right)^{-1} \tag{9.58}$$

Note that $x_{0|0}$ and $Qx_{0|0}$ refer to t_0. Therefore, for the epoch immediately following the initialization, the predicted state and predicted covariance matrix have to be transferred over a longer interval than is usually the case.

9.4.2 Model Validation

The above filter produces optimal estimates of the state vector with well defined statistical properties. The state estimates are unbiased, are Gaussian distributed and have minimum variance within the class of linear unbiased estimators. It is important to realize, however, that optimality is only guaranteed as long as the assumptions underlying the mathematical model hold. Misspecifications in the model will invalidate the results of estimation and thus also any conclusion based on them. It is therefore of importance to have ways to verify the validity of the working hypothesis H_0.

An important role in model testing is played by the predicted residual. The predicted residual is defined as the difference between the actual system output and the predicted output, based on the predicted state:

$$v_k = y_k - A_k x_{k|k-1} \tag{9.59}$$

Under the working hypothesis H_0, the predicted residual is Gaussian distributed with mean zero and covariance matrix:

$$E\{v_k v_l^*\} = Q_{v_k} \delta_{kl} \tag{9.60}$$

where:

$$Q_{v_k} = (Q_{y_k} + A_k Q_{x_{k|k-1}} A_k^*) \tag{9.61}$$

This knowledge of the distribution of the predicted residual under H_0 enables us to test the validity of the assumed mathematical model.
The following two hypotheses are considered:

$$H_0 : v \sim N(0, Q_v) \text{ and } H_a : v \sim N(\nabla v, Q_v) \tag{9.62}$$

with v the vector of predicted residuals up to epoch k. The vector ∇v will be parametrized as:

$$\nabla v = C_v \nabla \tag{9.63}$$

where C_v is a $(\sum_{i=1}^{k} m_i)$-by-b matrix and ∇ is a vector of dimension b. The matrix C_v is assumed to be known and of full rank b, and the vector ∇ is assumed to be unknown. The appropriate test statistic for testing H_0 against H_a reads then:

$$T = v^*Q_v^{-1}C_v[C_v^*Q_v^{-1}C_v]^{-1}C_v^*Q_v^{-1}v \tag{9.64}$$

The testing procedure for the real-time validation of GPS code and carrier observations consists of the following three steps:

1. Detection: An overall model test is performed to diagnose whether an unspecified model error has occurred.

2. Identification: After detection of a model error, identification of the potential source of the model error is needed. This implies a search among the candidate hypotheses for the most likely alternative hypothesis and their most likely time of occurrence.

3. Adaptation: After identification of an alternative hypothesis, adaptation of the recursive filter is needed to eliminate the presence of biases in the state vector

Detection. The objective of the detection step is to test the overall validity of the mathematical model H_o. In general, a distinction is made between local and global validity of the model. Here we will restrict ourselves to giving a description of the procedure for local detection of model errors.
To test the local validity of the model, we consider the following two hypotheses:

$$H_0^k : v_k \sim N(0, Q_{v_k}) \text{ and } H_a^k : v_k \sim N(C_{v_k}\nabla, Q_{v_k}) \tag{9.65}$$

In order to test the overall validity of the local hypothesis H_o^k, the mean ∇v_k of v_k under H_a^k should remain completely unspecified. This implies mathematically that the matrix C_v should be chosen as a square and regular matrix. By restricting T of (9.64) to time k, the invertible matrix C_v gets eliminated and the appropriate uniformly most powerful invariant (UMPI)-test statistic for local detection follows as:

$$T_{LOM}^k = \frac{v_k^* Q_{v_k}^{-1} v_k}{m_k} \tag{9.66}$$

with m_k the possibly time varying number of observables. This UMPI-test statistic, which has expectation one, is now used to perform a local overall model test for detecting unspecified model errors in the local null hypothesis H_0^k. The local overall model (LOM)-test reads therefore as follows: an unspecified local model error is considered present at time k if and only if:

$$T^k_{LOM} \geq F_\alpha(m_k, \infty, 0) \tag{9.67}$$

where $F_\alpha(m_k, \infty, 0)$ is the upper α probability point of the central F-distribution with m_k,∞ degrees of freedom.

Identification. The next step after detection is the identification of the most likely alternative hypothesis. As with detection, identification is based on the test statistic (9.64). For identification, candidate alternative hypotheses need to be specified explicitly. This specification is non-trivial and probably the most difficult task in the process of quality control. It depends to a great extent on experience and oncs knowledge of the system at hand. Here we will restrict ourselves to local identification of model errors in the measurement model. The following class of local alternative hypotheses is considered:

$$H_a^k : y_k = A_k x_k + C_k \nabla + e_k \tag{9.68}$$

This class of alternative hypotheses can be seen to model a slip in the mean of the vector of observables at time k. As such it can accommodate instrumental biases, sensor failure, and one or more outliers in the data. It is therefore particularly suited for the integrity monitoring of GPS data. The dimension of the b-vector ∇ in (9.68) depends on the alternative hypothesis considered and can range for identification purposes from 1 to m_k. In case of outlier-identification, b equals the number of assumed outliers. Here we will consider the case b=1, that is, the case of a single model error.
Since the local alternative hypothesis H_a^k of (9.68) is restricted to the measurement model, we have $C_v=C_k$. With b=1, the vector ∇ reduces to a scalar and the matrix C_k reduces to a vector, which will be denoted by the lower case kernel letter c_k. Restricting T of (9.64) to time k and taking the square root, we get a test statistic, which takes the form:

$$t^k = \frac{c_k^* Q_{v_k}^{-1} v_k}{(c_k^* Q_{v_k}^{-1} c_k)^{1/2}} \tag{9.69}$$

This is the local slippage (LS)-test statistic for the identification of a single local model error. Identification proceeds now as follows: the test statistic t^k is computed for each of the candidate one-dimensional alternative hypotheses. The alternative hypothesis for which $|\, t^k \,|$ is at a maximum is considered the one that contains the most likely model error. The likelihood of the most likely model error can be tested as follows: if

$$|t^k| \geq N_{\frac{1}{2}\alpha_0}(0, 1) \tag{9.70}$$

then the corresponding most likely model error can be considered likely enough to have occurred.
This procedure works well in case only one model error is present. However, in general multiple model errors may be present and we have to use an alternative procedure to identify them. This alternative procedure is characterized by a recursive filter, which operates in backward mode, see De Jong [1994].

Adaptation. After identification of the most likely alternative hypothesis, adaptation of the recursive filter is needed to eliminate the presence of biases in the filtered state of the system. In order to be able to adapt the filter, we first need an estimate of the identified model error ∇. The best linear unbiased estimator of the b-vector ∇ under H_a^k can be computed directly from the predicted residuals and reads as:

$$\hat{\nabla}^k = (C_k^* Q_{v_k}^{-1} C_k)^{-1} C_k^* Q_{v_k}^{-1} v_k \tag{9.71}$$

with corresponding covariance matrix:

$$Q_{\hat{\nabla}_k} = (C_k^* Q_{v_k}^{-1} C_k)^{-1} \tag{9.72}$$

Note that the estimator of a single model error can be computed directly from the LS-test statistic of equation (9.69) as:

$$\hat{\nabla}^k = \frac{t^k}{(c_k^* Q_{v_k}^{-1} c_k)^{1/2}} \tag{9.73}$$

The adapted filtered state at time k is given by:

$$x_{k|k}^a = x_{k|k} - K_k C_k \hat{\nabla}^k \tag{9.74}$$

with corresponding covariance matrix:

$$Q_{x_{k|k}^a} = Q_{x_{k|k}} + K_k C_k Q_{\hat{\nabla}_k} C_k^* K_k^* \tag{9.75}$$

The Minimal Detectable Bias (MDB). The Minimal Detectable Bias (MDB) is defined as the size of the model error ∇ that can be detected with probability γ with the one-dimensional test statistic (9.69). The quantity γ is called the power of the statistical test. The power is defined as the probability of rejecting H_0 when an alternative hypothesis H_a is true. The power depends on the chosen level of significance α (probability of false alarm), the number of degrees of freedom b, and the non-centrality parameter λ of the corresponding test statistic:

$$\gamma = \gamma(\alpha, b, \lambda) \tag{9.76}$$

The power γ is a monotonic function in α and λ, and a monotonic decreasing function in b. Since λ depends on the assumed model errors in H_0, the power function (9.76) can be used to determine how well particular model errors can be detected with the associated test. A low probability corresponds with poor detectability.
Instead of using the power function (9.76), we propose using the inverse power function:

$$\lambda = \lambda(\alpha, b, \gamma) \tag{9.77}$$

The rationale for using the inverse power function is that in practical applications one is usually much more interested in the size of the model error that can be detected with a certain probability γ than in the power γ itself. For chosen reference probabilities α_0 and γ_0 the non-centrality parameter λ_0 is given by:

$$\lambda_0 = \lambda(\alpha = \alpha_0, b, \gamma = \gamma_0) \tag{9.78}$$

Finally, the MDB ∇ is defined as:

$$\nabla = \sqrt{\frac{\lambda_0}{c_{v_k}^* Q_{v_k}^{-1} c_{v_k}}} \tag{9.79}$$

Values for the MDB are given in De Jong [1996, 1997] and Jin [1996b] for various models and input parameters. According to De Jong [1996] the MDB of the L1 and L2 code are 1.0 and 1.4 meter respectively. The carrier MDB's were in the range of 0.01 meter for 1 second data, up to 0.03-0.05 meter for 30 second data. The standard deviations of the L1 and L2 observations were 0.25 and 0.35 meter for code, 0.0019 and 0.0024 for the carrier. The model was quite conservative; a power spectral density of 10^{-9}-10^{-8} m^2/s^3 was assumed for the ionosphere, multipath was taken constant, and S_k has not been modelled at all. The MDB's for the carrier measurement given in Jin [1996b] are of the same order of magnitude. The MDB's for the code measurement are somewhat larger; they depend mainly on the standard deviation of the code measurement and the power spectral density of the bias terms. The carrier MDB is not very sensitive for the choice of model parameters. With four observables, and 1 second sample interval, the MDB for the carrier is always in the range of 1-2 cm. Also when

Anti-Spoofing is on.

Recursive processing and validation strategies. After the filter has been initialized, recursive processing and quality control of the GPS observations may begin. First the detection step, using the LOM test, is carried out. If this test is rejected alternative hypotheses are formulated, against which the null hypothesis (no model errors present) is tested. The alternative hypotheses are formulated as follows: an error is present in one or more of the original (untransformed) observations. The errors may be either integer cycle slips in the carrier measurements or outliers in both the carrier and code observations. Three types of alternative hypotheses are tested:

1) Error in a code observation
2) Error in a phase observation
3) Cycle slip in a phase observation

In the first two cases the c-vectors are restricted to the epoch at hand. The c-vectors for the original observations are very simple, an example is

$$\mathbf{c} = (0\ 0\ 1\ 0)^*$$

for a hypothesis specifying an error in the third observation. However, since the filter is usually specified in terms of the transformed observations the c-vectors have to be transformed as well. So for the specification of the alternative hypothesis $\mathbf{c}'=\mathbf{T}\mathbf{c}$ must be used instead.

For cycle slips the global identification and adaptation procedure of chapter 7 must be used. Using only local identification it is impossible to distinguish between the second an third type of hypotheses. This means that c-vectors must be constructed which span two or more epochs. It also means that the identification of cycle slips lags one or more epochs behind the first occurrence. Instead of the global identification and adaptation procedure a different procedure has been used by De Jong [1994].

In De Jong [1994] identification and estimation of model errors is done in two steps. In the first step the errors are identified and estimated. It is assumed that carrier errors consist of an integer, non-stochastic cycle slip and/or a stochastic fractional part. The estimated phase errors are rounded to the nearest integer, resulting in the non-stochastic cycle slips. These cycle slips are then subtracted from the predicted residuals. In the second step, the identification and estimation procedure is carried out again, using the corrected residuals. The errors detected in this second step (if any) will be free of cycle slips and consist of the stochastic code and fractional carrier errors. If model errors were identified and estimated in this second step, the updated state vector and covariance matrix has to be adapted. This approach is only allowed when the MDB's of the carrier measurement are smaller than a half cycle. Fortunately, this is almost always the case with 1 second data.

Jin [1996] uses a partial re-initialization of the filter when an outlier in the carrier observation is detected. So, in this implementation no distinction is made between slips and outliers.

The model used for the least squares initialization has not sufficient redundancy to detect all model errors. Only after the first filter epoch are the MDBs of the carrier observations small enough and is it possible to also detect small cycle slips. Therefore, if the first LOM test after filter initialization is rejected, the filter initialization is repeated, but shifted by one epoch.

9.4.3 Implementation Aspects and Choice of Models

Until now we have discussed the general principles by which we will do the integrity monitoring. We have not yet discussed the number of parameters in the state vector components, or their power spectral density, or even which of the parameters we are actually going to model.

The time dependent parameters were sorted in the previous section in order of decreasing variability. S_k has the largest time variations; even after subtracting an a-priori range, S_k-S_k^0, can be as large as several hundreds of kilometres. S_k-S_k^0 contains the effects from:

- orbit errors and errors in the a-priori position, which are very smooth,
- receiver clock errors, especially for receivers without accurate clocks, this may give short periodic fluctuations and jumps due to clock steering effects
- satellite clock errors, mainly S/A clock dither, with periodic variations of 100-200 meters with periods of 5-7 minutes
- tropospheric delays

Most effects are relatively smooth, with the possible exception of the receiver clock contribution. According to Jin [1996] S_k can be modelled using three parameters, i.e. m+1=3. For the power spectral density $q^{(3)}$ a value of 10^{-5} m^2/s^5 was chosen if an external clock has been used. In case an external clock was not available, the power spectral was chosen to be 0.1 m^2/s^5. In De Jong [1994, 1996] S_k is not modelled at all.

The second largest effect is the ionospheric delay I_k; typical values are in the range of 1-50 meters, with a significant diurnal and satellite elevation dependent variation. For the ionospheric model, usually two parameters, i.e., m+1=2, are sufficient. For the spectral density $q^{(2)}$ a value of 10^{-9} ~ 10^{-8} m^2/s^3 is realistic.

The code multipath, with constitute the main part of $b_{P1,k}$ and $b_{P2,k}$, is the next largest effect. The variability is in the range of several decimeters up to a few meters. For the multipath model as well, two parameters, i.e., m+1=2 are sufficient. For the spectral density $q^{(2)}$ a value of 10^{-5} - 10^{-3} m^2/s^3 seems realistic.

These values are based on Jin [1995]. These values are actually larger than the power spectral density for the ionosphere parameters, so its assumed that the ionospheric delay, which is based on phase data, is much smoother. The multipath and instrumental effects on the phase observations, $\bar{b}_{P1,k}$ and $\bar{b}_{P2,k}$, are the smallest systematic effects. Typical values are centimetres and subcentimetres (except for some effects such as antenna rotation, etc.). These effects are usually lumped with other terms, and will not be modelled explicitly, except for the six observation case where it is assumed to be constant.

In De Jong [1994] the number of parameters for the multipath model may be zero. In this case multipath is not modelled at all (assumed to be zero). In general one can say that the greater the number of ionosphere free, ionospheric and/or multipath parameters one wants to estimate, the more epochs of data are required for the initialization.

The input parameters for the recursive filter also consists of the standard deviations of the original code and carrier observations (P1, P2, C/A and L2⊗L1). For most geodetic receivers the values are roughly:

0.25 ~ 1.00 m - P1, P2 and C/A code
0.01 cycles - P1, P2 and C/A phase

In order for the quality control procedure to produce meaningful results, the values for the number of model parameters, power spectral density and standard deviation of the observations, should be as realistic as possible. If the standard deviations are too small, the model will be too optimistic and too many observation errors will be detected. On the other hand, if the standard deviations are too large, the model will be too pessimistic and it is not possible to detect small model errors. The models, as they have been implemented in the monitoring software developed in Delft, use a satellite elevation dependency in the precision of the observations. See the example in Figure 9.3.

The expected value of the LOM-test statistic is equal to one. If some of the model parameters have been chosen too optimistically this test statistic will in general be greater than one; if the parameters are too pessimistic, it will in general be smaller than one. Thus, to tune the model parameters, the LOM-test statistic is a useful tool. Since the LOM-test statistic is computed from the predicted residuals, they can be used for the fine tuning of individual parameters as well. An example has already been given in Figure 9.3, where the predicted residuals were used to estimate the elevation dependency of the standard deviation of the code measurements. The individual points in these figures are the root mean square of the predicted residual, computed over 10 minute batches, and plotted versus elevation. The histogram in Figure 9.4 shows the histogram of the predicted residuals of the L1 phase measurement for one satellite from the same receiver (TurboRogue SNR-8000).

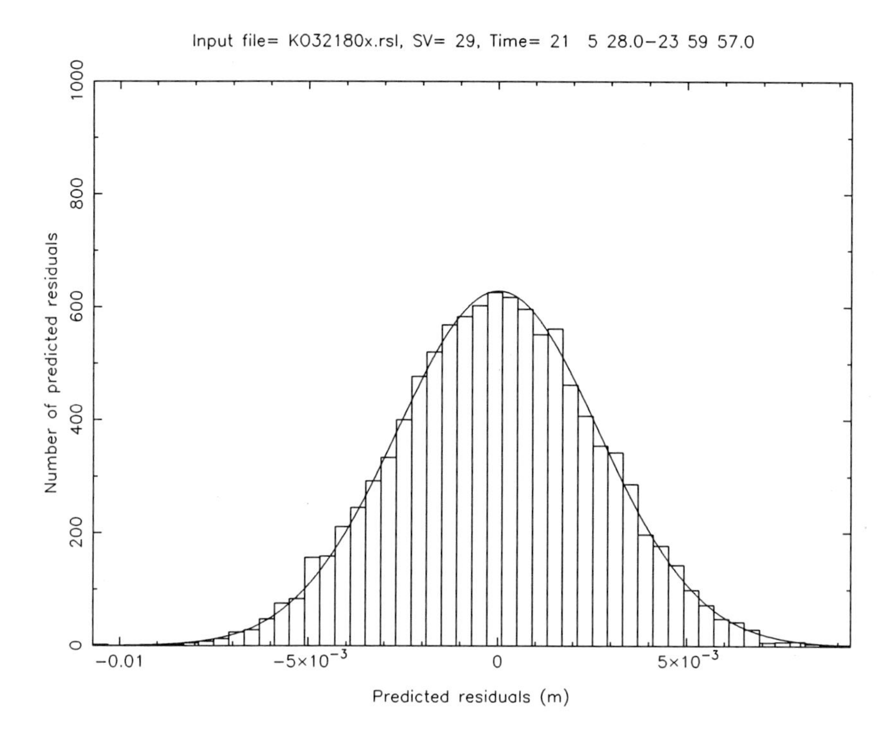

Figure 9.4. Histogram of carrier predicted residual.

9.4.4 Computation of D-GPS Corrections

Differential corrections, and differential correction rate, can be computed with (9.23) when both S_k and I_k are modelled. However, when computing differential corrections proper care must be taken of the phase ambiguities. As long as we did integrity monitoring, the constant phase ambiguities were irrelevant, because the dynamical models we use are not influenced by a constant offset. The phase ambiguities may be estimated using equation (9.22), using the filtered states instead of the derived observations. When one uses these ambiguity estimates, the differential correction is not any more a function of phase measurements only. However, as time progresses, the ambiguity estimate, and hence the differential correction, will become more precise.

In case S_k cannot be represented by a dynamical model, like in De Jong [1994, 1996], it becomes difficult to compute the differential correction rate. This may happen when the receiver clock is not smooth. Even when S_k is not modelled, it can be considered as a free observation, and since it is correlated with the other observations it may be given a correction after each recursive filtering step, i.e.

$$\Delta S_k = Q_{12} Q_{22}^{-1} \begin{pmatrix} \Delta I_k \\ \Delta b_{P1,k} \\ \Delta b_{P2,k} \end{pmatrix} \tag{9.80}$$

with $\mathbf{Q}_{22}$ the 3-by-3 submatrix of $\mathbf{Q}_{y'}$ corresponding to I_k, $b_{P1,k}$ and $b_{P2,k}$, and $\mathbf{Q}_{12}$ the 1-by-3 submatrix with the co-variances of these parameters with S_k. However, something can still be done considering the clock noise is common to all satellites. Therefore, differential corrections may be changed by arbitrary numbers, provided this is done for all satellites in the same way. A practical application to this concept is to model single differences of S_k between satellites, instead of S_k itself, introduce a new parameter for the clock which follows the actual clock smoothly, and add the difference of the smoothed and actual clock to the differential correction. The basis for this procedure is formed by the multiple channel approach of the next section.

9.4.5 Multiple Channel Approach

Until now the integrity monitoring has been based on a single channel dynamical model. This approach worked well for dual frequency receivers, especially in case A-S is off. However, the single channel approach does not work with single frequency receivers. Also when A-S is on, the single channel approach may not work well with some of the cross correlation dual frequency receivers.

The problem with the cross correlation receivers is caused by the fact that the cross correlation technique introduces correlations between the dual frequency observations. The ionospheric delay I_k, which has the 'strongest' dynamical model, is now fully determined by the cross correlation measurement, and a cycle slip or outlier on L1 will hardly affect the estimate of I_k. The other way round: it is not possible to find cycle slips or outliers on L1 with only a dynamical model for I_k. Therefore the quality control for L1 phase and code observations depends on the other dynamical models, and in case S_k is not modelled, only the dynamical model for $b_{P1,k}$ is available for this purpose. Therefore, in case of A-S on, it is important to be able to use a dynamical model for S_k. This means that the sample rate cannot be too high. It also means that we might need a smoothly running (external) receiver clock.

The receiver clock is the term which is usually responsible for most of the jitter on S_k. However, since the receiver clock error is the same for all channels, it will be eliminated from single differences of S_k between satellites. The single differences of S_k contain only satellite clock error, geometric range differences and a small tropospheric error, which are very smooth over a period of several sample intervals. Therefore, it makes sense to use a dynamical model for single differences of S_k instead of S_k itself. This is the approach taken by De Jong [1997] in case of the single frequency model, but it works also well in case of dual frequency receivers. The implementation of this model is more difficult than the single channel model, especially because changes in the satellite configuration should be dealt with. For more information on the implementation see De Jong [1997].

9.5 ACTIVE GPS REFERENCE SYSTEMS

An active GPS reference system consists of a number of Active GPS control stations operating in a network mode and a computing centre. Data from the control stations is collected at a computing centre, and the data is also distributed from there. At the computing centre the data can be checked before distribution. This can be done more rigorously than before at the individual control stations, because at the computing centre a baseline or network adjustment is possible, instead of the procedure discussed in section 9.4. Usually, this step is done in post-processing, typically once per day, or once per hour, depending on when the GPS data is received from the control stations. However, this processing step could be done in near real-time as well.

One of the benefits of the network processing is the data validation and cycle slip repair. Compared to the reference station software, which at best can detect cycle slips, a full cycle slip repair is now feasible. The processing may also provide other parameters, like station coordinates, which may be used to monitor the stability of the control network, or atmospheric parameters, which may be used for meteorology. An active GPS reference system may also add extra functionality to the data in the form of special services for users. Services which may be implemented are for instance: a data service, a single point correction service, a virtual reference station service or a data processing service.

The *data* service delivers GPS data of the individual reference stations, after they have gone through the above described cycle slip and data validation step. The purpose of this service is to give reliable data. A user may download data from one or more reference stations, and use this data for his own processing. The user has full responsibility for the processing. If data from a single station is downloaded only limited quality control is possible, although this problem is mitigated to some extend by the data validation at the computing centre, and the precision of these results are a function of the distance to the reference station. To obtain the best possible precision the user should download data from all reference stations. However, this will complicate his processing seriously, and actually only few users have the desire to do this. For the majority of users this is not a realistic option. For them a *virtual reference station* service may be more useful.

The virtual reference station is a novel concept, which is currently being implemented in the Dutch Active GPS Reference System. The *virtual reference station* service provides data for a non-existent, virtual, reference station. Instead of downloading data from several reference stations, the user downloads data of a single non-existent reference station in the neighbourhood of his project. The virtual station data is not data from a real receiver, although it is computed from real GPS observations of the active GPS control stations. The basic idea is that the data from the virtual reference station is as close as possible to data of a real receiver at the virtual reference station location. This implies that in the virtual station data atmospheric errors, for the virtual site, are introduced. On the other hand, all other errors (e.g., orbit errors and multipath errors) are eliminated as much as possible. Because virtual reference station data is computed from several real stations, errors will anyhow tend to average out. Therefore, the purpose of a virtual reference station is to give data with properties of a non existing station close to the user's location, but let the user benefit from data of all GPS reference stations, without burdening him with a complicated processing. The benefits for the user are that he can use standard, commercially available software, using a limited amount of data from the active GPS control system (one virtual reference station for the duration of his measurement) and obtain a precision and reliability comparable to a multistation approach.

Virtual reference station data is computed using the following procedure. First a network adjustment with GPS data from all reference stations is done. This results in adjusted GPS observations and atmospheric parameters. After the adjustment all data is consistent. Following the adjustment several corrections are applied to the GPS data:

- A geometric correction (re-location) to the phase and code data to make the user's software believe that the station is at a different location. To compute this correction (predicted) precise ephemerides can be used instead if broadcast ephemerides. This will mitigate the effects of orbit errors.
- A correction for the antenna phase centre, and elevation and azimuth dependent phase centre variations, to make the user's software believe that an other antenna is used. If the user selects antenna type he is actually using he does not have to bother himself about phase centre corrections.
- Atmospheric corrections: The ionosphere correction is based on a regional ionosphere model, estimated from the GPS data, or on interpolation using all GPS reference stations. The correction for the troposphere is based on the results from the network adjustment. Atmospheric models are used to 'interpolate' the correction to the virtual station site.

Before the user can download the virtual reference station data, the antenna type and receiver type he is using, and a position in his area of operation, has to be specified.

The *single point correction* service is based on a different concept. The idea is that the computing centre provides satellite clock corrections in addition to the

IGS orbits. This would easily give sub-meter positioning accuracies, but a better precision is possible when the clock corrections are tailored for a specific area. Actually, taking the virtual reference station data, and subtracting the topocentric distance computed from the orbit data, would just give such corrections.

The idea behind the *data processing* service is that the user, instead of downloading control station data, uploads his own data and that this data is processed by the computing centre. The user can then later retrieve the results (coordinates for his points). So, instead of downloading data and processing this by himself, he can upload his data to a central processing facility, where the data is processed, and retrieve the results. This is by far the most advanced, and for the user the easiest, service, but it provides little control and can be a logistical burden.

9.6 SUMMARY AND CONCLUSIONS

In this chapter an introduction was given to the theory, implementation and application of Active GPS control stations. An active GPS control station consists of a permanently operating dual frequency GPS receiver, a computer and communication facilities. The computer performs an important role in the management of the station. It is used to store the data for considerable periods of time, distribute this data to a computing centre and other users, watch over the integrity of the data, compute differential corrections, and remotely control the receiver. Some of these functions may be performed by a special computing centre, with the advantage that quality control and data distribution tasks can be performed better at the computing centre than at a stand alone active GPS control station.

The integrity monitoring, computation of differential corrections, and data decimation, were based on a regular and invertible transformation of the GPS observations. This transformation splits the original observations into its principle components: the combined range and clock terms, an ionospheric term, and one or more bias terms which mainly depend on multipath. The dynamical behaviour of these terms can be modelled using a simple dynamical model, based on polynomials. The validation can then be based on the predicted resi-duals, computed from a recursive (Kalman) filter. Validation consists of three steps: detection, identification and adaptation. Identification of the most likely model has been based on alternative hypotheses, specified through c-vectors, for outliers in the code and phase data, and cycle slips. The recursive (Kalman) filter can also be used for the computation of differential GPS corrections.

Acknowledgement

Parts of section 9.4 were edited from De Jong [1994]. Figures 9.3 and 9.4 are from Jin [1996b].

References

Abe, Y., H. Tsuji (1994), A nationwide GPS array in Japan for geodynamics and surveying. *Geodetical Info Magazine*, Vol. 8, No. 10, pp. 29-31.

Allen Osborne Associates (1995): *User manual TurboRogue™ SNR8000, SNR8100 and SNR-12 RM GPS Receivers*. Allen Osborne Associates, Westlake Village, CA, October 1995.

Beutler, G., I.I. Mueller, R. Neilan (1994), The international GPS Service for Geodynamics (IGS): Development and Start of Official Service on 1 January 1994, *Bulletin Géodésique,* 68, pp. 43-51.

Bock, Y., S. Shimada (1990), Continuously monitoring GPS networks for deformation measurements. In: *Global Positioning System: An Overview.* Edited by Y. Bock and N. Leppard, Springer-Verlag, Berlin, pp. 337-377.

Bruyninx, C., W. Gurtner, A. Muls (1996), The EUREF permanent GPS network. In: *Proceedings of the symposium of the IAG subcommission for Europe (EUREF)*, Ankara, 1996, pp. 123-130.

Delikaraoglou, D., H. Dragert, J. Kouba, K. Lochhead, J. Popelar (1990), The development of a Canadian GPS active control system: status of the current array. In: *Proceedings Second International Symposium on Precise Positioning with GPS, GPS'90,* Ottawa, pp. 191-202.

Frohlich, M. (1994), The high precise permanent positioning service (HPPS) in north Germany - Tests and first results. *Proceedings DSNS 94 - Third International Conference on Differential Satellite Navigation Systems,* London, 7 pp.

Jin, X.X., H. van der Marel, C.D. de Jong (1995), Computation and Quality Control of Differential GPS corrections. In: *Proceedings of ION GPS-95*, The Institute of Navigation, Alexandria, VA, pp. 1071-1079.

Jin, X.X., C.D. de Jong (1996a), The relationship between satellite elevation precision of GPS code observations. *Journal of Navigation*, Vol. 49(2), pp. 253-265.

Jin, X.X. (1996b), *Theory of carrier adjusted DGPS positioning approach and some experimental results*. PhD-thesis, Delft University of Technology, Delft University Press.

Jong, C.D. de (1994), *Real-time quality control of dual frequency GPS observations*. Internal publication of the Delft Geodetic Computing Centre, 1994, 59 p.

Jong, C.D. de (1996), Real-time integrity monitoring of dual-frequency GPS observations from a single receiver. *Acta Geod. Geoph. Hung.*, Vol. 31 (1-2), pp. 37-46.

Jong, C.D. de (1997), *Principles and applications of permanent GPS arrays.* PhD-thesis, Technical University of Budapest, Hungary, Delft University Press.

Hedling, G., B. Jonsson (1995), SWEPOS - A swedish network of reference stations for GPS. In: *Proceedings DSNS95 - Fifth International Conference on Differential Satellite Navigation Systems,* Bergen, 8 pp.

Pesec, P., E. Erker (1996), Concept and design of the remotely controlled permanent geodynamic (EUREF) GPS-network in Austria. *Proceedings of the Symposium of the IAG Subcommission for Europe (EUREF),* Ankara, 1996, pp. 131-134.

Quek, S.H., M. Craymer, R.B. Langley, D. Parkhill, B. Arseneau, D. McArthur, K. Lochhead (1989), Development of a GPS Active Control Point Station. *Journal of Surveying Engineering,* Vol. 115, No. 1, pp. 46-55.

Salzmann, M.A. (1993), *Least squares filtering and testing for geodetic navigation applications.* PhD-thesis, Delft University of Technology, Delft, 209 pp.

Teunissen, P.J.G. (1990), An integrity and quality control procedure for use in multi sensor integration. In: *Proceedings ION GPS-90*, Colorado Springs, 19-21 September 1990, pp. 513-522

Young, W., Y. Bock, S. Marquez (1995), Economics of GPS and CORS. *Proceedings ION GPS-95*, Palm Springs, pp. 105-113.

10. SINGLE-SITE GPS MODELS

Clyde C. Goad
Department of Geodetic Science and Surveying, The Ohio State University, 1958 Neil Avenue, Columbus OH 43210-1247 U.S.A.

10.1 INTRODUCTION

While the major use of GPS to most geodesists involves the use of two or more receivers in interferometric mode, it is very important to keep in mind the reason GPS was developed in the first place — to determine at an instant the location of a soldier, ship, plane, helicopter, etc. without any equipment other than a single GPS receiver and antenna. This is often referred to as absolute positioning. Without satisfying this requirement, there would be no GPS. Thus it is important that some time and effort be spent in the study of single-site modeling. In this chapter, the processing techniques using the pseudorange measurement are discussed. Also the combination of pseudorange and carrier phase are introduced.

10.2 PSEUDORANGE RELATION

Much of the groundwork has already been done. From Chapter 5 we have been exposed to pseudoranges that were designed by the planners of the GPS system of satellites for recovery of single site coordinates. In particular, we review equation (5.25):

$$\begin{aligned} P_i^k(t) = {} & \left\| (\boldsymbol{r}^k(t-\tau_i^k) - d\boldsymbol{r}^k(t-\tau_i^k)) - (\boldsymbol{r}_i(t) + d\boldsymbol{r}_i(t)) \right\| + \\ & I_i^k + T_i^k + c[dt_i(t) - dt^k(t-\tau_i^k)] + \\ & c[d_i(t) + d^k(t-\tau_i^k)] + dm_i^k + e_i^k \end{aligned} \tag{5.25}$$

Here (5.25) will be rewritten dropping those terms that can be computed or estimated by others and thus removed from each measurement, i.e., the measurement can be "corrected." These include satellite center of mass offset, tropospheric refraction, ionospheric refraction. Also for simplicity, multipath will be ignored. Thus (5.25) can be simplified to

$$P_i^k(t) = \left\| \boldsymbol{r}^k(t-\tau_i^k) - \boldsymbol{r}_i(t) \right\| + c[dt_i(t) - dt^k(t-\tau_i^k)] + e_i^k \tag{10.1}$$

Here it is seen that (10.1) is nonlinear in satellite and receiver coordinates and linear in clock offsets. For single-site positioning, a model for the satellite clock offset is contained in the navigation message and looks as follows:

$$dt^k(t) \approx a_0 + a_1(t - t_{oc}) + a_2(t - t_{oc})^2 \tag{10.2}$$

where a_0, a_1, a_2 are polynomial coefficients and t_{oc} is the reference time (time of clock) for the coefficients. Specifically a_0 is the offset at time t_{oc}, a_1 is the drift rate at t_{oc}, and a_2 is half the clock acceleration at t_{oc}. The idea in providing (10.2) to users is that, while admittedly it is only a prediction of clock behavior, it should be fairly precise since high-quality oscillators (mostly cesium) are used in the GPS satellites. Activation of Selective Availability (SA) will intentionally degrade this modeling. Cesium oscillators should be good to 10^{-13} (or equivalently one part in 10^{13}). That is the standard deviation or

$$\sigma\left(\frac{\Delta f}{f}\right) \approx 10^{-13}.$$

With these coefficients, the satellite clock offset can be computed and removed from (10.1) to yield the desired pseudorange model

$$P_i^k(t) = \left\| \boldsymbol{r}^k(t - \tau_i^k) - \boldsymbol{r}_i(t) \right\| + c\,dt_i(t) + e_i^k \tag{10.3}$$

The traditional way of solving (10.3) is to use a Newton-Raphson iteration. Using this technique, one must first obtain an initial guess of the receiver position and clock offset. The difference between an actual observation and what is calculated using the guessed values is a measure of the goodness of the guess. Assuming that the function behaves linearly (described by first derivatives), corrections can be computed assuming that sufficient measurements exist to solve for corrections to all unknowns (partial matrix has full column rank). However, before attempting to implement these techniques, one must first be able to compute the expected measurement value based on the current guess. Here the coordinate system used is very important. Normally one thinks of the vectors $\boldsymbol{r}^k$ and $\boldsymbol{r}_i$ as being inertial, but the navigation message provides parameters that allow one to compute coordinates in an Earth-centered, Earth-fixed system. Since this set of coordinates is attached to the rotating Earth's frame, some corrections are required — commonly called the "Earth rotation correction."

10.2.1 Calculation of the Distance Term When Using ECF Coordinates

First, the orientation of the Greenwich meridian must be defined. This has already been discussed in Chapter 1. At any time t, we will refer to the Greenwich sidereal angle as θ_t. Generally speaking, $\theta_t = \theta_0 + \omega t$ where ω is the mean earth spin rate and θ_0 is the value of θ_t at $t = 0$.

Now let's define the rotation matrix

$$R_3(\theta_t) = \begin{bmatrix} \cos\theta_t & \sin\theta_t & 0 \\ -\sin\theta_t & \cos\theta_t & 0 \\ 0 & 0 & 1 \end{bmatrix}$$

$R_3(\theta_t)$ is defined such that

$$\boldsymbol{r}_{\mathrm{ECF}}(t) = R_3(\theta_t)\ \boldsymbol{r}_{\mathrm{I}}(t)$$

or equivalently

$$\boldsymbol{r}_{\mathrm{I}}(t) = R_3^T(\theta_t)\ \boldsymbol{r}_{\mathrm{ECF}}(t) \tag{10.4}$$

when $\boldsymbol{r}_{\mathrm{I}}$ is a vector expressed in an inertial system, and $\boldsymbol{r}_{\mathrm{ECF}}$ is the same vector but expressed in an earth-centered fixed system.

From (10.3) the term between the vertical bars is the distance:

$$\rho = \left\| \boldsymbol{r}^k(t-\tau_i^k)_{\mathrm{I}} - \boldsymbol{r}_i(t)_{\mathrm{I}} \right\|$$

where ρ stands for distance, and subscript I denotes the choice of inertial coordinates. We can substitute for the inertial vectors using (10.4) to yield the following:

$$\rho = \left\| R_3^T(\theta_{t-\tau_i^k})\boldsymbol{r}(t-\tau_i^k)_{\mathrm{ECF}} - R_3^T(\theta_t)\boldsymbol{r}_i(t)_{\mathrm{ECF}} \right\|$$

Realizing that the first rotation can be expressed as two separate ones, we get

$$\rho = \left\| R_3^T(\theta_t)R_3^T(-\omega\tau_i^k)\boldsymbol{r}^k(t-\tau_i^k)_{\mathrm{ECF}} - R_3^T(\theta_t)\boldsymbol{r}_i(t)_{\mathrm{ECF}} \right\|$$

The common rotation $R_3^T(\theta_t)$ does not change the vector length, so the above is rewritten

$$\rho = \left\| R_3(\omega\tau_i^k)\boldsymbol{r}^k(t-\tau_i^k)_{\mathrm{ECF}} - \boldsymbol{r}_i(t)_{\mathrm{ECF}} \right\| \tag{10.5}$$

where we also use $R_3^T(-\omega\tau_i^k) = R_3(\omega\tau_i^k)$. So from (10.5) we see that when using coordinates in an ECF system, one must rotate the satellite position vector about the 3-axis an amount equal to the angular rotation of the earth in the time it takes the signal to travel from the satellite to the receiver. The height of a GPS satellite is about 20,000 km, thus the signal transit time is about 66 ms. The earth rotates 15 arcsec/s, so the angular displacement of the earth about its rotation axis during signal travel is roughly 1 arcsec. So if ECF coordinates are used and the correction is not applied, then the recovered station coordinate will be biased by about one arcsec. in longitude. So now (10.3) can be rewritten as

$$P_i^k(t) = \rho(t, t - \tau_i^k) + c\, dt_i(t) + e_i^k \tag{10.6}$$

and one should substitute (10.5) for the distance term when using ECF coordinates. It is also important to notice that the $\boldsymbol{r}_i(t)_{\mathrm{ECF}}$ is a function of time. Here station motion due to gravitationally induced tides, loading, crustal motion, displacements due to earthquakes, etc. must be considered.

10.2.2 Linearization

The linearized form of (10.6) has already been derived in Section 5.4.2. Actually, it included terms we have already neglected here, but then we were advised to assume that their errors were zero. Actually, this is just a statement that we did the best job we could in preparing the data for the adjustment process. Repeating (5.80), we see the following:

$$\Delta P_i^k = -(\boldsymbol{u}_i^k)^T \Delta \boldsymbol{r}_i + c\, \Delta dt_i' + \nabla_i^k \tag{5.80}$$

Equation (5.80) represents the linearization of (10.6). The $\Delta P_i^k = P_i^k$ (observed) – P_i^k (calculated). Each of these right-side entries represents a single real number. The "observed" value is delivered to us by the GPS receiver. It is the observed pseudorange measurement appropriately corrected for satellite clock error, tropospheric refraction, etc. The "calculated" term is the value we expect based on the best guess of station coordinates, clock states, etc. If the guesses are good, then ΔP_i^k will be small. The reader is cautioned, however, that the $\boldsymbol{u}_i^k$ unit vector should use the components as dictated in (10.5). That is, the satellite coordinates must be rotated by $R_3(\omega \tau_i^k)$ prior to their use if ECF coordinates are used.

If, however, the choice of coordinate system is inertial, then both satellites and stations exhibit continuous motion, and their locations must be computed for the respective transmit or receive time. More will be said about this in later chapters. Here we shall continue to concentrate on traditional positioning techniques as provided by the GPS broadcast message parameters.

For all pseudorange measurements at an epoch or instant of time, we can "stack" them to form a system of equations as follows:

$$[\Delta \boldsymbol{P}]_{m\times 1} = A_{m\times 4} \begin{bmatrix} \Delta \boldsymbol{r}_i \\ \Delta dt_i' \end{bmatrix}_{4\times 1} + \boldsymbol{e}_{m\times 1} \tag{10.7}$$

where $[\Delta \boldsymbol{P}]$ is the "stacked" vector of observed-computed pseudorange values, A is a matrix with each row composed of (row) vectors

$$\begin{bmatrix} \boldsymbol{u}_i^k \\ c \end{bmatrix}^T,$$

and $\boldsymbol{e}$ is a vector representing random errors present in the observed pseudoranges. The sizes of the matrices are given; the number of pseudoranges at an epoch is denoted by the letter "m". Thus (10.7) reminds us of the familiar notation

$$y = Ax + \boldsymbol{e} \tag{10.8}$$

with least-squares normal equations

$$(A^T \Sigma^{-1} A)\hat{x} = A^T \Sigma^{-1} y \tag{10.9}$$

where E($\boldsymbol{e}$)=0 and Σ= E($\boldsymbol{ee}^T$). Assuming the normal matrix ($\mathbf{A}^T\Sigma^{-1}\boldsymbol{A}$) to be of full rank, one can get

$$\hat{x} = (A^T \Sigma^{-1} A)^{-1} A^T \Sigma^{-1} y \tag{10.10}$$

Substituting from (10.7), we get

$$\begin{bmatrix} \hat{\Delta \boldsymbol{r}_i} \\ \Delta dt'_i \end{bmatrix} = (A^T \Sigma^{-1} A)^{-1} A^T \Sigma^{-1} [\Delta \boldsymbol{P}] \tag{10.11}$$

Normally the pseudoranges are assumed to have independent errors with $\boldsymbol{e} \sim (0, \sigma^2 \mathrm{I})$ meaning the vector $\boldsymbol{e}$ has zero mean and variance matrix $\sigma^2\mathrm{I}$. If the assumption is true, then (10.11) simplifies to

$$\begin{bmatrix} \hat{\Delta \boldsymbol{r}_i} \\ \Delta dt'_i \end{bmatrix} = (A^T A)^{-1} A^T [\Delta \boldsymbol{P}] \tag{10.12}$$

An interesting application that is used in GPS surveying quite often is to use all the pseudorange data from all epochs to estimate a single antenna location and clock offsets during the period of stationarity. Suppose that at each epoch i one has the following:

$$\left[\Delta \boldsymbol{P}_i\right] = A_i \left[\Delta \boldsymbol{r}\right] + b_i \Delta dt'_i + \boldsymbol{e}_i \tag{10.13}$$

where A_i is the i-th epoch partial derivative matrix with respect to station coordinates, $b_i = [c\,c \ldots c]^T$, a vector composed of the constant speed of light, c.

As before, we now "stack" epochs similar to the stacking of measurements

$$\begin{bmatrix} [\Delta \boldsymbol{P}_1] \\ [\Delta \boldsymbol{P}_2] \\ \vdots \\ [\Delta \boldsymbol{P}_n] \end{bmatrix} = \begin{bmatrix} A_1 \\ A_2 \\ \vdots \\ A_n \end{bmatrix} [\Delta \boldsymbol{r}] + \begin{bmatrix} b_1 & 0 & 0 & \cdots & 0 \\ 0 & b_2 & 0 & \cdots & 0 \\ \vdots & \vdots & \vdots & \vdots & \vdots \\ 0 & 0 & 0 & \cdots & b_n \end{bmatrix} \begin{bmatrix} dt_1' \\ dt_2' \\ \vdots \\ dt_n' \end{bmatrix} + \begin{bmatrix} \boldsymbol{e}_1 \\ \boldsymbol{e}_2 \\ \vdots \\ \boldsymbol{e}_n \end{bmatrix} \tag{10.14}$$

The least-squares normal equation system becomes

$$\begin{bmatrix} b_1^T b_1 & 0 & \cdots & 0 & b_1^T A_1 \\ 0 & b_2^T b_2 & \cdots & 0 & b_2^T A_2 \\ \vdots & \vdots & \vdots & \vdots & \vdots \\ A_1^T b_1 & A_2^T b_2 & \cdots & A_n^T b_n & \sum_i A_i^T A_i \end{bmatrix} \begin{bmatrix} dt_1' \\ dt_2' \\ \vdots \\ \Delta r \end{bmatrix} = \begin{bmatrix} b_1^T [\Delta P_1] \\ b_2^T [\Delta P_2] \\ \vdots \\ \sum_i A_i^T [\Delta P_i] \end{bmatrix} \tag{10.15}$$

The position corrections can be found using Gaussian elimination:

$$[\hat{\Delta \boldsymbol{r}}] = \left[\sum_i (A_i^T A_i - A_i^T b_i (b_i^T b_i)^{-1} b_i^T A_i) \right]^{-1} \cdot \left[\sum_i (A_i^T \boldsymbol{y}_i - A_i^T b_i (b_i^T b_i)^{-1} b_i^T [\Delta \boldsymbol{P}_i]) \right] \tag{10.16}$$

Back substitution can be used to calculate the least-squares estimate of clock offsets if they are desired. Should the clock terms change "smoothly" from epoch to epoch, then one could consider modeling the clock drift by smooth functions such as polynomials, splines, etc. to simplify (10.13)–(10.16). However, the above approach is advised because manufacturers have been known to introduce clock jumps or change the rate of the clock in order to keep data sampling between any two receivers within a specified tolerance. Such jumps in clock state or one of its derivatives will invalidate a model that expects smoothly changing values.

Also apparent is that in the stationary mode, no longer is one required to collect four measurements per epoch. Any number of measurements could be used so long as the reduced normal matrix is regular. Clearly, two or more measurements per epoch are required for data during that epoch to provide information about position. If only one measurement is available, it would be used only to estimate the clock offset at that epoch.

10.2.3 Equivalence of the Linear Gauss-Markov Models With and Without Nuisance Parameters

The technique of bias or nuisance parameter elimination from the system of the observation equations is widely used in the solution of numerous surveying and geodetic problems. One of them is the subsequent removal of the receiver and transmitter clock offsets or integer ambiguities by forming consecutive differences of the GPS observables. Another example is the elimination of the orientation unknown in the problem of a terrestrial network adjustment. The elimination scheme, simple, fast and effective, requires nontrivial theoretical validation so that the estimates of the nonstochastic parameters, common to both systems, original and reduced, will coincide. As Schaffrin and Grafarend [1986] have proved, elimination of the nuisance parameter vector η from the partitioned linear Gauss-Markov model described as follows:

$$E\{\boldsymbol{Y}\} = \boldsymbol{A}\xi + \boldsymbol{B}\eta, \qquad D\{\boldsymbol{Y}\} = \boldsymbol{P}^{-1}\sigma^2,$$

leads to the system that provides the least-squares solution for ξ identical to the estimate obtained from the original system, under the condition that the covariance matrix for the reduced system is modeled properly. $\boldsymbol{A}$ and $\boldsymbol{B}$ are design matrices, such that $rk(\boldsymbol{A})+ rk(\boldsymbol{B}) = rk[\boldsymbol{A}, \boldsymbol{B}]$, thus column spaces for $\boldsymbol{A}$ and $\boldsymbol{B}$ are complementary, so that separability of both groups of nonstochastic unknown vectors ξ and η is assured. The reduced system is obtained by finding an $n \times (n\text{-rk}(\boldsymbol{B}))$ matrix $\boldsymbol{R}$ of maximum column rank such that:

$$\boldsymbol{R}^T\boldsymbol{B} = 0 \text{ and } rk(\boldsymbol{R})+rk(\boldsymbol{B})=n$$

where n is a number of rows in $\boldsymbol{A}$ and $\boldsymbol{B}$. Thus the new, $\boldsymbol{R}$-transformed Gauss-Markov model can now be characterized as follows:

$$E\{\boldsymbol{R}^T\boldsymbol{Y}\} = \boldsymbol{R}^T\boldsymbol{A}\xi \text{ and } D\{\boldsymbol{R}^T\boldsymbol{Y}\} = \boldsymbol{R}^T\boldsymbol{P}^{-1}\boldsymbol{R}\sigma^2$$

so the bias vector η is not present in the reduced system; note that the dispersion matrix of the new model is also $\boldsymbol{R}$-transformed according to the law of error propagation.

10.2.4 Searching

We can take advantage of the Schaffrin-Grafarend theorem in a search if no reasonable guess is available for use in (10.7). A search algorithm can be employed. Here some very important information is available. For example, if the minimum of four pseudorange measurements is available, then the transmit times are known and thus the latitude, longitude, and height of each satellite are known at these transmit times from evaluation of the ephemeris using broadcast parameters. Thus one could average the latitudes and the longitudes of the satellite positions to determine a hemisphere for consideration. These average values of latitude and longitude could be used to secd a search of receiver locations. But what about the clock values? Clearly one does not want to include clock offsets in the search algorithm since these values are not bounded. Let us look again at (10.6). This time, however, we will rewrite it in terms of all measurements. For discussion purposes, let us assume that the minimum of four pseudorange measurements is available. The following discussion is also clearly valid for any greater number of available pseudoranges. Thus we group four pseudorange measurements according to (10.6) as follows:

$$\begin{aligned} P_1^1(t) &= \rho_1^1(t-\tau_1^1,t)+c\,dt_1(t)+e_1^1 \\ P_1^2(t) &= \rho_1^2(t-\tau_1^2,t)+c\,dt_1(t)+e_1^2 \\ P_1^3(t) &= \rho_1^3(t-\tau_1^3,t)+c\,dt_1(t)+e_1^3 \\ P_1^4(t) &= \rho_1^4(t-\tau_1^4,t)+c\,dt_1(t)+e_1^4 \end{aligned} \tag{10.17}$$

Now the idea is to somehow eliminate the term $c\,dt_1(t)$. Since we are only interested in finding a reasonable guess of position, then why try to find a reasonable value of $dt_1(t)$ during the search process? One way to avoid a search that includes $dt_1(t)$ is to generate another set of relations that eliminate this term analytically. Here differencing can be used. Probably the simplest is to subtract the first equation in (10.17) from the next three. Sequential differencing will also work. Doing so we get

$$\begin{aligned} P_1^{2,1}(t) &= \rho_1^2 - \rho_1^1 + e_1^2 - e_1^1 \\ P_1^{3,1}(t) &= \rho_1^3 - \rho_1^1 + e_1^3 - e_1^1 \\ P_1^{4,1}(t) &= \rho_1^4 - \rho_1^1 + e_1^4 - e_1^1 \end{aligned} \tag{10.18}$$

where functional arguments have been dropped since what is needed to evaluate them is obvious, and a pair of superscripts has been used to denote differencing between satellites. For example, $P_1^{2,1}(t) = P_1^2(t) - P_1^1(t)$. Also shown in (10.18) is the equivalence between (10.17) and hyperbolic positioning. The solution to each equation is the locus of all points whose difference in distance from the two satellites is a constant. This surface is a hyperboloid, and thus the phrase hyperbolic positioning is used. The right side of (10.18) can be evaluated only knowing the station-satellite geometry. Measurement errors will be ignored here.

A search could proceed as follows: Using the averaged satellite latitude and longitude as a start point (the pole of a hemisphere), search the half sphere where the test points are at the center of a tesseral bounded by distance and azimuth boundaries. The search space would look as follows:

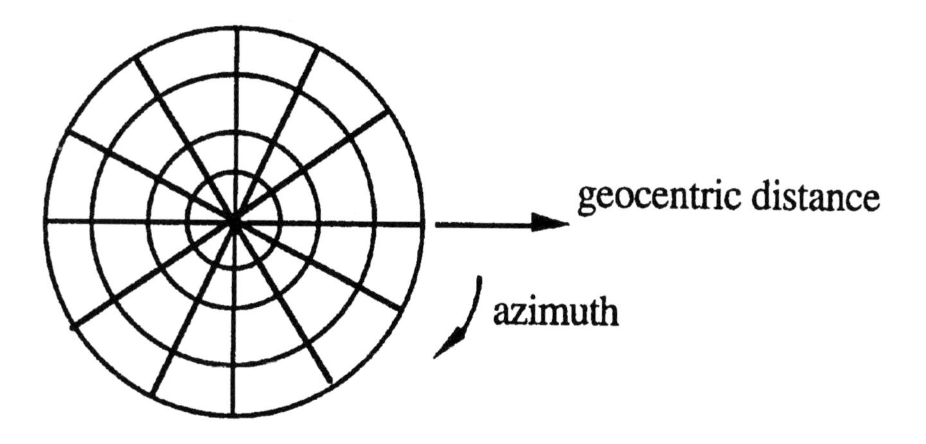

Figure 10.1.

The center is the averaged latitude and longitude, say, ϕ_0, λ_0. Then a recursive scheme can be used to generate the latitudes and longitudes at the centers of equi-angular blocks, say, every 10° of geocentric distance and azimuth. There should be no need to search more than a hemisphere. Goodness of fit can be defined as the sum of squares of residuals at each test point, for example.

One test point should fit the data better than any other. Then this point is used to seed another test, but now at higher resolution, say, at 1°. This time a smaller area is searched. This solution then seeds a 0.1° search area. And so the search continues until a desired resolution is achieved. And this search has used differenced measurements that eliminated the clock term as a consideration.

Once the search has resolved the position to some level of acceptance, then probably an iterative Newton-Raphson procedure would be employed to find the final solution. One schooled in weighted least squares will note that this is not a "proper" weighting scheme. But here the idea was only to find a reasonable guess, so compromises in algorithms are allowed. This is an application of the Schaffrin-Grafarend [1986] theorem that states that under certain conditions a reduction in the number of measurements accompanied by a reduction in an equal number of unknown parameters does not alter the estimate of the other unknowns if the proper transformed measurement covariance matrix is utilized. Here we fail (on purpose to save time) to use the proper weighting of measurement residuals, but the idea behind the Schaffrin-Graffarend theorem is the same — reducing the number of measurements and an equal number of unknowns without altering the validity of the estimates obtained with this reduced data set. This technique is simple, effective, and robust. If used infrequently, then its use is justifiable. But because of the time required, it probably should not be used more than once per data set.

10.3 DIRECT SOLUTION OF POSITION AND RECEIVER CLOCK OFFSET — BANCROFT'S SOLUTION (NO A PRIORI INFORMATION REGARDING POSITION)

Till now, solution of the single-station positioning problem has used traditional Taylor series expansion techniques associated with nonlinear solutions. For final solutions, a point of expansion must be provided so that all elements of the Taylor expansion can be computed. Just how to find such a priori values (guesses) for station position and clock offsets is not always so obvious. We have just seen that a global search that minimizes sequentially differenced pseudoranges, which removes receiver clock offsets from consideration, will lead to a good guess of station position. Substitution of the position guess into the original pseudorange equations will then allow one to solve for the clock offset.

But though searching is a robust technique, it is also time consuming. So the quest for an analytical solution is worthwhile. Also, analytical solutions generally allow for more understanding of the overall geometrical aspects of the positioning problem. Fortunately for us, Bancroft [1985] has provided such a solution.

Although Bancroft's solution is noniterative in itself, the recovery of position and clock offset does require at least one iteration. This iteration is required since if a position guess is not available, then corrections for at least tropospheric refraction cannot be made because the satellite's elevation angle is not known, and this information is required for calculation of the correction.

In the first iteration such corrections can be ignored because they are small (but definitely not negligible). Using the first iteration's solution, the data can be appropriately corrected to cast the mathematical relations into the form that Bancroft solved. As before, it is assumed that the satellite clock correction polynomial is adequate to remove this effect prior to data processing. The reader is also directed to other papers dealing with further discussion of Bancroft's solution by Abel and Chaffee [1991a, 1991b, 1992] and Chaffee and Abel [1992].

10.3.1 The Solution

The Lorentz inner product is defined as follows:

$$\langle \boldsymbol{g}, \boldsymbol{h} \rangle \overset{\text{def}}{=} \boldsymbol{g}^T M \boldsymbol{h}, \quad \boldsymbol{g}, \boldsymbol{h} \in R^4, \quad M_{4\times4} = \begin{bmatrix} I_{3\times3} & 0 \\ 0 & -1 \end{bmatrix}. \tag{10.19}$$

We now look at a single pseudorange relation appropriately corrected as mentioned above,

$$P^i = \sqrt{(x^i - x)^2 + (y^i - y)^2 + (z^i - z)^2} + c \cdot dt \tag{10.20}$$

recognizing that $c \cdot dt$ is a scalar (c is the vacuum speed of light); let $b = c \cdot dt$.
Rewriting (10.20) as

$$P^i - b = \sqrt{(x^i - x)^2 + (y^i - y)^2 + (z^i - z)^2}\ ,$$

and squaring both sides, yields

$$\begin{aligned} P^{i^2} - 2P^i b + b^2 &= (x^i - x)^2 + (y^i - y)^2 + (z^i - z)^2 \\ &= x^{i^2} - 2x^i x + x^2 + y^{i^2} - 2y^i y + y^2 + z^{i^2} - 2z^i z + z^2 . \end{aligned} \qquad (10.21)$$

Grouping terms, one gets

$$\left[x^{i^2} + y^{i^2} + z^{i^2} - P^{i^2}\right] - 2\left[x^i x + y^i y + z^i z - P^i b\right] = -\left[x^2 + y^2 + z^2 - b^2\right]$$

or more compactly

$$\frac{1}{2}\left\langle \begin{bmatrix} \boldsymbol{r}^i \\ P^i \end{bmatrix}, \begin{bmatrix} \boldsymbol{r}^i \\ P^i \end{bmatrix} \right\rangle - \left\langle \begin{bmatrix} \boldsymbol{r}^i \\ P^i \end{bmatrix}, \begin{bmatrix} \boldsymbol{r} \\ b \end{bmatrix} \right\rangle + \frac{1}{2}\left\langle \begin{bmatrix} \boldsymbol{r} \\ b \end{bmatrix}, \begin{bmatrix} \boldsymbol{r} \\ b \end{bmatrix} \right\rangle = 0\ . \qquad (10.22)$$

One copy of equation (10.22) exists for each pseudorange measured. Assume there are four such pseudoranges that provide sufficient information to resolve receiver position and clock error. Let the matrix

$$B = \begin{bmatrix} x^1 & y^1 & z^1 & P^1 \\ x^2 & y^2 & z^2 & P^2 \\ x^3 & y^3 & z^3 & P^3 \\ x^4 & y^4 & z^4 & P^4 \end{bmatrix}$$

where x^i, y^i, z^i are the coordinates of the i-th satellite at transmission time and P^i is the measured pseudorange to satellite i. Then the four pseudorange relationships can be expressed as

$$\alpha - BM\begin{bmatrix} \boldsymbol{r} \\ b \end{bmatrix} + \Lambda\tau = 0 \qquad (10.23)$$

where

$$\Lambda = \frac{1}{2}\left\langle \begin{bmatrix} \boldsymbol{r} \\ b \end{bmatrix}, \begin{bmatrix} \boldsymbol{r} \\ b \end{bmatrix} \right\rangle, \ \tau = \begin{bmatrix} 1 \\ 1 \\ 1 \\ 1 \end{bmatrix},$$

and α is a 4×1 vector with

$$\alpha_i = \frac{1}{2}\left\langle \begin{bmatrix} \boldsymbol{r}^i \\ P^i \end{bmatrix}, \begin{bmatrix} \boldsymbol{r}^i \\ P^i \end{bmatrix} \right\rangle.$$

Solving (10.23) for $\begin{bmatrix} \boldsymbol{r} \\ b \end{bmatrix}$,

$$\begin{bmatrix} \boldsymbol{r} \\ b \end{bmatrix} = MB^{-1}(\Lambda\tau + \alpha). \tag{10.24}$$

Substituting (10.24) into (10.23) (for both $\begin{bmatrix} \boldsymbol{r} \\ b \end{bmatrix}$ and $\Lambda = \frac{1}{2}\left\langle \begin{bmatrix} \boldsymbol{r} \\ b \end{bmatrix}, \begin{bmatrix} \boldsymbol{r} \\ b \end{bmatrix} \right\rangle$),

and realizing that $\langle M\boldsymbol{g}, M\boldsymbol{h}\rangle = \langle \boldsymbol{g}, \boldsymbol{h}\rangle$, yields

$$\left\langle B^{-1}\tau, B^{-1}\tau\right\rangle\Lambda^2 + 2\left[\left\langle B^{-1}\tau, B^{-1}\alpha\right\rangle - 1\right]\Lambda + \left\langle B^{-1}\alpha, B^{-1}\alpha\right\rangle = 0. \tag{10.25}$$

Since (10.25) is a quadratic equation in Λ, its solution yields potentially two locations in space (substituting the solutions for Λ into (10.24)), one of which is the desired solution.

Equation (10.25) yields the solution to the case where exactly four pseudorange measurements are available. But most of the time, five or more measurements are available, and one should use all the available measurements if possible. A modification of (10.23) is possible to achieve this goal.

When (10.23) contains more than four measurements, then one could multiply by B^T to reduce the number of equations to four,

$$B^T\alpha - B^T B\, M\begin{bmatrix} \boldsymbol{r} \\ \boldsymbol{b} \end{bmatrix} + B^T\Lambda\tau = 0. \tag{10.26}$$

Following the same logic as given above, one gets the following:

$$\left\langle (B^TB)^{-1}B^T\tau,(B^TB)^{-1}B^T\tau\right\rangle\Lambda^2+2\left[\left\langle (B^TB)^{-1}B^T\tau,(B^TB)^{-1}B^T\alpha\right\rangle-1\right]\Lambda \\ +\left\langle (B^TB)^{-1}B^T\alpha,(B^TB)^{-1}B^T\alpha\right\rangle=0. \tag{10.27}$$

It is clear that this incorporates all measurements in a "least-squares" sense. That is, the coefficients of the quadratic polynomial are now minimum L_2 norm values, but the overall solution is not the usual least-squares solution. Equation (10.27) still requires the solution of a quadratic, thus some decision about which root to choose is also required.

This can easily be done by comparing the two solutions to the original pseudorange measurements. The solution to choose will agree with these original measurements in the case of four satellites and will be very close to the individual pseudoranges when more than four pseudorange measurements are used.

For example, pseudorange data were collected on March 8, 1994, at Columbus, Ohio. At one epoch, five pseudoranges were collected that allowed for five different four-measurement solution combinations to be analyzed according to (10.25). The results are as follows:

Table 10.1.

Combination		*x (m)*	*y (m)*	*z (m)*
1	$\Lambda+$	-776901.10	7011222.27	-6354587.74
	$\Lambda-$	595035.50	-4856359.62	4078237.14
2	$\Lambda+$	-1303230.47	4642879.48	-5190159.12
	$\Lambda-$	595037.19	-4856354.13	4078234.98
3	$\Lambda+$	861372.66	6727309.29	-4550450.65
	$\Lambda-$	595030.92	-4856358.96	4078232.20
4	$\Lambda+$	-1061927.87	5079711.95	-2948006.26
	$\Lambda-$	595036.73	-4856356.87	4078229.49
5	$\Lambda+$	-1970270.71	11580605.17	-8385168.84
	$\Lambda-$	595038.09	-4856367.65	4078239.22

Clearly the solution $\Lambda-$ is common in all combinations.

10.4 DILUTION OF PRECISION

Returning now to (10.11), one is often concerned with the geometrical strength of the solution, which is represented by the matrix of partial derivatives with respect to Cartesian coordinates and clock offset, A. However, it is not intuitive nor all that instructive to examine the values in the matrix A. One excellent measure is the inverse of the least-squares normal matrix $(A^T\Sigma^{-1}A)$. Should the proper measurement covariance matrix Σ have been chosen, then the inverse matrix would be the solution vector's covariance matrix. At each epoch this contains ten different linearly independent numbers, so it requires that one convey too many numbers to judge the geometric quality. And even if one used this information, a knowledge of $\Sigma=\sigma^2 I$ must also be factored into the argument. That is, $(A^T\Sigma^{-1}A)$ represents not only the geometrical strength, but also a combination of geometry and measurement precision.

Another consolidation (and also a reduction in information) might be to look at the trace of $(A^T\Sigma^{-1}A)^{-1}$. This quantity is the same regardless of which coordinate orientation one chooses, but does not allow one to judge the shape of the variance ellipsoid. However, it does allow us to convey in only one number some information about geometry — the sum of all the four variances, $\sigma_x^2+\sigma_y^2+\sigma_z^2+\sigma_{c\Delta t}^2$. But still measurement precision is included. To eliminate measurement precision and to isolate a quantity that is a function of only geometry, the Geometric Dilution of Precision is defined as follows:

$$\mathrm{GDOP} = \frac{\sqrt{\mathrm{trace}\ (A^T\Sigma^{-1}A)^{-1}}}{\sigma}$$

In case $\Sigma=\sigma^2 I$, as is usually accepted, the above simplifies to

$$\mathrm{GDOP} = \sqrt{\mathrm{trace}\ (A^T A)^{-1}} \tag{10.28}$$

Desirable values of GDOP are in the neighborhood of [0, 5] (units are m/m!).If the analyst is interested only in position, and not the clock, then only the diagonal terms involving position need to be included in the summation. We define then the Position Dilution of Precision to be

$$\mathrm{PDOP} = \frac{\sqrt{\sigma_x^2+\sigma_y^2+\sigma_z^2}}{\sigma}$$

where $\sigma_x^2, \sigma_y^2, \sigma_z^2$ are the first three diagonal elements of $(A^T\Sigma^{-1}A)^{-1}$ when the ordering of unknowns is $x, y, z, c\Delta t$.

We note that $\sigma_x^2+\sigma_y^2+\sigma_z^2 = \sigma_E^2+\sigma_N^2+\sigma_U^2$; that is, if the covariance matrix is transformed from x, y, z to E, N, U (east, north, up), then the trace is unaffected. This transformation requires the usual law of variance propagation. Let $P_{x,y,z,c\Delta t}=(A^T\Sigma^{-1}A)^{-1}$, then

$$P_{E,N,U,c\Delta t} = \begin{pmatrix} R & 0 \\ 0 & 1 \end{pmatrix} P_{x,y,z,c\Delta t} \begin{pmatrix} R^T & 0 \\ 0 & 1 \end{pmatrix},$$

where

$$R = \begin{pmatrix} -\sin\lambda & \cos\lambda & 0 \\ -\sin\phi\cos\lambda & -\sin\phi\sin\lambda & \cos\phi \\ \cos\phi\cos\lambda & \cos\phi\sin\lambda & \sin\phi \end{pmatrix}$$

and ϕ, λ are geodetic latitude and longitude respectively.

Now that $\sigma_E^2, \sigma_N^2, \sigma_U^2$ can be computed from $P_{x,y,z,c\Delta t}$, we can define the Horizontal Dilution of Precision as

$$\text{HDOP} = \frac{\sqrt{\sigma_E^2 + \sigma_N^2}}{\sigma},$$

and the Vertical Dilution of Precision as

$$\text{VDOP} = \frac{\sqrt{\sigma_U^2}}{\sigma} = \frac{\sigma_U}{\sigma}.$$

If one is interested in time, then TDOP = $\sigma_{c\Delta t}/\sigma$. So now we have GDOP^2 = PDOP^2 + TDOP^2 = HDOP^2 + VDOP^2 + TDOP^2. Normally the cofactor matrix $(A^TA)^{-1}$ is used so that division by the measurement standard deviation, σ, is not required.

The several DOPs can be computed based on either anticipated or actual satellite coverage. It is usually better to use the almanac rather than broadcast ephemeris for calculations of anticipated coverage. Using anticipated satellite measurements, one can "test the water" to see which part of the day will yield the best results. Clearly, some satellite configurations are better than others, and knowing the best coverage is important information to anyone using the GPS system.

These DOP calculations are generally available in all survey planning software products.

10.5 COMBINING PHASE AND PSEUDORANGE FOR SINGLE-SITE DETERMINATIONS

As already given in Chapter 5, phase and pseudorange measurements are similar. For convenience, the previously given relations are presented here. First, the pseudorange relation given in (5.23) is repeated:

$$\begin{aligned} P_i^k(t) = \rho_i^k(t, t-\tau_i^k) + I_i^k + T_i^k + dm_i^k + \\ + c[dt_i(t) - dt^k(t-\tau_i^k)] + c[d_i(t) - d^k(t-\tau_i^k)] + e_i^k \end{aligned} \qquad (5.23)$$

Now the phase relation (5.32) is repeated:

$$\Phi_i^k(t) = \rho_i^k(t, t-\tau_i^k) - I_i^k + T_i^k + \delta m_i^k + c[dt_i(t) - dt^k(t-\tau_i^k)] + \\ + c[\delta_i(t) + \delta^k(t-\tau_i^k)] + \lambda[\phi_i(t_0) - \phi^k(t_0)] + \lambda N_i^k + \varepsilon_i^k \tag{5.32}$$

The common and dissimilar terms are discussed in 5.1.2. Focusing on the task at hand, to incorporate phase with pseudoranges for more precise single-site modeling, let us disregard some of the terms.

The terms to be ignored are multipath and delays. Thus now (5.23) and (5.32) are rewritten as (10.29) and (10.30) respectively:

$$P_i^k(t) = \rho_i^k(t, t-\tau_i^k) + I_i^k + T_i^k + c[dt_i(t) - dt^k(t-\tau_i^k)] + e_i^k \tag{10.29}$$

$$\Phi_i^k(t) = \rho_i^k(t, t-\tau_i^k) - I_i^k + T_i^k + c[dt_i(t) - dt^k(t-\tau_i^k)] + \\ + \lambda[\phi_i(t_0) - \phi^k(t_0)] + \lambda N_i^k + \varepsilon_i^k \tag{10.30}$$

Now the notation ρ^* will be introduced to simplify the previous two relations. Define ρ^*, N^* as follows:

$$\rho^* = \rho_i^k(t, t-\tau_i^k) + T_i^k + c[dt_i(t) - dt^k(t-\tau_i^k)] \\ N^* = [\phi_i(t_0) - \phi^k(t_0)] + N_i^k \tag{10.31}$$

Using (10.31) we can now substitute into (10.29) and (10.30) to obtain

$$P_i^k(t) = \rho^* + I_i^k + e_i^k \tag{10.32}$$

$$\Phi_i^k(t) = \rho^* - I_i^k + \lambda N^* + \varepsilon_i^k \tag{10.33}$$

10.5.1 Single Frequency Smoothing

Equations (10.32) and (10.33) can be used to smooth the noisy pseudoranges with the precise but biased phases. Clearly, however, the ionospheric terms, I_i^k, cause problems should only single frequency measurements be available. The problem is associated with a lack of information to recover the offset in the ionospheric terms. That is, (10.32) – (10.33) do allow for the change in ρ^* and in I_i^k from one epoch to the next. Differencing (10.32) and (10.33) between two successive times yields the following:

$$\Delta P_i^k(t) = \Delta\rho^* + \Delta I_i^k + \Delta e_i^k$$
$$\Delta\Phi_i^k(t) = \Delta\rho^* - \Delta I_i^k + \Delta\varepsilon_i^k \tag{10.34}$$

Under the assumption that Δe_i^k and $\Delta\varepsilon_i^k$ are zero, then we have two equations in two unknowns — one measurement at the metre level, the other at the millimetre level. But what about I_i^k and λN^* at some epoch? Here there is no real reprieve. One could combine the I_i^k and ρ^* at an epoch and also I_i^k and λN^* so as to estimate these linear combinations. The data would then support the estimation of changes from epoch to epoch as shown in (10.34). Choosing the reference epoch at maximum elevation where the ionosphere's mapping function is a minimum is reasonable. Compromises must be made when only single frequency data are available.

The real information then is in the second equation of (10.34) which gives a millimetre "constraint" on the behavior of ρ^* and I^*. One could either use (10.32) and (10.33) while estimating the needed offsets, or the Schaffrin-Grafarend theorem can be applied to temporal differences with the proper modeling of the measurement covariance matrices that will not be diagonal if such differences are used.

10.5.2 Dual Frequency Smoothing

Having dual frequency data changes the situation dramatically. Now writing the four available (dual-frequency) measurement relations, we have

$$\begin{aligned}
P_{1i}^{\;k}(t) &= \rho^* + I_i^k + e_{1i}^{\;k} \\
P_{2i}^{\;k}(t) &= \rho^* + (f_1/f_2)^2\, I_i^k + e_{2i}^{\;k} \\
\Phi_{1i}^{\;k}(t) &= \rho^* - I_i^k + \lambda_1 N_1{}^* + \varepsilon_{1i}^{\;k} \\
\Phi_{2i}^{\;k}(t) &= \rho^* - (f_1/f_2)^2\, I_i^k + \lambda_2 N_2{}^* + \varepsilon_{2i}^{\;k}
\end{aligned} \tag{10.35}$$

The reader is cautioned here that ρ^* is not distance and N^* is not an integer!

It is seen that the right side of (10.35) contains only four nonstochastic parameters — ρ^*, I_i^k, $N_1{}^*$, $N_2{}^*$. For the sake of being easily recognized, we shall call ρ^* the ideal pseudorange for it represents what the pseudoranges (10.32) would be if the ionospheric effect were zero.

Now we rewrite (10.35) into a more usable form:

$$\begin{bmatrix} P_1 \\ P_2 \\ \Phi_1 \\ \Phi_2 \end{bmatrix} = \begin{bmatrix} 1 & 1 & 0 & 0 \\ 1 & (f_1/f_2)^2 & 0 & 0 \\ 1 & -1 & \lambda_1 & 0 \\ 1 & -(f_1/f_2)^2 & 0 & \lambda_2 \end{bmatrix} \begin{bmatrix} \rho^* \\ I \\ N_1{}^* \\ N_2{}^* \end{bmatrix} + \begin{bmatrix} e_1 \\ e_2 \\ \varepsilon_1 \\ \varepsilon_2 \end{bmatrix}. \tag{10.36}$$

where subscripts denoting station identifiers and superscripts denoting satellite identifiers have been omitted since here we are not combining data across stations or satellites.

Now for a quick analysis of (10.36), it is seen that data from only one epoch are sufficient to recover all parameters on the right assuming that the errors are zeros and that the design matrix is regular (which it is).

Here a sequential filter or batch least-squares algorithms can be constructed to take advantage of the unchanging character of N_1* and N_2*. Each new epoch adds four new measurements but only two new unknowns. The usual implementation is either a Bayes or Kalman filter.

The key element here is to use the average of all pseudoranges to identify the $N*$ values. Once these values are available, then the ideal pseudoranges, $\rho*$ values, can be obtained from the millimetre level phases.

Convergence of the $N*$ standard deviations behaves like $1/\sqrt{n}$ as shown in Figure 10.2, where n is the number of epochs. Euler and Goad [1991] showed that $N_1* - N_2*$ is determined much better than either N_1* or N_2*. This will be used again later.

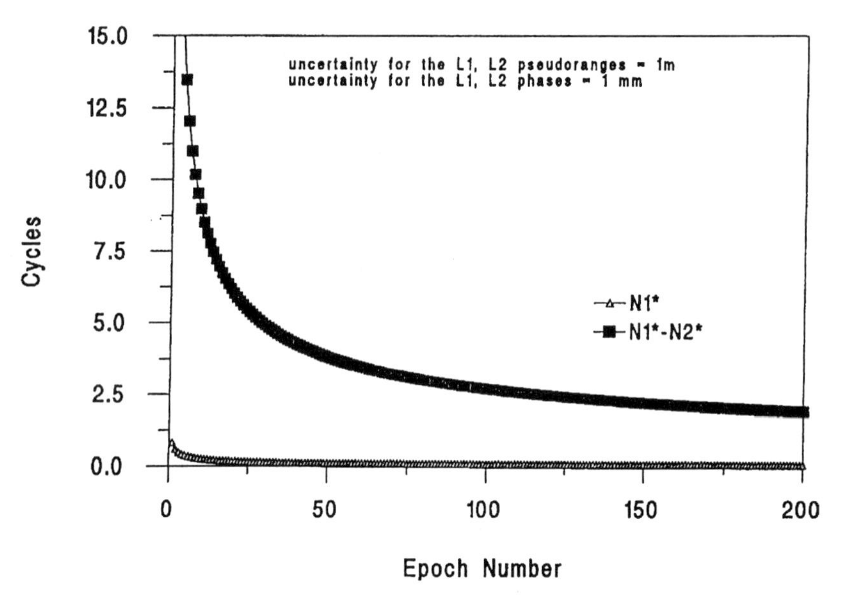

Figure 10.2. Standard deviation of the estimates of the N_1* and $N_1*- N_2*$ biases.

10.5.3 Discussion

Again it is worth reviewing what was assumed to obtain the benefits of combining phase and pseudorange. That is, that multipath was zero and the pseudorange and phase delays are either zero or have been accommodated through calibration.

The delays can be determined, but the user cannot totally control multipath. Antenna characteristics can be a major contributor to multipath sensitivity. And, of course, the reflective environment is the source of the multipath signals reaching the GPS receiver antenna.

Once the optimal estimate of ρ^* has been obtained, then the job of estimating position continues. That is, ρ^* on a one-way basis must be combined with other one-way measures to recover position using the usual techniques.

By far the most troublesome event is a situation of loss of lock on the GPS signal by the receiver. This loss of lock can be caused by many different possibilities. Physical blockages clearly can cause a break in signal tracking. Other occurrences are electronic in nature where unexpected large signal variations fall outside some predetermined range (bandwidth). Large and sudden changes in the ionospheric effects have been known to cause receivers to lose lock. The occurrences usually happen at the peaks of the eleven-year solar cycles.

Whatever the cause, one must verify that lock is or is not maintained. In the case of a loss, then once tracking is re-acquired, the implementation of (10.36) must react accordingly recognizing that new parameters N_1^* and N_2^* are to be estimated. This can be quite problematic. For example, what happens if the residuals in phase are at the four-centimetre level? Did multipath cause it? Or did the receiver lose lock by exactly one cycle and the parameters adjust accordingly?

Also, if a real-time result is required, no new tracking is available to be used in determining maintenance of lock or not. The reader is directed to Chapter 8 for an in-depth discussion of these topics.

Actually the N_1^*, N_2^* parameters can be classified as nuisance parameters. That is, their presence is needed only to be able to use the Φ_1 and Φ_2 measures. Another possibility is to use the Schaffrin-Grafarend theorem and eliminate the N^* values through differencing over time (epochs). That is, the data processing begins using only P_1, P_2 at the initial epoch. At the next and all following epochs, in addition to P_1 and P_2, one differences the previous phase measurements with those measured at the current epoch. Simple differencing then removes the two unknowns N_1^* and N_2^* and is accompanied by a reduction in an equal number of measurements, so the estimation of ρ^* and I values is guaranteed. (Actually it is not quite this straightforward, but the proper conditions discussed earlier are indeed satisfied.) Cycle slips are identified by large residuals in the differenced phase measurements between epochs. Such implementations must model the statistical correlation between the (differenced) measurements however. This complication leads most investigations to the more traditional modeling and thus the necessity of estimating N_1^* and N_2^*.

However this discussion shows the information content in the measurement process. Highly precise phase change measures can then be very useful in smoothing the pseudoranges in single site determinations.

10.6 SUMMARY

In this chapter the various processing techniques involving pseudoranges or pseudoranges/carrier phases for single-site (absolute) determinations were introduced. Due to the nature of the broadcast and precise orbits, these ECF orbits require an earth rotation correction in the usual processing step. Once completed, then either linearized or analytical solutions are possible. If the standard linearized approach is used, then one can use Newton-Raphson iteration or a search technique.

The differencing of pseudoranges at an epoch to eliminate the receiver clock term reveals the hyperbolic nature of the measurement process.

A study of tracking geometry can be made using the least-squares cofactor matrix ($A^T A$). Various DOPs can be computed based on the desired goals. Millimetre carrier phase measurements offer substantial improvement to the pseudorange processing. This improvement is much more complete in the case of dual-frequency tracking.

References

Abel, J. S. and J. W. Chaffee (1991a), Integrating GPS with Ranging Transponders, *Proceedings 1991 Institute of Navigation National Technical Meeting*, Phoenix AZ, Jan.

Abel, J. S. and J. W. Chaffee (1991b), *Existence and Uniqueness of GPS Solutions. IEEE Trans. Aerosp. and Elec. Systems*: AES-27 (6), 960-967.

Abel, J. S. and J. W. Chaffee (1992), Geometry of GPS Solutions, *Proceedings 1992 Institute of Navigation National Technical Meeting*, San Diego CA, Jan.

Bancroft, S. (1985), *An Algebraic Solution of the GPS Equations*, *IEEE Trans. Aerosp. and Elec. Systems*: AES-21 (7), 56-59.

Chaffee, J. W. and J. S. Abel (1992), The GPS Filtering Problem, *Proceedings 1992 PLANS*, Monterey CA, Mar.

Euler, H. J. and C. C. Goad (1991), *On Optimal Filtering of GPS Dual Frequency Observations Without Using Orbit Information*, *Bulletin Géodésique*, 65, 2, 130–143.

Schaffrin, B. and E. Grafarend (1986), *Generating Classes of Equivalent Linear Models by Nuisance Parameter Elimination—Applications to GPS Observations*, *manuscripta geodaetica*, 11, 262-271.

11. SHORT DISTANCE GPS MODELS

Clyde C. Goad
Department of Geodetic Science and Surveying, The Ohio State University, 1958 Neil Avenue, Columbus OH 43210-1247 U.S.A.

11.1 INTRODUCTION

Early on in Bossler et al. [1980] it was recognized that receivers that measure the (reconstructed) carrier phase differences between the satellite and receiver precisely would allow for precise recovery of baselines. This observation can be made through the study of equation (5.33). The key ingredients, at least at the introductory stage, are geometry and clock states. It is hoped that other contributors such as troposphere, ionosphere, multipath, and noise are very small or can be removed through calibration and modeling.

Both clocks and geometry remain either as systematically large in the case of geometry or capricious in the case of the clock states. While the effect of geometry is the desirable signal we want to exploit, one must, at the same time, eliminate the contribution of clock drift. This can be done either through modeling, Goad [1985] or through the use of physical differencing, Goad and Remondi [1983]. While the elimination of the clock states requires two satellite measurements to remove the receiver clock offset, or two receivers to remove the satellite clock offset, we see that double differences then are the natural quantities that are sensitive to only geometry and not clocks. This has already been shown in section 5.2.5. However, here it should be emphasized that the differencing can be done in the modeling rather than using a difference of measurements. Most, but not all, investigations choose to difference measurements. Here both single differences eq. (5.51) and double differences eq. (5.58) will be discussed in the context of estimating short baselines. So first we must address the concept of short.

11.2 SHORT DISTANCE GPS MODELS

Defining the concept of "short" baselines is not so easy however. Let us consider more carefully the ionosphere for example. The activity of the ionosphere is known to depend greatly on the eleven-year cycle of sunspot activity. So when the sunspot activity is low, then the ionosphere is not so active, and the effect on microwave signals from GPS satellites is similar over a wider area than when the sunspot activity is increased. In 1983 when the sunspot activity was low, newly introduced single frequency phase-measuring GPS receivers provided phase measurements which allowed for integer identification up to distances of 60 km. At the maximum of the most recent sunspot activity in 1990–1991, integers were difficult to identify, at times, over 10 km distances.

The residual troposphere (i.e., what is left after a model has been applied) also starts to decorrelate at about 15 km. So here we shall define a baseline to be short in the sense of normal surveying tasks. That is, most of the time surveyors will use GPS receivers to replace conventional angle and distance measuring chores where such is not very cost effective relative to the cost of using GPS.

Most of the time, this will involve the use of GPS where visibility with theodolites and EDMs or total stations is not possible. This could be quite short at the level of hundreds of metres and longer. Theoretically, there is not an upper limit in distance, but practically speaking most surveyors' project areas will be limited to a few tens of kilometres unless such projects involve the mapping of roadways, aqueduct systems, etc. Thus it is clear that relative to the ionosphere, most surveying projects can ignore the contribution of the ionosphere.

However, one should always be on guard to the potentially devastating contribution of "aggravated" ionospheric activity should techniques be in use which depend on a total cancellation of the ionosphere.

One might conclude then that for local surveying projects the less expensive single-frequency receivers should be the receivers of choice. However, this is not necessarily the case. We shall see that dual frequency technology does indeed allow for some extensions in data processing. As a matter of fact, such has already been shown in section 10.5.2.

11.2.1 Double Difference Schemes

As mentioned earlier, most investigators choose to form double differences rather than process undifferenced measurements or even single-differenced combinations. So now we must revisit equation (5.57). Here it is rewritten with the minor substitution of ρ in place of the magnitude notation:

$$\Phi_{ij}^{kl} = \rho_i^k(t, t-\tau_i^k) - \rho_i^l(t, t-\tau_i^l) - \rho_j^k(t, t-\tau_j^k) + \rho_j^l(t, t-\tau_j^l) \\ - I_{ij}^{kl} + T_{ij}^{kl} + \delta m_{ij}^{kl} + \lambda N_{ij}^{kl} + \varepsilon_{ij}^{kl} \tag{11.1}$$

Additionally we shall assume that over short distances the I_{ij}^{kl} is zero, that the T_{ij}^{kl} can be modeled, and that the multipath is small. With these additional considerations, (11.1) is now further reduced to

$$\Phi_{ij}^{kl} = \rho_i^k(t, t-\tau_i^k) - \rho_i^l(t, t-\tau_i^l) - \rho_j^k(t, t-\tau_j^k) + \rho_j^l(t, t-\tau_j^l) + \\ + \lambda N_{ij}^{kl} + \varepsilon_{ij}^{kl} \tag{11.2}$$

Equation (11.2) (or its counterpart in terms of cycles) is by far the most used formulation for short baseline reductions. Some additional corrections may also be needed since there is probably no possibility that both receivers collect measurements at exactly the same instant. This, however, does not cause too many problems. Similarly, the clock offsets need to be considered. After linearization as described in section 5.4.1, (11.2) is then rewritten in the form

$$\Delta\Phi_{ij}^{kl} = \left[(u_j^k)^T - (u_j^l)^T\right]\Delta r_j + \lambda \Delta N_{ij}^{kl} + \varepsilon_{ij}^{kl} \tag{11.3}$$

where the left side is the observed measurement minus that calculated based on the best guess of the position of the j-th station and the ambiguity N_{ij}^{kl}. (It is assumed that the i-th station's location is known.) The u's are the usual direction cosines. Thus (11.3) is linear in Δr_j and ΔN_{ij}^{kl}. Let us denote the vector $b^T = [(u_j^k)^T - (u_j^l)^T, \lambda]$, so now (11.3) can be simplified to

$$\Delta\Phi_{ij}^{kl} = b^T \begin{bmatrix} \Delta r_j \\ \Delta N_{ij}^{kl} \end{bmatrix} + \varepsilon_{ij}^{kl} \tag{11.4}$$

Stacking all measurements from all epochs results in the following:

$$\left[\Delta\Phi\right] = B \begin{bmatrix} \Delta r_j \\ \Delta N_1 \\ \Delta N_2 \\ \Delta N_3 \\ \vdots \end{bmatrix} + \left[\varepsilon\right] \tag{11.5}$$

where now all the different, but linearly independent, ambiguities are listed as ΔN_1, ΔN_2, etc.

Normally the baseline (vector) and ambiguities are estimated using the technique of least squares. That is, the best guess of the ambiguities and baseline are those values which minimize the sum of squares of measurement discrepancies once the estimated quantities' contributions are removed. In such implementations, one generally treats the ambiguities as real-valued parameters. These estimates then take on a (real) value which makes the measurement residual sum of squares a minimum. To the extent that common mode contributions to the measurements cancel, then the real-valued estimates of the ambiguities tend toward integer values. The classic case for such easy identification of integer-valued ambiguity estimates is when the baseline is short.

Not all possible difference combinations should be generated however. Theoretically, only those combinations of double differences which are linearly independent offer new information to a data reduction. A linearly dependent combination is one which can be obtained by linearly combining previously used double differences. For example, consider the following possible double differences:

$$\Phi_{9,12}^{6,18}$$

$$\Phi_{9,12}^{6,20}$$

$$\Phi_{9,12}^{18,20}$$

The last double difference can be obtained by a combination of the first two as follows: $\Phi_{9,12}^{18,20} = \Phi_{9,12}^{6,18} - \Phi_{9,12}^{6,20}$.

In other words, once $\Phi_{9,12}^{6,18}$ and $\Phi_{9,12}^{6,20}$ have been used, no new information is contained in $\Phi_{9,12}^{18,20}$. Thus such linearly dependent data should not be considered. If n represents the number of receivers and s the number of satellites being tracked at a data sampling epoch, the maximum number of linearly independent combinations is $(n-1)(s-1)$. For the simple case of just two receivers, then the generation of linearly independent data is not so difficult. But when the number of receivers is greater than two, the task of generating the maximum number of linearly independent measurements in order to gain the maximum amount of information possible is not so trivial. Goad and Mueller [1988] have addressed this problem in detail.

Since there are usually several ways one can combine data to form independent observables, then perhaps there are advantages of some schemes over others. Distance between receivers is one such consideration. Let's consider the case of three ground receivers (A, B, C) as given in Figure 11.1 below.

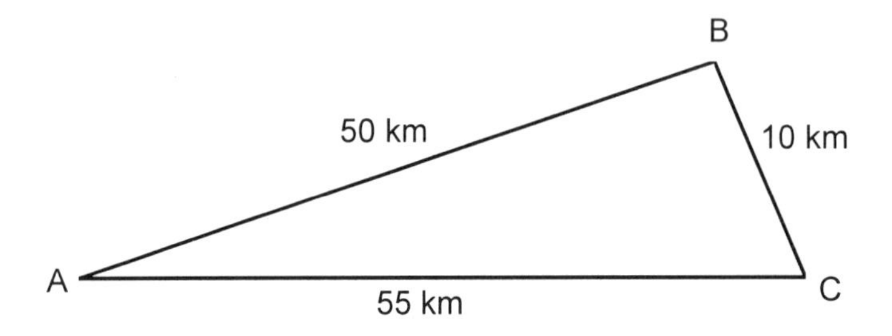

Figure 11.1. Possible geographical distribution of satellite receivers.

Here there are three possible baselines, only two of them being linearly independent. Which two should be chosen? Now it is appropriate to discuss those contributions which were ignored in the generation of equation (11). These include such items as tropospheric and ionospheric refraction, multipathing, arrival time differences, orbit error, etc. Two of these unmodeled contributions are known to have errors that increase with increasing distance between receivers — orbit error and ionospheric refraction. (Tropospheric refraction does also, but only to a limit of, say, 15–50 km). Now back to the figure above. Since we now realize that a more complete cancellation of unmodeled errors occurs for the shorter baseline and thus the use of equation (11.2) is more justified, one should definitely choose the baseline $\overrightarrow{BC}$ as one of the two independent lines. Although not so drastically different, one might as well choose $\overrightarrow{AB}$ as the other independent baseline since it is slightly shorter than the baseline $\overrightarrow{AC}$.

While 50 km approaches the limit of what we choose to call a short baseline length, it is not uncommon to collect data over such baseline lengths, especially if one must connect to an already established network location. What is emphasized here is to use the shortest baselines when possible. This argument assumes all things to be equal such as data collection interval, similar obstructions, etc. In the end, common sense should dictate which baselines to process.

Some always choose to process every possible baseline in order to compare results. This is probably a good idea, but one should be cautioned regarding the repeated use of the same data when the correlations between the solutions are ignored. Overly optimistic statistical confidences will result.

11.2.2 Dynamic Ranges of Double Differences

It is important to understand the impact of baseline length on double difference measurements. To do so, Figure 11.2 was generated using all available phases collected between Wettzel, Germany, and Graz, Austria, on January 15, 1995. The baseline length is 302 km. This length is definitely longer than what would qualify as being short. But the only ingredients that change in the measurements are the distance components, so the dynamic range should scale proportionally with distance.

First we notice that the range is in the neighborhood of 10^6 cycles. Many double differences vary by only one-tenth this amount, but several overhead passes do show large changes. Next, we note that the noise on the phases is of the order of 10^{-2} cycles, or maybe even less. So for 300 km lengths, the signal-to-noise values are in the range of $10^6 : 10^{-2}$ or one part in 10^8. This large range explains why integer ambiguities need not be determined for very long lines. But since we can scale these values to shorter lines, we can generate the following table:

Line Length (km)	S/N Ratio
300	10^8
30	10^7
3	10^6
0.3	10^5

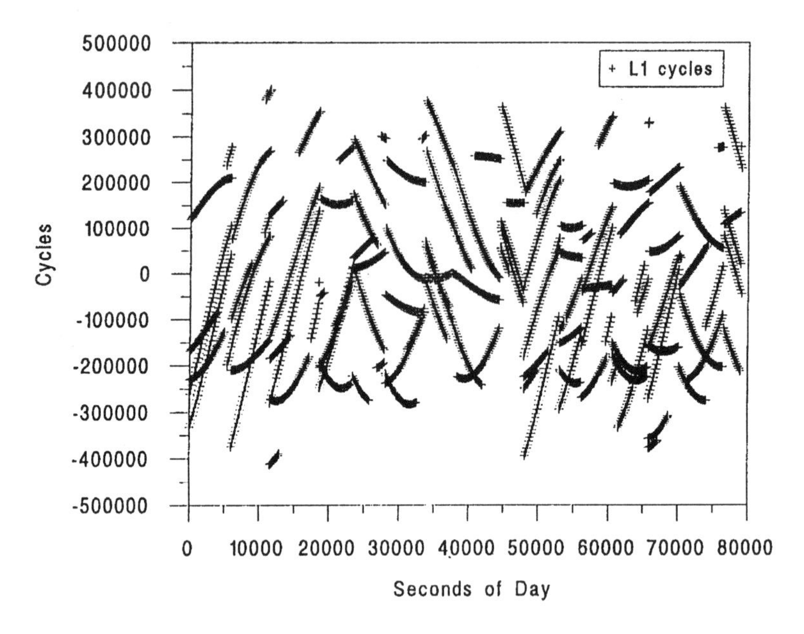

Figure 11.2. Double differences measured on January 15, 1995, Wettzell — Graz (302 km).

That is, as the baseline length decreases, noise plays an increasing role in the determination of location. Clearly, converting the system of measurements from biased ranges to unbiased ranges has a major beneficial impact on vector determination over short lines. Thus our goal for short baseline determination is to find the integer ambiguities.

11.2.3 Use of Pseudoranges

All that has been discussed regarding the use of phases can be extended to pseudoranges with the exception that all pseudorange ambiguities are zero. Thus (11.5) can be used to analyze pseudoranges as:

$$[\Delta P] = \begin{bmatrix} b_1^T \\ b_2^T \\ b_3^T \\ \vdots \\ b_n^T \end{bmatrix} [\Delta r_j] + [\varepsilon] \tag{11.6}$$

which excludes any ambiguity considerations. The above can be very useful when only meter-level precision is needed. Such applications normally go by the descriptor "differential positioning." Normally differential positioning is needed for such applications as navigation, near-terminal guidance, GIS applications, etc. Both smoothing and real-time (filtering) applications use (11.6).

11.2.4 Dual-Frequency Solutions

For a given set of two stations and two satellites contributing to double-difference measurements, one can consider the L_1 and L_2 measures if the receivers collect dual frequency measurements. Now (11.2) is rewritten as follows for each of the two frequencies, including the ionospheric contributions, and where the time arguments in the distance terms have been dropped:

$$\Phi_{ij}^{kl}(L_1) = \rho_i^k - \rho_i^l - \rho_j^k + \rho_j^l + I_{ij}^{kl} + \lambda_1 N_{ij}^{kl}(L_1) + \varepsilon_{ij}^{kl}(L_1) \tag{11.7}$$

$$\Phi_{ij}^{kl}(L_2) = \rho_i^k - \rho_i^l - \rho_j^k + \rho_j^l + (f_1/f_2)^2\, I_{ij}^{kl} + \lambda_2 N_{ij}^{kl}(L_2) + \varepsilon_{ij}^{kl}(L_2) \tag{11.8}$$

These have already been discussed in Chapter 5, but here it is reviewed in the more traditional way. The inclusion of the ionospheric terms complicates the situation quite a bit. We can combine the two double-difference measurements to eliminate the I_{ij}^{kl} terms. The idea is the same as before with pseudoranges. The idea is to choose coefficients α_1 and α_2 such that the following conditions are satisfied:

$$\begin{aligned} \alpha_1 + \alpha_2 &= 1 \\ \alpha_1 + (f_1/f_2)^2\, \alpha_2 &= 0 \end{aligned} \tag{11.9}$$

The first condition yields a combination which looks like the original L_1 relation, and the second condition ensures that the ionosphere is removed. The resulting solution is

$$\alpha_1 = f_1^2 \Big/ (f_1^2 - f_2^2) \approx 2.5457\,, \quad \alpha_2 = -f_2^2 \Big/ (f_1^2 - f_2^2) \approx 1.5457$$

Using the above, we can then combine L_1 and L_2 phase measures to yield

$$\begin{aligned} \Phi_{ij}^{kl}(no\ ion) &= \alpha_1 \Phi_{ij}^{kl}(L_1) + \alpha_2 \Phi_{ij}^{kl}(L_2) \\ &= \rho_i^k - \rho_i^l - \rho_j^k + \rho_j^l + \alpha_1 \lambda_1 N_{ij}^{kl}(L_1) + \alpha_2 \lambda_2 N_{ij}^{kl}(L_2) + \\ &\quad + \alpha_1 \varepsilon_{ij}^{kl}(L_1) + \alpha_2 \varepsilon_{ij}^{kl}(L_2) \end{aligned} \tag{11.10}$$

From the above, one sees that the integer nature of the double-difference N values is destroyed and the errors are amplified. If σ_1 and σ_2 represent the standard deviations of the L_1 and L_2 errors respectively and the errors are uncorrelated, then the $\sigma_{\text{no ion}} = \sqrt{(\alpha_1\sigma_1)^2 + (\alpha_2\sigma_2)^2}$.

Thus if in fact one can justify the fact that I_{ij}^{kl} does indeed equal zero, then generating the ion-free combination may actually amplify errors. Thus for short baselines where dual-frequency receivers are used, solutions using (11.10) can be used to judge whether the assumption of no double difference ionosphere is valid. However, once verified, the optimal solution could be one which processes the L_1 and L_2 phases individually to minimize the propagation of error. However finding the integer values of the ambiguities is equivalent to isolating the baseline to within 1/4 wavelength. The wavelengths of the L_1 and L_2 frequencies is 19 cm and 24 cm respectively. More discussion of these techniques can be found in Chapter 8.

11.2.5 Other Combinations of Dual-Frequency Phases

Other combinations can be considered. One class of combinations is

$$\phi_{ij}^{kl}(\beta_1,\beta_2) = \beta_1\phi_{ij}^{kl}(L_1) + \beta_2\phi_{ij}^{kl}(L_2)$$

where β_1 and β_2 are integers. Here, such linear combinations are guaranteed to produce new measurements with integer ambiguities.

The classic cases are wide- and narrow-lane combinations. For the wide-lane combination, choose $\beta_1 = +1$, $\beta_2 = -1$. So now the effective wavelength is given as

$$\frac{1}{\lambda_w} = \frac{1}{\lambda_1} - \frac{1}{\lambda_2} = \frac{1}{86\text{ cm}}$$

Thus the wide-lane combination almost quadruples the separation of integer solutions in space. The narrow-lane combination, $\beta_1 = +1$, $\beta_2 = +1$, has the opposite effect and is not used often. However, if one can confirm the value of the wide-lane combination (N_1-N_2). then the oddness or evenness is the same for the narrow-lane combination as for the wide-lane combination. Table 11.1 gives many of the possible combinations when integer coefficients have different signs. Other combinations appear to be useful, but the user is cautioned against the possibility of amplifying noise.

11.2.6 Effect of the Ionosphere and Troposphere on Short Baselines

The effect of the ionosphere on baselines most often appears to cause a shortening of the length, Georgiadou and Kleusberg [1988]. This reduction amounts to 0.25

ppm per one meter of vertical ionospheric delay. This is due to the impact of a layer of charged particles which, over short baselines, tends to be similar over the region of interest. Georgiadou and Kleusberg also showed that the data collected at only one dual frequency GPS receiver in an area can be used to infer this systematic effect on nearby single-frequency receiver data.

Table 11.1. Generated wavelengths (m) $\lambda = \dfrac{1}{(a/\lambda_1) - (b/\lambda_2)}$

		b									
		1	2	3	4	5	6	7	8	9	10
	1	.86	−.34	−.14	−.09	−.07	−.05	−.04	−.04	−.03	−.03
	2	.16	.43	−.56	−.17	−.10	−.07	−.06	−.04	−.04	−.03
	3	.09	.13	.29	−1.63	−.21	−.11	−.08	−.06	−.05	−.04
	4	.06	.08	.11	.22	1.83	−.28	−.13	−.09	−.06	−.05
a	5	.05	.06	.07	.10	.17	.59	−.42	−.15	−.09	−.07
	6	.04	.04	.05	.07	.09	.14	.35	−.81	−.19	−.11
	7	.03	.03	.04	.05	.06	.08	.12	.25	−14.65	−.24
	8	.03	.03	.03	.04	.05	.06	.07	.11	.19	.92
	9	.02	.03	.03	.03	.04	.04	.05	.07	.10	.16
	10	.02	.02	.02	.03	.03	.04	.04	.05	.06	.09

Brunner and Welsch [1993] have commented on the effect of the troposphere on height recovery. Heights are normally more problematic due to our inability to observe satellites in the hemisphere below us. Using a cutoff of, say, 15°, as suggested by Brunner and Welsch, to combat the deleterious effect of multipath and also refraction amplifies the problem even more.

Brunner and Welsch estimate that the differential height recovery uncertainty is of the order of three times the effect of differential tropospheric delay. Thus a delay error of only one centimeter results in a relative height error of 3 cm. Also, because of the various possible profiles for a given pressure, temperature, and relative humidity measurements at the surface, "actual meteorological observations at GPS sites together with conventional height profiles has often produced disappointing results" according to Brunner and Welsch. Davis [1986] showed that for VLBI analyses tropospheric mapping function errors seriously impact the estimate of the vertical component of position when the minimum elevation sampled drops below 15°.

11.3 USE OF BOTH PSEUDORANGES AND PHASES

It should now be obvious that for the most precise surveying applications the recovery of the ambiguities is required. Using the approach discussed earlier, the separation of the geometrical part (baselines) and the ambiguities requires some time to pass in order to utilize the accumulated Doppler. One major consequence of this approach is that the integer ambiguities are more difficult to identify with increasing baseline length due to the decoupling of unmodeled error sources such as tropospheric refraction and orbital errors. The same is true for the ambiguity search.

With the introduction of affordable receivers collecting both dual-frequency pseudoranges and phases, this laborious approach might be "laid to rest" if sufficient noise reduction can occur with the tracking of the precise pseudoranges. Techniques utilizing the P-code pseudoranges will now be discussed.

For some time now, the ability to use readily the pseudoranges in addition to dual frequency phase measurements to recover widelane phase biases has been well known, Blewitt [1989]; Euler and Goad [1991].
Here the simultaneous use of all four measurements (phases and pseudoranges from both L_1 and L_2 frequencies) will be visited. It will be shown that the four-measurement filter/smoother can be generated numerically from the average of two three-measurement filters/smoothers. Each of the three-measurement algorithms can be used to provide estimates of the widelane ambiguities provided that some preprocessing can be performed to reduce the magnitude of the L_1 and L_2 ambiguities to within a few cycles of zero. However, such a restriction is not required for the four-measurement algorithm.

11.3.1 A Review

To aid in the understanding of these techniques, a review is presented using the notation of Euler and Goad [1991]. First the set of measurements available to users of receivers tracking pseudoranges and phases on both the L_1 and L_2 frequency channels at an epoch is given mathematically as follows:

$$P_1 = \rho^* + I + \varepsilon_{R_1} \tag{11.11a}$$

$$\Phi_1 = \rho^* - I + N_1\lambda_1 + \varepsilon_{\phi_1} \tag{11.11b}$$

$$P_2 = \rho^* + (f_1/f_2)^2 I + \varepsilon_{R_2} \tag{11.11c}$$

$$\Phi_2 = \rho^* - (f_1/f_2)^2 I + N_2\lambda_2 + \varepsilon_{\phi_2} \tag{11.11d}$$

In equations (11.11a – d), the ρ^* stands for the combination of all nondispersive clock-based terms, or in other words the ideal pseudorange; the dispersive ionospheric contribution at the L_1 frequency is I (theoretically a positive quantity) with group delays associated with pseudoranges and phase advances associated with the phases. The two phase (range) measurements include the well known integer ambiguity contribution when combined in double difference combinations. And finally all measurements have noise or error terms, ε.

The equations (11.11a – d) can be expressed in the more desirable matrix formulation as follows:

$$\begin{bmatrix} P_1 \\ \Phi_1 \\ P_2 \\ \Phi_2 \end{bmatrix} = \begin{bmatrix} 1 & 1 & 0 & 0 \\ 1 & -1 & \lambda_1 & 0 \\ 1 & (f_1/f_2)^2 & 0 & 0 \\ 1 & -(f_1/f_2)^2 & 0 & \lambda_2 \end{bmatrix} \begin{bmatrix} \rho^* \\ I \\ N_1 \\ N_2 \end{bmatrix} + \begin{bmatrix} \varepsilon_{P_1} \\ \varepsilon_{\phi_1} \\ \varepsilon_{P_2} \\ \varepsilon_{\phi_2} \end{bmatrix} \qquad (11.12)$$

Here in equation (11.12) it is readily apparent that in the absence of noise, one could solve the four equations in four unknowns to recover ideal pseudorange, instantaneous ionospheric perturbations, and the ambiguities. Even though the noise values on phase measurements are of the order of a millimeter or less, we know that the pseudorange noises vary greatly from receiver to receiver. L_1 C/A-code pseudoranges have the largest noise values, possibly as high as 2-3 m. This is due to the relatively slow chip rate of 1.023 MHz. P-code chip rates are ten times more frequent which suggest noises possibly as low as 10–30 cm. Obviously to determine ambiguities at the L_1 and L_2 carrier frequencies $(\lambda_1 \cong 19\text{cm}, \lambda_2 \cong 24\text{cm})$, low pseudorange noise values play a critical role in the time required to isolate either N_1 or N_2, or some linear combination of them. In a least-squares smoothing algorithm, Euler and Goad [1991] showed that the worst and best combinations of L_1 and L_2 ambiguities are the narrow lane (N_1+N_2) and widelane (N_1–N_2) combinations, respectively. With 20 cm pseudorange uncertainties, the widelane estimate uncertainty approaches 0.01 cycles while narrow lane uncertainties are at about 0.5 cycles. These should be considered as limiting values since certain contributions to equations (11.11) and (11.12) were not included such as multipath and higher-order ionosphere terms, with multipath by far being the more dominant of the two.

The beauty of equation (11.12) lies in its simplicity and ease that one can implement a least-squares algorithm to obtain hopefully the widelane ambiguity values. Once the widelane bias is obtained, the usual ion-free combination of (11.11b) and (11.11d) yield biases which can be expressed as a linear combination of the unknown L_1 ambiguity and the known widelane ambiguity. Knowing the values of the widelane ambiguity makes it much easier then to recover the L_1 ambiguity. However, not knowing either ambiguity, and even knowing the baseline exactly is a situation in which very possibly the analyst will be unable to recover the integer values for N_1 and N_2.

Other factors, in addition to multipath which have been mentioned, which could influence in a negative way the use of equation (11.11) would be the non-simultaneity of sampling of pseudorange and phase measurements within the receiver or a smoothing of the pseudoranges using the phase (or Doppler) information which attempts to drive down the pseudorange noise but then destroys the relations (11.11a–d). Notice that theoretically no large ionosphere variations or arbitrary motions of a receiver's antenna negate the use of equations (11.11) or (11.12). Thus after sufficient averaging, widelane integer ambiguities can be determined for a receiver/antenna, say, involved in aircraft tracking or the tracking of a buoy on the surface of the sea. For many terrestrial surveys, once sufficient data have been collected to recover the widelane ambiguity, no more would be required except where total elimination of the ionosphere is required such as for orbit determination and very long baseline recoveries. For these situations, both L_1 and L_2 integers are desired and geometry changes between satellite and ground receivers are required unless the baseline vectors are already known. The technique of using such short occupation times along with the four-measurement filter to recover widelane ambiguities is known as "Rapid Static Surveying." Again, one must be aware that unmodeled multipath can be very detrimental value when very short occupation times are utilized.

An example of the use of equation (11.12) in a least-squares algorithm is illustrated in Figure 11.3. Here four measurements, P_1, P_2, Φ_1, Φ_2 were collected every 120 seconds at the Penticton, Canada, tracking station. Although the integer nature of the ambiguity can only be identified after double differencing, the one-way measurements (satellite-to-station) can be smoothed separately and the biases combined later to yield the double difference ambiguities. The figure shows the difference between the linear combination involving P_1, P_2, Φ_1, Φ_2 to yield the wide-lane ambiguity on an epoch-by-epoch basis with the estimated values. The reader will notice that individual epoch values deviate little from the mean or least-squares estimate; the rms. of these values is 0.06 cycles. The three-measurement combinations will be discussed in the next section.

Table 11.2 shows the estimates of the double difference ambiguities formed from the combination of one-way bias estimates between Canadian locations Penticton and Yellowknife which are 1500 km apart. The integer nature of the widelane values is clearly seen while the similar integer values of the L_1 and L_2 bias values cannot be identified. Clearly, in the processing steps, an integer close to the originally determined bias value has been subtracted from the corresponding phase measurements in an attempt to keep the double difference ambiguities close to zero. This was not a requirement of the four-measurement technique however.

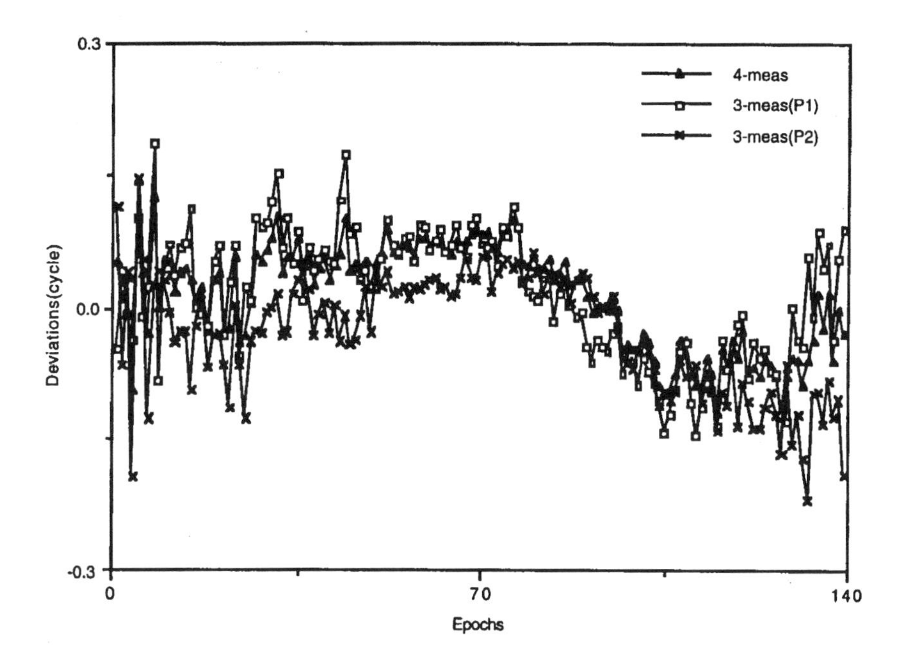

Figure 11.3. Deviations from mean values of the four- and three-measured combinations, Rogue receiver, Penticton, Canada, day 281, 1991, SV14.

Table 11.2 Estimated values of the N_1, N_2, and wide-lane (N_1–N_2) double difference ambiguities.

Sat	Sat	N_1	N_2	N_1–N_2
2	3	-0.162	-1.177	1.105
2	6	-0.284	-1.250	0.966
2	11	-0.002	-1.044	1.042
2	12	-0.539	-1.497	0.957
2	13	-0.450	-1.396	0.947
2	14	1.544	-0.542	2.086
2	15	-1.492	-2.562	1.070
2	16	0.174	-0.877	1.051
2	17	0.035	-1.015	1.051
2	18	-0.335	-1.382	1.047
2	19	0.905	-0.119	1.024
2	20	-0.214	-1.197	0.983
2	21	0.078	-0.984	1.063
2	23	-0.253	-1.316	1.063
2	24	-0.787	-2.735	1.948

11.3.2 The Three-Measurement Combinations

Here the derivations of the two three-measurement combinations are presented. First, one must use the two phase measurements, (11.11b) and (11.11d). Next choose only one of the two pseudorange measurements, P_1 or P_2. Let us choose to examine the selection of either by denoting the chosen measurement as P_i where i denotes either 1 or 2 for the L_1 or L_2 pseudorange respectively. To simplify the use of the required relations, the equations (11.11a–d) are rewritten as follows:

$$P_i = \rho^* + I \cdot \left(\frac{f_1}{f_i}\right)^2 + \varepsilon_{p_i} \tag{11.13a}$$

$$\Phi_1 = \rho^* - I + N_1\lambda_1 + \varepsilon_{\phi_1} \tag{11.13b}$$

$$\Phi_2 = \rho^* - (f_1/f_2)^2 I + N_2\lambda_2 + \varepsilon_{\phi_2} \tag{11.13c}$$

The question to be answered is: What is the final combination of N_1 and N_2 after eliminating the ρ^* and I terms in eq. (11.13a–c)? The desired combinations can be expressed as follows:

$$aP_i + b\Phi_1 + c\Phi_2 = dN_1 + eN_2 + a\varepsilon_{p_1} + b\varepsilon_{\phi_1} + c\varepsilon_{\phi_2} \tag{11.14}$$

where $d = b\lambda_1$, $e = c\lambda_2$. In order to assure the absence of the ρ^* and I terms, the a, b and c coefficients must satisfy the following:

$$a + b + c = 0 \tag{11.15a}$$

$$a\left(\frac{f_1}{f_i}\right)^2 - b - \left(\frac{f_1}{f_2}\right)^2 c = 0 \tag{11.15b}$$

One free condition exists. Since it is desirable to compare the resulting linear combinations of N_1 and N_2 to widelane combination, we choose arbitrarily to enforce the following condition:

$$d = b\lambda_1 = 1 \tag{11.15c}$$

Solving (11.13a–c) with i=1,2 yields the two desired three-measurement combinations with noise terms omitted:

$$-1.2844P_1 + 5.2550\Phi_1 - 3.9706\Phi_2 = N_1 - 0.9697N_2, \text{ for } i = 1 \tag{11.16a}$$

$$-1.0321P_2 + 5.2550\Phi_1 - 4.2229\Phi_2 = N_1 - 1.0313\,N_2, \text{for } i = 2 \tag{11.16b}$$

In practice, the coefficients in (11.16a) and (11.16b) should be evaluated to double precision. The errors in the above combinations are dominated by the pseudorange errors which depend on the receiver characteristics as discussed earlier. But when compared to even the most precise GPS pseudoranges, the phase uncertainties are orders of magnitude smaller. Thus the error in the combination (11.16a) in cycles is equal to 1.28 times the uncertainty of P_1 (in meters). Similarly the combination (11.16b) is equal in cycles to 1.03 times the uncertainty in P_2 (in meters). As with the four-measurement combination, averaging can be used to reduce the uncertainty of the estimated combination. Also the two three-measurement combinations possess almost all the desirable characteristics as the four-measurement combination. The same restrictions also apply. For example, simultaneity of code and phase is required; multipath is assumed not to exist; and filtering of the pseudoranges which destroys the validity of (11.13a–c) is assumed not to be present.

One situation does require some consideration — the magnitudes of N_1 and N_2. That is, in the four-measurement combination the identification of the widelane ambiguity is not hindered by large magnitudes of either N_1 or N_2. However, if either of the two three-measurement combinations differ from the widelane integers by 3% of the N_2 value, this difference could be very large if the magnitude of N_2 is large. Thus some preprocessing is required. For static baseline recovery, this is probably possible by using the estimated biases from the individual widelane and ion-free phase solutions. Using these ambiguity estimates, the L_1 and L_2 phase measurements can be modified by adding or subtracting an integer to all the one-way phases so that the new biases are close to zero. With near-zero L_1 and L_2 ambiguities, the magnitude of the 0.03 N_2 deviation from the widelane integer should be of no consequence in identifying the integer widelane value.

Also it appears that the average of the two, three-measurement combinations is equal to the four-measurement combination. This is not the case identically, but again with small L_1 and L_2 ambiguities, it is true numerically.

To illustrate the power in the three-measurement combinations, the data collected on the Penticton-Yellowknife baseline are used to estimate all three combinations. Table 11.3 shows the resulting estimates (the last column will be discussed later). It is clear that all three combinations round to the same integer values. Also apparent is that the numerical average of each of the three-measurement estimates equals the four-measurement estimate. Again this is due to the preprocessing step to ensure that ambiguities are close to zero.

The figure shows deviations of the one-way (satellite-station) means from the epoch-by-epoch values. The noise levels appear to be small for all the combinations. Large scatter is noted at lower elevation angles when the satellite rises (low epoch numbers) and sets (large epoch numbers). A cutoff elevation angle of 20° was used in the generation of the figure. Also as seen by Euler and Goad [1991], an increase in deviations with the model can be seen at the lower elevation angles. The obvious question is whether this is due to multipath.

Table 11.3. Four-measurement and two three-measurement double difference ambiguity estimates over the Penticton-Yellowknife baseline.

Sat	Sat	N_1-N_2	N_1−1.03 N_2	N_1−0.97 N_2	N_1−1.283 N_2
2	3	1.105	1.052	0.980	1.350
2	6	0.966	1.009	0.933	1.324
2	11	1.042	1.074	1.011	1.337
2	12	0.957	1.003	0.912	1.380
2	13	0.947	0.996	0.909	1.348
2	14	2.086	2.106	2.069	2.261
2	15	1.070	1.149	0.992	1.796
2	16	1.051	1.078	1.025	1.300
2	17	1.051	1.082	1.020	1.338
2	18	1.047	1.090	1.005	1.438
2	19	1.024	1.030	1.015	1.094
2	20	0.983	1.020	0.947	1.321
2	21	1.063	1.086	1.033	1.303
2	23	1.063	1.105	1.023	1.520
2	24	1.948	2.033	1.865	2.723

Clearly if one has all four measurement types, the four-measurement combinations would be used. However, at very little extra effort, all three combinations can be computed possibly helping to identify potential problems in either the P_1 or P_2 measurements.

11.3.3 Anti-Spoofing?

Under certain assumptions about Y-code structure (anti-spoofing turned on), a receiver can compare the two Y-codes and obtain an estimate of the difference between the two precise pseudoranges (P_1 – P_2). For this tracking scenario equation (11.13a) is replaced with

$$P_{1-2} = I\left[1-\left(f_1/f_2\right)^2\right] + \varepsilon_{R_1} - \varepsilon_{R_2} \tag{11.17}$$

Imposing the same restrictions as before on the coefficients a, b, and c, the following is obtained where again the error terms are ignored:

$$\frac{P_{1-2}}{\lambda_1} + \frac{\Phi_1}{\lambda_2} - \frac{\Phi_2}{\lambda_1} = N_1 - 1.2833\, N_2 \tag{11.18}$$

The recovery of N_1−1.283 N_2 using differences in pseudoranges from the Penticton-Yellowknife baseline are given in the last column of Table 11.3. Here, assuming the magnitude of N_2 to be less than or equal to 3, the values of N_1 and N_2 appear to be identifiable in some cases. Using an orbit to recover the ion-free double difference biases can also be of major importance for those cases where the integer values still remain unknown to within one cycle. In any event, some concern is warranted when one is required to use these measurements. Since $1/\lambda_1 = 5.25$, an amplification of the pseudorange difference uncertainty over the individual pseudorange uncertainty of $\sqrt{2} \times 5.25 = 7.42$ is present assuming that the pseudorange difference uncertainty is only $\sqrt{2}$ larger than either the L_1 or L_2 individual pseudorange uncertainties. This is far from the expected situation, so clearly some noisy, but unbiased, C/A-code pseudorange data are highly desirable. The usefulness of these data types when AS is operating is an open question and no definitive conclusions can be obtained until some actual pseudorange differences and C/A-code pseudoranges are available for testing.

11.3.4 Ambiguity Search

With the rapid improvement of personal (low cost) computers, a technique introduced by Counselman and Gourevitch [1981] is now being pursued by some investigators. In essence, it is a search technique which requires baseline solutions to have integer ambiguities. Two techniques have evolved — one which searches arbitrarily many locations in a volume and one that restricts the search points to those locations associated with integer ambiguities. Or to put it another way, one "loops" over all locations in a volume, or one loops over possible integer ambiguity values which yield solutions within a given volume. The explanation of this technique requires only the use of eq. (11.2). A sample location in space is chosen (arbitrarily). It then can be used to calculate the distances (ρ terms) in eq. (11.2) and if it is the actual location of the antenna, then that which is left after removing the ρ terms should be an integer. All measurements to all satellites at all epochs will exhibit this behavior. Locations which do not satisfy this requirement then cannot be legitimate baselines. The beauty of this search technique is that cycle slips (losses of lock) are not a consideration. That is even if the ambiguity changes its integer value, such an occurrence has no impact on the measure of deviation from an integer.

The volume search technique is the easiest to envision and the most robust. A suitable search cube, say one meter on a side, is chosen and each location in a grid is tested. Initial search step sizes of 2–3 cm are reasonable. Once the best search point is found, a finer search can be performed to isolate the best fitting baseline to, say, the mm level. Although the most robust, this volume search can be quite time consuming. An alternative is to choose the four satellites with the best PDOP, and test only those locations which are found from assuming that their ambiguities are integers. That is, one "loops" on a range of ambiguities rather than all locations in the test cube. Such a scheme is much faster, but can suffer if the implied test locations are in error due to unmodeled contributions to the measurements used to seed the search. Effects which can cause such errors are multipathing, ionosphere, etc.

In either case the key to minimizing computer time is to restrict the search volume. One such way is to use differential pseudorange solutions if the pseudoranges are of sufficient quality. Here P-code receiver measurements are usually superior to those which track only the C/A codes. However, some manufacturers are now claiming to have C/A code receivers with pseudorange precision approaching 10 cm. Of course, success can only be obtained if the initial search volume contains the location of the antenna within it. So one now has to contend with competing factors: the search volume needs to be as large as possible to increase the probability that the true location can be found, but then the search volume must be small enough to obtain the estimate in a reasonable amount of time. Clearly, the better the available pseudoranges and the greater the number of satellites being tracked, the better such a search algorithm will work.

These search techniques can also be used even when the antenna is moving. But in this case one needs to assume that no loss of lock occurs for a brief time so that the search can be performed on ambiguities. This then allows for the different epochs to be linked through a common ambiguity value since there is no common location between epochs of a moving antenna (unless the change in position is known which could be the case if inertial platforms are used).

As computers become even more powerful and if receivers can track pseudoranges with sufficient precision and orbits are known well enough, even baselines over rather long distances can be determined using these techniques.

In the end, due to the required computer time, one probably would not use these search techniques to determine the entire path of an airplane or other moving structure, but they could be very useful in providing estimates of integer ambiguities in startup or loss of lock situations.

11.3.5 Nonstatic / Quasi-Static Situation

Such techniques as kinematic, rapid static, stop-and-go, and pseudo-kinematic/ pseudo-static have received much attention for several years now. All techniques try to optimize the recovery of ambiguities and recognize that the integer ambiguities are applicable regardless of whether the antenna is moving or not. Here an attempt to clarify the types is given.

(a) Pseudo-Static / Pseudo-Kinematic. Here one realizes that the recovery of ambiguities over unknown baselines requires station/satellite geometry to change. While one waits for new geometries, the antenna can be taken to nearby locations and the data can be analyzed later. Thus returning to the original baseline and collecting more data for a short period, one then attempts to recover the integer bias. If successful, then the same integers can be used to obtain positions of intermediate locations visited, possibly, only once.

(b) Stop-and-Go. Stop-and-go applications are simply based on two items. The first is to know the ambiguity and the second is to occupy the desired mark for a short while to take advantage of averaging. Lock is maintained between occupations to be able to use the already known ambiguity.

(c) Rapid Static. In rapid static mode, the four-measurement filter is used to recover quickly the ambiguities during, say, a visit the order of five minutes or so when geometry is favorable. The receiver is turned off during travel to conserve batteries.

(d) Kinematic. Here one needs to know location of the antenna while moving. To take full advantage of the data one should either determine integers before moving, or track a sufficient number of satellites to enable integer identification eventually even though in motion.

11.3.6 Fast Ambiguity Resolution

We have already discussed using pseudoranges to enable fast recovery of ambiguities. When precise pseudoranges are not present, but dual frequency phases are tracked, then search techniques can be utilized.

For very local surveys, such as airports, construction sites, etc., one can use antenna swap techniques already discussed in section 5.5.3. Here the swap data can be used with the original data to recover ambiguities almost immediately. Also, known baselines can be occupied to allow for almost instantaneous integer ambiguity recovery.

11.4 DISADVANTAGES OF DOUBLE DIFFERENCES

Double difference data types have some major advantages. Of course the most prominent is the fact that the bias is theoretically an integer. Other advantages are simplicity of the model, cancellation of time varying biases between L_1 phase, L_2 phase, and pseudorange measurements, etc.

But along with these pluses come some minuses. One major consideration is how to define the solution biases. They are necessarily associated with particular satellites for a given baseline. For example, suppose one starts tracking and the common satellites observed by two stations are satellites 9, 11, 12, 18, and 23. From these measurements one could define the following sets of biases:

Set 1	Set 2	Set 3
9–11	9–11	9–11
9–11	11–12	12–18
9–18	12–18	11–23
9–23	18–23	18–23

Obviously Sets 1 and 2 have some easily recognizable algorithm in their generation, while Set 3 does not. But Set 3 is just as legitimate as Sets 1 and 2. So some algorithm should play a role. Now if one chooses Set 1, what happens when an epoch of data is encountered and there is no data from satellite 9? Clearly we have a potential problem. The approach most program designers choose is to assign a bias to each satellite and then to constrain one of the satellites to have a bias of exactly zero. So with this design, any order of satellites can be handled.

While such implementations can handle missing epochs, eventually the satellite with defined zero bias will set, and then that satellite can no longer be used as a reference. So now a changing reference should be accommodated. For any data processing scheme that can handle long arcs (more than a day?), some emphasis on how to define ambiguities is needed.

Another consideration is that the least-squares algorithm requires a proper covariance matrix to be utilized. If there is common tracking of three or more satellites at an epoch, then two or more double differences can be generated, and thus the covariance matrix will not be diagonal (uncorrelated). Thus one must compute this covariance matrix and then accumulate properly the normal matrix and absolute column (right side) in which case a matrix inversion is needed, or one must decorrelate (whiten) the set of measurements using a technique like Gram-Schmidt orthogonalization. Scaling the measurement vector by the inverse of the Cholesky factor of the covariance matrix accomplishes this task also. Properly processing data from several baselines simultaneously complicates matters to an even greater extent.

11.4.1 Single Differences

Processing single, rather than double, differences is an alternative, especially if one limits the processing to only one baseline at a time. In this case all measurements have diagonal covariance matrices, which is a major advantage. But other issues now must be accommodated.

For example, if only phase data are processed, then at least one of the satellite biases cannot be separated from the clock drift which now must be estimated also. Adding pseudoranges can help in this regard, but with the addition of pseudoranges one must make sure that there is no bias between the pseudorange tracking channels and the phase tracking channels. If a bias exists, then it must be accommodated.

This is a major concern because designers of GPS geodetic quality receivers try to ensure that all is stable when generating double differences. Small time varying biases which are common to all L_1 phase, and common to all L_2 phase channels have no effect on double differences. This is considered acceptable since double differences are unaffected by their presence. But should one choose to process single differences instead; accommodating these biases is an important consideration.

11.5 SEQUENTIAL VERSUS BATCH PROCESSING

Regardless of which measurement modeling is chosen, there could be major advantages to choosing a sequential processing algorithm (Bayes or Kalman) over the traditional batch least-squares algorithm. This advantage deals with the ease with which one can add or delete parameters. For example, if a parameter is to be redefined, such as clock drift in case of single difference processing, the variance can be reset to a large number during the prediction of the next state to accommodate a redefinition of parameters. While the clock states are obvious for these actions, such could be useful when cycle slips occur and biases must be redefined. Such redefinition of unknowns happens quite frequently. Adding new unknowns to a batch algorithm is not a very practical thing to do since the least-squares normal matrix increases as the square of the number of unknowns.

Also, in case of a Kalman implementation, the covariance matrix is carried along rather than its inverse (the normal matrix), so identification of integer biases could be aided since the covariance matrix is needed to judge the quality of the recovery.

If loss of lock occurs, then in a sequential implementation one must decide about whether a bias is integer or not. If successful, a constraint can be imposed before a new bias is estimated in the place occupied by the old bias. More about these concepts and ways to discover integer biases is given in Chapter 8.

11.6 NETWORK ADJUSTMENT— THE FINAL STEP

After the individual baseline vectors have been estimated using least-squares techniques, then these should form the pieces of a puzzle that must be fitted together to form a network. For the person responsible for the final product, this is the point where the true test of the surveying efforts is measured. If the vectors do not fit, no product can be delivered.

So it is very important that network adjustment programs provide the analyst with the proper tools to find and fix problems. The importance of this step cannot be overemphasized. For with almost every project there will be problems. For example, even though the GPS data are reduced without problems, if the wrong station name is used, then the software cannot properly connect the vectors together. Another potential problem is for the field operator to enter the wrong height of antenna above the ground marker, the "h.i." Here again the GPS data can process without problems, but then the vector will not connect to the other vectors. How is one led to the identification of the problem without the user having to go to extreme measures himself to discover them. Many simple procedures and choices can be part of any network adjustment software to enable the detective work to proceed easily. For example, look at Table 11.4, which is generated from a reduction of a very small network of GPS baseline vectors in the State of New Jersey.

This little example shows how the user can be helped by displaying information in an organized way. First of all, one notices that the observed baseline vectors are displayed in projection coordinates, not Cartesian. Thus one can immediately spot potential heightening problems as being distinct from horizontal positioning problems. Here one would normally be given the choice of which projection system to use. For example, in most countries there are adopted standards such as Gauss Krueger, Mercator, Lambert, Stereographic, UTM, etc. Other countries such as Denmark have very nonstandard systems. So then it is incumbent on the designers of the network software to incorporate these projections that are really very useful.

Another interesting characteristic seen in Table 11.4 is that the baseline vectors have been sorted by length Obviously one will generally not have control of the way the computer "reads in" baseline solution files. This is especially true if wild card names, such as *.SOL, can be used. Thus even though inputs may be random, outputs can be organized in ways to aid the analyst.

Another very important tool is the use of a (statistical) edit function. One should not be required to remove manually those measurements that do not connect within their statistical confidence region. For example, in Table 11.4, the "*" character denotes that baseline 15 has residuals (the first number of the pair inside parentheses) that are large compared to its estimated uncertainty. The second number, the normalized residual, is the so-called tau statistic, which is expressed in units of standard deviations. Here a cutoff of 3.0 standard deviations causes this baseline to be rejected. An automatic rejection option is considered a necessity when processing even a moderate number of baseline vectors.

Many other options and outputs can be considered when designing a network adjustment program. The following is a list of possible questions one can ask when considering buying or designing such a program:

1. Are vector correlations used?
2. Will the program handle correlated (multiple) baselines?
3. Can one choose the units such as feet, meters, degrees, gons, etc.?
4. Is there an upper limit on the number of baselines or number of unknowns?
5. When the number of unknowns is large, how much time does the program require?
6. Does the program provide plots of baselines, error ellipses, histograms, etc.?
7. Does the program identify "no check" baselines?
8. Does the program detect and accommodate multiple networks? (This happens quite often!)
9. Does the program accommodate geoid models so that mean sea-level heights can be used with GPS baseline vectors?
10. For projection coordinates, does the program provide scale and convergence values?
11. Can one input as a priori coordinates, location in projection coordinates?

12. What happens in the case of singularities? Will the program terminate abnormally?

The above is just a sampling of what to ask. When considering such programs, checking with those experienced in network adjustments is highly suggested. Leick [1995] gives an extensive discussion of adjustment issues in his Chapters 4 and 5. Another excellent text on parameter estimation is Koch [1988].

Table 11.4 . Output of measurements and residuals in projection coordinates from a network adjustment.

Measurements (State Plane Vectors, Ellipsoidal Heights) Iteration Number 3
State Plane Zone: New Jersey (NY East) (1983)

Vector No.	Northing (*m*)	(*vn*,	*v'n*) (*m*)	Easting (*m*)	(*ve*,	*v'e*)(*m*)	*dh* (*m*)	(*vdh*,	*v'dh*) (*m*)
1	–358.922	(+.001,	1.2)	-394.266	(+.002,	1.7)	–.968	(+.000,	.1)
2	–358.924	(–.001,	1.7)	-394.269	(–.001,	1.8)	–.967	(+.000,	.2)
3	5023.305	(+.010,	.7)	10664.560	(–.006,	1.0)	–12.853	(+.020,	.7)
4	5023.292	(–.004,	.4)	10664.568	(+.002,	.2)	–12.881	(–.007,	.3)
5	–5382.221	(–.002,	.2)	–11058.830	(+.004,	.6)	11.884	(–.022,	.7)
6	–5382.219	(–.001,	.1)	–11058.838	(–.004,	.3)	11.917	(+.011,	.5)
7	–6197.064	(–.030,	2.0)	–12392.559	(+.000,	.0)	1.202	(+.008,	.3)
8	–11220.307	(+.022,	1.1)	–23057.113	(+.012,	.4)	14.100	(+.034,	.7)
9	–11579.228	(+.023,	1.1)	–23451.381	(+.012,	.4)	13.138	(+.039,	.8)
10	–11579.244	(+.008,	.4)	–23451.399	(–.006,	.4)	13.035	(–.064,	.9)
11	13102.875	(–.010,	.6)	–22807.412	(+.030,	1.1)	236.069	(–.033,	.7)
12	12743.953	(–.009,	.5)	–23201.678	(+.033,	1.2)	235.104	(–.031,	.7)
13	12743.965	(+.003,	.1)	–23201.706	(+.004,	.2)	235.067	(–.068,	.8)
14	7720.662	(–.005,	.2)	–33866.270	(+.006,	.2)	248.001	(–.006,	.1)
15	7720.668	(+.001,	.0)	–33866.399	(–.122,	3.5)	248.188	(+.180,	1.3)*
16	1523.628	(–.005,	.1)	–46258.777	(+.058,	1.1)	249.228	(+.028,	.3)

v= residual, *v'*= normalized residual, * = automatically edited, *dh* = ellipsoidal height differences

11.7 SUMMARY

By far the most precise results using GPS receivers is in interferometric mode where the millimeter-level carrier phases are available. In the short distance mode, cancellations are quite complete when generating differences of measurements that offers the user certain opportunities.

The most used differencing scheme is one that differences between receiver and satellite — double differencing. Such schemes eliminate explicitly both receiver and satellite clock offsets. Over short distances, the Doppler difference is so small that resolution of the double difference integer ambiguity is very much desired to extract the maximum information from the unbiased (double differenced) phase ranges. Pseudoranges can play a major role in identifying these integers if the noise is not too large. Various schemes that utilize available dual-frequency data were discussed.

Because there are no observations to satellites in the hemisphere below the horizon, degradation by a factor of three in vertical recoveries compared to horizontal components are usual. Using cutoff angles to combat refraction that is especially difficult to model at low elevation angles tends to amplify this problem.

The filtering/smoothing of both dual frequency phases and pseudoranges, especially when the noise or the pseudoranges is low, can be quite beneficial in resolving the integer ambiguities so needed in resolving short baselines. These same techniques can be quite useful in studying the characteristics of different manufacturers' tracking performance and also multipathing.

The final step to any surveying project using GPS is the network adjustment. Here is where all hidden problems tend to be discovered. Having a network adjustment computer program that anticipates such problems and thus presents the results in ways that lead to identification of these problems can be of tremendous value to the analyst.

References

Blewitt, G. (1989), Carrier Phase Ambiguity Resolution for the Global Positioning System Applied to Geodetic Baselines up to 2000 km, *Journal of Geophysical Research*, 94 (B8), 10187–10203.

Brunner, F. K. and W. M. Welsch (1993), Effect of Troposphere on GPS Measurements, *GPS World*, January.

Bossler, John D., Clyde C. Goad, Peter L. Bender (1980), Using the Global Positioning System (GPS) for Geodetic Positioning, *Bulletin Géodésique*, 54, 553–563.

Counselman, C.C. and S. A. Gourevitch (1981), Miniature Interferometric Terminals for Earth Surveying: Ambiguity and Multipath with Global Positioning System, *IEEE Transactions on Geosciences and Remote Sensing*, 19, 4, 244–252.

Davis, J. L. (1986), Atmospheric Propagation Effects on Radio Interferometry, Air Force Geophysics Laboratory, AFGL-TR-86-0234, April.

Euler, Hans-Juergen and Clyde C. Goad (1991), On Optimal Filtering of GPS Dual Frequency Observations Without Using Orbit Information, *Bulletin Géodésique*, 65, 2, 130–143.

Georgiadou, Y. and A. Kleusberg (1988), On the effect of ionospheric delay on geodetic relative GPS positioning, *manuscripta geodaetica*, 13, 1–8.

Goad, Clyde C. and Achim Mueller (1988), An Automated Procedure for Generating an Optimum Set of Independent Double Difference Observables Using Global Positioning System Carrier Phase Measurements, *manuscripta geodaetica*, 13, 365–369.

Goad, Clyde C. (1985), Precise Relative Position Determination Using Global Position System Carrier Phase Measurements in a Nondifference Mode, *Positioning with GPS—1985 Proceedings*, Vol. 1, 593–597, NOAA, Rockville, Maryland.

Goad, Clyde C, and Benjamin W. Remondi (1984), Initial Relative Positioning Results Using the Global Positioning System, *Bulletin Géodésique*, 58, 193–210.

Koch, Karl-Rudolf (1988), *Parameter Estimation and Hypothesis Testing in Linear Models*, New York: Springer Verlag.

Leick, Alfred (1995), *GPS Satellite Surveying*, 2nd ed., New York, John Wiley & Sons.

12. MEDIUM DISTANCE GPS MEASUREMENTS

Yehuda Bock,
Cecil H. and Ida M. Green Institute of Geophysics and Planetary Physics, Scripps Institution of Oceanography, University of California, San Diego, 9500 Gilman Drive, La Jolla, California, 92093-0225 USA

12.1 INTRODUCTION

In this chapter we discuss GPS geodesy at medium distances which in some ways is the most challenging. As indicated in Chapter 1, GPS can be considered a global geodetic positioning system providing nearly instantaneous position with 1–2 cm precision with respect to a consistent terrestrial reference frame. Nevertheless, as we shall see in this chapter, the relation between intersite distance and geodetic precision is still important, particularly for integer-cycle phase ambiguity resolution. Currently, the main sources of error at medium distances are due to tropospheric and ionospheric refraction of the GPS radio signals.

We generalize the discussion to derivatives of precise geodetic positioning at regional scales. In particular, what information can we garner about the Earth from very accurate GPS position measurements? This leads us to a discussion of continuous GPS (CGPS) networks for monitoring crustal deformation and atmospheric water vapor, geodetic monumentation as a significant source of low-frequency error, and integration of the new technique of interferometric synthetic aperture radar (INSAR) with CGPS.

12.1.1 Definition of Medium Distance

Distance enters into GPS measurements when one or more stations are to be positioned relative to a base station(s) whose coordinates are assumed to be known, or more generally when a geodetic network is to be positioned with respect to a set of global tracking stations. Phase and pseudorange observations are differenced between stations and satellites (i.e., doubly-differenced or an equivalent procedure) to cancel satellite and station clock errors. The degree that between-station differences eliminate common-mode errors due to ionospheric, tropospheric and orbital effects depends to a large extent on the baseline distance.

It is not possible to define precisely a range of distances that can be called "medium." One can define, however, a lower limit as the shortest distance at which residual ionospheric refraction, tropospheric refraction, and/or orbital errors between sites are greater than total high-frequency site and receiver specific errors (i.e., receiver noise, multipath, antenna phase center errors). Another definition is the shortest distance at which dual-frequency GPS measurements start to make a difference in position accuracy compared to single-frequency measurements.

Likewise, one can define an upper limit as the minimum distance at which dual-frequency phase ambiguity resolution is no longer feasible or reference frame/orbital errors are the dominant error source. Nevertheless if we were required to define medium distance by a range of distances we might use the following: $10^1 - 10^3$ km.

12.1.2 Unique Aspects of Medium Distance Measurements

The primary distinguishing element of medium distance GPS measurements is the relationship between dual-frequency ambiguity resolution and ionospheric refraction. This is particularly the case for estimation of the horizontal position components. For the vertical component, the effects of tropospheric refraction and antenna phase center errors are most significant.

Highest geodetic (horizontal) precision requires ambiguity resolution in static, kinematic and dynamic GPS applications. Ionospheric refraction is the limiting factor in dual-frequency phase ambiguity resolution (in the absence of precise dual frequency pseudorange; if available, the limiting factors are then multipath and receiver noise). Although ionospheric effects can be canceled by forming the ionosphere-free linear combination of phase, any source of noise which is dispersive will be amplified. Let us express the ionosphere-free combination of phase as:

$$\phi_c = (\frac{1}{1-g^2})(\phi_1 - g\phi_2); \quad g = \frac{1227.6}{1575.42} = \frac{60}{77} \tag{12.1}$$

Assuming that measurement errors in both bands are equal and uncorrelated,

$$\sigma_{\phi_c}^2 = (\frac{1}{1-g^2})^2 (1+g^2)\,\sigma^2 = 10.4\,\sigma^2 \tag{12.2}$$

so that forming the ionosphere-free linear combination magnifies dispersive errors by a factor of about 3.2. For short distances, residual ionospheric errors are negligible compared to instrumental error, particularly multipath. Therefore, it is preferable to analyze L1 and L2 (if available) as independent observations. Ambiguity resolution is then straightforward because the L1 and L2 ambiguities can be determined directly as integer values.

At medium distances it is necessary to measure both L1 and L2 phase and to form the ionosphere-free linear combination. Let us express a simplified model for this combination as

$$\phi_c = \rho + \frac{1}{1+g} n_1 - \frac{g}{1-g^2}(n_2 - n_1) = \rho + 0.56 n_1 - 1.98(n_2 - n_1) \tag{12.3}$$

The reason why we use the L2-L1 ("wide-lane") ambiguity, n_2-n_1, is because of its longer wavelength (86 cm) compared to the ("narrow-lane") L1 ambiguity, n_1 (19 cm). The ionosphere-free observable has a *non-integer* ambiguity which is a linear combination of the integer-valued L1 and L2-L1 ambiguities. If we are able to resolve n_2-n_1 to its correct integer value, prior to forming this observable and losing the integer nature of the ambiguities, then this term can be moved to the left-hand side of (12.3)

$$\bar{\phi} = \rho + 0.56\, n_1; \quad \bar{\phi}_c = \phi_c + 1.98\,(n_2 - n_1) \tag{12.4}$$

so that the remaining *integer-valued* ambiguity is just n_1 scaled by 0.56 and thus with an narrower ambiguity spacing of 10.7 cm. This observable is free of ionospheric refraction effects so that if the remaining errors can be kept within a fraction of 10.7 cm, then the narrow-lane ambiguities can be resolved as well. Once the narrow-lane ambiguities are resolved the (double difference) phase observable becomes a (double difference) range observable,

$$\bar{\bar{\phi}} = \rho; \quad \bar{\bar{\phi}} - 0.56\, n_1 + 1.98\,(n_2 - n_1) \tag{12.5}$$

Resolving wide-lane phase ambiguities is primarily limited by ionospheric refraction which increases in proportion to baseline distance (as noted above in the absence of precise pseudorange; if available, the limiting factors are multipath and receiver noise). If the wide-lane ambiguities cannot be resolved, narrow-lane ambiguity resolution is futile. Once wide-lane ambiguity resolution is achieved, ionospheric effects can be eliminated as in (12.4). Narrow-lane ambiguity resolution is then limited by orbital and reference frame errors, multipath and receiver noise.

12.1.3 Types of Medium Distance Measurements

Medium distance surveys usually fall under one of the following classifications:

- *Field Campaigns.* A geodetic network is surveyed over a limited period of time by a number of roving receivers according to a fixed deployment and observation schedule. The network may be observed periodically (e.g., once per year) to estimate deformation, for example, e.g. Feigl et al. [1993]. These surveys may be static, kinematic and/or dynamic. In general, the number of stations occupied significantly exceeds the number of receivers used to occupy them.

- *Continuous arrays.* A network of GPS stations observes continuously for an extended period of time, e.g. Bock et al. [1997]. On a global scale, the growing network of GPS tracking stations (section 1.7) provides access to a consistent terrestrial reference frame and data for the computation of precise satellite ephemerides and Earth orientation parameters. On a regional scale, continuously monitoring GPS stations provide base measurements for field surveys, and "absolute" ties to the global reference frame. Furthermore, they provide enhanced temporal resolution and the ability to better characterize the GPS error spectrum than field campaigns. This leads to another mode:

- *Multimodal surveys.* Continuous arrays have begun to drastically alter the way field GPS surveys are conducted. Under the multimodal occupation strategy, Bevis et al. [1997], field receivers are positioned with respect to a continuous array backbone which provides base data and a consistent reference frame. Compared to campaign surveys, fewer receivers need be deployed (as few as one receiver) and there is more flexibility regarding observation scenarios and logistics.

The most straightforward application of medium distance GPS is geodetic control whether with non-active monumented geodetic stations, active control stations (continuous GPS), or a combination of both. Monitoring of geodetic positions over time brings us into the realm of geodynamics and crustal deformation, including such phenomena as tectonic plate motion, intraplate deformation, volcanism, post-glacial uplift, variations in sea level, land subsidence, and land sliding. Active GPS stations provide important calibration and control for other types of instrumentation such as synthetic aperture radars, altimeters, and aerial mappers (photogrammetry).

12.2 GPS MODELS AT MEDIUM DISTANCES

In this section we present linearized observation equations for dual-frequency carrier phase measurements following the development of Bock et al. [1986], Schaffrin and Bock [1988], [Dong and Bock [1989] and Feigl et al. [1993]. We construct an orthogonal complement to the ionosphere-free phase observable which includes a weighted ionospheric constraint. This formulation provides a convenient framework for phase ambiguity resolution at medium distances.

12.2.1 Mathematical and Stochastic Models

Linearized observation equations with ionospheric constraints. The linearized double difference carrier phase observation equations in their simplest general form can be expressed as

$$\mathbf{Dl} = \mathbf{DAx} + \mathbf{v} \tag{12.6}$$

where $\mathbf{D}$ is the double difference operator matrix which maps at each observation epoch the carrier phase measurements to an independent set of double differences, Bock et al. [1986]; $\mathbf{l}$ is the observation vector, $\mathbf{A}$ is the design matrix, $\mathbf{x}$ is the vector of parameters, and $\mathbf{v}$ is the double difference residual vector. We construct two orthogonal linear combinations of the $\mathbf{l}_1$ and $\mathbf{l}_2$ observation vectors

$$\mathbf{l}_{c1} = \mathbf{l}_1 - \left(\frac{g}{1-g^2}\right)(\mathbf{l}_2 - g\mathbf{l}_1) \tag{12.7}$$

$$\mathbf{l}_{c2} = \mathbf{l}_1 + \frac{1}{2g}(\mathbf{l}_2 - g\mathbf{l}_1) - \left(\frac{1+g^2}{2g^2}\right)\mathbf{l}_{ion} \tag{12.8}$$

where $\mathbf{l}_{c1}$ is the familiar ionosphere-free linear combination and $\mathbf{l}_{ion}$ is a pseudo observation (weighted constraint) of the ionosphere (arbitrarily in the L1 band)

$$\mathbf{l}_{ion,1} = \mathbf{l}_{ion,1} + \mathbf{v}_{ion,1} \tag{12.9}$$

with stochastic model (expectation and dispersion, respectively)

$$E\left\{\begin{bmatrix} \mathbf{v}_1 \\ \mathbf{v}_2 \\ \mathbf{v}_{ion} \end{bmatrix}\right\} = \begin{bmatrix} \mathbf{0} \\ \mathbf{0} \\ \mathbf{0} \end{bmatrix} \tag{12.10}$$

$$D\left\{\begin{bmatrix} \mathbf{v}_1 \\ \mathbf{v}_2 \\ \mathbf{v}_{ion} \end{bmatrix}\right\} = \begin{bmatrix} \sigma_1^2 \mathbf{D}\mathbf{C}_1\mathbf{D}^T & \mathbf{0} & \mathbf{0} \\ \mathbf{0} & \sigma_2^2 \mathbf{D}\mathbf{C}_2\mathbf{D}^T & \mathbf{0} \\ \mathbf{0} & \mathbf{0} & \sigma_{ion}^2 \mathbf{D}\mathbf{C}_{ion}\mathbf{D}^T \end{bmatrix} \tag{12.11}$$

Schaffrin and Bock [1988] demonstrated that a zero constraint on the ionosphere (i.e., assuming zero residual ionospheric effects with an absolute constraint) reduces the model to the simple case of independent L1 and L2 observations (short distance GPS model). On the other hand, assuming zero residual ionospheric effects but with an infinite error reduces the model to the ionosphere-free formulation (long distance GPS model). The observation equations for the two new observables can be written as

$$\begin{bmatrix} \mathbf{D}\mathbf{l}_{c1} \\ \mathbf{D}\mathbf{l}_{c2} \end{bmatrix} = \begin{bmatrix} \mathbf{D}\mathbf{A}_{c1} \\ \mathbf{D}\mathbf{A}_{c2} \end{bmatrix}\mathbf{x} + \begin{bmatrix} \mathbf{v}_{c1} \\ \mathbf{v}_{c2} \end{bmatrix} \tag{12.12}$$

The parameter vector $\mathbf{x}$ is partitioned into $\mathbf{x}_m$ which includes all non-ambiguity parameters, $\mathbf{n}_1$, the narrow-lane L1 ambiguity vector (19-cm wavelength), and $\mathbf{n}_2$-$\mathbf{n}_1$, the wide-lane L2-L1 ambiguity vector (86-cm wavelength). The observation equations can now be expressed as

$$\begin{bmatrix} \mathbf{Dl}_{c1} \\ \mathbf{Dl}_{c2} \end{bmatrix} = \begin{bmatrix} \mathbf{D}\tilde{\mathbf{A}}_1 & \frac{1}{1+g}\mathbf{D} & -\frac{g}{1-g^2}\mathbf{D} \\ \mathbf{D}\tilde{\mathbf{A}}_1 & \frac{1+g}{2g}\mathbf{D} & \frac{1}{2g}\mathbf{D} \end{bmatrix} \begin{bmatrix} \mathbf{x}_m \\ \mathbf{n}_1 \\ \mathbf{n}_2 - \mathbf{n}_1 \end{bmatrix} + \begin{bmatrix} \mathbf{v}_{c1} \\ \mathbf{v}_{c2} \end{bmatrix} \tag{12.13}$$

where the coefficient matrix for the non-ambiguity parameters is distinguished by a tilde. The stochastic model is

$$E\left\{\begin{bmatrix} \mathbf{v}_{c1} \\ \mathbf{v}_{c2} \end{bmatrix}\right\} = \begin{bmatrix} \mathbf{0} \\ \mathbf{0} \end{bmatrix} \tag{12.14}$$

$$D\left\{\begin{bmatrix} \mathbf{v}_{c1} \\ \mathbf{v}_{c2} \end{bmatrix}\right\} = \sigma_0^2 (1+g^2) \begin{bmatrix} \mathbf{d}_{11} & \mathbf{d}_{12} \\ \mathbf{d}_{21} & \mathbf{d}_{22} \end{bmatrix} \tag{12.15}$$

where

$$\mathbf{d}_{11} = \frac{1}{(1-g^2)^2} \mathbf{D}\mathbf{C}_\phi \mathbf{D}^T \tag{12.16}$$

$$\mathbf{d}_{22} = \frac{1}{4g^2} (\mathbf{D}\mathbf{C}_\phi \mathbf{D}^T + \frac{\sigma_{ion}^2}{\sigma^2} \frac{(1+g^2)}{g^2} \mathbf{D}\mathbf{C}_{ion} \mathbf{D}^T) \tag{12.17}$$

$$\mathbf{d}_{12} = \mathbf{d}_{21} = \mathbf{0} \tag{12.18}$$

and σ_0^2 is the variance of unit weight.

Cofactor matrices. Schaffrin and Bock [1988] constructed cofactor matrices $\mathbf{C}_\phi$ and $\mathbf{C}_{ion}$ for the dual-frequency phase measurements and ionospheric constraints, respectively, so that the propagated double difference cofactor matrices $\mathbf{D}\mathbf{C}_\phi\mathbf{D}^T$ and $\mathbf{D}\mathbf{C}_{ion}\mathbf{D}^T$ would reflect the nominal distance dependent nature of GPS measurement errors, i.e.,

$$\sigma^2 = a^2 + b^2 s_{ij}^2 \tag{12.19}$$

(for stations *i* and *j*), and be at least positive semi-definite.

For a single double difference observation the cofactor matrices take the general form

$$\mathbf{C}(\alpha, \beta, \delta) = \begin{bmatrix} \beta^2 & 0 & \alpha\beta^2 \text{sech}\,\delta & 0 \\ 0 & \beta^2 & 0 & \alpha\beta^2 \text{sech}\,\delta \\ \alpha\beta^2 \text{sech}\,\delta & 0 & \beta^2 & 0 \\ 0 & \alpha\beta^2 \text{sech}\,\delta & 0 & \beta^2 \end{bmatrix} \tag{12.20}$$

where sech is the hyperbolic secant function[1] and

$$\beta = \beta(a,b);\ \alpha = \alpha(a,b);\ \delta = \delta(s) \tag{12.21}$$

With the availability of very precise orbits it is possible to ignore the distance dependent measurement error term in medium distance analysis so that

$$\mathbf{C}_\phi(a) = \begin{bmatrix} a^2 & 0 & 0 & 0 \\ 0 & a^2 & 0 & 0 \\ 0 & 0 & a^2 & 0 \\ 0 & 0 & 0 & a^2 \end{bmatrix} \tag{12.22}$$

For a single double difference

$$\mathbf{D}\mathbf{C}_\phi\mathbf{D}^T = 4a^2 \tag{12.23}$$

Furthermore, we can ignore the constant term for the ionospheric constraints (a=0) so that

$$\mathbf{C}_{ion}(\bar\beta, \delta) = \begin{bmatrix} \bar\beta^2 & 0 & \bar\beta^2 \text{sech}\,\delta & 0 \\ 0 & \bar\beta^2 & 0 & \bar\beta^2 \text{sech}\,\delta \\ \bar\beta^2 \text{sech}\,\delta & 0 & \beta^2 & 0 \\ 0 & \bar\beta^2 \text{sech}\,\delta & 0 & \bar\beta^2 \end{bmatrix};\ \bar\beta = \bar\beta(b) \tag{12.24}$$

For a single double difference

$$\mathbf{D}\mathbf{C}_{ion}\mathbf{D}^T = 4\bar\beta^2(1 - \text{sech}\,\delta) \tag{12.25}$$

Note in (12.24) that when $\delta = 0$ (zero baseline) there is perfect correlation and as the baseline increases in length intersite correlations decrease. Suitable values for (12.25) are

$$\bar\beta^2 = 0.3[\text{b x } 10^4 \text{mm}]^2 \tag{12.26}$$

---

[1]The hyperbolic secant function is defined as sech(δ) = 2/[exp(-δ) + exp(δ)]

$$\delta = 0.56\,\lambda \tag{12.27}$$

where b is expressed in parts per million and λ is the baseline length in units of arc length. Other cofactor matrices (see Chapter 11, for example) can also be incorporated into the general stochastic model (12.14)–(12.15).

Ambiguity mapping. The weighted least squares normal equations can be expressed simply as

$$\begin{bmatrix} \mathbf{N}_{11} & \mathbf{N}_{12} \\ \mathbf{N}_{21} & \mathbf{N}_{22} \end{bmatrix} \begin{bmatrix} \mathbf{x}_m \\ \mathbf{n} \end{bmatrix} = \begin{bmatrix} \mathbf{u}_1 \\ \mathbf{u}_2 \end{bmatrix} \tag{12.28}$$

where

$$\mathbf{n} = \begin{bmatrix} \mathbf{n}_1 \\ \mathbf{n}_2 - \mathbf{n}_1 \end{bmatrix} \tag{12.29}$$

Explicit expressions for the elements of the normal equations, suitable for computer coding, can be found in Schaffrin and Bock [1988].

When transforming undifferenced carrier phase measurements into double differences, it is necessary to choose a linearly independent set of L1 and L2 ambiguities. Otherwise the normal equations (12.28) will be rank deficient. Below we apply a general mapping operator **B** which constructs an independent set of double difference ambiguity parameters such that

$$\begin{bmatrix} \mathbf{N}_{11} & \tilde{\mathbf{N}}_{12} \\ \tilde{\mathbf{N}}_{21} & \tilde{\mathbf{N}}_{22} \end{bmatrix} \begin{bmatrix} \mathbf{x}_m \\ \mathbf{Bn} \end{bmatrix} = \begin{bmatrix} \mathbf{u}_1 \\ \tilde{\mathbf{u}}_2 \end{bmatrix} \tag{12.30}$$

where

$$\tilde{\mathbf{N}}_{12} = \mathbf{N}_{12}\bar{\mathbf{B}}^T = \tilde{\mathbf{N}}_{21}^T \tag{12.31}$$

$$\tilde{\mathbf{N}}_{22} = \bar{\mathbf{B}}\mathbf{N}_{22}\bar{\mathbf{B}}^T \tag{12.32}$$

$$\tilde{\mathbf{u}}_2 = \bar{\mathbf{B}}\mathbf{u}_2 \tag{12.33}$$

$$\bar{\mathbf{B}} = (\mathbf{B}\mathbf{B}^T)^{-1}\mathbf{B} \tag{12.34}$$

The preferred mapping is to choose those baselines that yield real-valued ambiguities with lowest uncertainties and minimum correlation structure (see discussion on *ambiguity decorrelation* in Chapter 8). Other mappings that have been used to order the ambiguities according to baselines with increasing length considering that the total GPS error budget increases with baseline length. Yet another is to choose base stations (and base satellites). In the latter two mappings, the elements of **B** include combinations of +1, -1 and 0; each row contains two +1's and two -1's, which map four phase ambiguities into one double difference ambiguity.

Least squares solution. The weighted least squares solution is computed by inverting (12.30)

$$\begin{bmatrix} \hat{\mathbf{x}}_m \\ \hat{\bar{\mathbf{n}}} \end{bmatrix} = \begin{bmatrix} \mathbf{N}_{11} & \tilde{\mathbf{N}}_{12} \\ \tilde{\mathbf{N}}_{21} & \tilde{\mathbf{N}}_{22} \end{bmatrix}^{-1} \begin{bmatrix} \mathbf{u}_1 \\ \tilde{\mathbf{u}}_2 \end{bmatrix} = \begin{bmatrix} \mathbf{Q}_{11} & \mathbf{Q}_{12} \\ \mathbf{Q}_{21} & \mathbf{Q}_{22} \end{bmatrix} \begin{bmatrix} \mathbf{u}_1 \\ \tilde{\mathbf{u}}_2 \end{bmatrix} \; ; \bar{\mathbf{n}} = \mathbf{Bn} \tag{12.35}$$

with

$$\hat{\mathbf{v}}^T\mathbf{P}\hat{\mathbf{v}} = \mathbf{l}^T\mathbf{Pl} - \begin{bmatrix} \mathbf{u}^T_1 & \tilde{\mathbf{u}}^T_2 \end{bmatrix} \begin{bmatrix} \hat{\mathbf{x}}_m \\ \hat{\bar{\mathbf{n}}} \end{bmatrix} \tag{12.36}$$

where **P** is the inverse of the dispersion matrix (12.15).

Sequential ambiguity resolution. An efficient algorithm for sequential ambiguity resolution uses the following relations to update the solution vector and the sum of residuals squared, see Dong and Bock [1989]

$$\hat{\mathbf{x}}_{new} = \hat{\mathbf{x}} + \mathbf{Q}_{12}\mathbf{Q}_{22}^{-1} (\bar{\mathbf{n}}_{0,fixed} - \bar{\mathbf{n}}_0) \tag{12.37}$$

$$(\mathbf{v}^T\mathbf{Pv})_{new} = \mathbf{v}^T\mathbf{Pv} + (\bar{\mathbf{n}}_{0,fixed} - \bar{\mathbf{n}}_0)^T \mathbf{Q}_{22}^{-1} (\bar{\mathbf{n}}_{0,fixed} - \bar{\mathbf{n}}_0) \tag{12.38}$$

$$\mathbf{Q}_{new} = \mathbf{Q}_{11} - \mathbf{Q}_{12}\mathbf{Q}_{22}^{-1}\mathbf{Q}_{21} \tag{12.39}$$

The parameter vector **x** contains all parameters except those few bias parameters $\bar{\mathbf{n}}_0$ that are fixed during each step of sequential ambiguity resolution (wide-lane and then narrow-lane).

Wide-lane ambiguity resolution. There are two basic approaches to wide-lane ambiguity resolution.

Ionospheric constraint formulation. The purpose of the ionospheric constraint formulation given above is to be able to resolve the 86-cm wavelength wide-lane ambiguities. We cannot do this directly using the ionosphere-free combination since the resulting ambiguities are no longer of integer values, as can be seen in (12.3). A reasonable constraint on the ionospheric "noise" ranging from b=1 to 8 parts per million in (12.26) facilitates an integer search in the space of wide-lane ambiguities (see also Chapter 8).

Precise pseudorange formulation. Wide-lane ambiguity resolution using the ionospheric constraint formulation is possible only when ionospheric effects on carrier phase are a small fraction of the 86-cm wavelength ambiguity. This is a function primarily of baseline length, but also station latitudes, time of day, season, and the sunspot cycle. Blewitt [1989] describes the use of precise dual-frequency pseudoranges in combination with carrier phase to resolve wide-lane ambiguities.

The simplified observation equations for phase and pseudorange, given in units of cycles of phase, can be expressed as

$$\phi_1 = -\frac{\rho}{c} f_1 + I_1 + n_1 + v_{\phi 1} \tag{12.40}$$

$$\phi_2 = -\frac{\rho}{c} f_2 + I_2 + n_2 + v_{\phi 2} \tag{12.41}$$

$$P_1 = -\frac{\rho}{c} f_1 - I_1 + v_{P1} \tag{12.42}$$

$$P_2 = -\frac{\rho}{c} f_2 - I_2 + v_{P2} \tag{12.43}$$

The wide-lane ambiguity can be computed at each observation epoch from these four equations such that

$$n_2 - n_1 = \phi_2 - \phi_1 + \frac{f_1 - f_2}{f_1 + f_2} (P_1 + P_2) + v_{(n_2 - n_1)} \tag{12.44}$$

This combination is solely a function of the phase and pseudorange observations and their combined measurement errors, and is independent of GPS modeling errors (e.g., orbits, station position, atmosphere). Thus, not only is it powerful for wide-lane ambiguity resolution but also for fixing cycle slips in undifferenced phase measurements, see Blewitt [1990]. However, the pseudoranges must be sufficiently precise to make this procedure successful, that is a small fraction of the 86-cm wide-lane ambiguity wavelength. This is not always the case, particularly in the anti-spoofing (AS) environment and for short observation spans.

Narrow-lane ambiguity resolution. The sole purpose of the ionosphere constraint formulation is to resolve wide-lane ambiguities which is limited either by pseudorange measurement noise or ionospheric refraction, or both. Once (and only if) the wide-lane ambiguities are resolved then the ionosphere-free observable is formed and resolution of (now integer-valued) narrow-lane ambiguities can proceed (12.4). The limiting error sources are then orbital and reference frame errors (which are distance dependent), tropospheric refraction, and multipath and receiver noise which are magnified in the ionosphere-free observable (12.2).

Four-step algorithm. Dong and Bock [1989] and Feigl et al. [1993] describe a 4-step procedure for sequential ambiguity resolution as follows[2]:

(1) All parameters are estimated using the ionosphere-free linear combination of carrier phase. Tight constraints are applied to the station coordinates to impose a reference frame.

(2) With the geodetic parameters held fixed at their values from step 1, wide-lane ambiguity resolution proceeds sequentially using either the ionospheric constraint formulation or precise pseudoranges (or both).

(3) With the wide-lane ambiguity parameters held fixed at the values obtained in step 2, the narrow-lane ambiguities and the other parameters are estimated using the ionosphere-free linear combination. Tight constraints are imposed on the station coordinates as before. Narrow-lane ambiguity resolution proceeds sequentially.

(4) With the wide-lane and narrow-lane ambiguities fixed to their integer values obtained in steps 2 and 3, the geodetic parameters are estimated using the ionosphere-free linear combination of, now unambiguous, double difference phase.

12.2.2 Estimated Parameters

Geometric parameters. The geometric term of the GPS model can be expressed as

$$\phi_i^k(t) = \frac{f_0}{c}[\rho_i^k(t, t-\tau_i^k(t))] = \frac{f_0}{c}\left\|\mathbf{r}^k(t-\tau_i^k(t)) - \mathbf{r}_i(t)\right\| \tag{12.45}$$

where $\mathbf{r}_i$ is the geocentric station position vector with Cartesian elements

$$\mathbf{r}_i(t) = \begin{bmatrix} X_i(t) \\ Y_i(t) \\ Z_i(t) \end{bmatrix} \tag{12.46}$$

and $\mathbf{r}^k$ is the geocentric satellite position vector, both given in the same reference frame.

---

[2] In practice an additional two solutions are generated during this procedure. These are very loosely constrained solutions in which the terrestrial reference frame is undefined (essentially free adjustments). The solutions (parameter adjustments and full covariance matrices) are available for subsequent network adjustment. This will be described in more detail in section 12.4.

The last two steps are:

(5) Step 1 is repeated but with loose constraints on all the geodetic parameters, with the ambiguity parameters free to assume real values.

(6) Step 4 is repeated with the ambiguities constrained to integer values but with loose constraints on all the geodetic parameters.

The equations of motion of a satellite can be expressed by six first-order differential equations, three for position and three for velocity,

$$\frac{d}{dt}(\mathbf{r}^k) = \dot{\mathbf{r}}^k \tag{12.47}$$

$$\frac{d}{dt}(\dot{\mathbf{r}}^k) = \frac{GM}{r^3}\,\mathbf{r}^k + \ddot{\mathbf{r}}^k_{\text{Perturbing}} \tag{12.48}$$

where G is the universal constant of attraction and M is the mass of the Earth. The first term on the right-hand side of (12.48) contains the spherical part of the Earth's gravitational field. The second term represents the perturbing accelerations acting on the satellite (e.g., nonspherical part of the Earth's gravity field, luni-solar effects and solar radiation pressure). In orbit determination or orbit relaxation (see section 12.3.2), the satellite parameters are the initial conditions of the equations of motion and coefficients of a model for nongravitational accelerations. These parameters can be treated deterministically or stochastically. The estimation of GPS satellite orbits are discussed in detail in Chapters 2 and 14.

Tropospheric refraction parameters. Estimation of tropospheric refraction parameters is an important element in modeling medium distance measurements, primarily for the estimation of the vertical component of station position. With continuous GPS networks with well coordinated positions, it is possible to accurately map tropospheric water vapor at each site.

The physics of the atmospheric propagation delay have been discussed in Chapter 3. Ionospheric refraction is dispersive (frequency-dependent) while tropospheric refraction is neutral (frequency-independent) at GPS frequencies. Tropospheric delay accumulated along a path through the atmosphere is smallest when the path is oriented in the zenith direction. For slanted paths the delay increases approximately as the secant of the zenith angle. It is typical to model the delay along a path of arbitrary direction as the product of the zenith delay and a dry and wet 'mapping function' which describes the dependence on path direction such that

$$\Delta L = \Delta L_h^0\, M_h(z) + \Delta L_w^0\, M_w(z) \tag{12.49}$$

where ΔL_h^0 is the zenith hydrostatic (dry) delay (ZHD), ΔL_w^0 is the zenith wet delay (ZWD), $M_h(z)$ and $M_w(z)$ are the hydrostatic and wet mapping functions, respectively, and z is the zenith angle. The total zenith delay is denoted here by ZND (zenith neutral delay). Various mapping functions have been described in Chapter 3 which take into account the curvature of the Earth, the scale height of the atmosphere, the curvature of the signal path, and additional factors. Usually it is assumed that the delay is azimuthally isotropic, in which case the mapping function depends on a single variable, the zenith angle[3]. For this class of model, signal delays are totally specified by the (time-varying) zenith delay. This allows us to introduce zenith delay parameter estimates for each station in a network. The simplest approach for retrieving the zenith delay is to assume that it remains constant (or piecewise linear) for one or more time intervals, and to estimate these values more or less independently. A more sophisticated approach utilizes the fact that the temporal variation of the zenith delay has exploitable statistical properties. The zenith delay is unlikely to change by a large amount over a short period of time (e.g. ten minutes). In fact the zenith delay can be viewed as a stochastic process, and the process parameters can be estimated using a Kalman filter (see section 12.4.4).

The ZHD has a typical magnitude of about 2.3 meters. Given surface pressure measurements accurate to 0.3 millibars or better, it is usually possible to predict the ZHD to better than 1 mm. The ZWD can vary from a few millimeters in very arid conditions to more than 350 mm in very humid conditions. It is not possible to predict the wet delay with any useful degree of accuracy from surface measurements of pressure, temperature and humidity. It is possible to estimate the wet delay using relatively expensive ground-based water vapor radiometers (WVR's). Alternate less expensive approaches include estimation of ZND from the GPS observations, or measurement of ZHD, using barometers, and estimation of the remaining wet delay as part of the GPS adjustment process. One advantage of decomposing the ZND in this way is that it enables the delay models to incorporate separate hydrostatic and wet mapping functions, thereby taking better account of the differing scale heights of the wet and hydrostatic components of the neutral atmosphere. This approach is highly advantageous in the context of VLBI, in which radio sources are tracked down to elevation angles as low as 5°. For GPS, it is typical to process only those observations collected from satellites with elevation angles greater than 15 degrees[4]. In this case, the wet and dry mapping functions differ only very slightly, and it is reasonable to lump the wet and hydrostatic delays together and use a single mapping function, thereby parameterizing the problem solely in terms of the total zenith delay. Once the ZND parameters have been estimated during the geodetic inversion, it is possible to estimate the ZWD by subtracting the ZHD from the ZND, where the ZHD is derived from surface pressure readings.

For GPS networks with interstation spacing of less than several hundred kilometers, the ZWD parameters inferred across the network contain large but highly correlated errors. In this case one may infer relative ZWD values across the network but not the absolute values. Remote stations can be included in the geodetic inversion, in which case the absolute values of the ZND parameters are readily estimated. Continuously operating GPS stations of the global GPS tracking

---

[3] Recently, there has been a considerable amount of research on estimating horizontal tropospheric gradients by GPS, e.g. Ware et al. [1997].

[4] With the development of new tropospheric mapping functions, e.g. Niell [1996] and improved antenna phase center models, GPS observations can now be processed more accurately at lower elevation angles.

network can be used for this purpose. Since precise surface pressure measurements are not available for most global tracking sites it is not possible to determine the hydrostatic delays at these sites. The total zenith delay can be estimated from each site and the hydrostatic delays are subtracted for those sites that are equipped with precise barometers to determine the ZWD. For more details on this approach see Duan et al. [1996].

12.3 ANALYSIS MODES

The expansion of the global GPS tracking network, the availability of highly precise satellite ephemerides, Earth orientation and satellite clock parameters, improvements in the terrestrial reference frame, the proliferation of continuous GPS arrays (at medium/regional scales), and technological advances in GPS software and hardware are changing the way medium-distance surveys are performed and analyzed. The capability of positioning a single receiver anywhere in the world with respect to the ITRF with centimeter-level, three-dimensional accuracy, and in nearly real time is becoming a reality.

In this section we describe several analysis modes that are suitable for medium distance GPS. The models described in section 12.2 are suitable for any of these modes.

12.3.1 Baseline Mode

The analysis mode with the longest history is the baseline mode. The first GPS network at medium distance (in this case at 10–20 km station spacing) was surveyed in 1983 in the Eifel region of West Germany in the baseline mode, using fixed broadcast ephemerides and single-frequency receivers, see, Bock et. al [1985]. In the simplest case, one GPS unit surveys at a well-coordinated base station and a second unit is deployed sequentially at stations with unknown coordinates. The baseline or the three-dimensional vector between the base station(s) to each unknown station is then estimated with postprocessing software using standard double difference algorithms. This is the way that most commercial GPS software packages have worked for many years and still do today. A standard network adjustment program (see section 12.4) is then used to obtain consistent estimates for the coordinates of the unknown stations with respect to the reference frame defined by the fixed coordinates of the base station(s).

All advanced GPS packages use the more rigorous session mode analysis described in the next section. However, with today's technology, the baseline mode has become an accurate and straightforward method for medium distance surveys. The IGS provides highly precise, timely, and reliable satellite ephemerides in standard SP3 format that can be read by all major GPS software packages. The base station can either be a continuous GPS site, a monumented (nonpermanent) geodetic station, or a temporary station, all of which can be coordinated with respect to ITRF with sufficient accuracy for relative positioning. Ambiguity resolution is usually successful for single baselines of distances up to several hundreds of kilometers, depending on the length of the observation span.

12.3.2 Session Mode

Session mode describes a variety of analysis techniques in which all data observed over a particular observation span (a session) are simultaneously analyzed. It was first introduced to treat inter-baseline correlations rigorously. Later, orbit estimation, in addition to station coordinate estimation, became part of session mode processing at regional, continental and global scales. Orbit estimation is often referred to as *orbit relaxation* at regional scales, see e.g., Shimada and Bock [1992], *fiducial tracking* at continental scales, see, e.g., Dong and Bock [1989]; Larson et al. [1991]; Feigl et al. [1993], and *orbit determination* at global scales, see, e.g., Lichten and Border [1987]. This different terminology is primarily a function of the type of constraints placed on the orbital parameters. Zenith delay estimation always accompanies orbit estimation. The simultaneous analysis of several sessions to improve ambiguity resolution and orbital estimation is referred to as *multi-session* mode.

Session mode with orbital estimation allows for a bootstrapping approach to ambiguity resolution in which ambiguities are searched sequentially over baselines of increasing length, see Blewitt [1989]; Counselman and Abbot [1989] and Dong and Bock [1989].

12.3.3 Distributed Session Mode

The proliferation of permanent global and regional continuous GPS networks, and the change in the nature of campaign-type surveys makes the traditional session mode analysis of regional, continental and global GPS data computationally prohibitive and unnecessary. In this section, we describe an implementation of a distributed processing scheme, see Blewitt et al. [1993] that divides the data analysis into manageable segments of regional and global data without significant loss of precision compared to simultaneous session adjustment. As described in section 12.3.1 the traditional baseline mode of processing is a viable and efficient procedure for medium-distance GPS. However, distributed session processing is a more rigorous procedure which retains the covariance structure between geodetic parameters, is computationally efficient, does not significantly sacrifice geodetic accuracy, and allows for straightforward integration of various type geodetic networks with respect to ITRF.

The simplest form of observation equation for GPS estimation was given by (12.6) with weighted least squares solution

$$\hat{\mathbf{x}} = (\mathbf{A}^T\mathbf{W}\mathbf{A})^{-1}\,\mathbf{A}^T\mathbf{W}\mathbf{l}\ ;\ \mathbf{W} = \mathbf{D}^T(\mathbf{D}\mathbf{C}_\phi\mathbf{D}^T)^{-1}\mathbf{D} \tag{12.50}$$

and (unscaled) covariance matrix

$$\mathbf{Q} = (\mathbf{A}^T\mathbf{W}\mathbf{A})^{-1} \tag{12.51}$$

Suppose that data from (e.g. two regional networks) are analyzed simultaneously with global data, to estimate station coordinates of the global and regional sites. We normalize the diagonal elements of the covariance matrix to unity by computing the correlation matrix of the form

$$\mathbf{C}_{all} = \begin{bmatrix} \mathbf{C}_g & \mathbf{C}_{r1,g} & \mathbf{C}_{r2,g} \\ \mathbf{C}_{g,r1} & \mathbf{C}_{r1} & \mathbf{C}_{r2,r1} \\ \mathbf{C}_{g,r2} & \mathbf{C}_{r1,r2} & \mathbf{C}_{r2} \end{bmatrix} \tag{12.52}$$

where $\mathbf{C}_g$, $\mathbf{C}_{r1}$, and $\mathbf{C}_{r2}$ are correlation matrices for the site coordinates of the global sites, region 1, and region 2, respectively, and $\mathbf{C}_{g,r1}$, $\mathbf{C}_{g,r2}$, $\mathbf{C}_{r1,r2}$ are cross-correlation matrices.

The structure of the parameter covariance matrix resulting from a simultaneous analysis of regional-scale networks and the global IGS network has been studied by Zhang [1996]. High correlations (≥ 0.8) are concentrated in the regional coordinates. Cross-correlations are low (< 0.3) among distinct regions, and between each region and the more globally distributed stations. Furthermore, cross correlations between different components at a particular site are low regardless of station distribution. Correlations among components are uniformly low between regions and global sites, in particular for the longitudinal and radial components. The radial components are weak (≤ 0.3) even within regional networks. The longitudinal components are more highly correlated (≤ 0.5), and the latitudinal components most correlated (≤ 0.8).

From these types of studies, it has been shown that the analysis of global and regional data can be distributed among several smaller manageable segments, without significant loss of geometric strength or precision, see Oral [1994] and Zhang [1996]. The distributed processing scheme, then, neglects a large fraction of weakly correlated cross covariances between regional and global coordinates, i.e.,

$$\mathbf{Q}_{all} = \begin{bmatrix} \mathbf{Q}_g & \mathbf{Q}_{g,r1} & \mathbf{Q}_{g,r2} & \mathbf{Q}_{g,r3} \\ & \mathbf{Q}_{r1} & \mathbf{0} & \mathbf{0} \\ & & \mathbf{Q}_{r2} & \mathbf{0} \\ \text{Sym} & & & \mathbf{Q}_{r3} \end{bmatrix} \tag{12.53}$$

However, the regional segments should contain at least 3 stations in common with the top-level global network, which are denoted by the off-diagonal submatrices in (12.53). An important component of this scheme is a top-level analysis of one or more global segments of 30–40 global tracking (IGS) stations, a manageable number with today's computing technology. Any number of solutions of regional segments can then be combined with the global segment(s) in a rigorous network adjustment of station positions (and velocities) with respect to a globally consistent reference frame (see section 12.4.1).

12.3.4 Point Positioning Mode

An efficient "point positioning" mode has been proposed by Zumberge et al. [1997]. It is similar to the baseline mode but uses satellite clock estimates (from global tracking data analysis) in place of between station differencing to eliminate satellite clock errors. Thus, one can point position a site with respect to ITRF using fixed IGS orbits, Earth orientation parameters, and satellite clock estimates. Station position, zenith delays, station clock, and phase ambiguity parameters are estimated using the ionosphere-free linear combination.
The advantages of this approach are

(1) It does not require fiducial (base station) data in the computation of position;
(2) In principle, it provides consistent accuracy worldwide. This should be the case when the global tracking network is more uniformly distributed; and
(3) It is a very efficient procedure and requires minimum computer processing time. However, it is not significantly more efficient than performing baseline mode processing, and is essentially equivalent to baseline mode with a fixed base station (without ambiguity resolution).

Additional limitations include:

(1) It does not directly allow for integer-cycle ambiguity resolution thus limiting horizontal precision to about 1 cm, compared to baseline mode with ambiguity resolution which is 2–3 times more precise, primarily in the east component. This is certainly a limiting factor for medium-distance measurements where ambiguity resolution is critical;
(2) It is more difficult to edit phase data in undifferenced phase measurements, particularly with AS turned on; and
(3) The analysis must be performed with the same processing software (i.e., models) as was used to compute the fixed orbits, Earth orientation parameters and satellite clocks.

12.3.5 Kinematic and Rapid Static Modes

Kinematic and rapid static modes have been discussed extensively in Chapters 5 and 13. At medium distances ionospheric refraction and its effect on successful ambiguity resolution is, of course, more severe than in static mode since the GPS unit is in motion between station occupations and, hence, more susceptible to cycle slips and losses of lock in the phase measurements. Furthermore, station occupations are shorter so that ambiguity resolution is even more difficult even in the absence of cycle slips, primarily because less time is available for averaging down multipath effects.

Genrich and Bock [1992] applied kinematic/rapid static techniques to simulate a continuous GPS baseline (network), and demonstrated it on a short baseline across the San Andreas fault in central California. The key to this approach is that once the phase ambiguities are resolved (in the same way as in the baseline/session mode), the baseline can be computed epoch by epoch. Cycle slips and rising of new satellites can be accommodated by on-the-fly ambiguity resolution techniques. Multipath effects which repeat from day to day (with an offset of about 4 minutes) can be nearly totally eliminated by stacking the daily positions (see section 12.4.5). This technique has been applied successfully at medium distances for detecting coseismic displacements associated with the January 1994 Northridge Earthquake and the January 1995 Kobe Earthquake. The key to this method at medium distances is that once the dual-frequency integer cycle ambiguities are resolved, one can use the ionosphere-free doubly differenced range measurements to estimate positions epoch by epoch with fixed external (IGS or other) precise orbits.

Multimodal surveys show great promise for regional kinematic and rapid static analysis, Bevis et al. [1997]. That is, regional continuous GPS arrays will anchor the field measurements, see e.g. Genrich and Bock [1997] and will provide maps and profiles of atmospheric refraction. Furthermore, ionospheric corrections should enhance ambiguity resolution at longer distances and hence horizontal precision. Tropospheric corrections (primarily for the wet component, see section 12.2.2) should enhance vertical precision.

12.3.6 Dynamic Mode

Dynamic measurements at medium distances are a greater challenge since the platform (e.g., an aircraft) is always in motion complicating phase initialization and reinitialization (see Chapter 13 for a detailed discussion). Mader [1992] applied an ambiguity function technique, Counselman and Gourevitch [1981], that is applicable to kinematic and rapid static measurements at medium distances. The ambiguity function is given by

$$A(x,y,z) = \sum_{k=1}^{K}\sum_{j=1}^{J}\sum_{l=1}^{2} \cos\{2\pi\,[\phi^{jkl}_{obs}(x_0,y_0,z_0) - \phi^{jkl}_{calc}(x,y,z)\,]\} \qquad (12.54)$$

where $\phi^{jkl}_{obs}(x_0,y_0,z_0)$ is the doubly differenced observed phase whose correct position is (x_0,y_0,z_0) and $\phi^{jkl}_{calc}(x,y,z)$ is the calculated doubly differenced phase at the initial position (x,y,z). The subscripts *j*, *k*, and *l* refer to satellite, epoch and frequency (L1 and L2), respectively. The difference term is the *a priori* phase residual. Phase ambiguity terms are neglected since they would cause an integer number of rotations leaving the ambiguity function unchanged; hence it is immune to cycle slips between epochs included in its estimation. It is clear that, for a particular satellite, epoch, and frequency the ambiguity function will have a maximum value of 1 when the phase residual is an integer or zero. This will occur when $(x,y,z) = (x_0,y_0,z_0)$, assuming that all other errors are negligible, and at all other positions where the difference in distance computed between this satellite and the reference satellite for double differencing is an integer number of wavelengths. The search algorithm must distinguish between these different *optima.* In order to find the correct position, the summation above is made over different satellites, epochs and frequencies which results in a combination of intersecting surfaces that will interfere constructively at the correct position and destructively at incorrect positions, with the correct position emerging as a recognizable peak as the ambiguity function is computed over a volume that includes its position. If there are a sufficient number of satellites (≥ 5) present at a given epoch, a unique solution may be obtainable from that one epoch. This is crucial in the dynamic modc. As described in the previous section, precise orbital information and corrections for ionospheric and tropospheric refraction, if available, could be of significant value in this aspect, although antenna multipath and unmodeled antenna phase center variations would still be a limiting factor.

12.4 NETWORK ADJUSTMENT

12.4.1 Free-Network Quasi-Observation Approach

The free-network quasi-observation approach is the standard and preferred approach to GPS network adjustment. It is probably the only practical approach to integrating GPS networks with other types of geodetic networks (see section 12.4.2) and maintaining a consistent terrestrial reference frame. The quasi-observation approach uses point-position, baseline- or session-adjusted GPS free network (or very loosely constrained) solutions including full covariance matrix information as observations to a standard weighted least squares adjustment. In the simplest and most straightforward case only the geodetic coordinate adjustments and their covariance matrices are included as quasi-observations. However, it is also possible and often preferable to include other GPS estimated parameters (e.g., orbits and Earth orientation parameters) in the network adjustment process. In any case, the reference frame is imposed at the final network adjustment stage by fixing (or tightly constraining) a subset of the station coordinates.

Linearized observation equations. We review this approach according to the development of Dong [1993] which follows the well known four-dimensional integrated geodesy approach, see e.g. Collier et al. [1988]. We ignore the gravitational potential term for this discussion.

The nonlinear mathematical model for the geodetic measurement can be expressed familiarly as

$$\mathbf{l}(t) = \mathbf{F}\{\mathbf{X}(\mathbf{a},t), \mathbf{h}(t)\} \tag{12.55}$$

where $\mathbf{X}(\mathbf{a},t)$ is the geocentric Cartesian position vector, whose time dependence is described by the parameters **a**, and **h**(t) are additional GPS parameters such as orbits, Earth orientation, and reference frame (translations, rotations, and scale). The linearized observation equations are

$$\delta\mathbf{l}(t) = \mathbf{A}[\Delta X_0 + (t - t_0)\Delta\dot{\mathbf{X}}_0] + \mathbf{B}\Delta\dot{\mathbf{X}}_0 + \mathbf{C}\Delta\mathbf{h}_0 + \mathbf{v} \tag{12.56}$$

where $\delta\mathbf{l}(t)$ is the observed minus computed value of the observable based on the *a priori* model, $\Delta\mathbf{X}_0$ is the adjustment of the *a priori* position vector, $\Delta\dot{\mathbf{X}}_0$ is the adjustment of the *a priori* station velocity vector, $\Delta\mathbf{h}_0$ is the adjustment of the *a priori* additional parameter vector,

$$\mathbf{A} = \frac{\partial \mathbf{F}}{\partial \mathbf{X}} \tag{12.57}$$

$$\mathbf{B} = \frac{\partial \mathbf{F}}{\partial \mathbf{a}} = \mathbf{A}\frac{\partial \mathbf{X}}{\partial \mathbf{a}} \tag{12.58}$$

$$\mathbf{C} = \frac{\partial \mathbf{F}}{\partial \mathbf{h}} \tag{12.59}$$

and **v** is the error term such that

$$E\{\mathbf{v}\} = \mathbf{0}\,;\, D\{\mathbf{v}\} = E\{\mathbf{v}\mathbf{v}^T\} = \sigma_0^2\,\mathbf{P}^{-1} \tag{12.60}$$

For the purposes of this discussion, we have assumed that station motion is linear in time, so that vector **a** in (12.55) includes only the site velocity $\dot{\mathbf{X}}$.

Observation equations for site coordinates and velocity. In a geocentric Cartesian reference frame, the observation equations for station coordinates and velocities are given by

$$\delta\mathbf{X} = \mathbf{L}_X \begin{bmatrix} \delta\mathbf{X}_0 \\ \delta\dot{\mathbf{X}}_0 \end{bmatrix} + \mathbf{L}_X \begin{bmatrix} \mu & \mathbf{0} \\ \mathbf{0} & \mu \end{bmatrix} \begin{bmatrix} \omega_X \\ \omega_{\dot{X}} \end{bmatrix} + \mathbf{L}_X \begin{bmatrix} \tau_X \\ \tau_{\dot{X}} \end{bmatrix} \tag{12.61}$$

$$\delta\dot{\mathbf{X}} = \mathbf{L}_{\dot{X}} \begin{bmatrix} \delta\mathbf{X}_0 \\ \delta\dot{\mathbf{X}}_0 \end{bmatrix} + \mathbf{L}_{\dot{X}} \begin{bmatrix} \mu & \mathbf{0} \\ \mathbf{0} & \mu \end{bmatrix} \begin{bmatrix} \omega_X \\ \omega_{\dot{X}} \end{bmatrix} + \mathbf{L}_{\dot{X}} \begin{bmatrix} \tau_X \\ \tau_{\dot{X}} \end{bmatrix} \tag{12.62}$$

where

$$\mathbf{L}_X = \begin{bmatrix} \mathbf{I}_3 + \frac{\partial \dot{\mathbf{X}}_0}{\partial \mathbf{X}_0}(t - t_0) & \mathbf{I}_3(t - t_0) \end{bmatrix} \tag{12.63}$$

$$\mathbf{L}_{\dot{X}} = \begin{bmatrix} \frac{\partial \dot{\mathbf{X}}_0}{\partial \mathbf{X}_0} & \mathbf{I}_3 \end{bmatrix} \tag{12.64}$$

$$\mu = \begin{bmatrix} 0 & -z_0 & y_0 \\ z_0 & 0 & -x_0 \\ -y_0 & x_0 & 0 \end{bmatrix} \tag{12.65}$$

$\mathbf{I}_3$ is the (3x3) identity matrix, ω_X is the rotation angle vector, $\omega_{\dot{X}}$ is the rotation angle rate vector, τ_X is the translation vector , $\tau_{\dot{X}}$ is the translation rate vector, and (x_0, y_0, z_0) are the *a priori* coordinates.

Observation equations for baseline and baseline rate vectors. Again, in a geocentric Cartesian reference frame,

$$\delta(\mathbf{dX}_{ij}) = \mathbf{L}_X \begin{bmatrix} \delta\mathbf{X}_{0_j} - \delta\mathbf{X}_{0_i} \\ \delta\dot{\mathbf{X}}_{0_j} - \delta\dot{\mathbf{X}}_{0_i} \end{bmatrix} \tag{12.66}$$

$$\delta(\mathbf{d\dot{X}}_{ij}) = \mathbf{L}_{\dot{X}} \begin{bmatrix} \delta\mathbf{X}_{0_j} - \delta\mathbf{X}_{0_i} \\ \delta\dot{\mathbf{X}}_{0_j} - \delta\dot{\mathbf{X}}_{0_i} \end{bmatrix} \tag{12.67}$$

Observation equations for episodic site displacements. Unknown episodic site displacements at a subset of sites are modeled as step functions in site position (e.g., coseismic displacements, antenna offsets, eccentricities). The observation equations are given by

$$\delta\mathbf{X}(t) = \mathbf{L}_X \begin{bmatrix} \delta\mathbf{X}_0 \\ \delta\dot{\mathbf{X}}_0 \end{bmatrix} + \sum_k [r_k(t, t_k)\delta\xi_k] \tag{12.68}$$

$$r_k(t,t_k) = \begin{matrix} -1 \\ 0 \\ 1 \end{matrix} \left\{ \begin{matrix} \text{if } (t < t_k < t_0) \\ \text{if } (t > t_k, t_k < t_0 \text{ or } t < t_k, t_k > t_0) \\ \text{if } (t > t_k > t_0) \end{matrix} \right\} \tag{12.69}$$

where t_k is the occurrence epoch, of the *k*-th event, and $\delta\xi_k$ is the site displacement vector from the *k*-th event.

Transformation to other coordinate frames. Although the geocentric Cartesian frame is conceptually simple, other frames are more intuitive for representation of position.

Geodetic Coordinates (ϕ, λ, h). For an ellipsoid with semimajor axis *a* and eccentricity '*e*' (see section 1.6.6)

$$\begin{aligned} X(t) &= [N + h(t)] \cos \phi(t) \cos \lambda(t) \\ X(t) &= [N + h(t)] \cos \phi(t) \sin \lambda(t) \\ Z(t) &= [N(1 - e^2) + h(t)] \sin \phi(t) \end{aligned} \tag{12.70}$$

where

$$N(t) = \frac{a}{\sqrt{(1 - e^2) \sin^2 \phi(t)}}$$

is the radius of curvature in the prime vertical.

Local topocentric coordinate frame. The local vector $\mathbf{dx}_{ij}(t)$ emanating from site *i* is transformed to the geocentric Cartesian frame by the rotation matrix **R** such that

$$\mathbf{dx}_{ij}(t) = \mathbf{R}_i(t)\mathbf{dX}_{ij}(t) = \mathbf{R}_i(t)[\mathbf{X}_j(t) - \mathbf{X}_i(t)] \tag{12.71}$$

$$\mathbf{R}_i(t) = \begin{bmatrix} -\sin\Lambda_i(t) & \cos\Lambda_i(t) & 0 \\ -\sin\Phi_i(t)\cos\Lambda_i(t) & -\sin\Phi_i(t)\sin\Lambda_i(t) & \cos\Phi_i(t) \\ \cos\Phi_i(t)\cos\Lambda_i(t) & \cos\Phi_i(t)\sin\Lambda_i(t) & \sin\Phi_i(t) \end{bmatrix} \tag{12.72}$$

$$\Phi_i - \phi_i = \xi_i \; ; \; \Lambda_i - \lambda_i = \frac{\eta_i}{\cos \phi_i}$$

where (Φ, Λ) are astronomic coordinates (latitude and longitude), and (ξ, η) are the north-south (positive south) and east-west (positive west) deflections of the vertical, respectively. For representing relative positions in terms of "north, east, and up" components, it is sufficient to substitute geodetic latitude and longitude for astronomic latitude and longitude, in (12.72).

12.4.2 Integration with Other Geodetic Measurements

SLR and VLBI. The observation equations for satellite laser ranging determinations of position and velocity are given by (12.61) and (12.62); very long baseline interferometry baseline and baseline rate determinations by (12.66) and (12.67). Similarity transformation parameters given in (12.61) and (12.62) can be estimated to correct for reference frame differences (see also section 1.6.7).

Conventional terrestrial observations. In crustal deformation studies with GPS, there are often older measurements available from triangulation, see e.g. Bibby [1982], leveling, and trilateration, see e.g. Lisowski et al. [1991]. The linearized observation equations for azimuth (α), vertical angle β and distance ℓ for local site i to site j are given by

$$\begin{bmatrix} \delta\alpha \\ \delta\beta \\ \delta\ell \end{bmatrix} = \begin{bmatrix} (\ell_0 \cos\beta_0)^{-1} & 0 & 0 \\ 0 & \ell_0^{-1} & 0 \\ 0 & 0 & 1 \end{bmatrix} \left[\mathbf{S}_0^T \mathbf{R}_0 \mathbf{L}_X \begin{pmatrix} \delta\mathbf{X}_{0_j} - \delta\mathbf{X}_{0_i} \\ \delta\dot{\mathbf{X}}_{0_j} - \delta\dot{\mathbf{X}}_{0_i} \end{pmatrix} \right] \tag{12.73}$$

where

$$\mathbf{S}_0 = \begin{bmatrix} \cos\alpha_0 & -\sin\alpha_0 \sin\beta_0 & \sin\alpha_0 \cos\beta_0 \\ -\sin\alpha_0 & -\cos\alpha_0 \sin\beta_0 & \cos\alpha_0 \cos\beta_0 \\ 0 & \cos\beta_0 & \sin\beta_0 \end{bmatrix} \tag{12.74}$$

$\mathbf{R}$ is given by (12.72), and the zero subscript indicates values calculated from *a priori* values.

12.4.3 Software Independent Exchange Format (SINEX)

A software independent exchange format (SINEX), see Blewitt et al. [1994], has been developed by the IGS (see section 1.7) for GPS solution files. A SINEX file includes parameter adjustments, full covariance matrices, and all necessary auxiliary information. The adoption of this format by the geodetic community (akin to the widespread use of the RINEX format) is facilitating the exchange of solutions and the rigorous integration of geodetic networks, from a variety of terrestrial and space geodetic measurements.

12.4.4 Estimation Procedures

Input to network adjustment. The basic input to a geodetic network adjustment are the adjusted parameter vectors and corresponding full covariance matrices from a set of separate GPS (and other geodetic) solutions. Stochastic and mathematical models are chosen (section 12.4.1), as well as an estimation method. The basic output is a consistent set of geodetic coordinates (and velocities) referred to some epoch in time, and possibly other parameters of interest.

Adjustment of Baseline Mode Solutions. Network adjustment of baseline-mode solutions is well known, e.g. Bock et al. [1985], and several software packages are available to perform this task. The observation equations are given in 12.4.1 and the parameters estimated are station coordinates (and similarity transformation parameters, if necessary). The reference frame is defined by fixing (or tightly constraining) at least one set of station coordinates in the network to define the origin (preferably in the ITRF). When only an origin is fixed, the solution is commonly referred to as a *minimum-constraint* solution. In practice, since external satellite ephemerides and Earth orientation parameters are fixed in the baseline analysis (e.g., to IGS and IERS values), scale and rotation are already defined, although these can be adjusted as well if mixing several types of geodetic solutions.

Adjustment of Session Mode Solutions. GPS medium distance (regional) solutions are obtained after ambiguity resolution as described in section 12.2, and may contain a variety of geodetic parameters (the simplest case being baseline-mode solutions). Generally, network adjustment of tightly constrained solutions results in relatively poor long-term position repeatability because of global errors, primarily reference frame and orbital model deficiencies. In principle, one could use a free network adjustment (i.e., inner constraint adjustment, see below). From a computational point of view, it is preferable to use a loosely constrained solution, loose enough not to bias the solution but tight enough to allow (Cayley) inversion of the normal equations. It is important, though, to iterate these solutions to convergence since GPS adjustments are highly nonlinear (in the orbit and station coordinate parameters). In addition, loosely constrained solutions allow flexibility in modifying the underlying terrestrial reference frame and combining GPS solutions with other geodetic solutions.

Adjustment of Global and Regional Solutions. With the availability of a robust global tracking network (the IGS, section 1.7), it is feasible to produce regularly (e.g., daily or weekly) global geodetic solutions (adjustments and full covariance information), estimating tracking station coordinates, satellite orbital elements, and Earth orientation parameters[5]. Then, regional GPS solutions can be adjusted conveniently in (distributed) session mode, with each solution adjusted in common with a continental-scale subset of the global stations (see example in section 12.5.1). Thus regional coordinates can be estimated with respect to the global terrestrial reference frame from a network adjustment of regional and global solutions. Another alternative is the point-positioning mode discussed in section 12.3.4.

---

[5]Such a daily global solution with full covariance information has been produced in GLOBK h-file format [Herring, 1994] at the Scripps Orbit and Permanent Array Center, in La Jolla, California, since November, 1991. These files are available on Internet via anonymous ftp (lox.ucsd.edu) and the WWW (http://lox.ucsd.edu); e-mail: pgga@pgga.ucsd.edu. The IGS Associate Analysis Centers of Type II, Blewitt et al. [1994] have begun producing global SINEX files (section 12.4.3) based on weekly global solutions.

Combination of solutions. Each loosely constrained solution (global and regional) has a very weak underlying reference frame. At the network adjustment stage we impose a consistent reference frame by applying tight constraints on a subset of station coordinates and velocities. Here we discuss three estimation options for network adjustment.

Inner Constraint Estimation. A convenient initial step is to perform the familiar inner constraint adjustment to assess the internal precision of the network. The normal equations are augmented by the inner constraints

$$\mathbf{CX} = \mathbf{0} \tag{12.75}$$

where

$$\mathbf{C} = [\mathbf{I\,I\,I} \ldots \mathbf{I}]_{(3\text{x}3\text{k})} \tag{12.76}$$

I is the (3x3) identity matrix and k is the number of stations. The inner constraint (free) adjustment is then

$$\hat{\mathbf{X}} = (\mathbf{A}^{\mathrm{T}}\mathbf{PA} + \mathbf{C}^{\mathrm{T}}\mathbf{C})^{-1}\mathbf{A}^{\mathrm{T}}\mathbf{Pl} \tag{12.77}$$

and covariance matrix

$$\Sigma_{\hat{\mathrm{X}}} = \sigma_0^2\left[(\mathbf{A}^{\mathrm{T}}\mathbf{PA} + \mathbf{C}^{\mathrm{T}}\mathbf{C})^{-1} - \mathbf{C}^{\mathrm{T}}(\mathbf{CC}^{\mathrm{T}})^{-1}(\mathbf{CC}^{\mathrm{T}})^{-1}\mathbf{C}\right] \tag{12.78}$$

where for this particular case

$$\mathbf{C}^{\mathrm{T}}(\mathbf{CC}^{\mathrm{T}})^{-1}(\mathbf{CC}^{\mathrm{T}})^{-1}\mathbf{C} = \frac{1}{\mathrm{k}^2}\begin{bmatrix}\mathbf{I} & \cdots & \mathbf{I}\\ \vdots & \ddots & \vdots\\ \mathbf{I} & \cdots & \mathbf{I}\end{bmatrix} \tag{12.79}$$

The inner constraint solution preserves the centroid determined by the *a priori* input coordinates. In practice this type of solution has little physical justification. Other more physically plausible, geometrically constrained solutions (outer coordinate solution and model coordinate solution) are reviewed by Segall and Mathews [1988].

Bayesian Estimation. Bayesian estimation is well suited for imposing weighted constraints on a subset of the station coordinates. The estimation model can be expressed as $(\mathbf{l},\mathbf{AX},\mathbf{Q}_{\mathrm{v}}\,\overline{\mathbf{X}}\,,\Sigma_{\overline{\mathrm{X}}})$ where $\overline{\mathbf{X}}$ and $\overline{\mathbf{V}}$ are random variables such that

$$\mathrm{E}\{\overline{\mathbf{X}}\} = \mathbf{X};\ \mathrm{D}\{\overline{\mathbf{X}}\} = \Sigma_{\overline{\mathrm{X}}} = \mathrm{E}\{(\overline{\mathbf{X}} - \mathbf{X})(\overline{\mathbf{X}} - \mathbf{X})^{\mathrm{T}}\} \tag{12.80}$$

$$E\{\mathbf{V}\} = \mathbf{0};\ D\{\mathbf{V}\} = \mathbf{Q}_v = E\{\mathbf{V}\mathbf{V}^T\} = \Sigma_\mathbf{l} \quad (12.81)$$

from which

$$E\{\mathbf{l}\} = \mathbf{AX};\ D\{\mathbf{l}\} = \mathbf{A}\Sigma_{\overline{\mathbf{X}}}\mathbf{A}^T + \Sigma_\mathbf{l} \quad (12.82)$$

We can construct an estimate that is unbiased by assuming that

$$E\{\hat{\mathbf{X}}\} = (\mathbf{GA} + \mathbf{G}_x)\mathbf{X} = \mathbf{X};\ \mathbf{G}_x = \mathbf{I} - \mathbf{GA} \quad (12.83)$$

and impose minimum variance such that

$$\mathrm{tr}\{\mathbf{G}\Sigma_\mathbf{l}\mathbf{G}^T + (\mathbf{I} - \mathbf{GA})\Sigma_{\overline{\mathbf{X}}}(\mathbf{I} - \mathbf{GA})^T\} = \min \quad (12.84)$$

which gives the result

$$\mathbf{G} = \Sigma_{\overline{\mathbf{X}}}\mathbf{A}^T(\mathbf{A}\Sigma_{\overline{\mathbf{X}}}\mathbf{A}^T + \Sigma_\mathbf{l})^{-1} \quad (12.85)$$

Assuming that $\Sigma_{\overline{\mathbf{X}}}$ is positive definite and $\mathbf{M} = \Sigma_{\overline{\mathbf{X}}}^{-1}$ then

$$\hat{\mathbf{X}} = \overline{\mathbf{X}} + (\mathbf{A}^T\mathbf{PA} + \mathbf{M})^{-1}\mathbf{A}^T\mathbf{P}(\mathbf{l} - \mathbf{A}\overline{\mathbf{X}}) = (\mathbf{A}^T\mathbf{PA} + \mathbf{M})^{-1}(\mathbf{A}^T\mathbf{Pl} + \mathbf{M}\overline{\mathbf{X}}) \quad (12.86)$$

and

$$\Sigma_{\hat{\mathbf{X}}} = \sigma_0^2(\mathbf{A}^T\mathbf{PA} + \mathbf{M})^{-1} \quad (12.87)$$

where σ_0^2 is the variance of unit weight. Equation (12.86) shows that the Bayesian estimate is just a correction term to the *a priori* model assumption, $\overline{\mathbf{X}}$.

*Kalman Filtering***.** The Kalman filter formulation is quite useful in combining many individual GPS solutions with respect to a consistent terrestrial reference frame. The Kalman filter is an extension of Bayesian estimation described in the previous section. Its advantages are that it can conveniently be applied in a sequential manner and that the parameters can be treated as stochastic processes. Sequential processing allows us to easily distinguish poor sessions and to diagnose problems in the input data.
Let us start from the linearized Gauss-Markov model $(\mathbf{l}, \mathbf{AX}, \mathbf{Q}_v)_t$ generalized in time as

$$\mathbf{l}_t = \mathbf{A}_t\mathbf{X}_t + \mathbf{V}_t \quad (12.88)$$

$$E\{\mathbf{V}_t\} = \mathbf{0};\ E\{\mathbf{V}_t\mathbf{V}_u) = \mathbf{0};\ D\{\mathbf{V}_t\} = E\{\mathbf{V}_t\mathbf{V}_t^T\} = (\mathbf{Q}_v)_t \quad (12.89)$$

where u is any epoch other than t. The dynamics of the parameters are given by the state transition equation

$$\mathbf{X}_{t+1} = \mathbf{S}_t\mathbf{X}_t + \mathbf{W}_t \tag{12.90}$$

where the state transition matrix $\mathbf{S}_t$ operates on the state of the system at epoch t to give the expected state at epoch t+1 and

$$E\{\mathbf{W}_t\} = \mathbf{0}\,;\ E\{\mathbf{W}_t\mathbf{W}_u) = \mathbf{0};\ D\{\mathbf{W}_t\} = E\{\mathbf{W}_t\mathbf{W}_t{}^T\} = (\mathbf{Q_X})_t \tag{12.91}$$

and for the cross covariances at epochs t and u (t≠u)

$$E\{\mathbf{V}_t\mathbf{W}_u) = E\{\mathbf{V}_t\mathbf{X}_u{}^T\} = E\{\mathbf{X}_t\mathbf{W}_u{}^T\} = \mathbf{0} \tag{12.92}$$

A deterministic (nonstochastic) parameter has by definition

$$W_t = 0 \tag{12.93}$$

The forward Kalman filter is performed sequentially and is given by

$$\hat{\mathbf{X}}^t_{t+1} = \mathbf{S}_t\hat{\mathbf{X}}^t_t \tag{12.94}$$

$$\mathbf{C}^t_{t+1} = \mathbf{S}_t\mathbf{C}^t_t\mathbf{S}^T_t + (\mathbf{Q_X})_t \tag{12.95}$$

where $\mathbf{C}$ is the covariance matrix, and

$$\hat{\mathbf{X}}^{t+1}_{t+1} = \hat{\mathbf{X}}^t_{t+1} + \mathbf{K}(\mathbf{l}_{t+1} - \mathbf{A}_{t+1}\hat{\mathbf{X}}^t_{t+1}) \tag{12.96}$$

$$\mathbf{C}^{t+1}_{t+1} = \mathbf{C}^t_{t+1} - \mathbf{K}\mathbf{A}_{t+1}\mathbf{C}^t_{t+1} \tag{12.97}$$

where $\mathbf{K}$ is the Kalman gain

$$\mathbf{K} = \mathbf{C}^t_{t+1}\mathbf{A}^T_{t+1}[(\mathbf{Q_V})_{t+1} + \mathbf{A}_{t+1}\mathbf{C}^t_{t+1}\mathbf{A}^T_{t+1}]^{-1} \tag{12.98}$$

Compare the Kalman gain matrix to matrix $\mathbf{G}$ in equation (12.85).

After all the observations have been processed by the filter, the resultant state yields the estimates of the nonstochastic parameters (if all the parameters are nonstochastic then the forward Kalman filter estimate reduces to the weighted least squares solution). The estimates of the stochastic parameters are determined sequentially by a backward or smoothing filter which is just the forward filter run with time in reverse, and then taking the weighted mean of the forward and backward runs such that

$$\hat{\mathbf{X}}_t^s = \hat{\mathbf{X}}_+ + \mathbf{B}(\hat{\mathbf{X}}_- - \hat{\mathbf{X}}_+) \tag{12.99}$$

$$\mathbf{C}_t^s = \mathbf{C}_+ - \mathbf{B}\mathbf{C}_+ \tag{12.100}$$

$$\mathbf{B} = \mathbf{C}_+(\mathbf{C}_- + \mathbf{C}_+)^{-1} \tag{12.101}$$

where the positive subscript indicates that the estimates are from the forward filter and the negative subscripts indicates that the estimates are from the backward filter. The superscript s indicates the smoothed estimate.

Deterministic parameters could include (a subset of) station coordinates and velocities. The coordinates and velocities of a subset of global stations could be tightly constrained to define the reference frame, for example. The stochastic parameters could be very loosely constrained (between successive days) and include regional site coordinates, satellite orbit, and Earth orientation (polar motion and UT1 rate) parameters.

12.4.5 Common-Mode Analysis of Adjusted Positions

Cancellation of common-mode errors by differencing of phase (and pseudorange) *observables* has always been an inherent part of GPS analysis whether performed explicitly (double differencing) or implicitly (epoch by epoch clock estimation). It is well known that between-stations differencing eliminates common-mode satellite clock errors, and between-satellites differencing eliminates common-mode station clock errors. The former also cancels common-mode atmospheric and orbital errors on short baselines. In this section we present the concept of cancellation of common-mode errors by differencing estimated *positions* (after network adjustment). There are two general classes of techniques that we discuss here: *stacking in time* and *stacking in space.*

It is preferable, of course, to eliminate or reduce systematic errors at the instrumentation level or at the estimation stage. For example, one could attempt to design a site that minimizes signal scattering or model this complicated signal site by site, see e.g. Elósegui, et al. [1995] and Jaldehag et al. [1996]. Stacking provides a powerful and complementary approach as we show below.

Stacking in time. The removal of daily multipath signatures by stacking in time has been described by Genrich and Bock [1992], and subsequently used by several investigators to enhance resolution of kinematically-determined station positions just before and after major earthquakes. The GPS satellites have semi-diurnal orbital periods so that the same satellite geometry essentially repeats over successive days, except for an approximately 3^m 56^s negative shift in time due to the difference between sidereal and universal time (see equation 1.22). Assuming that site characteristics have not changed, multipath signatures will be highly correlated from day to day and will be manifested as relatively low-frequency noise superimposed on higher-frequency measurement noise. The strong day to day correlation makes it possible to suppress this noise by subtracting the low-pass-filtered signature that is evident during the first observation session from the time series of subsequent days (shifted in time by $3^m\ 56^s$).

Stacking in space. Continuous GPS networks and multimodal techniques allow us to take advantage of common-mode position estimate signatures in medium distance (regional) networks resulting from global systematic errors which we refer to as *stacking in space*. Zhang [1996] demonstrated that there exist strong statistical correlations among absolute coordinate estimates of regional networks, and weak correlations between the coordinate estimates of regional and global networks. Any global errors are expected to propagate in a similar manner to all regional site estimates, in particular satellite orbit, Earth orientation and reference frame errors. A comparison of the raw position time series estimated by a continuous GPS array in southern California (see section 12.5.1) reveals that there is indeed a systematic bias in each coordinate component which is distinguishable from site-specific effects, Wdowinski et al. [1997]. These biases are attributed to global errors. Based on this experience, a stacking technique has been developed to eliminate common-mode signatures in the time series of daily positions, computed by network adjustment with respect to the ITRF, Bock et al. [1997] and Wdowinski et al. [1997].

The stacking algorithm isolates and removes the common systematic signatures in the position time series, component by component, and can be summarized as follows:

1. *Detrending.* Remove a linear trend from the raw time series. When a discontinuity is present in the time series (e.g. due to an antenna switch) detrending is performed separately for data before and after the discontinuity.
2. *Correcting for discontinuities.* The magnitude of discontinuities can be estimated by averaging the detrended positions for a few weeks before and after the discontinuity and differencing the positions. The filtered series (see step 6) provides more sensitivity to correcting the discontinuities than the raw or cleaned series (see step 3).
3. *Outlier Detection.* Clean the series of data outliers.
4. *Stacking.* Sum and average the cleaned series to determine a common-mode bias. Problematic sites can be excluded from the stacking, for example low-frequency fluctuations in the time series due to monument and/or site instability.
5. *Filtering.* Remove the common-mode bias from the cleaned series. The bias is subtracted from all series, including the data from sites that are not used in the stacking.
6. *Correcting for discontinuities.* Same as step 2 but for the filtered series.

This sequence of steps is iterated until there are no more detected outliers and residual offset estimates are less than a specified level (e.g., sub-mm). If a particular filtered time series is seen to have non-linear trends, it is removed from the stack and steps 1-4 are repeated. The station velocities are determined at step 1. They are minimally affected by the stacking procedure.

The stacking algorithm is a powerful technique for isolating site specific signatures, whether due to "noise" (e.g., site instability or multipath) or "signal" (e.g., postseismic displacements or interseismic strain variations). The stacking signature can also be removed from roving receivers within the region, being careful to match total observation time with the continuous trackers. One could also apply a combination of time and space stacking for continuous GPS networks. An application of the stacking in space algorithm is given in the next section.

12.5 CASE STUDIES

12.5.1 Accuracy of GPS Position Estimates at Medium Distances

It is well know that GPS measurements are inherently capable of providing three-dimensional positions with sub-millimeter accuracy over short distances. Over medium distances error sources such as atmospheric (tropospheric and ionospheric) refraction, and orbital and reference frame errors will degrade position accuracy as some function of the distance between sites. Site-specific errors will also degrade GPS position accuracy, for example, geodetic monument instability. It is important to be able to assess the short- and long-term accuracy of GPS positions over medium-distance scales. For crustal deformation and related applications it also critical to be able to determine the accuracy of quantities derived from repeated GPS position estimates, for example site velocities and episodic site displacements. The availability of longer and longer time series of GPS positions estimated from data collected by regional continuous GPS arrays, such as the southern California Integrated GPS Network (SCIGN), allows us to better assess the accuracy of GPS positions at medium-distance scales, and hence, quantities derived from these positions.

Colored noise in geodetic data. Experience with other types of continuous geodetic measurements indicates that instability of geodetic monuments, Karcz et al. [1976] introduces significant temporal correlations in these data. Expansive clays in near-surface rocks can cause monument motions comparable to the measurement precision of GPS, complicating the detection of tectonic phenomenon. The power spectra for monument displacements shows the ground to undergo random-walk-like motion. Frequent two color geodimeter (distance) measurements, see Langbein et al. [1987] and Langbein and Johnson [1995] exhibit power spectra that rise at low frequencies in proportion to f^{-2}, which is characteristic of random walk processes. The random walk noise in these data ranges from 1.3 to 4.0 $mm/yr^{0.5}$, and is attributed to monument instability , see Langbein et al. [1990] and Langbein and Johnson [1997]. Johnson and Agnew [1995] performed simulations with synthetic data to investigate the effect of random walk on the estimation of station velocities by continuous geodetic measurements. They conclude that to achieve better estimates of station velocity from continuous GPS measurements, compared to campaign measurements, monument instability must be held to a small fraction of the measurement system precision.

To understand the effect of colored noise in GPS position estimates on derived quantities consider linear regression of an equally-spaced time series of GPS positions for the purpose of estimating site velocity, Zhang et al. [1997]. In the case of white noise (WN), velocity uncertainty σ_r is proportional to the magnitude of white noise a_{WN}, and inversely proportional to the total observation interval T, and the square root of the number of measurements N according to

$$\left(\sigma_{\hat{r}}\right)_{WN} = \frac{a_{WN}}{T}\sqrt{\frac{12\,(N-1)}{N^2+N}} \simeq \frac{2\sqrt{3}\,a_{WN}}{N^{1/2}T} \quad (\text{for } N >> 2) \tag{12.102}$$

This is the basis for the familiar $N^{1/2}$ increase in accuracy sometimes assumed for continuous GPS measurements compared to less frequent field measurements. The magnitude of the white noise component is limited by the inherent GPS system measurement error which can be as small as 1 mm or less, Genrich and Bock [1992]. In the case of random walk noise (RWN)

$$\left(\sigma_{\hat{r}}\right)_{RWN} = \frac{b_{RWN}}{T^{1/2}} \tag{12.103}$$

That is, velocity uncertainty is proportional to the magnitude of random walk noise b_{RWN}, and inversely proportional to only the square root of the total time span T. It is *independent* of the number of observations. The presence of random walk motion in continuous GPS measurements of magnitudes 1.3-4 mm/yr$^{0.5}$ seen in longer series of conventional geodetic measurements by Langbein and Johnson [1997] would significantly degrade site velocity uncertainties as can be easily be seen by comparing (12.102) and (12.103).

Regional continuous GPS arrays. In the last few years there has been a growing interest in measuring crustal deformation at tectonic plate boundaries by remotely controlled, continuously monitoring arrays (section 12.1.3) in which GPS units are deployed permanently and unattended over highly stable geodetic monuments. The first continuous GPS network was established in 1988 in the Kanto-Tokai region of central Japan, Shimada et al. [1990]. In 1990, the first 4 stations of the Southern California Permanent GPS Geodetic Array (PGGA) were established as a pilot project to demonstrate the feasibility and effectiveness of continuous GPS, see Bock [1991] and Lindqwister et al. [1991]; today the PGGA, see Bock et al. [1997] is the regional component of the Southern California Integrated GPS Network with more than 45 stations operational (Figure 12.1) and an expansion to 250 stations to be accomplished by the end of 1998, Prescott [1996] and Bock and Williams [1997].

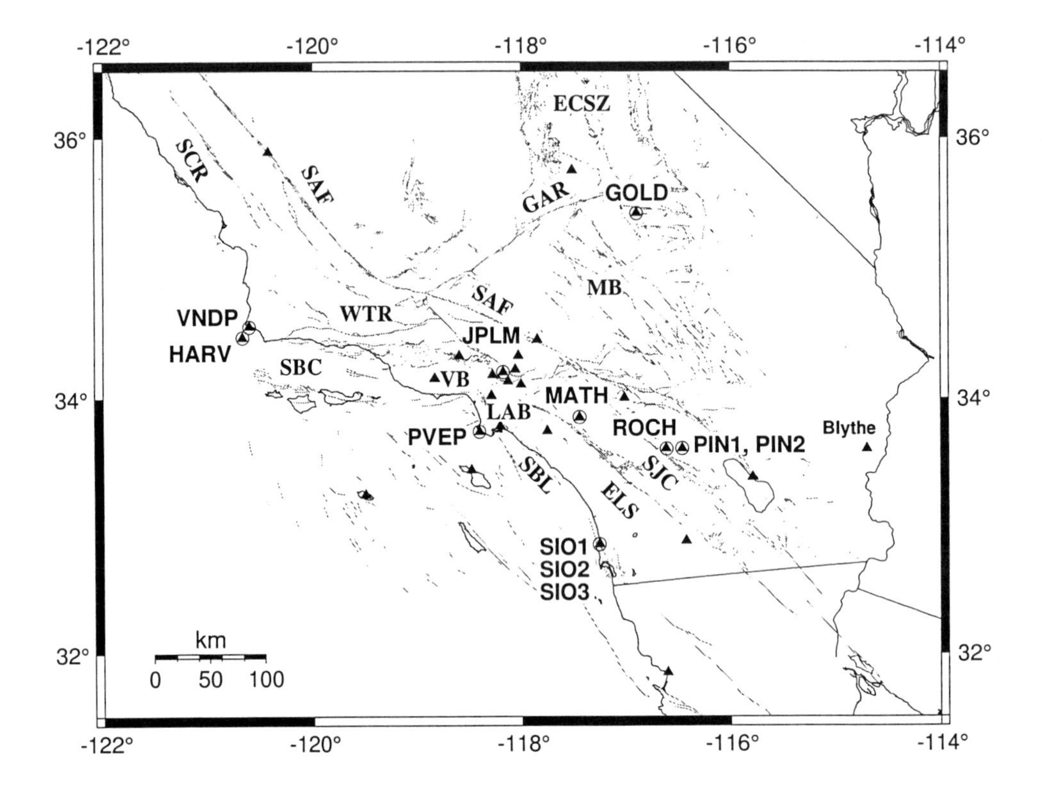

Figure 12.1. Sites of the Southern California Integrated GPS Network (SCIGN) *circa* mid-1996, including the regional Permanent GPS Geodetic Array (PGGA) sites and Dense GPS Geodetic Array (DGGA) in the Los Angeles basin. PGGA sites operational between the 1992 Landers and 1994 Northridge earthquakes are denoted by circumscribed triangles and their four-character codes. Other SCIGN sites are denoted by triangles only. Tectonic features: ECSZ, Eastern California Shear Zone; ELS, Elsinore fault; GAR, Garlock fault; LAB, Los Angeles basin; MB, Mojave Block; SAF, San Andreas fault; SBC, Santa Barbara Channel; SBL, Southern Borderlands; SCR, Southern Coast Ranges; SJC, San Jacinto fault; VB, Ventura basin; WTR, Western Transverse Ranges. Adapted from Bock et al. [1997].

Other regional continuous GPS networks can be found, for example, in western Canada, Dragert and Hyndman [1995], northern California, King et al. [1995], Japan, Tsuji et al. [1995], and Scandinavia, Jaldehag et al. [1996].

Continuous GPS provides temporally dense measurements of surface displacements induced by crustal deformation processes including interseismic, coseismic, postseismic and aseismic deformation, and the potential for detecting anomalous events such as preseismic deformation and interseismic strain variations. Within several months of the start of regular operations, the PGGA recorded far-field coseismic displacements induced by the June 28, 1992 (M_W=7.3), Landers earthquake. see Blewitt et al. [1993] and Bock et al. [1993], the largest magnitude earthquake in California in the past 40 years and the first one to be recorded by a continuous GPS array. Only nineteen months later, on January 17, 1994, the PGGA recorded coseismic displacements, see Bock [1994], for the strongest earthquake to strike the Los Angeles basin in two decades, the (M_W=6.7) Northridge earthquake. Fortuitously, the IGS became operational only several weeks before the Landers earthquake, providing access to a consistent global terrestrial reference frame and data for the computation of precise satellite ephemerides and Earth orientation, IGS [1993, 1995]. The IGS/PGGA synergism provides a rigorous way of computing positions of regional sites with respect to a global reference frame, rather than the traditional approach of computing intraregional relative site positions (baselines). It then becomes possible to position one or more roving GPS receivers with respect to the regional anchor provided by a continuous GPS array, so that logistically complex field campaigns with many receivers are no longer necessary for monitoring crustal deformation, Bevis et al. [1997]. Furthermore, continuous arrays are able to generate other classes of geophysical information, for example integrated water vapor, see e.g. Bevis et al. [1992] and Duan et al. [1996] and ionospheric disturbances, Calais and Minster [1995].

Stable GPS monuments were especially designed for use in the PGGA network, Wyatt et al. [1989] to minimize low-frequency colored noise in data, described in the previous section. One such monument consists of two parts:

(1) a ground-level base that is anchored at depth (~10 m), laterally braced, and decoupled from the surface (~3 m), as much as possible; and

(2) an antenna-mount (~1.7 m tall) that can be precisely positioned on this base (Figure 12.2a). The removable antenna mount can also support other types of measurements such as leveling rods and EDM reflectors. A modified version was also constructed at consisting of one vertical and four obliquely braced metal pipes anchored at depth (~10 m), decoupled from the surface, and welded together at their point of intersection about 1.5 m above the surface (Figure 12.2b). This type of monument was successful in attenuating seasonal displacements of 2-color geodimeter reflector monuments at Parkfield, California, to less than 1 mm compared to annual variations of up to 12 mm exhibited by earlier shallow, vertically driven monuments, Langbein et al. [1995]. Another strategy used to monument PGGA sites was anchoring stainless steel rods in exposed bedrock or massive concrete structures (Figure 12.2c).

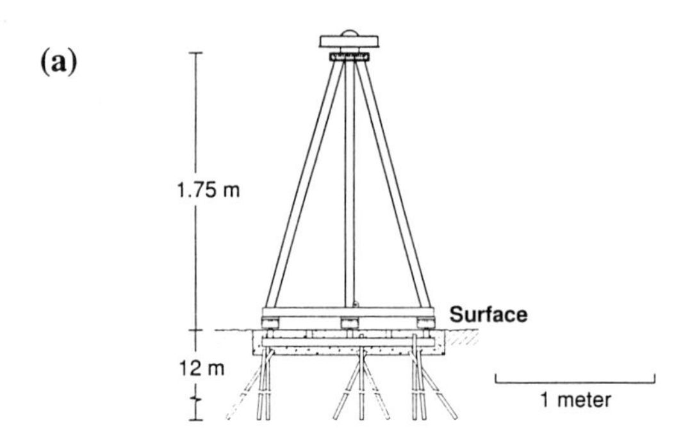

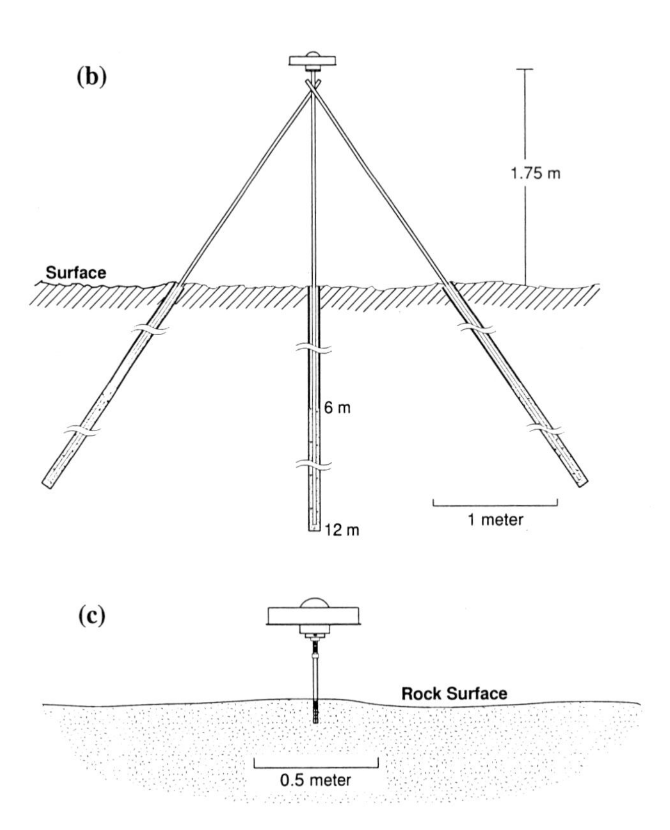

Figure 12.2. Monument designs for SCIGN: (a) deeply anchored monument with concrete base and removable antenna mount, constructed at Piñon Flat Observatory (PIN1), Scripps (SIO1, SIO3), and Vandenberg (VNDP), (b) deeply anchored monument of quinqunx design, constructed at Piñon Flat Observatory (PIN2), (c) typical bedrock anchored monument, constructed at Lake Mathews (MATH) and Pinemeadow (ROCH). Adapted from Bock et al. [1997].

Time series analysis of continuous GPS data. Zhang et al. [1997] analyzed time series of daily position estimates for ten PGGA sites for the 19-month period between the Landers and Northridge earthquakes. An example of raw and filtered position time series (demeaned) for north, east and vertical components of PGGA sites at Goldstone (GOLD) and JPL (JPLM) are shown in Figure 12.3 (these time series have been filtered using the algorithm outlined in section 12.4.5). The circled points in the raw time series are considered outliers and are deleted from further analysis. The triangulated points indicate days when anti-spoofing (AS) was activated. The bottom panel shows the time history of the common-mode bias removed daily from the raw time series (cleaned of outliers) of each site to construct the filtered time series. It is clear, for example, that the north component of GOLD exhibits temporal correlations (this may be due to the fact that it is positioned at an unguyed 25 m tower). Zhang et al. [1997] investigated four noise models for the filtered time series.

The first three are a white noise (WN) model, a white noise plus flicker noise model (WN+FN), and a white noise plus random walk model (WN+RWN). In the frequency domain, the spectral index of white noise is zero (a flat spectrum), the spectral index of flicker (1/f) noise is one, and the spectral index of random walk ($1/f^2$) noise is two. Of these three models, the WN+FN model best fits these data. This analysis is performed in the time domain using a maximum likelihood estimation (MLE) algorithm, see Langbein and Johnson [1997] which is currently implemented only for noise models with integer-valued spectral indices.

Zhang et al. also investigated in the frequency domain a fourth model which assumes that the noise can be described by a single-parameter power law process of the form

$$S_x(\omega) = S_0(\omega / \omega_0)^{-\kappa} \tag{12.104}$$

where ω is spatial or temporal frequency, S_0 and ω_0 are normalizing constants, and κ is the spectral index, Mandlebrot and Van Ness [1968]. The larger k is in (12.104), the smoother is the process and the more dominant are the lower frequencies. Geophysical phenomena often approximate processes with spectral index in the range $1< \kappa <3$, see e.g. Agnew [1992] and Davis et al. [1994], and are termed "fractal random walk", Mandlebrot and Van Ness [1968]. For classical Brownian motion ("random walk"), $\kappa = 2$. An example of a fractal random walk process is "Kolmogorov turbulence" with $\kappa = 5/3$, Kolmogorov [1941], which plays an important role in GPS atmospheric errors at medium scales, e.g. Treuhaft and Lanyi [1985]. Processes with spectral index larger than 1 are nonstationary. "Fractal white noise" is defined within the range $-1< \kappa <1$, Mandlebrot [1983]. These processes are stationary or independent of time. Classical white noise has a spectral index of 0 and flicker noise has a spectral index of 1. Zhang et al. found that a fractal white noise model with a spectral index of about 0.4 also fits the 19-month PGGA time series.

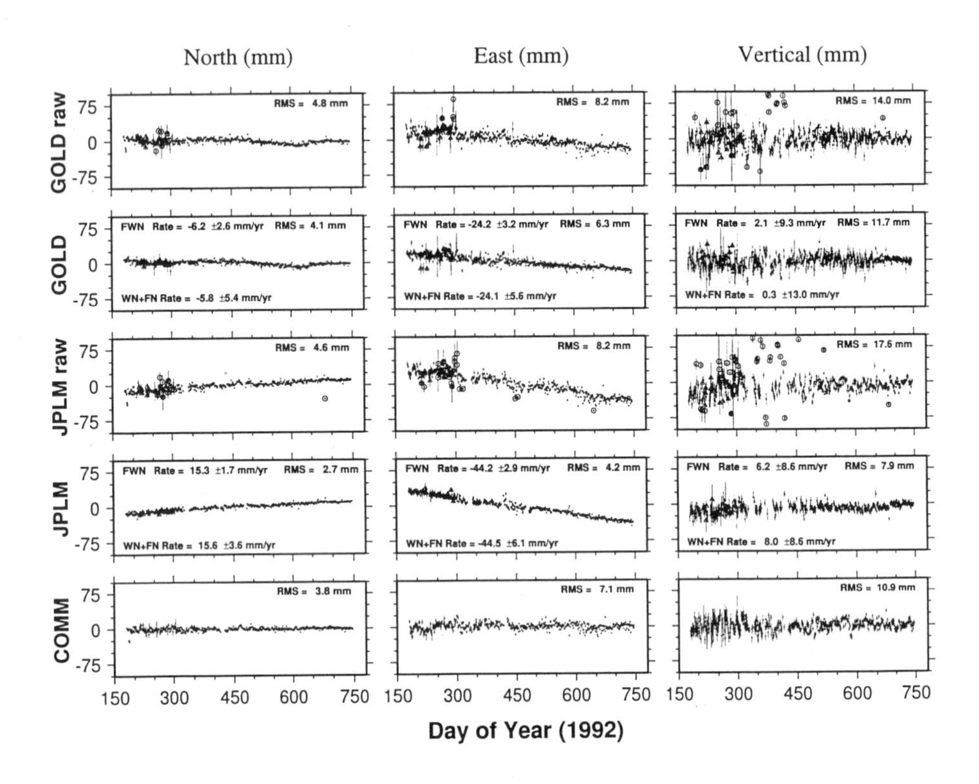

Figure 12.3. Demeaned raw and filtered position time series for north, east, and vertical components of sites at Goldstone (GOLD) and JPL (JPLM) for the 19-month period between the Landers and Northridge earthquakes. The circled points in the raw time series are considered outliers and are deleted from further analysis. The triangulated points indicate days when GPS anti-spoofing (AS) was activated. Listed for each filtered time series is the rate computed by linear regression and its uncertainty (95% confidence interval), under the fractal white noise (FWN) model and the white noise plus flicker noise (WN+FN) model, Zhang et al. [1997], and the weighted RMS scatter of the detrended time series. The bottom panel shows the common-mode bias removed daily from the raw time series (cleaned of outliers) of each site to construct the filtered time series. Adapted from Bock et al. [1997].

Estimation of site velocity. Shown in Figure 12.4 are the horizontal velocities of the PGGA sites fit to the 19-month position data, with 95% confidence ellipses, for both the FWN model and the WN+FN model. It is clear that the choice of model will affect the significance of the estimated velocities. and influence the tectonic interpretation. Bock et al. [1997] compared velocities at four PGGA site locations with those reported by Feigl et al. [1993], which were derived from an independent set of GPS and VLBI measurements collected over nearly a decade prior to the Landers earthquake (Figures 12.5 and 12.6). The velocity differences for three sites 65-100 km from the earthquake's epicenter are of order 3-5 mm/yr and are systematically oriented clockwise, at angles of 20-45°, with respect to the corresponding directions of coseismic displacement. The fourth site, 300 km from the epicenter, shows no velocity difference. These velocity changes cannot be attributed to a simple reference frame misalignment, for example, a horizontal rotation, and the directions of coseismic displacements are only very weakly dependent on the chosen reference frame. The statistical significance of these observations is complicated by an incomplete knowledge of the noise properties of the time series since two possible noise models fit the PGGA data equally well. Under the FWN model velocity differences for all three sites are statistically different at the 99% significance level. The WN+FN model results in significance levels of only 94%, 43%, and 88%.

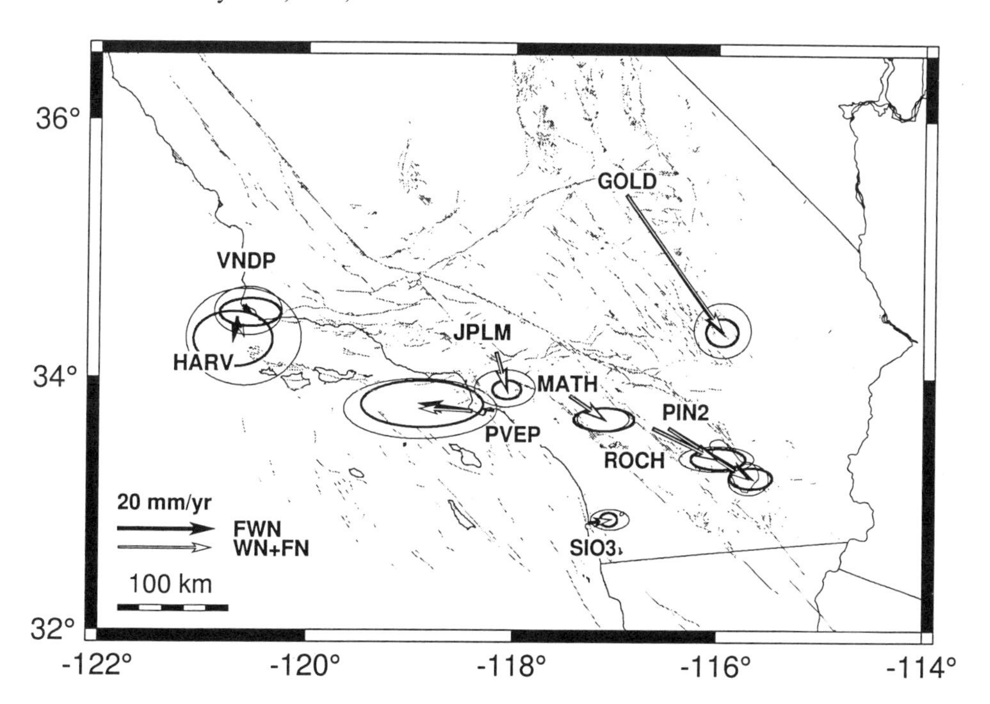

Figure 12.4. Observed 19-month PGGA horizontal site velocities (mm/yr) relative to the NUVEL-1A Pacific pole of rotation for the fractal white noise (FWN) and the white noise plus flicker noise (WN+FN) models. Estimates of north and east components are assumed to be uncorrelated. Error ellipses are 95% confidence. Adapted from Bock et al. [1997].

If these signals turn out to be of tectonic origin, this will be the first time that postseismic deformation of this duration has been observed over such a large distance from the rupture zone of a major strike-slip earthquake, and may have important implications for earthquake research. This case study demonstrates that assessing the geodetic accuracy of GPS measurements at medium distance is critical for the proper interpretation of derived quantities, in this case site velocities.

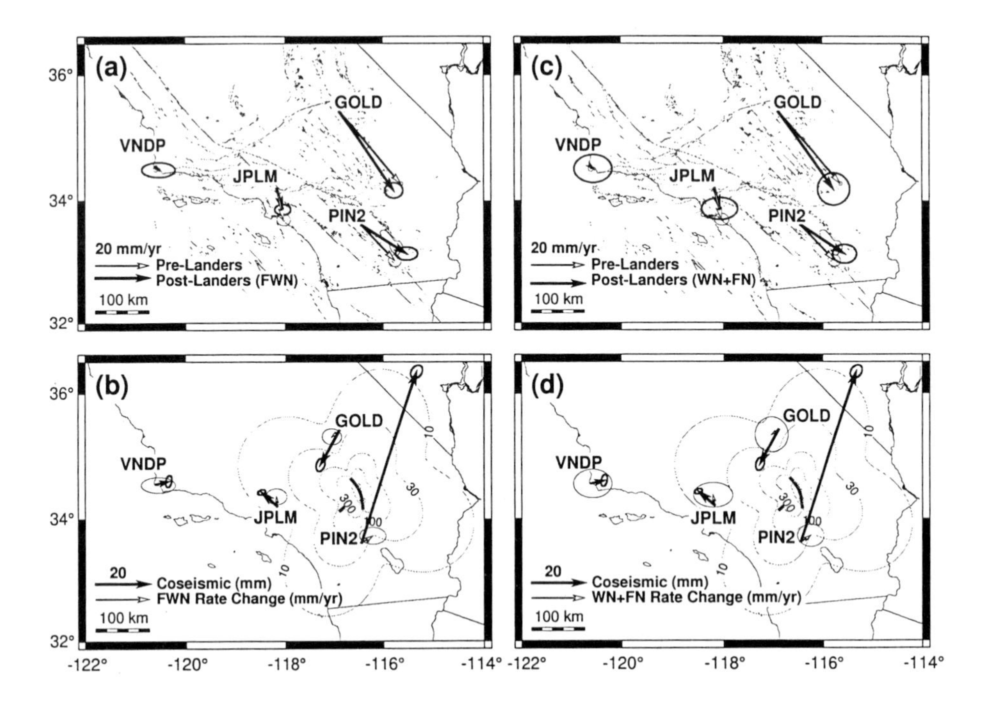

Figure 12.5. Change in regional deformation rates induced by the 1992 Landers earthquake. All ellipses are 95% confidence. (a) and (c) Pre-Landers site velocities estimated by Feigl et al. [1993] and post-Landers PGGA velocities estimated by Zhang et al. [1997], for the fractal white noise (FWN) and white noise plus flicker noise (WN+FN) models. Velocities are in units of millimeters per year and are with respect to the NUVEL-1A Pacific plate. (b) and (d) PGGA observed coseismic displacements induced by the Landers earthquake, the trace of the earthquake's surface rupture (arc segment), the Big Bear earthquake's subsurface trace (line segment), and model coseismic contours are adapted from Wdowinski et al. [1997]. Observed and contoured model coseismic displacements are in units of millimeters. Post-Landers velocity changes are the result of differencing the post-Landers velocities from the pre-Landers velocities, for the FWN and WN+FN models. Adapted from Bock et al. [1997].

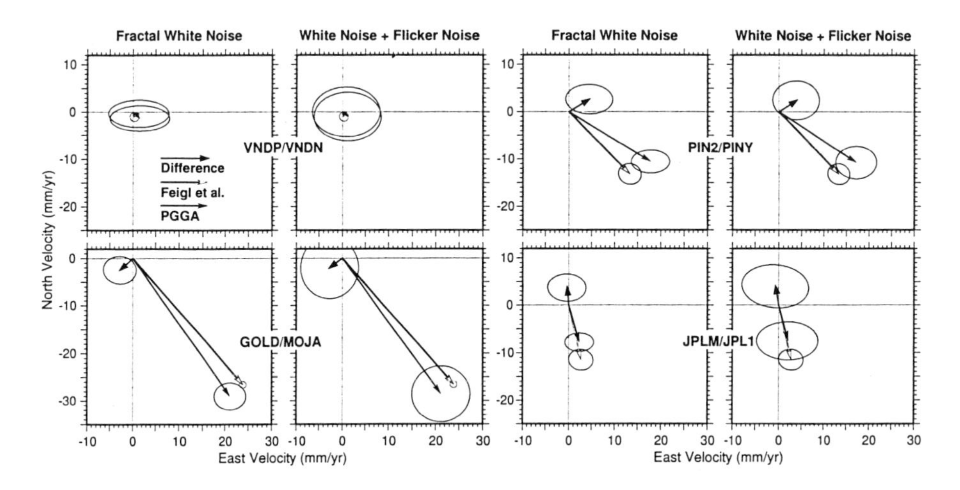

Figure 12.6. Site-by-site 95% confidence regions for the pre-Landers velocity vectors, Feigl et al. [1993], post-Landers velocity vectors, Zhang et al. [1997], and velocity differences under the fractal white noise and the white noise plus flicker noise models for the PGGA time series. Velocities are in units of millimeters per year and are with respect to the NUVEL-1A Pacific plate. Adapted from Bock et al. [1997].

12.5.2 GPS STORM Experiment for Mapping Atmospheric Water Vapor

In the next case study we demonstrate that regional continuous GPS arrays can also be applied to the determination of atmospheric water vapor, which has important applications in real-time "nowcasting," weather, and long-term climate studies, Bevis et al. [1992].

Description. A 30-day field experiment called "GPS/STORM" was mounted in May 1993 in order to demonstrate the feasibility of retrieving PW from GPS observations, Bevis et al. [1992] and Duan et al. [1996]. Dual frequency GPS observations were collected for 22 hours each day at six stations in Oklahoma, Kansas and Colorado (Figure 12.7). At four of these sites the GPS receivers were colocated with water vapor radiometers. Precisely calibrated barometers were available at all six sites, in addition to radiosonde observations.

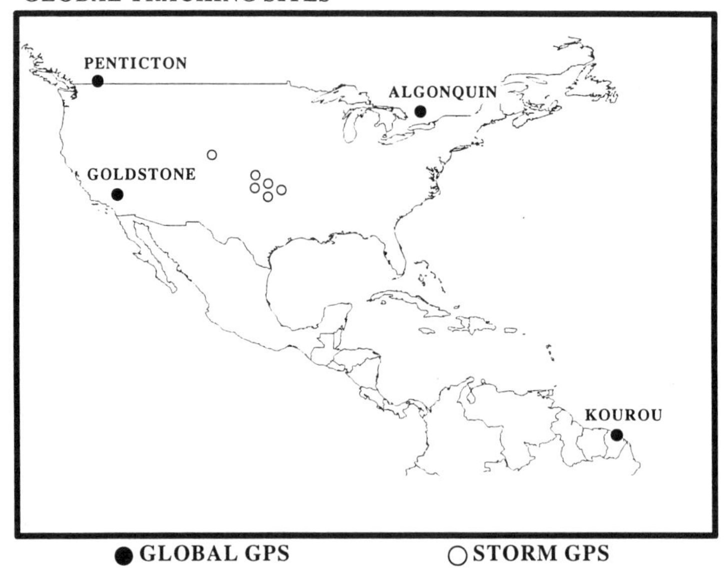

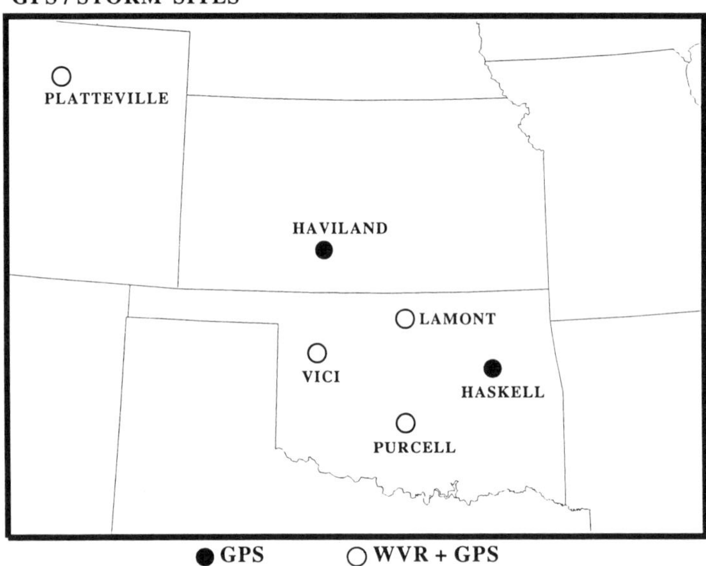

Figure 12.7. Station coverage for the GPS/STORM experiment. Top map shows the 5 IGS stations that provided fiducial control for regional data analysis. Bottom map shows the location of 5 stations in Oklahoma, Kansas, and Colorado, where precipitable water (PW) estimates were derived using the GPS meteorology technique. Open circles indicate that water vapor radiometers were also deployed at the sites for an independent determination of PW. Adapted from Duan et al. [1996].

Estimation of precipitable water. Having estimated the ZWD history at a site (see section 12.2.2) it is possible to transform this time series into an estimate of the precipitable water (PW). PW is defined as the length of an equivalent column of liquid water and can be related to ZWD by

$$PW = \Pi \ ZWD \tag{12.105}$$

where the constant of proportionality Π is given by

$$\Pi = \frac{10^6}{\rho R_v[\frac{k_3}{T_M} + k_2']} \tag{12.106}$$

where ρ is the density of liquid water, R_v is the specific gas constant for water vapor, and T_M is a weighted mean temperature of the atmosphere defined as

$$T_M = \frac{\int (P_v / T)\, dz}{\int (P_v / T^2)\, dz} \tag{12.107}$$

$$k_2' = k_2 - mk_1 \tag{12.108}$$

$$m = \frac{M_w}{M_d} \tag{12.109}$$

where T is temperature, m is the ratio of the molar masses of water vapor and dry air, and the integrations occur along a vertical path through the atmosphere. The physical constants are from the formula for atmospheric refractivity

$$N = k_1 \frac{P_d}{T} + k_2 \frac{P_v}{T} + k_3 \frac{P_v}{T^2} \tag{12.110}$$

where P_d and P_v are the partial pressures of dry air and water vapor, respectively. The time-varying parameter T_M can be estimated using measurements of surface temperature or numerical weather models with such accuracy that very little noise is introduced during the transformation 12.105. That is, the uncertainty in the PW estimate derives almost entirely from the uncertainty in the earlier estimate of ZWD, Bevis et al. [1994].

The GAMIT software, see King and Bock [1995], was used to analyze the GPS/STORM data using the algorithms described in section 12.2.1. Four remote stations (Figure 12.8) were incorporated into the analysis to establish the link to the global reference frame (in this example, ITRF93). The 30 days of observations were analyzed one day at a time. Using the GLOBK software by Herring [1995], the daily solutions were combined with daily solutions of 32 globally distributed stations produced by the Scripps Orbit and Permanent Array Center (SOPAC), using the Kalman filter algorithm described in section 12.4.4. This step provided precise (sub-centimeter) geocentric positions for the six GPS/STORM stations.

The daily GAMIT solutions were then repeated, tightly constraining the positions of all ten stations. ZND parameters were estimated under the assumption that they behave as a first-order Gauss-Markov process. The process correlation time was set to 100 hours, the process standard deviation was set to 2.5 mm, and the ZND was estimated every 30 minutes at each station. The CfA zenith delay mapping function, Davis, et al. [1985], was used. The ZND estimates produced by GAMIT were then used to estimate PW. The ZWD histories at the six GPS/STORM sites were recovered from the ZND estimates by subtracting the ZHD time series computed using surface pressure measurements. The ZWD estimates were then transformed into PW estimates using (12.105).

The GPS-derived PW solutions thus obtained were compared with the WVR-derived PW solutions at the four stations with colocated WVRs and GPS receivers. A representative segment of the time series acquired at site Purcell is shown in Figure 12.8. A weighted root-mean-square deviation between the WVR and GPS time series was computed for each station, with the results falling in the range 1.15–1.45 mm of PW.

These results illustrate a significant advantage of estimating PW from GPS observations. The measurements of PW provided by WVRs are virtually useless during brief but not rare episodes in which these instruments are wetted by rainfall or dew. Note that the GPS solutions do not suffer from this problem. Additionally GPS receivers are robust, all-weather devices requiring minimal levels of maintenance, and they are far cheaper than WVRs.

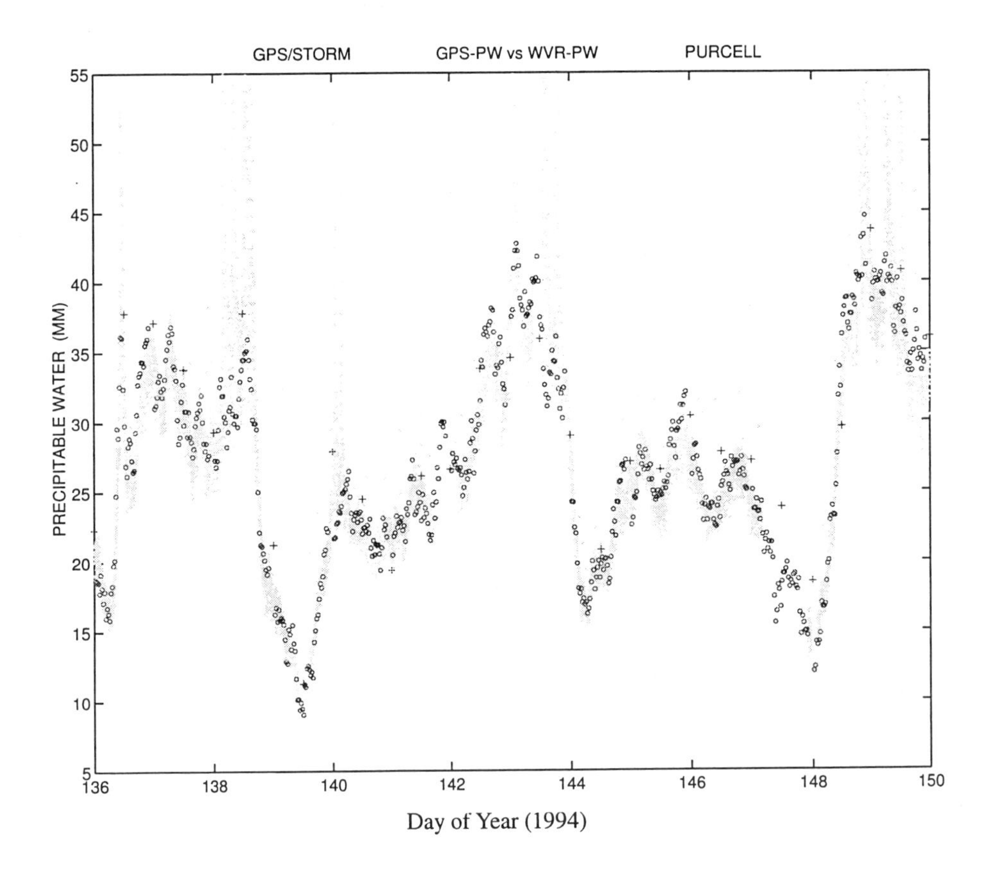

Figure 12.8. Precipitable water from GPS meteorology and water vapor radiometry. Plot of PW estimates over a two-week period at station Purcell during the GPS/STORM experiment. The open circles include radiosonde measurements. Note the extremely high (and erroneous) values from WVR during brief but not rarer episodes in which the instruments are wetted by rainfall or dew. Note that the GPS solutions do not suffer from this problem. This experiment demonstrated that PW can be estimated with GPS at about 1-2 mm accuracy. Adapted from Duan et al. [1996].

12.5.3 Integrated Satellite Interferometry

In this section we discuss the integration of two satellite interferometric techniques, continuous GPS (CGPS) and interferometric synthetic aperture radar (INSAR), as an important development in crustal deformation monitoring at regional scales.

Radar phase difference maps for satellite images acquired before and after the 1992 Landers earthquake in southern California provided, for the first time, a spectacular contour map of coseismic deformation, see e.g. Massonnet et al. [1993] and Zebker et al. [1994]. However, INSAR provides very dense spatial resolution, its accuracy is less than that of GPS. We demonstrate this in Figure 12.9 where Landers earthquake coseismic displacements estimated at 4 PGGA site locations, see Bock et al. [1993], are compared with the interferogram determined by Massonnet et al. [1994]. The GPS analysis yields estimates of absolute coseismic displacement with mm accuracy more than 150 km from the surface rupture zone, but cannot compete with the spatial coverage provided by INSAR (even if there had been hundreds of GPS sites). On the other hand, interferometric fringes are restricted to a smaller region about the Landers rupture zone than the coseismic displacements observed by the PGGA. This is a consequence of having assumed zero deformation at the outer edges of the interferogram (to reduce satellite orbit error), when clearly this is not the case. Furthermore, the nearly instantaneous satellite images are more affected by atmospheric refraction errors than the continuous GPS measurements.

Developing a way to harness CGPS and INSAR to function effectively as a single geodetic instrument could provide an extremely detailed and accurate picture of crustal deformation. Fortunately, the two techniques are quite complementary as summarized in Table 12.1. To take one point, as described in section 12.5.1 one of the great difficulties in high precision GPS measurements is monumentation. INSAR, although it measures to the least stable part of the ground (the surface), could potentially overcome this problem because of the spatial averaging inherent in this technique (each value being the average over a 30x30m square for ERS-2 imagery, for example). On the other hand, it seems doubtful, without some additional source of geodetic control, that INSAR could make stable measurements over periods of years and contribute to measurements of interseismic deformation which are currently considered to be at the extreme limit of the technique's resolution.

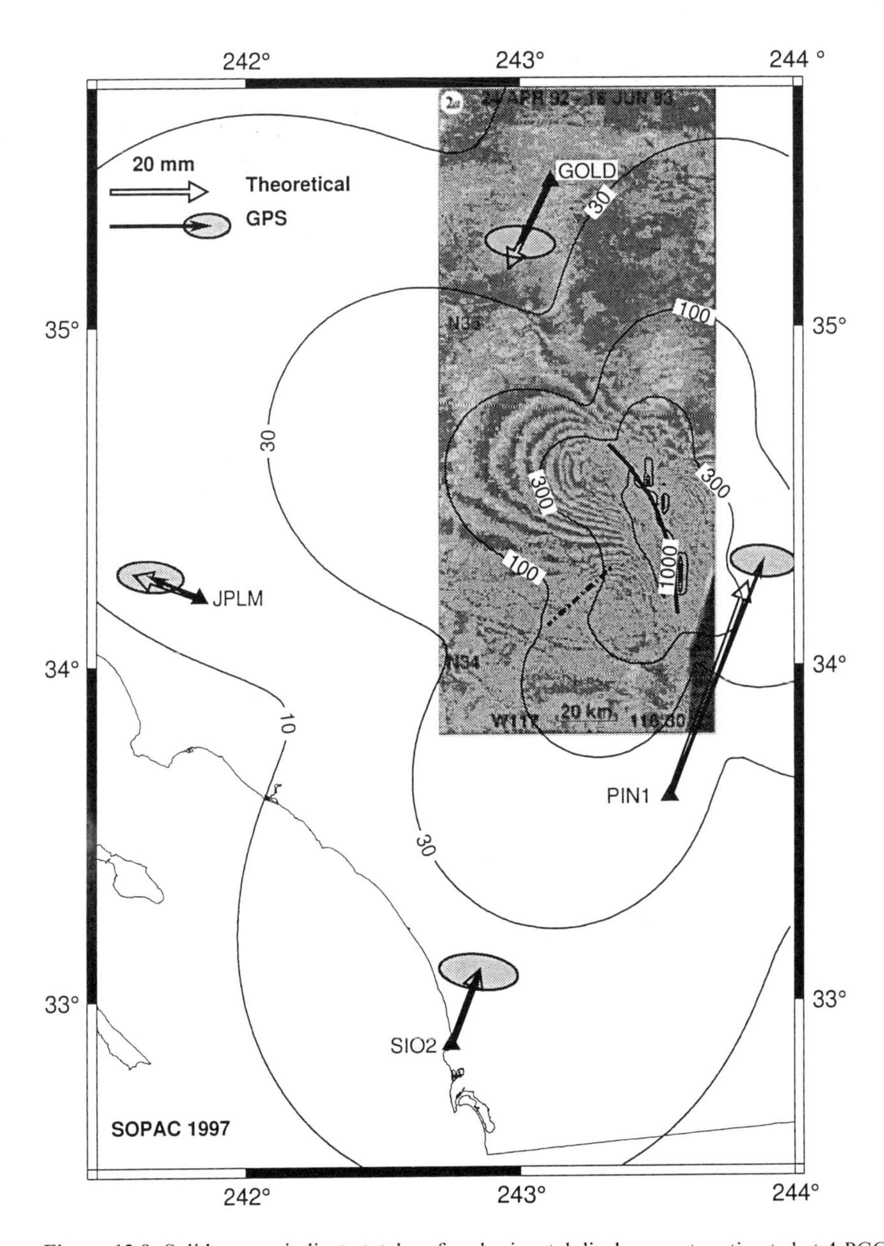

Figure 12.9. Solid arrows indicate total surface horizontal displacements estimated at 4 PGGA sites operating during the Landers earthquake. Blank arrows and contour lines show the magnitude (in mm) of the horizontal displacements predicted by a dislocation model that assumes 7 linear segments describing the rupture geometry of the Landers earthquake. The heavy line denotes the surface trace of the Landers rupture, the dashed line is the Big Bear earthquake's subsurface trace. Use of interferogram is courtesy of D. Massonnet. Adapted from Bock and Williams [1997].

Table 12.1. Comparison of CGPS and INSAR

	Continuous GPS	**Interferometric SAR**
Strengths	High temporal densitiy 3-D vector measurement Global tracking network Map atmospheric delays	High spatial density Remotely sensed No monumentation required No siting difficulties
Weaknesses	Limited spatial density Siting difficulties Antenna phase centers Signal multipath Stable monumentation	Limited temporal density 1-D Scalar measurement Atmospheric refraction Baseline (orbital) errors Image decorrelation Processing intensive

CGPS data can help reduce two of the major error sources in determining ground displacements with INSAR:

(1) "Baseline" or orbital errors. Inaccuracies in the estimated SAR satellite orbits can introduce long wavelength quadratic phase distortions that can mask the similar long wavelength component of interseismic deformation. Knowledge of the precise position of a GPS site in a SAR image can be used to provide 3-D control on the spacecraft orbit and improve the accuracy of the ground displacement measurements. Critical to this problem is the number and optimum position of the GPS sites in each image. It has been shown that several well-identified points are required in each image to properly correct for orbital errors , e.g. Zebker et al. [1994].

(2) Troposphere propagation errors. Spatial and temporal variations in the line-of-sight integrated water vapor content of the atmosphere, if not properly accounted for, can produce artifacts in the radar displacement fields with magnitudes of up to several centimeters. Significant phase variations can appear in SAR interferograms that are due entirely to changes in two-way atmospheric delay of the radar signals, Zebker et al. [1997] and Bock and Williams [1997]. Continuous GPS data can be used to map the spatial and temporal distribution of water vapor and correct the delays in the radar data. Again, the position and number of GPS sites is crucial for the successful reduction of propagation errors in the image.

Using CGPS data from the Los Angeles basin, Bock and Williams [1997] have shown that more than 90% of the troposphere signal can recovered by CGPS with station spacing of 10-20 km (Figure 12.10). Outside of the basin, where the station separation is much greater, the percentage drops to a minimum of about 40% at the outermost GPS sites (Figure 12.1). The standard deviations of the observed-minus-predicted residuals interpolated over the region are within the observational accuracy of the GPS ZND measurements themselves (generally about 6-8 mm in total delay). In the Los Angeles basin where station density is highest the standard deviations are about 4 mm.

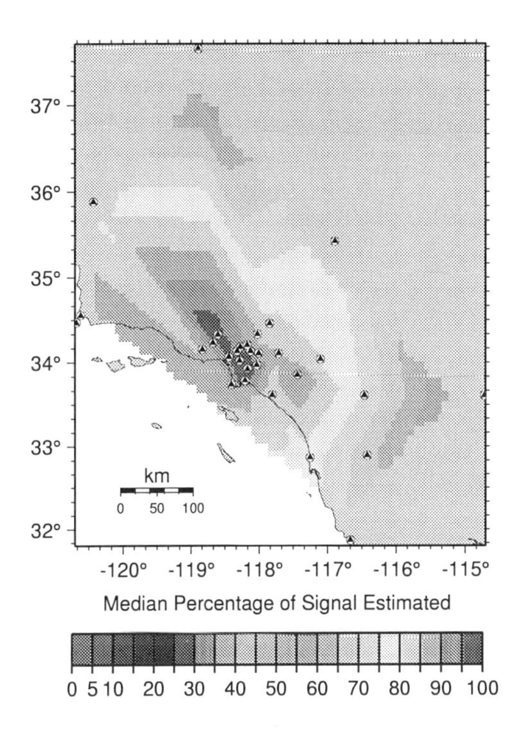

Figure 12.10. Use of CGPS to calibrate troposphere refraction errors, as a way of improving INSAR images. Shown are median percentage of semi-hourly tropospheric delay signals recoverable by the current array of CGPS sites (triangles) in southern California (see Figure 12.1). Adapted from Bock and Williams [1997].

12.6 SUMMARY

We reviewed in section 2 the GPS mathematical and stochastic models for medium distance measurements. We concentrated on the main parameters of interest at medium distances: station coordinates and tropospheric zenith delay parameters.

In section 3, we reviewed the various analysis modes used today for medium distance GPS. We stressed that accurate processing of data by a variety of modes was being simplified by the availability on the Internet of global and regional array solutions with full covariance information.

In section 4, we described the free-network quasi-observation approach to network adjustment of session-mode solutions, exploiting the full covariance information available today. We provided observation equations for site coordinates and velocity, baseline coordinates and velocities, and episodic site displacements. We indicated a scheme to integrate VLBI and SLR space geodetic solutions (with full covariance information), as well as classical terrestrial geodetic measurements. We reviewed several estimation algorithms for network adjustment including free adjustment, Bayesian estimation and Kalman filtering. We described algorithms to remove common-mode systematic effects from regional station position estimates, both temporally and spatially. We reviewed the physical models behind estimation of tropospheric water vapor by continuous GPS networks.

In section 5, we presented two case studies using much of the material covered in sections 2-4. The first case study used data from the Permanent GPS Geodetic Array in southern California to estimate regional postseismic deformation after the 1992 M_w=7.3 Landers earthquake. The second example, showed how precisely positioned continuous GPS networks could be used to track atmospheric water vapor using the same procedures outlined in this chapter. Finally, we discussed a new development at medium distance scales, the integration of continuous GPS and interferometric synthetic aperture radar.

Acknowledgements

I would like to thank Bob King and Paul Tregoning for their critical comments on an earlier version of this chapter, Peng Fang, Simon Williams, Jie Zhang, Jeff Dean, Chris Roelle, and Myra Medina of the Scripps Orbit and Permanent Array Center (SOPAC), Hadley Johnson, Tom Herring, Danan Dong, and my colleagues at IGS and SCIGN. Didier Massonnet allowed use of the Landers earthquake interferogram. Significant portions of this chapter have been extracted from Schaffrin and Bock [1985], Dong and Bock [1989], Dong [1993], Duan et al. [1996], Bock et al. [1997], Zhang et al. [1997], and Bock and Williams [1997], listed in the references. Supported by grants from the National Aeronautics and Space Administration, the Southern California Earthquake Center, the U.S. Geological Survey, and the U.S. National Science Foundation.

References

Argus, D. F. and R. G. Gordon (1991), No-Net-Rotation model of current plate velocities incorporating plate rotation model NUVEL-1, *Geophys. Res. Lett.*, *18,* 2039-2042.

Beutler, G. and E. Brockmann, eds. (1993), *Proceedings of the 1993 IGS Workshop*, International Association of Geodesy, Druckerei der Universitat Bern.

Beutler, G., P. Morgan, and R. E. Neilan (1993), International GPS Service for Geodynamics: Tracking satellites to monitor global change, *GPS World*, February issue, 40-46.

Beutler, G., I. I. Mueller, and R. E. Neilan (1994), The International GPS Service for Geodynamics (IGS): Development and start of official service on January 1, 1994, *Bull. Géodésique*, *68*, 39-70.

Bevis, M., S. Businger, T. Herring, C. Rocken, R. Anthes and R. Ware (1992), GPS meteorology: Remote sensing of atmospheric water vapor using the Global Positioning System, *J. Geophys. Res. 97*, 15,787-15,801.

Bevis, M., S. Businger, S. Chiswell, T. Herring, R. Anthes, C. Rocken and R. Ware (1994), GPS Meteorology: Mapping zenith wet delays onto precipitable water, *J. Appl. Met. 33*, 379-386.

Bevis, M., Y. Bock, P. Fang, R. Reilinger, T.A. Herring, J. Stowell, and R. Smalley (1997), Blending old and new approaches to regional GPS geodesy, *Eos Trans. AGU, 78*, p. 61.

Bibby, H.M. (1982), Unbiased estimate of strain from triangulation data using the method of simultaneous reduction, *Tectonophysics*, *82*, 161-174.

Blewitt, G. (1989), Carrier phase ambiguity resolution for the Global Positioning System applied to geodetic baselines up to 2000 km, *J. Geophys. Res., 94*, 10,187-10, 283.

Blewitt, G. (1990), An automatic editing algorithm for GPS data, *Geophys. Res. Lett.*, *17*, 199-202.

Blewitt, G. (1993a), Advances in Global Positioning System technology for geodynamics investigations: 1978-1992, in Contributions of Space Geodesy to Geodynamics: Technology, D. E. Smith and D. L. Turcotte, eds., *Geodynamics Series 25*, Amer. Geophys. Union.

Blewitt, G., M. B. Heflin, K. J. Hurst, D.C. Jefferson, F.H. Webb and J.F. Zumberge (1993b), Absolute far-field displacements from the June 28, 1992, Landers earthquake sequence, *Nature*, *361,* 340-342.

Blewitt, G., Y. Bock and G. Gendt (1993c), Regional clusters and distributed processing, Proc. IGS Analysis Center Workshop, J. Kouba, ed., Int. Assoc. of Geodesy, Ottawa, Canada, 61-92.

Bock, Y. (1991), Continuous monitoring of crustal deformation, *GPS World*, 40-47, June.

Bock, Y. Z., Crustal deformation and earthquakes (1994), *Geotimes*, *39*, 16-18.

Bock, Y., R. I. Abbot, C.C. Counselman, S. A. Gourevitch, and R.W. King (1985), Establishment of three-dimensional geodetic control by interferometry with the Global Positioning System, *J. Geophys. Res.*, *90*, 7689-7703.

Bock, Y., and S. Williams (1997), Integrated satellite interferometry in southern California, *Eos Trans. AGU*, 78, pp. 293,299-300.

Bock, Y., R. I. Abbot, C.C. Counselman, S. A. Gourevitch, and R.W. King (1986), Interferometric analysis of GPS phase observations, *Manuscripta Geodaetica*, *11*, 282-288.

Bock, Y., J. Zhang, P. Fang, J.F. Genrich, K. Stark and S. Wdowinski (1992), One year of daily satellite orbit and polar motion estimation for near real time crustal deformation monitoring, *Proc. IAU Symposium No. 156*, Developments in Astrometry and their Impact on Astrophysics and Geodynamics, I. I. Mueller and B. Kolaczek, eds., Kluwer Academic Publishers, 279-284.

Bock Y., D.C. Agnew, P. Fang, J.F. Genrich, B.H. Hager, T.A. Herring, K.W. Hudnut, R.W. King, S. Larsen, J.B. Minster, K. Stark, S. Wdowinski and F.K. Wyatt (1993), Detection of crustal deformation from the Landers earthquake sequence using continuous geodetic measurements, *Nature*, *361*, 337-340.

Bock, Y., P. Fang, K. Stark, J. Zhang, J. Genrich, S. Wdowinski, S. Marquez (1993), Scripps Orbit and Permanent Array Center: Report to '93 Bern Workshop, Proc. 1993 IGS Workshop, G. Beutler and E. Brockmann, eds., Astronomical Institute, Univ. of Berne, 101-110.

Bock Y., S. Wdowinski, P. Fang, J. Zhang, S. Williams, H. Johnson, J. Behr, J. Genrich, J. Dean, M. van Domselaar, D. Agnew, F. Wyatt, K. Stark, B. Oral, K. Hudnut, R. King, T. Herring, S. Dinardo, W. Young, D. Jackson, and W. Gurtner (1997), Southern California Permanent GPS Geodetic Array: Continuous measurements of crustal deformation between the 1992 Landers and 1994 Northridge earthquakes, *J. Geophys. Res.*, 102, 18,013-18,033.

Boucher, C., Z. Altamimi, and L. Duhem (1994), Results and analysis of the ITRF93, *IERS Tech. Note 18*, Int. Earth Rotation Service, Observatoire de Paris, Paris, France.

Calais, E. and J.B. Minster (1995), GPS detection of ionospheric perturbations following the January 17, 1994, Northridge earthquake, *Geophys. Res. Lett.*, *22*, 1045-1048.

Collier, P., B. Eissfeller, G. W. Hein and H. Landau (1988), On a four-dimensional integrated geodesy, *Bull. Geod.*, *62*, 71-91.

Counselman, C.C. and S. A. Gourevitch (1981), Miniature interferometer terminals for Earth surveying: ambiguity resolution and multipath with the Global Positioning System, *IEEE Trans. on Geoscience and Remote Sensing*, *GE-19*.

Counselman, C.C. and R. I. Abbot (1989), Method of resolving radio phase ambiguity in satellite orbit determination, *J. Geophys. Res.*, *94*, 7058-7064.

Davis, J. L., T.A. Herring, I.. Shapiro, A. E. Rogers, G. Elgered (1985), Geodesy by radio interferometry: Effects of atmospheric modeling on estimates of baseline length, *Radio Sci. 20*, 1593-1607.

Davis J. L., T.A. Herring, and I. I. Shapiro (1991), Effects of atmospheric modeling errors on determination of baseline vectors from very long baseline interferometry, *J. Geophys. Res.*, *96*, 643-650.

DeMets, C., R. G. Gordon, D. Argus and S. Stein (1990), Current Plate Motions, *Geophys. J. Int.*, *101*, 425-478.

DeMets, C., R. G. Gordon, D. Argus and S. Stein (1994), Effects of recent revisions to the geomagnetic reversal time scale on estimates of current plate motions, *Geophys. Res. Lett.*, *21*, 2191-2194.

Dixon, T. H. (1991), An introduction to the Global Positioning System and some geological applications, *Reviews of Geophysics*, v. 29, 249-276.

Dong, D. and Y. Bock (1989), Global Positioning System network analysis with phase ambiguity resolution applied to crustal deformation studies in California, *J. Geophys. Res.*, *94*, 3949-3966.

Dong, D. (1993), The horizontal velocity field in southern California from a combination of terrestrial and space-geodetic data, Doctoral Dissertation, Massachusetts Institute of Technology.

Duan, J. et al. (1996), Remote sensing of atmospheric water vapor using the Global Positioning System, *J. Applied Meteorology*, *35*, 830-838.

Elósegui, P., J. L. Davis, R. T. K. Jaldehag, J. M. Johansson, A. E. Niell and I. I. Shapiro (1995), Geodesy using the Global Positioning System: The effects of signal scattering on estimates of site position, *J. Geophys. Res.*, *100*, 9921-9934.

Fang, P. and Y. Bock (1995), Scripps Orbit and Permanent Array Center Report to the IGS, 1994 Annual Report, International GPS Service for Geodynamics, J. F. Zumberge, R. Liu and R. E. Neilan, eds., IGS Central Bureau, Jet Propulsion Laboratory, Pasadena, 213-233.

Feigl, K.L. (1991), Geodetic measurement of tectonic deformation in central California, Doctoral Dissertation, Massachusetts Institute of Technology.

Feigl, K. L., et al. (1993), Space geodetic measurement of crustal deformation in central and southern California, 1984-1992, *J. Geophys. Res.*, *98*, 21,677-21,712.

Genrich, J.F. and Y. Bock (1992), Rapid resolution of crustal motion at short ranges with the Global Positioning System, *J. Geophys. Res.*, *97*, 3261-3269.

Genrich, J. F., Y. Bock, and R. Mason (1997), Crustal deformation across the Imperial Fault: Results from kinematic GPS surveys and trilateration of a densely-spaced, small-aperture network, *J. Geophys. Res.*, *102*, 4985-5004.

Georgiadou, Y. and A. Kleusberg (1988), On carrier signal multipath effects in relative GPS positioning, *Manuscripta Geodaetica*, *13*, 172-179.

Heflin, M. B., W. I. Bertiger, G. Blewitt, A. P. Freedman, K. J. Hurst, S. M. Lichten, U. J. Lindqwister, Y. Vigue, F.H. Webb, T. P. Yunck, and J.F. Zumberge (1992), Global geodesy using GPS without fiducial sites, *Geophys. Res. Lett.*, *19*, 131-134.

Herring, T.A., J. L. Davis, J. L. and I. I. Shapiro (1990), Geodesy by radio interferometry: The application of Kalman Filtering to the analysis of very long baseline interferometry data, *J. Geophys. Res. 95*, 12,561-12,583.

Herring, T.A. (1997), Documentation of the GLOBK Software v. 4.0, Mass. Inst. of Technology.

Hudnut, K. W. (1995), Earthquake geodesy and hazard monitoring, *Rev. of Geophys. Suppl.*, 249-255.

Hudnut, K. W. et al. (1994), Coseismic displacements of the 1992 Landers earthquake sequence, *Bull. Seismol. Soc. Amer.*, Special Issue on the Landers earthquake sequence, *84*, 625-645.

Hudnut, K. W. et al. (1996), Coseismic displacements of the 1994 Northridge, California, earthquake, *Bull. Seismol. Soc. Amer.* , Special Issue on the Northridge earthquake, S19-S36.

International GPS Service for Geodynamics (1993), *Proceedings of the 1993 IGS Workshop*, International Association of Geodesy, Beutler, G. and E. Brockmann, eds., Druckerei der Universitat Bern.

International GPS Service for Geodynamics (1995), 1994 Annual Report, J. F. Zumberge, R. Liu and R. E. Neilan, eds., IGS Central Bureau, Jet Propulsion Laboratory, Pasadena.

King R.W. and Bock, Y. (1997), Documentation of the GAMIT GPS Analysis Software v. 9.5, Mass. Inst. of Technology and Scripps Inst. of Oceanography.

Jaldehag, R. T. K., J. M. Johansson, J. L. Davis, and P. Elósegui (1996), Geodesy using the Swedish permanent GPS network: Effects of snow accumulation on estimates of site positions, *Geophys. Res. Lett.*, *23*, 1601-1604.

Johnson, H. O. and F. K. Wyatt (1994), Geodetic network design for fault-mechanics studies, *Manuscripta Geodaetica*, *19*, 309-323.

Johnson, H. O. and D. C. Agnew (1995), Monument motion and measurements of crustal velocities, *Geophys. Res. Lett.*, *22*, 2905-2908.

Karcz, I., L. J. Morreale, and F. Porebski (1976), Assessment of benchmark credibility in the study of recent vertical crustal movements, *Tectonophysics*, *33*, T1-T6.

Kolmogorov, A. N. (1941), Local structure of turbulence in an incompressible fluid for very large Reynolds number, *Doklad Akad. Nauk SSSR*, *30*, 299-303.

Langbein, J. O., M. F. Linker, A. F. McGarr, and L. E. Slater (1987), Precision of two-color geodimeter measurements: results from 15 months of observations, *J. Geophys. Res.*, *92*, 11,644-11,656.

Langbein, J. O., R. O. Burford and L. E. Slater (1990), Variations in fault slip and strain accumulation at Parkfield, California: Initial results using two-color Geodimeter measurements, 1984-1988, *J. Geophys. Res.*, *95*, 2533-2552.

Langbein, J. O., F. Wyatt, H. Johnson, D. Hamann, and P. Zimmer (1995), Improved stability of a deeply anchored geodetic monument for deformation monitoring, *Geophys. Res. Lett.*, *22*, 3533-3536.

Langbein J. and H. Johnson (1997), Correlated errors in geodetic time series: Implications for time-dependent deformation, *J. Geophys. Res.*, *102*, 591-604.

Larson, K.M. and D.C. Agnew (1991), Application of the Global Positioning System to crustal deformation, 1. precision and accuracy, *J. Geophys. Res.*, *96*, 16,547-16,566.

Larson, K.M., F.H. Webb and D.C. Agnew (1991), Application of the Global Positioning System to crustal deformation, 2. The influence of errors in orbit determination networks, *J. Geophys. Res.*, *96*, 16,567-16,584.

Leick A. (1990), GPS Satellite Surveying, John Wiley and Sons, New York.

Lichten, S. M. and J. S. Border (1987), Strategies for high precision Global Positioning System orbit determination, *J. Geophys. Res.* *92*, 12,751-12,762.

Lindqwister, U., G. Blewitt G., J. Zumberge and F. Webb (1991), Millimeter-level baseline precision results from the California permanent GPS Geodetic Array, *Geophys. Res. Lett.*, *18*, 1135-1138.

Lisowski, M, J. C. Savage and W. H. Prescott (1991), The velocity field along the San Andreas fault in central and southern California, *J. Geophys. Res.*, *96*, 8369-8389.

Mader, G. L. (1992), Rapid static and kinematic Global Positioning System solutions using the ambiguity function technique, *J. Geophys. Res.*, *97*, 3271-3283.

Mandlebrot, B. (1983), *The Fractal Geometry of Nature*, W. H. Freeman, San Francisco, 1983.

Mandlebrot, B., and J. Van Ness (1968), Fractional Brownian motions, fractional noises, and applications, *SIAM Rev.*, *10*, 422-439.

Massonnet, D., K. Feigl, M. Rossi, and F. Adragna (1994), Radar interferometric mapping of deformation in the year after the Landers earthquake, *Nature*, *369*, 227-230.

Massonnet, D., W. Thatcher, and H. Vadon (1996), Detection of postseismic fault-zone collapse following the Landers earthquake, *Nature*, *382*, 612-616.

Murakami, M., M. Tobita, S. Fujiwara, T. Saito, and H. Masaharu (1996), Coseismic crustal deformations of 1994 Northridge, California, earthquake detected by interferometric JERS 1 synthetic aperture radar, *J. Geophys. Res.*, *101*, 8605-8614.

Murray, M. H. (1991), Global Positioning System measurement of crustal deformation in central California, Doctoral Dissertation, Woods Hole Oceanographic Institution and Massachusetts Institute of Technology.

Niell, A. E. (1996), Global mapping functions for the atmospheric delays at radio wavelengths, *J. Geophys. Res.*, *101*, 3227-3246.

Oral, M. B. (1994), Global Positioning System (GPS) measurements in Turkey (1988-1992): Kinematics of the Africa-Arabia-Eurasia plate collision zone, Doctoral Dissertation, Massachusetts Institute of Technology.

Peltzer, G., P. Rosen, F. Rogez, and K. Hudnut (1996), Postseismic rebound in fault step-overs caused by pore fluid flow, *Science*, *273*, 1202-1204.

Papoulis, A. (1965), Probability, random variables, and stochastic processes, McGraw-Hill Book Company.

Prescott, W. H. (1996), Satellites and earthquakes: A new continuous GPS array for Los Angeles, Yes, It will radically improve seismic risk assessment for Los Angeles, *Eos Trans. AGU*, 77, p.417.

Priestley, M. B. (1981), *Spectral Analysis and Time Series*, Academic, San Diego, Calif.

Rocken, C., R. Ware, T. VanHove, F. Solheim, C. Alber, J. Johnson, M. Bevis and S. Businger (1993), *Geophys. Res. Lett.* *0***2**, 2631.

Savage, J. C., and J. L. Svarc (1997), Postseismic deformation associated with the 1992 Mw=7.3 Landers earthquake, southern California, *J. Geophys. Res*., 102, 7565-7577.

Schaffrin, B. and Y. Bock (1988), A unified scheme for processing GPS dual-band observations, *Bulletin Geodesique*, *62*, 142-160.

Schupler, B. R., R. L. Allhouse, and T. A. Clark (1994), Signal characteristics of GPS user antennas, *Navigation*, *41*, 277-295.

Segall, P. and M. V. Mathews (1988), Displacement calculations from geodetic data and the testing of geophysical deformation models, *J. Geophys. Res.*, *93*, 14,954-14,966.

Shen, Z-K., D. D. Jackson, Y. Feng, M. Cline, M. Kim, P. Fang and Y. Bock (1994), Postseismic deformation following the Landers earthquake, California, June 28, 1992, *Bull. Seismol. Soc. Amer.* , Special Issue on the Landers earthquake sequence, *84*, 780-791.

Shimada et al. (1990), Detection of a volcanic fracture opening in Japan using Global Positioning System measurements, *Nature*, *343*, 631-633.

Shimada, S. and Y. Bock (1992), Crustal deformation measurements in Central Japan determined by a GPS fixed-point network, *J. Geophys. Res.*, *97*, 12,437-12,455.

Tralli D. M., T.H. Dixon and S. A. Stephens (1988), Effect of wet tropospheric path delays on estimation of geodetic baselines in the Gulf of California using the Global Positioning System, *J. Geophys. Res. 93*, 6545-6557.

R. N. Treuhaft and G. E. Lanyi (1987), *Radio Sci. 22*, 251.

Tsuji, H., Y. Hatanaka, T. Sagiya and M. Hashimo (1995), Coseismic crustal deformation from the 1994 Hokkaido-Toho-Oki earthquake monitored by a nationwide continuous GPS array in Japan, *Geophys. Res. Lett. 22*, 1669-1673.

Ware, R., C. Alber, C. Rocken, and F. Solheim (1997), Sensing integrated water vapor along GPS ray paths, *Geophys. Res. Lett.*, 24,417-420.

Wessel, P. and W. H. F. Smith (1991), Free software helps map and display data, E*OS, Trans. AGU*, *72*, pp. 445-446.

Wdowinski, S., Y. Bock, J. Zhang, P. Fang, and J. Genrich (1997), Southern California Permanent GPS Geodetic Array: Spatial Filtering of Daily Positions for Estimating Coseismic and postseismic displacements Induced by the 1992 Landers earthquake, *J. Geophys. Res.*, 102, 18,057-18,070.

Wyatt, F. (1982), Displacements of surface monuments: horizontal motion, *J. Geophys. Res.*, *87*, 979-989.

Wyatt, F. (1989), Displacements of surface monuments: vertical motion, *J. Geophys. Res.*, *94*, 1655-1664.

Wyatt F., K. Beckstrom, and J. Berger (1982), The optical anchor — a geophysical strainmeter, *Bull. Seismol. Soc, Amer.*, *72*, 1701-1715.

Wyatt, F. K., H. Bolton, S. Bralla, and D. C. Agnew (1989), New designs of geodetic monuments for use with GPS, *Eos Trans. AGU*, *70*, 1054-1055.

Wyatt, F. K., D. C. Agnew and M. Gladwin (1994), Continuous measurements of crustal deformation for the 1992 Landers earthquake sequence, *Bull. Seismol. Soc. Amer.*, Special Issue on the Landers earthquake sequence, *84*, 768-779.

Zebker H. A., C. L. Werner, P. A. Rosen, and S. Hensley (1994), Accuracy of topographic maps derived from ERS-1 interferometric radar, *IEEE Trans. Geosci. Rem. Sens.*, *32*, 823-836.

Zebker, H. A., P. A. Rosen, and S. Hensley (1997), Atmospheric effects in interferometric synthetic aperture radar surface deformation and topographic maps, *J. Geophys. Res.*, *102*, 7547-7564.

Zhang, J. (1996), Continuous GPS measurements of crustal deformation in southern California, PhD dissertation, University of California, San Diego.

Zhang, J., Y. Bock, H. Johnson, P. Fang, J. Genrich, S. Williams, S. Wdowinski and J. Behr (1997), Southern California Permanent GPS Geodetic Array: Error analysis of daily position estimates and site velocities, *J. Geophys. Res.*, 102, 18,035-18,055.

13. LONG-DISTANCE KINEMATIC GPS

Oscar L. Colombo
University of Maryland/NASA's Goddard S.F.C., Code 926
Greenbelt, Maryland 20771, U.S.A.

13.1 INTRODUCTION

13.1.1 Why Long-Range?

The subject of this chapter is very precise navigation using the Global Positioning System (GPS) of navigation satellites, over distances of up to several thousand kilometres from the nearest reference receiver, particularly when the GPS data is *post-processed.* As the distances among receivers increase, sources of error that have almost the same effects when the receivers are a short distance apart, and cancel out in differential and interferometric GPS positioning (e.g., using single, double, or triple-differenced data), become different enough not to cancel out any longer. These errors now must be estimated and corrected out, or filtered, to determine the vehicle trajectory accurately. Because it is a more complex problem, the reliability and accuracy of long-range navigation, at this early stage of development, are not as good as for similar applications of GPS navigation over distances of a few tens of kilometres. One source of error that cannot be differenced away over long baselines is ionospheric refraction, limiting the ability to deal with carrier phase ambiguities even when using expensive dual-frequency receivers. Such problems aside, the experience obtained so far shows that decimetre-level positioning can be a practical proposition with many important uses, for distances of more than 1000 km. (For 2-3 metre-level, real-time navigation with pseudo-range, the closely related idea of Wide Area Differential GPS Augmentation concept, is being tested on a continental scale for future use in civil and commercial aviation.)

The main reasons for developing precise long-range navigation with GPS are:

(1) To assist large-area surveys with remote-sensing instruments on board ships or aircraft, when they require geographical data registration accurate to one decimetre or better. Such remote sensing includes: airborne altimetry and laser depth sounding, photogrammetry without ground control over large regions, SAR interfrometry, as well as ship-borne sonar interferometry. Such surveys tend to have mostly scientific objectives, such as mapping topography or bathymetry, and monitoring surface shape changes over time. Other types of remote sensing may benefit also from position-derived velocity (synthetic aperture radar and sonar), or acceleration corrections (e.g., airborne gravimetry). Among the main uses for

large-area surveys are in ocean and polar studies, hydrographic charting, and photogrammetric mapping of inaccessible or inhospitable regions.

Very precise survey data are usually *post-processed* (they are processed after all the observations have been collected), and so this chapter deals with techniques that are suitable for post-processing. However, many of the ideas discussed here are also useful for real-time processing.

(2) To improve GPS navigation at all distances, developing methods that can be used reliably in most applications, because they are effective at the very limits allowed by the technology.

Generally speaking, the kinematic methods described here are adaptations of *static* techniques developed during the eighties and early nineties for precise surveys of very large networks, primarily to monitor crustal-tectonic movements. Such methods are described in detail elsewhere in this book, and here it will be shown how they can be adapted for kinematic use.

Methods for relative kinematic positioning using, primarily, GPS carrier phase data, have become increasingly precise, fast, and reliable in recent years. The main reasons for this have been: (a) better receivers, computers, and communications hardware, and (b) improved data analysis methods, particularly in dealing with phase ambiguities.

13.1.2 Vehicle Dynamics

Depending on the type of vehicle, one may distinguish three main cases:

Very low dynamics. The position can be predicted with an error of less than 1/2 cycle of the wide-lane (43 cm) over a period of some minutes. Examples: Static receivers; receivers on board space- or earth-stabilized satellites.

Low dynamics. The vehicle is subject to unpredictable acceleration changes that often make predicting its position to within 1/2 wide-lane impossible after some tens of seconds. Examples: Receiver on a boat at anchor in still waters; on a balloon in calm air.

Moderate dynamics. Position can be predicted within 1/2 wide-lane only a few seconds ahead. Examples: Large aircraft or ship cruising under calm conditions.

High dynamics. The unpredictable changes in acceleration are often too large for the position to be predicted to better than 1/2 wide-lane in one second. Examples: Receiver on an airplane flying through strong turbulence; on a car suddenly accelerating, breaking, turning, or changing lanes.

In general, by understanding the nature of the vehicle's movement, or *dynamics*, one becomes able to put strong *a priori* constraints on the positions to be determined along the trajectory. Such strong constraints should be applied with great caution, because using the wrong dynamics may produce very bad results (with very good formal precision). To avoid this may require laborious and expensive modeling of the forces acting on the vehicle, and this task can often be too complicated to be carried out successfully.

A static receiver has the simplest and best understood dynamics: the coordinates are constant, except for such predictable things as earth tides. At the opposite extreme is a dirt bike racing across rugged terrain. In this and in many other practical cases, it is simplest and safest to assume "high dynamics". The most common form of high-dynamics kinematic GPS is known as "*no dynamics*". This means the coordinates of the vehicle are treated as completely unpredictable, and given large *a priori* uncertainties. This is the only type of vehicle dynamics considered in this chapter.

Notice that vehicles with similar type of dynamics may move at very different speeds (and heights): a fixed station on the ground and an artificial satellite moving at about 7 km/s along a low earth orbit (hundreds of km above the ground), both have "very low dynamics", because any unpredictable acceleration changes are, for both, at the micro-gravity level (except during orbital manoeuvers).

13.1.3 Multiple Reference Stations

In differential kinematic GPS there is a moving receiver on a vehicle, and fixed receivers at one or more known locations, or *reference stations*.

For *long-range* applications it is preferable to have *several reference stations* all of them accurately positioned in a common reference frame, which should be also the frame of the GPS satellite ephemerides. At distances of more than 700 km, receivers see less satellites simultaneously. The roving receiver may see satellites S1, S2, S3 in common with station A, and satellites S1, S3, S4, and S5 in common with station B. Using both stations together, there are five satellites in common view among vehicle and stations, but only three or four using either station alone. That could make the difference between being able to navigate precisely or not, since usually a minimum of five satellites is required for precise long-range positioning.

Using several fixed reference stations to position a roving receiver is the same as solving for the coordinates of one station in a geodetic network, while keeping the coordinates of all the others fixed to precisely known *fiducial values*. In the case of kinematic GPS, the most convenient type of network is shaped as a *star*, with the rover at the centre (at least topologically, if not geographically).

Multiple reference stations provide obvious redundancy in hardware (if one receiver breaks down, there are others left to finish the job), and less obvious redundancy in information. For example, each receiver at a different location on the ground "see" different components of the errors in the GPS satellite orbits, and this multiplicity of information helps filter out those errors better than with a single receiver.

13.1.4 Data Corrections

Data pre-processing consists in detecting and editing out bad data, correcting some flaws in the data (such as cycle-slips), and making corrections to remove certain well-understood effects of natural or artificial origin, such as "dry" tropospheric refraction, the slow periodical movement of the markers of reference stations due to the Earth's body tides, and the different positions of the centre of each antenna and its marker or, for a GPS satellite, the phase centre of the transmitting antenna array, and the centre of mass of the spacecraft (the actual point whose position and velocity are given by the satellite ephemerides). When double-differencing data, relativistic effects and phase wind-up due to antenna rotation are canceled out to the point that it is usually safe to ignore them.

13.1.5 Ambiguities: Fix or Float?

Fixing the ambiguities. In kinematic applications of differential GPS, the elimination of phase ambiguities is likely to be successful as long as the distance to the nearest reference receiver is less than 10 - 30 km, depending on location, time of day, and the age of the 11-year sun-spot cycle. Over short distances, ambiguities can be calculated exactly, as integer number of cycles of the L1 and L2 carriers. Usually, this can be done in a few minutes, even while the vehicle and its receiver are moving (on-the-fly ambiguity resolution). As a result, relative kinematic positioning using the unambiguous carrier phase, accurate to a few centimetres, becomes possible in real time, assuming a good satellite distribution across the sky (one with a low value of Position Dilution Of Precision, or *PDOP*).

The trouble with the ionosphere. The ionosphere is made up of several layers of ionized gas enveloping the Earth, produced mostly by the ultra-violet radiation of the Sun acting on the upper atmosphere. The mean height of these layers is about 400 km, they are irregular in density, and can be imagined as formed by cloud-like "blobs" of higher ion concentration, that expand, thicken, or dissipate more or less quickly, covering areas some hundreds of kilometres across.

For short distances between receivers, the signals coming to them from the same satellite must cross the same "blobs", so they arrive with refraction-induced delays so similar that they cancel each other out when single or double-differencing the data; that is not the case for long distances.

The effect of ionospheric refraction is a delay in the envelope of the signal, and an advance in the phase of the carrier. As a result, refraction affects pseudo-range and phase measurements to the same extent, but with opposite signs: Phase measurements of distance are always "short", while pseudo-range measurements are always "long".

Ionospheric refraction for decimetre radio waves varies pretty much as the inverse of frequency squared. As a result, it is not the same for L1 and L2, and there is no way of eliminating this refraction effect from the ambiguity resolution.

This makes it quite different from tropospheric refraction, which is almost constant across the radio-wave spectrum.

Above the ionosphere proper, there is a more tenuous, and more uniform, distribution of ions, or plasma, trapped inside the space known as the magnetosphere, where most of the Earth's magnetic field is confined by its interaction with the charged particles streaming out of the Sun (solar wind). GPS signals travel more than 19,000 km through this plasma. Though the plasma is extremely rarefied, its cumulative effect over the long signal path adds several decimetres to the "ionospheric" refraction effect, and that is important when determining the trajectories of satellites with on-board GPS receivers.

The extent and predictability of the ionospheric refraction varies according to the thickness and degree of turbulence of the ionosphere. The worst places, from the point of view of GPS users, are near the equator and in polar regions (particularly under the aurora ovals, ring-shaped regions centered roughly on the magnetic poles, where the charged particles that cause the Northern and Southern Lights stream in with high energy and density along magnetic field lines, hit atmospheric gas molecules, and excite them to emit light). The worst times are during the night; the worst epoches are those times when the number of sun-spots is high, or during magnetic storms caused by solar flares. The Sun is a variable star, and sun-spots peak and then fall in their numbers over a period of 11 years. The next maximum in the number of sun-spots is expected around 2001, with a fast buildup likely during the years 1998-2000 (the solar cycle is not perfectly periodic, or even always observable; the last time it failed to happen coincided with the "The Little Ice Age", a century-long interval of very cruel winters and cold summers, centuries ago).

Beyond 20 or 30 km between receivers, the effect of ionospheric refraction becomes an increasing impediment to ambiguity resolution. One has to solve or eliminate additional unknowns representing ionospheric refraction in the observation equations, without being able to increase the number of data and, thus, of equations (one may include the pseudo-range as data, but pseudo-range is a poor source of reliable information for this purpose, because of possible long-period multipath biases).

To some extent, the ionospheric effect is predictable in time and interpolable in space, so observations from all receivers available within a given area might be used to predict what the ionospheric refraction at any given time and location might be, and then this prediction could be used to correct the phase measurements, making them "ionospheric-free". Variants of this idea have been used successfully in static applications, but the adaptation to kinematic positioning is not simple and needs further study.

Floating the ambiguities. The ionospheric effect normally cancels out at the sub-centimetre level when forming the ion-free combination, or L3, of the L1 and L2 phases or pseudoranges:

$$L3 = L1\,(+/-)\,\frac{\lambda_1^2}{(\lambda_2-\lambda_1)(\lambda_2+\lambda_1)}\,(L1-L2),\ \text{with}\ \frac{\lambda_1^2}{(\lambda_2-\lambda_1)(\lambda_2+\lambda_1)} \approx 1.5457 \qquad (13.1)$$

where the "+" corresponds to phase and the "-" to pseudorange.

However, L3 is about three times noisier than either L1 or L2, and the L1 and L2 ambiguities $N1$ and $N2$ create a *bias* in L3 equal to

$$\frac{\lambda_1\lambda_2}{2(\lambda_2-\lambda_1)}(N1-N2)+\frac{\lambda_1\lambda_2}{2(\lambda_2+\lambda_1)}(N1+N2)\ \text{metres,}$$
$$\text{with}\ \frac{\lambda_1\lambda_2}{2(\lambda_2-\lambda_1)} \approx 0.43,\ \frac{\lambda_1\lambda_2}{2(\lambda_2+\lambda_1)} \approx 0.06 \qquad (13.2)$$

The minimum difference between biases caused by different combinations of $N1$ and $N2$ is about 3 mm. This is the "wavelength" of the L3 phase, but since it is smaller than the standard deviation of the noise in L3, the usual approach is to treat the biases as being real-valued unknowns. This is known as "*floating the ambiguities*" (as opposed to "fixing the ambiguities" to their exact integer values). Many observing epoches are needed to estimate the ion-free biases with enough accuracy to achieve sub-decimetre positioning (for that, the biases should be known to within a few centimetres). An observing period of one hour or longer may be needed (but the availability of even a few resolved ambiguities can speed this up quite considerably). Clearly, estimating floated biases on long baselines can be a much slower process than fixing the corresponding L1 and L2 integer ambiguities over shorter baselines, so floating is a technique used mostly in post-processing, as opposed to real-time.

Loss of lock. This is a situation associated with poor reception, very fast acceleration changes, and shadowing of satellite signals by obstacles in their way. The receiver cannot keep track on the change in the phase of the received signal, so a sudden jump in ambiguity takes place. One should distinguish the case where the change in phase ambiguity can be easily found and corrected *(a resolvable cycle-slip),* so there is no need to add extra ambiguity parameters, and the case where this is not possible and a new ambiguity must be estimated following the loss of lock *(a complete loss of lock).*

Every time a receiver completely looses lock on the signal from a satellite, it becomes necessary to start estimating a completely new ambiguity parameter. Frequent loss of lock may prevent estimating the ambiguities for long enough to achieve the accuracy needed for good position determination, before the process has to be re-started. Consequently, it is important to adopt all possible measures during a survey to minimize the chance of losing lock at any time, particularly if the distances are such that the ambiguities have to be floated.

Single-difference ambiguities and changes in reference satellites. A commonly encountered problem occurs when changing the reference satellite, whose data is used in all simultaneous double-differences, from one that is setting to another high enough to stay in view for a long time yet. At that epoch, all ambiguities and L3 biases experience large discontinuities equal to the ambiguity of the old reference satellite minus that of the new. One could start estimating all ambiguities again, but that would weaken the solution greatly and quite unnecessarily. A much better approach is to have *two* ambiguities or biases assigned to each double-difference, one for each "between-receivers" single-difference, and to make the ambiguity or bias for the single-difference of the reference satellite an unknown common to all double-differences. These single-difference biases are not discontinuous across a change in reference satellite. To avoid making the problem ill-conditioned, at least one single-difference bias should be initially fixed to zero.

13.1.6 Errors in GPS Orbits, in Reference Station Coordinates and in Tropospheric Corrections

As in long-baseline static GPS for geodetic applications, uncertainties associated with tropospheric refraction, stations, and orbits can be handled by adding error states to the Kalman filter formulation (the exact equivalent to adding nuisance unknowns to a batch adjustment). Imperfect tropospheric refraction corrections are taken into account by including zenith distance error states, one per station, with "partials" derived from the same atmospheric model used to make the corrections. At each site, the refraction varies over time and is usually modelled as a random-walk. There is also some spatial correlation between different sites, particularly when they are closer than 50 km.

Station coordinate errors add 3 unknowns per station, their "partials" the direction cosines of the vectors between the GPS satellites and each station.

In the case of orbit error, the unknowns are the initial errors in satellite position and velocity according to the known ephemerides (e.g., GPS broadcast ephemerides, IGS SP3 orbits, etc.) or any equivalent set of six parameters. The "partials", or coefficients of those unknowns in the observation equations, can be obtained from linearized orbital dynamics theory. They can be approximated by sums of simple functions of time such as sines and cosines of the true anomaly (the orbital angle between the perigee and satellite), plus some constants, and some linear functions of time.

When filtering out nuisance unknowns, such as orbit errors, refraction, etc., the main point *is not* to find the exact values of such parameters, but to get rid of their unwanted effects. For example: unless a network of GPS stations extends around the world, with stations north and south of the equator, it cannot be used to determine the orbit errors with uniform accuracy along the whole orbit. But the same errors estimated within a regional network, however wrong individually, usually represent quite well the combined effect of all those errors on the data.

13.1.7 Satellites in Common View

While remote-sensing instruments such as gravimeters, cameras, or altimeters, usually operate side-by-side with GPS in the same vehicle, their dependence on the world outside is vastly different. The remote-sensing hardware may work entirely on its own, and usually is just as accurate and reliable pretty much everywhere in the world, at any time. The GPS receiver, on the other hand, is only the most visible part of a complex, world-wide system of satellites, tracking stations, radio links, and so forth, and this global system *does not* work equally well everywhere, under all circumstances, at any time. This simple fact may baffle people used to self-contained remote-sensing instruments. It is a point that has serious practical implications, particularly for large-area surveys requiring long-range kinematic GPS.

To be able to position one receiver relative to others with any accuracy at all, it is necessary to have a sufficient number of satellites in common view of all receivers. This means a minimum of five satellites (four to solve for the instantaneous coordinates and clock error of the receiver, and another one to have enough observations to solve also for biases, etc.)

Generally speaking, in most of the world, satellite/site intervisibility is not a big problem with less than 700 km between receivers.

The longer the separation between receivers, the less the number of satellites in common view. Moreover, the number of satellites that are useful are less than those in common view: to minimize refraction and multipath effects, only those satellites are used that appear above a minimum elevation on the horizon at each receiver. This minimum elevation normally is chosen between 15 and 18 degrees. The actual choice is a fine balancing act: even a small increase of 1 degree usually decreases the number of satellites above the elevation cutoff, and increase the PDOP. But below 15 degrees, signal multipath and errors in tropospheric refraction correction become worse very quickly.

One way of extending the range within which one may reasonably expect to have a sufficient number of simultaneous observations at most epoches is to have more navigation satellites than just those in the GPS constellation. GLONASS is one such set of satellites which could double the number if fully incorporated in the navigation. The main obstacles at present are the scarcity and cost of good combined GPS and GLONASS receivers, and uncertainty as to the future of the Russian space program. Other satellite systems may incorporate GPS-like transmitters to add their signals to those of the original constellation. An example of this is INMARSAT.

A more distant possibility is the termination of the use of selected availability, or S/A (see next section), by the US military in control of GPS. That would greatly reduce the need for differential satellite clock corrections and, in post-processing, one could use the very precise estimates of the satellite orbits and of their clock errors distributed by such global organizations as the International GPS Service for Geodynamics, or *IGS*, and do single-receiver positioning (using between-satellites single-differences to eliminate the receiver clock). However, present plans call for S/A to remain active until 2006.

The requirements of satellite inter-site observability includes having a satisfactory sky-distribution with a low PDOP. That becomes increasingly less likely beyond +/- 55 degrees of latitude (equal in absolute value to the inclination of the orbits) because, closer to the poles, no GPS satellite can reach the zenith and are all confined to progressively lower elevations. Things are worst inside the polar regions, where the circumpolar "hole" in the distribution of GPS satellites in the sky is always above the observer.

When, as a result of poor satellite distribution, or insufficient numbers, kinematic positioning accuracy is lost, also lost is the ability to detect and correct cycle slips reliably, and to perform other necessary calculations correctly.

Therefore, careful pre-survey planning includes making sure that enough reference receivers will be available at well-distributed sites in or around the area to be surveyed, to secure adequate intervisibilibity most of the time.

13.1.8 Unfriendly Skies

Since late 1993, precise GPS positioning has benefited from the complete deployment of the constellation of satellites. On the other hand, since February of 1994, this progress has been hampered by military restrictions on the accuracy of the signals and of the orbital information openly available to everyone. The restrictions are achieved by satellite clock dithering and broadcasting less precise ephemerides, and are known collectively as "selected availability", or "S/A". Moreover, a secret modulation has been added to the signal, to help distinguish true GPS transmissions from those that might be broadcast by an enemy during a confrontation, to confuse the GPS navigation of US and allied troops. This is known as "anti-spoofing", or "A/S". The effect of S/A can be overcome using differential, or interferometric positioning, and that is one reason why such positioning is the main subject of this book. Differencing phases between receivers and satellites eliminates the effect of clock errors and of S/A's satellite clocks dithering. A mathematically equivalent approach is to use undifferenced observations (from the same arrangement of moving and fixed receivers), and to solve at every epoch, without *a priori* constraints, for unknown errors in the clocks of all the receivers and of all the satellites in view.

The lack of accuracy in the GPS broadcast ephemerides under S/A is circumvented in various ways. For precise post-processing, it is particularly useful to have access via the *Internet* to the precise orbits distributed by the IGS. In any case, differential techniques greatly reduce the effect of the errors through cancellation over relatively short inter-receiver baselines.

A/S, for its part, affects the quality of the measurements made by the receivers, particularly their pseudo-ranges, and this cannot be ameliorated by differential techniques. With the modulation on, the correlation of the received L2 signals with the receiver-generated P-code is noisier and more easily corrupted by the superposition of the directly-received satellite signals with their reflections off nearby objects (i.e., multipath). Efforts to mitigate the problems created by A/S have concentrated on improving the receivers' ability to measure the C/A code pseudo-range on L1, which is not subject to A/S modulation, although it is intrinsically less precise than P-code pseudo-range. A promising development along these lines are advanced correlators that exploit the power of modern digital processors to do very elaborate computations in real time, as in the multipath estimating delay lock-loop, or MEDDL technique developed by R. van Nee while a doctoral candidate in electrical engineering at the Delft University of Technology.

However, better processing based on the C/A code cannot directly improve the L2 pseudo-range, since there is no L2 C/A code. Nevertheless, some commercial receivers for general civilian use make quite good pseudo-range measurements in L1 and L2, even if still inferior to P-code based L1 and L2 pseudo-ranges. Therefore, great emphasis has been given to the use of differenced carrier phase as the primary data type for very precise positioning, with the pseudo-range getting only limited use (such as finding initial, rough estimates for the position of the vehicle, or making a preliminary guess of each L3 bias by subtracting the pseudo-range from L3).

All things considered, the non-military users of GPS have adapted successfully to conditions under S/A and A/S. Nowadays the relative position accuracy obtained from GPS is about the same whether those measures are on or off. (On the other hand, single-receiver, or "absolute" positioning is severely affected when they are on.)

13.2. DATA ANALYSIS

13.2.1 Data Pre-Processing

"Pre-processing" GPS receiver data means:

(a) Deleting bad observations, after they fail one or more simple quality tests (e.g., for low signal/noise ratio).

(b) Finding and, if possible, correcting cycle-slips in the carrier-phase observations.

(c) Calculating the receiver clock errors to within 1-2 microseconds and, if necessary, the approximate receiver coordinates in single-receiver solutions based on the pseudo-range (accurate to 100- 200 metres under S/A).

(d) Forming single or double-differences, as needed for the subsequent data analysis.

Some of these operations can be conducted at various stages in the data processing, starting with simple scans to check for bad data in each individual receiver observations file.

Scanning the data files. Bad, or suspicious, data can be detected by looking at the signal-strength indicator, rejecting observations where this is below 4 (in RINEX files). The ionospheric observable (L1-L2) for a particular satellite signal (or *channel*) is likely to follow closely a straight line over a few seconds. In a short time interval with three or more epoches in it, the first and last ones may be used to fit a straight line to the data. If the residuals are too large at the intermediate points, after that line is subtracted from the data, all observations in this interval, from the channel being tested, may be discarded. If the data was recorded while the receiver was *stationary*, parabolas can be fitted to 3 or more successive values of the phase in L1, and in L2. These parabolas can be used to predict the next values of the L1 and L2 phases. If the actual phase for either frequency differed from its predicted value by more than 1/2 cycle, then a cycle slip might have been found. Its size, an integer number of cycles of L1 or L2, would be determined by the size of the discrepancy between the actual and the predicted phase.

This approach is limited by the presence of S/A to checking intervals only a few seconds long. For longer intervals and a more thorough testing, it is better to do a further scan of a file of double-differences (once these are formed), which are not seriously affected by S/A.

Cycle-slip detection. One can identify three separate cases (all data are supposed to have been double-differenced):

(1) *Static case.* The present value of a double-difference can be predicted from its earlier ones, and a sufficiently large discrepancy between its true and its predicted value treated as the potential indicator of a cycle-slip. To predict that value it is preferable to use a simple Kalman filter. In practice, that means one filter for each double-difference and frequency, each filter running in parallel with all the others. This is a channel-by-channel (where now "channel" means the same as "double-difference"), or "*single channel*", approach.

(2) *Quasi-Static case.* This may involve an artificial satellite, its movement relative to the GPS satellites almost as predictable as that of a fixed receiver. (For most artificial satellites, at least when no manoeuvers are taking place, the accelerations of the vehicle and its antenna can be calculated using mathematical models that are accurate to better than 10^{-8} of G, the normal acceleration of gravity). Or it could be that the short-term position changes in a more conventional vehicle are already known from separate measurements (e.g., from an inertial navigation unit), so their effects can be corrected out of the double-differences. In either case, the changes over time in the double-differences are, or can be made, smooth enough to resemble the static case. Then the detection/correction of cycle-slips can be done "single channel", as in the static case.

(3) *Kinematic case.* This is the most difficult case, since "single channel" procedures generally do not work very well, or at all, in the moderate and high dynamics cases. The later is characterized by frequent changes in acceleration that are impossible to predict, and large enough to produce departures from a constant-acceleration parabola of more than 1/2 cycle of L1, L2, or some combination of both (e.g., the wide-lane), in the interval between two consecutive epoches (a

step-wise acceleration change of 0.1 G during a 1-second interval translates into a discrepancy of 0.5 m, or more than 1/2 wide-lane cycles).

Here it is necessary to use "*multichannel*" methods. In essence, one has to solve for the unknown *changes* in coordinates from one epoch to the next, using double-difference changes (or *triple-differences*) as observations. With more than three simultaneous triple-differences (more than four satellites in common view), checking the residuals of that solution for outliers might reveal the presence of cycle-slips. If there is only one cycle-slip present, the identification is straightforward. Otherwise, an elaborate series of tests is often used to determine if there are really several of them, and in which channels, and of what sizes. One limitation of this idea is that the uncertainty in the initial position of the moving receiver, and unpredictable changes in ionospheric refraction, are sources of error that grow with time, so this method may not be accurate enough to find cycle slips at the end of gaps in the data of more than a few minutes.

Since the ionospheric observable (L1-L2) is fairly smooth over short intervals, single-channel checks based on it are still possible, and can complement the "multi-channel" approach.

However, there are some combinations of simultaneous cycle-slips in L1 and L2 (4-3, 5-4, 9-7 cycles of L1 and L2, and their multiples) where the change in (L1-L2) is too small to detect in the presence of noise, so these can be found only by multi-channel checks.

In "multi-channel mode", solving for the change in position is subject to inaccuracies in the data, and so it depends on the geometric strength indicated by the PDOP. Just as for short-baseline ambiguity resolution, there should be at least 5 satellites in common view. If conditions are not favorable, the cycle slips may go undetected or, worse perhaps, a non-existing cycle-slip might be "found", and then "corrected", effectively planting a nasty systematic error in what was perfectly good data.

Redundant hardware: multiple on-board receivers. One way to improve the reliability of the results is to use additional hardware to get redundant information. A helpful practice is to use two or more receivers simultaneously on the vehicle. The baseline between two such receivers is always short, so the ionospheric and tropospheric refraction, orbit errors, etc., are virtually the same at each receiver and cancel out in their double-differences. All satellites visible from one receiver are in common view with the rest (unless obstructed by some part of the vehicle), and the cancellation of tropospheric refraction allows the use of very low elevation satellites (as long as multipath is not a serious problem); so one can get a good PDOP at most times. In consequence, detecting cycle-slips in the double-differences between pairs of on-board receivers is usually a much more reliable operation than finding them between an on-board receiver and a distant reference station. When a cycle-slip happens in an on-board receiver, then that cycle slip also appears in any double-difference between that receiver and any fixed station observing the same satellite. So detecting cycle-slips between the on-board receivers means finding cycle-slips in the navigation baselines. Of course, two or more on-board receivers may experience cycle-slips at the same time, and all that can be detected is their difference. As long as that difference is not zero, its detection flags down very clearly which double-differences are affected, and when, so further analysis can be focused on those channels and epoches.

Synthesized antenna and vehicle orientation. Other uses of multiple on-board receivers are: (a) creating a synthetic antenna at a convenient location inside the vehicle; (b) determining the orientation of the vehicle in the same reference frame as the trajectory.

(a) If the relative positions of the antennae in a reference frame attached to the vehicle are known from a preliminary survey, then with four or more receivers one can form a linear combination of their phases equal to the phase at any given point in their neighborhood. This is a virtual, synthetic antenna that can be "put" anywhere on board, including places where it is impossible to mount a real antenna. Such a place could be, for example, a reference mark on an altimeter or a camera, making it unnecessary to correct for the offset between a real antenna and the instrument's mark.

As a simple example, consider two real antennae: A1 and A2, receiving signals from the same satellite with phases ϕ_1 and ϕ_2. Along the straight line passing between the two antennae, the phase f varies almost linearly with distance. If B is any point on this line within 20 metres of A1 and A2, then an antenna at B would receive signals whose phase, to better than 0.001 cycle (given the great distance of the satellite to both antennae), should be:

$$\phi = a\,\phi_1 + b\,\phi_2 \tag{13.3}$$

where $a = d_1 / d$ and $b = d_2 / d$, d being the distance between A1 and A2, d_1 the distance of B from A1, and d_2 the distance of B from A2. Either d_1, d_2, or both, can be greater than d. Of course, the accuracy with which ϕ can be calculated is limited by how well d, d_1, and d_2 are known from a previous survey of the vehicle.

With two real antennae, the virtual antenna can be anywhere on the same line; with three, anywhere on the same plane; with four or more, anywhere in the surrounding space, or even in the interior of the vehicle.

(b) With the coordinates of the real antennae already known in the frame of the vehicle, and their relative coordinates estimated from the observed phases, in the earth-fixed (navigation) frame, one can determine the instantaneous transformation from vehicle to navigation frame. This transformation defines the spatial orientation of the vehicle in the earth-fixed frame, usually within a few minutes of arc.

Undetected cycle-slips. Cycle-slips left undetected (false negatives) or introduced by mistake (false positives) can be a serious problem in long-range kinematic GPS, where ambiguities are floated. This is often a slow process. (In short-range kinematic, the possibility of repeatedly and quickly fixing the ambiguities can be used to detect cycle slips by looking for changes in those ambiguities from one fixing to the next.)

The cycle-slip is an extremely systematic error, with ill-effects that are likely to build up over time causing a progressively larger position error, as shown in Figure 13.1 (from a simulation of an airborne survey, assuming all the L1 and L2 ambiguities are correctly resolved up to the time of the cycle-slip).

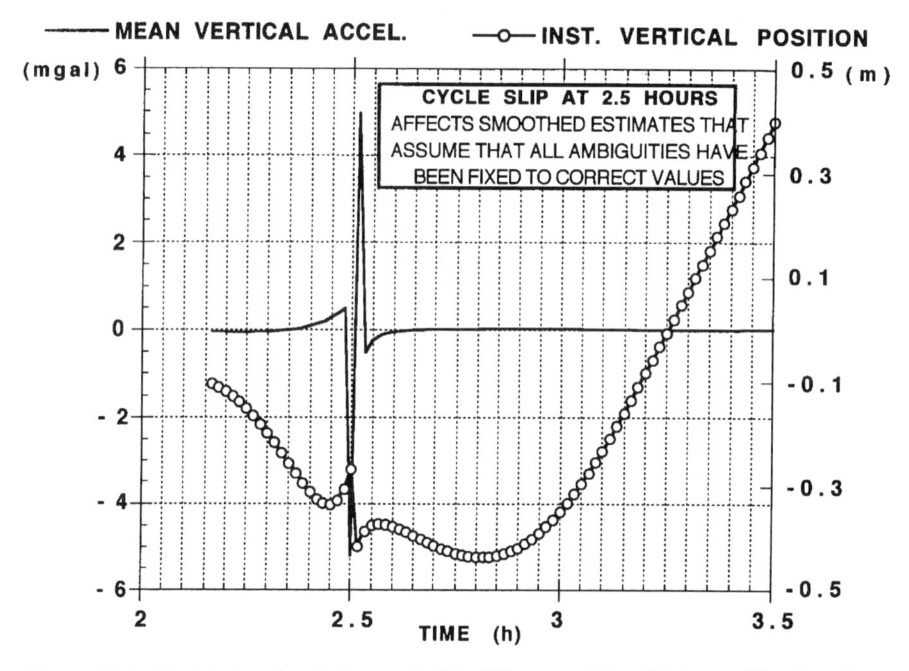

Figure 13.1. Simulated cycle-slip in one double-difference at time 2.5 hours. The plots show the effect on GPS-based Kalman filter / smoother estimates of airplane height and vertical acceleration (the acceleration obtained by numerically differencing the height twice).

The cycle-slip happens at the point of discontinuity in the plot, but its effect appears both before and after it, because this post-processed solution is from a Kalman filter and smoother algorithm that uses all the data to get the position at any given epoch, as explained in the next section.

One way of mitigating the effects of possible errors in cycle-slip detection is to assign an *a priori* uncertainty, or variance, to the estimated corrections. The affected L3 biases have their variances increased at the time of the presumptive cycle-slip, when the correction starts to be applied. The additional uncertainty makes the results less accurate, but if this is done only a few times, it can be quite helpful.

Using more than one reference station cuts down the odds of making costly mistakes, since unlikely combinations of simultaneous cycle-slips in separate receivers can be ruled out as false alarms caused by some other data problems. For example: it is highly unlikely that the same cycle-slip will happen at the same time in different reference receivers far from each other, so detecting it in more than one baseline either indicates a cycle-slip in the common roving receiver, or else it is a false alarm. Also, with more than one baseline it is likelier that at least five satellite channels will not be affected by cycle-slips, and can be used to test the others in a multi-channel solution.

The difficulty of the task is best understood if one considers that a long-range survey may produce many hours of data, from many receivers, at the rate of once per second or higher. Each single phase observation, at every epoch, may have a cycle-slip, and has to be tested for it. To have a reasonable chance of making no mistakes in the whole run, the likelihood of any one of hundreds of thousands of tests being wrong must be extremely low. In any human activity there is always the implicit gamble that things are going to work out all right. To develop better techniques for cycle-slip detection is to improve the odds that this bet will pay off (and that the failure to correct a few lowly cycle-slips will not ruin an important survey, possibly wasting a great deal of effort, money, and good will).

13.2.2 Final Data Analysis (Post-Processing)

The Kalman filter. In its various forms, Kalman filter algorithms are efficient and convenient ways of doing linear, bayesian least-squares (b.l.s.) estimation in an important class of problems involving large numbers of observations. Such problems appear often in navigation. For example: using high-rate observations from a moving GPS receiver, find the coordinates of tens of thousands of successive antenna positions over a period of many hours, while also floating the L3 biases, estimating orbit errors, etc. With the arrival of the Kalman filter, in the early 60's, solving such problems finally became practical. Since then, this method has found widespread application in such as: navigation, digital communications, industrial engineering, machine control, etc. Even with the primitive and slow computers of the 60's, this technique could be implemented with reasonable run times and (even then) with *realistic* amounts of memory. That often meant *very* small amounts of memory: early hardware components were several orders of magnitude larger in size than they are today, and in some military and in space applications, the space available for the navigation computer was often very small.

For long-range GPS navigation, the number of calculations in the Kalman filter can be much greater than for short-range, due to the often large number of nuisance unknowns. Calculations and information flow should be kept to a minimum, to make computing time reasonably brief and the size of some temporary data files manageably large, and to avoid excessive accumulation and propagation of arithmetic round-off errors.

Kalman filtering consists of a sequence of bayesian least square solutions, each with one more observation than the previous one, so at every epoch the estimates of vehicle coordinates and nuisance parameters are based on all data up to that observation. In consequence, the filter is well-suited for real-time applications. For post-processing, it is a convenient first step in a two-step solution, at the end of which all of the data have been used to estimate the unknowns, fully exhausting the information available. That second step, the smoother, is explained later in this section.

After processing (or "assimilating") a new observation, the Kalman-filter finds new estimates of the unknowns and their updated variance-covariance matrix, by combining that observation with the previous estimates in a small bayesian adjustment. Since, at each epoch, the results are partly based on those of the previous one, the algorithm is said to be *recursive*. As more data are used to estimate the same unknowns, the estimates tend to improve. That is not always true, because the introduction of *process noise* increases the *a priori* covariances of the unknowns from one epoch to the next, and tends to raise the uncertainty of subsequent results (for example, on loosing lock, it may be necessary to start estimating new biases with large initial uncertainties). Also, changes in the geometry (PDOP) of the observations may cause results to get worse. In many cases, though, the accuracy of the results tends to improve over time. The filter is then said to *converge*. How fast it converges depends on the strength of the data (signal to noise ratio and receiver/satellite geometry). For a while, the results may fluctuate wildly, as shown in Figure 13.2 (a simulated airplane flight where there is a complete loss of lock in all channels at the point marked by a spike in the formal precision of the airplane height), but eventually become quite good. If one is using GPS data, the main reason for a delay in getting good vehicle coordinates is the time it takes the filter to make good estimates of the L3 biases.

If the vehicle is close enough to a reference station to fix ambiguities, these can be entered into the Kalman filter solution as if they were very precise, uncorrelated observations of the L3 biases.

The smoother. Once the filter has processed all the data, at the final epoch, one may have very good estimates of all L3 biases. Assuming no intervening cycle-slips, one could "go back in time" and use these well-determined final estimates of the biases to correct the data affected by them at all earlier epoches. This should improve results along the whole trajectory, possibly making all of it about as well-determined as its very last point, as shown in Figure 13.2. This is the idea at the heart of *smoothing*. That procedure uses all the data available to solve the b.l.s at each and every epoch, not just all the data up to that epoch, as in Kalman filtering.

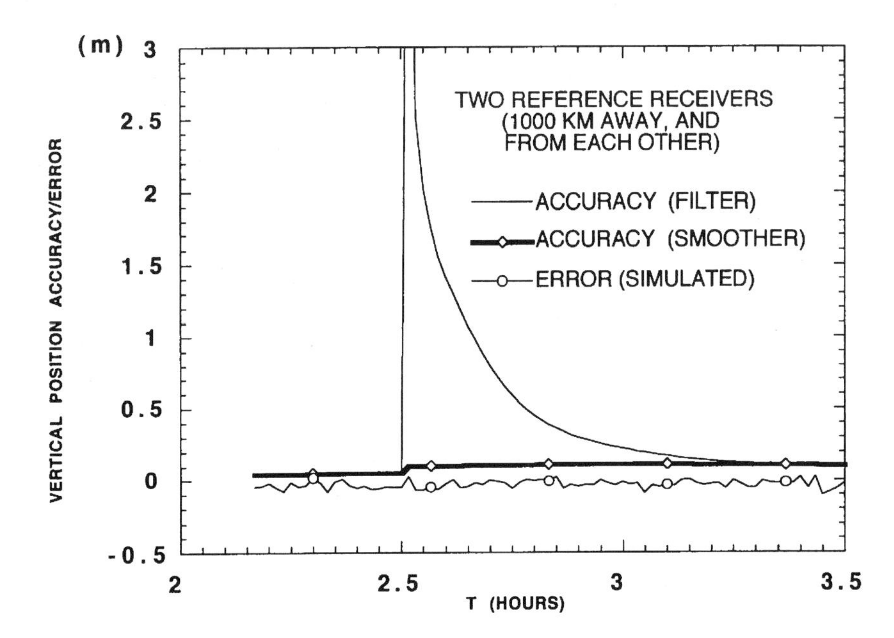

Figure 13.2. Filter and smoother vertical accuracy and vertical error in smoother solution (simulation). Airplane flying more than 800 km away from either of two reference stations that operate simultaneously. Complete loss of lock at 2.5 hours. (Also shown, the simulated height error after smoothing.)

The data is analyzed in the smoother in reverse sequence, starting from the *last* epoch and proceeding "backwards in time" towards the first, so smoothing (as defined here) can only be done in post-processing. Kalman filtering alone can be equivalent to filtering+ smoothing if the unknown states are constants (e.g., static receiver coordinates, initial state errors in the orbits, L3 biases). Their last estimate is their best estimate for all time, so going back to earlier epoches to improve their results is pointless. Unknowns of this type are said to be *unsmoothable*, since their estimates cannot be improved by smoothing.

To understand how the smoother works, consider its simplest and best known form (if not the most useful one in practice): the backwards, or reverse-time K-filter. The filter can assimilate the data in a sequence that it is not necessarily in the "natural", "past-to-present" order. In particular, one can run the filter "backwards", or "future-to-present", and get an estimate of the present error states based on all the future data, as well as the variance-covariance matrix of such estimates. Combining at each epoch, in a weighted average, this "future-based" estimate with the "past-based" estimate of a forward running Kalman-filter, one gets the equivalent of a batch solution based on all the data collected during the survey.

Stability. Recursive filtering/smoothing propagates arithmetic round-off from epoch to epoch and, under some conditions, the errors may accumulate and corrupt the solution. To make this less likely, a number of measures may be taken. The simplest are: *symmetrization* of the variance-covariance matrix of the unknown error states (making pairs of matrix elements symmetrical with respect to the main diagonal equal to the average of their calculated values); *regularization*, (adding very small positive quantities to the diagonal elements of the state variance-covariance matrix; this is particularly necessary with constant error states, such as L3 biases); and *pre-whitening*, where *correlated* simultaneous observations are de-correlated by multiplying them and the corresponding observation equations by the inverse of the Cholesky factor of their noise matrix. This allows the use of computationally thrifty filter algorithms that assimilate one de-correlated observation at the time, without necessarily making inexact and time-consuming matrix inversions at every single epoch.

More elaborate approaches have to do with the actual sequencing of arithmetic operations, the choice of error-state dynamics to reduce calculations, and the use of a type of transformation known as *factorization*, that converts the problem into a new one that is less susceptible to round-off error accumulation.

Finally, there is data *batching*, and *compression* (see next two sections), or the use as observations of successive data averages formed over many epoches, to calculate slow-varying states while greatly reducing the number of recursive steps in the filter/smoother solution.

13.2.3 Cutting back Calculations

The basic Kalman Filter algorithm can be implemented in many different ways. In general, the more observations are processed, the longer it takes to arrive to the final result, the greater the loss in accuracy caused by arithmetic round-off propagation, and the larger the scratch files' space on hard disk needed for communication between filter and smoother. There are a few common sense measures that can be taken to reduce these problems. By cutting back the number of calculations, round-off propagation becomes manageable without using complicated filter schemes (e.g., with factorized variance-covariance matrices; see Bierman's book for detailed information on factorized forms of filters and smoothers)

Eliminating unknowns. An example of this is the elimination of clock error parameters by double-differencing. In fact, this is equivalent to implicitly solving for the eliminated parameters as freely adjusted unknowns, but without actually taking any time to calculate their values. Of course, if one wished to know those values, additional calculations would be needed.

Near-diagonal state transition matrix. With a thoughtful choice of states, the state transition matrix can often be diagonal, with most non-zero elements equal to 0 or 1, or, at worst, it might be block-diagonal, with blocks of very small dimension. Multiplying full (nxn) matrices, requires about $2n^3$ floating-point calculations. If the state transition matrix is diagonal, with 0 and 1 elements, the same matrix multiplication involves setting to 0, or else leaving unchanged, certain rows and columns. This takes a very small fraction of the time it would if the state transition matrix were a full matrix. Choosing (a) the "no-dynamics" (or white noise) form for the vehicle coordinates, (b) the natural "constant" form for the ambiguities, (c) a random walk for the tropospheric refraction, and so on, one ends up with a diagonal state transition matrix that has mostly 1's and 0's in the main diagonal.

Batching. States that vary slowly enough may be regarded as constant over a number k of observing times. In that case the state transition matrix equals the unit matrix and process noise is absent for all k epochs. This is equivalent to treating these k consecutive observations as if they were part of a *batch* of simultaneous observations. Batching results in fewer calculations because: (1) the unit state transition matrix is diagonal, with all non-zero elements equal to 1, and (2) the process noise variance update is not computed or used in the stochastic updates.

Compression. The data can be averaged over separate, consecutive intervals, to estimate slow-varying nuisance parameters with much fewer data updates than using the original instantaneous measurements. The idea is discussed further in the next section.

13.2.4 Data Compression

In a filter-smoother solution there are two filters that run in opposite directions in time, the results being the weighted averages of those from each filter. To average the backward and forward-filter results at any epoch during the backward run, one needs to get, from some storage device, the results of the forward run for that epoch. This means storing the state estimates, their variance-covariance matrix, and other useful information (such as the matrix of the observation equations) for *all* epoches. For long runs with high data-rates, and with up to 100 states being estimated at any given time, the size of those files in a long-range kinematic solution can become very large and present a serious storage problem. The time needed to compute the filter-smoother solution, and the numerical stability of the results after many recursive steps, also give cause for concern. All these problems can be taken care in a rather elegant and effective way by using *data compression.* Some error states may vary slowly enough to be nearly constant for significant periods of time, and one may take advantage of this to reduce the number of computations and storage. Less frequent filter updates, using data averages (compressed data) rather than instantaneous values, may be sufficient to get results that are very close to the optimal ones based on instantaneous data.

Compression: Static case. This case is easier to explain. In very precise *static* surveys with VLBI, satellite laser ranging, or GPS, the unknown states are either constant (station coordinate errors), or vary quite slowly and regularly (satellite orbit errors). The data can be *compressed* by averaging over several minutes, and the averaged data can then be used instead of the full-rate data with very little loss of information. Virtually the same results can be obtained with much fewer data updates: if the data rate is q observations per minute, and the duration of the compression interval is m minutes, then the number of updates decreases by a factor of $q \times m$. If observations are taken once per second (as it is often done in kinematic GPS), then averaging over 1 minute would result in a reduction of some 60 times in computer operations, round-off propagation, running time, and in the size of the data files needed to pass information from the filter to the smoother.

More concretely, let the equations for the observations at epoch "j" be, in matrix form:

$$y(j) = H(j)\, x(j) + n(j),$$

where "H" is the matrix of "partials", or measurement's matrix, "y" is a vector of simultaneous observations, "x" is the vector of error states, and "n" is the noise.

Averaging over "k" epoches, and assuming that the "$x(j)$" are all constant and equal to "$x(i)$" for all epoches in the "ith" compression interval:

$$\frac{1}{k_i} \sum_{j=1}^{k_i} y(j) = \frac{1}{k_i} \sum_{j=1}^{k_i} H(j)\, x(i) + \frac{1}{k_i} \sum_{j=1}^{k_i} n(j) \tag{13.4}$$

Σ is the operator that "sums over k epoches in the interval i". This average can be calculated easily by taking advantage of the gradual change with time of the "partials", the elements of matrix "H". For several minutes these can be approximated by low-degree polynomials, such as parabolas. Three points are needed to find the coefficients of a parabola, so one could use the value of each partial at the beginning, near the middle, and at the end of each compression interval, to find three coefficients $a(l,m,i)$, $b(l,m,i)$, $c(l,m,i)$ that, in any given interval "i" starting at time "$t_0(i)$", satisfy (approximately):

$$h_{lm}\, t(j) = a(l,m,i)\left\{\left(t(j)\text{-}t_0(i)\right)^2\right\} + b(l,m,i)\ \left\{t(j)\text{-}t_0(i)\right\} + c(l,m,i) \tag{13.5}$$

where "$h_{lm} t(j)$" is the element of H in row "l" and column "m" at time "$t(j)$", or epoch "j". The average over the compression interval "i" is:

$$\frac{1}{k_i}\sum_{j=1}^{k_i} h_{lm}t(j) = \frac{1}{k_i}\,[a(l,m,i)\sum_{j=1}^{k_i}\left(t(j)\text{-}t_0(i)\right)^2 + b(l,m,i)\sum_{j=1}^{k_i}\left(t(j)-t_0(i)\right)+c(l,m,i)] \tag{13.6}$$

The last term in (13.4) is the noise in the averaged data, and it requires a corresponding noise covariance matrix in order to implement the b.l.s solution correctly. Because of poorly understood correlation in GPS observations, largely caused by multipath, the noise matrix for the averaged data are empirically determined assuming that observations are strongly correlated over 10 or 15 seconds. So (if all observations are equally accurate), instead of multiplying the variance-covariance matrix for single-epoch noise by the inverse of the square-root of the number of observations k, one may use some modified law, such as the inverse of the square root of $k/10$ or of $k/15$ (if "k" is in seconds).

Length of the compression intervals. Experience with the use of compression for kinematic GPS shows that one may use averages as long as 7 minutes with little loss of accuracy. To avoid problems combining averages with different numbers of data (because some satellite observations may have gaps), and given that the statistics of the data noise are not that well understood, it is best to leave out all data from those epoches where some observations are missing, using only averages based on more than a minimum percentage (e.g., 60%) of epoches with complete data. The occasional lack of sufficient data to form such averages in a compression interval, the rising or setting (below the elevation cutoff) of some satellites, the beginning of long gaps in some channels, the starting of new ambiguities (or the relaxing of old ones suspect of harboring cycle-slips), etc., represent natural boundaries where the averaging intervals must stop. So, in practice, successive compression intervals may have different lengths. The rules for starting and ending a compression interval can be quite elaborate, in order to cope with the many inconvenient oddities of real data.

Setup intervals. In the case of multi-satellite data such as GPS double-differences, practical considerations when using data compression dictate a certain rigidity in the way the double-differences are formed (which satellites' data are used from which receivers). The double-differences are formed in the same way at all epoches within a certain period of time, followed by another period with a new, fixed choice of double-differences, and so on. These periods, which may last from a few minutes to a few hours, are not compression intervals (they may contain many compression intervals) but *setup intervals*, with a fixed selection of satellites and stations peculiar to each one of them. The rules for deciding when to end a setup interval and begin the next can be quite elaborate, but are susceptible to automation, at least to the extent that scanning the data with a subroutine may be enough to produce a good "first draft" of the setup, that can then be easily and quickly modified and improved by hand, if necessary.

Compression: kinematic case. In kinematic GPS some error states, such as the vehicle coordinates, may vary extremely fast. The solution is to compress the data anyway, and then estimate the actual values of the slow-varying states, and the *average values* of the fast-varying states (e.g., vehicle coordinates). The estimates of the slow states alone are used to correct the full-rate data, in order to eliminate as much as possible the effect of those states from them. The corrected full-rate data can then be processed epoch by epoch, in a series of separate, non-recursive, small, free (unconstrained) least-square adjustments, with only the vehicle coordinates at each epoch for unknowns. This last part of the procedure is, in fact, the usual algorithm for short-baseline navigation after all ambiguities have been resolved and eliminated from the problem. The extra calculations needed to estimate the many nuisance error states peculiar to long-range navigation, take place in the first stage, the analysis of the compressed data. This "long-range" computing overhead is greatly reduced by compression, to the point that filtering 100 error states may account for 20% or less of the total computing needed to post-process a multi-baseline, multi-hour survey, with observations taken every second.

Iteration. Depending on how the *a priori* position of the vehicle at a given epoch has been found in order to linearize the estimation problem at that epoch, the uncertainty in the *a priori* coordinates may range from 200 metres (single receiver under S/A, with poor geometry) to tens of metres (using double-differenced pseudo-ranges), to less than one metre (low dynamics vehicle). In the first two cases, one or two iterations of the least squares solution may be needed to eliminate nonlinear effects, and also to get a sufficiently accurate vehicle height to make good tropospheric corrections. In the case of long-range navigation, this iteration cannot be limited to the vehicle coordinates. All unknowns, nuisance parameters and vehicle coordinates alike, must be iterated together. The way to do that is to repeat the whole K-filter/smoother procedure, using points along the trajectory estimated in one iteration, as the *a priori* positions for the next (but always using the same *a priori* accuracy!).

Comparing each iterated trajectory with the previous one, one may stop iterating when the r.m.s. of the discrepancies between them becomes a small fraction of the data noise. For L3 phase double-differences, the noise is some 2 cm (r.m.s.), so a discrepancy of 1 mm (r.m.s.) or less indicates convergence.

The use of iteration offers a special advantage in the case of compressed data. In the expressions for the averaged observation equations, the coordinates of the moving vehicle are treated as if they were constants. This seems absurd, until one realizes that what is being estimated is not the coordinates themselves, but first order corrections to their *a priori* values. These corrections will be mainly long-period errors in the previous iteration, caused by imperfectly estimated orbit, tropospheric refraction, and L3 bias errors. The faster the actual changes in position, the likelier they are to be resolved at the very first iteration, and so to be automatically subtracted out of the problem in later iterations. If the compression intervals are not too long, the dx, dy, dz corrections to the iterated coordinates of the vehicle are likely to be nearly constant within each interval, making the kinematic case as amenable to compression as the static one, even if this is not obvious at first sight. As in the static case, the coefficients in the observation equations corresponding to the fast-varying states are the averages of the instantaneous "partials".

Interpolation of corrections. Accuracy of the full-rate solution. Two practical problems with compression are:
(1) The slow error states "x" may be constant over the whole period, but some of their partials are not constant. Therefore, the data corrections for "slow" states (except L3 biases) are time-varying, and have to be calculated at every observation epoch, to be applied to the full-rate data. In the author's software this is done by calculating their values at the start of each compression interval, and interpolating them at the full-rate epoches with cubic splines (except for the biases, that may vary abruptly from interval to interval, if there is loss of lock, but should be constant within each properly chosen interval and require no interpolation there).
(2) The variances and co-variances of the instantaneous vehicle coordinates cannot be obtained directly from the simple, epoch-by-epoch series of small least squares adjustments with the corrected, full-rate data. The uncertainties in the estimated data corrections would have to be brought into the formulation of the problem. That is difficult to implement efficiently, and rigorously. A simple way out is to process a single epoch of data after each step with compressed data. The "single epoch" (or "instantaneous") data step, also updates the full error state vector and covariance matrix. The 3x3 sub-matrix corresponding to the variances and covariances of the coordinates of the vehicle can be shown to be the correct one for the full-rate solution at that epoch. Moreover, the accuracy of the full-rate solution changes so slowly that the same 3x3 sub-matrix closely reflects the accuracy of the vehicle coordinates throughout the interval. This idea is relatively easy to implement, but means doubling the number of steps in the filter/smoother algorithm. In general, with modern fast PC's and workstations, the algorithm can be implemented so this doubling is not a problem.

Effect of compression on accuracy. The "slow" states are not exactly constant, and the approximation of the "partials" with low-degree polynomials (such as parabolas) not quite exact. The longer the compression interval, the less correct those two assumptions become, so one may expect a progressive deterioration in results with increasingly long intervals. The experience of the author indicates that such deterioration is not significant for intervals of up to 7 minutes; beyond that, it becomes noticeable very quickly. In general, a compression interval of 5 minutes is a good choice; with one-second receiver sampling rates, this means a compression rate of 300, and a reduction of 150 times in the number of steps and in the size of the intermediate filter/smoother scratch files, even when one has doubled the steps by including an instantaneous data update after each compressed data update, as explained above.

For a more comprehensive discussion of compression in GPS navigation, in particular the use of "augmented state spaces", and the equivalence of compressed and full-rate kinematic solutions, see the author's 1992 paper on this subject, listed in the References.

Stop-and-go kinematic, rapid-static, station revisiting. All these variants of the basic GPS positioning techniques can be implemented also in long-range kinematic GPS, and entirely within the framework of the ideas presented here. The author has found no problems implementing (and using) such features in his own long-range navigation software. It is possible to switch back and forth from static to kinematic modes by merely changing the state transition matrix corresponding to the vehicle coordinates from a "constant" to "white noise" form and viceversa.

13.3 TESTING THE ACCURACY OF LONG-RANGE KINEMATIC GPS

13.3.1 Approach

To test the accuracy of long-range navigation, first an accurate control trajectory is computed using nearby stationary receivers (< 20 km from the vehicle) as reference stations. The data are L1 and L2 carrier phase double-differences with all their ambiguities properly resolved, if possible by static initialization prior to the start of the trip, as well as just after it. Such trajectory is usually good to a few centimetres, so it can be used as "truth", or control, to verify the positioning of the same vehicle relative to very distant receivers.

For the short and long-baseline results to be comparable, all fixed receivers must have their coordinates precisely determined in a common reference frame. This can be done by making geodetic ties between those receivers and the International GPS Service for Geodynamics (IGS) stations in the same area, whose coordinates in the International Terrestrial Reference Frame (ITRF) are known with great accuracy.

13.3.2 First Case: Boat Experiment, Sydney Harbour, Australia

Setup. In August of 1994 a test was conducted during which a small motor-boat was positioned with GPS while cruising Sydney Harbour. This was done in cooperation with many colleagues throughout Australia and in New Zealand, the main organizers being Dr. Chris Rizos, the late Mr. Bernd Hirsch, and their co-workers at the School of Geomatic Engineering of the University of New South Wales, in Sydney. There were two stationary GPS receivers in Fort Denison, which is on a small island near the Opera House, another receiver, 6 km away, was atop a tower in the building housing the School of Geomatic Engineering of the University of New South Wales (UNSW). These were the local receivers. The distant ones were in Brisbane (750 km), Adelaide (1100 km), Hobart (1100 km), and Wellington (2200 km), as shown in Figure 13.3. The receivers were Trimble SSE's and Leicas RS299's. ITRF coordinates for the distant receivers and for UNSW were obtained from a geodetic solution done at the University of Canberra. The boat traveled the 12 km from Denison Island to Manly Cove, and back, at an average speed of 10 km/hour. Occasionally, it was rocked by waves and by the wakes of large passenger ferries. Because of the length of some baselines, precise IGS orbits were used, and their errors estimated as part of the filter-smoother navigation solution. At times, the PDOP was > 10.

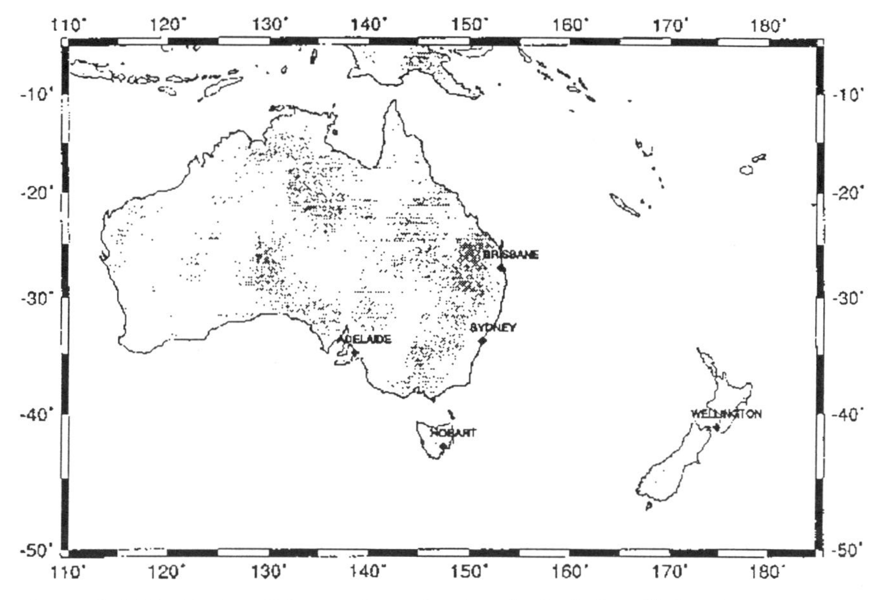

Figure 13.3. Distribution of distant reference stations for the Sydney Harbour test.

Results. In normal operations, an aircraft or ship might leave an airfield or dock and come back to it at the end of its trip. There would be a reference receiver near the runway or berth, allowing to resolve the phase ambiguities exactly at the start and at the end of the kinematic survey. The trajectory estimates might be strengthened considerably by resolving those ambiguities, even if all satellites concerned are present only for a minor part of the survey period. To simulate this situation, data from nearby UNSW was included for the first hour and also for the last ten minutes of the solution, with all ambiguities resolved. For the remaining time, the boat was positioned only relative to distant sites. In the case shown in Figure 13.4, these sites are Adelaide and Wellington (1100 km and 2200 km) used simultaneously. The plots in the figure are those of the vertical and horizontal discrepancies between the short-range "truth" or control trajectory and the long-range solution in the period when only the distant sites were used. The position discrepancy is 12 cm (peak), 7 cm (r.m.s.), and the drift is less than 5 cm/hour. Hours of data from several receivers were analyzed in less than 30 minutes in a commonly used type of workstation (HP 9000-735), showing that this is a practical approach for use in large-area surveys.

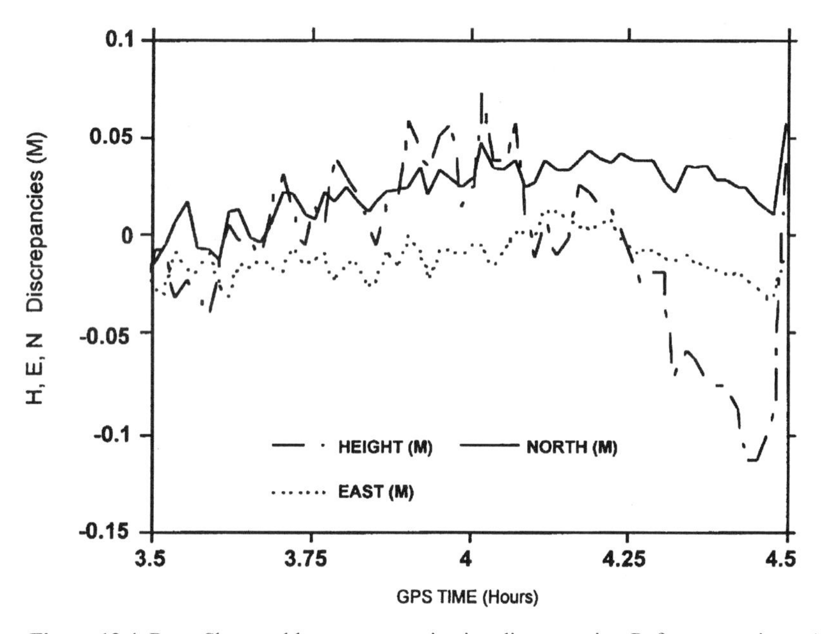

Figure 13.4. Boat: Short and long-range navigation discrepancies. Reference stations: Adelaide (1100 km away) and Wellington (2200 km)

Orbital decay. The reference stations were positioned by colleagues at the University of Canberra with ties to the IGS stations in Australia, in a simultaneous adjustment of orbits and site coordinates, with IGS and test data. Unfortunately, the adjusted orbits were lost, and it become necessary to use SP3 orbits from the IGS instead. These orbits were not quite in the same de facto reference frame as the test stations; there was something like a 0.1 arc second misalignment between them, resulting in a rotation of the estimated orbit planes relative to the fiducial stations. This was the equivalent of introducing large (up to 20 m) orbit errors in most satellites in view, so it was necessary to allow the SP3 orbits to adjust to the same extent, relaxing their a priori uncertainties to 10 m per coordinate and 1 mm/s per velocity component. Such a large orbit adjustment was not without complications. At first, calculations were done with data from the remote stations only, without the early/late contribution from the UNSW data described in the previous paragraph.

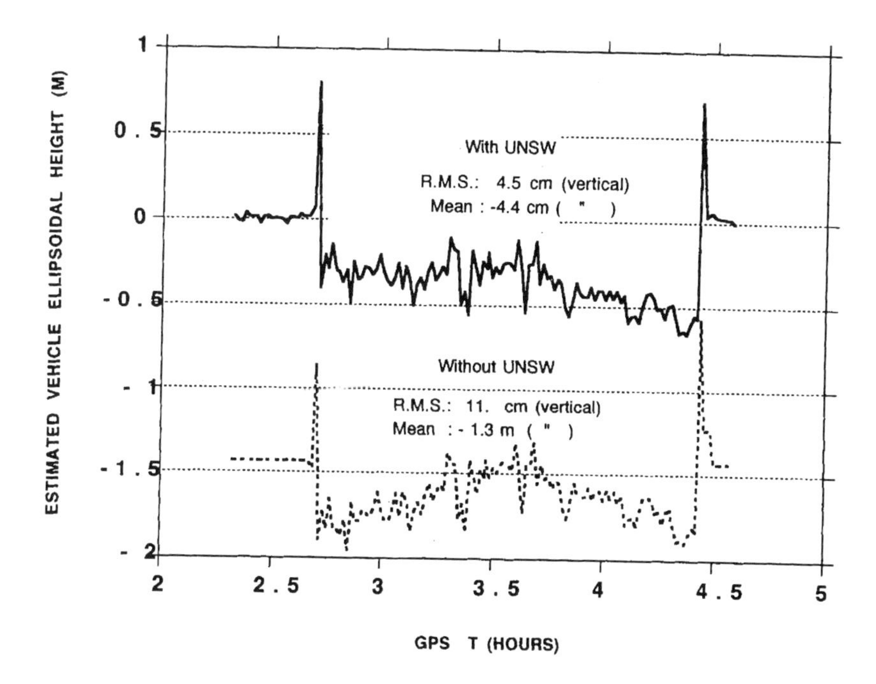

Figure 13.5. Estimated vehicle height with Adelaide (1200 km) and Wellington (2300 km) as fiducial reference stations: (a) with, and (b) without UNSW (6 - 13 km). Shown (inset numbers): r.m.s. and bias of the discrepancies in boat height between estimates (a) or (b) and the short-baseline control trajectory. At the beginning and at the end the rover's antenna rests on the same tripod, at the pier, so the height is the same. The two "spikes" are periods when the antenna was risen to be taken to and from the boat. In between, the signature of waves while the antenna was on the boat (heights plotted every 2 minutes).

The results were strange: the estimated long-range trajectory was shifted almost parallel to its short-range counterpart by several metres during the whole session. Once the problem became better understood, the idea of adding local data from UNSW at the start and at the end, with all the ambiguities resolved, was tried successfully. In this case, the stronger data at the beginning and the end helped separate the initial position of the vehicle from the orbit error and the floated L3 biases. Without the local data, and with loose a priori constraints on both the orbits and the ambiguities, these three types of unknowns cannot be separated well, because their slow-varying partials are nearly linear combinations of each other, at least over a period of three hours. On the other hand, the changes in position were much better resolved. The terms in the observation equations consisting of the products between partials for the boat coordinates, and their "white noise" states (i.e., "no dynamics" kinematic), represented a much higher-frequency portion of the signal spectrum, making it easier to separate position change from the rest.

Repeating the same calculations with the broadcast ephemerides, instead of SP3 orbits, and allowing them to adjust with the same large a priori uncertainties, the results were quite similar to those with the SP3's in each case (with and without UNSW). This suggests that "quick look" solutions can be done on the field immediately after returning to base station, without having to wait days or weeks for the release of the precise IGS ephemerides.

13.3.3 Second Case: Airplane Flight off Adelaide, Australia

Setup. In December of 1995, an ORION P-3 from the Naval Research Laboratory (NRL), in the USA, flew one of several tests carrying the Australian laser depth sounder (LADS) off the coast of Adelaide, in South Australia. During this flight, GPS data was collected aboard the airplane and on the ground, near the airfield. The GPS receivers were dual-frequency Ashtech Z-XII's, and the coordinates of the reference site were found using an SP3 orbit file for that day computed at the U.S. National Geodetic Survey (NGS). The author is thankful to Dr. John Brozena and to Mr. James Jarvis, both at NRL, for sharing their data with him.

The airplane flew out over the ocean, reaching a maximum distance of 270 km from the fixed receiver before coming back. Orbit and tropospheric correction errors were adjusted simultaneously with the trajectory. During the whole flight, which lasted almost 4.5 hours, the exact integer ambiguities for all the satellites used were known. Both at the start and at the end it was possible to fix those ambiguities on the fly.

Floating versus fixing. The trajectory was found again, this time with all the ambiguities floated, as if the vehicle had always been too far away from the reference site to resolve them exactly on the fly. Both trajectory calculations were made using the same SP3 orbits with which the coordinates of the reference site had been determined.

When compared, the trajectory found floating all the ambiguities turned out to be in good agreement with the one calculated with all the ambiguities resolved on the fly. In this case, the 3-dimensional position discrepancy had an r.m.s. of about 5 cm, a peak value of 10 cm, and a shift, or mean vector offset, of 2 cm. The Height, North and East components of the discrepancies between the trajectories are all plotted in Figure 13.6. The orbits only adjusted their initial conditions within a few decimetres of their initial values, and by amounts that were always less than one a posteriori sigma, so the calculated corrections for them were "in the noise".

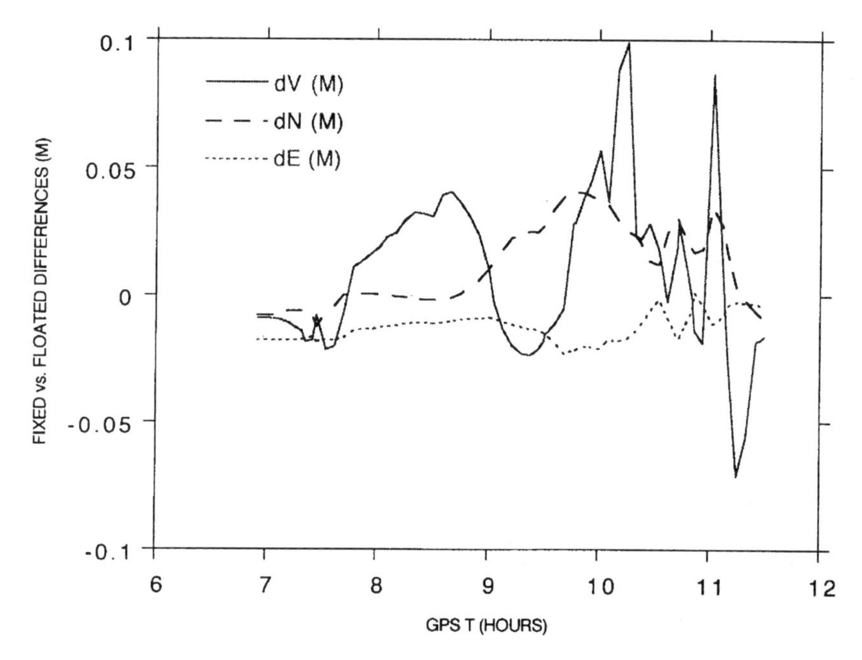

Figure 13.6. Height, North, East discrepancies between two estimates of the trajectory of an airplane calculated with (a) all ambiguities fixed, and (b) all ambiguities floated. Flight off Adelaide, Australia, December 21, 1995. (Maximum distance from fixed receiver near airfield: 270 km.)

In brief: it made little difference whether the ambiguities were floated or fixed. This is a very encouraging result, at least for "medium-range" applications where the vehicle is never close enough to a reference station to fix ambiguities, and post-processing is a viable option.

Floating or fixing? In the case just shown, thanks to careful flying that avoided sharp turns with banking angles in excess of 15 degrees, there were very few cycle-slips.

In cases where the ride is not as smooth, frequent loss of lock would produce frequent re-starting of the L3 bias estimates in the Kalman filter, without enough time to reach decimetre accuracy in vehicle position before the next loss of lock comes about. In such case, even after smoothing, the overall accuracy could be much poorer than if all the ambiguities had been resolved. Therefore, floating is a viable way of dealing with ambiguities only under conditions where the occurrence of frequent and unresolvable cycle-slips is unlikely.

13.4 CONCLUSIONS

Early tests suggest that it is both possible and practicable to position a vehicle with sub-decimetre accuracy, more than 1000 km away from any reference station (e.g., to one part in 10,000,000, or better) using differential kinematic GPS, at least in post-processing.

These long-range kinematic results resemble, both in accuracy and in baseline length, much earlier static ones that, in the mid-eighties, first showed that differential GPS positioning could be better than 1 part in 10^7 for fixed stations and baselines of more than 1000 km. Since then, many advances have pushed the accuracy and reliability of such ultra-precise static surveys to the point where an international effort such as the IGS continuously runs a global network of GPS observatories, and its member organizations routinely produce orbits, earth orientation parameters, and station coordinates accurate to 1 - 3 parts in 10^8 of baseline length.

Significant advances can be expected also for long-range kinematic GPS. Perhaps not so much in accuracy, but certainly in terms of reliability and versatility. Problems that remain unanswered today seem to have mostly to do with increasing the immunity of the results to cycle slips, and developing better methods for dealing with phase ambiguities beyond a few tens of kilometres from a reference site.

Some of the future progress will be driven by better and cheaper hardware, and some by more powerful analysis algorithms and software. The general availability of additional navigation satellites from systems such as GLONASS could go a long way to making possible 10-centimetre navigation anywhere, at any time, by increasing the number of spacecraft in common view over very long distances. The increase in sun-spot activity that is expected over the next few years could try severely the techniques now in use, most of which have been developed into their present form during the years of quiet sun now coming to and end.

As it continues to develop and mature, precise, long-range kinematic GPS will become a valuable tool for the scientific interpretation of altimetry, SAR, gravimetry, laser depth-sounding, or photogrammetry, as these are used to survey the oceans, great deserts, jungles, and polar regions of the Earth.

References

Bierman, G.J. (1977), Factorization Methods for Discrete Sequential Estimation, Academic Press, New York.

Brown, R.G. (1983), Introduction to Random Signal Analysis and Kalman Filtering, John Wiley and Sons, New York

Brozena, J.M., Mader, G.L., and Peters, M.F. (1989), Interferometric GPS: Three-dimensional Positioning Source for Airborne Gravimetry, J. Geophys. Res., 94, B9. 12153-12162, U.S.A.

Colombo, O.L. (1989), The Dynamics of GPS Orbits and the Determination of Precise Ephemerides, Journal of Geophys. Res., 94, B7, 9167-9182, U.S.A.

Colombo, O.L. (1991a), Errors in Long Distance Kinematic GPS, Proceedings ION GPS-91, 4th International Tech. Meeting Inst. of Nav. Satellite Div., Albuquerque, New Mexico.

Colombo, O.L. (1991b), Airborne Gravimetry, Altimetry, and GPS Navigation Errors, Proceedings I.A.G. Symposium on the Determination of the Gravity Field from Space and Airborne Methods, Vienna.

Colombo, O.L. (1992), Precise, long-range aircraft positioning with GPS: The use of data compression, Proceedings VI International Symposium on Satellite Positioning, Columbus, Ohio, March, 1992.

Colombo, O.L., Rizos, C., and B. Hirsch (1995), Long-Range Carrier Phase DGPS: The Sydney Harbour Experiment. Proceedings of DSNS 95, from the 4th International Conference on Differential Satellite Navigation Systems. April 1995, Bergen, Norway.

Gelb, A., editor (1974), Applied Optimal Estimation, (11th printing, 1989), M.I.T. Press.

Hatch, R. (1991), Instantaneous Ambiguity Resolution, Proceedings KIS-90 Symposium, Springer-Verlag, N. York.

Hermann, B.R., Evans, A.G., Law, C.S., and B.W. Remondi (1995), Kinematic On-the-Fly GPS Position Relative to a Moving Reference. Navigation, Vol. 42, No. 3, Fall 1995 issue, p. 487-501.

Herring, T.A., Davis, J.L., Shapiro, I.I. (1990), Geodesy by Radio Interferometry: The Application of Kalman Filtering to the Analysis of VLBI Data, J. Geophys. Res., 95, B8, p. 12561-12581, U.S.A.

Krabill, W.B., and Martin, C. (1987), Aircraft Positioning Using GPS Carrier Phase Data, Navigation, Spring Issue, U.S.A..

Loomis, P. (1989), A Kinematic GPS Double-differencing Algorithm, Proceedings 5th International Symposium on Satellite Positioning, Las Cruces, N.M., N. Mexico State U., p. 611-620.

Lucas, J.R. (1987), Aerotriangulation without Ground Control, Photogramm. Eng. Remote Sensing, 53, p. 311-314, U.S.A.

Maybeck, P.S. (1979), Stochastic Models, Estimation, and Control, Academic Press, New York.

Remondi, B.W. (1991), Recent Advances in Pseudo-kinematic GPS, Proceedings, Ottawa.

Zumberge, J.F., Liu, R., and R.E. Neilan, Editors (1995), IGS, International GPS Service for Geodynamics, 1994 Annual Report, JPL Publication 95-18 9/95.

14. THE GPS AS A TOOL IN GLOBAL GEODYNAMICS

G. Beutler, R. Weber, E. Brockmann, M. Rothacher, S. Schaer, A. Verdun
Astronomical Institute, University of Berne, Sidlerstrasse 5, CH-3012 Berne, Switzerland

14.1 INTRODUCTION

Until a few years ago it was believed that the GPS would never play an important role in Global Geodynamics. There was a general consensus that the Global Terrestrial Reference Frame and the Celestial Reference Frame would be uniquely defined by VLBI, that the geocenter and the Earth's potential would be defined essentially by Laser Observations. It was believed that GPS would play a decisive role in the densification of the Terrestrial Reference Frame, a role as an interpolation tool for the other more absolute space techniques so to speak.

This view of affairs had to be modified considerably in consideration of the success of the International GPS Service for Geodynamics (IGS). The contributions of the IGS and its Analysis Centers to the establishment of

- polar motion (x and y components),
- length of day (or, alternatively the first time derivative of ΔUT), and
- the IERS Terrestrial Reference Frame (ITRF)

became more and more accurate and reliable with the duration of the IGS experiment (test campaign in summer 1992, IGS Pilot Service (1 November 1992 - 31 December 1993), official service since January 1, 1994)).

Today the IGS products play an essential role for the Rapid Service Subbureau of the IERS; the contribution is getting more and more weight also in the IERS Central Bureau's analyses. Temporal resolution and the timeliness of the IGS analyses are unprecedented, the consistency is comparable to that of the other space techniques.

Unnecessary to say that GPS actually plays a decisive role for the densification of the ITRF. As a matter of fact, the IGS at present organizes a densification of the ITRF through regional GPS networks (see proceedings of the 1994 IGS workshop in December 1994 in Pasadena, Zumberge [1995a]).

Should we thus conclude that GPS is on its way to take out VLBI and SLR as serious contributors to global geodynamics? The answer is a clear *no*! Let us remind ourselves of the limitations of the GPS:

- GPS is *not* capable of providing absolute estimates of ΔUT. Length of day estimates from relatively short data spans may be summed up to give a ΔUT curve referring to a starting value taken from VLBI. SLR (Satellite Laser Ranging), as every technique in satellite geodesy *not* including direction measurements with respect to the inertial reference frame suffers from the same problems. GPS gives valuable contributions in the domain of periods between 1 and 40 days.

- So far, GPS has given *no* noteworthy contributions to the establishment of the Celestial Reference Frame. Below, we will see that contributions of a kind comparable to the length of day are actually possible.
- GPS, as a radio method, suffers from the limitations due to the wet component of tropospheric refraction. VLBI has the same problems, SLR is in a much better situation in this respect. Compared to either VLBI or GPS the SLR measurements are *absolute* in the sense that ground meteorological data are sufficient to reduce the tropospheric correction to a few millimeters for SLR established ranges. This fact gives SLR a key role for the calibration of troposphere estimates as they are routinely performed in GPS and VLBI analyses. The fact that some (at present two) GPS satellites have Laser reflectors clearly underlines this statement.
- GPS is an interferometric method. Highest accuracy is achieved in the *differences* of measurements taken (quasi-) simultaneously at different points of the surface of the Earth. The issue of common biases when analyzing data taken at sites which are separated by 500 - 2000 km *only* is not completely resolved. A combination of different space techniques (of SLR and GPS in this particular case) will help to understand and resolve the problems. This aspect does in particular affect the estimated station heights.
- GPS makes extensive use of the gravity field of the Earth as it was established by SLR. If GPS receivers on low Earth orbiters become routinely available there will also be GPS contributions in this area. Modeling of non-gravitational forces for such satellites (which certainly will *not* be *canon ball satellites*) still will be problematic, however.

These aspects should be sufficient to remind ourselves that space geodesy does not take place *in a single spectral line*. A combination of all methods is mandatory and will eventually give most of the answers we would like to have in geodynamics.

14.2 THE PARTIAL DERIVATIVES OF THE GPS OBSERVABLE WITH RESPECT TO THE PARAMETERS OF GLOBAL GEODYNAMICS

Let us remind ourselves of the transformation between the Earth fixed and the celestial coordinate systems (section 2.2.2 eqn. (2.21)) and apply it to the coordinates of a station on the surface of the Earth $(\mathbf{R}'')$ and in space ($\mathbf{R}$):

$$\mathbf{R}'' = R_2(-x) \cdot R_1(-y) \cdot R_3(\theta_a) \cdot N(t) \cdot P(t) \cdot \mathbf{R} \tag{14.1a}$$

$$\mathbf{R} = P^T(t) \cdot N^T(t) \cdot R_3(-\theta_a) \cdot R_1(y) \cdot R_2(x) \cdot \mathbf{R}'' \tag{14.1b}$$

where θ_a is the Greenwich apparent siderial time:

$$\theta_a = \theta_m(UTC + \Delta UT) + \Delta\psi \cdot \cos\varepsilon \qquad (14.2)$$

where θ_m is the mean sidereal time in Greenwich at observation time
ε is the apparent obliquity of the ecliptic at observation time
$\Delta\psi$ is the nutation in longitude at observation time
ΔUT is UT1-UTC

We refer to Chapter 2, eqn. (2.21) for the other symbols in eqn. (14.1).

To the accuracy level *required for the computation of the partial derivatives* of the GPS observable with respect to the parameters of interest, we may approximate the nutation matrix as a product of three infinitesimal rotations:

$$N(t) = R_1(-\Delta\varepsilon) \cdot R_2(\Delta\psi \cdot \sin\varepsilon) \cdot R_3(-\Delta\psi \cdot \cos\varepsilon) \qquad (14.3)$$

Introducing this result into equation (14.1b) we obtain the following *simplified* transformation equation *which will be used for the computation of the partial derivatives* only:

$$\mathbf{R} = P^T(t) \cdot R_1(\Delta\varepsilon) \cdot R_2(-\Delta\psi \cdot \sin\varepsilon) \cdot R_3(-\theta_m) \cdot R_1(y) \cdot R_2(x) \cdot \mathbf{R}'' \qquad (14.4)$$

The global geodynamic parameters accessible to the GPS are all contained in eqn. (14.4). It is our goal to derive expressions for the partial derivatives of the GPS observable with respect to these parameters.

Neglecting refraction effects and leaving out range biases (ambiguities), we essentially observe the slant range d between the receiver position $\mathbf{R}''$ resp. $\mathbf{R}$ at observation time t and the GPS satellite position $\mathbf{r}''$ resp. $\mathbf{r}$ at time t-d/c (where c is the velocity of light). This slant range may be computed either in the Earth-fixed or in the celestial (inertial) coordinate system. Let us use the celestial reference frame subsequently:

$$d^2 = (\mathbf{r} - \mathbf{R})^T \cdot (\mathbf{r} - \mathbf{R}) \qquad (14.5)$$

Let p stand for one of the parameters of interest (e.g., a polar wobble component x or y, (ΔUT) or one of the nutation parameters). From eqn. (14.5) we easily conclude that

$$\frac{\partial d}{\partial p} = -\mathbf{e}^T \cdot \frac{\partial \mathbf{R}}{\partial p} \qquad (14.6)$$

where $\mathbf{e}$ is the component matrix of the unit vector pointing from the receiver to the satellite:

$$\mathbf{e} = (\mathbf{r} - \mathbf{R})/d \tag{14.6a}$$

In eqn. (14.6) we assumed that the partial derivative of the satellite position **r** with respect to the parameter p is zero. In view of eqn. (2.22) this is not completely true – but the assumption is good enough for the computation of partial derivatives.

Equations (14.6) and (14.4) allow it to compute the partial derivatives in a very simple way. Let us explain the principle in the case of the derivative with respect to $\Delta\varepsilon$:

$$\frac{\partial \mathbf{R}}{\partial \Delta\varepsilon} = P^T(t) \cdot \frac{\partial}{\partial \Delta\varepsilon}\left(R_1(\Delta\varepsilon)\right) \cdot R_2(-\Delta\psi \cdot \sin\varepsilon) \cdot R_3(-\theta_m) \cdot R_1(y) \cdot R_2(x)\mathbf{R}''$$

where: $\frac{\partial}{\partial \Delta\varepsilon}\left(R_1(\Delta\varepsilon)\right) = \begin{pmatrix} 0 & 0 & 0 \\ 0 & 0 & 1 \\ 0 & -1 & 0 \end{pmatrix}$

Retaining only terms of order zero in the small angles x, y, $\Delta\psi$, $\Delta\varepsilon$ we may write for eqn. (14.6):

$$\frac{\partial d}{\partial \Delta\varepsilon} = -\mathbf{e}'^T \cdot \begin{pmatrix} 0 \\ R' \\ -R_2' \end{pmatrix} = -e_2' \cdot R_3' + e_3' \cdot R_2' = \frac{1}{d} \cdot \left(R_2' \cdot r_3' - R_{3'} \cdot r_2'\right) \tag{14.7a}$$

where the prime "′" denotes a coordinate in the system referring to the true equatorial system of observation time. In the same way we may compute the partial derivative with respect to the nutation correction in longitude and with respect to ΔUT:

$$\frac{\partial d}{\partial \Delta\psi} = -\frac{1}{d} \cdot \sin\varepsilon \cdot \left(R_3' \cdot r_1' - R_1' \cdot r_3'\right) \tag{14.7b}$$

$$\frac{\partial d}{\partial \Delta UT} = -\frac{1}{d} \cdot \left(R_1' \cdot r_2' - R_2' \cdot r_1'\right) \tag{14.7c}$$

The partials with respect to the components x and y of the pole formally look similar as those with respect to the nutation terms. This time, however, the appropriate coordinate system is the Earth fixed system.

$$\frac{\partial d}{\partial x} = \frac{1}{d} \cdot \left(R_3'' \cdot r_1'' - R_1'' \cdot r_3''\right) \tag{14.7d}$$

$$\frac{\partial d}{\partial y} = \frac{1}{d} \cdot \left(R_2'' \cdot r'' - R_3'' \cdot r_2'' \right) \tag{14.7e}$$

We recognize on the right hand side of eqns. (14.7a-e) the components of the vectorial product of the geocentric station vector with the geocentric satellite vector. The components refer to different coordinate systems, however.

In the above formulae we assumed that all the parameters are small quantities. This is not too far away from the truth. But, let us add, that we might have given more correct versions for the above equations by writing e.g. the nutation matrix as a product of the matrix due to the known a priori model and that due to the unknown small correction. The principle of the analysis – and the resulting formulae – are similar. The above equations are good enough for use in practice.

Until now we assumed that the unknown parameters x, y, etc. directly show up in the above equations. It is of course possible to define *refined* empirical models, e.g., of the following kind:

$$x := x_0 + x_1 \cdot \left(t - t_0 \right) \tag{14.8}$$

Obviously the partial derivatives with respect to our *new* model parameters have to be computed as

$$\frac{\partial d}{\partial x_i} = \frac{\partial d}{\partial x} \cdot \frac{\partial x}{\partial x_i} \quad , \quad \frac{\partial x}{\partial x_0} = 1 \quad , \quad \frac{\partial x}{\partial x_1} = \left(t - t_0 \right) \tag{14.9}$$

Models of type (14.8) are of particular interest for those parameters which are not directly accessible to the GPS (i.e., for ΔUT and nutation parameters).

14.3 GEODYNAMICAL PARAMETERS NOT ACCESSIBLE TO THE GPS

As opposed to VLBI analyses we always have to solve for the orbital elements of all satellites in addition to the parameters of geodynamic interest in satellite geodesy. This circumstance would not really matter, *if we would observe the complete topocentric vector to the satellite* and not only its length or even – as in GPS – its length biased by an unknown constant (initial phase ambiguity).

Let us first formally prove that it is *not* possible to extract ΔUT from GPS observations in practice. This is done by showing that the partial derivatives with respect to ΔUT and with respect to the right ascension of the ascending node are (almost) linearly dependent. Let us assume at present that the orbit is Keplerian (i.e., we neglect all perturbations). Let us furthermore assume that we refer our orbital elements to the true equatorial system at the initial time t of our satellite arc. We may now write the component matrix $\mathbf{r}$ in this equatorial system (compare eqn. (2.8)) as

$$\mathbf{r} = R_3(-\Omega) \cdot \mathbf{r}^*$$

where $\mathbf{r}^*$ are the coordinates of the satellite in an equatorial system which has its first axis in the direction of the ascending node.

The partial derivative of $\mathbf{r}$ with respect to the r.a. of the ascending node may be computed easily:

$$\frac{\partial \mathbf{r}}{\partial \Omega} = \begin{pmatrix} -\cos\Omega & -\sin\Omega & 0 \\ \sin\Omega & -\cos\Omega & 0 \\ 0 & 0 & 0 \end{pmatrix} \cdot \mathbf{r}^* = \begin{pmatrix} -r_1 \\ r_2 \\ 0 \end{pmatrix}$$

The partial derivative of d with respect to the r.a. of the node is thus computed as (compare eqn. (14.6))

$$\frac{\partial d}{\partial \Omega} = \mathbf{e}^T \cdot \frac{\partial \mathbf{r}}{\partial \Omega} = \frac{1}{d} \cdot (R_1 \cdot r_2 - R_2 \cdot r_1) \tag{14.10}$$

A comparison of eqn. (14.10) and (14.7c) clearly proves the linear dependence of the two equations. Of course one might argue that neither of equations (14.10) and (14.7c) are completely correct. For a refined discussion of this problem we would have to consider perturbations in eqn. (14.10) and we would have to take into account the partial derivative of the satellite vector with respect to ΔUT in eqn. (14.7c). We resist this temptation and just state that *in practice it is not possible to estimate* ΔUT *using the usual GPS observables.*

One can easily verify on the other hand that it is possible without problems *to solve for a drift in* ΔUT by adopting a model of the type (14.9):

$$\Delta\mathrm{UT} = d\mathrm{UT}_0 + (d\mathrm{UT}_0)^{(1)} \cdot (t - t_0) \tag{14.11}$$

Thanks to this time dependence, the parameters $(d\mathrm{UT}_0)^{(1)}$ and the r.a. of the node(s) are *not* correlated. This demonstrates that the length of day may very well be estimated with the GPS. This drift parameter would be correlated with a first derivative of the node. But, because we assume that the force model is (more or less) known, there is no necessity to solve for a first derivative of the node. On the other hand, a good celestial mechanic would have no problems to introduce a force (periodic, in W-direction) which would perfectly correlate with $(\mathrm{dUT}_0)^{(1)}$ (?).

What we just showed for ΔUT in essence is also true for the nutation terms: GPS has no chance whatsoever to extract these terms. It is very well possible on the other hand to extract the first derivatives for these parameters. The formal proof follows the same pattern as in the case of ΔUT but it is somewhat more elaborate because two orbital parameters, r.a. of the ascending node and inclination, are involved.

Let us also point out one particular difficulty when going into the sub-diurnal domain with the estimation of the pole parameters x and y. A diurnal signal in polar wobble of the form

$$x = \mu \cdot \cos(\theta + \phi)$$
$$y = \mu \cdot \sin(\theta + \phi)$$

may as well be interpreted as a constant offset in nutation (depending only on ϕ and μ). This fact is well known; it is actually the justification to introduce the ephemeris pole and *not* the rotation pole for the definition of the pole on the surface of the Earth and in space, Seidelmann [1992]. When using simple empirical models of type (14.8) in the subdiurnal domain in GPS analyses, we will run into difficulties even if we do not solve for nutation parameters because we have to solve for the orbit parameters (which in turn, as stated above, are correlated with the nutation parameters). It is thus no problem to generate any diurnal terms of the above type using GPS! Sometimes these terms are even interpreted.

Let us conclude this section with a positive remark: GPS is very well suited to determine the coordinates x and y of the pole – provided that the terrestrial system (realized by the coordinates of the tracking stations) is well defined. Within the IGS this is done by adopting the coordinates and the associated velocity field for a selected number of tracking sites from the IERS, Boucher et al. [1994].

14.4 ESTIMATING TROPOSPHERIC REFRACTION

Tropospheric refraction is probably the ultimate accuracy limiting factor for GPS analyses (as it is for VLBI). The total effect is about 2.3 m in zenith direction, the simplest mapping function (not even taking into account the curvature of the Earth's surface) tells us that the correction $dr(z)$ at zenith distance z is computed as $dr(z) = dr/\cos(z)$.

This means that we are looking at an effect of about 7 m at $z = 70°$, a frightening order of magnitude if we remind ourselves that we are actually trying to model the GPS observable with millimeter accuracy. The situation is critical in particular in global analyses, because, in order to get a good coverage we have to allow for low elevations.

It would be the best solution if the tropospheric zenith correction would be provided by independent measurements. To an accuracy level of a few centimeters this is actually possible using surface meteorological data. Much better corrections (better than 1 cm?) are provided by water vapor radiometers.

But even in this case the corrections available are not of sufficient quality to just apply and forget the effect in GPS analyses. The conclusion for global applications of the GPS is thus clear: one has to solve for tropospheric refraction corrections for each site.

Two methods are used today in global applications of the GPS

(1) Estimation of site- and time- specific tropospheric zenith parameters. A priori constraints may be introduced for each parameter, constraints may also be applied for the differences between subsequent parameters.

(2) The tropospheric zenith correction is assumed to be a random walk in time with a power spectral density supplied by the user (see below). In this case the conventional least squares approach has to be replaced by a Kalman filter technique.

Let us briefly remind ourselves of the principal difference between sequential least squares techniques and Kalman Filter techniques. Let us assume that the set of observation equations at time t_i reads as

$$A_i \cdot \mathbf{x} - \mathbf{y}_i = \mathbf{v}_i \tag{14.12}$$

where A_i is the first design matrix, $\mathbf{x}$ is the parameter array, $\mathbf{y}_i$ is the array containing the terms *observed-computed*, and $\mathbf{v}_i$ is the residuals array for epoch t_i.

In the conventional least squares approach we just compute the contribution of the observations (14.12) to the complete system of normal equations. If we are interested in a solution at time t_i using all observations available up to that time, we may use the algorithms developed in sequential adjustment calculus (the roots for such procedures go back to C.F. Gauss, the motivation was to save (human) computation time at that epoch). These algorithms allow us to compute the best estimate $\hat{\mathbf{x}}_{i+1}$ and the associated covariance matrix Q_{i+1} at time t_{i+1} using all the observations up to time t_{i+1} in a recursive way:

$$\begin{aligned} \hat{\mathbf{x}}_{i+1} &= \hat{\mathbf{x}}_i + K \cdot \left(\mathbf{y}_{i+1} - A_{i+1} \cdot \hat{\mathbf{x}}_i\right) \\ Q_{i+1} &= Q_i - K \cdot A_{i+1} \cdot Q_i \end{aligned} \tag{14.13}$$

where the gain matrix K is computed as

$$K = Q_i \cdot A_{i+1}^T \cdot \left(\mathrm{cov}(v_i) + A_{i+1} \cdot Q_i \cdot A_{i+1}^T\right)^{-1} \tag{14.13a}$$

Kalman estimation on the other hand allows for a stochastic behavior of the parameter vector:

$$\mathbf{x}_{i+1} = \mathbf{x}_i + \mathbf{w}$$

where $\mathbf{w}$ is the vector of random perturbations affecting the parameters in the time interval between subsequent observations. The optimal estimation using the same set of observations at time t_i looks quite similar as in the conventional least squares case. The difference consists of the fact that the variance-covariance matrix has to be propagated from time t_i to time t_{i+1} and that it contains an additional term:

$$Q_{i+1}^{i} = Q_i + W \tag{14.14}$$

where W is the covariance matrix of the random perturbations vector $\mathbf{w}$.

We assume W to be a diagonal matrix with

$$W_{ii} = \phi_i \cdot dt \tag{14.14a}$$

where ϕ_i is the power spectral density for the stochastic parameter no. i, dt is the time interval between the previous and the current observation epoch.

From now onwards the Kalman solution follows the same pattern as the conventional sequential least squares solution: The *Kalman gain matrix K* is computed in analogy to eqn. (14.13a):

$$K = Q_{i+1}^{i} \cdot A_{i+1}^{T} \cdot \left(\mathrm{cov}(v_i) + A_{i+1} \cdot Q_{i+1}^{i} \cdot A_{i+1}^{T}\right)^{-1} \tag{14.14b}$$

The best estimate of $\mathbf{x}$ using all observations up to time t_{i+1} and the covariance matrix associated with it read as

$$\hat{\mathbf{x}}_{i+1} = \hat{\mathbf{x}}_i + K \cdot \left(\mathbf{y}_{i+1} - A_{i+1} \cdot \hat{\mathbf{x}}_i\right) \tag{14.14c}$$
$$Q_{i+1} = Q_{i+1}^{i} - K \cdot A_{i+1} \cdot Q_{i+1}^{i}$$

Let us add that in both, the conventional least squares approach and in the Kalman Filter approach, it is possible to perform a parameter transformation – in principle after each observation epoch:

$$\mathbf{x}_{i+1} = S_i \cdot \mathbf{x}_i + \mathbf{w} \tag{14.15}$$

where $\mathbf{w}$ would simply be a zero array in the least squares case. The only difference again consists of the computation of the propagated variance covariance matrix. Eqn. (14.14) has to be replaced by

$$Q_{i+1}^{i} = S_i \cdot Q_i \cdot S_i^{T} + W \tag{14.15a}$$

More information about sequential adjustment vs. sequential filter estimates may be found in Beutler [1983]. Also, there are many good textbooks on Kalman filtering (see, e.g., Gelb [1974]).

Both approaches, conventional least squares estimates and Kalman estimates have their advantages and disadvantages. Let us list a few characteristics:

- The Kalman approach may be considered to be more general because sequential adjustment is contained in it (in the absence of stochastic parameters).
- Least squares generally is more efficient (as far as computer time is concerned) because epoch specific solutions only have to be performed if they are actually required by the user.
- Kalman techniques have the problem of the initialization phase: the matrix Q_i has to be known initially.
- If the values of the stochastic parameters at epoch t_i are of interest we have to take into account that the Kalman estimates at time t_i are not optimal because they do not take into account the measurement at times t_k, $k > i$. This may be problematic for the stochastic parameters in particular. Theoretically *optimal smoothing* would solve the problem. The technique is time-consuming, on the other hand. In practice a backwards Kalman filter step may be added, see e.g. Herring et al. [1990], a technique which is also quite elaborate.
- A general Kalman scheme is very flexible. In principle stochastic properties may be assigned to each parameter type, many different error sources actually showing up in practice may be dealt with separately. The problem *only* consists of specifying appropriate variances for all these stochastic variations.
- In conventional least squares algorithms there are no stochastic parameters. The effects which are described by one stochastic parameter in the case of a Kalman filter have to be described by many (certainly more than one) parameters (e.g., by one or more polynomials) in the conventional approach. If there is a high frequency component in the effect to be modeled (as it supposedly is the case for tropospheric refraction) this noise has to be interpreted as measurement noise, and the observations have to be weighted accordingly.
- The effect of an increased number of parameters in the case of conventional least squares adjustment may be reduced considerably by making use of the fact that for one specific observation time there is only one parameter of this type active.

This list of characteristics might be made considerably longer. It is a fact, however, that in practice the results of both methods are of comparable quality, *provided* the same statistical assumptions are made (to the extent possible) in both cases. It is our impression that in practice the differences between methods are philosophical in nature.

Those IGS processing centers using approach (2) usually set up between 1 and 12 troposphere parameters per station and day. The consequences for the other parameters (those of interest to geodynamics and geodesy) seem to be rather small. We refer to section 14.8.2 for examples.

14.5 MISCELLANEOUS ORBIT MODELING

As mentioned in Chapter 2 the attitude of GPS space vehicles is maintained by momentum wheels using the information from horizon finders (to let the antenna array point to the center of the Earth) and from Sun-sensors (to guarantee that the y-axis is perpendicular to the direction satellite $\rightarrow$ Sun). Obviously during eclipse seasons attitude control is problematic because the Sun-sensors do not see the Sun if the satellite is in the Earth's shadow. Figure 14.1 illustrates the situation.

According to Bar-Sever [1994], before 6 June, 1994 the rotation about the Z-axis was rather arbitrary during the time of the eclipse, after the exit from the Earth's shadow the satellite was rotated with maximum angular velocity around the Z-axis to get back to the theoretical position. The maximum angular velocity is about 0.12 °/s for GPS satellites. Depending on the actual position of the Y-axis at the end of the eclipse up to about 30 minutes were necessary to bring the Y-axis back to the nominal position. After June 6, 1994 the rotation about the Z-axis during the eclipse phase is more predictable (for most Block II satellites): they rotate at maximum speed with known sense of rotation. The result at first sight is not much better, however: because the maximum rotation rate is not really the same for all satellites, again the Y-axis may be in an arbitrary position after the shadow exit. The advantage of the new attitude control resides in the fact that a determination of the motion during eclipse is more easily possible.

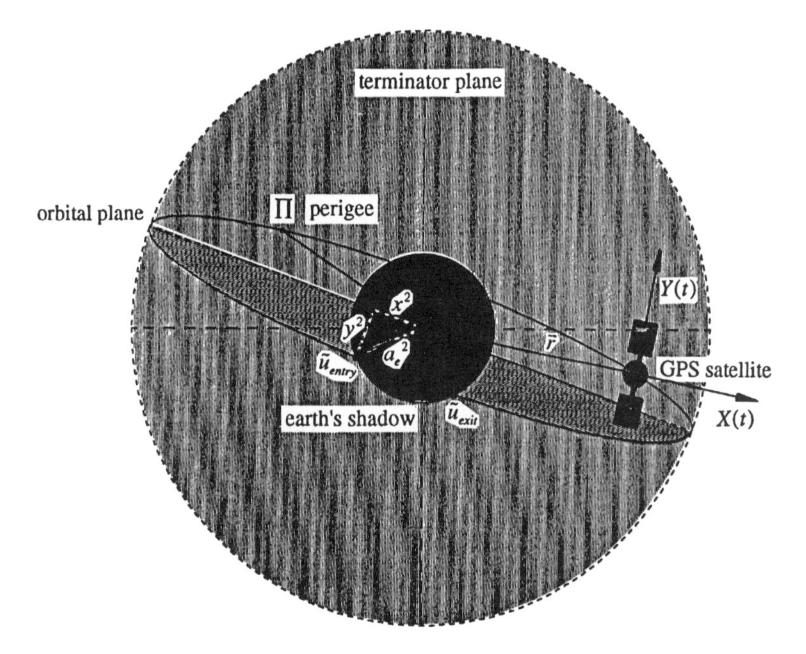

Figure 14.1. Satellite orbit as seen from the Sun.

Two effects should be distinguished: (a) the geometrical effect caused by the rotation of the phase center of the antenna around the satellite's center of mass, and (b) a dynamical effect due to radiation pressure caused mainly by a (possible) serious misalignment of the space vehicle's Y-axis. In principle it is possible to determine the attitude during eclipse seasons using the geometric effect and to apply the dynamical effect afterwards.

There is also a simpler *standard corrective action*, however: one just removes the data covering the time interval of the eclipse plus the first 30 minutes after shadow exit (to get rid of the geometric effect), and one allows for impulse changes in given directions (e.g., in R, S, and W directions, see eqn. (2.30)) at (or near) the shadow exit times. The resulting orbit is continuous, but there are jumps in the velocities at the times of the pulses. The partial derivatives of the orbit with respect to these *pseudo-stochastic pulses* may be easily computed using the perturbation equations (2.30) to relate an impulse change at time t to corresponding changes in the osculating elements at any other time. More information about this technique may be found in Beutler et al. [1994] and the explicit formulae for the partial derivatives are given by Beutler et al. [1995].

The obvious alternative to the introduction of *pseudo-stochastic pulses* for Kalman-type estimators is to declare some of the orbit parameters as stochastic parameters (horribile dictu for a celestial mechanic!) with appropriate (very small) values for the corresponding power spectral densities; the technique is that outlined in the preceding section. Again, both methods lead to comparable result. For a description of stochastic orbit modeling techniques we refer to Zumberge et al. [1993].

There are more arguments for setting up pseudo-stochastic pulses in practice under special circumstances. So-called *momentum dumps* (deceleration of the momentum wheels) at times require small impulse changes performed by the thrust boosters, but there are also other abnormal satellite motions. In practice one just reports *modeling problems* for certain satellites for certain time spans. In the orbit combination performed by the IGS Analysis Center Coordinator impulse changes are set up, if all analysis centers have consistent modeling problems for particular satellites more or less at the same time. This, e.g., often is the case for PRN 23. For an example we refer to Kouba et al. [1995].

Beutler et al. [1994] showed that the Rock4/42 radiation pressure models are not sufficient for long arcs (of 1-4 weeks). An alternative model describing radiation pressure by nine parameters was developed and tested. The radiation pressure was decomposed into three directions, namely the z-direction (pointing from the Sun to the satellite), the y-direction (identical with the space vehicle solar panels axis, the Y-axis) and the x-direction (normal to the z- and y-directions). The parameters are defined as:

$$x(t) = x_0 + x_c \cdot \cos(u + \phi_x)$$

$$y(t) = y_0 + y_c \cdot \cos(u + \phi_y) \qquad (14.16)$$

$$z(t) = z_0 + z_c \cdot \cos(u + \phi_z)$$

where u is the argument of latitude of the satellite at time t. The conventional radiation pressure model just optimizes the parameters z_0 and y_0, whereas all nine parameters (x_0, y_0, z_0, x_c, y_c, z_c, ϕ_x, ϕ_y, ϕ_z) are adjusted in the new approach.

Figure 14.2a shows the radial- along track-, and out of plane residuals R, S, and W for PRN 19 of a seven days orbit fit using the Rock4/42 model (and adjusting the two conventional radiation pressure parameters); seven consecutive orbit files of the CODE processing center of the IGS were used as pseudo-observations.

Figure 14.2b gives the residuals using the same data sets but the new radiation pressure model (14.15) instead of the Rock4/42 model. All nine parameters in eqns. (14.15) were adjusted.

Figure 14.2c finally proves that actually PRN 23 had modeling problems in week 787. The orbit is completely unsatisfactory without setting up pseudo-stochastic pulses somewhere in the middle of the arc, Kouba et al. [1995].

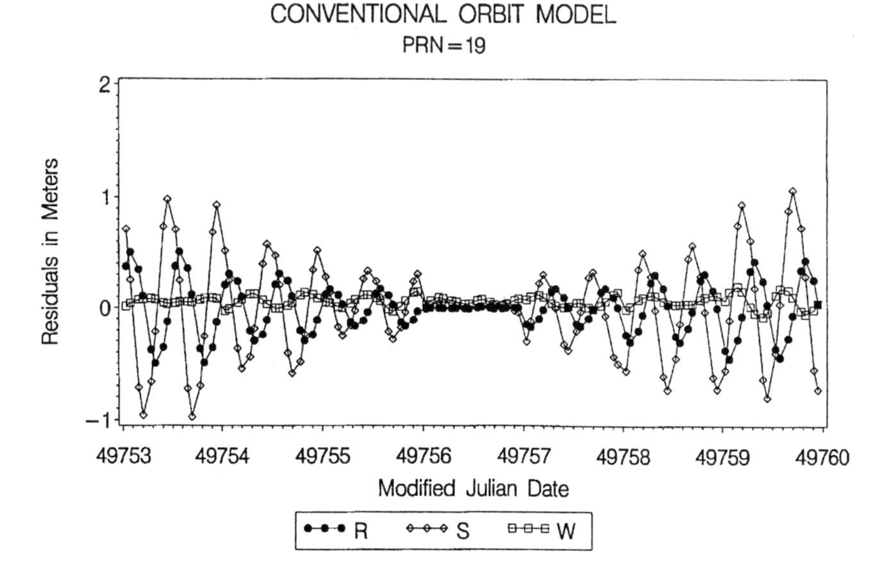

Figure 14.2a. Residuals in radial (R), along track (S), and out of plane (W) direction for PRN 19 using the RPR Model in the IERS Standards. Week 787, 7 files of the CODE Analysis Center used.

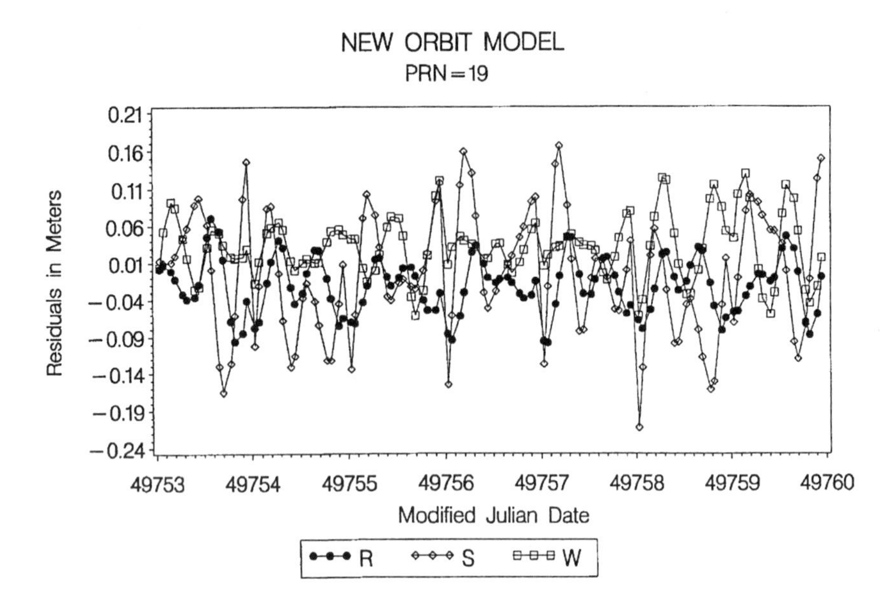

Figure 14.2b. Residuals in radial (R), along track (S), and out of plane (W) direction for PRN 19 using the RPR Model (14.15). Week 787, 7 files of the CODE Analysis Center used.

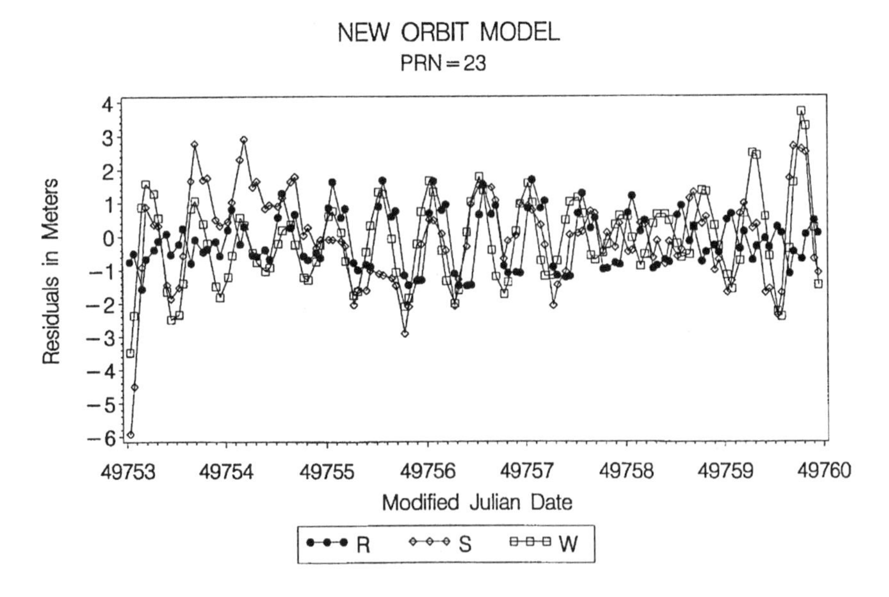

Figure 14.2c. Residuals in radial (*R*), along track (*S*), and out of plane (*W*) direction for PRN 23 using the RPR Model (14.15). Week 787, 7 files of the CODE Analysis Center used.

14.6 SATELLITE- AND RECEIVER- CLOCK ESTIMATION

Four of the seven IGS Analysis Centers, namely EMR, ESA, GFZ, and JPL produce satellite- and receiver- clock parameters in their analyses. These centers include satellite clock estimates in the precise orbit files (i.e., one clock estimate is available every 15 minutes for each satellite). The CODE and the NGS Analysis Centers do not produce satellite clock estimates, but they include the Broadcast-clocks into their precise ephemerides files. No clocks are produced or reported by SIO.

Why this inhomogeneous treatment of clocks by the IGS, where most of the other aspects seem to be so organized? The answer resides in the different processing *philosophies*: those centers producing clock information use so-called *zero difference* procedures, i.e. they essentially solve for one clock parameter for each station and each epoch, whereas the other centers analyze differences between measurements. It was shown in previous chapters that the clocks need to be known only with a modest accuracy (microseconds instead of fractions of nanosecond) if double differences are analyzed.

Solving for clock parameters really makes sense in global analyses like those performed by the IGS: some of the receiver clocks in the network are of excellent quality (hydrogen masers are, e.g., driving the receivers at Algonquin, Fairbanks, Wettzell, etc). The service to the IGS user community by including clock estimates is considerable: The clocks in the ephemerides files may, e.g., be used to produce excellent single point solutions (decimeter accuracy) using code measurements in very remote areas. Also the implications for time transfer in the (sub-) nanosecond domain are obvious.

It would in principle be easy to produce clock solutions for double difference processing schemes, too. The precise code files, possibly together with the phase files, might be re-processed under the assumption that all parameters (orbits, coordinates, troposphere) except satellite- and receiver-clocks are known from the double difference solution. Such clock solutions would be of a quality comparable to that of the centers using zero difference approaches.

The clock solutions were the only IGS products seriously affected by the Anti-Spoofing AS (turned on permanently basis since end of January, 1994). The reported accuracies today are again of the order of few nanoseconds. The next generation of receivers will allow for even better clock estimates.

14.7 PRODUCING ANNUAL SOLUTIONS

The IGS Analysis Centers turn out one solution for every calendar day. Apart from the NGS all centers base their daily products on more than one day of observations. At CODE we use e.g. three full days of observations. Consequently the satellite orbits made available by CODE through the IGS data centers correspond to the center portion (day) of overlapping three days arcs.

Satellite orbits clearly are day- or arc-specific. The same is true for ambiguity parameters, Earth rotation parameters, and troposphere parameters. Station coordinates, on the other hand, are general parameters in the sense that they show up in all the daily solutions. Each daily solution may be considered as an (independent) estimate of one and the same set of three coordinates (for each station). This statement would be completely true if the Earth were a rigid body. On the accuracy level reached today we have to take into account the motion of the stations. Consequently we have to write each station position $\mathbf{R}''(t)$ at time t as a function of station position and velocity at time t_0:

$$\mathbf{R}_i'' = \mathbf{R}''(t_i) = \mathbf{R}_0'' + (t_i - t_0) \cdot \mathbf{V}_0'' \quad , \quad i = 1,2,\ldots,n \tag{14.17}$$

Provided the part of the normal equation system corresponding to the coordinates $\mathbf{R}''(t)$ of all stations is stored for each day i (let us assume that all the other parameters are pre-eliminated), eqn. (14.17) makes it easy to set up a new normal equation system combining all the daily systems. The new normal equation system does not contain n different sets of coordinates but only one set of coordinates and velocities corresponding to time t_0 as unknowns for each station.

Such stacking procedures are standard in geodesy and need no further explanation. Equation (14.16) demonstrates that variable transformations are possible to a certain extent after the daily solutions. This is an important aspect, because an actual reprocessing starting from the observation equations is virtually impossible in GPS. In this respect SLR and VLBI are in a much better position.

All IGS processing centers developed such stacking capabilities. Usually these procedures include also the *daily* parameters – it would thus theoretically be possible to generate, e.g., new sets of orbits referring to a *new* edition of the ITRF. Nobody does that, because normally the differences are very small and barely noticeable in practice. The procedure usually is rigorously performed for the Earth rotation parameters.

Beutler et al. [1995] showed that it is possible to generalize such techniques to produce a solution combining the normal equation systems from n consecutive days (not overlapping), where the n one-day-arcs are replaced by one n-day-arc for each satellite. The procedure is very flexible and much less time consuming than an actual re-processing. The technique is used in routine production since autumn 1994 at CODE.

14.8 RESULTS

The results presented in this section stem from the CODE Analysis Center. Let us point out that other IGS Analysis Centers produce results of comparable quality. Many figures and the corresponding results are extracted from the CODE contribution to the 1994 IGS Annual Report (in preparation).

14.8.1 Earth Rotation Parameters

Figure 14.3 shows the motion of the ephemeris pole on the surface of the Earth in the Earth-fixed system. There is an obvious improvement in the accuracy of the estimates. Today the accuracy of our daily pole coordinates are believed to be of the order of about 0.2-0.3 mas.

Figures 14.4a,b show the *x* and *y* estimates and the best fitting curves with 8 parameters (offset, drift, periodic term with annual period and with Chandler period; each periodic term characterized by amplitude, phase angle, and period). We have to point out that the time period is still rather short for such analyses. Nevertheless the Chandler period was estimated with 445 days and the annual period with about 336 days. The rms of the fit is 6 mas in both cases.

Figure 14.5a shows the length of day (LoD) estimates before, Figure 14.5b after removal of the terms due to the fixed body tides. These LoD values show an excellent agreement with the values derived from VLBI. It is allowed to conclude

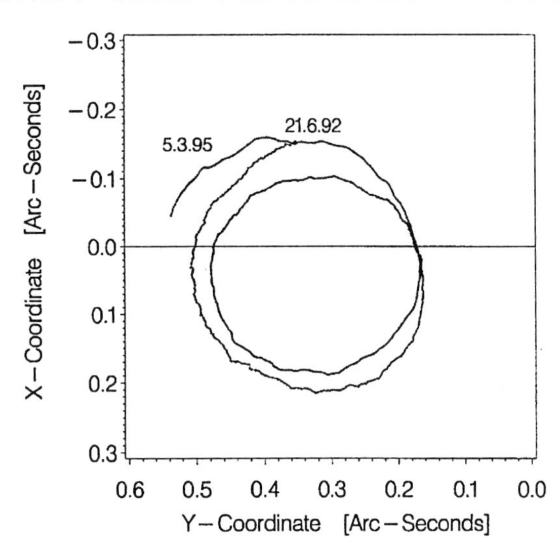

Figure 14.3. Polar motion 21 June 1992 - 5 March 1995 as produced by the CODE Analysis Center of the IGS.

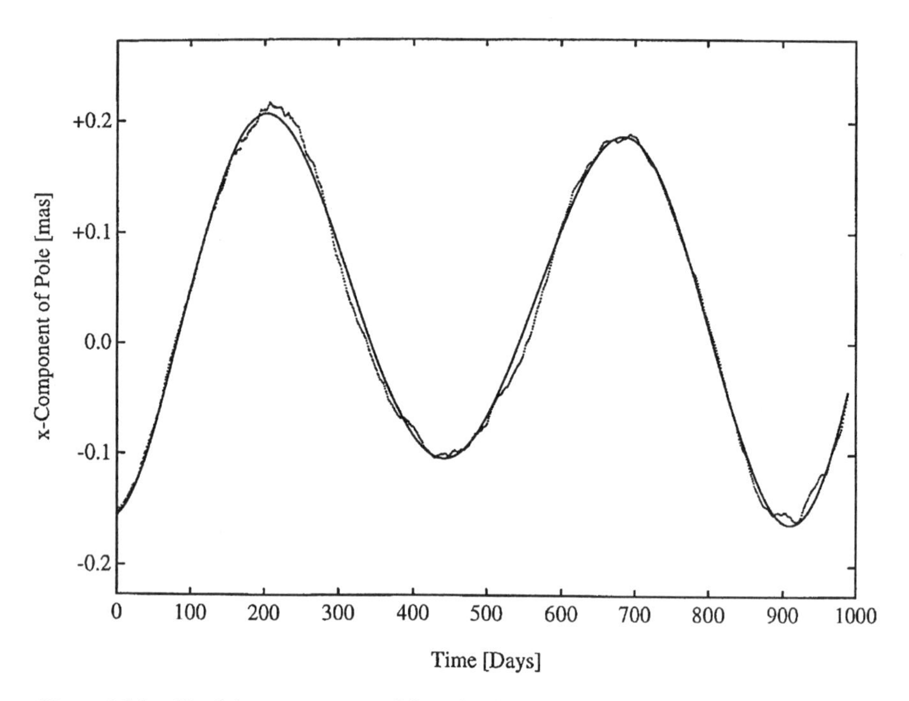

Figure 14.4a. Fit of the x component of the pole (CODE data, 8 parameters adjusted).

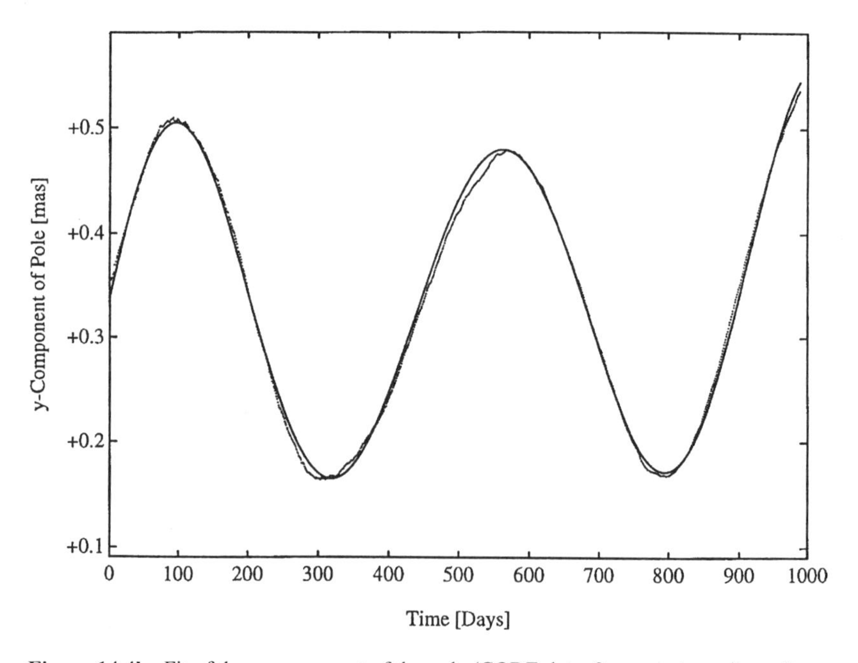

Figure 14.4b. Fit of the y component of the pole (CODE data, 8 parameters adjusted).

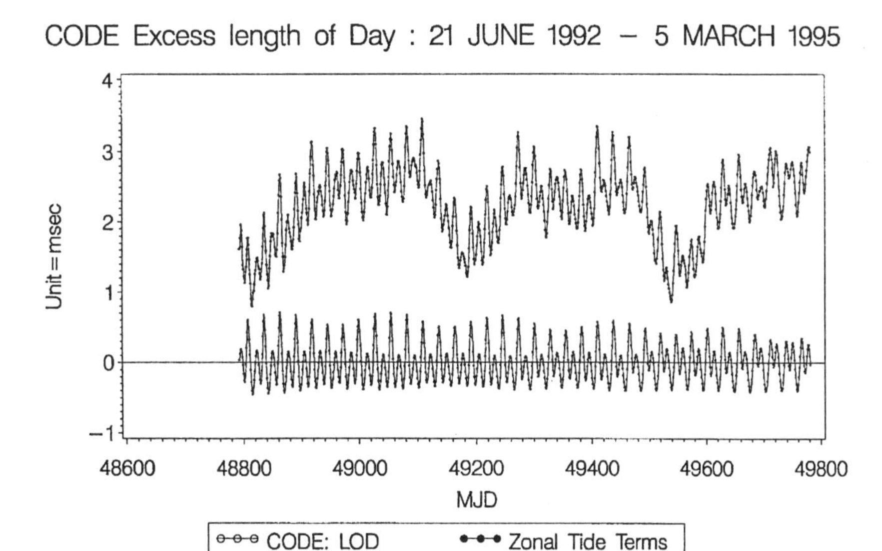

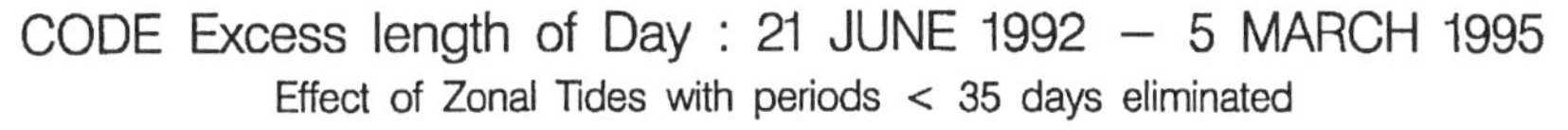

Figure 14.5a. CODE LoD estimates.

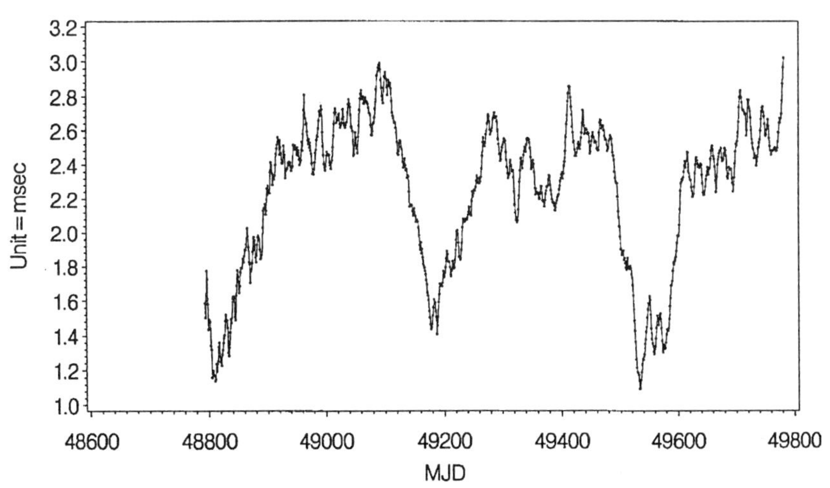

Figure 14.5b. CODE LoD estimates after removal of zonal tides.

that GPS will be an important contributor to the LoD series in future. We believe that the CODE LoD estimates are good to about 0.03 msec/day rms.

The IERS is much more interested on the other hand to have directly ΔUT values available. We already pointed out in section 14.1 that the GPS – as every satellite method *not* including direction measurements – is not able to measure ΔUT directly. It is possible, on the other hand, to sum up the LoD estimates and to produce a ΔUT curve relative to a VLBI-defined initial epoch. The question is simply after what time the GPS derived values start to deviate significantly from the VLBI-curve.

Figure 14.6 shows the result of two such GPS reconstructions relative to the VLBI-curve. In the reconstruction (a) the ΔUT drifts were extracted from one day arcs, in case (b) from three-day-arcs. Obviously the arc length plays an essential role! From Figure 14.6 we also conclude that GPS might be used very well to interpolate ΔUT between – let us say – monthly values established by VLBI.

The difference between the one day and the three day estimates is remarkable. It seems that the change in the reference frame on January 1, 1994 resp. 1995 (change of coordinates and associated velocities of the tracking stations to the ITRF 92 then to the ITRF 93) was of vital importance for the daily estimates (?!).

As pointed out in the introduction to this chapter the GPS so far gave no contribution to the motion of the pole with respect to the celestial reference frame. In section 14.3 we pointed out that GPS is not able to measure nutation directly, but we also found that, as in the case of ΔUT, it should be possible to extract the first derivative of the motion of the celestial pole using the GPS.

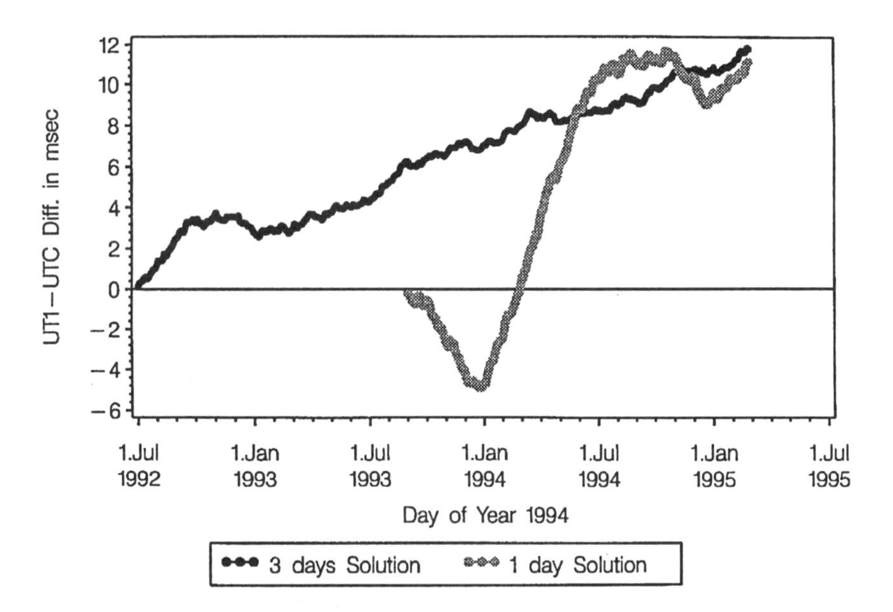

Figure 14.6. ΔUT estimates from CODE one resp. three days solutions relative to VLBI solutions (from the IERS).

At CODE we are routinely solving for drifts in $\Delta\varepsilon$ and in $\Delta\psi$ since January 1, 1994. The accuracy of these daily drifts corresponds to the ΔUT estimates, it is of the order of 0.3 mas/day. These rms values are of course relatively big compared to the expected signals, because we refer our estimates to the IAU model 1980 of nutation. Ideally we should (a) see essentially the same frequencies as VLBI in the spectrum of our estimated drifts and (b) get the same order of magnitude when estimating the relevant terms. We have to take into account that our time interval (approximately 1.2 years) still is very short compared to that available to the VLBI. Figures 14.7a and 14.7b show a frequency analysis of the drifts in $\Delta\varepsilon$ and in $\Delta\psi$ over the time interval of 1.2 years. The $\Delta\varepsilon$-spectrum shows the maxima roughly at the expected places. The corresponding curve for the nutation in longitude is somewhat less convincing – but again the growing time base will cure many problems. We are convinced that the GPS will give essential contributions in the frequency domain between 1 and 60 days in future.

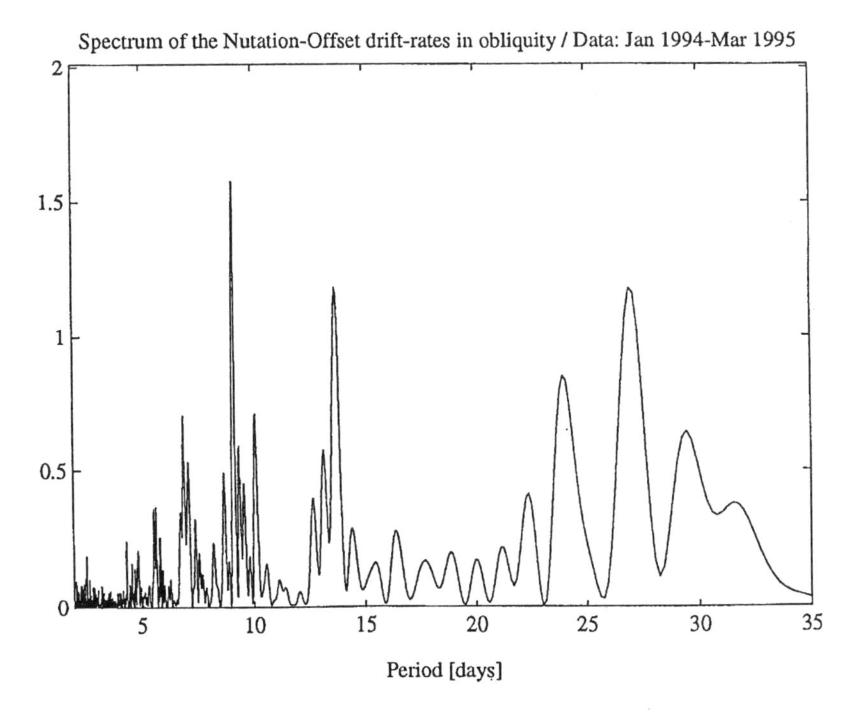

Figure 14.7a. Frequency analysis of the drifts in $\Delta\varepsilon$ as estimated by the CODE processing center.

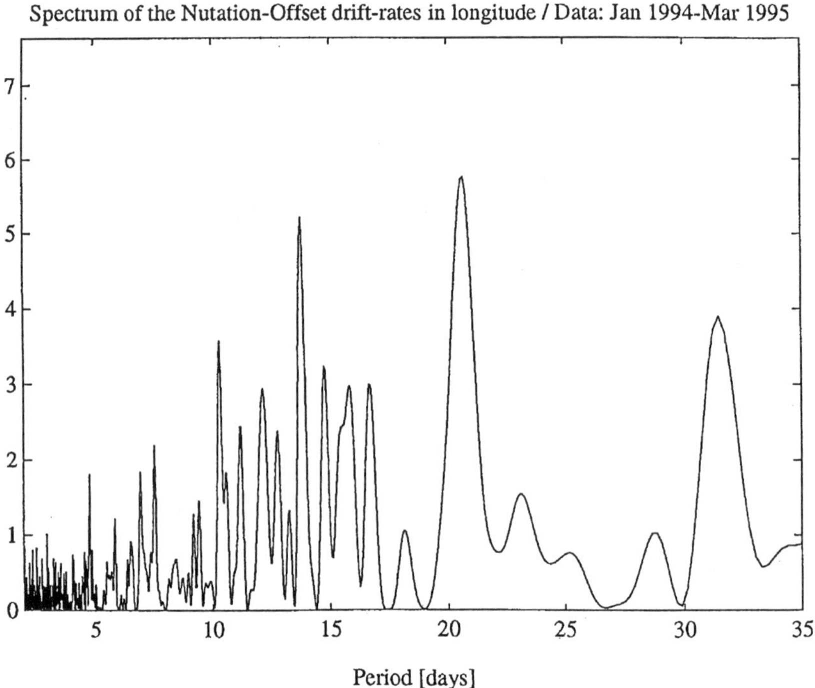

Figure 14.7b. Frequency analysis of the drifts in $\Delta\psi$ as estimated by the CODE processing center.

14.8.2 Troposphere Parameters

Troposphere parameters have to be estimated in GPS global analyses for each daily solution for each station. In the CODE solutions produced for the IGS we introduce one tropospheric zenith delay parameter for each time interval of six hours and for each station. In Figure 14.8a we show the values estimated from GPS for the station of Zimmerwald *and* the tropospheric zenith corrections computed from surface meteorological data gathered at the Zimmerwald observatory (pressure, temperature, and humidity are measured and recorded every 15 minutes). Figure 14.8b contains the corresponding information for the station of Wettzell (about 500 km away from Zimmerwald). It is encouraging to see that, statistically speaking, the GPS derived values and the values derived from the met-sensors are very highly correlated. The mean values of the differences *GPS-Sensor* agree to within 1.5 cm, the rms of the difference is about 2 cm in both cases. This agreement, on the other hand is clearly *not* sufficient to rely on surface met data in Global GPS analyses.

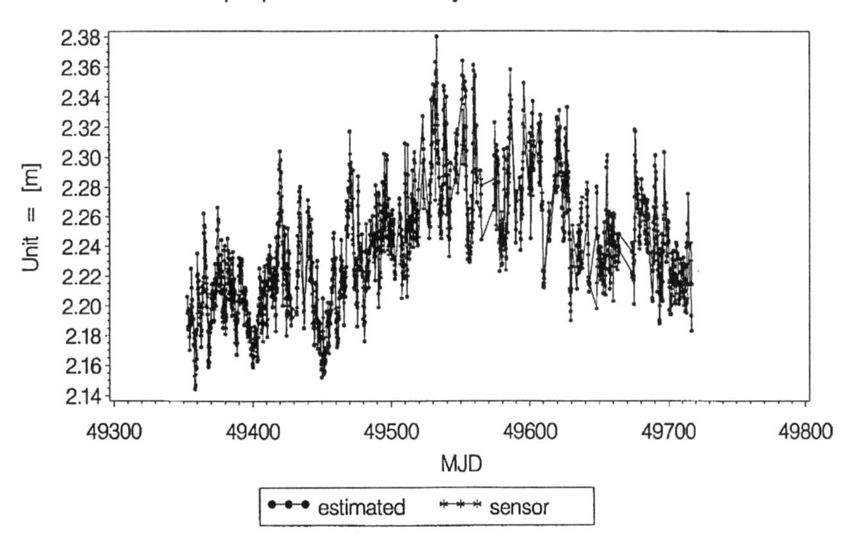

Figure 14.8a. Tropospheric zenith delay estimated in GPS processing and calculated from surface met data for the station Wettzell.

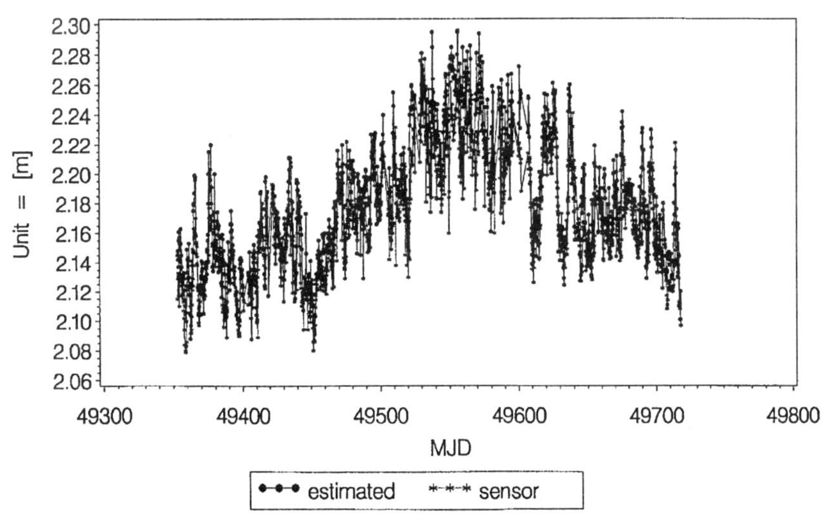

Figure 14.8b. Tropospheric zenith delay estimated in GPS processing and calculated from surface met data for the station Zimmerwald.

The quality is of interest on the other hand for meteorologists. If precise temperature- and pressure- measurements are available at the stations, the wet component of tropospheric refraction may be reconstructed by subtracting the dry component using surface met from the GPS estimates of the total tropospheric refraction. The total precipitable water vapor content of the atmosphere may then be computed from the reconstructed wet tropospheric refraction. This in turn is a decisive quantity for weather forecasts! Figures 14.9a,b show such reconstructed wet tropospheric delays.
According to Bevis et al. [1992] these tropospheric delays have to be divided by about a factor of six to obtain the precipitable water content of the atmosphere.

This particular application of the GPS is still very young. We are convinced that this branch should be systematically explored and that the IGS should start collecting ground met data of excellent quality as soon as possible.

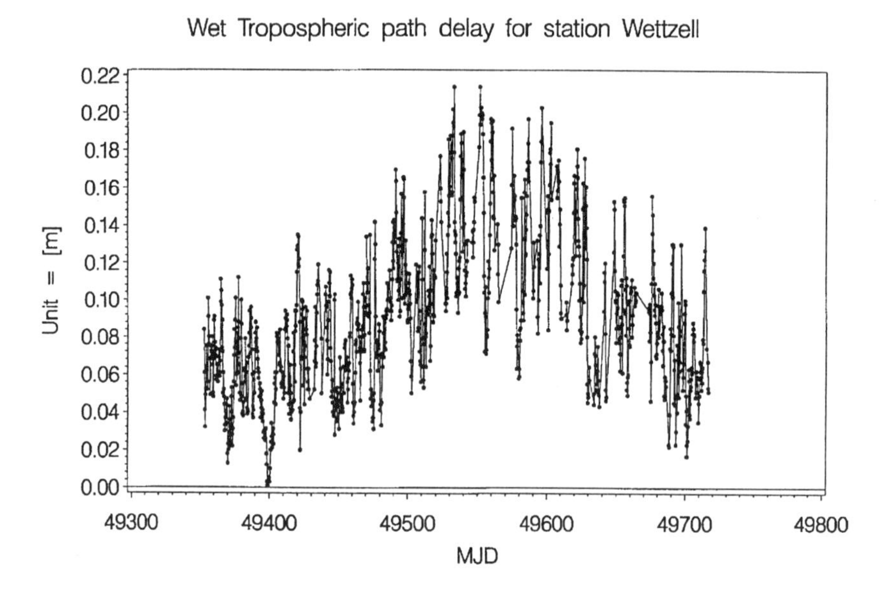

Figure 14.9a. Reconstructed wet tropospheric path delay for Wettzell (Year 1994).

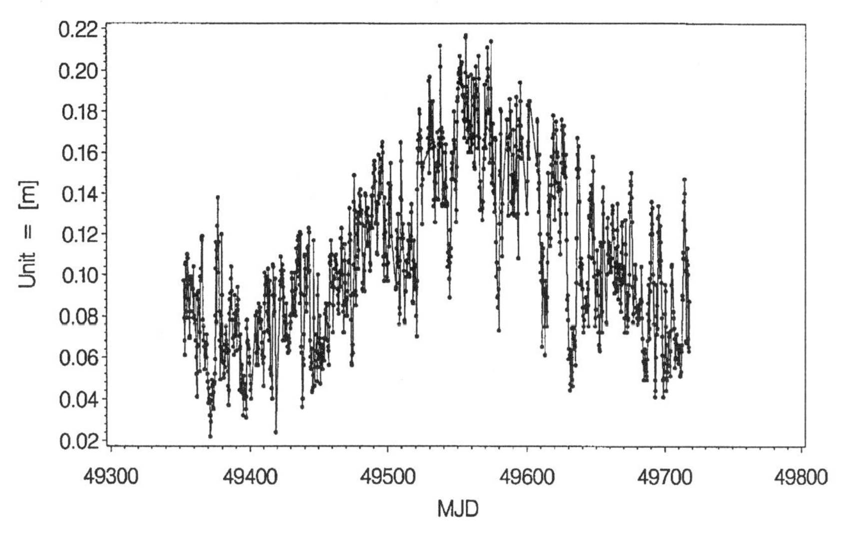

Figure 14.9b. Reconstructed wet tropospheric path delay for Zimmerwald (Year 1994).

14.8.3 Station Coordinates and Velocities

Some of the stations have to be assumed as known (or their coordinates are closely constrained) in the daily solutions of the IGS Analysis Centers. The IGS makes sure that its Analysis Centers use essentially the same terrestrial frame. At present the ITRF93, see Boucher et al. [1994], is used within the IGS by adopting the ITRF93 coordinates and velocities for the 13 stations listed in Table 14.1.

Table 14.1: Stations kept fixed in daily IGS analyses.

Stations kept fixed in the ITRF93			
153	KOSG	13504M003	Europe
154	MADR	13407S012	Europe
156	TROM	10302M003	Europe
157	WETT	14201M009	Europe
351	HART	30302M002	Africa
451	ALGO	40104M002	North America
452	FAIR	40408M001	North America
453	GOLD	40405S031	North America
454	KOKB	40424M004	Hawaii
458	YELL	40127M003	North America
461	SANT	41705M003	South America
551	TIDB	50103M108	Australia
552	YAR1	50107M004	Australia

When combining daily solutions the coordinates and velocities of Table 14.1 have to be estimated in addition to the coordinates of all other stations used in the daily solutions. Such combined solutions usually are called *free network solutions*. This expression is not entirely correct because completely free solutions lead to singularities. It is the responsibility of the IERS to define the terrestrial reference frame when combining the final results of different analysis centers using different techniques. The analysis centers contributing to the ITRF have to make sure that their contributions allow the adoption of the system conditions:

- no net rotation
- no translation
- no scale

for any set of stations the IERS may wish to select in the combined IERS solution. Today essentially VLBI, SLR, GPS contribute to the definition of the ITRF. The French DORIS system is about to start contributing.

Until now, the free network solutions of IGS analysis centers were not compared with the same intensity as e.g. the orbits or the Earth rotation parameters. It was in fact the IERS which compared the annual solutions before producing a combined solution. At the IGS workshop in December 1994 it was decided that in future specialized Associate Analysis Centers will compare and combine these individual IGS solutions on a weekly basis (at least initially). The result might be a combined IGS coordinate and velocity set, which in turn might be considered as the GPS/IGS contribution to the definition of the ITRF. For more details we refer to Zumberge [1995a]. It should be pointed out, however, that the actual definition of the ITRF is a very delicate task asking for the contributions of all space techniques. The IERS clearly has the responsibility to implement this definition.

GPS derived station coordinates and velocities at present are made available by some of the IGS processing centers (e.g., by JPL and SIO).

Figure 14.10 gives the result of a combination of 23 months of data gained at the CODE processing center. It is a very *loose* solution indeed: The coordinates of the stations in Table 14.1 show no net translation and rotation with respect to the ITRF93, and the velocity of the station Wettzell was kept fixed on the ITRF93 values. In Figure 14.10 we can see the GPS derived velocities (arrows) and the official ITRF93 velocities. These velocities seem to be quite well established on the Northern hemisphere, the agreement is not so good in the South. The longer time base really favors the ITRF velocities which today are essentially established by VLBI and SLR. From Figure 14.10 we also conclude, on the other hand, that the GPS contribution starts to become significant in this domain, too.

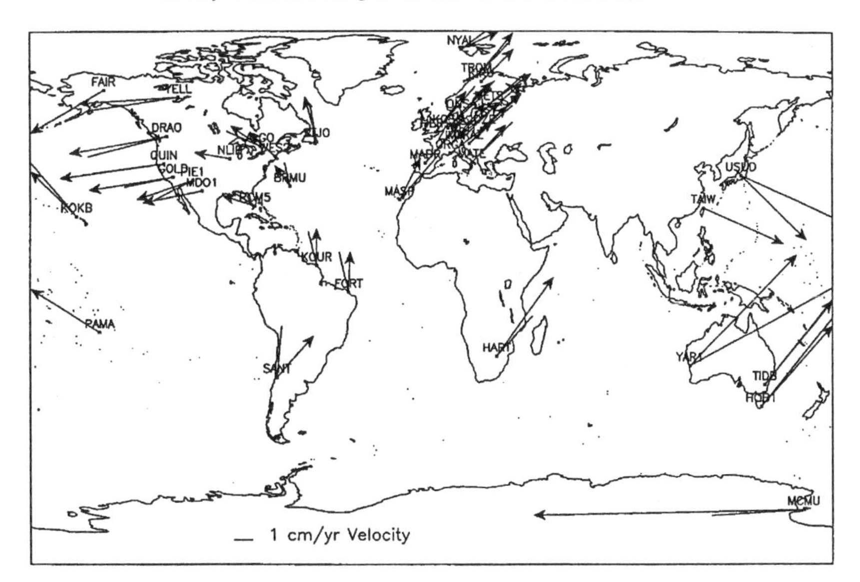

Figure 14.10. Station velocities based on 23 months of CODE results.

14.8.4 The Impact of Ambiguity Resolution on Global GPS Analyses

In regional and global applications ambiguity resolution becomes more and more difficult with the increasing size of the network. Mervart et al. [1994] and Mervart [1995] developed a technique to resolve the ambiguities in the baseline mode even on very long baselines using highly accurate orbits and coordinates of the IGS. A fair percentage of ambiguities may be safely resolved using this technique. What is the benefit?

Ambiguity resolution plays a key role if only short data spans are available. In regional and global applications the effect of ambiguity resolution is less spectacular – just because the *ambiguities free* results are already excellent.

This fact is underlined by Figure 14.11 which shows the rms of Helmert transformation of ambiguities fixed resp. free solution with respect to the *true* coordinates of a European network consisting of 13 stations (BRUS, KOSG, MADR, ONSA, WETT, GRAZ, JOZE, ZIMM, MASP, METS, TROM, MATE, NYAL) using data spans of different lengths (1 hour to 24 hours). Obviously for such applications the coordinate quality becomes comparable after about 8 hours.

Do we have to conclude that ambiguity resolution is unimportant in big permanent tracking networks? Not quite. Whereas the impact is small for the north-south and for the height components the improvement is important in the east-west component. The east-west repeatability of 14 daily solutions was improved by about a factor of two (from about 4 - 8 mm in the ambiguities free to about 2 - 4 mm in the ambiguities fixed case).

Mervart [1995] also reports that ambiguity resolution does significantly strengthen the orbital elements (the semimajor axis, inclination and r.a. of ascending node, and the radiation pressure parameters in particular), whereas only marginal changes could be seen in the troposphere parameters.

An ambiguities resolved solution is being produced in parallel to the officially released solution since October 1994. It will be analyzed in the near future.

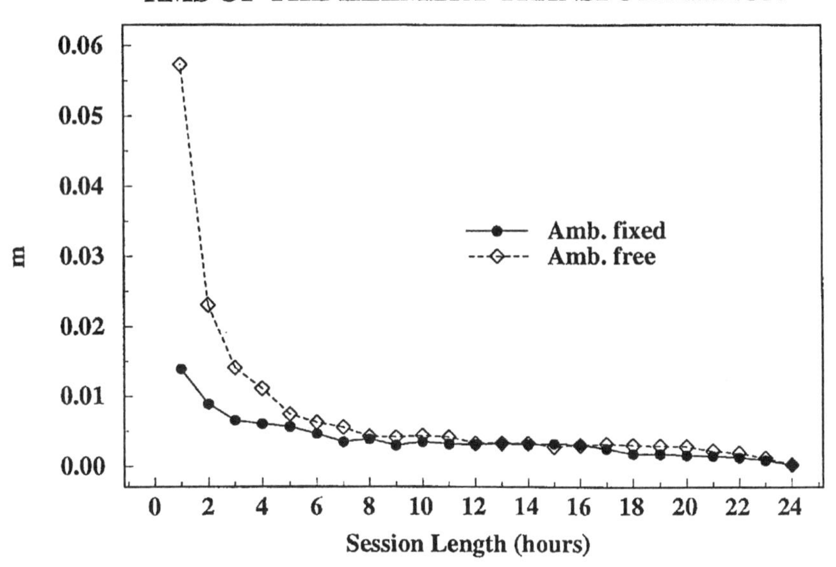

Figure 14.11: rms of 7 parameter Helmert transformations of ambiguities fixed resp. free solutions using short data spans in a European network of 13 stations with respect to a *true* solutions (combining 14 days of observations). Stations BRUS, KOSG, MADR, ONSA, WETT, GRAZ, JOZE, ZIMM, MASP, METS, TROM, MATE, NYAL involved (taken from Mervart [1995])

14.9 SUMMARY AND CONCLUSIONS

In section 14.1 we compared the GPS contribution to that of the other space techniques (VLBI and SLR). We concluded that all techniques give a significant contribution and that only from a combination of all techniques we may expect an answer to all relevant questions in the field of global geodynamics.

In section 14.2 we derived simple expressions for the partial derivatives of the GPS observable with respect to the parameters of geodetic interest. In section 14.3 we saw that it is possible to extract the x and y components of the pole using the GPS (problems only exist if we are moving towards the subdiurnal domain). We showed on the other hand that only the time derivatives of ΔUT and of the nutation terms are accessible to the GPS observable.

In section 14.4 we introduced two different ways of taking into account tropospheric refraction, namely Kalman filter techniques and the conventional technique (introducing time and stationspecific troposphere parameters). We pointed out that in practice both methods lead to results of comparable quality.

In section 14.5 we briefly touched the possibility of (pseudo-) stochastic orbit modeling. Again we made the distinction between conventional and Kalman-type approaches.

In section 14.6 we pointed out that a network like that of the IGS is also well suited to extract receiver and satellite clock information. We concluded that this information is beneficial to the user community and that IGS Analysis Centers using the double difference processing approach should start producing time information, too.

In section 14.7 we gave some clues how the normal equation systems which are produced on a daily basis by the IGS processing centers may be rigorously combined to long-term (e.g., annual) solutions. The establishment of such techniques is of particular importance, because in GPS it is virtually impossible to actually reprocess long global time spans from scratch.

The chapter was concluded with some results of a typical IGS processing center.

Acknowledgements

The author wishes to express his gratitude to Dr. Robert Weber and Dipl.-Astr. Andreas Verdun for their invaluable assistance in editing and writing this chapter. It should be mentioned in particular that Dr. Robert Weber is responsible for analyzing the CODE earth orientation and earth rotation parameters, the nutation parameters, in particular. Section 14.8.1 is based to a large extent on his research results.

Dipl.Ing. Elmar Brockmann developed the theoretical foundations and wrote the programs to combine daily solution to monthly, annual, etc., solutions. Stacking procedures as those briefly mentioned in section 14.7 are the essential tools for his Ph.D. thesis. Dr. Leos Mervart studied the impact of ambiguity resolution on global GPS analyses in his Ph.D. thesis. Section 14.8.4 contains some of his essential findings.

Many of the examples and techniques (the extended orbit modeling techniques in particular) presented are extracted from results acquired by the CODE processing center of the IGS or describe techniques used by CODE. Let me cordially thank Dr. Markus Rothacher, who is the head of CODE, for making the results available and for assisting me to describe the procedures. CODE stands for Center for Orbit Determination in Europe, a joint venture of four European institutions (Astro-nomical Institute University of Bern (Switzerland); Federal Office of Topography (Switzerland); Institute for Applied Geodesy (Germany);

Institut Geographique National (France)); and IGS stands for International GPS Service for Geodynamics.

Last but not least, I would like to thank Ms. Christine Gurtner for the actual typing of the manuscript. Her contribution was essential for the timely completion of the manuscript.

References

Bar-Sever, Y.E. (1994), *"New GPS Attitude Model."* IGS Mail No. 591, IGS Central Bureau Information System.

Beutler, G., (1983), *"Digitale Filter und Schaetzprozesse."* Mitteilung No.11 der Satellitenbeobachtungsstation Zimmerwald, Druckerei der Universitaet Bern.

Beutler, G., E. Brockmann, W. Gurtner, U. Hugentobler, L. Mervart, M. Rothacher, A. Verdun (1994), *"Extended Orbit Modelling Techniques at the CODE Processing Center of the IGS: Theory and Initial Results."*, Manuscripta Geodaetica, Vol. 19, pp. 367-386.

Beutler, G., E. Brockmann, U. Hugentobler, L. Mervart, M. Rothacher, R. Weber (1995), *"Combining Consecutive One-Day-Arcs into one n-Days-Arc."*, Submitted for publication to Manuscripta Geodaetica, October 1994.

Bevis, M., S. Businger, T.A. Herring, Ch. Rocken, R.A. Anthens, R.H. Ware (1992*), "GPS Meteorology: Remote Sensing of Atmospheric Water Vapor using the Global Positioning System."*, Jounal of Geophysical Research, Vol. 97, pp 15'787-15801.

Boucher, C., Altamimi, Z., L. Duhem (1994), *"Results and Analysis of the ITRF93."*, IERS Technical Note, No. 18, October 1994, Observatoire de Paris.

Gelb, A. (1974). *"Applied Optimal Estimation."* MIT Press, Cambridge, Mas.

Herring, T.A., J.L. Davis, I.I. Shapiro (1990), *"Geodesy by Radio Interferometry: the Application of Kalman Filtering to the Analysis of Very Long Baseline Interferometry Data."*

Kouba, J., Y. Mireault, F. Lahaye (1995), "Rapid Service IGS Orbit Combination - Week 0787." *IGS Report No 1578*, IGS Central Bureau Information System.

Mervart, L., G. Beutler, M. Rothacher, U. Wild (1994), *"Ambiguity Resolution Strategies using the Results of the International GPS Service for Geodyanamics (IGS)."*, Bulletin Géodesique, Vol. 68, pp. 29-38.

Mervart, L. (1995), *"Ambiguity Resolution Techniques in Geodetic and Geodynamic Applications of the Global Positioning System."*, PhD Thesis, University of Bern, Druckerei der Universität Bern.

Seidelmann, P.K. (1992), *"Explanatory Supplement to the Astronomical Almanach."*, University Science Books, Mill Valley, California, ISBN 0-935702-68-7.

Zumberge, J.F., D.C. Jefferson, G. Blewitt, M.B. Heflin, F.H. Webb (1993), "Jet Propulsion Laboratory IGS Analysis Center Report, 1992.", *Proceedings of the 1993 IGS Workshop*, Druckerei der Universität Bern.

Zumberge, J.F., R. Liu (1995a), "Densification of the IERS terrestrial reference frame through regional GPS networks". Workshop Proceedings of the IGS Workshop, November 30 - December 2, 1994, Pasadena 1994. Available through the IGS Central Bureau.

Zumberge, J. F. , Liu, R., Neilan, R. E. (eds.)(1995b), IGS 1994 Annual Report. IGS Central Bureau, JPL, CalTech, September 1, 1995.

15. ATMOSPHERIC MODELS FROM GPS

Alfred Kleusberg
Institut für Navigation, Universität Stuttgart, Geschwister-Scholl-Str. 24D, 70174 Stuttgart, Germany

15.1 INTRODUCTION

As discussed in earlier chapters of this monograph, atmospheric refractivity affects GPS measurements by introducing ionospheric and tropospheric delays. If not properly corrected or removed otherwise, these delays will cause position errors. This point of view treats the effects of atmospheric refraction on GPS measurements as a nuisance. The complementary point of view is that the refraction effects in GPS measurements contain useful information accumulated while passing through the atmosphere. In this scenario, the ionospheric and tropospheric delays are seen as remotely sensed data related to atmospheric parameters, with the possibility to recover some or all of these parameters through proper GPS data analysis.

This chapter contains three main sections and a summary at the end. Section 15.2 provides a general understanding of the distribution of atmospheric refractivity in the troposphere and in the ionosphere. The term *troposphere* is used in this chapter interchangeably with the term *neutral atmosphere*. This section also reviews the relation between refractivity and parameters describing the chemistry of the atmosphere. Section 15.3 introduces and discusses the retrieval of vertical refractivity profiles using radio signal occultation methods, and the resulting profiles of atmospheric parameters. Section 15.4 gives an overview of methods for determining the refractivity in the ionosphere and the related models for electron density.

15.2 DISTRIBUTION OF REFRACTIVITY IN THE ATMOSPHERE

The basic relations between refractivity N and atmospheric parameters were derived in Chapter 3. Within the neutral atmosphere this relation is described in an approximation known as the Smith and Weintraub equation (c.f. equation (3.24)):

$$N = 77.6\frac{P}{T} + 3.73 \cdot 10^5 \frac{e}{T^2} \qquad (15.1)$$

Here P denotes atmospheric pressure in units of mbar, e is the partial water vapor pressure in units of mbar, and T is the air temperature in Kelvin. For a more detailed discussion of the relation between refractivity and atmospheric parameters and better approximations, we refer to chapter 3. The first term on the right hand side of equation (15.1) is called the *hydrostatic* delay (often referred to as the *dry delay*), the latter one is called the wet delay. The atmospheric parameters are obviously functions of position, with the primary variability in the vertical direction.

Several models for the vertical profile of tropospheric refractivity have been developed in the past. Such models are valuable for the correction of GPS and other satellite measurements, and additionally they can be used to provide a general overview of the distribution of refractivity in the neutral atmosphere. One such model was developed by Hopfield [1969]. Figure 15.1 shows the wet and dry components of refractivity, and resulting total refractivity, as a function of height. The profiles were calculated from the Hopfield model based on surface meteorological data of 1013 mbar atmospheric pressure, 20 mbar water vapor pressure, and a temperature of 288 K. The refractivity amounts to about 363 at sea level, and decreases rapidly with height, reaching zero level at a height of 42 km. A GPS signal passing vertically through this model atmosphere would suffer a delay of 8.4 ns, equivalent to a range error of 2.5 m.

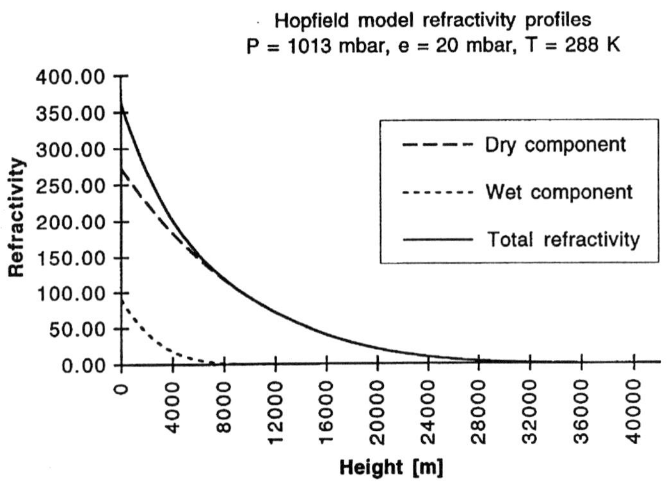

Figure 15.1. Refractivity in the neutral atmosphere.

Obviously, the bulk of the tropospheric refractivity is accumulated in the very low part of the lower atmosphere. This becomes even more obvious, if vertically integrated refractivity is plotted against height, see Figure 15.2. It can be seen that 90% of tropospheric refraction are accumulated in the lowest 15 km of the atmosphere; 99% are accumulated in the lowest 25 km. Although these numbers are based on a refractivity model and not on true refractivity profiles, they nevertheless give a good impression of the general structure of refractivity distribution in the neutral atmosphere.

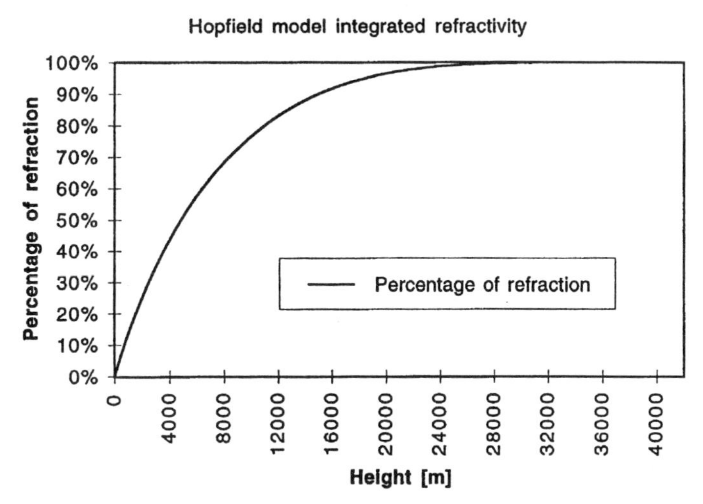

Figure 15.2. Cumulative percentage refraction in the neutral atmosphere.

Within the ionosphere, the relation between the phase refractive index n_ϕ and atmospheric parameters is described by the simplified and approximated Appleton-Hartree formula (c.f. equation 3.42)

$$n_\phi = 1 - 40.3\frac{N_e}{f^2} \tag{15.2}$$

The group refractive index n_P as applicable to the pseudorange measurements is obtained as (c.f. equation (3.44)):

$$n_P = 1 + 40.3\frac{N_e}{f^2} \tag{15.3}$$

Here f is the carrier frequency of the signal propagating through the ionosphere in units of Hz, and N_e is the electron density in units of m^{-3}. Introducing a refractivity N for the ionosphere similar to the neutral atmosphere as the deviation of the refractive index from unity, scaled by the factor 10^6, we obtain

$$N = 40.3 \cdot 10^6 \frac{N_e}{f^2}. \tag{15.4}$$

Electron density in the upper atmosphere is undergoing variations on a number of different time scales. One of the major temporal variations of the ionosphere follows the 11 year solar activity cycle by Hargreaves [1992]. This cycle goes hand in hand with the number of observed sunspots, and is often represented in terms of the so called *sunspot numbers.* Figure 15.3 shows the last four solar cycles as presented by Komjathy et al. [1995].

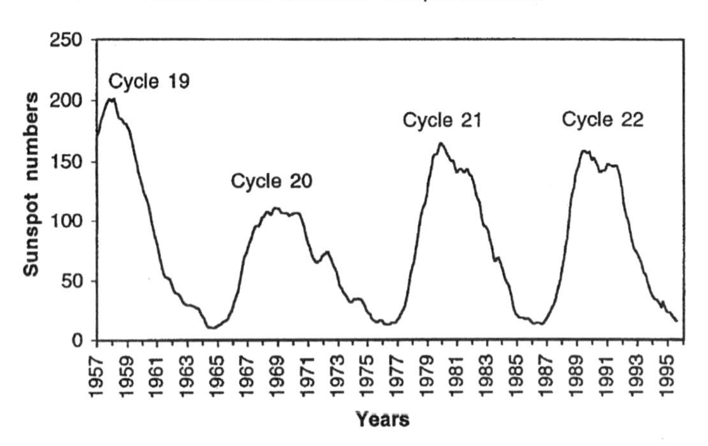

Figure 15.3. The 11 year solar cycle of sunspot numbers; after Komjathy et al. (1995).

Several models for the spatial and temporal variations of the ionosphere including the distribution of the electron density have been developed in the past. One of these models is the *International Reference Ionosphere 1995*, [*IRI-95*], the latest version of the series of IRI models by Bilitza [1993]. IRI represents a smoothly varying mean ionosphere for undisturbed conditions. We will use IRI-95 data to illustrate the main features of the distribution of electron density and the corresponding variations in refractivity.•

The amount of influence of the solar activity onto electron density in the upper atmosphere is clearly visible in Figure 15.4, showing the refractivity (calculated from equation (15.4)) for the altitude of maximum electron density. The electron density values are derived from IRI-95 for January 1 of each year. Comparing to Figure 15.3, the two curves are not only in-phase, but also the relative sizes of the peak values are retained.

The cyclic variation of the refractivity in Figure 15.4 is also reflected in the difference between the two vertical electron density profiles of Figure 15.5. The January 1, 1990 profile for a period of high solar activity shows a peak refractivity value of 27 at an altitude of about 350 km; the profile for January 1, 1995 reaches refractivity values up to 8 at a much lower altitude of about 200 km. GPS L1 signals propagating vertically through these reference ionospheres would be

• The figures showing IRI plots were created with data obtained from the IRI web site at http://nssdc.gsfc.nasa.gov/space/model/models/iri.html

affected by signal delays equivalent to 11.2 m (January 1990) and 2.1 m (January 1995).

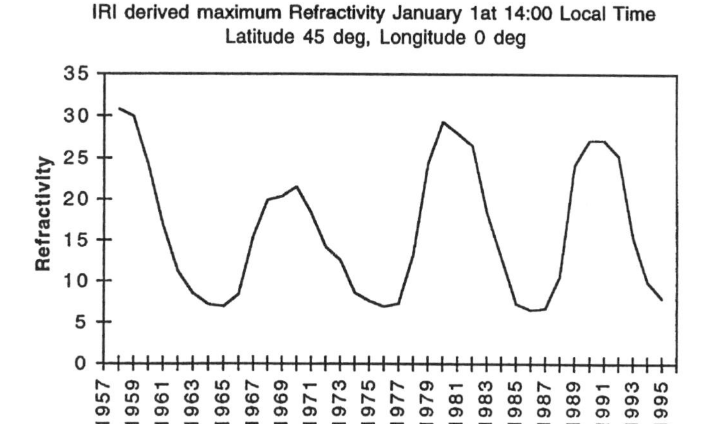

Figure 15.4. Maximum refractivity variation during four solar activity cycles.

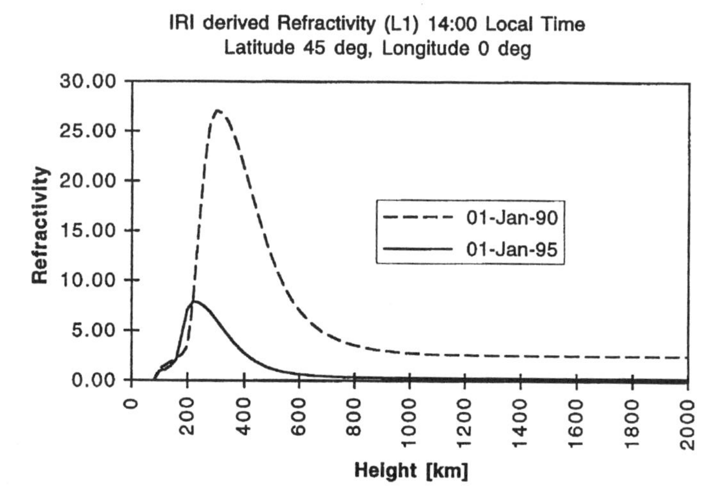

Figure 15.5. Vertical refractivity profiles for periods of high and low sunspot numbers.

At this point it is also interesting to briefly compare the broad features of the ionospheric refractivity profiles with the vertical refractivity in the neutral atmosphere as displayed in Figures 15.1 and 15.2. It is obvious that even during periods of high solar activity the ionospheric refractivity remains an order of magnitude lower than the tropospheric refractivity. However, whereas in the neutral atmosphere the refractivity is concentrated in a thin shell of a few tens of km thickness, the bulk of the ionospheric refractivity is spread throughout an area several hundreds of km thick.

In addition to the 11 year cycle related to the radiation output of the sun, there are two more significant cycles in ionospheric refractivity related to radiation intake in the upper atmosphere of the earth. The first of these is the seasonal variation, reflecting the annual motion of the earth around the sun. Typically, electron density levels are in the winter higher than in the summer. This *seasonal anomaly* is somewhat unexpected since radiation intake is actually higher in the summer months, see Hargreaves [1992].

Figure 15.6 shows mid IRI derived northern latitude vertical refractivity profiles for January 1 and July 1 of the year 1990, a year of high solar activity. The peak density value in the summer is about one half of the winter value. In this particular example the altitudes of the peak densities are almost the same. A GPS L1 signal propagating vertically through the reference ionosphere for July 1 would be affected by a signal delay equivalent to 6.9 m as compared to 11.2 m on January 1.

The third main ionospheric activity cycle results from the rotation of the earth. The rotation meant here is with respect to the sun (not the stars), having therefore a period of a solar day. Typically, the electron density will follow the incident solar radiation with some delay, having a maximum in the early afternoon (Local Time) and reaching a minimum after mid-night. Figure 15.7 shows IRI derived vertical refractivity profiles for 2:00 and 14:00 Local Time in northern mid latitudes for January 1, 1995, a time of low solar activity. Peak refractivity values at 2:00 are about one quarter of the value for 14:00. A GPS L1 signal propagating vertically through the reference ionosphere at 2:00 would be affected by a signal delay equivalent to 0.6 m as compared to 2.1 m at 14:00 Local Time. A characteristic feature is also the disappearance of electrons from the lower layers of the ionospheric region.

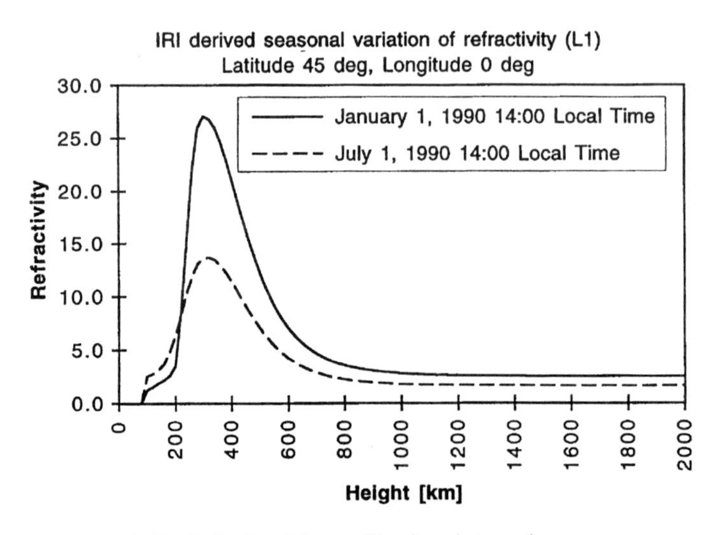

Figure 15.6. Vertical refractivity profiles for winter and summer.

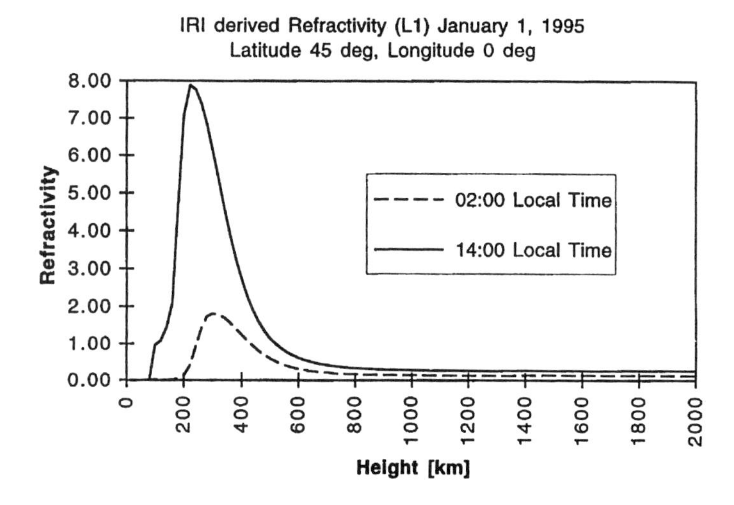

Figure 15.7. Vertical refractivity profiles for early morning and early afternoon.

The diurnal variation of the electron density is to a large degree an artifact of the point of view chosen to represent the ionospheric variability for a fixed point of the earth. Since the diurnal variation results primarily from the changing levels of solar radiation received in the higher atmosphere, the main features of the ionosphere will remain at a fixed orientation with respect to the sun. In other words, looking at refractivity profiles for two different locations along a geodetic parallel at the same local solar time, most of the differences seen will disappear. In a coordinate system based on latitude and local time (or latitude and hour angle of the sun), the diurnal temporal variations shown in Figure 15.6 will be appearing as spatial variations of the ionosphere.

Once more we will look back to the refractivity of the neutral atmosphere discussed earlier. Comparing the refractivity shown in Figure 15.7 for 2:00 Local Time to the tropospheric refractivity shows the following relation. The maximum ionospheric refractivity for a GPS L1 signal during a low in the solar cycle and a low in the diurnal cycle is less than 1% of the refractivity at the surface of the earth. Also, during such periods the signal delays caused by the ionosphere can be considerably smaller than tropospheric delays.

15.3 VERTICAL PROFILES OF REFRACTIVITY FROM RADIO SIGNAL OCCULTATION

Several satellite missions to the planets of the solar system have been used in the past to study the composition of the planetary atmospheres by means of radio occultation experiments, see Fjeldbo et al. [1971]. Such methods are based on deriving the refractive index of the atmosphere from perturbations of radio signals transmitted by the space probes while the satellite is moving around the planet. In the early years of GPS, Yunk et al. [1988], proposed to use GPS measurements obtained with a receiver on a Low Earth Orbit (LEO) satellite to study the earth's atmosphere in a similar manner.

15.3.1 Principles of Refractivity Recovery from Signal Occultation

In this section we will derive the principles of the recovery of the refractive index of the atmosphere from radio occultation experiments, following primarily the ideas developed by Fjeldbo et al. [1971]. A more comprehensive description can be found in Melbourne et al. [1994] and Hoeg et al. [1995].

In the absence of a refractive medium, a signal transmitted at point T and received at point R would propagate along the straight line path indicated by the unit vector $\mathbf{e}$ in Figure 15.8. In the presence of the atmosphere, the signal will propagate along a curved path with tangent lines in the direction of $\mathbf{e}_t$ at the transmitter and $\mathbf{e}_r$ at the receiver. As a result, there will be an excess Doppler shift Δf observed at R, which is often referred to as the *atmospheric frequency perturbation*, see Fjeldbo et al. [1971]. Δf can be expressed as a function of the three unit vectors shown in the figure, and the velocities of the transmitter, $\mathbf{v}_t$, and the receiver, $\mathbf{v}_r$, according to

$$\Delta f = \frac{f}{c}\left[\left(\mathbf{v}_t \cdot \mathbf{e}_t - \mathbf{v}_r \cdot \mathbf{e}_r\right) - \left(\mathbf{v}_t \cdot \mathbf{e} \;\; - \mathbf{v}_r \cdot \mathbf{e} \;\right)\right]$$

$$\Delta f = \frac{f}{c}\left[\mathbf{v}_t \cdot \left(\mathbf{e}_t - \mathbf{e}\right) - \mathbf{v}_r \cdot \left(\mathbf{e}_r - \mathbf{e} \;\right)\right], \tag{15.5}$$

where we have used (·) to indicate the scalar product of two vectors. Under the assumption of a spherically symmetric atmosphere, the atmospheric frequency perturbation can be related to the angle of intersection between the two tangent lines, γ.

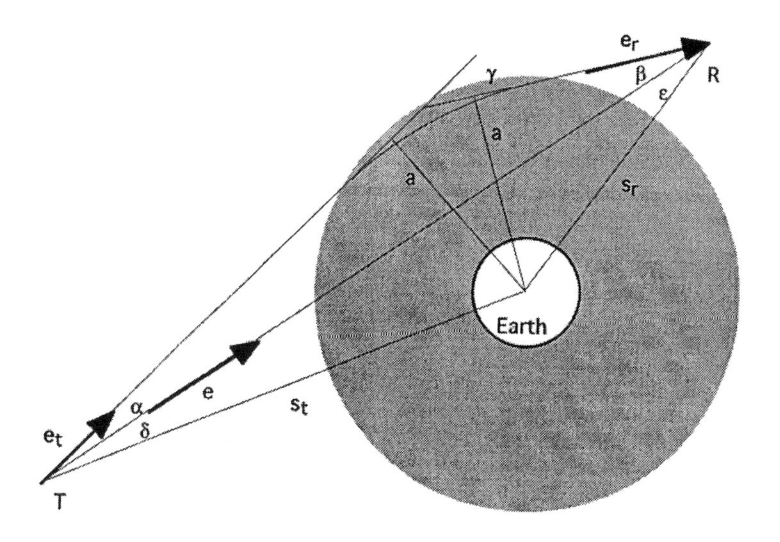

Figure 15.8. Geometry of signal occultation in the atmosphere.

Given that the positions and the velocities of the transmitter and receiver are known with respect to a geocentric coordinate system, the unit vector **e**, the distances s_t and s_r, and the angles δ and ε can be derived. This leaves on the right hand side of equation (15.5) two unknown quantities: the angles α and β between the unit vectors **e** and $\mathbf{e}_t$, and **e** and $\mathbf{e}_r$, respectively. Symmetry requires that the ray path tangent lines have equal distance, a, to the geocentre. This gives rise to the equation

$$s_t \cdot \sin(\alpha + \delta) = s_r \cdot \sin(\beta + \varepsilon) \tag{15.6}$$

which together with equation (15.5) can be solved for the angles α and β. The angle γ is then calculated as the sum of α and β, and the distance of the ray path tangent lines from the geocentre (the asymptote miss distance) is equal to

$$a = s_r \cdot \sin(\beta + \varepsilon)\,. \tag{15.7}$$

As the geometry of the transmitter and receiver changes, the ray path will descend into the atmosphere. This will yield a sequence of atmospheric frequency perturbations, which can be related to a sequence of angles γ and distances a, which can be expressed by

$$\gamma = \gamma(a)\,. \tag{15.8}$$

In the next step, the variation of this angle will be related to the refractive index of the atmosphere, which is assumed to be spherically symmetric in the region under consideration. Figure 15.9 shows a close-up of the resulting symmetric path bending geometry. The distance between the geocentre and the point of closest approach of the ray is denoted by r_0, any other point along the ray path can be identified by its polar coordinates r and ϕ.

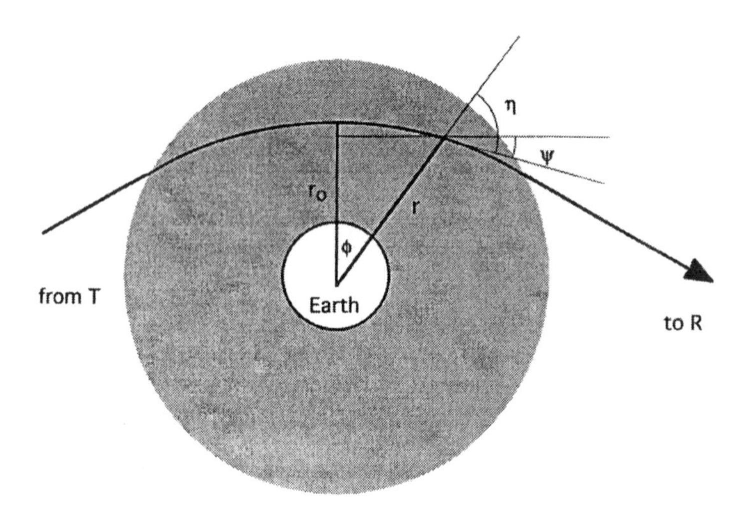

Figure 15.9. Geometry of symmetric ray path bending.

The angle ψ between the ray path tangent at a point (ϕ, r) and the line perpendicular to r_0 is the change in ray path direction between the point of closest approach and the point (ϕ, r). Given the symmetry assumption for the ray path, then the total angular deflection angle g (cf. Figure 15.8) is given by

$$\gamma = 2 \int_{r=r_0}^{\infty} d\psi \tag{15.9}$$

From the sum of the three angles in the plane triangle the relation between the differentials of ϕ, η, and ψ can be established.

$$d\psi = d\eta + d\phi \tag{15.10}$$

To establish the differential $d\eta$ and $d\phi$, we observe the following differential relation in the plane triangle:

$$r\frac{d\phi}{dr} = \tan\eta \tag{15.11}$$

A second relation can be found from the *formula of Bouger*, see Born and Wolf [1965, p. 123] which states that for all points along a particular ray path in a spherically symmetric refractive medium the triple product of the refractive index n, the distance r, and the sine of the angle h is constant.

$$n \cdot r \cdot \sin\eta = const. \tag{15.12}$$

The size of the constant on the right hand side of the equation can be easily found by considering a point of the ray path outside of the refractive medium. By definition, there the refractive index is unity, and the constant is found to be equal to the asymptote miss distance, a discussed above. Using this constant, we can derive from the formula of Bouger

$$\tan\eta = \frac{a}{\left[(nr)^2 - a^2\right]^{1/2}} \tag{15.13}$$

which can be set equal to the left hand side of equation (15.11) to yield an expression for $d\phi$

$$d\phi = \frac{a}{\left[(nr)^2 - a^2\right]^{1/2}} \cdot \frac{dr}{r}. \tag{15.14}$$

Finally, the differential of Bouger's formula can be formulated:

$$d\eta = -\frac{a\left(n + r\frac{dn}{dr}\right)}{\left[(nr)^2 - a^2\right]^{1/2}} \cdot \frac{dr}{nr}. \tag{15.15}$$

Equations (15.14) and (15.15) combined with equations (15.9) and (15.10) then yield a relation between the total angle of refraction γ, the asymptote miss distance a, and the vertical profile of the refractive index $n(r)$.

$$\gamma = -2a \int_{r=r_0}^{\infty} \frac{a}{\left[\left(n(r)\cdot r\right)^2 - a^2\right]^{1/2}} \cdot \frac{1}{n(r)} \cdot \frac{dn}{dr} \cdot \frac{dr}{r} \tag{15.16}$$

This is a *Volterra type integral equation*, the solution of which, the natural logarithm of the vertical profile of the refractive index, is obtained through an integral transform, see Fjeldbo et al. [1971].

$$\ln(n(a)) = \frac{1}{\pi} \int_{x=a}^{\infty} \frac{\gamma(x)}{\left(x^2 - a^2\right)} dx \tag{15.17}$$

If the function $\gamma(x)$ is known from the observations of atmospheric frequency perturbations, equations (15.5) - (15.8), then equation (15.17) can be used to compute a vertical profile of the refractive index. The kernel of the integral is unbounded for $x = a$. This is unfavorable from a numerical point of view, and can be overcome by further evaluating the right hand side of equation (15.17) through integration by parts. Noting that

$$\left[\gamma(x)\cdot \ln\left(\frac{x}{a} + \left(\frac{x^2}{a^2} - 1\right)^{1/2}\right)\right]_a^{\infty} = 0\,, \tag{15.18}$$

we obtain as the final result

$$n(a) = \exp\left\{-\frac{1}{\pi}\int_a^{\infty} \ln\left(\frac{x}{a} + \left(\frac{x^2}{a^2} - 1\right)^{1/2}\right)\frac{d\gamma}{dx}dx\right\}. \tag{15.19}$$

Comparing to equation (15.17) it is obvious, that the integral kernel is now bounded for $x = a$, and that the vertical derivative of the angle γ is now needed instead of γ itself. The angle γ and its derivative are zero for an asymptote miss distance larger than the upper limit of the atmosphere, putting thereby an upper limit on the integration domain.

15.3.2 The GPS/MET Experiment

On April 3, 1995 the OrbView-1 satellite (also known as MicroLab) was launched into a near circular orbit of about 740 km with an inclination of 70°. Apart from other sensors this satellite carries instrumentation for the GPS/MET experiment, see Hajj et al. [1995]. This instrumentation consists basically of a dual frequency GPS receiver tracking the signals from up to eight GPS satellites.

The antenna points towards the negative direction of velocity of the MicroLab satellite and allows to track signals from GPS satellites while the ray path descends through the atmosphere. Given that precise positions and velocities of the GPS satellites are available, the measurements from non occluded GPS signals, together with measurements from a ground based receiver are used to determine the position, velocity and clock drift of the GPS/MET instrumentation, c.f. Figure 15.10.

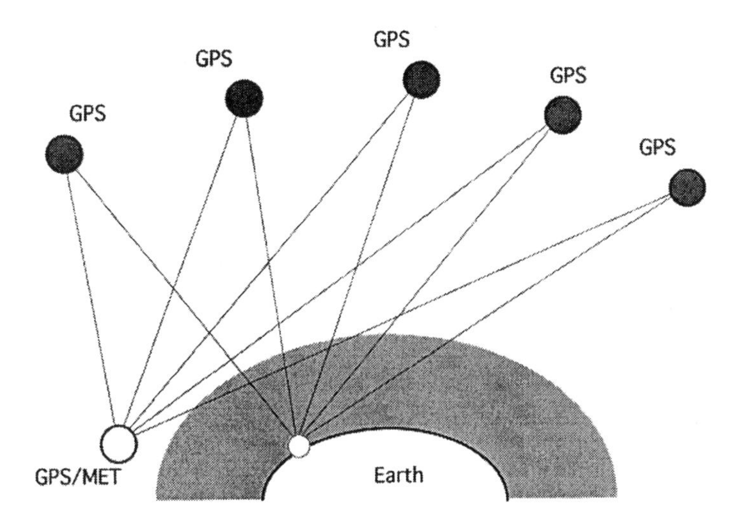

Figure 15.10. Configuration of GPS/MET and GPS satellites.

These positions and velocities determine the right hand side of equation (15.5) up to the two unit vectors $\mathbf{e}_t$ and $\mathbf{e}_r$. The GPS/MET phase measurements to the GPS satellite in occultation together with the corresponding ground based measurements and the GPS/MET clock drift determine the atmospheric frequency perturbation, i.e. the left hand side of equation (15.5).

The primary purpose of the GPS/MET experiment is to monitor the lower neutral atmosphere. In that context, the contribution of the ionospheric refractivity to the signal bending and the corresponding atmospheric frequency perturbation is considered a distortion which needs to be removed. Since the GPS/MET receiver provides dual frequency measurements, it is possible to derive bending angles as a function of ray path asymptotes for both signals L1 and L2 according to equation (15.8):

$$\gamma_1 = \gamma_1(a_1), \quad \gamma_{21} = \gamma_2(a_2) \tag{15.20}$$

These can be combined to (approximately) remove the ionospheric effect, based on the dispersive character of the ionosphere for the GPS frequencies. According to Hajj et al. [1995] the combination

$$\alpha(a) = \frac{f_1^2}{f_1^2 - f_2^2} \cdot \alpha_1(a_1) - \frac{f_2^2}{f_1^2 - f_2^2} \cdot \alpha_2(a_1) \tag{15.21}$$

with $\alpha_2(a_1)$ being the L2 bending function interpolated to the L1 ray path asymptote. Subjecting this corrected ray bending to the mathematical procedure of equation (15.19) gives a profile of the refractive index, which can be directly converted into the refractivity of the neutral atmosphere as described by equation (15.1).

In the upper part of the neutral atmosphere where the contribution to refractivity from water vapor is negligible small (cf. Figure 15.1), equation (15.1) reduces to the hydrostatic delay term

$$N = 77.6 \frac{P}{T} \tag{15.22}$$

which is linearly related to the density of air given by the equation of state for dry air, see, e.g. Hardy et al. [1993].

$$\rho = \kappa \frac{P}{T} \tag{15.23}$$

with a proportionality factor κ. Under these conditions (i.e. dry air), the profile of air density then can be converted

$$\frac{dP}{dh} = -g\rho \tag{15.24}$$

where h denotes height and g is the gravity acceleration. Once the profile of atmospheric pressure has been obtained from equation (15.24), the profile of temperature can be derived from either equation (15.22) or (15.23).

The situation is more complicated in the lower neutral atmosphere, where a significant contribution to the atmospheric refractivity results from the second term on the right hand side of equation (15.1) , the so called wet delay term. The profile of refractivity obtained from GPS/MET occultation data cannot separate between the dry and wet delay terms, and therefore cannot determine simultaneously profiles of atmospheric pressure, temperature and water vapor pressure. In this case, additional information provided by other types of instrumentation is needed.

15.4 MODELS FOR THE IONOSPHERE

Dual frequency GPS receivers provide a direct measurement of integrated electron density. We may recall from chapter 5 the model for the inter frequency difference of pseudorange and carrier phase measurements, repeated here for convenience in a simplified manner.

$$P_{i,2}^{k} - P_{i,1}^{k} = C_{i,p}^{k} + \left(I_{i,2}^{k} - I_{i,1}^{k}\right) + e_{i}^{k} \tag{15.25}$$

$$\Phi_{i,2}^{k} - \Phi_{i,1}^{k} = C_{i,\Phi}^{k} - \left(I_{i,2}^{k} - I_{i,1}^{k}\right) + \varepsilon_{i}^{k} \tag{15.26}$$

Here the left hand side of the equations are the inter frequency differences between simultaneous pseudorange (P) or carrier phase (Φ) measurements. Subscript i identifies the receiver, superscript k identifies the satellite, and subscripts 1, 2 identify the frequency.

The model for the pseudorange difference consists of the term C_p, the inter frequency ionospheric group delay difference (I_2-I_1), and an error term e. The first term results from delay differences of the L1 and L2 signal in the satellite and receiver hardware and is assumed to be constant over time for a particular satellite and receiver pair, even for different passes of the same satellite. The last term contains primarily pseudorange measurement noise and multipath errors.

The model for the carrier phase difference consist of the term C_Φ, the negative inter frequency ionospheric group delay difference, and the error term ε. The first term consists here of the sum of hardware delays and linear combinations of carrier phase ambiguities (cf. equation (3.49)). This term remains constant as long as a receiver keeps lock to the signals; any phase discontinuity (cycle slip) will introduce a discontinuity into this term. Typically, the term assumes different values for measurements of subsequent passes of the same satellite. The last term contains the carrier phase measurement noise and multipath errors.

Both equations (15.25) and (15.26) relate GPS measurement to signal delays resulting from the ionosphere, and both equations are free of clock errors, tropospheric delays, and the satellite to receiver range. They are therefore well suited to recover information about the ionospheric refractivity.

In chapter 5 we derived the ionospheric delay as the integral over the refractive index minus unity along the signal path. Using the refractive index for pseudorange measurements as given by equation (15.3) we obtain:

$$I_1 = \frac{40.3}{f_1^2} \int_{path} N_e ds\,, \quad I_2 = \frac{40.3}{f_2^2} \int_{path} N_e ds \tag{15.27}$$

Combining these equations with (15.25) and omitting the satellite and receiver indices, we get the following four equivalent representations of the pseudorange difference in terms of ionospheric parameters.

$$\frac{f_2^2}{f_1^2 - f_2^2}\left(P_2 - P_1\right) = \frac{f_2^2}{f_1^2 - f_2^2} C_p + I_1 + e \tag{15.28}$$

$$10^6 \frac{f_2^2}{f_1^2 - f_2^2}\left(P_2 - P_1\right) = 10^6 \frac{f_2^2}{f_1^2 - f_2^2} C_p + \int_{path} N_1 ds + e \tag{15.29}$$

$$\frac{f_1^2 f_2^2}{40.3\left(f_1^2 - f_2^2\right)}\left(P_2 - P_1\right) = \frac{f_1^2 f_2^2}{40.3\left(f_1^2 - f_2^2\right)} C_p + TEC + e \tag{15.30}$$

$$\frac{f_1^2 f_2^2}{40.3\left(f_1^2 - f_2^2\right)}\left(P_2 - P_1\right) = \frac{f_1^2 f_2^2}{40.3\left(f_1^2 - f_2^2\right)} C_p + \int_{path} N_e ds + e \tag{15.31}$$

The first of these equations relates the pseudorange measurements to a constant and the ionospheric delay of the L1 signal (in units of distance). The second equation relates the pseudorange measurements to the integral along the signal path over L1 refractivity N_1, as defined by equation (15.4). The third equation gives a relation between the measurements and the Total Electron Content (TEC) along the signal path, and the last one relates the inter frequency differences of pseudoranges to the integral along the signal path over the electron density.

Equations (15.29) and (15.31) contain the three-dimensional structure of the ionosphere in terms of refractivity and electron density, respectively. Therefore they are in principle suited to recover a three-dimensional model of the ionosphere from GPS pseudorange measurements. Equations (15.28) and (15.30) contain integrals of ionospheric parameters along the signal path. Under certain assumptions, these equations are useful to recover two-dimensional models of the ionosphere. A similar set of equations can be written for the inter frequency difference of carrier phase measurements.

15.4.1 Three-Dimensional Models for the Ionosphere from GPS Data

The refractivity N_1 of the ionosphere at the L1 frequency can be described by a four-dimensional model

$$N_1(\mathbf{x}, t) = \sum_{k=1}^{K} \alpha_k(t) \cdot F_k(\mathbf{x}) \tag{15.32}$$

where $\mathbf{x}$ is a representation of three-dimensional position, t is a time parameter, F_k, k=1, K are a set of appropriately chosen base functions of position, and α_k are a set of K coefficients describing the ionosphere. As described in section 15.2, the ionosphere undergoes considerable variability on various time scales. The shortest cycle of systematic variation is the diurnal one, cf. Figure 16.7. It is caused by the effect that the predominant features of the ionosphere rotate around the earth synchronous with the apparent motion of the sun. This means, that the coefficients α_k are undergoing cyclic diurnal variations, if $\mathbf{x}$ is a representation of position in terms of geographic longitude, latitude and altitude.

This systematic diurnal variability can be eliminated through the use of sun fixed longitude χ instead of geographic longitude λ, or equivalently the hour angle of the sun. In this rotating coordinate system, the variation of the coefficients α_k is smaller over short time intervals of several hours, and the may be replaced by constants.

$$N_1(\varphi, \chi, h) = \sum_{k=1}^{K} \alpha_k \cdot F_k(\varphi, \chi, h) \tag{15.33}$$

Building a three-dimensional model for the ionosphere from GPS measurements now means determining the coefficients α_k from a combination of equations (15.29) and (15.33). For the sake of clarity and simplicity, we will assume in the following that the coefficients C_p in equation (15.29) have been determined beforehand, and a corrected measurement of the type

$$10^6 \frac{f_2^2}{f_1^2 - f_2^2}\left(P_2 - P_1 - C_p\right) = \int_{path} N_1 ds + e \tag{15.34}$$

is available. Then the integrant can be replaced by equation (15.33), and exchange of integration and summation yields the expression

$$10^6 \frac{f_2^2}{f_1^2 - f_2^2}\left(P_2 - P_1 - C_p\right) = \sum_{k=1}^{K} \alpha_k \int_{path} F_k(\varphi, \chi, h) ds + e \tag{15.35}$$

This is a linear observation equation for the unknowns α_k. If a number of globally distributed dual frequency receivers provide a total of L measurements l for the time interval under consideration, these measurements can be assembled into a system of observation equations as follows.

$$\begin{bmatrix} l_1 \\ \bullet \\ \bullet \\ \bullet \\ l_L \end{bmatrix} = \begin{bmatrix} D_{11} & \bullet & \bullet & D_{1K} \\ \bullet & \bullet & \bullet & \bullet \\ \bullet & \bullet & \bullet & \bullet \\ \bullet & \bullet & \bullet & \bullet \\ D_{L1} & \bullet & \bullet & D_{LK} \end{bmatrix} \cdot \begin{bmatrix} \alpha_1 \\ \bullet \\ \bullet \\ \alpha_K \end{bmatrix} + \begin{bmatrix} e_1 \\ \bullet \\ \bullet \\ \bullet \\ e_L \end{bmatrix} \tag{15.36}$$

The coefficients in the design matrix are computed as integrals over the base functions along the individual ray paths according to:

$$D_{ij} = \int_{path\#i} F_j(\varphi, \chi, h)\,ds \tag{15.37}$$

In principle, least squares normal equations could be formed from the system of observation equations (15.36) in order to determine the coefficients α_k. Unfortunately, the normal equation matrix proves to be close to singularity, making an inversion all but impossible. This singularity can be understood from the following geometrical considerations illustrated with a simple model for the ionosphere as shown in Figure 15.11.

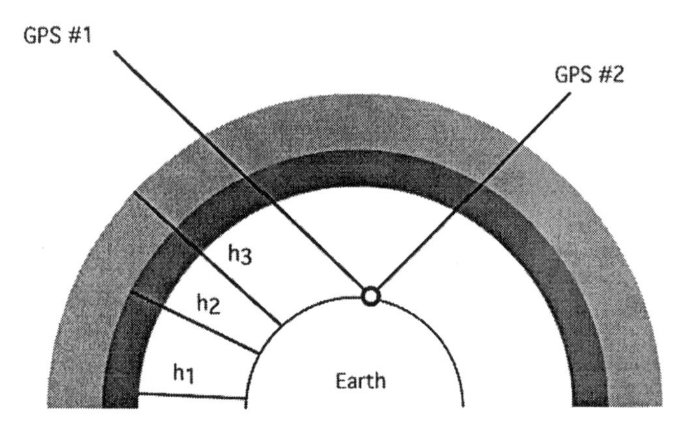

Figure 15.11. Model for a simplified ionosphere.

The model for the ionosphere shown in this figure consists of two spherical shells with uniform but different refractivity within each shell. The lower shell extends in height between h_1 and h_2, the second shell from h_2 to h_3. Within the lower shell, refractivity is equal to N_l; within the higher shell it is equal to N_h. Then according to equation (15.33) refractivity is represented as

$$N_1(h) = \alpha_1 F_1(h) + \alpha_2 F_2(h) \tag{15.38}$$

with α_1 equal to N_l, α_2 equal to N_h, and

$$\begin{aligned} F_1(h) &= 1 \; for \; h_1 < h^2 h_2, \quad zero \; otherwise \\ F_2(h) &= 1 \; for \; h_2 < h^2 h_3, \quad zero \; otherwise \end{aligned} \tag{15.39}$$

The elements of the design matrix can be computed from equation (15.37) and (15.39). In this case, these elements are obviously the signal path lengths within the two spherical shells. In particular, we obtain:

$$D_{11} = \int_{path\#1} F_1(h)ds = \int_{h_1}^{h_2} \frac{dh}{\cos z_{11}}, \quad D_{12} = \int_{path\#1} F_2(h)ds = \int_{h_2}^{h_3} \frac{dh}{\cos z_{12}} \tag{15.40}$$

$$D_{21} = \int_{path\#2} F_1(h)ds = \int_{h_1}^{h_2} \frac{dh}{\cos z_{21}}, \quad D_{22} = \int_{path\#2} F_2(h)ds = \int_{h_2}^{h_3} \frac{dh}{\cos z_{22}} \tag{15.41}$$

Here $\cos z_{ij}$ is the cosine of the signal zenith angle along the i-th ray path within the j-th shell. As can be seen from Figure 15.11, the signal zenith angle does not change significantly from the lower to the higher shell. Thus $\cos z_{i1} \approx \cos z_{i2}$, and $D_{12} \approx \kappa D_{11}$, $D_{22} \approx \kappa D_{21}$, where κ is equal to the ratio of the shell thicknesses. Obviously, the second column of the resulting design matrix is then approximately equal to the first column times a constant factor.

Adding more observations to other satellites or from other receiver locations does not remove this nearly linear dependence of the two columns of the design matrix. Changing the base functions for the vertical electron density distribution does not help either. The problem is caused primarily by the poor capability of the model to distinguish between electron density contributions from different altitudes, as long as all measurements are made on transionospheric signals with ground based receivers.

It is then perhaps appropriate to revisit space borne GPS measurements like the ones discussed in section 15.3. Figure 15.12 shows the same simple electron density distribution discussed above. However, now one ionospheric delay measurement is made on an transionospheric signal with a ground based receiver, the other measurement comes from a space borne receiver. As can be seen in the figure, the signal received at the LEO satellite passes only through the upper part of the simple model ionosphere. Writing now the observation equations for the two measurements according to equation (15.36), the design matrix will contain in the first row the elements D_{11} and D_{12} as described by equation (15.40). The second signal does not pass through the lower shell of the model ionosphere, therefore D_{21} is zero. The fourth element D_{22} is equal to the length of the signal path in the upper shell, i.e. the distance d indicated in the figure. Obviously, the new design matrix has full column rank and the electron densities within the two shells can be determined.

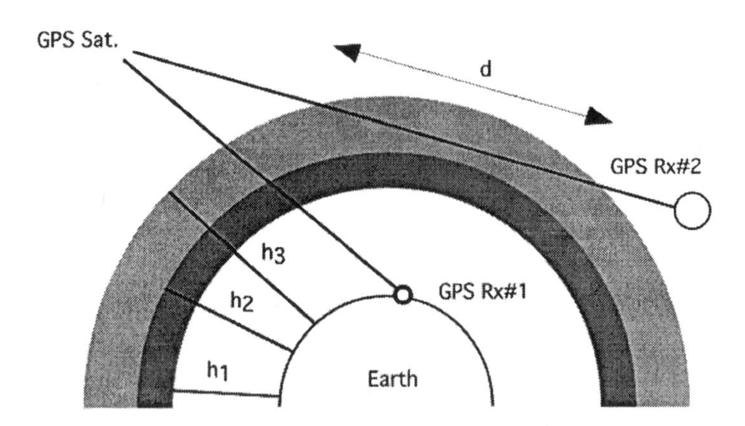

Figure 15.12. Space borne GPS measurements of ionospheric delay.

Space based ionospheric delay measurements from a LEO satellite together with ground based delay observations apparently allow the determination of a three-dimensional model of electron density. It can even be shown that horizontal variations in electron content are the primary contribution resulting from ground based observations, whereas the vertical variations are contributed by the space based delay measurements.

This should come to no surprise after reading section 15.3. It was shown there that vertical profiles of refractivity could be extracted from space based GPS measurements. There the discussion was restricted to the neutral atmosphere, with the contribution of the ionosphere eliminated, cf. equation (15.21). The occultation method outlined in section 15.3 can also be applied to the ionosphere region, see Hajj et al. [1993]. Since the assumption of spherical symmetry required in the occultation method is not truly valid in the ionosphere, ground based delay measurements can complement the space based measurements by providing a description of the asymmetry reflected in the horizontal electron density variation.

15.4.1 Two-Dimensional Models for the Ionosphere from GPS Data

As discussed in the previous section, ground based ionospheric delay measurements are not very sensitive to the vertical structure of electron density within the ionosphere. It then seems to be appropriate to describe such delays as resulting from a two-dimensional ionospheric model without vertical extension. A popular choice for a two-dimensional model ionosphere is an infinitesimally thin spherical shell (a spherical layer) with a layer density equal to the vertically integrated electron density, the Vertical Total Electron Content (VTEC), as shown in Figure 15.13.

$$VTEC(\varphi,\lambda) = \int N_e(\varphi,\lambda,h)dh \qquad (15.42)$$

Typically, the height h of the shell is selected to coincide with the height of maximum electron density at about 350 km, cf. section 15.2.

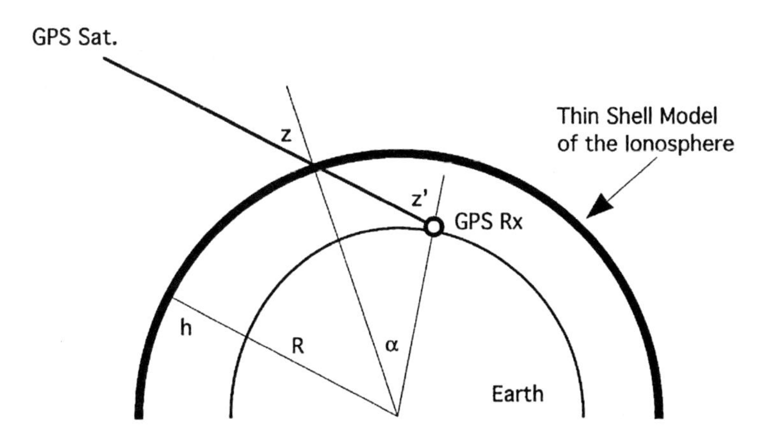

Figure 15.13. Two-dimensional thin shell model for the ionosphere.

The thin shell is pierced by the GPS signal with a zenith angle z which can be computed from the height of the spherical shell and the signal zenith angle z' at the receiver, see, e.g. Georgiadou and Kleusberg [1988]

$$z = \sin^{-1}\left(R\sin z'/(R+h)\right) \qquad (15.43)$$

where R is the radius of the earth (in spherical approximation). The distance of the *subionospheric point*, the point on the earth having the same coordinates as the signal piercing point at the shell, is obtained from

$$\alpha = z' - z \qquad (15.44)$$

and the geographic coordinates of the subionospheric point are calculated from. e.g. Hakegard [1995]

$$\begin{aligned} \varphi &= \sin^{-1}(\cos\alpha \sin\varphi' + \sin\alpha \cos\varphi' \cos A') \\ \lambda &= \lambda' + (\sin\alpha \sin A'/\cos\varphi) \end{aligned} \qquad (15.45)$$

where φ', λ' are latitude and longitude of the receiver, and A' is the signal azimuth at the receiver location. The signal passing through the spherical shell will suffer an ionospheric delay proportional to the TEC along the signal path through the model ionosphere.

$$TEC(\varphi,\lambda) = VTEC(\varphi,\lambda) / \cos z \tag{15.46}$$

in general, the layer density $VTEC(\varphi,\lambda)$ will have spatial and temporal variations. As discussed in section 15.4.1, a major systematic variation results from the rotation of the earth with respect to the sun. It was also discussed there that the ionosphere exhibits a much more static pattern, if it is represented in terms of the sun fixed longitude χ rather than geographic longitude. If represented in terms of geographic latitude and sun fixed longitude, *VTEC* may be considered constant for short intervals of time (several hours), and an appropriate model can be written as

$$VTEC(\varphi,\chi) = \sum_{k=1}^{K} \alpha_k F_k(\varphi,\chi) . \tag{15.47}$$

$F_k(\varphi,\chi)$ are appropriately chosen base functions, and α_k are the (constant) coefficients of the model for *VTEC*. Equations (15.46) and (15.47) then may be used to replace the *TEC* in equation (15.30).

$$\frac{f_1^2 f_2^2}{40.3\left(f_1^2 - f_2^2\right)}\left(P_2 - P_1\right) = \frac{f_1^2 f_2^2}{40.3\left(f_1^2 - f_2^2\right)} C_p + \frac{1}{\cos z} \sum_{k=1}^{K} \alpha_k F_k(\varphi,\chi) + e \tag{15.48}$$

This equation can be seen as an observation equation modelling the observed inter frequency difference of pseudoranges on the left hand side in terms of a constant hardware delay term C_p, and a total of K constant coefficients for the base functions $F_k(\varphi,\chi)$.

One equation (15.48) can be written for each satellite-receiver pair per observation epoch. Provided that several hours of measurements from a globally distributed network of dual frequency receivers are available and that base functions are chosen appropriately, the resulting equation system will have a design matrix with full column rank. This system then can be solved through inversion after forming least squares normal equations.

The result of this process is a set of hardware delays, and the coefficients of the thin layer model represented through the chosen base function, usually *surface spherical harmonics, see,* e.g. Heiskanen and Moritz [1967]. In a variation of this procedure, Schaer et al. [1996] present VTEC results based on globally distributed carrier phase measurements. Yet another variation is to use *geomagnetic latitude* rather than geographic latitude to represent the layer density. This approach was followed by Brunini and Kleusberg [1996] who estimate vertical ionospheric delay of the L1 signal rather than VTEC. Figure 15.14 is taken from this article.

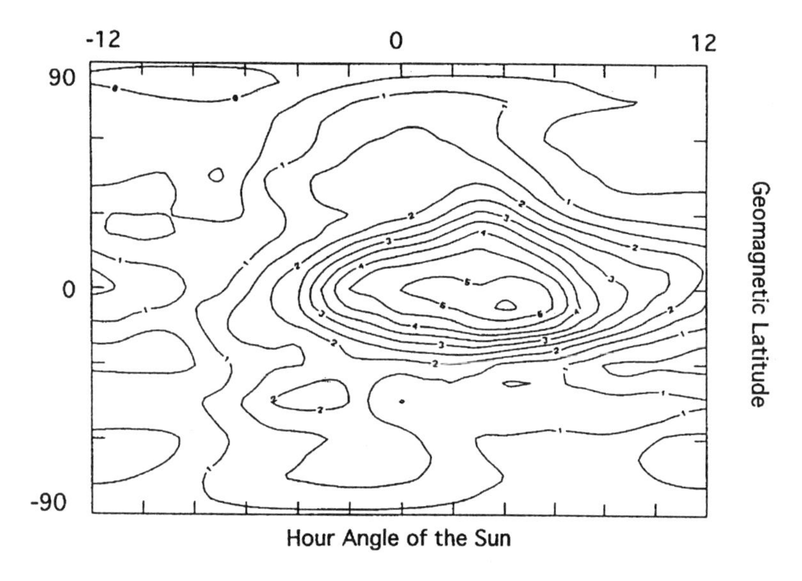

Figure 15.14. Estimates of vertical delay at the L1 frequency.

This figure shows results for estimated L1 delays based on 10 hours of dual frequency pseudorange data from 28 globally distributed stations of the network established by the International GPS Service for Geodynamics, IGS; Beutler et al. [1994]. The start of the 10 hour data interval is 16:00 UT on 28 October 1995. This is a period of low solar activity, as can be seen from Figure 15.3.

The results are based on the thin shell model described above with surface spherical harmonics up to degree 12. It can be seen that the delay estimates range up to about 5.5 m, with a maximum concentrated in equatorial regions for hour angles of the sun between 2 and 6 h. Also apparent is that the delay estimates dip just below zero in the upper left part of the plot; since the delay is non-negative by definition, this reflects the inaccuracies of the data and the modeling approach.

So far we have discussed global thin shell models based on spherical harmonics as base functions. Other representations are possible and have been successfully implemented. Mannucci et al. [1993] estimated the parameters of a global TEC representation where the thin shell is subdivided into 1280 tiles of triangular shape, with the TEC represented by linear functions within each tile.

Models for TEC and/or ionospheric delay have also been introduced at the regional and local level. At the regional level, these efforts concentrate on the provision of ionospheric corrections for Wide Area Augmentation System (WAAS), see, e.g. Engler et al. [1995]; Stull and Van Dierendonck [1995]; Conker et al. [1995]; Hakegard, [1996] and GPS Active Control Systems, e.g. Gao et al [1994]. In these cases, regionally defined base functions or grids are used to represent the TEC or delay values. Especially in the context of WAAS, also the real-time aspects of delay estimation are of importance.

15.5 SUMMARY AND CONCLUSIONS

In the second section of this chapter, we have discussed the refractive properties of the neutral atmosphere and the ionosphere, and how parameters describing refraction are related to the propagation of GPS signals. Existing models approximating the refractivity in various regions of the atmosphere were used to get a general overview of refraction affecting GPS measurements.

The third section provided an introduction in the determination of atmospheric refractivity from radio signal occultation methods. It was seen that vertical profiles of refractivity can be estimated from measurements of atmospheric frequency perturbations obtained with a GPS receiver onboard a LEO satellite. The GPS/MET experiment was introduced, and procedures for deriving atmospheric pressure and temperature for dry air from refractivity were described. Although this method is primarily designed to determine refractivity in the neutral atmosphere, it is also capable of evaluating vertical profiles of electron content in the ionosphere.

Three-dimensional and two-dimensional ionospheric models from GPS were the topic of the fourth section. It was shown that the vertical structure of the ionosphere cannot be determined well from ground based GPS measurements alone. Various types of global two-dimensional ionosphere models based on a thin shell approximation and methods to determine the coefficients of such models were shown.

References

Beutler, G., Mueller, I.I. and R. Neilan (1994), The International GPS Service for Geodynamics (IGS): Development and Start of Official Service on 1 January 1994, *Bulletin Geodesique* 68, 43-51.

Bilitza, D. (1993), *International Reference Ionosphere - Past, Present, Future. Solar-Terrestrial Predictions IV*, NOAA Environmental Research Laboratories, Boulder, CA. Vol. 3, 313-337.

Born, M. and E. Wolf (1965), *Principles of Optics*. Third Edition, Pergamon Press.

Brunini, C. and A. Kleusberg (1996), Mapas Globales de Retardo Ionosferico Vertical a Partir de Observaciones GPS. *Proceedings IV Intern. Congress Earth Sciences*, Santiago, Chile.

Conker, R., El-Arini, M.B., Albertson, T., Klobuchar, J. and P. Doherty (1995), Development of Real-Time Algorithms to Estimate the Ionospheric Error Bounds for WAAS. *Proceedings ION GPS-95*, 1247-1258.

Engler, E., Sardon, E. and D. Klähn (1995), Real Time Estimation of Ionospheric Delays. *Proceedings ION GPS-95*, 1183-1191.

Fjeldbo, G., Kliore, A.J., and V.R. Eshleman, (1971), The Neutral Atmosphere of Venus as Studied with the Mariner V Radio Occultation Experiments. *Astronomical Journal* 76(2), 123-140.

Gao, Y., Heroux, P. and J. Kouba (1994), Estimation of Receiver and Satellite L1/L2 Signal Delay Biases Using Data from CACS. *Proceedings KIS-94*, Banff, 109-117.

Georgiadou, Y. and A. Kleusberg (1988), On the Efeect of Ionospheric Delay on Geodetic Relative GPS Positioning. *Manuscripta Geodaetica* 13, 1-8.

Hajj, G., Ibanez-Meier, R. and E.R. Kursinski (1993), Ionospheric Imaging from a Low Earth Orbiter Tracking GPS. *Proceedings ION GPS-93*, 1315-1322.

Hajj, G., Kursinski, E.R., Bertiger, W., Leroy, S., Romans, L. and J.T. Schofield (1995), Sensing the Atmosphere from a Low-Earth Orbiter Tracking GPS: Early Results and Lessons from the GPS/MET Experiment. *Proceedings ION GPS-95*, 1167-1174.

Hakegard, O.P. (1995), *A Regional Ionospheric Model for Real-Time Predictions of the Total Electron Content in Wide Area Differential Satellite Navigation Systems*. Dr.ing thesis. Norwegian Institute of Technology, Trondheim.

Hargreaves, J. K. (1992), *The Solar-Terrestrial Environment*. Cambridge University Press.

Heiskanen, W.A. and H. Moritz (1967), *Physical Geodesy*. Freeman and Co., San Francisco.

Hoeg, P., Hauchecorne, A., Kirchengast, G., Syndergaard, S., Belloul, B., Leitinger, R., and W. Rothleitner, (1995), Derivation of Atmospheric Properties Using a Radio Occulation Technique. *Danish Meteorological Institute Scientific, Report 95 - 4*, Copenhagen, Denmark.

Hopfield, H. S. (1969), Two-Quartic Tropospheric Refractivity Profile for Correcting Satellite Data, *Journal of Geophysical Research* 74 (18), 4487-4499.

Komjathy, A., R.B. Langley, and F. Vejrazka (1995). A Comparison of Predicted and Measured Ionospheric Range Error Corrections, *Eos Trans*, AGU, 76(17), Spring Meeting. Suppl., S87.

Mannucci, A.J., Wilson, B.D. and C.D. Edwards (1993), A New Method for Monitoring the Earth's Ionospheric Total Electron Content Using the GPS Global Network. *Proceedings ION GPS-93*, 1323-1332.

Melbourne, W.G., Davis, E.S., Duncan, C.B., Hajj, G.A., Hardy, K.R., Kursinski, E.R., Meehan, T.K., Young, L.E., and T.P. Yunck, (1994), The Application of Spaceborne GPS to Atmospheric Limb Sounding and Global Change Monitoring. *Jet Propulsion Laboratory Publication 94 - 18*, Pasadena, CA, USA.

Schaer, S., Beutler, G., Rothacher, M., and T. Springer (1996), Daily Global Ionosphere Maps Based on Carrier Phase Data Routinely Produced by the CODE Analysis Center. Paper pres. *IGS AC Workshop*, Silver Springs, MD, USA, March 19-21.

Stull, C. and A.J. Van Dierendonck (1995), Test Results of Wilcox Electric's Ionospheric Monitoring Network. *Proceedings ION GPS-95*, 1219-1228.

Yunk, T.P., Lindal, G.F. and C.-H. Liu (1988), The Role of GPS in Precise Earth Observation. Proc. *IEEE PLANS'88*.

16. THE ROLE OF GPS IN SPACE GEODESY

Gerhard Beutler
Astronomical Institute, University of Berne, Sidlerstrasse 5,
CH-3012 Berne, Switzerland

16.1 INTRODUCTION

GPS may be considered a fully mature space geodetic technique in 1997. Some people even would say that GPS is *the dominant* technique in geodesy and geodynamics. Certainly, when reading this book, one must be impressed with the achievements of GPS in geodesy and geodynamics.

In this chapter we look more closely at the role of ground based GPS in Space Geodesy and we discuss interactions with the other space geodetic techniques. In the second part of this Chapter we will study in particular possible interactions between SLR and GPS.

Let us repeat here four essential characteristics to assess the role of GPS in Space Geodesy (compare Introduction to Chapter 14).

- GPS is a *satellite geodetic technique*. There are other satellite geodetic techniques, in particular *SLR (Satellite Laser Ranging)* and, to some extent, *LLR (Lunar Laser Ranging)*. An almost "died-out species" are astrometric observations to artificial satellites.

- GPS is an *interferometric technique*. Highest accuracy may only be achieved using two or more receivers operating simultaneously. *VLBI (Very Long Baseline Interferometry)* is an interferometric technique, too.

- GPS is a *satellite microwave technique*, *i.e.*, signals are transmitted on microwave carriers through the Earth's atmosphere. There are other such techniques like, *e.g.*, the French DORIS system, the German PRARE System, and, last but not least, the Russian GLONASS System.

- GPS allows for an unlimited number of receivers on and near the Earth surface (up to Low Earth Orbiters).

The first three of the above characteristics show that GPS shares important characteristics with other well established techniques, in particular with VLBI and with SLR. We thus expect that GPS-only analyses will reveal some of the strengths of those techniques, but it will also suffer from the rather problematic aspects of these methods. From the theoretical point of view we expect that by combining results from different techniques it should be possible to eliminate (or essentially reduce) the weaknesses of individual contributors and to come up with solutions reflecting the strength of all contributors. The production of such combinations, if done rigorously from the mathematical point of view, is an ambitious task.

In this Chapter we will briefly review the situation in Space Geodesy in 1997 (Section 16.2), look at the performance of the major Space Geodetic Systems in Section 16.3, address the issue of comparing or combining techniques in Section 16.4, and finally focus on *SLR observations to GPS satellites* in Section 16.5. The findings will be summarized in Section 16.6.

16.2 SPACE GEODESY IN 1997

Three types of systems have to be considered in Space Geodesy today:

- **SLR and LLR:** Satellite Laser Ranging and Lunar Laser Ranging measure the distance to artificial satellites and to the moon with (sub-) cm accuracy. SLR and LLR are operating in the *optical band* of the electromagnetic spectrum.

- **VLBI:** Very Long Baseline Interferometry measures the distance difference between VLBI telescopes as seen from *Quasars* with (sub-) cm accuracy by correlating the Quasar signals received by different telescopes. VLBI is operating in the *microwave band.*

- **Satellite microwave systems:** The satellites are emitting signals in the microwave band which are received (or retransmitted) by receivers (transponders) on or near the Earth's surface. GPS, DORIS, GLONASS, PRARE are systems available today.

Coordination is vital in Space Geodesy. Three major organizations try to achieve this goal:

- **IERS:** The International Earth Rotation Service is responsible for defining and maintaining the conventional terrestrial and celestial reference frame, and for determining the transformation parameters between them. The IERS uses results from all space geodetic techniques for that purpose.

- **IGS:** The International GPS Service for Geodynamics is responsible for making available GPS data from its global network, for producing and disseminating high accuracy GPS orbits, Earth rotation parameters, station coordinates, atmospheric information, *etc.*

- **CSTG:** The Commission on International Coordination of Space Techniques for Geodesy and Geodynamics is responsible to develop links between groups engaged in space geodesy and geodynamics, coordinate their work, elaborate projects implying international cooperation, etc.

The first two organizations are *scientific services in support of geodesy and geodynamics*, the latter is an *IAG Commission* and a *COSPAR Subcommission.*

The *International Earth Rotation Service (IERS)* was established in 1987 by IAU and IUGG and started operation on January 1st, 1988. It replaced the international Polar Motion Service (IPMS) and the Earth Rotation Section of the Bureau International de l'Heure (BIH); the activities of BIH on time are continued at the Bureau International des Poids et Mesures (BIPM). IERS is a member of the *Federation of Astronomical and Geophysical Data Analysis Services (FAGS).*

The *International GPS Service for Geodynamics (IGS)* officially became an IAG Service on January 1, 1994. Previously the *IGS Oversight Committee* organized the *1992 IGS Test Campaign (21 June - 23 September 1992)* and subsequently the *IGS Pilot Service (1 November 1992 - 31 December 1993).* Since January 1, 1996 the IGS is a FAGS service, as well.

The *Commission on International Coordination of Space Techniques for Geodesy and Geodynamics (CSTG)* was formally established at the XVII General Assembly of the *IUGG (International Union of Geodesy and Geodynamics)* in Canberra in 1979. CSTG is Commission VIII in Section II of *the International Association of Geodesy (IAG)* and Subcommission B.2 in *COSPAR.* CSTG consists of *Subcommissions and Projects.* The Subcommissions are more permanent while the Projects are of a temporary nature. The CSTG structure is periodically re-defined at the *IUGG General Assemblies*, *i.e.*, every four years. Obviously the task of CSTG can only be performed through a close collaboration with the two space geodetic services IERS and IGS.

Despite all the fantastic achievements in Space Geodesy over the last twenty years it is felt by many colleagues that there are some signs of a crisis in Space Geodesy. Like in all branches of science, **funding is a problem** in Space Geodesy, today. We are all suffering from the same disease! Obviously, VLBI and SLR/LLR are expensive techniques. It was always clear that there would be an enormous pressure on both techniques *if* new, and cheap techniques should be developed allowing results of the same kind and of comparable quality as in SLR and VLBI. Well, GPS is a cheap technique, provided the space segment is assumed free of charge.

It was already mentioned in Chapter 14 that GPS was originally thought to serve as the key tool to densify the ITRF, but that its contributions were successfully extended to other fields of global geodynamics. It is also clear that the GPS contributions will become even more significant in future when *spaceborne applications of the GPS* will start to play a significant role.

The **key question** is, however, *whether or not GPS is capable of replacing the expensive SLR and VLBI techniques in every respect.* It is clear that IERS, CSTG, and IGS have to deal with these questions considering the new situation created by the IGS. Essentially, a **new equilibrium of techniques in Space Geodesy** has to be sought. In order to find answers to these questions we have to review the contributions made by different techniques in the past.

16.3 PERFORMANCE OF THE MAJOR SPACE TECHNIQUES IN THE RECENT PAST

One way of looking at the performance of space techniques in the recent past is to study the IERS annual reports. Table 16.1 gives the estimated accuracy of individual, technique-specific series for the years 1993-94 (from Charlot [1995]).

We observe that all techniques contribute to the establishment of *x*- and *y*-components of polar motion. We also see, that according to the IERS, all techniques perform roughly on the same accuracy level.

The picture is quite different when looking at the establishment of *UT1-UTC*. Clearly, the VLBI contribution is superior to the satellite geodetic contributions stemming from SLR and GPS. This is due to the fact that the latter methods are only capable of establishing the *length of day* (or the first time derivative of *UT1-UTC*). The UT values given in Table 16.1 are as a matter of fact summed-up length of day values referring to some initial point which was established by VLBI. Also, offsets and drifts were subtracted from the satellite geodetic series prior to the comparison. It may thus be stated that the *long term stability* of *UT1-UTC* is *uniquely established through VLBI.*

Table 16.1. RMS of contributing EOP Determinations to 1993-94 C04 Series.

Analysis Center	Technique	RMS RELATIVE TO IERS C04 SERIES					
		X [mas]	Y [mas]	UT1 [0.1 ms]	dΨ [mas]	dε [mas]	LOD [0.1 ms]
GSFC	**VLBI**	0.32	0.27	0.09	0.10	0.08	
NOAA	**VLBI**	0.24	0.30	0.10	0.25	0.10	
USNO	**VLBI**	0.20	0.16	0.10	0.19	0.08	
CODE	**GPS**	0.33	0.36	0.45			0.33
EMR	**GPS**	0.35	0.35	0.45	-	-	
GFZ	**GPS**	0.35	0.26		-	-	0.69
JPL	**GPS**	0.34	0.34		-	-	
CSR	**SLR**	0.20	0.22	0.51	-	-	

We see furthermore that the satellite geodetic methods *SLR* and *GPS* do *not* contribute to the establishment of the *nutation* in longitude and in latitude. As we saw in Chapter 14 it is theoretically possible for these techniques to contribute to nutation in the same sense as they do to *UT1-UTC*: SLR and GPS are sensitive to the first (and higher) time derivatives of precession and nutation.

Although this does not show up in Table 16.1 it should be mentioned that satellite geodetic methods do *not* contribute to the establishment of the International Celestial Reference Frame (ICRF). Today, VLBI is the only technique capable of providing this system.

Table 16.1 gives a somewhat optimistic picture of the situation in space geodesy. That the situation is very problematic today (that we are already in a crisis) is illustrated by the following facts:

- Only VLBI and GPS routinely contribute estimates based on one day of observations to the IERS time series. SLR values are based on three days of data.

- The regular *VLBI contribution to polar motion* is essentially **one observation session of 24 hours** of the five station network *Wettzell, Greenbank, Fairbanks, Kokee, Fortaleza* **per week!**

- *UT1-UTC* is established by analyzing the VLBI baseline Wettzell-Greenbank on a daily basis!

- GPS delivers *one data point* for the *x*- and *y*-coordinates of the pole and for *length of day* (or the *UT1-UTC drift*) on a daily basis.

There is thus a certain danger that IERS results will rely more and more on one technique only (namely GPS) for the generation of its products.

There are of course other results in Space Geodesy than those reflected in Table 16.1. Let us mention in particular the positions and velocities of the space geodetic observatories. These contributions to the *ITRF (International Terrestrial Reference Frame)* are analyzed each year by the ITRF section of the IERS Central Bureau (see, *e.g.*, Boucher et al. [1994]).

Other important space geodetic results are related to the Earth's gravity field. SLR was the key contributor in the past. Even in future, when there will be dedicated gravity missions (e.g., GRACE (NASA) and/or GOCE (ESA)), SLR, due to its long time series, will continue playing an essential role. The important *SLR* contributions according to Eanes and Watkins [1995] are:

- Establishment of non time-varying geopotentional models,
- Establishment of constraints on secular changes of low degree zonal harmonics,
- Anelastic solid Earth response at the 18.6 year period,
- Interannual, seasonal, intraseasonal gravity field changes,
- Selected ocean tide harmonics, tidal energy dissipation,
- High resolution of the mean sea surface and gravity anomalies,
- Degree 1 harmonics (motion of the ITRF with respect to the geocenter).

The above contributions are mainly due to the coordinated tracking of the *"cannon ball" satellites Lageos I, Lageos II*, and *Starlette*.

Table 16.2. Contributions of VLBI, SLR and GPS to geodynamics.

Technique	Contributions		
	Unique	**Common**	**Supporting**
VLBI	Cel. Frame UT1-UTC Precession, Nutation	Polar Motion Terr. Frame	GPS, SLR (ICRF) GPS, SLR
SLR	Gravity Model and Temp. Var. Center of Mass Orbit Det. (Passive Sats)	Polar Motion High Freq. UT1-UTC Terr. Frame	GPS (Grav. Model) GPS
GPS **DORIS** **PRARE** **GLONASS** ...	Densific.of ITRF Orbit Det. (Spaceborne GPS)	Polar Motion High Freq. UT1-UTC High Freq. ICRF Terr. Frame Center of Mass	VLBI and SLR through Colloc.of Sites (SLR)

The individual achievements of space techniques are summarized in Table 16.2, where the attempt is made to distinguish between *unique contributions*, *i.e.*, results which cannot be achieved (with a comparable level of accuracy and/or efficiency) by any other space technique and *common contributions*, *i.e.*, results which are established through a combination of different space geodetic methods. Moreover we specify in Table 16.2 which other technique is depending most on a given space technique.

From Table 16.2 we conclude that both, SLR and GPS, rely on VLBI for the definition of the Celestial Reference Frame, which is the natural system for the formulation of the equations of motion. Moreover, SLR and GPS rely on VLBI for the establishment of precession, nutation, and *UT1-UTC* as essential elements of the transformation between the Celestial and Terrestrial Reference Frame.

Table 16.2 also tells us that VLBI and GPS heavily rely on SLR for the definition of the geocenter (where one may of course argue that VLBI does not need to have access to the geocenter), and that SLR and VLBI fully rely on GPS for collocation, *i.e.*, for the geometric tie between all space geodetic observatories.

In summary we draw the following conclusions from Table 16.2:

- There is a clear necessity for *all* major space geodetic techniques, *i.e.*, for VLBI, SLR/LLR, and satellite microwave systems.
- The International organizations dealing with Space Geodesy, in particular IERS, IGS and CSTG have to define a *New Concept* for the co-existence of these techniques.

16.4 SPACE GEODESY: THE FUTURE

Based on the facts and developments given in sections 16.2 and 16.3, and we conclude that there is a *clear demand* for

- an *International Space Geodetic Network* which is in essence the combination of all existing technique specific networks *and* for the
- *combination of space geodetic analysis* in order to make full use of the complementarity of space geodetic results.

The two issues were addressed several times in the past and promising developments are underway right now. Although it is not yet possible to present "final" solutions it is worthwhile to briefly address the two issues.

The International Space Geodetic Network

The necessity for a better coordination of observations in Space Geodesy was recognized a long time ago. IUGG Resolution Number 8 (1991 IUGG General Assembly in Vienna) *urges that relevant organizations, agencies, and Member Countries should review the geographical distribution of those geophysical stations under their control which make continuous Earth and space observations, and should jointly locate as many of these stations as is practical, so that data gathering on a global basis by ground and space-based measurements be optimized.*

A first attempt to follow this resolution was made at the XXIst IUGG General assembly with a further resolution asking for the establishment of a *Fundamental Reference and Calibration Network (FRCN)*. The basic idea was to create a new network of "superstations" which would contain, if possible, all space geodetic plus all "normal" geodetic and geophysical instruments, Mueller et al. [1996]. The resolution was *not* accepted by IUGG in 1995.

The idea was subsequently taken up by the IAG Executive Committee at its meeting in Copenhagen in November 1995, which created an IAG/CSTG Working Group to come up with *the scientific and logistic rationale for an IUGG Fundamental Reference and Calibration Network (FRCN), not replacing but supporting the existing individual networks and to present a report to the IAG Executive Committee at its meeting in spring 1997,* Beutler, Mueller, Schwarz, [1996]. CSTG, IGS, IERS, and all other (than IAG) IUGG Associations were represented in this group.

The working group organized several meetings, including a one-day workshop in Paris (October 1996). The findings are contained in Drewes [1997]. The final conclusion of the working group is *that there are no compelling scientific reasons to identify stations as FRCN stations*. It was argued instead that an *International Space Geodetic Network (ISGN) as a combination of the stations of all space geodetic techniques already exists and that this network is well suited to take over most of the tasks that were considered for the FRCN.*

The IAG Executive Committee, at its Meeting of April 28-29, accepted the report and gave the task of the establishment of such a network to CSTG and IERS.

16.4.2 Combination of the Analysis

Combining the results stemming from different space techniques is a problem that routinely has to be addressed by the *IERS*. The IERS evolved in a remarkable way since it started its official activities in 1988. In particular it has accepted two new techniques as official IERS techniques, namely the GPS technique in 1993 (organized by the IGS) and the DORIS technique in 1996.

The IERS team 1995-1999 with Prof. Christoph Reigber as Chairman and Dr. Martine Feissel as Director of the IERS Central Bureau found it appropriate to initiate a *thorough external review* of IERS activates. This review took place at the *1996 IERS Workshop in Paris* in October 1996 and is fully documented in Reigber and Feissel [1997]. The Workshop focused on a *review of the IERS mission and structure* and on various *options to extend the IERS activities* (*e.g.*, to include tide gauges, to improve the vertical datum, to study Earth rotation dynamics and fluids).

The workshop came up with many recommendations with the goal to make the IERS even more efficient and relevant for Earth sciences than it is already now. Let us briefly comment two of these resolutions. For the entire set of recommendations we refer to Reigber and Feissel [1997].

Recommendation 1: Each Coordinating Center (CC) should strengthen overall control of observations, quality control, and data processing for products of interest to the IERS (coordinates, EOP series, *etc.*) in accordance with the IERS Terms of Reference. Combinations of individual solutions into technique specific products are strongly encouraged.

Recommendation 3: Combination of products from the various CCs shall be performed by at least two IERS Components such as Sub-Bureaus, specialized Analysis Centers, Central Bureau Sections. The Central Bureau should disseminate the state-of-the-art combined solutions *and* provide the official IERS products. The criteria for the selection or production of this official solution will be determined by the IERS Directing Board.

Recommendation 1 essentially draws the conclusions from the experiences gained through the development of the *International GPS Service for Geodynamics (IGS)*. It proved to be very efficient when *one* organization takes care of collecting *and* processing data, *and* of validating the results gained by its analysis centers. Attempts are underway to implement similar structures in other techniques, as well. (The DORIS technique, as the youngest contributor to the IERS automatically chose this approach).

Recommendation 3 directly addresses the combination issue, where it is assumed that the Coordinating Centers for the individual techniques make available technique specific combined solutions, already. Whereas in the past such multi-technique combinations were performed in a rather simple way (essentially by computing weighted averages of individual techniques) more elaborate approaches are feasible today thanks to the availability of powerful computers and internet connections. More explicitly it was recommended at the 1996 IERS workshop

- to foster the collaboration of *new* research groups interested in combining the results stemming from all different techniques contributing to the IERS,
- to continue resp. maintain unique series of *official IERS products* for all types of IERS users,
- and that the IERS Directing Board is responsible for selecting resp. giving the directions for generating these official products.

That rigorous combination of results stemming from different space techniques must lead to better results was also recognized within CSTG. The coordination of activities and efforts in these domains is the main objective of the CSTG Project *Coordination and Combination of the Analysis in Space Geodesy* and the Subcommission *Precise Satellite Microwave Systems* which were both established at the XXIst IUGG General Assembly in Boulder in summer 1995.

The approach of the CSTG Project on the *Coordination and Combination of the Analysis in Space Geodesy* is (or was, initially) intimately related to the *IGS project on the densification of the ITRF using regional GPS analyses,* Zumberge and Liu [1995]. This *IGS project* may be summarized as follows:

- The project officially started on September 3, 1995. In a first phase (till mid 1996) the seven IGS Analysis Centers *COD, EMR, ESA, GFZ, JPL, NGS, and SIO* produced so-called *free network solutions* which may subsequently be combined into a unified IGS coordinate solution. The AC contributions are in the *SINEX format* (see below) which are delivered at weekly intervals.

- Three IGS *Global Network Associate Analysis Centers (GNAACs)* are combining the individual contributions every week:

  - J.P.L. (Jet Propulsion Laboratory) with Mike Heflin.
  - M.I.T. (Massachussetts Institute of Technology) with Tom Herring,
  - N.C.L. (University of Newcastle) with Geoff Blewitt and Phil Davies,

- On January 11, 1996 a *Call for Participation* was issued for *IGS Regional Associate Analysis Centers (RNAACs)* to perform regional analyses using IGS global products, and to produce in turn weekly SINEX files to be combined by the GNAACs every week (IGS-mail message No. 1178). About 15 organizations responded to the call and are contributing smaller or larger regional networks to the IGS project on a weekly basis.

- The solutions prepared by these RNAACs are combined into weekly solutions by the above mentioned GNAACs, the result being weekly coordinate sets in the best possible IGS realization of the ITRF.

- The results of the three GNAACs are in turn analyzed by the IGS Analysis Center Coordinator. For more information we refer to the IGS Annual reports for 1995 and 1996.

Of central importance was the development of *SINEX*, the *S*oftware *IN*dependent *EX*change format, which allows it to combine solutions prepared by different agencies on the normal equation system level.

It was recognized that SINEX (or a similar format) is *the key* for combining results from different techniques, as well. Assuming that strong technique specific organizations (as those mentioned in IERS Resolution 1 above) make available their results in such a format, it would then be relatively easy to come up with results based on different space geodetic techniques. This task of generalizing the SINEX format to accommodate results based on different space geodetic observations was officially given to the CSTG project at the 1996 Paris Workshop.
The effort is ongoing and hopefully converging to a useful tool.

The CSTG Subcommission *Precise Satellite Microwave Systems* aims at coordinating activities of various groups dealing with satellite microwave systems, in particular GPS, GLONASS, DORIS, and PRARE. From the geodetic point of view GPS (organized within the IGS) and DORIS may be considered operational. We believe that GLONASS, which is operational for navigational applications will play an important role in future, as well. The subcommission will have to coordinate its activities with the IGS.

16.5 SLR OBSERVATIONS TO GPS SATELLITES

Two GPS Block-IIa satellites, namely PRNs 5 and 6 have one Laser reflector array on board. The reflectors are of the same type as those used on all Russian GLONASS satellites.

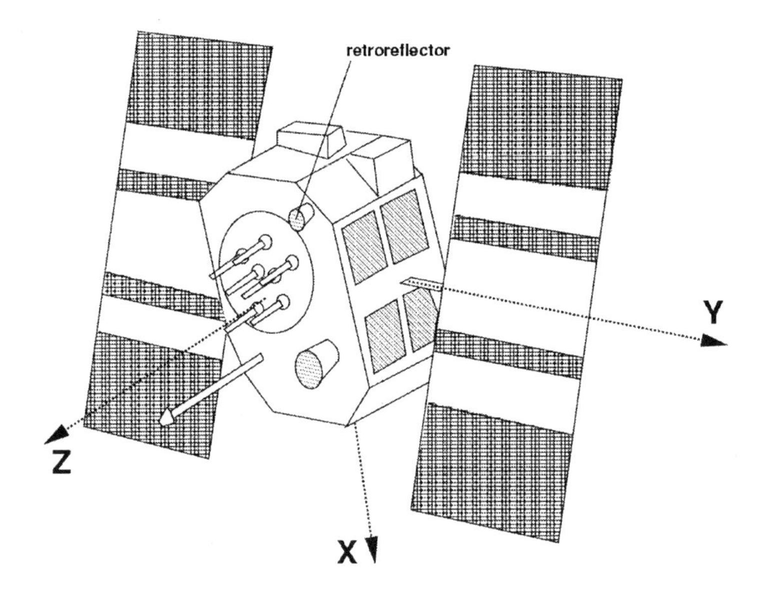

Figure 16.1. Schematic View of a GPS Satellite

The approximate location of the reflector array may be seen in Figure 16.1, the coordinates relative to the center of mass may be found in Degnan, J., E. Pavlis [1994]. PRN 5 was launched in August 1993, PRN 6 in March 1994.

GPS satellites are difficult targets for the SLR community due to the distance of the satellite of more than 20'000 *km*, and due to the fact that the mentioned GPS satellites carry only one corner-cube array. Day-time observations, due to the bad *Signal/Noise ratio*, are particularly difficult. Therefore, only few SLR observatories are capable of ranging to these targets routinely.

It is also worth mentioning that the observation geometry is "always bad": the angle between the observatory and the geocenter, as seen from the GPS satellite, never exceeds $\approx 13^{\circ}$. The situation is of course considerably improved if several observatories are ranging simultaneously to the same GPS satellite.

16.5.1 Motivation of Ranging to GPS Satellites

From a "philosophical point of view" it is always interesting and important to observe one and the same process using (essentially) different methods. Let us give a number of reasons why SLR observations to GPS satellites are of interest:

- SLR *with a demonstrated ranging accuracy of 1-2 cm* offers a unique opportunity for an independent check of the GPS/IGS orbit quality with a demonstrated *consistency* of 5-7 cm (rms per coordinate).

- Due to the fact that (for obvious reasons) the on-board *sun sensors* cannot "see" the sun when the satellite is eclipsed through the Earth, GPS satellites behave in a strange way (not according to the theoretical attitude control as described e.g., in Fliegel et al. [1992]) when they are eclipsed by the Earth. The actual rotation of the satellites may be directly monitored provided several SLR stations are tracking the satellite simultaneously. (Determinations using GPS are possible and are performed e.g., by Bar-Sever et al [1995] , but they are much more elaborate).

- IGS ephemerides refer to the center of mass of the satellite. In order to produce these ephemerides one has to know (in the satellite fixed system) the exact (<1 *cm*) location of the phase center (actually the L3 phase-center !) of the satellites. This information is (probably) not available with the required accuracy, at present. One might hope that a common analysis of SLR observations and GPS phase observations would help to determine this location.

- SLR observations are much less sensitive to the wet component of tropospheric refraction. This leads to major differences in processing.

- SLR observations are unbiased range observations, GPS observations may be considered as biased range observations. Even in the best possible case, when all double difference ambiguities are resolved, the (GPS derived) ranges are biased by a *satellite* and a *receiver clock* error.

Let us focus from now on the latter two aspects, namely on tropospheric refraction and on the impact of clock biases.

The delay of the signals due to the dry component of the troposphere is similar in both domains, **the delay due to the wet component, however, is significantly different in the optical and in the microwave band**. The Marrini Murray model for optical observations computes the *wet tropospheric zenith delay* as (see McCarthy [1996]):

$$\Delta r_{wet} \approx 0.00041 \cdot e \tag{16.1}$$

where e is the water vapour pressure.

The Saastamoinen model for microwave observations computes the *wet tropospheric zenith delay* as (see Saastamoinen [1972]):

$$\begin{aligned} \Delta r_{wet} &\approx 0.0022733 \cdot \left(\frac{1225}{T(Kelvin)} + 0.05 \right) \cdot e \\ &\approx 0.0095 \cdot e \end{aligned} \tag{16.2}$$

where e is the water vapour pressure. We used a temperature of T=273°+20° to compute the above numerical value.

If we compare the two equations (16.1) and (16.2) giving the wet tropospheric zenith delay in the optical and in the microwave band, we obtain:

$$\frac{\Delta r_{wet}(micro)}{\Delta r_{wet}(opt)} \approx 67 \tag{16.3}$$

The ratio (16.3) leads us to the following conclusions:

- Typically the wet component of the tropospheric zenith delay varies between 5 cm and 40 cm (very wet and humid conditions) *for GPS observations.*

- The corresponding variation for SLR observations lies only between 0.1 cm and 0.6 cm.

- About 80 % of the wet troposphere delay may be modeled by using *surface pressure, temperature, and humidity* as established on the station.

We thus conclude that *for SLR observations the effect is safely modeled* using surface met data, but that for **GPS observations surface met values are far from being sufficient** to account for the wet path delay. When processing GPS data, tropospheric refraction has to be modeled, instead.

These considerations also lead to the conclusion that SLR might serve as a **Calibration Technique** for Microwave Systems like GPS, DORIS, GLONASS, PRARE. This role was actually played by SLR several times in the past (satellite missions).

One may argue, of course, that the wet path delay may be (and actually is) modeled in GPS analyses. One should not forget, however, that modeling considerably weakens the GPS solutions: when looking at the partial derivatives of the GPS observable *w.r.t.* the relevant parameters we observe strong correlations between *troposphere estimates, station heights, and clock parameters* in GPS analyses. The situation is even more complicated because *elevation dependent phase center variations* have to be established in addition (at least) for each GPS antenna type. (The *Bermuda Triangle* troposphere, station heights, GPS antennas is the final destination for many GPS analysts ...).

In order to get some insight into the impact of the peculiarities of GPS analyses, which force us to solve for station height corrections Δh *and* for troposphere zenith corrections Δr_t, let us assume that we use the GPS observable ρ *uniquely* to determine these two parameters (orbits, horizontal positions, clocks, ambiguities, *etc.* are assumed known).

To our knowledge, Santerre [1989, 1991], was the first analyst to study systematically the problem of correlations between station heights, clock parameters, and tropospheric zenith delays. The findings of this *in-depth analysis* are still valid today. Below we present a simplified study along the lines of Rock Santerre.

The *observation equation* for a zero difference observation made at zenith distance z may then be written as:

$$-\cos z \cdot \Delta h + \frac{1}{\cos z} \cdot \Delta r_t - (\rho - \rho_0) = v \tag{16.4}$$

where v is the residual of the observation. Assuming that we have a uniform satellite distribution between a minimum zenith distance z_0 and a maximum zenith distance z_{max} this gives a normal equation matrix $N = A^T P A$ of the following kind

$$N = \kappa \cdot \begin{pmatrix} \int_{z_0}^{z_{max}} \sin^2 z \cos^2 z \, dz & -\int_{z_0}^{z_{max}} \sin^2 z \, dz \\ -\int_{z_0}^{z_{max}} \sin^2 z \, dz & \int_{z_0}^{z_{max}} \sin^2 z \cos^{-2} z \, dz \end{pmatrix} \tag{16.5}$$

where κ is a positive constant of no interest for us. The inverse of this equation (in essence) is the variance-covariance matrix of our estimated parameters:

$$N^{-1} = \tilde{\kappa} \cdot \begin{pmatrix} \int_{z_0}^{z_{max}} \sin^2 z \cos^{-2} z \, dz & \int_{z_0}^{z_{max}} \sin^2 z \, dz \\ \int_{z_0}^{z_{max}} \sin^2 z \, dz & \int_{z_0}^{z_{max}} \sin^2 z \cos^2 z \, dz \end{pmatrix} \tag{16.6}$$

This matrix may be used to compute the ratio σ_1/σ_0, which we define as the ratio of the *rms errors* with and without simultaneous estimation of one parameter for the tropospheric zenith correction, and the *correlation coefficient* of the parameters Δh and Δr_t. The numerical results, as a function of the maximum zenith distance z_{max} are given for a site at the equator ($z_0 = 0$) and one at the pole $z_0 = 45$ in Table 16.3.

Table 16.3. Ratio of mean errors of height *with* and *without* estimation of a troposphere zenithdelay and correlations between station heights and troposphere parameters.

z_0	z_{max}	σ_1/σ_0	Corr	z_0	z_{max}	σ_1/σ_0	Corr
0	60	2.79	+0.93	45	60	5.08	+0.98
0	65	2.29	+0.90	45	65	3.48	+0.96
0	70	1.90	+0.85	45	70	2.54	+0.92
0	75	1.59	+0.78	45	75	1.93	+0.86
0	80	1.35	+0.67	45	80	1.51	+0.75
0	85	1.15	+0.50	45	85	1.22	+0.57

We see that merely due to the fact that we had to introduce *one* tropospheric zenith delay parameter, our height estimates were weakened by a factor of about 2 when using a maximum zenith distance of $z_{max} = 70°$. We also observe a rather pronounced *correlation between troposphere and height estimates.*

The situation may be considerably improved by *increasing the maximum zenith distance* z_{max}. The draw-back is an increased level of *multipath effects.* Also one is more exposed to phase center variations which tend to become less well known at lower elevations (see also remarks below).

Let us become one step more realistic and introduce a receiver clock parameter $c \cdot \Delta t$ into the observation equations (we are thus identifying the GPS observable ρ with the biased range). A single observation equation then reads as:

$$-\cos z \cdot \Delta h + \frac{1}{\cos z} \cdot \Delta r_t - c \cdot \Delta t - (\rho - \rho_0) = v \tag{16.7}$$

Making the same assumptions as above it is easy to deduce the variance-covariance matrix (this is left to the reader) and to produce Table 16.4 giving the ratio of rms errors and the correlations between heights and troposphere parameters, but this time assuming that we have to solve for (only one) station clock parameter in addition to the tropospheric zenith delay and the height component.

The numbers in Table 16.4 are frightening, indeed. They should be interpreted carefully, however: We compare the performance of the *real GPS* with that of a *hypothetical optical GPS.* Nevertheless, the statement made above that *SLR measurements may be considered as calibration measurements for GPS* is not completely "academic" in view of Tables 16.3 and 16.4.

Correlations of the kind given in Table 16.4 are actually encountered in practice, *and* they occur *independently* on whether one processes data in the *zero, single, or double difference mode* and independently on the actual procedure chosen to model tropospheric refraction. Table 16.5 giving the correlations *and* the height rms errors for one endpoint of a baseline (actually Kootwijk-Wettzell on day 109 of year 1997) using one day of data underlines this fact.

Table 16.4. Ratio of mean errors of height *with* and *without* estimation of a troposphere zenith delay *and* a clock parameter and correlations between station heights and troposphere parameters.

z_0	z_{max}	σ_1/σ_0	Corr	z_0	z_{max}	σ_1/σ_0	Corr
0	60	31.23	-0.985	45	60	113.32	-0.996
0	65	20.64	-0.976	45	65	52.29	-0.991
0	70	13.78	-0.964	45	70	27.11	-0.983
0	75	9.23	-0.943	45	75	14.99	-0.967
0	80	6.13	-0.907	45	80	8.52	-0.937
0	85	3.94	-0.830	45	85	4.78	-0.868

Table 16.5 also indicates that despite all correlations the quality of GPS determined heights is not too bad, after all. Even with an elevation cut-off angle of 20^0 a *formal accuracy* of about 2 *mm* is promised.
Experience tells, that the formal parameter errors are too optimistic by about a factor of 3. This is confirmed by Figure 16.2, which shows the repeatability of GPS established coordinate series in the case of the permanent Onsala GPS site. One can see that the height error is dominant, the rms being of the order of 5-6 *mm*. The horizontal coordinates are determined better by a factor of about three than the height component (the result stems from a regional European analysis as performed by the CODE processing center of the IGS).

Let us once more address the correlations seen in Tables 16.4 and 16.5. It is clear that the height rms errors (and the corresponding ratios) already account for these correlations. One might thus argue that the situation is not so bad after all. One should keep in mind, however, that these rms errors are realistic only *if the stochastic and the deterministic model of the parameter estimation problem are both correct.* If an effect remains unmodeled, the correlation may have a devastating influence on GPS determined heights. This is illustrated by the following example.

Table 16.5. Height rms and correlations between height and troposphere parameter for Kootwijk-Wettzell in the double difference mode (coordinates and troposphere for Wettzell fixed on "true" values).

z_{max}	Height rms (*mm*)	Corr
60	5.1	-0.986
65	3.2	-0.986
70	2.1	-0.964
75	1.4	-0.944
80	1.0	-0.903
85	0.8	-0.872

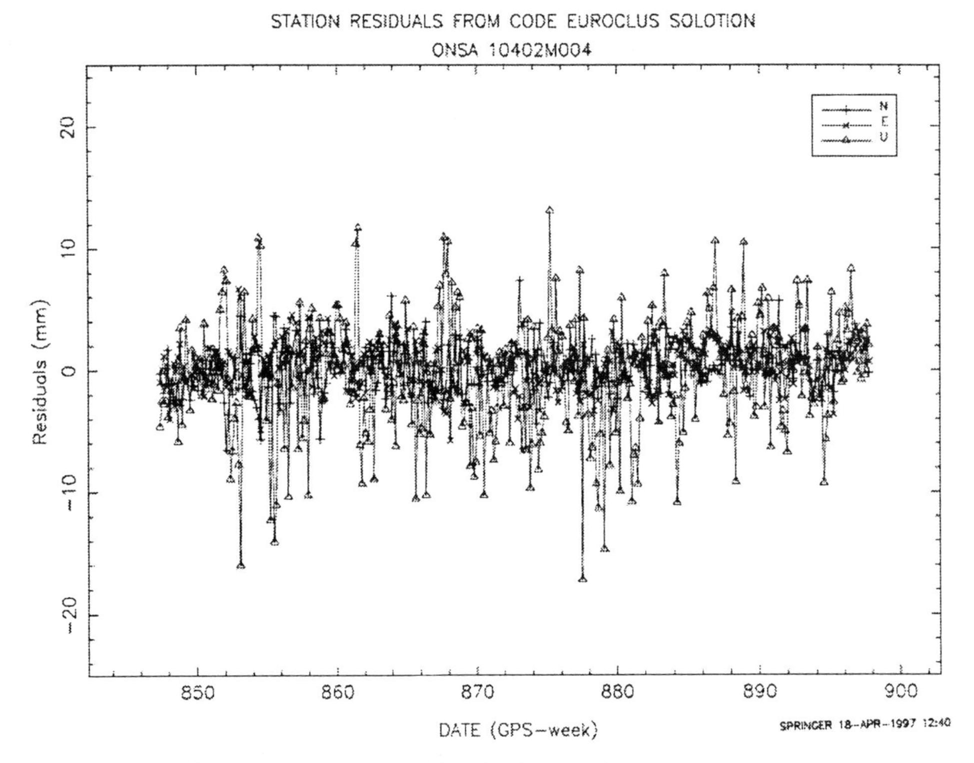

Figure 16.2. Coordinate Repeatability of Onsala Site Coordinates, Daily European network Solution (Minimum Elevation 15°, station Velocity removed)

Since 1992 a permanent Trimble SST receiver was installed at the Zimmerwald SLR observatory. Its data were regularly processed by the CODE processing center of the IGS since 21 June 1992 and it became quite clear after a while that there was a systematic difference of about 10 *cm* (!) between the GPS- and the SLR-determined station heights for Zimmerwald. The usual explanation of an imperfect local tie could be ruled out rather quickly. Depending on the point of view one was speaking of an SLR- or a GPS- height bias.

The issue could be resolved, when an Ashtech receiver (for a different purpose) was running in parallel to the Trimble receiver for a time interval of about three months in 1992 in Zimmerwald. The bias miraculously disappeared, when the Ashtech instead of the Trimble data were used in the daily CODE solutions. The effect was *not* seen when the local (very short) Trimble-Ashtech baseline was processed in the *standard way*, *i.e.*, introducing only the coordinates of one receiver (*but no troposphere biases*) as unknown. The height difference between the two receivers corresponded within a couple of millimeters to ground truth. *When artificially introducing a troposphere bias for one endpoint of the short baseline, the height "jumped" by about 10 cm and the troposphere estimate indicated rather pronounced climatic differences between the two endpoints of the baseline*!

The example shows the importance of using independent measurement systems in Space Geodesy (and elsewhere). The particular example triggered indepth studies of *antenna-type specific height variations*. It became clear soon that the explanation for the "Zimmerwald height problem" was explained by substantial differences in the elevation dependent phase center variations of the two antenna types.

The problem was studied subsequently by different authors using different methods. We mention in particular *anechoic phase chamber measurements* where the attempt is made to establish the phase center variations in a laboratory environment and *GPS antenna calibration campaigns*, where antennas of different type are brought to a test range, where all antenna locations are known within 1-2 millimeters using classical (non-GPS) geodetic methods. The latter experiments allow to model the phase center differences between different antenna types as a function of elevation (and if necessary azimuth). Such experiments eventually led to the *establishment of an IGS file containing the phase center corrections for all known antenna types*. For more information we refer to Rothacher et al. [1996a, 1996b].

16.5.2 Actual Observations and Results

Since the launch of the PRN 5 and PRN 6 a number of SLR observatories successfully tracked the two satellites despite the fact that both targets are difficult for the SLR observatories. These stations may be found in Figure 16.3. *Graz, Haleakala, Monument Peak, Herstmonceux, Orroral, Yarragadee, MLRS (Fort Davis), Wettzell, Potsdam, MacDonald, Riga, Komsomolsk, and Greenbelt* were the principal contributing observatories.

Degnan and Pavlis [1994] and Pavlis and Beard [1996] describe the first analyses that were performed using SLR observations to GPS satellites. Essentially the attempt was made to derive high accuracy orbits based on SLR observations only and to compare these orbits to the IGS orbits (after a system transformation). The critical point in such an approach has to be seen in the scarcity of SLR data. Long arcs (14 days or longer) had to be formed in order to obtain a stable SLR orbit. It is known on the other hand that the force model becomes a delicate issue if arcs of this length are considered Beutler et al. [1994] if an accuracy below the decimeter is aimed at. Pavlis and Beard [1996] could conclude, however, that GPS-derived and SLR-measured ranges did *not show big inconsistencies*. An agreement between IGS orbits and SLR distances on the 10 *cm* rms level could be established rather quickly.

It was only normal that the IGS community became interested in the SLR observations. It seemed that a regular SLR tracking of GPS satellites and an inclusion of SLR observations into the GPS analyses might be beneficial to the IGS community for *calibration purposes*. It was also clear, however, that such a development would only take place if the number of SLR passes would significally increase. This was why the IGS Governing Board asked the SLR community with a letter dated January 29, 1996 for more frequent SLR tracking of the two GPS satellites equipped with reflectors.

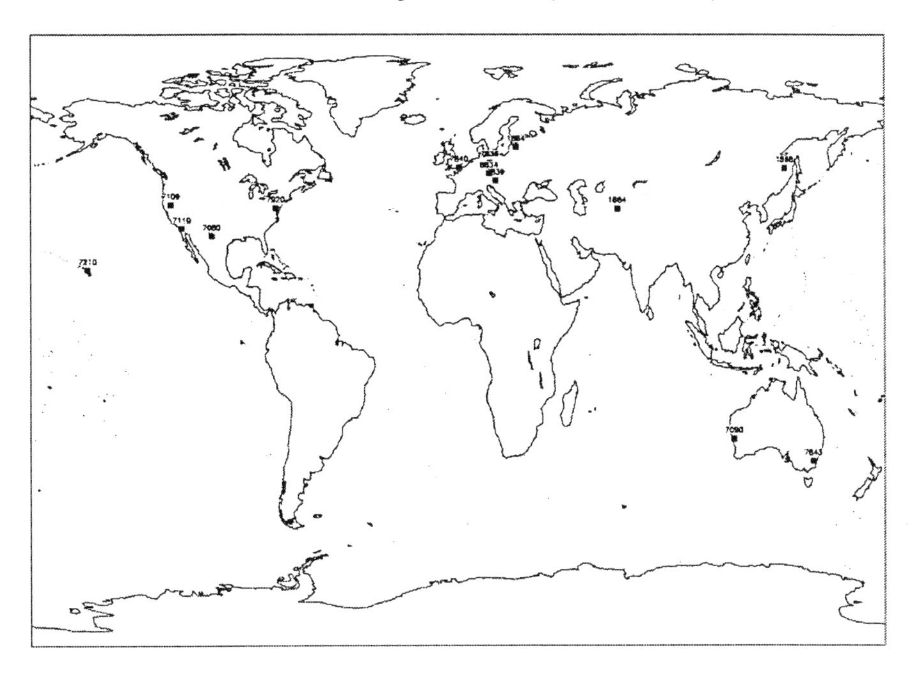

Figure 16.3. Stations of the International SLR Network actually Tracking GPS Satellites.

In June 1996 the SLR Community reacted by publicly announcing a coordinated tracking campaign for the time period between October 20 and December 15, 1996. The success of the campaign was not overwhelming. It was confirmed that GPS satellites are extremely difficult targets for the current SLR network which gives highest priority to tracking the LAGEOS satellites. There is some hope, however, that the situation might improve in future.
Nevertheless this particular activity and the (nevertheless) considerable number of SLR observations to GPS satellites accumulated since late 1993 gave rise to some analyses performed by groups interested in both, SLR and GPS. Let us mention in particular the analysis performed by Zhu, Reigber, and Kang [1996], the overview given by Watkins and Bar-Sever [1996] at the AGU Fall Meeting in 1996, and the analysis presented by Springer et al. [1996] at the same meeting.

Let us first give an overview of the various kinds of analyses performed, give an impression of the quality of SLR observations, and conclude with the description of some results. The following types of analyses were performed so far:

- **(1): Direct check of IGS orbits**
  Let

  i observation number
  t'_i reflection time of Laser pulse at observation time for observation i
  ρ_i the SLR observation at time t'_i
  $\boldsymbol{r}_{sat}(t'_i)$ the position of the satellite at time t'_i
  $\boldsymbol{R}_{obs}(t'_i)$ the(hopefully known) station position.

Then the differences

$$\Delta_i = \rho_i - |\boldsymbol{r}_{\text{sat}}(t'_i) - \boldsymbol{R}_{\text{obs}}(t'_i)| \quad , \quad i = 1,2,\ldots \tag{16.8}$$

are analyzed without trying to incorporate the SLR observations into an orbit adjustment process. A mean value for these differences and the associated rms error are usually estimated.

- **(2): Site estimates using SLR observations *and* IGS orbits:**

  The differences (16.8) were used as the terms *observed-computed* in a parameter estimation process where only the station positions (and possibly some bias terms) of the Laser observatories were estimated. A direct tie of the SLR network into the IGS network may be performed using this method.

- **(3): Orbit determination using only SLR observations:**

  The SLR observations ρ_i are used to generate orbits for the satellites *PRN5, PRN6*. These orbits are then compared to the IGS-orbits or to the orbits of individual analysis centers.

- **(4): Combined SLR- and GPS orbits:**

  Orbits are computed using *SLR- and GPS-observations* in the same adjustment process. Attempts are made to establish the quality of these combined orbits, *e.g.*, by looking at discontinuities at arc boundaries.

From Springer et al. [1996] we include three Figures which should give an impression of the quality of SLR observations. Figure 16.4 gives an overview of all (accepted) SLR observations of the time period from 1995 to early 1997 considered in this study. More precisely the differences (16.8) using CODE orbits and known station positions (taken from IGS and SLR analyses) are given in this Figure. We see at once that the agreement is quite good in general, we also see that a *small bias* between SLR measured distances and GPS-derived ranges seems to exist.

Figure 16.4 makes us believe that the rms of SLR observations is of the order of 10 *cm*. That this not the case is shown by Figures 16.5 and 16.6. Figure 16.5 shows two passes of PRN5 as seen from two different observatories. The actual measurement noise is of the order of a few millimeters, the slope seen in one case may be either due to inconsistent station coordinates, an (along track) error in the reference ephemeris, or a time bias in the SLR observation. It seems quite clear, however, that the SLR observations could very well contribute to the orbit quality.

Figure 16.5 is typical for satellite passes corresponding to non-eclipsing orbits. If the satellite is entering into the Earth shadow things look different due to the attitude problem addressed in Section 14.5. Figure 16.6 gives an example. We see "residuals" of up to half a meter. It is clear in this case that these excursions are due to a modeling deficit in the CODE orbits, because the *irregular attitude control* (see Section 14.5) is not (yet) taken into account in the CODE analyses. It is also clear from this figure that SLR observations are perfectly suited to study this effect in detail for PRNs 5 and 6.

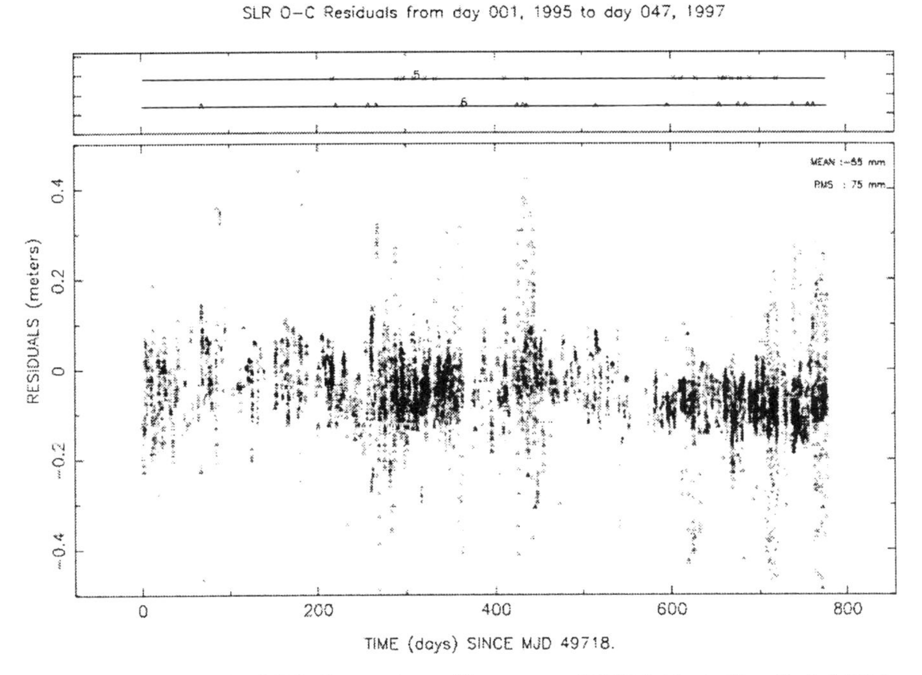

Figure 16.4. Residuals of SLR Observations with respect to CODE Orbits in Time Period 1995-early 1997.

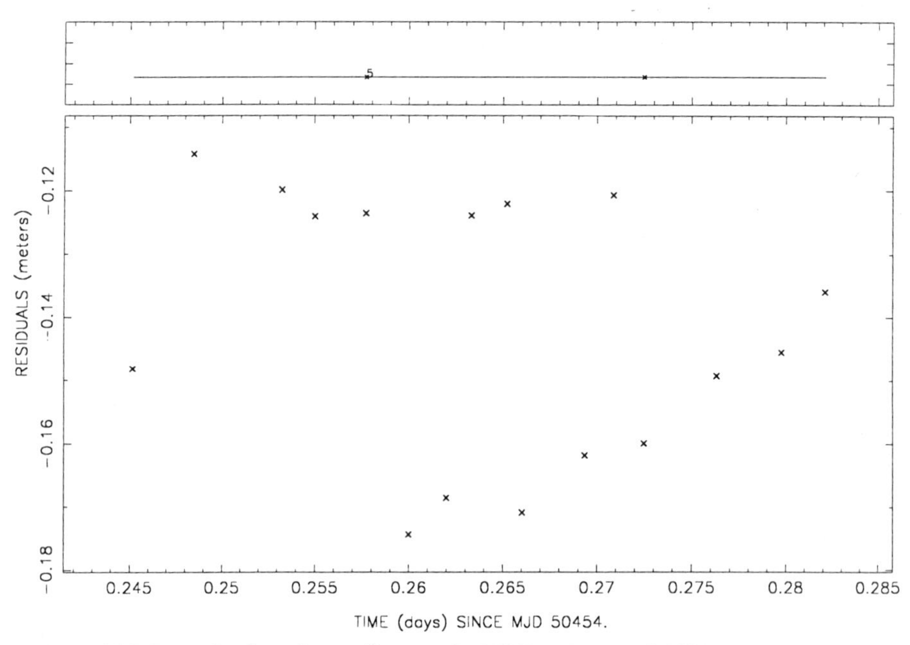

Figure 16.5. Example of two "normal" passes for PRN5 on January 6, 1997

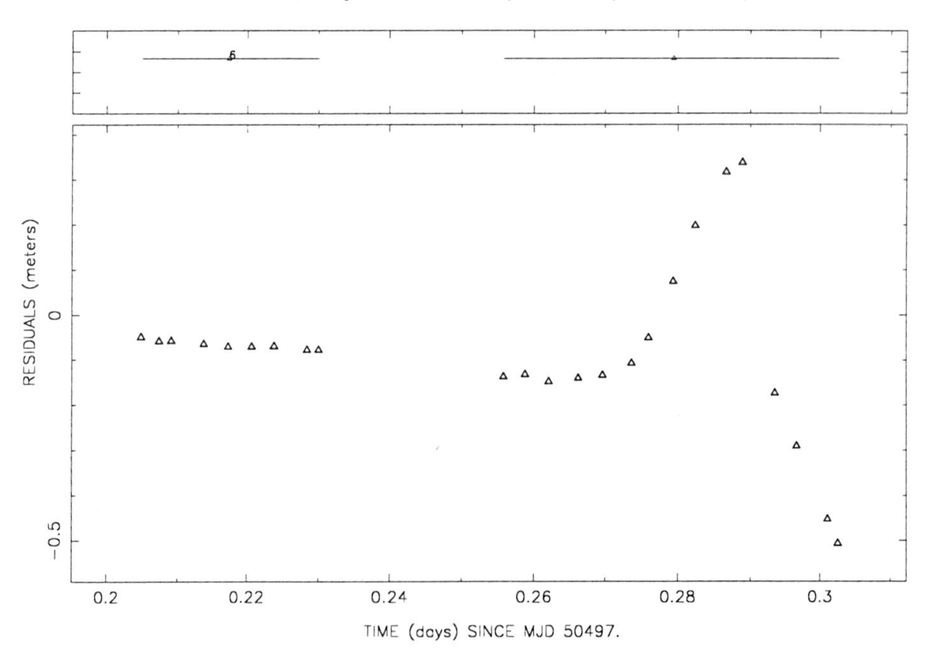

Figure 16.6. Example for an PRN 6, which was observed during Eclipse from the Maui Observatory.

Let us now briefly summarize the three most recent studies processing SLR data. Watkins and Bar-Sever [1996] analyze SLR observations performed by *Haleakala, Monument Peak, Yaragadee, Wettzell, Herstmonceux, Graz, Ft. Davis, Greenbelt, Orroral, and Quincy* in 1995 and 1996. Essentially an **analysis of type (1)** is performed. They use ITRF94 values for the station coordinates and the following orbits: (a) IGS orbits, (b) CODE orbits, and (c) JPL orbits. The distinction is made between *full sun* and eclipsing orbits. A bias of -39 *mm*, -45 *mm*, and -33 *mm* is reported for the IGS, the CODE, and the JPL orbits for the full sunlight orbits. Very similar numbers for eclipse seasons, as well. The data from 1995 and 1996 were analyzed separately. The biases seem to be about one centimeter smaller in 1995. Special attention was paid to the eclipse orbits. It was confirmed that the attitude modeling as performed by Bar-Sever et al [1995] is quite successful. JPL orbits (which take this effect into account) agree somewhat better with the SLR observations in these cases. The authors see a potential in future for ambiguity resolution, and for the collocation of the SLR and the IGS networks. Their main conclusions are that the GPS-derived radial orbit error does not exceed 50 *mm* for the "better" IGS Analysis Centers, that the three dimensional orbit error does probably not exceed 150 mm, and that SLR tracking may provide an interesting way to get improved frame ties.

Zhu, Reigber, Kang [1996] were (to our knowledge) the only ones to perform an **analysis of type (2)**. They report an agreement with the corresponding SLR coordinates (after a suitable transformation which may probably be considered as the best system transformation between the IGS and the SLR network) at the 10 *mm* level. They also produced **analyses of type (4)**, *i.e.*, a combination of SLR and GPS observations to come up with combined orbits for the entire set of 24 GPS satellites. Studying the overlap regions of consecutive arcs they claim an improved orbit quality for PRNs 5 and 6, and a marginal improvement for the other satellites. They argue that even a small number of SLR observations to GPS satellites would clearly lead to an improved set of orbits. All SLR observations of good quality were used in this analysis.

Springer et al. [1996] reported on the generalization of the **Bernese Software Version 4** to process *SLR Observations to GPS Satellites* (and zero difference code and phase observations). The implementation was performed to allow for the highest degree of flexibility and modularity: SLR and GPS observations are first treated *separately* by the *same* parameter estimation program. The *Normal Equation System (NEQS)* pertaining to one day of SLR and GPS observations are then stored separately on files (NEQ-files). Subsequently *any number of NEQ-files for consecutive days* may be combined to generate *n-day-solutions* using only SLR, only GPS, or using both, SLR *and* GPS NEQ-files. This procedure allows it to perform studies of all the types mentioned above. As a first result a study of **type (1)** was presented. Essentially the same biases as in the study presented by Watkins and Bar-Sever [1996] were confirmed. The effort is ongoing and an inclusion of SLR observations into the IGS analyses at CODE is considered.

In summary we may state that the analyses performed so far have shown that SLR observations to GPS satellites *actually are of great interest*: Comparing the GPS only and the GPS/SLR combined solutions reveals differences on the level of up to 5-10 cm rms for the entire GPS satellite system. Zhu et al [1996] even claim that the SLR observations improve the misclosures at the arc boundaries. Such results are promising and one may hope that SLR/GPS combined solutions will be produced regularly in future at some of the IGS analysis centers.

Should a GLONASS tracking network with good dual-band receivers be deployed in future (probably as a subset of the IGS network) the impact of SLR observations on microwave satellites might become even more important: All GLONASS satellites carry SLR reflectors, and the signal-to-noise ratio is much better than for GPS satellites. Moreover, with *Etalon-1 and Etalon-2* there are two cannon-ball satellites available in GLONASS-type orbits. SLR observations would thus be of greatest use for the determination of the *non-gravitational forces* acting on GLONASS satellites. We also believe that the impact will be more than marginal on the future improvement of the GPS force model, too.

16.6 SUMMARY

In this Chapter we reviewed the role of GPS within Space Geodesy. It became clear in Sections 16.1 and 16.2 that GPS actually plays an increasingly important role thanks to the excellent and rather inexpensive equipment. Global GPS activities are very well coordinated by the IGS. Space geodetic activities involving several techniques are coordinated by IERS, CSTG and the IGS.

A review of the contributions of the mature space techniques in Section 16.3 clearly showed that there is a necessity for all space techniques.

In Section 16.4 we gave arguments for the formal creation of an *International Space Geodetic Network*, where GPS has the obvious role of a collocation instrument. We also argued that analyses correctly combining several space geodetic methods are badly needed today. We presented examples for such developments within IERS and CSTG.

In Section 16.5 we looked into a particular experiment, namely SLR observations of the two GPS satellites PRN 5 and PRN 6 equipped with Laser reflectors. We first discussed the motivation for such studies, where we saw that the interest goes far beyond a purely “philosophical one”. Using a very simple theoretical model we showed that due to the necessity to solve for clock and troposphere parameters in GPS analyses there must be strong correlations between the station height, the troposphere parameters, and the clock parameters, which leads to a considerably reduced accuracy of the station heights (and the other parameters). We gave evidence that correlations of the type postulated actually exist in practice. Due to the fact that SLR does not suffer from such problems -- is it neither necessary to solve for clock parameters nor for troposphere parameters – SLR promises to play a significant role as a calibration technique for GPS.

The section was concluded with a discussion of some of the recent results of the SLR/GPS experiment. All investigators concluded that the consistency of GPS-derived distances and SLR measurements is surprisingly good, a range bias of only about 3-5 *cm* between the two systems was reported. The yet unexplained bias seems to be significant and reasonably constant in time. All investigators agree that even with few SLR observations routinely available there is a considerable interest for GPS (microwave) applications. Improvement of GPS orbits, improvement of ambiguity resolution and constraining of troposphere parameters were mentioned as key issues. Zhu et al. [1996] also used the SLR data to derive directly a transformation between the IGS and the SLR network.

We believe that the results achieved so far are interesting and encouraging. It is clear that an improved SLR tracking scenario of GPS satellites would lead to even better results. We also argued that SLR observations might become even more important if some time in the future a GLONASS tracking network should be built up.

Acknowledgements

Part of the material presented here is taken from an overview paper by Rothacher and Beutler which was presented at the 197 Vienna EGS Meeting and which should be published late 1997. The author wishes to thank Tim Springer for making available his SLR related work and for his help in the preparation of this manuscript, Markus Rothacher and Stefan Schaer for their contributions related to antenna phase center variations and tropospheric refraction, and, last but not least, to Leos Mervart for a careful review of this chapter and for preparing the manuscript in the required form.

References

Bar-Sever, Y., J. Anselmi, W. Bertiger, E. Davis (1995), *Fixing the GPS Bad Attitude: Modeling GPS Satellite Yaw During Eclipse Seasons*. Proceedings of the ION National Technical Meeting, Anaheim, CA, January 1995.

Beutler, G., E. Brockmann, W. Gurtner, U. Hugentobler, L. Mervart, M. Rothacher, A. Verdun (1994), *Extended Orbit Modelling Techniques at the CODE Processing Center of the IGS: Theory and Initial Results*. Manuscripta Geodaetica, Vol. 19, pp. 367-386.

Beutler, G., I.I. Mueller, K.P. Schwarz (1996), *Memorandum for the Record "stablishment of an IAG/CSTG Working Group: IUGG Fundamental Reference and Calibration Network"* CSTG Bulletin No. 12, pp. 75-94, Deutsches Geod„tisches Forschungsinstitut, Abt. I - Theoretische Geodäsie.

Charlot, P.(ed.) (1995), *Earth Orientation, Reference Frames and Atmospheric Excitation Functions submitted for the 1994 IERS Annual Report. VLBI, LLR, GPS, SLR, DORIS, and AAM Analysis Centers.*, IERS Technical Note 19, Observatoire de Paris, September 1995.

Drewes, H., Beutler, G.(1997), Report of the IAG/CSTG Working *Group "IUGG Fundamental Reference and Calibration Network"*. In: Beutler, G. Drewes, H. Hornik, H. (eds.): CSTG Bulletin No. 13, Progress Report 1997, Munich, Deutsches Geod„tisches Forschungsinst., Abt. 1, 1997. pp. 20-35.

Eanes, R.J., M.M. Watkins (1995), *The Contributions of Satellite Geodesy to Geodynamics.*, EOS Transactions G32A-3, Vol. 76, No. 17, April 25, 1995, Supplement, p. S88, 1995 AGU Spring Meeting in Baltimore.

Boucher, C., Z. Altamimi, L. Duhem (1994), *Results and Analysis of the ITRF93*, IERS Technical Report No. 18, Observatoire de Paris.

Degnan, J., E.C. Pavlis (1994), *Laser Ranging to GPS Satellites with Centimeter Accuracy.* GPS World, September 1994, pp. 62-70.

Fliegel, H.F., T.E. Gallini, E.R. Swift (1992), *Global Positioning System Radiation Force Model for Geodetic Applications.* Journal of Geophysical Research, Vol. 97, No B1, pp. 559-568.

McCarthy D.D. (ed.), (1996), *IERS Conventions (1996)*, IERS Technical Note 21, July 1996, Observatoire de Paris.

Mueller, I.I., Montag, H., Reigber, Ch., Wilson, P. (1995), *An IUGG Network of Fundamental Geodetic Reference Stations: Rationale and Recommendations*, CSTG Bulletin No. 12, pp. 75-94, Deutsches Geod,,tisches Forschungsinstitut, Abt. I -- Theoretische Geodäsie.

Pavlis, E.C., R.L. Beard (1996), *The Laser Retrofelector Experiment on GPS 35 and 36.* in GPS Trends in Precise Terrestrial, Airborne, and Spaceborne Applications, IAG Symposia, No. 115, pp. 154-158, Springer Verlag, Berlin.

Reigber, Ch., M. Feissel (1997), *IERS Missions, present and future, Report on the 1996 IERS Workshop.* IERS Technical Note No 22, Observatoire de Paris.

Rothacher, M., S. Schaer, L. Mervart, G. Beutler (1996a), *Determination of Antenna Phase Center Variations using GPS Data.* Proceedings, 1995 IGS Workshop "special Topics and New Directions", pp. 205-220, IGS Central Bureau.

Rothacher, M., S. Schaer (1996b), *Antenna Phase Center Offsets and Variations Estimated for GPS Data.* Proceedings, 1996 IGS Analysis Center Workshop, pp. 321-246.

Saastamoinen, J. (1972), *Atmospheric Correction for the Troposphere and Stratosphere in Radio Ranging of Satellites.* The use of artificial satellites for geodesy, Geophysical Monograph 15, American Geophysical Union, Washington, D.C., pp. 247-251

Santerre, R. (1989), *GPS Satellite Sky Distribution: Impact on the Propagation of Some Important Errors in Precise Relative Positioning*, Technical Report No. 145, Dept. Of Surveying Engineering, University of New Brunswick.

Santerre R. (1991), Impact of GPS satellite sky distribution, *Manuscripta Geodaetica*, Vol. 16, pp. 28-53

Springer, T., G. Beutler, E. Brockmann, M. Rothacher (1996), *Orbit Improvement of GPS Satellites using GPS and SLR Observations.* EOS Transaction, Vol. 77, No. 46, p. F135.

Tscherning, C.C. (1992), The Geodesist's Handbook 1992. *Bulletin Géodésique, Vol. 66, No. 2*, Springer-Verlag.

Watkins M.M., Y.E. Bar-Sever (1996), *SLR Contributions to GPS.* EOS Transaction, Vol. 77, No. 46, p. F130.

Zhu, S.Y., Ch. Reigber, Z. Kang (1996), *Apropos Laser Tracking to GPS Satellites.*

Zumberge, J.F., R. Liu (1995), *Densification of the IERS Terrestrial Reference Frame through Regional GPS Networks.* Workshop Proceedings of the IGS Workshop, November 30-December 2, 1994, Pasadena 1994. Available through the IGS Central Bureau.

Printing: Mercedesdruck, Berlin
Binding: Buchbinderei Lüderitz & Bauer, Berlin